THE ORIGINS OF ORDER

THE ORIGINS OF ORDER

Self-Organization and Selection in Evolution

Stuart A. Kauffman

University of Pennsylvania
and
The Santa Fe Institute

New York   Oxford
Oxford University Press
1993

Oxford University Press

Oxford   New York   Toronto
Delhi   Bombay   Calcutta   Madras   Karachi
Kuala Lumpur   Singapore   Hong Kong   Toyko
Nairobi   Dar es Samaalm   Cape Town
Melbourne   Auckland   Madrid

and associated companies in
Berlin   Ibadan

Published by Oxford University Press, Inc.,
200 Madison Avenue, New York, New York 10016

Oxford is a registered trademark of Oxford University Press

Library of Congress Cataloging-in-Publication Data
Kauffman, Stuart A.
The origins of order : self-organization and selection in
evolution / Stuart A. Kauffman.
p.   cm.
Includes bibliographical references (p.   ) and index.
ISBN 0-19-505811-9
ISBN 0-19-507951-5 (pbk.)
1. Life—Origin.   2. Self-organizing systems.   3. Molecular
evolution.   4. Evolution—Philosophy.   I. Title.
QH325.K39   1993
577—dc20   91-11148

9 8 7 6 5 4

Printed in the United States of America
on acid-free paper

To my daughter
Merit Leslie Kauffman,
whom I honor and love still,

and

Ethan and Elizabeth,
who survive her

Preface

This book is an attempt to focus attention on new themes in developmental and evolutionary biology. It is, in fact, an attempt to include Darwinism in a broader context. The central themes are easily stated. Simple and complex systems can exhibit powerful self-organization. Such spontaneous order is available to natural selection and random drift for the further selective crafting of well-wrought designs or the stumbling fortuity of historical accident. Yet no body of thought incorporates self-organization into the weave of evolutionary theory. No research program has sought to determine the implications of adaptive processes that mold systems with their own inherent order. Yet such must be our task. And more as well, for some systems adapt readily, whereas others are so badly disrupted by minor modifications that adaptive improvement by random mutation and selection can hardly occur. Darwin simply assumed that such improvement was possible. One might have thought, more than a century later, that we would understand the construction requirements which permit complex systems to adapt. But we do not. Nor do we understand the extent to which selection can achieve systems able to adapt successfully. This book explores these broad themes.

Like many other books by scientists, this one is ineluctably autobiographical. It witnesses one mind's sense of mystery. The famous physicist Wolfgang Pauli is said to have remarked that the deepest pleasure in science comes from finding an instantiation, a home, for some deeply felt, deeply held image. I share that odd sense. In my own case, over thirty years ago, walking in Hanover, New Hampshire, as an undergraduate at Dartmouth College, I looked into a bookstore window and realized that, someday, I would write a book filled with points mysteriously connected with arrows. Like cubist paintings, with their quixotic surfaces and angles meeting, dispersing, fanning, and fading into the background, edging polygons and polyhedra with currents of meaning, the points and arrows in my yet-to-be-written book hinted at spontaneous order. In subterranean ways this image works in me still. The greater mystery, after all, is not the answers that scientists contrive, but the questions they are driven to pose. Why? Why this question rather than another? Why this search, hope, dispair, rather than another? Why this ill-lit, nil understood, hobo path? And why the outrageous confidence, born of no evidence, to tred it? I do not know. But I know that this sense is not rare. What a strange pleasure it is to seek.

I am glad to thank students, friends, and colleagues who have taken the time to read parts of the manuscript as it evolved. In particular, I mention Pere Alberch, Jeanette Alexander, Philip Anderson, Vahe Bedian, Richard Burian, James Crow, Manfred Eigen, Warren Ewens, Doyne Farmer, Marjory Grene, Scott Gilbert, Brian Goodwin, John Holland, Jeremy Knowles, Thomas LaBean, David Lane, Simon Levin, David Margolin, Jay Mittenthal, John Miller, Sandy Mitchell, Harold Morowitz, Leslie Orgel, Norman Packard, Rob Page, Richard Palmer, Alan Perelson, Rudy Raff, Bob Richardson, Irwin Rose, Peter Schuster, Daniel Stein, Martin Weigert, and Edward Weinberger. Their help, however, has not eliminated all the defects which the reader will encounter. My editors at Oxford University Press, Judith May and William Curtis, added their efforts. Carole Gan and Elizabeth Kauffman were patient and painstaking with the bibliography. Jeanette Alexander drafted many of the final figures. Recent close collaborative work has been carried out with Richard Bagley, Lloyd Clark, Brian Goodwin, Sonke Johnson, Alan Perelson, and Edward Weinberger. I am grateful to them all.

In the past five years, I have had the pleasure to be associated closely with the Santa Fe Institute. I am pleased to acknowledge both the institute and the University of Pennsylvania.

Authors know that books are not easily written. This one has been no exception. Yet writing a scientific book can be like writing a novel. Ideas, like characters, once loose upon a page harbor their own lives, follow their own unsuspected paths, mature in unforeseen ways, and mingle with their own logic. If useful, they have progeny.

Santa Fe, N.M.   S.A.K.
February 1993

Contents

Themes

The title of this book, *Origins of Order: Self-Organization and Selection in Evolution*, states the book's task: to answer the question, What are the sources of the overwhelming and beautiful order which graces the living world? To presume to ask such a question is also to know one must not presume to succeed. Questions such as this must ever be asked anew as each generation comes to perceive new ways of ordering its view of life.

One view, Darwin's, captivates us all: natural selection and the great branching tree of life, spreading from the major phyla to the minor genera and species, to terminal twigs, to curious humans seeking their place. Darwin and evolutionism stand astride us, whatever the mutterings of creation scientists. But is the view right? Better, is it adequate? I believe it is not. It is not that Darwin is wrong, but that he got hold of only part of the truth. For Darwin's answer to the sources of the order we see all around us is overwhelmingly an appeal to a single singular force: natural selection. It is this single-force view which I believe to be inadequate, for it fails to notice, fails to stress, fails to incorporate the possibility that simple and complex systems exhibit order spontaneously. That spontaneous order exists, however, is hardly mysterious. The nonbiological world is replete with examples, and no one would doubt that similar sources of order are available to living things. What is mysterious is the extent of such spontaneous order in life and how such self-ordering may mingle with Darwin's mechanism of evolution—natural selection—to permit or, better, to produce what we see.

Biologists have not entirely ignored the spontaneous emergence of order, the occurrence of self-organization. We all know that oil droplets in water manage to be spherical without the benefit of natural selection and that snowflakes assume their evanescent sixfold symmetry for spare physicochemical reasons. But the sheer imponderable complexity of organisms overwhelms us as surely as it did Darwin in his time. We customarily turn to natural selection to render sensible the order we see, but I think the answer to our questions about the origins of order is broader. We already have some inkling of the kinds of spontaneous order which may bear on biological evolution, and I believe we must make the most profound assessment of such self-organization. We must look in any direction that seems profitable because whatever spontaneous order may abound is available for evolution's continuing uses.

What makes the present stage of biological science so extraordinary is that molecular biology is driving us to the innermost reaches of the cell's ultimate mechanisms, complexity, and capacity to evolve. At the very same time, work in mathematics, physics, chemistry, and biology is revealing how far-reaching the powers of self-organization can be. These advances hold implications for the origin of life itself and for the origins of order in the ontogeny of each organism. One major theme of this book is an effort to link recent work in molecular biology with these new insights into spontaneous order in complex systems. Union of the two streams of insight promises to transform our understanding. The order inherent in the busy complexity within the cell may be largely self-organized and spontaneous rather than the consequence of natural selection alone.

Yet our task is not only to explore the sources of order which may lie available to evolution. We must also integrate such knowledge with the basic insight offered by Darwin. Natural selection, whatever our doubts in detailed cases, is surely a preeminent force in evolution. Therefore, to combine the themes of self-organization *and* selection, we must expand evolutionary theory so that it stands on a broader foundation and then raise a new edifice. That edifice has at least three tiers:

• We must delineate the spontaneous sources of order, the self-organized properties of simple and complex systems which provide the inherent order evolution has to work with *ab initio* and always.

• We must understand how such self-ordered properties *permit, enable,* and *limit* the efficacy of natural selection. We must see organisms in a new light, as the *balance found,* the *collaboration achieved,* when natural selection acts to further mold order which preexists. In short, we must integrate the fact that selection is not the sole source of order in organisms.

• We must understand which properties of complex living systems *confer* on the systems their capacities to adapt. For Darwin simply assumed that the accumulation of advantegeous mutations was possible, and yet the capacity to do so is not self-evident. Some systems can hardly adapt at all. Indeed, we must investigate the possibility that selection itself achieves the kinds of organisms which can adapt successfully. Therefore, we must also wonder whether there may be characteristic features so deeply requisite for the capacity to adapt in a coevolutionary process that their presence in organisms is itself a lawlike consequence of selection operating on complex coevolving systems.

While these points hardly seem contentious, it is no secret that we have, as yet, no theory which embodies them. Physics has its examples of remarkable order, but no use for natural selection. Biologists are secretly aware that selection must be working on systems which to one degree or another exhibit order by themselves. D'Arcy Thompson (1942) told us so with eloquence years ago, but we have not troubled to think through the implications. How strange, yet therefore how inviting, that we may one day bring ourselves to see life in a new light.

The major parts of the book discuss the following topics.

The introduction, Chapter 1, outlines our contemporary view of organisms, order, and evolution. Here we have been persuaded by Monod's (1971) evocative phrase, "Evolution is chance caught on the wing." And we are equally persuaded by Jacob's (1983) view that evolution "tinkers together contraptions." Here broods our sense of organisms as ultimately accidental and evolution as an essentially historical science. In this view, the order in organisms results from selection sifting unexpected useful accidents and marshaling them into improbable forms. In this view, the great universals of biology—the genetic code, the structure of metabolism, and others— are to be seen as frozen accidents, present in all organisms only by virtue of shared descent. The quiet sense that spontaneous order is everywhere present is itself not central to this view. Hence it is not stressed, not investigated, not integrated.

The first part of the book, Chapters 2 through 6, examines the power and limits of selection when acting on complex systems exhibiting spontaneous order, explores our first examples of self-organization, and proposes that the evolutionary marriage of self-organization and selection is itself governed by law: Selection achieves and maintains complex systems poised on the boundary, or edge, between order and chaos. These systems are best able to coordinate complex tasks and evolve in a complex environment. The typical, or generic, properties of such poised systems emerge as potential ahistorical universals in biology.

None can doubt Darwin's main idea. If we are to consider the implications of spontaneous order, we must certainly do so in the context of natural selection, since biology without it is unthinkable. Therefore, we must understand how selection interacts with systems which have their own spontaneously ordered properties. At a minimum, we must wonder whether selection in sufficiently powerful to obviate any inherent order in life's building blocks. If so, the order seen might reflect selection's dictates alone. Thus Chapters 2 to 4 consider the character of adaptive evolution under strong natural selection on mountainous "fitness landscapes," with high mountain tops representing peaks of fitness and ridges and deep valleys representing low fitness. We shall in fact find critical limits to the power of selection: As the entities under selection become progressively more complex, selection becomes less able to avoid the typical features of those systems. Consequently, should such complex systems exhibit spontaneous order, that order can shine through not because of selection, but despite it. Some of the order in organisms may reflect not selection's success, but its failure.

Much of the discussion in Chapters 2 to 4 focuses on adaptation in *sequence spaces,* such as among possible DNA or protein sequences, where we can conceive of evolution as carrying out *adaptive walks* toward peaks that represent how well proteins perform specific catalytic or ligand binding tasks. Consideration of the evolution of proteins able to carry out new catalytic functions, in turn, leads to the abstract concept of a catalytic task space. Among the implications of such a space is that about 100 million roughed-in enzymes might constitute a universal enzymatic toolbox able to catalyze almost any reaction. The immune repertoire of about 100 million may already be a first example of such a universal set. This possibility is not merely abstract, for Chapter 4 leads us toward practical implications as well. It is now possible to use genetic-engineering techniques to generate extremely large numbers of random or quasi-random DNA sequences, hence very large numbers of random or quasi-random RNA sequences and quasi-random proteins. Thus it is possible to explore sequence spaces for the first time. I believe this exploration will lead in the coming decades to what might be called "Applied Molecular Evolution" with very great medical and industrial implications, such as rapid evolution of new drugs, vaccines, biosensors, and catalysts.

Chapter 5 seeks the principles of construction in "parallel-processing" integrated systems of elements that allow the systems to adapt their behavior in a complex environment. We find two themes: First, the emergence of profound spontaneous order. Second, a bold hypothesis that the target of selection is a characteristic type of adaptive system poised between order and chaos. The unexpected spontaneous order is this: Vast interlinked networks of elements behave in three broad regimes—ordered, chaotic, and a complex regime on the frontier between order and chaos. The spontaneous order of the ordered regime foretells much of the order seen in aspects of developmental biology. The bold hypothesis states construction requirements which permit complex systems to adapt optimally through accumulation of useful mutations, even in a coevolutionary context where an adaptive move by one "player" distorts the fitnesses and the fitness landscapes of the coevolving partners. Ordered systems, particularly those near the edge of chaos, have the needed properties.

In Chapter 6, we see that the same construction requirements find echos at higher levels, such as whole ecosystems. Here the problem is to understand how such systems are coupled so that members coevolve successfully and how selection itself may achieve such coupling. Again, such ecosystems can behave in three broad regimes— ordered, complex, and chaotic. Again, remarkably, coevolving systems may optimize their capacity to coevolve by mutually attaining the edge of chaos.

The second and third parts of the book discuss other major examples of powerful self-ordering. In each case, the spontaneous order appears so impressive that it would be shortsighted to ignore the possibility that much of the order we see in the biological world reflects inherent order.

In the second part, Chapters 7 to 10, I discuss the origin of life. It requires no more words than this phrase to remember that we do not now know how life may have started. Any discussion is at best a body of ideas. The central problem is this: How hard is it to obtain a self-reproducing system of complex organic molecules, capable of a metabolism coordinating the flow of small molecules and energy needed for reproduction and capable of further evolution? Contrary to all our expectations, the answer, I think, is that it may be surprisingly *easy*. To state it another way, I want to suggest that we can think of the origin of life as an expected *emergent collective property* of a modestly complex mixture of catalytic polymers, such as proteins or catalytic RNA, which catalyze one another's formation. I believe that the origin of life was not an enormously improbable event, but law-like and governed by new principles of self-organization in complex webs of catalysts. Such a view has many implications. Among them, the template-replicating properties of DNA and RNA are not essential to life itself (although these properties are now essential to *our* life). The fundamental order lies deeper, the routes to life are broader.

Further, I suspect that the same principles of self-organization apply to the emergence of a protometabolism. I suggest that the formation of a connected web of metabolic transformations arises almost inevitably in a sufficiently complex system of organic molecules and polymer catalysts. This view implies that, from the outset, life possessed a certain inalienable holism. It also suggests that almost any metabolic web, were life to evolve again, would have a very similar statistical structure. Thus I find myself wondering if the web structure of a metabolism may reflect not the contingent consequences of this particular history of life, but some underlying ordering principles in biology.

These ideas are generalized in Chapter 10 to a new class of "random grammar" models which exhibit functional integration and transformation in coevolving systems, ranging from prebiotic chemical systems with protoorganisms to the emergence of mutualism and antagonism between simple organisms to similar features of economic and cultural systems. Grammar models are new testbeds for the locus of law in deeply historical sciences such as biology.

The third part, Chapters 11 to 14, examines the "genetic program" which controls cell differentiation during development of the adult from the fertilized ovum, and the machinery which yields ordered morphologies. The main intent is to suggest that many highly ordered features of ontogeny are not the hard-won achievements of selection, but largely the expected self-organized behaviors of these complex genetic regulatory systems.

The problem of cell differentiation, the focus of Chapters 11 to 13, is one of the two most basic issues in developmental biology. Different cell types—nerve, muscle, liver parenchymal—arise and *differentiate* from earlier cell types during development and, ultimately, in a human, form several hundred cell types. Each cell in a human's body contains essentially the same genetic instructions as all other cells. Those instructions include the structural genes coding for about 100 000 different proteins. Cell types differ because different subsets of genes are "active" in the different cell types. The activation and repression of genes is itself controlled by an elaborate regulatory network in which the products of some genes switch other genes on or off. More generally, expression of gene activity is controlled at a variety of levels, ranging from the gene itself to the ultimate protein product. It is this web of regulatory circuitry which orchestrates the genetic system into coherent order. That circuitry may comprise thousands of molecularly distinct interconnections. In evolution, the very circuitry is persistently "scrambled" by various kinds of mutations, as is the "logic" of the resulting developmental program.

In Chapters 11 to 13, I try to show that such properties as the existence of *distinct cell types,* the *homeostatic stability of cell types,* the *number of cell types* in an organism, the *similarity in gene expression patterns* in different cell types, the fact that development from the fertilized egg is organized around *branching pathways of cell differentiation,* and many other aspects of differentiation are all consequences of properties of self-organization so profoundly immanent in complex regulatory networks that *selection cannot avoid that order.* All aspects of differentiation appear to be properties of complex parallel-processing systems lying in the ordered regime. These properties may therefore reflect quasi-universal features of organisms due not to selection alone, but also to the spontaneous order of the systems on which selection has been privileged to act.

Chapter 14 treads D'Arcy Thompson's ground and considers the second fundamental problem in developmental biology: morphology. The actual morphologies of organisms must also be viewed as a collaboration between the self-ordered properties of physicochemical systems together with the action of selection. Oil droplets *are* spherical in water because that is the lowest energy state. The membrane of a cell, a bilipid structure, forms spherical closed surfaces because that is its lowest energy state. Other aspects of spatial order in organisms reflect *dissipative structures* rather like whirlpools, which require a continuous flow of matter and energy to maintain the form. Thus the genome's capacity to generate a form must depend on very many physicochemical processes constituting a panoply of developmental mechanisms beyond the sheer capacity of the genome to coordinate the synthesis of specific RNA and protein molecules in time and space. Morphology is a marriage of underlying laws of form and the agency of selection. The task is to find the laws and hallow the marriage.

I should make it clear that there are many fundamental problems in evolution and development which I have made no attempt to discuss. Most notably, the study of evolution has focused and will continue to focus on analysis of branching phylogenies, with related debates about the tempo and mode of evolution and the roles of natural selection and drift in the evolutionary process. In the best sense, this tradition studies this history of life. My aim in this book, nowhere in opposition to the familiar tradition, is to examine some new directions in which the occurrence of spontaneous order underpins this history of life.

I should also stress that, while the book is finished, it is not a finished book. Some of the subjects are familiar and can be discussed with a modest sense of completion. Others, however, constitute new areas of thought and investigation. Premises and conclusions stand open to criticism. If useful, I hope they are open to improvement.

THE ORIGINS OF ORDER

CHAPTER 1

Conceptual Outline of Current Evolutionary Theory

There are times in any science when one senses that a transformation to deeper understanding is pressing upward in some as yet poorly articulated form. We may be in such a period in biology. The recent achievements in molecular biology have brought us close to the molecular basis of evolution and ontogeny. Yet the organization which stands revealed is puzzling in its vast complexity. It is the hallmark of our present understanding of organisms to see this complexity as a result of historical accident, due to random mutations and to natural selection, which cobbles together jury-rigged contraptions.

It is the aim of this chapter to explore the genesis of the contemporary view of organisms as mixtures of contraption and design under the aegis of natural selection. I begin with the pre-Darwinian Rational Morphologists, who believed they were examining an unchanging, fixed set of species and sought *ahistorical laws of form.* This look at the Rational Morphologists leads us into the onset of the theory of evolution, with its branching phylogenies and Darwin's central concept of natural selection. Evolution, while destroying the idea of fixed species, simultaneously swept away the impetus to seek ahistorical laws of organic form. Further, the concept of branching phylogenies brought with it an emphasis on "population thinking" rather than the kind of Platonic "typology" that underwrote the Rational Morphologists' efforts. Evolution, branching phylogenies, and natural selection in turn engendered our current picture of organisms as opportunistic, ad hoc solutions to design problems wrought by selection and random genetic drift. The second section of the chapter considers familiar and less familiar criticisms of our present view, setting the stage for the rest of the book.

THE EMERGENCE OF THE NEO-DARWINIAN SYNTHESIS

The Rational Morphologists and Laws of Form

Publication of Darwin's book (1859) changed the agenda of biological science. So deep has the transformation been, that the serious purposes of pre-Darwinian biologists have receded from our collective scientific mind, not merely by passage of time but also by passage from relevance. Yet the aims of pre-Darwinian biology were fully sensible and are important to recall here in order to trace the development of contemporary evolutionary theory and place it in a broader context.

Late-eighteenth- and early-nineteenth-century biology was faced with the concept of *fixed, unchanging species.* It is difficult for us now to grasp the questions which arose naturally given this fundamental assumption. If organisms do not literally evolve into one another, as we now believe, if they are fixed forever, what account are we to make of the obvious and profound similarities among organisms? Granted that those similarities were taken to be God's work, nevertheless one wished to suppose a rational creator (Newton's laws were taken as signs of God's law immanent on earth). The overwhelming similarities between organisms, later seen as evidence of common descent by Darwin, seemed instead to harbor the possibility that there might be underlying simple laws of form which a rational mind might apprehend. It was entirely rational for such biologists to focus on comparative anatomy. For such Rational Morphologists as Goethe, Cuvier, and Geoffroy St. Hillaire, despite their differences (Appel 1987), the guiding interest was a search to find some underlying logic or laws which would let us understand similar organisms as variations on some simple mechanisms that generate living forms.

Behind this search was a coherent philosophical tradition. Webster and Goodwin (1982) note that the benchmark view of organisms in the Enlightenment was set by Kant, who undertook to distinguish organisms from mechanical devices. For Kant, organisms were fundamentally self-reproducing, and therefore *self-organizing wholes.* In a mechanical device, the parts exist only *for* one another in that each is the condition of the others' functions toward a common functional end. In contrast, in an organism, the "parts" exist both for one another and *by means* of one another. For Kant, an organism "is that in which everything is both a means and end" (Webster and Goodwin 1982; for a contrasting account, see Lenoir 1982). While this point of view might seem mere common sense, we shall see that it has dwindled from an operant role in contemporary biology. Within the Kantian view, and more deeply embedded in the tradition of science from Newton, lay the ideal of accounting for the diversity of superficially heterogeneous phenomena on the basis of relatively few underlying universal principles.

Comparative anatomy leads naturally to a systematic taxonomy. One result of this natural progression, Linnean taxonomy, remains with us today. It is important to be clear that the intellectual aims of a systematic taxonomy were not merely to categorize organisms into nested clusters, but to discover the *natural orderings* and groupings among all living things. If similarities in morphology were to reveal what the proper laws of form might be, then proper clustering of organisms into a system was critical to seeing the proper similarities, hence finding the laws. To this end, Linnean taxonomy sought knowledge of the natural order, the "Plan of Creation," linking natural kinds in a hierarchical pattern. The aim was to find a kind of inherent necessity, the logic behind the apparent diversity.

The Rational Morphologists' research program to find that logic lay in the supposition, held first by Cuvier and Geoffroy and then by Owen, that organisms were built up by combinatorial variations of a small number of principles. The purpose of the research was to discover those principles. Hans Driesch (1892, 1894, 1908, 1914), the brilliant experimentalist, understood better than later writers the aims of the Rational Morphologists, when noting their goal compared with that of their Darwinian successors: they "sought to construct what was typical in the varieties of form into a system which should not be merely historically determined, but which should be intelligible from a higher and more rational standpoint." This purpose can be seen in, for example, the work of Geoffroy and later Owen on the homologies between the various forms of the vertebrate limb. All vertebrate limbs were seen as members of a common series, transformable one into another by either deletion of or alteration in the character or in the connections of one or more elements. The same effort can be seen in Reichert's work (see de Beer 1958) on the homology between the reptile jaw and the mammalian middle ear, where invariant relations can be found despite marked functional alterations. Homology between organisms and within organisms appeared to point toward a rather small number of common patterns, recurrently deployed in different species.

The concept of laws of form, which seems alien to us as biologists reared in the Darwinian tradition, is immediately comprehensible if we consider an analogy with crystals. All possible space groups for conceivable regular crystals—tetrahedrons, dodecahedrons, icosahedrons, and so forth—are now well known. Crystallographers have succeeded in finding laws of form ordering diverse and complex morphologies in terms of combinations of only a few underlying principles. Prior to the advent of evolutionism and branching phylogenies, biologists confronted by temporally fixed species were perfectly sensible in hoping for just those kinds of laws.

Four major conceptual strands lead from the Rational Morphologists to contemporary biology. Darwin's theory is the first. The other three are Mendel's discovery of the basic laws of transmission genetics, Weismann's concept of the continuous germ plasm from which each organism grows, and the advent of population genetics in the effort to account for Darwinian evolution by selection acting on single genes. We trace briefly each of these major strands as they lead to the current neo-Darwinian Synthesis.

Evolutionism, Branching Phylogenies, and Darwin

The onset of evolutionism brought with it the concept of branching phylogenies. The branching image, so clear and succinct, has come to underlie all our thinking about organisms and evolution. The roots of the branching idea are intriguing. From Greek and medieval sources came the concept of a linear, unbranched Chain of Being or, later, the Scala Naturae, linking lower to higher forms of life (Lovejoy 1936). As Stanley (1979) points out, Cuvier, an arch antievolutionist, played an unwitting role in clipping links in the one-dimensional chain of being. In recognizing four major waves of creation, corresponding to the Precambrian, Paleozoic, Mesozoic, and Cenozoic, during which increasing complex but fixed species were formed, Cuvier recognized four major branches in the Chain of Being. Leibniz and Geoffroy, as well as others before Darwin, considered branching phylogenies, but with Darwin the image snaps into focus.

With the onset of full-blown evolutionism and Darwin's outlook based on branching phylogenies, the very notion that biology might harbor ahistorical universal laws other than "chance and necessity" has become simple nonsense. Darwin's ascension marks a transition to a view of organisms as ultimately accidental and historically contingent. Our purposes have become analysis of branching evolutionary paths and their causes on one hand, and reductionistic unraveling of the details of organismic machinery accumulated on the long evolutionary march on the other.

Darwin's development of the theory of evolution by natural selection after reading

Malthus's theory of population limitation is a familiar story. New conceptual systems such as Darwin's theory emerge as intellectual wholes. Once in place, the logical structure of the system inevitably begets a coherent set of questions. Given the view that species evolve into one another, then members of one species must somehow give rise to members of another species. It follows that members of the second species must somehow derive as variants of members of the first. A focus on variants emerges. Speciation must then rest on conversion of variation arising within one species to variation between species, hence from variation within some population to variation between populations. Thus the advent of Darwin heralds, as Mayr notes (1982), the transition from the typological thinking of the Rational Morphologists to the population thinking of contemporary biology.

The nub of the theory of evolution is as much branching phylogenies as natural selection. Natural selection is the force which "wedges new gaps in the economy of nature," yielding the "well-marked varieties" which are, for Darwin, species. Successive, unending, natural wedging yields branching speciation to form the tree of life. It is now branching speciation, the product of "descent with modification," which enjoins the systematists to harmonize their classes with the pattern of branching: Species, genera, families, order, and the other higher taxa cluster hierarchically because they express the pattern of branching speciation, not because they reflect a timeless and universal Plan of Creation.

Branching speciation generates further natural questions: Who branched from whom? What was the common ancestral species? What is the branching pattern among a given set of species? Did a given species evolve from one or several ancestral populations? Do similarities among organisms in related species reflect evolutionary homologies due to the propinquity of their descent from a common ancestor? Or are such similarities "analogies" arising by parallel adaptive selection in independent branches of organisms? How rapidly does phenotypic evolution within one species occur? What are the tempo and mode of branching evolution? Even though natural selection as the driving "force" underlying adaptation and speciation has been doubted by many biologists, questions such as these, based on branching phylogenies, have remained the unchallenged core of evolutionary theory.

A curious, logically unnecessary, but powerfully influential feature of Darwin's thinking was that the variation within one species which paved the way for emergence of well-marked varieties constituting two species was of an indefinite range. The idea that variations could occur in virtually any direction, an idea which dominates in Darwin's work despite attention to correlations among traits under selection, has had important conceptual consequences. It follows that selection alone can discriminate which new variants will be found in later generations. Here is one root of our current idea that selection is the sole source of order in the biological world.

Mendel's Atoms of Heredity

The second fundamental strand in our contemporary view stems from Mendel's discovery of transmission genetics. Consider Mendel as though he were a serious atomist, for atomism in chemistry, with its remarkable ratios of small whole numbers, was in the air. Imagine asking oneself whether there might be permanent, unalterable atoms of heredity. Like chemical atoms, such atoms of heredity might form many new combinations. Given two parents and the observation that offspring often resemble both parents, it becomes plausible that the offspring receive atoms of heredity from each parent. Each sexual organism has two parents, four grandparents, and

2^N ancestors in the Nth prior generation. If atoms of heredity pass into a member of the present generation from all prior generations, then two possibilities immediately arise. Either all the atoms from the past cumulate in each generation, which leads to an overabundance of atoms of heredity, or some form of steady state is reached in which each organism has a constant number of the hoped-for hereditary atoms. In the latter case, on average, the organism must receive a twofold dose from its two parents and, to maintain the steady state, had best pass along half of its twofold dose to each offspring. In particular, the simplest hypothesis must be that, for each property of the organism, it receives one atom from its mother and one from its father, then passes along one or the other to each offspring. But then the wonderful consequence: If atoms of heredity are truly permanent and can combine and reemerge, their passage through first- and second-generation offspring should be traceable, and simple analysis suggests that small-whole-number ratios of effects should be found. Mendel's peas.

It is not overly important whether Mendel actually thought in just this pattern (Olby 1979). Rather, this intellectual reconstruction, also attributed to Mendel by Fisher (1930), helps exhibit the historical context and natural reasoning by which Mendel's discovery harmonizes with the ideas of his time.

Mendel was obviously no modern academic scientist. He sent his work to some of the best botanists of his day, even to Darwin, with almost no response. He made little further effort to attract attention. The story of the rediscovery of Mendel's laws at the turn of the century and their linking to the observed movements of chromosomes is part of the folklore of biology. Interestingly, it is one of the clear cases where a theory predated and helped decisively to identify a functionally important cellular component. Recognition that chromosomes might be the carriers of hereditary information rested in no small part on the parallels between their behavior in meiosis and that required by Mendel.

Weismann's Doctrine of the Germ Plasm: Toward the "Genetic Program"

For Kant and the Rational Morphologists, as we noted earlier, an organism was a self-organizing entity whose parts existed for one another and by means of one another. That is, an organism was seen as both a structural and a functional whole. Early in the nineteenth century, this Enlightenment ideal was challenged by the idea of historical or developmental science (Cassier 1950). The question became, How are organisms generated, both in the history of the species and in the history of the individual?

In their tentative reconstruction of the intellectual transitions that occurred during the early nineteenth century, Webster and Goodwin (1982; and see Lenoir 1982 for a more detailed account of German biology of the period) note that the capacity of mechanical concepts to account for organismic development was an abject failure in the eighteenth century. Therefore, some nonmaterial controlling agency might be at work. The conceptual lineage may run something along these lines: Kant, in the *Critique of Judgment,* discusses the necessity of using teleological principles if biological organization is to be made intelligible. The possibility of a teleological cause has analogies to human planned action, where the idea of the effect becomes a condition of the cause. Reified, the "Idea" becomes a distinct cause separate from mechanical causes, either transcendent or immanent. As Goodwin and Webster point out, in this development of German Romanticism the organism is seen for the first time in modern biology as a dualistic entity—a material substance which is reactive to and expressive of a central directing agency, the Idea. The earlier view of an organism as a self-generating material whole is gone.

In our own time, this dualism has not been lost but only washed of its transcendent character. It finds a material form in Weismann's doctrine of the continuity of the germ plasm passed from parent to offspring. In each offspring, the germ cells were to control the development of the organism but not themselves be part of the body, or soma. Weismann argued persuasively that the problem of inheritance is not a question of how the structure of the parent is transmitted to the offspring, but is about growth and development. Offspring resemble parents because both are the results of identical processes of growth and development, located in a distinct structure, the germ cell, which contains a specific substance with a "highly complex structure," the germ plasm, that has "the power of developing into a complex organism" (Weismann 1895, 1904). In turn, germ cells are derived not from the body of the individual but from parent germ cells, whose substance has remained in perpetual continuity from the first origin of life (Weismann 1895, 1904). The organism therefore is not a self-organizing whole but an expression of the commands of the central directing agency of the germ plasm.

The theme of a central directing agency mediated by the germ plasm has grown into the concept of the genome controlling development. We now conceive of the central directing agency as a genetic blueprint, or *genetic program.* The concept of a genetic program continues to play a profound role in the thinking of developmental and evolutionary biologists. I will discuss it briefly below and in great detail in Chapter 12.

The Core Paradigm and Population Thinking

Our contemporary view of biology is truly the marriage of the views of Darwin, Mendel, and Weismann. With the rediscovery of Mendelian transmission genetics and the tentative identification of chromosomes as the significant subcellular entities which constitute the active substances in Weismann's germ plasm, passing from generation to generation and directing the growth of each organism, the central conceptual structure is in place. That structure culminates in contemporary molecular biology, for the natural questions which emerge are familiar ones. What complex chemical substances are the carriers of Mendelian genes? How do genes accomplish their transmission? How do such genes become expressed as traits in the offspring? Answering these questions has led to the elucidation of the structure of DNA and the genetic code, the expression of structural genes as specific proteins playing catalytic or structural roles in the developing organism, and the concepts of regulatory genes and "cybernetic" genetic regulatory circuits governing patterns of gene expression. It is here that the program of reductionism in biology has been most profoundly successful.

The complement to this reductionistic strand in cellular biology has been the insertion of population biology in evolutionary theory. Mayr (1982) is correct in noting that the triumph of contemporary biology was attained by replacing typological thinking by population thinking. The approval expressed by Mayr is not only positive, in favor of the analysis of population dynamics, but also negative, in favor of cleaving away nonsensical medieval ideas about "types" based on an outmoded essentialism. As noted, population thinking has been central to the analysis of branching phylogenies in Darwinian terms, where speciation is the emergence of well-marked varieties from the variation present within an initial population. And in particular, population thinking is central to the emergence of population genetics.

Population genetics is the fourth major strand leading to the current neo-Darwinian Synthesis. It is interesting that population genetics grew out of recognition, by the early decades of this century, that Mendel's laws of particulate inheritance solve the deep problem Darwin faced because of his concept of blending inheritance: loss of variability in the population. According to blending inheritance, offspring blend the traits of their parents much as two colors—yellow and blue, say—blend into a single color—green, in this case—thereby losing the variability of two original colors. In Mendel's theory, the persistent atoms of heredity combine and reshuffle in each generation and pass unchanged to subsequent generations, permitting variability to be maintained. This basic result was obtained by the famous mathematician Hardy at Cambridge University, as well as by Weinberg independently.

While the Hardy–Weinberg result helped explain Darwin's theory, it did not suffice, for scientists still had to show that selection could lead to evolution as Darwin imagined. It is the effort to establish this point which led to the foundations of contemporary population genetics. The question posed by Fisher, a young Cambridge mathematician, was whether a single mutant gene which conferred a very slight selective advantage over the normal, or wild-type, gene and which arose in a single individual in a breeding population could spread throughout a population by virtue of the selective advantage it offered. If it could be shown that slight selection was adequate to substitute a more favorable for a less favorable allele, this would sustain Darwin's theory that slight selective advantages acting over long times could mold and remold populations (Fisher 1930). It should be clear that this question is genuinely the basic question of population genetics. The minimal microevolutionary step is conceived as the substitution of one allele for another in a population. General analysis of the conditions under which such a substitution can be expected to occur has formed the core of the powerful body of theory now available. This body of theory, united with paleontological evidence and experimental transmission and developmental genetics, forms the core of the contemporary neo-Darwinian Synthesis.

The Neo-Darwinian Synthesis and the Neutralists

The neo-Darwinian Synthesis (exemplified by Dobzhansky 1937, 1970; Simpson 1944, 1950, 1953; Mayr 1942, 1982; Mayr and Provine 1980; Stebbins 1950), drawing on Fisher (1930), Haldane (1932), and Wright (1931, 1932) (see Provine 1971, 1986), has been extremely successful. It has sought to understand the conditions under which slightly advantageous mutant alleles at one or several genes might invade a population. Analysis has concerned the influence of population size, the selective advantage of the allele, and the relative fitness of homozygotes and heterozygotes for the different alleles of a gene. The Synthesis, in its population-genetics arm, has examined the effects of linkage of many genes on one chromosome and the effects of recombination between maternal and paternal chromosomes, which tends to disrupt favorable combinations of alleles at different loci on the same chromosome. The conditions favoring the emergence of sex in evolution have been studied. All the vast panoply of phylogenetic variation, the tempo and mode of evolution seen in the paleographic record was, it was hoped, to be brought under the umbrella of natural selection acting on individuals, substituting one allele for another over evolutionary time.

Despite its wide range and apparent success, the Synthesis has come under attack.

Undoubtedly the most successful attack has challenged the neo-Darwinian view that most or all evolutionary change is driven by selective advantage. Sewall Wright, one of the three pioneers of population genetics, emphasized the role in evolution of small peripheral populations that are largely isolated from the main body of a species (1931, 1932). In small enough isolates, the particular sample of the gene pool normally shared by the entire population which might be present in the isolate can be sharply divergent from the average of the entire pool. In this way, a gene which has a low frequency in the population as a whole can have, by chance, a high frequency in the isolate. Consequently, the evolutionary dynamics can lead to that allele's displacing all alternative alleles of the gene simply by random chance or, more formally, by random drift. (For the roles of geographical isolation and founder effects, see also Mayr 1942.)

The *neutral theory* (Kimura 1983) is the strongest statement of the role assigned to random drift. In its strongest formulation, the neutral theory asserts that all or most evolution at the molecular level is due to random drift among *selectively neutral genetic variants,* even in large populations. This theory has unleashed an enormous amount of energy in an attempt to understand the extent to which the evolutionary substitution of one allele by another is, or is not, due to selective differences or to chance fluctuations. It is not the purpose of this book to enter into the debate between the neutralists and the selectionists. However, some of the material in Chapters 2 and 3 on the mountainous structure of fitness landscapes—itself an idea introduced by Wright—bears on these difficult issues.

In the next section I will close the chapter with further criticisms of the neo-Darwinian Synthesis. In the remainder of this section, however, I want to sketch some correlates of the Synthesis.

Four Conceptual Correlates of the Neo-Darwinian Synthesis

None of the four correlates discussed here need be taken as requisite to the logical structure of the Synthesis. Yet each is pervasive and plays a powerful role in guiding our thinking. More important, the first three of these patterns of thought harmfully constrain our thinking. Not surprising, of course, each contains an element of truth.

• The first correlate is that we have come to see selection as the sole source of order in organisms. This may slightly overstate our common view, but overstatement is meant to help clarify.

• The second is that we have come to think of the development of an organism as though that development were controlled by a "genetic program." Our problem, I shall argue, is that we wrongfully identify the concept of a genetic program with current serial algorithms. A better image of the genetic program—as a *parallel distributed regulatory network*—leads to a more useful theory.

• The third correlate is that we tend to see organisms, perhaps excessively, as ad hoc contraptions cobbled together by evolution.

• The fourth has been and remains a major useful research paradigm in evolution. It is the concept of developmental constraints in evolution. This theme is broached here and is the major subject of Chapter 14, on morphogenesis.

Selection as the Sole Source of Order.   Since Darwin we have come to view selection as the overwhelming, even the sole, source of order in organisms. Natural selection operating on gratuitous random mutations is the sieve that retains order and lets chaos pass into oblivion. This phrase is no understatement of our world view; it is its heart. No idea derivative from Darwin lies deeper in our minds than this: myriad mutations, selection sifting. Here rebels the "Creation Scientist," here cavil many, but here is the core.

The routes to and roots of the acceptance of this core belief in selection are interesting. Darwin was himself inheritor of the tradition of Natural Theology, a tradition in which organisms were considered to have been constructed by the agency of God. This tradition focused on the design of organisms, their intimate meeting, and the matching of their traits to their environments, all as evidence of a higher purpose and intelligence. Such a tradition is uninterested in the earlier Kantian idea of an organism as a self-organized whole, which, even if true, hardly accounts for the organism's fit to its environment. Powerfully inimical to the theological consequences, Darwin's notion of natural selection can be enthroned in God's stead as the creative agency. The conceptual structure is already in place; only a kind of governmental revolution and regicide are needed. Selection slips into place as an agency creating order from chaotic variation. (The change in governments has not been without detractors with a sense of humor. It was noted that making selection a creator of order was parallel to creating an apple tree by cutting away its branches, a point not without its merits.)

If we take selection as the *sole source of order,* it is because we have come to suppose that without selection there could be only chaos. The roots for our continuing strong adherence to this core Darwinian view are probably twofold. First, as already noted, for Darwin and many biologists, the initial idealization that variation can occur in any direction, known by all biologists to be literally false, is too useful a simplification to give up. Even should there be some constraints, the open texture of possibilities looms so vast that the idealization was and remains a reasonable starting point for thinking. Further, since many variations lead to loss of function, some selective processing maintaining function must be supposed. Even deeper than this set of ideas, which have long appealed to biologists, lies the second root for our belief in the Darwinian view: our present intuitions derived from the development of the science of statistical mechanics, the idea of entropy, and the second law of thermodynamics, which states that, without work, the internal disorder of a system always increases. Intuition based on the second law, I think, has now become the deeper root. Left to themselves, systems are inherently disordered and unstructured. Should some order be created and the ordering efforts cease, then lapse to disorder will occur. Therefore, selective "work" is necessary to achieve and maintain order.

Monod's (1971) book captures the role of selection in maintaining order in the biological world, but it also explores evolutionary freedom. For example, Monod notes that metabolic regulation is achieved by modification of allosteric enzymes through binding of the regulatory molecule at a site other than the site of enzyme activity. The consequence is that, from the chemical point of view, the regulatory molecule need bear no resemblance to the substrates or products of the enzyme. This freedom finds its expression in the capacity of selection to construct rational metabolic circuits in which the terminal product of a pathway can feed back to the first unique step on that pathway and inhibit the first enzyme if the level of the terminal product becomes too high. This leads to arbitrariness, freedom to vary without constraints, selection, and design. ·

The same freedom occurs at the level of genetic regulation. An example is the famous lactose operon in the bacterium *E. coli.* Control of synthesis of the messenger RNA coding for the proteins involved in lactose metabolism is mediated by binding of a repressor protein to the DNA *operator site* blocking transcription. Removal of the repressor from the operator occurs by a metabolic derivative of lactose, allolactose, which binds to the repressor. No chemical necessity requires that it be allolactose which binds the repressor protein. The genetic-circuit feedback might have been controlled by some other metabolite in the cell. Therefore, Monod argues that the structure of regulatory circuitry is arbitrary from the chemical point of view. Nor is the *logic* of the genetic circuit constrained. The lactose operon works by repressing transcription of the lactose-metabolizing genes through a protein called the repressor, which is constantly being produced. Transcription is activated by "derepression" when allolactose is present. But genetic regulatory circuits with different "logical behavior" occur. Some genes are normally inactive and positively regulated to switch on in the presence of a specific inducer. Again, the inducer need bear no chemical similarity to the chemicals upon which the products of the regulated genes subsequently act.

The chemical freedom to construct arbitrary genetic circuits is but one expression of evolutionary freedom. That freedom then relies upon selection to construct *useful* genetic circuits.

The "Genetic Program": Sequential Turing Machine or Parallel Computing Network?   Discovery of the operon helped introduce cybernetic phraseology into biology, and the image of genes turning one another on and off is now coupled to the central notion of a genetic program underlying ontogeny. Since Boole formulated logic in his binary laws, it has become clear that with a small number of Boolean functions—in particular, "No," "Or," and "If"—it is possible to derive all of logic. A critical implication of the completeness of "No," "Or," and "If" was the discovery by Turing of universal computing systems. Such systems are able to carry out any algorithmically specifiable computation. This universal computing capacity is the logical basis of general-purpose computers, which can perform, in principle, any well-specified algorithm. Thus the fact that regulatory relations in metabolic pathways and in control interactions among the genes themselves can be arbitrary and gratuitous from the chemical point of view has been taken to imply that the cybernetic system within an organism—in other words, the genetic program of the organism—can in principle be rearranged to compute any algorithm, to behave in any way.

From universal computation we have been led to our loose contemporary idea of a genetic program underlying the ontogeny of each organism. The genome is pictured as a kind of computer specifying the structure of the chemical elements, the RNA, and proteins and the regulatory interactions among those components. Like a familiar computer program, the genetic program unfolds coherently in development (see, for instance, Apter 1966). The intellectual strands leading back to the idea of development controlled by a central directing agency are clear.

Two features of programs in universal computers fit with exquisite precision into our view of organisms. First, reprogramming allows the computer to compute any arbitrary algorithm. Hence in principle we find ourselves thinking that with adequate reprogramming the genome could generate an arbitrary diversity of organisms. Second, most programs are devastatingly fragile to minor variations in the instructions. This strengthens our supposition that organisms are precise and must be maintained against chaotic degradation by selection. Conversely, we know this is an overstatement; organisms can suffer substantial variation and still function. Somehow genetic programs must be both free to vary widely and buffered against catastrophic failure for many minor changes. As we shall see, the probable answer to this seeming paradox lies in the fact that the term "genetic program," if it points to anything, points to a parallel-processing genetic regulatory network which can exhibit self-organized buffered behavior.

Organisms as "Locked-In" Historical Contingencies: Rube Goldbergs.   It is no accident that we have come to view organisms as historical accidents. Nor is this view due merely to the recent advent of the Neutral theory; rather, it is rooted in the utter blindness and gratuity of mutations, their arbitrary randomness with respect to prospective usefulness.

Current theory affords us only two ways to think about any phenotype—from protein structure to limb morphology: random drift and selection. To selection we couple some insight into design: Wings must meet aerodynamic criteria, legs must confront the law of the lever, and so forth. But within enormous latitudes, we understand the branching flow of phylogenies as either natural selection or drift acting upon myriad mutations, most of which are harmful, such that the more or less rare successes are accumulated in any phylogenetic lineage. Evolution is thus seen as an opportunist, remolding hard-won successes for novel uses. As a consequence, the results of an evolutionary flow have an historically contingent character, a somewhat ad hoc accumulation of accidental successes culled over the eons.

The coordinated, gratuitous, ad hoc character of organisms is captured by humorist Rube Goldberg, whose wonderful cartoons appeared in the post–World War II period. Consider the design of a device to ring a bell: Father in the wing chair allows tea to fall from his cup to the floor, inciting the cat to run through the resultant puddle and leave a trail of conducting moisture to the terminal of a strategically located battery. The current loop is completed, activating a small motor, which moves a lever pulling a string releasing a pendulum latched to the shelf and allowing it to swing against the gong hung from the Tiffany lamp nearby. Beyond the charm of his style, Goldberg's ad hoc machinery demonstrates a basic principle. Once the components are assembled and once the system works, the system is an integrated whole. Removing or sharply changing any component will probably lead to failure. That is, solutions, once found, are more or less locked in.

This theme in contemporary biology is not implicit; it is fully explicit. The common current view of the origin of the genetic code holds that the particular code now used, and its minor variants, rather than being preordained by chemical affinities, are a "frozen accident" (Crick 1968). We might have had different codes, but once our code is in place, altering any single codon might be expected to alter so many proteins that its effects would be lethal. Similar arguments are presented for the chirality of metabolites and amino acids. The choice of dextro- or levorotary, once made in evolution as an historical accident breaking a symmetry, was frozen in place. The well-known evolutionary deployment of the jawbones of reptiles to the middle ear of mammals exemplifies the theme of gradual molding of a hard-won success. Our sensitivity to the frozen accidents of evolutionary history finds its expression in the familiar sense of biology as a deeply historical discipline.

There is no doubt that our awareness of historical contingency is proper. The question we must address is whether there might be statistical order within such historical processes. A loose analogy makes this point. Imagine a set of identical round-topped hills, each subjected to rain. Each hill will develop a particular pattern of rivulets which branch and converge to drain the hill. Thus the particular branching pattern will be unique to each hill, a consequence of particular contingencies in rock placement, wind direction, and other factors. The particular history of the evolving patterns of rivulets will be unique to each hill. But viewed from above, the statistical features of the branching patterns may be very similar. Therefore, we might hope to develop a theory of the statistical features of such branching patterns, if not of the particular pattern on one hill.

Evolutionary biologists are well aware of this kind of analogy and have constructed many such statistical theories with respect to branching phylogenies (Gould, Gilinsky, and German 1987), ecosystem stability (May 1973), and other areas.

Evolutionary Constraints in the Darwinian–Mendelian Tradition.   Biologists have recognized at all times that evolution is constrained (Bateson 1894; J. M. Smith, Burian, et al. 1985), and the supposition that variation might occur in any direction has been well criticized. Even though Darwin considered internal correlations within the organism, such that selection for one feature would pull along other, correlated features, he did stress the capacity for variation in almost any direction, and it is fair to say that much of late-nineteenth-century and early-twentieth-century biology was concerned with refuting this claim. If evolution is branching phylogenies converting variation within populations to variation between populations, then evidently a thorough analysis of the patterns of variability thrown up by any population will reveal the avenues available to evolution. Constraints on those patterns of variation are constraints on branching phylogenies.

The question under analysis in asking whether constraints exist in evolution should be made explicit. It is to ask whether there may be restrictions in generating *neighboring* forms or organisms, given that the process is starting at a specific point— in a given species or in a specific member of the species. Restated, the major idea of a constraint in our tradition is a *local* constraint on transitions between neighboring forms.

Analysis of phylogenies in terms of such local constraints has a rich and successful history. Two of the major successful strands are considerations of heterochrony and allometries.

Heterochronic studies posit that changes in the time of onset of particular developmental pathways in an organism, changes in the rate of progression of such developmental processes, and changes in their time of cessation can change the resulting organism. If, for instance, limb development were initiated earlier in one species and lasted longer in that species than its neighbors, limbs in the first species would be expected to be longer. Such alterations in the timing of growth and differentiation are clearly a way to deform one organism to a closely *neighboring* organism. More dramatic alterations can be achieved as well, as when, for example, a juvenile form distinct from the adult form becomes sexually mature. The well-known example of the axolotl, in which the juvenile form is sexually mature without undergoing metamorphosis, comes to mind.

Recent studies in this tradition have been carried out by Alberch and his colleagues (Alberch 1980, 1981, 1982; Alberch and Alberch 1981; Alberch and Gale 1983; Alberch, Gould, et al. 1979) on several varieties of newts, comparing the interspecific variation in loss of digits with the order within one species with which digits are lost when development of the limb is slowed. The patterns of variation are highly similar, strongly supporting the thesis that evolution has made use of the internal patterns of variation and achieved the transformation of digits in neighboring species by playing on developmental rates.

Allometric transformations, studied by Thompson (1942) in his analysis of coordinate transformations mapping one form into a family of neighboring forms, and named and studied in detail by Huxley (1932), are in the same tradition as analysis of heterochrony. It is interesting that the Darwinian idea that *absence* of change requires no selection but reflects simple heredity reemerges in the analysis of allometric patterns, but in a hidden form. Consider a series of species with correlated alteration in the size of two elements, perhaps body weight and antler spread. It may be discovered that a single allomorphic transformation, giving the rate of increase of antler spread with respect to change in body weight, yields a curve onto which all the species fall but with some scatter about the curve. Then the analysis will often be interpreted to mean that the allomorphic transformation reflects internal constraints, while the deviations from the curve reflect selection. Note, then, that this interpretation supposes that no selection is needed to ensure the constraint itself. As was Darwin's hope, the constraint persists by some sort of inertia.

Hints of a Structuralist Paradigm

The analysis of patterns in neighboring organisms which we have discussed can be based on the idea of the organism as generated by a developmental algorithm. Then our purpose becomes one of assessing what the generative algorithm—more properly, the developmental mechanism—is and consequently what the truly neighboring organisms are. Consider phyllotaxis, conveniently seen in pine cones and sun flowers. The scales, as is well known, form in double spirals which radiate from a center, one clockwise, the other counterclockwise. The surprising feature is that the number of spirals in one direction is related to the number in the other direction as two adjacent numbers in the Fibbonacci series 1, 1, 2, 3, 5, 8, 13, 21, 34. . . .

A number of models to account for this double-spiral pattern have been suggested; an example by G. J. Mitcheson (1977) is attractive. Mitcheson suggests that tight packing of scale primordia on the conical meristem suffices to generate the observed phyllotactic series. The issue here is not the adequacy of his model, but the mode of analysis it exemplifies. First, given the presumptive developmental mechanism, we have an algorithm to generate a given form and, by modifying parameters of the model, a way of computing and predicting neighboring forms. Thus the developmental mechanism predicts a family of forms it will generate. Second, unlike the cases of heterochrony and allomorphic transformation just noted, where a continuous gradation of neighboring forms occurs, here true neighbors are adjacent pairs in the Fibbonacci series: either 8-13 or 13-21. Thus *true neighboring morphologies in evolution reflect transformations to neighboring forms in the family of forms generated by the underlying developmental mechanisms.*

Goodwin and Trainor (1983) have stressed the fact that analysis in terms of developmental mechanisms and the families of forms they can generate is a structuralist pattern. As with the pre-Darwinian Rational Morphologists, one attempts to find an underlying algorithm, or law of form, which relates a *family* of different morphologies as members of a *series which can transform into one another.* In such a view, the organism cannot vary equally in any direction but is strongly biased by the underlying developmental algorithm to vary among a defined set of neighboring morphologies. This point of view has been expressed by a number of authors (Alberch 1980, 1982; Alberch, Gould, et al. 1979; Oster and Alberch 1982).

The case of phyllotaxis is just one of a number to which we will return in Chapter 14. In thinking about the evolution of organisms, we would dearly like to know the extent to which orderliness in such developmental programs can have origins largely *independent of selection,* and what roles selection has in maintaining or furthering that order.

ENLARGING THE FRAMEWORK

In the previous section, I sketched the genesis of the main strands in the Darwinian–Mendelian paradigm which led to the neo-Darwinian Synthesis. In the present section, I cover three general topics.

The first topic is a brief review of familiar criticisms of the neo-Darwinian position. These critiques are largely parenthetical to the main issues of this book and are included merely to orient the general reader to the present position of the neo-Darwinian view.

The second topic is the difficulty that evolutionary theory has in accounting for the *evolutionary origin of complex "wholes."* The specific example I discuss is the origin of life itself. Creditable arguments by respected scientists have led to the unfortunate conclusion that we cannot exist. In my view, these arguments fail by failing to utilize the self-organized collective properties of simple and complex systems. Instead, these arguments hope to find answers to origins in the length of time available for evolutionary search. The origin-of-life argument I examine is important precisely because plausible calculations demonstrate that insufficient time has elapsed for life to have originated by chance without such self-organization. This single example therefore is a harbinger of an alternative hope: If we understand self-organization, many such origin questions in evolution may become understandable. This brief discussion also serves to introduce the entire origin-of-life issue examined in later chapters.

The third topic introduces part of the framework I want to use to think about the relation between selection and self-organization. The essential idea is simple. It is to think of selection as acting on systems that spontaneously exhibit some particular form of order that is *typical of an entire class of similar systems,* called an *ensemble.* Selection can be thought of as moving a "population cloud" to particular parts of that ensemble. The *balance* between the self-organized properties typical in the ensemble and selection then depends upon the extent to which selection can move the population cloud to parts of the ensemble which no longer exhibit the typical order. The critical point we shall find is this: *In sufficiently complex systems, selection cannot avoid the order exhibited by most members of the ensemble.* Therefore, such order is present not because of selection but *despite* it. This implies that the kinds of collective self-organization which we will discuss in the remainder of the book can reasonably be expected to account for some of the order exhibited by organisms.

Familiar Criticisms

A number of criticisms against Darwinism and the neo-Darwinian Synthesis have been raised. I discuss next eight such issues to help orient the general reader.

Is the Concept of Natural Selection Circular?   First among the criticisms is the concern that the theory is circular, a mere tautology: the survival of the survivors. In its neo-Darwinian reformulation, in which the microevolutionary event is a change in gene frequencies, the worry is restated as the circularity of defining natural selection as the differential reproductive success of genotypes and then defining the fitness of a gene as its average probability, over all genetic backgrounds, of being propagated to the next generation. This restatement again leads to the possibility of circularity, survival of the survivor.

The criticism has been answered by attempts to show that fitness is logically distinct from the tautology and has reference to design traits in the organism (Sober 1984). It is of some interest, however, to consider a different response—that such circularity is normal at the core of many sciences. For example, a very similar circularity is found at the core of Newtonian mechanics. Newton's second law of motion states that $F = ma$. Acceleration is independently defined as measurably changing velocity. But force is taken as a primitive, and inertial mass is defined as the proportionality constant relating force to acceleration. Mass has no definition independent of force; a force is that which, acting on a mass, accelerates it. The couplet of terms is defined in a circle; each requires the other. In fact, this circularity lay behind Poincaré's conception of fundamental laws as definitional conventions.

The reasons for such circularity are, in fact, not strange and are related to a necessary holism in science pointed out by the philosopher W. V. Quine, who noted that no hypothesis confronts the "world" alone (1961). Instead, it confronts the world as part of an entire world view of linked hypotheses plus statements about the experimental situation. Given a negative experimental result, something must be rejected. Either some hypothesis is wrong, or a description of the experimental situation is incorrect. But, as Quine pointed out, the choice of which hypothesis to reject is a *free one*. As he notes, we can salvage the hypothesis we want. Different choices of which hypothesis to reject impinge on the whole web of hypotheses and laws. To maintain coherence in that web, we typically choose to salvage a central *circularly interdefined cluster* of hypotheses. In effect, we treat this central cluster as unfalsifiable by insisting that it remain true. But this circularity is not merely definitional, with no grip on the world, so long as the circle—or better, web—is reasonably large. Thus although a detailed argument in Quine's direction of conceptual "holism" has not to my knowledge been worked out for Darwinism, it seems reasonable that the sense of circularity in Darwinian theory is no more malign than in other areas of science.

Panselection. An important argument has developed about panselectivist interpretations of Darwin (Gould and Lewontin 1979). The conceptual basis of this interpretation is that, in order to be present in organisms at all, any trait must be under direct or indirect positive selection. The issue is this: The idea of natural selection implies that advantageous variant features will tend preferentially to be passed on if there is heritable variance for those features. But Darwin and many others have noted that traits are *correlated;* therefore, a deleterious trait might persistently be selected in evolution because of its coupling to some positively useful trait or traits. This argument appears to sustain the logical coherence of a panselectionist position. Among the best arguments against this view is the entire body of evidence amassed by the Neutralist school (Kimura and Ohta 1971; Kimura 1983). As noted on page 10, the Neutralist claim is that many features, at least at the molecular level, persist because they are selectively neutral.

Restricted Selection. The more restricted interpretation of Darwinism has been that some but not all features of organisms are present as a result of positive selection. But this sensible application of Darwinisn has always faced the difficulty of discriminating which features were present even if they were not favored by selection, and has always faced the danger of facile constructions of "just-so" stories plausibly positing a use for a feature in the face of no possible tests at all. This has truly been one of the major problems in evolutionary biology, for most biologists, myself included, would deny a panselectivist view but hold to the claim that many aspects of organisms are present due to past or continuing selection. The overall problem here is great. We do not doubt that, for example, the eye has been selected to enable distant vision because such vision makes a significant contribution to organismic fitness, but in innumerable concrete cases it has proved elusive to establish what the function of a given structure or property is. If selection alone is to account for the order we find and if we cannot genuinely determine that which is selected, then our account of the order in organisms stands in peril of weakening into a formal explanation whose validity can rarely be ascertained in concrete cases.

The Abundance of Variation.   The unexpected abundance of variation in the amino acid sequences of proteins segregating in natural populations has posed certain problems for the Darwinian view of evolution. The supposition that most mutants are harmful has been correlated with the assumption that organisms must be highly precise entities in order to survive. In turn, this assumption has been linked to the widespread belief earlier in the century that organisms are nearly homozygous for wildtype genes; that is, that organisms will have the same normal copy of each gene on the maternal and paternal chromosomes. Thus the famous geneticist H. J. Muller (1950a, 1950b) reasoned that members of a species would be nearly homozygous and was concerned with genetic purity and the accumulation of defective mutants in human populations. With the advent of protein gel electrophoresis and the discovery of electrophoretic protein variants due to amino acid substitutions in the proteins (Hubby and Lewontin 1966; Lewontin and Hubby 1966), it became possible in the 1960s to measure protein variants in the population. Very large variability was found (Lewontin 1974). A large fraction of the genes segregating in a population have more than one allele, and many individuals carry different alleles on their maternal and paternal chromosomes (in other words, many individuals are heterozygotes) at a large fraction of their genetic loci. These studies are now being extended to the DNA level, where widespread polymorphism has similarly been described.

High levels of variability lead to the question of how so much variability is maintained. For Neutralists, it can reflect the birth rate of new mutants, their rate of random drift through the population, and their rate of either fixation in or loss from the population (Ewens 1979; Kimura 1983). For those who maintain a selectionist view, maintenance of high levels of variability is taken to reflect selection in favor of the heterozygotic condition at each such locus (Lewontin 1974). Whether such selection is biologically plausible and adequate to account for the observed variability has been sharply debated (Kimura 1983).

One implication of the discovery of so much variability in a given population is that the idea that organisms must be precise in order to function at all needs to be rethought. As we shall see in Chapters 11 through 14, the variability uncovered in enzyme polymorphisms is beginning to be extended to the cybernetic aspects of the control systems by which genes mutually regulate one another's activities. It therefore becomes important to assess how much regulatory variability exists and, even more profound, how perhaps imprecise genetic programs function adequately. This topic is returned to in those chapters, where the stable self-organized structures and behaviors of genetic regulatory systems are examined.

Epistemological Adequacy of Population Genetics.   The epistemological adequacy of population genetics has been questioned. At its very inception, population genetics began by ignoring the organism and ascribed fitness to a given gene. The necessity for making this simplification in order to construct a theory was clear and in fact acceptable, given the initial question for which the theory was constructed: Could a gene which conferred a slight selective advantage spread through a population and displace the former wild-type gene? While it may or may not be sensible to discuss the average fitness of a gene against a range of genetic backgrounds, the entire body of theory can be considered to be epistemologically incomplete. This issue was cogently raised by the population geneticist R. Lewontin in his Jesup Lectures (1974), where he pointed out that at least the following are needed: (1) a theory mapping the genotype to the phenotype; (2) a theory relating phenotype to fitness; and (3) a theory relating changes in phenotype to resulting changes in genotype. With these three theories in place, explanatory closure might be attained for both phenotypic and genotypic evolution under the drive of either selection or drift. The problem, of course, has lain with the difficulty of obtaining theories which map the genotype to the phenotype and the phenotype to the genotype. These mappings, and the implications of self-organization for them, are discussed in Chapters 11 through 14.

Missing Phenotypes.   An interesting criticism leveled at evolutionary theory has been the "emptiness" of phenotypic space. The puzzle seems to be this: In the linear Chain of Being, each being is linked to those above and below it, there are no gaps, and the economy of nature is filled, an idea expressed in the doctrine of Plentitude. As S. M. Stanley (1979) remarks, Darwin is presumably thinking of this gapless state in his own initial image of selection, wedging places in the filled economy of nature where new species can shoulder into the crowded bustle. But very many conceivable useful phenotypes do not exist. Why not, and how can Darwinian theory account for their absence? Missing phenotypes appear to imply that failure to achieve so many useful phenotypes might be due to unknown constraints that limit the effectiveness of natural selection.

I confess that I believe the emptiness of phenotypic space is filled with red herrings. Forget for a moment the possibility of constraints in development. Imagine that we can describe the phenotype of an organism by measuring some large number of continuous metric characters, each corresponding to an *axis* in a phenotypic space. Then each organism's phenotype is a single point in phenotypic space, and a species is some cloud of points in the space. Phenotypic evolution in one species corresponds to the cloud's moving in some trajectory across phenotypic space. Similarly, branching phylogenies are recorded as branching trajectories in this phenotypic space.

Under the null hypothesis that no constraints at all exist, the branching pathways through space taken by this process constitute a random-branching walk in a high-dimensional space. The typical property of such a walk in a high-dimensional space is that most of the space is empty. Thus the emptiness of phenotypic space itself reveals nothing with respect to constraints on the evolutionary process. Natural selection may be critical in evolution, but it may not have to wedge very hard to create open vistas in morphospace. This comment, of course, does not imply that phylogenetic evolution is such a random walk.

Phenotypic Stasis.   Evolutionary stasis, the long-term persistence of form or feature, is a particularly serious problem. Despite the fact that evolutionary biologists have been familiar for years with living fossils, such as the opossum, the problem of stability of species form, or stasis, has recently come to the foreground in the debate between phyletic gradualism and the theory of punctuated equilibrium. The classical neo-Darwinist position favors phyletic gradualism, in which small phenotypic changes accumulate slowly in a species (Simpson 1944; see Levinton 1988 for a general discussion). The punctuated equilibrium theory holds that most phenotypic differences occur at speciation but that species are remarkably stable in phenotype thereafter (Eldredge and Gould 1972).

Whatever the merits of the general case made by the advocates of punctuated equilibrium (see criticisms by Charlesworth, Lande, and Slatkin 1982, and Levinton 1988), these advocates have properly focused on the fact that phenotypic stability over millions of years, typically 3 to 6 million for the average species, is not at all rare. The problem which must be answered is how such stability is achieved. If mutations continue to accumulate, as evidenced by the reasonably steady rate of nucleotide and amino acid substitution, what holds phenotypes in typical form? For the neo-Darwinian, the answer is that powerful selection maintains the normal form. The simplest argument is that, since stable species live in stable environments, selection favors the established normal form.

The difficulty with this plausible argument, as with so many other arguments in evolution, lies in assessing its degree of plausibility. It is difficult to confirm that stable species live in "sufficiently stable environments" to account for stasis. The environment of a species is not merely its physical environment but also the many other species that constitute its prey, its predators, its competitors. It is at least unclear that stability of environment is the norm. If the average species exists for 3 to 6 million years and if each species interacts significantly with N other species in its way of life and if those species also either alter or become extinct, then on average the species aspect of the niche changes N times during the duration of each species. How stable does an environment have to be to support phenotypic stability?

Are stable features obviously maintained by normalizing selection? This is extremely hard to assess, since in general field tests for normalizing selection are not feasible. Species of trilobites appear to have persisted virtually unaltered in morphology over tens of millions of years, with the exception of the addition of a single row of ommatidia to their compound eyes. It is certainly plausible that these trilobites were a maximally adapted phenotype in a stable environment. But is it true? If stasis is more common than rare, then a selectionist argument is left explaining a dominant feature of the evolutionary record in terms of a sufficiently constant environment, which is often unmeasurable, and in terms of normalizing selection for the established form, which is also rarely testable. One need not be a disciple of Karl Popper, who holds that theories must be refutable if they are to be scientific at all, to be slightly anxious.

Macroevolutionary Challenges.   The centerpiece of the neo-Darwinian Synthesis is that natural selection, acting upon individual variations within a population to substitute one allele for another, is the major force driving adaptive evolutionary change. In this view, speciation and the patterns of phylogenies from species to the higher taxa all express natural selection acting at the level of individual organisms. Yet students of the major trends at higher taxonomic levels have recently begun to doubt that selection at the level of the individual organism is the only factor accounting for patterns among higher taxa. Suppose, for example, that some species are prone to speciate, while others are not. If the propensity to speciate is itself heritable, then the former group will ultimately branch into a large number of species, and the latter will not. This difference in tendency to speciate may or may not be related to fitness at the level of the individual, but may dramatically affect the patterns seen within the corresponding genera and families. Further, some authors have suggested that species-level selection may play a role in such macroevolutionary processes (Stanley 1979). This set of problems seems to me to be important. If these challenges to the

Synthesis are correct, then the patterns of macroevolutionary change may be partially cut off or screened from the action of natural selection at the level of individual fitness variations.

The Origin of Complex "Wholes" and the Problem of Adequate Time

One of the purposes of an examination of self-organization in complex systems is the hope that spontaneous order will help account for origin problems in evolution. Everywhere in thinking about evolution, one confronts the question of how hard it may have been to "find" a particular structure or property. Such problems appear most trying when the structure or property in question requires the concerted action of a large number of constituents. It is here, I believe, that consideration of self-ordering may prove most useful. In contrast to the hope that insight into spontaneous order may help us to understand origins, typical thinking in evolutionary biology is couched in terms of the notion of an adequate time for the eventual "discovery" of the useful property. I discuss next an interesting case concerning the origin of life. My purpose is not only to introduce the origin-of-life topic but also to give a clear example of the difficulties which often arise from reliance on adequate search time.

No less a scientist than George Wald, 1967 Harvard Nobel laureate for his work in the chemistry of vision, published an article in *Scientific American* in 1954, based on the idea of adequate search time (quoted from Shapiro 1986): "One has only to contemplate the magnitude of this task to concede that spontaneous generation of a living organism is impossible. Yet we are here—as a result, I believe, of spontaneous generation." Wald goes on to argue that, with very many trials, the unthinkably improbable becomes virtually assured: "Time is in fact the hero of the plot. The time with which we have to deal is of the order of two billion years. What we regard as impossible on the basis of human experience is meaningless here. Given so much time, the impossible becomes possible, the possible probable, and the probable virtually certain. One has only to wait: time itself performs the miracles."

This line of argument has been sharply criticized, but the critique only leads to deeper problems. I should stress that I find the critique below inadequate and shall return to this problem in detail in Chapter 7. Robert Shapiro, in *Origins: A Skeptic's Guide to the Creation of Life on Earth* (1986), is not the first to attack Wald's view, but Shapiro's argument suffices to exemplify the point. He calculates the number of "trials" which can have occurred in the history of the earth, the probability of success per trial, and hence the overall probability of success. To calculate the number of trials, "We will need to know two items, the length of time needed for a single trial, and the number of trials that can take place simultaneously. . . . The bacterium *E. coli* replicates in twenty minutes, . . . let us presume, however, that a simpler bacterium than *E. coli* is involved and estimate one minute as the time for a trial . . . (thus) 5×10^{14} minutes were available." Shapiro then purposefully overestimates the space available for trials by assuming an ocean 10 kilometers deep and a volume of 1 cubic micrometer per trial, yielding 10^{36} simultaneous trials. Over 2 billion years, this yields 2.5×10^{51} trials.

Shapiro continues with an effort to calculate the odds of attaining, by chance, something like *E. coli*. He begins with an argument by Sir Fred Hoyle and N. C. Wickramasinghe (1981). Rather than estimate the chances for obtaining an entire bacterium, these authors try to calculate the chances of obtaining a functioning enzyme. They begin with the set of the 20 amino acids which are used to construct enzymes. If the amino acids were selected at random and arranged in random order, what would be the chances of obtaining an actual bacterial product? For a typical enzyme with 200 amino acids, the probability is obtained by multiplying the probability for each amino acid, 1 in 20, together 200 times, yielding 1 in 20^{200}. Since more than one sequence of amino acids might provide enzymes with proper function, Hoyle and Wickramasinghe estimate that the chance of obtaining an enzyme of the appropriate type at random was "only" 1 in 10^{20}. But to duplicate a bacterium, one would have to assemble *2000 different functioning enzymes.* The odds against this would be 1 in 10^{20} multiplied 2000 times, or *1 in $10^{40\,000}$*. As Shapiro points out, it is clear why Hoyle and Wickramasinghe gave up on spontaneous generation, since the likelihood of the event was comparable to the chances that "a tornado sweeping through a junkyard might assemble a Boeing 747 from the materials therein."

Against Wald's world enough and time, Shapiro says that with only 10^{51} possible trials, the odds of success—1 in $10^{40\,000}$—are vastly too improbable to have happened. Life, on this argument, cannot have arisen spontaneously.

The arresting feature of this example is not the apparent improbability of success but how typical the apparent failure is (Eden 1967; Schutzenberger 1967). The same sense of mystery surrounds the origin of a coupled metabolism, of the genetic code, of tissue organization. The general feature of each of these mysteries is that each exhibits, in one form or another, the evolutionary emergence of a mutually necessary set of processes. Each time we confront the evolutionary emergence of such a whole, whose parts are mutually necessary to one another, there is a tendency to reason along with Shapiro. What is the chance of obtaining the first part, of obtaining the second part—and since each is useless without the rest, what is the chance of obtaining them jointly? Where the odds can be estimated, even crudely, the joint probability is always very low. Typically, one cannot even estimate the odds of obtaining any single part.

Yet it is clear that Shapiro's argument is flawed. Having calculated the probability of obtaining a protein with some particular catalytic activity as 10^{20}, he then argues that an organism would require a set of *2000 enzymes for 2000 particular reactions.* It is this requirement for one particular set of coupled enzymatic activities which yields the overwhelmingly poor odds he calculates. We should instead be concerned with the probability of finding *any one* of possibly very many properly coupled sets of enzymatic activities which might constitute a living proto-organism. I will suggest in Chapter 7 that, viewed in this way, the origin of life was a quite probable consequence of the collective properties of catalytic polymers. More generally, I suggest throughout this book that many properties of organisms may be probable emergent collective properties of their constituents. The evolutionary origins of such properties, then, find their explanation in principles of self-organization rather than sufficiency of time.

A Framework to Think About Selection and Self-Organization

The task of enlarging evolutionary theory would be far from complete even if we could show that fundamental aspects of evolution and ontogeny had origins in some measure reflecting self-organizing properties of the underlying systems. The present paradigm is correct in its emphasis on the richness of historical accident, the fact of drift, the many roles of selection, and the uses of design principles in attempts to characterize the possible goals of selection. Rather, the task must be to include self-organizing properties in a broadened framework, asking what the effects of selection and drift will be when operating on systems which have their own rich and robust self-ordered properties. For in such cases, it seems preeminently likely that what we observe reflects the interactions of selection processes *and* the underlying properties of the systems acted upon.

In the remainder of the chapter, I want to introduce a straightforward framework in which to begin to think about the relation between selection and self-organization. The framework has direct analogies with the well-understood example of statistical mechanics and can be described quite informally. In Chapter 3, I will take up the problem in considerable detail.

Consider a gas at thermodynamic equilibrium, confined to a box. Statistical mechanics is constructed from Newtonian mechanics, and Newton's laws of motion apply to each molecule of the gas. The position of each molecule in the box can be described by three spatial coordinates. Similarly, the momentum of each molecule can be described by three coordinates showing how fast the molecule is moving as a projection of its velocity onto the three spatial coordinates. Therefore, each molecule's position and momentum at any instant can be described by six coordinates. If there are N molecules in the box, then $6N$ coordinates specify the positions and momenta of all N molecules at one instant.

It is convenient to conceive a $6N$-dimensional phase space, each axis of which represents one coordinate among the $6N$ specifying the current positions and momenta of the molecules. Then the present state of the *entire N molecules* can be visualized as a *single point* in this $6N$-dimensional space. Furthermore, over time the gas molecules move and collide with one another, thereby changing position and momentum. Consequently, over time the point representing the entire system moves through a *trajectory* in its phase space. The entire phase space represents all possible combinations of positions and momenta of the N molecules in the box. This set of all possible combinations is the ensemble of possible states of the gas.

Statistical mechanics is built up from analysis of this ensemble of possibilities. Any specific combination of positions and momenta is as unlikely as any other, since each is a single point in phase space. However, the *volume* of phase space which corresponds to a certain *class* of states may be much smaller than the volumes which correspond to other classes of states. For example, one class corresponds to all the ways the N molecules might be distributed within a specified small distance from one specific corner of the box. Another class corresponds to all the possible ways the N molecules might be distributed such that any volume representing $\frac{1}{1000}$ of the total volume of the box contains roughly the same number of molecules as all the other such volumes. Obviously, this class of nearly homogeneous distribution of the gas molecules corresponds to a volume of phase space that is vastly larger than the volume which corresponds to the case where all the molecules are confined near one specific corner of the box.

The statistical idea of entropy depends upon the hypothesis that the system's trajectory will wander aimlessly, or ergodically, in phase space. Therefore, over long periods of time, the probability that the point representing the system is in any one small region equals the probability that it is in any other small region. Consequently, the system is more likely to spend time in classes of states represented by large volumes in phase space than in those represented by small volumes. The idea that entropy increases is restated in statistical mechanics as the idea that the system leaves classes corresponding to small volumes in phase space and is overwhelmingly likely to be found in those large volumes which represent the near-homogeneous distribution.

Now consider Maxwell's demon. (I shall take the demon as an analogue of natural selection below.) The box is partitioned into left and right boxes by a wall with a flap valve. The demon opens the valve to allow faster molecules to pass from the left to the right box, and to allow the slower molecules to pass from the right to the left box. The faster molecules accumulate in the right box, raising its temperature and pressure with respect to the left box. Then, due to the pressure and temperature differential, useful work can be extracted from the pair of boxes as the system relaxes back to thermodynamic equilibrium. Maxwell's purpose in constructing his demon was to show that the second law of thermodynamics might be violated—a possible violation which has been resolved by noting that the operation of the valve requires information, whose energetic price compensates the later useful work. My point in reintroducing Maxwell's small colleague is rather to note that, as the demon acts, the pressure in the right box increases and opposes his efforts. If the demon is sufficiently powerful, he may succeed in separating all the faster molecules into the right box and slower molecules into the left box. But if the demon is finite, and rather weak, then he will succeed in shifting only a few of the faster molecules into the right box before the increased back pressure balances his efforts. In this latter case, the system will come to rest at a steady state, displaced away from thermodynamic equilibrium by the demon's efforts, but perhaps close enough to thermodynamic equilibrium that the statistical features of the equilibrium gas distributions will be shifted only slightly. Then the statistically robust features seen at equilibrium will remain good predictors of the features found in the presence of the demon.

In the examples of self-organization in complex systems to be discussed below, we shall in each case find that it is natural to first consider an ensemble of all possible systems and then characterize the *typical, average,* or, more generally, *generic* features of such systems. It will become natural to think of evolution as exploring such an ensemble, as mutations drive populations through neighborhood volumes of the ensemble. I shall want to say that selection is analogous to Maxwell's demon, for selection may attempt to pull the evolving population toward properties which are rare in the ensemble, but as it does so, the "back pressure" of mutations toward the statistically typical properties of the ensemble will increase. Thus if selection is a sufficiently weak force with respect to the mutational processes, the evolutionary process will come to rest at an equilibrium modestly displaced from the average properties of the underlying ensemble. But then those robust generic properties will serve as good predictors of properties actually found. In short, if selection is operating on systems with strongly self-organized properties that are typical of the ensemble being explored, then those properties simultaneously are the proper *null hypotheses* concerning what we would expect to find in the *absence* of selection and may be good predictors of what we will observe even in the presence of *continuing* selection. In brief, if selection can only slightly displace evolutionary systems from the generic properties of the underlying ensembles, those properties will be widespread in organisms not *because* of selection, but *despite* it.

The onset of "population thinking" in biology led to the loss of Platonic archetypes. I suggest that incorporation of self-organization into evolution may lead to new archetypes which are to be ensemble average properties. In principle, this development of new archetypes allows a marriage between population thinking and the search for underlying laws.

The possibility that generic properties of the ensembles in which evolution is occurring can be used to predict widespread and persistent features of organisms depends upon selection's *failing to escape those properties.* As we shall see, two quite different *limitations* on the capacity of selection to pull an evolving population to arbitrary regions of such an ensemble shall require analysis. In the first, selection is simply too weak in the face of mutations to hold a population at small volumes of the ensemble which exhibit rare properties; hence typical properties are encountered instead. In the second, even if selection is very strong, the population typically becomes trapped on suboptimal peaks which do not differ substantially from the average properties of the ensemble. As we shall see, both limitations tend to become more powerful as the *complexity* of the entities under selection increases. That is, there is a strong tendency in complex systems for selection to be unable to avoid the typical properties of the class of systems in which evolution is occurring.

But it will turn out that the marriage between selection operating on complex systems and self-organized properties promises to be yet more subtle. Under favorable circumstances, selection may be able to change the kinds of entities upon which it operates, hence change the ensemble of systems being explored, in such a way that the two limitations noted above are mitigated. In particular, the general strong tendency for adaptive processes under strong selection to become trapped on ever "lower" local optima as complexity increases can be mitigated. The various ways in which this limitation can be mitigated suggest lawlike universals in the ways complex systems adapt.

Just as Darwin's introduction of the ideas of evolution and natural selection created a web of questions which naturally arise, so an effort to include the emergent self-organizing properties typical of large ensembles of systems in evolutionary theory must provoke a resonant set of questions and consequences. Not the least of these is an interesting epistemological implication. If we should find it possible to account for, explain, predict widespread features of organisms on the basis of the generic properties of underlying ensembles, then we would not need to carry out in detail the reductionistic analysis of organisms in order to explain some of their fundamental features. As the physicist explains ice formation as a typical phase transition in a general class of systems, so we might explain aspects of organisms as typical of their class. Physics has ahistorical laws and no selection. Biology since Darwin is unthinkable without selection, but may yet have universal laws.

SUMMARY

This chapter has traced the outlines of our current view in evolutionary biology from its pre-Darwinian precursors. The major themes are encapsulated by Monod's lovely dictum "evolution is chance caught on the wing." This phrase captures the sense of freedom, of accident, of historical contingency, of design, of selection, of drift. Tracing the evolution of evolutionary theory has led us to see that the pre-Darwinians, considering fixed species with resemblances to one another, were fully reasonable in seeking laws of form which might account for such related morphologies. The rational sense of that enterprise is caught by remembering that crystallographers account for all possible crystals as members of a modestly large family of possible forms based on underlying laws of symmetry and structure.

Darwinism—the joint advent of evolving species arrayed in branching phylogenies and of natural selection as the force driving adaptation—undermined the previous search for laws of form. In place of this search, we have focused on design criteria which we hope selection has chanced to achieve. We have come to think of selection as essentially the only source of order in the biological world. If "only" is an overstatement, then surely it is accurate to state that selection is viewed as the overwhelming source of order in the biological world. It follows that, in our current view, organisms are largely ad hoc solutions to design problems cobbled together by selection. It follows that most properties which are widespread in organisms are widespread by virtue of common descent from a tinkered-together ancestor, with selective maintenance of the useful tinkerings. It follows that we see organisms as overwhelmingly contingent historical accidents, abetted by design.

The entire neo-Darwinian paradigm has been subject to serious criticisms. The most persistent critics have been the neutralists, who argue that much of evolution at the molecular level is selectively neutral. At the macroevolutionary level, two issues—(1) the problem of morphological stasis and punctuated equilibrium, and (2) species selection—have challenged neo-Darwinism.

Nevertheless, neo-Darwinism has largely withstood these attacks, both because no serious biologist doubts the historical fact of evolution and because few thinking biologists can contemplate the vertebrate eye and not harbor the conviction that natural selection is sometimes a dominant factor in evolution. If the tradition is under attack, it is only at the margins: Some, not all, evolution may be neutral. At the macroevolutionary level, perhaps species selection rather than individual selection plays a role in the formation of higher taxa.

My own aim is not so much to challenge as to broaden the neo-Darwinian tradition. For, despite its resilience, that tradition has surely grown without seriously attempting to integrate the ways in which simple and complex systems may spontaneously exhibit order. Since we shall see in a number of examples that such spontaneous order occurs, we must not be surprised if evolutionary theory must expand to embrace these facts.

PART I

Adaptation to the Edge of Chaos

The first part of this book, Chapters 2 to 6, stalks answers to new questions: What kinds of complex systems can evolve by accumulation of successive useful variations? Does selection achieve complex systems able to adapt? Are there lawful properties characterizing such systems? The overall answer may be that complex systems constructed such that they are poised on the boundary between order and chaos are the ones best able to adapt by mutation and selection. Such poised systems appear to be best able to coordinate complex, flexible behavior and best able to respond to changes in their environment. I suggest that selection does achieve and maintain such poised systems. Further, beyond the selective molding of individual adaptive systems, there are provocative, promising indications that linked coevolving complex systems are led by selection, as though by an invisible hand, to form ecosystems whose members mutually attain the edge of chaos. Here all may sustain the highest expected fitness, even while avalanches of coevolutionary changes propagate through the ecosystem, ringing out old species and ringing in new ones.

These are new issues in our understanding of the evolution of life. Darwin told us that adaptive evolution occurs by gradual accumulation of useful variants but failed to tell us what kinds of systems can evolve successfully by random variation and selection for fitter variants. It is remarkably easy to lay our minds around this set of issues. Most readers will be familiar with programs for contemporary sequential processing computers. The issue is this: How readily might a process of random mutation and selection operating on the instructions in a computer program succeed at attaining a complex program to carry out a desired computation? It is clear that evolution of useful sequential programs is very difficult for several reasons, the most obvious being that almost all random alterations in the code wreak dramatic changes in the computation being performed.

Adaptive evolution occurs largely by the successive accumulation of minor variations in phenotype. The simple example of computer programs makes it clear that not all complex systems are graced with the property that a minor change in system structure *typically leads to a minor change in system* behavior. *In short, as their internal structure is modified, some systems change behavior relatively smoothly, some relatively radically. Thus we confront the question of whether selective evolution is able to "tune" the structure of complex evolving systems such that they evolve readily.*

The variability in behavior as the structure of a system is altered can be pictured as characterizing the ruggedness of a fitness landscape. In Chapters 2 and 3 we shall discuss the concept of such landscapes. In particular, we shall focus on the simple case of protein molecules in "sequence spaces," where each protein is located next to a large number of other proteins which differ from it at only one amino acid position. The capacity of each protein to carry out some specified function allows us to define the fitness landscape over protein space with respect to this function. Here fitness peaks represent either local or global optima for such a function. As we shall see, such landscapes range from smooth and single-peaked to very rugged and multipeaked.

The character of adaptive evolution depends on the structure of such fitness landscapes. Most critically, we shall find that, as the complexity of the system under selection increases, selection is progressively less able to alter the properties of the system. Thus, even in the presence of continuing selection, complex systems tend to remain typical members of the ensemble of possibilities from which they are drawn. We shall find in this book many examples of spontaneous order present in entire ensembles of complex systems. Thus if selection, when operating on complex systems which spontaneously exhibit profound order, is unable to avoid *that spontaneous order, that order will "shine through." In short, this theme, central to our concerns, states that much of the order in organisms may be spontaneous. Rather than reflecting selection's successes, such order, remarkably, may reflect selection's failure.*

Analysis of the structure of fitness landscapes and its implications occupies Chapters 2 and 3. Chapter 4 considers adaptive evolution in the space of possible proteins. Here we find that it is possible to construct statistical models of molecular fitness landscapes which apply to the rapid protein evolution seen in maturation of the immune response. But protein evolution to carry out progressively different catalytic functions drives us to conceive of an abstract catalytic task space. Among the surprising implications of this view is the possibility that a finite number of enzymes can carry out all enzymatic tasks: A universal enzyme toolbox is possible. Remarkably, about 100 000 000 protoenzymes may well constitute such a universal toolbox, and the immune repertoire of about 100 000 000 variant antibody molecules is probably also a universal toolbox. These possibilities lead me, in Chapter 4, to discuss what might become a vastly important new arena of biotechnology: applied molecular evolution. I believe we are now crossing the threshold into an era where it shall become possible to evolve biopolymers that act as vaccines, drugs, enzymes, biosensors, and so forth, serving a wide range of practical medical and other functions.

Chapters 5 and 6 present evidence for adaptation to the edge of chaos both within single complex systems and between the linked members of a coevolving system. Chapter 5 investigates the emergence of spontaneously ordered behavior in parallel-processing systems and elucidates the three broad regimes which occur in such systems: ordered, complex, and chaotic. Here we encounter our first powerful example of spontaneous order. Contrary to intuition, even randomly constructed networks of elements which turn one another on and off according to complex rules can exhibit extremely ordered behavior. This ordered regime is characterized by the formation of a large connected set of elements of the system that freeze into fixed activity states. This frozen component percolates (spans) across the system and leaves behind isolated unfrozen islands free to vary activities in complex ways. The chaotic regime corresponds to the case where the frozen component does not percolate across the system. Rather, the unfrozen component of elements spans the system, leaving behind isolated frozen islands embedded in the fluctuating sea. Transition back and forth between these two regimes corresponds to a phase transition at which the frozen component begins to melt and the fluctuating sea begins to coalesce; this phase-transition region is the complex regime. The most complex but controllable behavior arises in parallel-processing systems poised in this complex regime on the boundary between order and chaos.

I hold that the exorbitant order of the ordered regime underlies the evolutionary emergence of order in ontogeny—that spontaneous order lies to hand, free, as it were, for selection's further molding. In particular, asking what form such molding may take and what laws might govern it leads us to the hypothesis that the target which selection achieves is complex systems poised in the complex regime on the boundary between order and chaos. Such systems, it begins to appear, harbor behavior which is the most flexible, complex, and adaptable. If so, we may have uncovered a universal in biology relating the mutual implications and interpretations, the true marriage, of self-organization and selection: Selection, in attaining complex systems, may build toward and sustain a characteristic poised order, an entire ensemble coursing back and forth along a high-dimensional boundary between order and disorder. Then further selection would be unable to avoid the typical features of this poised ensemble, whose generic features would emerge as additional potential biological universals.

Chapter 6 examines coevolution. Here we find evidence of a selective metady-

namics which may lead coevolving systems jointly to the edge of chaos. I ask in this chapter what the implications for coevolution may be of the ruggedness of the fitness landscapes of each of the partners, and of how much an adaptive move by one partner deforms the landscapes of others. We are led to provocative results. If landscapes are too smooth compared with the landscape deformation caused by partners, then each partner can hardly respond by increasing fitness as other partners move. The entire system is chaotic. Sustained fitness is low because the landscape of each partner is drastically altered, typically casting the partner to low fitness, by the moves of its coevolving partners. If landscapes are too rugged compared with the deformation caused by others, then all partners rapidly freeze into fixed but poor compromise phenotypes. Again, sustained fitness is low. Just at the boundary between frozen order and chaotic wandering, just at the edge of chaos when some, but not all, partners cease changing and form a percolating frozen component, leaving isolated islands of partners that continue to coevolve and change, the coevolving system attains a structure where all partners attain the highest expected sustained fitness. Adaptation, through a selective metadynamics altering landscape structure and landscape deformation, again attains the edge of chaos.

In summary, Part I investigates the mutual implications of self-organization and selection for adaptive evolution. We ask what the conditions are within and between evolving entities which permit adaptive evolution, and whether the attainment of those conditions is itself a lawful consequence of selection's operation. The tentative answer, to be held as a working hypothesis at this stage, is that at levels within organisms which must coordinate complex tasks, within evolving populations, and within coevolving systems, selection attains a near-universal poised state hovering between unexpected, profound spontaneous order and the incoherence of chaos. Borrowing a culminating phrase from my colleagues N. Packard and C. Langton, life exists at the edge of chaos.

CHAPTER 2

The Structure of Rugged Fitness Landscapes

In this chapter, I begin discussing in detail the central conceptual framework of the book: to examine the relation between selection and self-organization, opening with a discussion of the structure of the rugged fitness landscapes underlying evolution. I shall, for the moment, consider such landscapes as fixed in structure. In reality, fitness landscapes deform in response to changes in the abiotic environment and in response to coevolution. In coevolutionary processes, the fitness of one organism or species depends upon the characteristics of the other organisms or species with which it interacts, while all simultaneously adapt and change. A critical difference between evolution on a fixed landscape and coevolution is that the former can be roughly characterized as if it were an adaptive search on a "potential surface," or "fitness surface," whose peaks are the positions sought. In coevolution, there may typically be no such potential surface, and the process is far more complex. These more complex situations are deferred to Chapter 6.

As Jacob pointed out in "Evolution and Tinkering" (1977a; see also Jacob 1983), adaptation typically progresses through small changes involving a *local* search in the space of possibilities. The paradigm is one of local hill climbing via fitter mutants toward some local or global optimum. Despite this transparent metaphor, such a process involves complex, combinatorial optimization. In such optimization searches, many parts and processes must become coordinated to achieve some measure of overall success, but conflicting "design constraints" limit the results achieved. One purpose of this chapter is to show that increasing levels of conflicting constraints make the landscape more rugged and multipeaked.

The hill-climbing framework is hardly new, for I borrow it with minor modifications directly from Wright (1931, 1932), who introduced the concept of a space of possible genotypes. In one version of his idea, each genotype has a "fitness," and the distribution of fitness values over the space of genotypes constitutes a *fitness landscape*. [Often Wright thought of the fitness of a given gene or genotype as a function of its frequency in the population (Wright 1931, 1932; Provine 1986). In this chapter, I shall use the simpler idea that each genotype can be assigned a fitness.] Depending upon the distribution of the fitness values, the fitness landscape can be more or less mountainous. It may have many peaks of high fitness flanked by steep ridges and precipitous cliffs falling to profound valleys of very low fitness. Or it may be, like the gentle Normandy countryside, smoothly rolling with low hills and gentle valleys.

In this framework, adaptive evolution in a population is a hill-climbing process. The population can be thought of as a tight or loose cluster of individuals located at different points in the landscape. Mutations move an individual, or its offspring, to neighboring points in the space, representing neighboring genotypes. Selection is reflected in differential reproduction by individuals with different fitness values. Therefore, over time the cluster of individuals representing the population will flow over the fitness landscape. In the simplest cases, the population will climb to and cluster about one of perhaps a large number of different fitness peaks (Crow and Kimura 1965, 1970; Ewens 1979; Gillespie 1983, 1984). In more complex cases, the cluster representing the population may spread widely across the landscape, passing via a rich web of ridges somewhat below the fitness peaks (Eigen 1985; Schuster 1986, 1987; Fontana and Schuster 1987; Kauffman and Levin 1987; Kauffman, Weinberger, and Perelson 1988; Kauffman 1989a; Eigen, Gardiner, et al. 1989). In cases that are still more complex, the population may drift down from the peaks and wander within a band of modest fitness altitudes virtually anywhere across the fitness landscape (Fontana and Schuster 1987; Kauffman, Weinberger, and Perelson 1988; Kauffman 1989a; Eigen, Gardiner, et al. 1989).

It is intuitive from this description that the behavior of an adapting population depends on how mountainous the fitness landscape is, on how large the population is, and on the mutation rate which moves an individual from one genotype to another genotype in the space. The flow of a population over a fitness landscape also depends on whether the population is sexual, where mating allows mixing of genotypes from distant points in the landscape in a new individual, or asexual.

Self-Organization and Selection

As I sketched in the last section of Chapter 1 and now describe in slightly greater detail, part of the relation between self-organization and selection depends upon how well selection can move an adapting population to arbitrary regions of the fitness landscape. If selection can move a population to virtually any region of the landscape, then selection is powerful enough to avoid any spontaneously ordered properties which most but not all entities on the landscape may exhibit. If selection is not able to move an adapting population to virtually any region of the landscape, then spontaneously ordered properties which are widespread in the landscape are very likely to be found in organisms even in the presence of continuing selection.

The general ideas will be clearer in the context of a concrete example which will occupy us in Chapter 12: the control of cellular differentiation. Among the most obvious features of cell differentiation is that less complex organisms possess fewer cell types than more complex organisms. For example, yeast has three cell types, the coelenterate hydra has 15 to 17, annelid worms have about 60, and humans about 250. A plot of the number of cell types in an organism as a function of the estimated number of genes in that organism shows a profound and simple relation: The number of cell types increases as about a square-root function of the number of genes (Figure 12.7). Why should such a remarkably simple relation exist across such a wide range of phyla? Yeast, hydra, annelides, and vertebrates diverged around 600 million years ago. It is hard to believe that such a relationship is a heritable remnant of a property derived from a primitive archean ancestor in the Cambrian or Precambrian as a result of propinquity of descent. Perhaps instead selection has directly opted for this property?

Perhaps not. I shall argue that this strong correlation across many phyla is an indirect and virtually inevitable property of the kinds of cybernetic gene-regulatory systems which control cell differentiation and ontogeny. We shall see that a very large ensemble of genetic cybernetic systems exhibit a wide range of spontaneously ordered properties. Among them, in most members of the ensemble the number of cell types the regulatory system can create is about the square root of the number of its genes. Thus this first example is a bellwether for all our questions about self-organization and selection. Here is a striking and ordered property which is *typical of most but not all* members of a vast ensemble of regulatory systems. Does its presence in organisms reflect selection or the fact that this ordered property is widely distributed in the ensemble and cannot be avoided by selection?

Throughout this book, it will prove convenient to conceive of an ensemble of possible "systems," each "next to" slightly different versions of itself. In the cases we shall examine, most but not all members of the ensemble exhibit some spontaneously ordered properties, which may or may not have anything to do with fitness. If an ordered property is selected, we might of course expect to see it in organisms. But if an ordered property is *not* under selection, or if it is even modestly selected against, might we still see it? The answer can be "yes." The conditions under which this can happen are the central issue. The point here is similar to imagining an ensemble of objects most of which are blue but vary in size, weight, and so forth. If selection acts on size and weight, will we see blue objects in the continuing presence of selection? There are two ways blue objects might persevere in the presence of selection:

1. If under the conditions of adaptation the population does *not* remain tightly clustered around single peaks of high fitness but instead wanders within some larger volume of the ensemble, the chances are high that most members of that volume exhibit the ordered property. This corresponds directly to the analogy discussed in the preceding chapter of selection as a weak Maxwell's demon. If the demon is weak, then the statistical distribution of the faster and slower gas molecules in the two boxes will not differ sharply from the distribution at thermodynamic equilibrium. Similarly, if selection is too weak to hold an adapting population in very small volumes of the ensemble, then even in the presence of continuing selection the population will almost certainly exhibit the "typical" ordered properties of most ensemble members. Hence I tend to say that such adapting systems exhibit order not *because* of selection but *despite* it.

2. The second fundamental way that adapting populations might continue to exhibit the spontaneous order typical of most members of the ensemble, *even if selection can hold the population within very small volumes* of the ensemble representing adaptive peaks, is that the vast majority of *the adaptive peaks remain typical of the ensemble as a whole.* This limitation tends to become powerful as landscapes become very rugged and multipeaked. In such landscapes, adaptation tends to become "trapped" on local peaks and thus cannot move long distances to rare regions of the space. If there are many peaks, most of them are likely to be in typical regions of the space. Therefore, if adaptation starts at some typical spot in the space of possibilities, it will become trapped on a local peak which is very likely to remain typical of the space as a whole. In terms of our imaginary example, blue objects will be found. But a related feature of very rugged landscapes will soon emerge: As many rugged landscapes become more multipeaked, the peaks *dwindle in altitude* to mere hills, then hummocks, then faint bumps. As the peaks fall ever lower, they necessarily become progressively more typical of the space as a whole. Thus even when selection is very powerful and can hold populations on

any accessible peak, the peak almost certainly exhibits properties typical of the entire space of possibilities. Thus if the space is the space of genetic regulatory systems and if member systems typically exhibit the property that the number of cell types is about the square root of the number of genes, that relation can be expected across phyla despite strong selection. In the simple image, blue objects will be found.

These two limitations constitute what I shall call two *complexity "catastrophes,"* for we shall see that one or the other must ultimately occur as the *complexity of the entities under selection increases.* This is the most important point of the theory discussed in this and the following chapter. As the complexity of entities increases, one or the other basic mechanism ultimately *limits the power of selection.*

The intuitive reasons for these limitations are not hard to see, and the *NK* model I introduce in this chapter makes the case clearly. In an adapting system of many parts, either those parts are fully independent of one another or they are coupled together. In the limiting case where the parts are independent, each part typically makes, to the overall function of the system, a contributuion which *decreases* in relative importance as the total number of parts in the system increases. For a system with a sufficient number of parts, the fitness loss due to mutational damage of one part becomes small. Therefore, the selective force tending to restore the damage becomes weaker than the mutational pressure tending to damage the part. In short, selection becomes too weak a force to hold an adapting population at adaptive peaks. The population flows down the adaptive hillside to the lowlands. This contention of mutational and selective forces leads, as we shall see, to a complexity catastrophe when the number of parts exceeds a critical value. Beyond that level of complexity, selection cannot climb to peaks or remain there.

At the opposite extreme, the parts are very richly coupled. But in this case common experience suggests that conflicting design constraints make it difficult to achieve overall success. As we shall soon see, such conflicting constraints lead to an adaptive landscape which becomes more multipeaked as the number of parts increases. Thus adaptation, which must search such rugged landscapes, tends to become trapped in very small regions of the space. Worse, due to the increasing numbers of conflicting constraints, the peaks become ever poorer compromises among those constraints, withering to mere bumps hardly better than chance agglomerations of the parts.

These investigations must brusquely suggest that adaptive evolution is bounded by the character of fitness landscapes. But that character in turn depends upon the entities which are evolving. Hence evolution can change the rugged structure of fitness landscapes and their impact on evolution by changing the adapting entities. Thus we shall ask what kinds of landscapes, in what conditions, allow adaptive evolution to be optimized.

FITNESS LANDSCAPES IN SEQUENCE SPACE

Sequence Space: The "Practical" Importance of a Theory of Adaptation on Rugged Landscapes

The framework just sketched has, as one overarching purpose, the aim of analyzing the relation between self-ordering and selection in complex systems. But there are more immediate reasons to develop a theory of adaptation on rugged fitness land-scapes. The same framework provides a crisp means of describing the selective evo-lution of, say, proteins or RNA molecules for specific functions. More generally, it applies to adaptive evolution in sequence spaces.

Before continuing, I must clarify what I mean by a fitness landscape. For an evo-lutionary biologist, "fitness" applies principally to an entire organism. It has com-ponents of fecundity, fertility, and other factors leading to reproductive success (Crow and Kimura 1965, 1970; Ewens 1973). These include complex issues such as the *frequency* of each genotype variant of the organism in the population, the *density* of each genotype variant in a region, and even the entire *ecosystem* with which each organism interacts (S. A. Levin 1978). Therefore, in the general context, it is difficult to assign a fitness to a gene or even to a genotype, since all these factors depend upon the other organisms in the population.

For the purposes of the present chapter, I shall use the term "fitness landscape" in a much more restricted sense to refer to any well-defined property and its distribution across an ensemble. For example, the capacity of each protein in protein space to catalyze a specific reaction under specified conditions is, in principle, a well-specified property. The velocity of the reaction catalyzed by each protein can then be *defined* as the fitness of that protein. Then the distribution of velocities across the space of proteins constitutes the fitness landscape with respect to that defined function. Adap-tive evolution with respect to that function is a search in protein space which attempts to optimize the capacity to catalyze that specific reaction. It is an entirely different issue whether optimization of any specific reaction velocity optimizes the overall fitness of the organism harboring the protein.

The concept of protein space was, to my knowledge, first introduced by Smith (1970). It has since been reinvented by a number of authors, including Ninio (1979), Eigen (1985, 1987), and Schuster (1986, 1987). Others (Borstnick et al. 1987; Kauff-man and Levin 1987; Kauffman et al. 1988; Kauffman 1989a, 1989b) have utilized Smith's initial idea as well. The idea is straightforward. Proteins are linear polymers comprising various combinations of 20 different amino acids. Because proteins have distinguishable carboxy and amino terminal ends, each polymer is oriented. The total number of proteins of a specific length N is just 20^N. Therefore, this set of all possible proteins of length N constitutes an ensemble. Furthermore, each protein can be mutated to other proteins by changing any amino acid at one position in the pro-tein to one of the 19 other possible amino acids. Therefore, for a protein length N, there are $19N$ "one-mutant" neighbor proteins. A *protein space* therefore is a high-dimensional space in which each point represents one protein and is next to $19N$ points representing all the one-mutant neighbors of that protein. The protein space therefore simultaneously represents the entire ensemble of 20^N proteins and keeps track of which proteins are one-mutant neighbors of each other.

Although it is difficult to draw a picture of such high-dimensional spaces, a sense of their structure can be captured by considering proteins with only two kinds of amino acids, say alanine and glycine, which can be represented by 1 and 0 (1 = ala-nine, 0 = glycine). Figure 2.1a shows all 16 possible peptide sequences of length 4 using these two amino acids. Each vertex corresponds to one of the 16 possibilities and is linked by a line to four other vertices representing the four other peptides which differ from the first by a single amino acid. The point to carry away from this picture is that all possible length-4 peptides made up of the two amino acids are rep-resented in this small peptide space and that each is "next to" those four others which differ from it by one amino acid in each possible position.

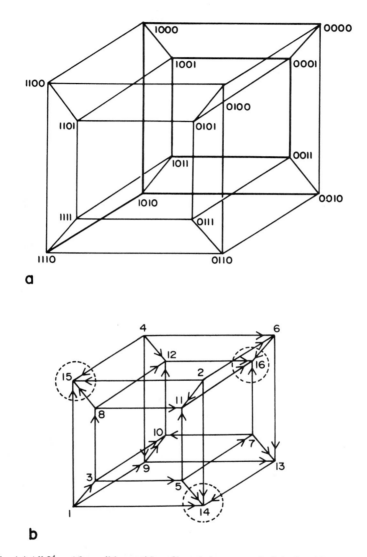

Figure 2.1 (*a*) All $2^4 = 16$ possible peptides of length 4 composed of alanine (1) and/or glycine (0), arranged as vertices on a four-dimensional Boolean hypercube. Each peptide is connected to its four one-mutant neighbors, accessible by changing a single amino acid from a 1 to a 0 or from a 0 to a 1. Thus, the hypercube represents this four-dimensional peptide space. (*b*) Each peptide has been assigned, at random, a rank-order fitness, ranging from the worst, 1, to the best, 16. Adaptive "moves" pass from any peptide to those of its one-mutant neighbors with higher fitness. Directions of such moves between adjacent peptides are shown by arrows from the less fit to the more fit. Peptides fitter than all one-mutant neighbors are local optima.

Smith's purpose in invoking protein space was to note that if adaptive evolution in general occurs by substitution of single amino acids, then evolution is a "walk" between adjacent vertices in protein space. Therefore, to improve function, any such *adaptive walk* in general must be a *connected* walk through a succession of adjacent vertices, all of which exhibit *improved function.*

The most straightforward version of Smith's idea is shown in Figure 2.1. Suppose I measure the capacity of these 16 peptides to carry out some specific function—for example, that of being bound by a specific cell receptor. I may define the affinity with which each peptide is bound as its fitness. In Figure 2.1b, I have arbitrarily assigned 16 affinity fitness values. To make inspection easier, I have rank-ordered the 16 peptides from worst (1) to best (16). An adaptive walk might begin with any peptide. The walk will "move" to a one-mutant neighbor only if the second peptide is fitter than the first. Then any adaptive walk starts at a peptide and passes via fitter one-mutant neighbors which improve fitness until a peptide is reached which is fitter than all its one-mutant neighbors. Any such peptide is a *local optimum* in peptide space.

Inspection of Figure 2.1b shows that three of the 16 peptides are local optima. If the adaptive walk starts at one of these three, it stops immediately. If the process begins at another peptide, then after one, two, or perhaps three improvement steps, the process reaches a local optimum and stops. This model is very useful because it immediately focuses attention on a number of obvious questions about such adaptive walks.

1. How many local optima exist in the space?
2. What is the average number of improvement steps on an adaptive walk to a local optimum?
3. What is the average number of mutants tried on an adaptive walk to a local optimum?
4. What is the ratio of the mutations accepted to those tried?
5. What is the average number of alternative local optima which can be reached from a peptide in the space? What is the maximum?
6. After each improvement step, the number of fitter one-mutant neighbors may change. At each local optimum, there are zero fitter one-mutant neighbors. Therefore, we might expect the average number of fitter one-mutatnt neighbors to *dwindle to zero* on an adaptive walk. How does it dwindle?
7. How many peptides can climb to the same local optimum?

In analyzing the statistical structure of fitness landscapes, I shall use the simple image of an adaptive walk via fitter one-mutant variants. I adopt this idealization in order to consider how mountainous such high-dimensional landscapes might be expected to be. Nevertheless, it is important to stress that this idealized image of walks constrained to pass only via fitter one-mutant neighbors corresponds to one plausible limiting case of the adaptive flow of a real population under the drives of mutation, selection, and recombination. Gillespie (1983, 1984) has shown that this constrained version of an adaptive walk corresponds to an adapting population in which the rate of finding fitter variants is very low compared with the fitness differentials between the less fit and the more fit allele. In such a limit, if the population begins entirely at the less fit allele, a single mutant will eventually encounter the fitter allele. Either that mutant dies out before leaving offspring, or a few of the fitter mutant type are produced. Once the number of fitter type produced is sufficient to reduce the chance fluctuation leading to their death, the fitter type rapidly takes over the entire population. Thus the entire population "hops" to the fitter neighboring genotype. Gillespie has shown that the entire adaptive process in this limit can be treated as a continuous-time, discrete-state Markov process. Each state corresponds to one genotype. The population hops as a whole with different probabilities to one or another of the fitter neighboring genotypes. The conditions required for Gillespie's limit are that the product of population size and mutation rate be low compared with the rate of finding fitter variants.

THE *NK* MODEL OF RUGGED FITNESS LANDSCAPES

The NK *Model of Random Epistatic Interactions*

I now introduce a simple formal model of rugged fitness landscapes, called the *NK* model. In this model, *N* refers to the number of parts of a system—genes in a genotype, amino acids in a protein, or otherwise. Each part makes a fitness contribution which depends upon that part and upon *K* other parts among the *N*. That is, *K* reflects how richly cross-coupled the system is. In the geneticist's term, *K* measures the richness of epistatic interactions among the components of the system.

Since the model is abstract and since it forms one of the conceptual backbones of this book, I should make clear why I believe analysis of it warrants detailed attention. The ruggedness of fitness landscapes with respect to catalytic or other protein functions are unknown, but knowable. Discovery of the structures of such landscapes is of the deepest importance. Although we do not yet know what the real landscapes are like, we may be able to develop some intuition for their typical, or statistical, structures by building simple models. That is, we need a kind of statistical mechanics for fitness landscapes to help us to understand their expected features. The *NK* model is meant to accomplish this. As the main parameters are altered, the model generates a *family* of increasingly rugged multipeaked landscapes.

A second reason to develop a formal model for the statistical structure of rugged fitness landscapes is that we want to predict and understand the structure of actual fitness landscapes in protein space and elsewhere. The *NK* model is the first effort in this direction. Perhaps surprisingly, given its simplicity, the model performs rather well when confronted by known adaptive landscapes in protein space.

The model can be interpreted as a model of genetic interactions, and it is in this genetic framework that I now introduce it.

One of the earliest population-genetic models focuses on haploid organisms with a single copy of each chromosome. Each chromosome has some number of distinct genes, and the chromosome set has a total of N distinct genes. Each gene may occur in more than one allele. In the simplest case, each gene can occur in two alleles (Crow and Kimura 1965, 1970; Ewens 1979). Then the haploid genotype has N genetic loci, each with two alleles; more generally, each locus might have some larger number of alleles A. In the first case, the total number of genotypes is 2^N; in the latter case, the total number is A^N.

The set of possible genotypes constitutes the ensemble. Each genotype is a one-mutant neighbor of all those genotypes accessible by mutating a single locus from one allele to another. In the N-locus, two-allele case, each genotype is a one-mutant neighbor of N other genotypes. In the A-allele-per-locus case, each genotype is the one-mutant neighbor of $N(A - 1)$ other genotypes. The number of one-mutant neighbors, which I will symbolize by "D," is the *dimensionality* of the genotype space and gives the number of directions in which each genotype can change to another neighboring genotype by a minimal alteration.

As I described in Chapter 1, population geneticists have been interested in whether selection can substitute a gene which confers a slight selective advantage on the organism and hence have talked in terms of the "fitness contribution" of each gene. To carry out their central analysis, therefore, they naturally ask whether selection can substitute one allele for a second allele when the first confers slightly higher fitness. In models with N loci each having two alleles, it is necessary to specify the fitness of the entire genotype of N genes given the fitness contribution of the allele present at each of the N genetic loci. The most idealized genetic models assume that each locus contributes to the overall fitness of the genotype *independently* of all the other loci. Therefore, given the fitness contribution of the allele at each locus, the fitness of the genotype is just the sum of the N independent fitness contributions divided by N, hence the average of those contributions. This central model is, not surprisingly, called the *N-locus, two-allele additive fitness model*.

The assumption that each gene contributes to overall fitness independently of all other genes is clearly an idealization. In a system with N genes, the fitness contribution of one or another allele of one gene may often depend upon the alleles of some of the remaining $N - 1$ genes. Such dependencies are called *epistatic interactions*, and their existence, which is well known, raises experimental and theoretical issues. The experimental issues concern how to measure the extent of epistatic interactions. I will not discuss this further. The theoretical problem is how to build useful models of epistatic interactions. One device commonly used has been to assume that genetic loci which interact can be represented by *multiplying* their fitness contributions (Franklin and Lewontin 1970; Lewontin 1974; Ewens 1979). Multiplication captures a kind of positive cooperativity. High fitness contribution by two epistatic loci requires that *both* have high fitness. If either locus has low fitness, then the product may be low even if the other has high fitness.

The problem with any such model is that the ways in which different alleles at the N loci might be coupled to one another epistatically to produce an overall fitness for each genotype might be extraordinarily complex. In general, we truly have almost no idea what those mutual influences on overall fitness might be. Take Mendel's peas. He found two alleles for seed color, yellow and green, and two alleles of a second gene for seed texture, rough and smooth. *A priori* we have no idea which of the four combinations of these traits will be of highest fitness, nor how changing from any one combination of traits to any other will affect fitness. If the fitness contribution of each gene is epistatically affected by a large number of other genes, the possible conflicting constraints among the complex web of epistatically interacting genes are both unknown and likely to be extremely complex. This complexity suggests that it might be useful to confess our total ignorance and admit that, for different genes and those which epistatically affect them, essentially arbitrary interactions are possible. Then we might attempt to capture the statistical features of such webs of epistatic interactions by assuming that the interactions are so complex that we can model the statistical features of their consequences with a random fitness function. This leads to the *NK* model.

Consider an organism with N gene loci, each with two alleles, 1 and 0. Let K stand for the average number of other loci which epistatically affect the fitness contribution of each locus. Thus the two main parameters of the NK model are the number of genes N and the average number of other genes K which epistatically influence the fitness contribution of each gene. A third parameter characterizes how the K genes are distributed among the N genes. In general, the sensible step to take with such models is to assess which parameters matter. It turns out that, to a very large extent, *only N and K* matter. The distribution of K among the N appears to be far less important. (For the moment, restrict A, a fourth parameter, to two.)

Having assigned to each locus i the K genes which impinge upon it (Figure 2.2a), it is necessary to assign fitness contributions to each gene in the context of the K genes which epistatically influence it. The fitness contribution of the allele at the ith locus depends upon itself (in other words, whether it is 1 or 0) and on the alleles, 1 or 0, at K other loci, hence upon $K + 1$ alleles. The number of combinations of these alleles is just 2^{K+1}. Since we have no idea what the effects of each such combination on the fitness contribution at the ith locus might be, let us *model* those effects by assigning to each of the 2^{K+1} combinations at random, a different fitness contribution drawn from the uniform distribution between 0.0 and 1.0. Therefore, the fitness contribution w_i of the ith locus is specified by a list of random decimals between 0.0 and 1.0, with 2^{K+1} entries (Figure 2.2b).

The fitness contribution of each allele at each gene in the context of the K other genes which impinge upon that gene must be specified. For each gene, its fitness contribution w_j is generated by random assignment on the 2^{K+1} allele combinations of the $K + 1$ genes which impinge upon it (Figure 2.2b).

Having assigned the fitness contributions, we may now define the *fitness of an entire genotype* as the average of the contributions of all the loci:

$$W = \frac{1}{N} \sum_{i=1}^{N} w_i \qquad (2.1)$$

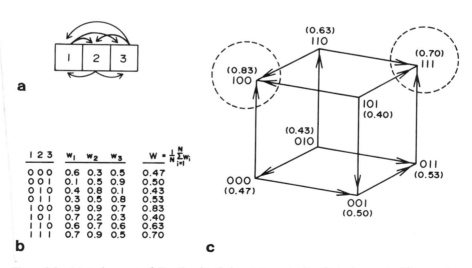

Figure 2.2   (*a*) Assignment of $K = 2$ epistatic inputs to each site. (*b*) Assignment of fitness values to each of the three genes with random values for each of the eight combinations of $K + 1$ alleles bearing on genes 1, 2, and 3. These fitness values then assign a fitness to each of the $2^3 = 8$ possible genotypes as the mean value of the fitness contributions of the three genes, as given in Equation 2.1. (*c*) Fitness landscape on the three-dimensional Boolean cube corresponding to the fitness values of the eight genotypes in (*b*). Note that more than one local optimum exists.

Given this definition, the *NK* model is fully specified. Therefore, the fitness of each possible genotype in the space has been assigned, and as a consequence, a fitness landscape over the genotype space has been created. Figure 2.2c shows such a fitness landscape for the eight possible genotypes available with three genes, each having two alternative alleles. In this case, I have assumed that the fitness contribution of each gene depends upon the allele at that gene and upon the alleles at the remaining two genetic loci. Note that the resulting landscape, like that in Figure 2.1, has more than one local optimum and that the questions raised on page 39 about adaptive walks in sequence space regarding Figure 2.1 can be posed with respect to this landscape as well.

The *NK* model is very similar to a famous and well-studied class of models which arises in statistical physics, called *spin-glasses* (Edwards and Anderson 1975; Sherrington and Kirkpatrick 1975; P. W. Anderson 1985; Binder and Young 1986; Stein, Baskaran, et al. 1987). Indeed, Anderson (private communication) has pointed out that the *NK* model is a form of spin-glass. This is important to us for two reasons. First, a feature of spin-glasses called *frustration* helps account for the multipeaked features of fitness landscapes. Second, there are profound similarities between the behaviors of a physical system in a complex potential surface at a finite temperature and an adapting population on a rugged fitness landscape at a finite mutation rate. In other words, the tools of statistical physics bear on population biology.

Spin-glasses are disordered magnetic materials in which the orientation of nearby magnetic dipoles may be either parallel or antiparallel. This "preference" for one polarity or the other varies sinusoidally in space. Thus whether two dipoles prefer to orient either in the same direction or in opposite directions and how much so depends upon their precise distance apart. Models of spin-glasses typically consist in two- or three-dimensional lattices of spins with each spin pointing either up or down. Each spin is coupled with some set of other spins which may be limited to its neighbors or may range outward to all other spins in the lattice. For each pair of coupled spins, their preference to point either in the same direction or in opposite directions and how strong that preference may be are decided at random but then fixed. Preferences correspond to energies. The entire spin-glass has N spins, each up or down; hence there are a total of 2^N possible configurations of spins. Each configuration has a total energy given by a Hamiltonian:

$$H = -\Sigma J_{ij}(s_i \times s_j) \qquad s_i, s_j = \pm 1$$

where s_i and s_j are the orientations (up or down) of the two spins, J_{ij} is the energy reflecting how strongly the two are coupled and hence prefer to be in the more favorable relative orientation, and the sum is over all coupled pairs of spins.

The similarity to the *NK* model is clear. In the *NK* model, each site makes a fitness contribution which depends in a random way on the "allele" at that site and on the "alleles" at K other sites. The fitness of a given genotype is the normalized sum of the randomly assigned fitness contributions of the N sites. Each genotype can change to $N(A - 1)$ one-mutant neighbors, and adaptive walks via fitter variants pass to local optima. In a spin-glass, each of the 2^N configurations has an overall energy, and each can change to N one-mutant neighbors by flipping a single spin to its opposite orientation. The configuration space of a spin-glass has an energy landscape which is directly analogous to a fitness landscape. At 0 K temperature, a spin-glass descends to local minima by flipping spins and passing to neighboring configurations—typically constrained to be one-flip neighbors—which are of successively lower energy. A main difference between the two models is that biologists like to think of adaptation climbing hills, while physicists like to think of systems minimizing energy.

Conflicting constraints in spin-glasses account for the rugged structure of their potential surfaces. Consider a set of four adjacent spins in a square-lattice spin-glass, where three of the four pairs prefer to point in the same direction, while the fourth prefers to point in the opposite direction. No arrangement of spin up and spin down around the square can satisfy all these constraints. Such a square is said to be *frustrated* (Anderson 1985). This frustration, which results from conflicting constraints, leads to a complex energy surface with very many potential minima. We shall find the direct analog in the *NK* model, for as *K* increases, the conflicting constraints lead to an ever more rugged multipeaked fitness landscape.

Many properties of the landscapes created by the *NK* model appear to be surprisingly robust and depend almost exclusively upon *N* and *K* alone. Therefore, it is important at the outset to explain which features are dependent on other aspects of the model.

One very sensitive feature of the model is the *range of fitness values* assigned to the space of genotypes. This range depends upon the assumption that the fitness value assigned to each of the A^{K+1} combinations of alleles influencing each allele *i* was drawn at random from the uniform interval between 0.0 and 1.0. I might instead have assigned values at random from different underlying distribution—for example, a peaked Gaussian distribution between 0.0 and 1.0 in which the random decimals are more likely to be near 0.5 than near 0.0 and 1.0, or a U-shaped distribution between 0.0 and 1.0 in which the random decimals are more likely to be near 1.0 or 0.0 than near 0.5. In the Gaussian case, this choice of range would tend to "squeeze" fitness values assigned to all possible genotypes closer to the mean of that distribution, 0.5. Use of the U-shaped distribution would tend to expand the deviation of fitness values assigned to all possible genotypes farther away from the mean fitness of the ensemble, 0.5. Since the actual fitness values assigned are sensitive to the choice of the underlying distribution used, I shall avoid properties of the *NK* model known to be sensitive to this choice.

The reasonably insensitive properties of the fitness landscapes generated by the *NK* model appear to include

- The number of fitness peaks in the genotype space
- The lengths of walks via fitter neighbors to fitness optima (equivalently, this is the number of accepted mutations on an adaptive walk)
- The total number of mutants tried before an optimum is reached
- The ratio of accepted to tried mutations on a walk
- The number of alternative optima to which one genotype can climb
- The number of genotypes which can climb to the same optimum
- The rate at which the fraction of fitter neighbors *dwindles* to zero along walks to fitness peaks
- The similarity of local optima

These properties are kinds of rank-order statistics. As *N* and *K* are changed, these statistical features of the corresponding more or less jagged fitness landscapes over genotype space alter. These features, however, are largely insensitive to the choice made for the underlying distribution—uniform, Gaussian, U-shaped, or other—and therefore to the range of fitness values in a landscape. This insensitivity rests on the fact that, for the moment, I shall count a neighboring genotype fitter than another even if the fitness difference is infinitesimal. Ignoring the actual fitness differences amounts to a kind of rank-ordering of the fitness of all possible genotypes. In this case, adaptive walks will pass from any genotype to any fitter genotype.

I should emphasize that, at present, all we can say is that these properties *appear* to be insensitive to the underlying distributions from which fitness values are assigned. This assertion is based on numerical investigations (Kauffman, Weinberger, and Perelson 1988). The extent of insensitivity warrants further study.

By focusing on the rank-order statistics of the NK family of landscapes, we achieve a class of models which yields substantial insight into the statistical structure of rugged fitness landscapes. However, the *flow* of an adapting population on such a landscape also depends critically on the fitness differences between adjacent genotypes in the space.

We turn next to an examination of landscape structure as a function of N and K. I discuss first the two extremes $K = 0$, which corresponds to the limit of a smooth landscape with a single fitness peak, and $K = N - 1$, which corresponds to a completely random landscape with very many peaks. Thereafter, I characterize the family of correlated landscapes which lies between these extremes.

K = *0 Corresponds to the Additive Genetic Model and Yields a Single-Peaked and Smooth Correlated Fitness Landscape: Fujiyama*

The first case to examine is the $K = 0$ limit, with the further condition that each gene has only two alleles. Then there are no epistatic interactions. We show now that the structure of this fitness landscape has a single global optimal genotype, that all other genotypes are suboptimal and can climb to the global optimum via fitter neighbors, and that all one-mutant neighbors have nearly the same fitness.

At each locus, by chance, either allele 0 or allele 1 makes the higher fitness contribution. Therefore, there is a special genotype having the fitter allele at each locus which is *the global optimum genotype.* Furthermore, any other genotype, which must of course have lower fitness, can be sequentially changed to the globally optimal genotype by successive flipping of each gene which is in the less favored allele to the more favored allele. Therefore, any such suboptimal genotype *lies on a connected pathway via fitter one-mutant variants to the global optimum.* It follows trivially that there are no optima other than the single global optimum. All other genotypes are below the global optimum and can climb to it.

In previous sections, I used the pictorial image of a rugged or smooth fitness landscape without defining the terms. A more precise term is the "correlation structure" of the fitness landscape. By this I mean how similar the fitness values of one-mutant neighbors in the space are. A *smooth* landscape is one in which neighboring points in the space have nearly the same fitness value. Knowing the fitness value of one point carries a lot of information about the fitness value of neighboring points. At the opposite extreme, a *maximally rugged* landscape is one in which the fitness values are entirely uncorrelated. Knowing the fitness at one point would then carry no information about the fitness of neighboring points. A variety of alternative measures can be used to characterize the correlation structure of a fitness landscape. I discuss one below.

The $K = 0$ additive model corresponds to a very smooth, highly correlated fitness landscape. This is clear because the fitness of one-mutant neighbors cannot differ by more than $1/N$. Therefore, for large N, the fitness of one-mutant neighbors is very similar.

Two other features of the $K = 0$ model with two alleles per locus are immediately understandable. If an adaptive walk starts anywhere and climbs via successively fitter one-mutant variants, then the number of fitter neighbors dwindles by 1 at each step upward. If the walk starts with a randomly chosen genotype, on average half the N loci are already in the more favored allele, the other half are in the less favored allele. Therefore, the expected number of steps to the optimum is just $N/2$. This implies that walk lengths to the global optimum increase linearly as N increases.

K = N — 1 Corresponds to Fully Random Fitness Landscapes

The largest possible value of K is $N - 1$. In this limit, each gene is epistatically affected by all the remaining genes. It is particularly easy to show that in this limit the resulting fitness landscape is entirely uncorrelated. In other words, the fitness value of one genotype gives no information about the fitness value of its one-mutant neighbors. As I show in this section, we can understand a number of quite surprising features of such extremely rugged fitness landscapes (Kauffman and Levin 1987; Weinberger 1988, 1991a; Macken and Perelson 1989). In particular, we shall see that

1. The number of local fitness optima is extremely large.
2. The expected fraction of fitter one-mutant variants dwindles by ½ at each improvement step.
3. The lengths of adaptive walks to optima are very short and increase only as a logarithmic function of N.
4. The number of mutants tried to reach an optimum is proportional to the dimensionality of the space.
5. The ratio of accepted to tried mutations scales is $\ln N/N$ for the two-allele case.
6. Any genotype can climb to only a small fraction of the local optima.
7. Only a small fraction of the genotypes can climb to any given optimum.

These seven features are all ordering properties of completely uncorrelated landscapes. Perhaps the most important implication of such landscapes, however, is this:

8. As the number of genetic loci N increases, the local optima fall toward the mean fitness of the space of genotypes.

As we shall soon see, this last feature carries over to a large class of rugged but correlated landscapes. This feature is so central that I shall call it a further kind of complexity catastrophe. It points to a fundamental restraint on adaptive selection. Conflicting constraints in complex systems limit the optimization of function possible. As we shall see, this limitation appears to be a very general problem for many classes of systems.

In the $K = N - 1$ limit, the fitness vector w_i for each gene i, $i = 1, 2, \ldots, N$, is a function of all $K + 1 = N$ genes. Consider any initial genotype among the 2^N genotypes with two alternative alleles at each locus. Alteration of the allele at any single locus affects each of the N genes, since that alteration alters the combination of the $K + 1 = N$ alleles which bear on the fitness of each gene. In turn, this second alteration alters the fitness contribution of each gene to a different randomly chosen value between 0.0 and 1.0. The fitness of the new one-mutant neighbor genotype is therefore a new sum of N random decimals between 0.0 and 1.0. Therefore, the new fitness value is *entirely uncorrelated* with the old fitness value. Since fitness values of one-mutant neighbors are entirely random with respect to one another, the $K = N - 1$ landscape is fully uncorrelated. Figure 2.2 is a concrete example.

The first point to stress is straightforward: since $K = 0$ corresponds to fully correlated smooth landscapes and $K = N - 1$ corresponds to fully uncorrelated rugged landscapes, it must be true that, as K increases, landscapes must change from smooth through a family of increasingly rugged landscapes to fully uncorrelated landscapes. *Increasing the richness of epistatic interations K increases the ruggedness of fitness landscapes.* Since increasing epistatic interactions simultaneously increases the number of conflicting constraints, increased multipeaked ruggedness of the fitness landscape as K increases reflects those increasingly complex mutual constraints.

The Rank-Order Statistics on K = N − 1
Random Landscapes

The Number of Local Optima Is Very Large.   We now calculate the expected total number of local optima with respect to one-mutant neighbors. In keeping with the hypothesis that walks must pass via fitter one-mutant neighbors regardless of how small the fitness differentials may be, it is convenient to rank-order all the genotypes from worst (1) to best (2^N). The probability P_m that any genotype is a local optimum is just the probability that it has higher rank-order than any of its N one-mutant neighbors:

$$P_m = \frac{1}{N + 1} \tag{2.2}$$

Since the total number of genotypes with two alleles per locus is 2^N, the expected total number of local optima with respect to one-mutant moves $M1$ is

$$M1 = \frac{2^N}{N + 1} \tag{2.3}$$

Therefore, the number of local optima is extremely large and increases almost as rapidly as the number of genotypes 2^N. This means that these extremely rugged landscapes are so rife with local optima that trapping on such optima is essential inevitable.

While I have considered only walks via fitter one-mutant neighbors, it is useful to calculate the number of local optima if walks can proceed by two-mutant, three-mutant, r-mutant neighbors. The denominator in Equation 2.3 is replaced by the total number of genotypes which can be reached in r or fewer mutations. This is just the cumulative binomial sum

$$\sum_{j=0}^{r} \binom{N}{j}$$

where $r = 1$ in Equation 2.3. Thus for any small value of r, as N increases, the number of genotypes increases exponentially, but the number of local optima with respect to walks via fitter r-mutant neighbors increases very rapidly as well.

It is easy to generalize the $K = N - 1$ model from two alleles per locus to an arbitrary number of alleles per locus A. The number of genotypes in the space is then A^N. As defined earlier, the *dimensionality* D of the space (the number of one-mutant neighbors to each genotype) is $N(A - 1)$. Substitution into Equation 2.3 by the number of genotypes in the numerator and the number of one-mutant neighbors in the denominator gives the expected number of local optima:

$$\frac{A^N}{D + 1}$$

The Expected Fraction of Fitter One-Mutant Neighbors Dwindles by ½ on Each Improvement Step.   The landscape is entirely uncorrelated. Let the adaptive walk begin from the lowest ranked genotype. All its D neighbors are fitter, with rank-orders spread randomly between 2 and A^N. The walk samples neighbors at random and moves to the first fitter one encountered. Since those fitter neighbors are spread uniformly in rank-order from just above the current genotype to the top, and since a random fitter neighbor is picked, on average, its rank-order lies *halfway to the top*. When the process moves to that neighbor, because it is expected to be halfway to the top, only half its one-mutant neighbors are still fitter. On average, each successive step jumps half the remaining distance to the top rank; hence at each step the expected number of fitter one-mutant neighbors dwindles by ½.

This argument replaces the mean of a family of such adaptive walks with a "mean walk." In short, on random landscapes the number of ways uphill increases rapidly. Recall, by contrast, that in $K = 0$ smooth landscapes, the number of ways uphill decreases only by 1 at each improvement step.

Walks to Local Optima Are Short and Vary as a Logarithmic Function of N.   In general, adaptive walks might begin anywhere. However, to obtain an upper bound on walk lengths, we consider walks which begin at the lowest ranked genotype. Because the walk steps halfway to the top at each step r, the expected *relative rank order x/T* at each step, where T is the top rank, is

$$\frac{x}{T} = \frac{2^r - 1}{2^r} \tag{2.4}$$

When the walk arrives at a genotype of relative rank-order x/T, that genotype is fitter than at least the one from which the process just left. Therefore the probability P_m that the newly encountered genotype is itself a local optimum is

$$P_m = \left(\frac{x}{T}\right)^{D-1} \tag{2.5a}$$

When x and T are small, this equation must be modified slightly to take account of the lack of replacement in calculating P_m (Kauffman and Levin 1987):

$$P_m = \frac{(x - 1)!(T - D - 1)!}{(T - 1)!(x - D - 1)!} \tag{2.5b}$$

Combining equations 2.4 and 2.5*a* allows us to calculate the probability P_l that an adaptive walk continues for l steps *without* encountering a local optimum:

$$P_l = \prod_{r=0}^{l} \left[1 - \left(\frac{2^r - 1}{2^r} \right)^{D-1} \right] \tag{2.6}$$

Each term in this product is 1 minus the probability that the current genotype reached, which has a relative rank-order given by Equation 2.4, is actually a local optimum, as given by Equation 2.5*a*. Hence each term is the probability that the current genotype is not a local optimum and therefore that the adaptive walk continues at least one more step.

As l increases, the product in Equation 2.6 decreases and eventually falls below 0.5. The value of l at which this occurs is the number of steps taken such that the walks will have arrested in half the trials, while the rest may continue. Thus Equation 2.6 yields an estimate of expected walk lengths before a local maximum arrests progress upward.

Equation 2.6 implies that the adaptive walks in uncorrelated landscapes are surprisingly short and tightly bounded. Note that if $r = \log_2 (D - 1)$, the corresponding term in P_l is

$$1 - \left(1 - \frac{1}{(D - 1)^{D-1}} \right)$$

which is extremely well approximated by $1 - 1/e = 0.63$. Moreover, if this term is the rth term in P_l, it is easily shown that the preceding terms are approximately $1 - 1/e^2 = 0.86, 1 - 1/e^4 = 0.98$. Thus there is very little probability of the process stopping more than one or two adaptive steps before $r = \log_2 (D - 1)$. To a high degree of accuracy, then, the adaptive walk will stop on average at the rth step when

$$r = \log_2 (D - 1) \tag{2.7a}$$

Equation 2-7*a* therefore shows that the expected lengths of adaptive walks r in uncorrelated landscapes are *short*, on the order of $\log_2$ of the number of neighbors of each entity in the space.

Weinberger (1988) and Macken and Perelson (1989) have carried out more detailed analysis of such walks, examining the entire distribution of walk lengths with similar results. Because adaptive moves which happen to step more than halfway to the top at each improvement are more likely to truncate an adaptive walk than moves which step less than halfway to the top are likely to lengthen the walk, Equation 2.7*a* is an overestimate of walk lengths to optima. Accounting for such fluctuations shows that adaptive walk lengths r are more nearly the natural logarithm of the number of one-mutant neighbors. This feature shows that, in highly rugged landscapes, there are so many peaks that the local ones which can be climbed from any point are very close in sequence space:

$$r \approx \ln(D - 1) \tag{2.7b}$$

The Expected Time to Reach an Optimum Is Proportional to the Dimensionality of the Space.   Consider a walk which begins at the lowest ranked genotype and climbs to a local optimum. Let the walk examine the D one-mutant neighbors sequentially, taking one unit of time for each examination. This corresponds to the waiting time for a mutational event but ignores the fact that such mutations do not examine all neighbors in order. The modification makes only a minor difference.

Since on average the adaptive walk steps halfway to the top rank at each improvement step, the expected waiting time to find a fitter variant *doubles* after each improvement step. This result is equivalent to the theory of records found in Feller's (1971) classical probability text. The first improvement step occurs after one moment; the second on average requires two moments; the third on average requires four moments. The expected number of improvement steps to reach a local optimum is $\log_2 (D - 1)$. Thus the expected waiting time t_{op} to reach that optimum is just

$$t_{op} = \sum_{l=0}^{\log_2(D-1)-1} 2^l \tag{2.8}$$

When $\log_2 (D - 1)$ is an integer, this series sums to $D - 1$. Therefore, t_{op} is proportional to the number of one-mutant neighbors or, equally, to the dimensionality of the space. The time to reach an optimum, of course, is also equivalent to the total number of mutants tried before reaching an optimum.

Macken and Perelson (1989) rederived these results and found a surprising additional fact: For large N and K, t_{op} is nearly independent of starting fitness.

The Ratio of Accepted to Tried Mutations Scales as ln **N/N** .   Since the length of an adaptive walk is the number of accepted mutations and the time is the number of tried mutations, the ratio of these is just ln N/N for the two-allele case. Macken and Perelson (1989) make the same point with the addendum that these results may be quite insensitive to starting fitness.

Any Genotype Can Climb to Only a Small Fraction of the Local Optima.   A maximal estimate of the number of branches to fitter variarnts which might emerge from the lowest ranked genotype can be obtained: D of its neighbors are fitter. On average, after a single improvement step, $(D - 1)/2$ (in other words, almost $D/2$) of the neighbors of that first-step variant are still fitter. After successive steps, on average $D/4$, $D/8, \ldots$ neighbors are fitter. Adaptive walks continue for about $\log_2 (D - 1)$ steps. The series $(D \times D/2 \times D/4 \times \ldots D/D)$ yields a gross upper bound on the expected number of alternative local optima accessible from the lowest ranked entity. This bound is

$$\frac{D^{\log_2 D}}{2^{(\log_2 D)(\log_2 D - 1)/2}} \tag{2.9a}$$

$$\frac{D^{\log_2 D}}{D^{(\log_2 D + 1)/2}} \tag{2.9b}$$

$$D^{(\log_2 D - 1)/2} \tag{2.9c}$$

This bound is an overestimate, since it ignores *convergence* of walks. Nevertheless, it suffices to establish the major point: Branching adaptive walks in uncorrelated landscapes reach only a small fraction of the total number of local optima. Thus for an uncorrelated genotype space using $A = 20$ alleles per locus and $N = 64$ genetic loci, an upper bound on the number of accessible local optima from the lowest ranked genotype is about 10^{14}; by contrast, the total number of local optima in the uncorrelated genotype space is about 10^{80}. Thus only 10^{14} among the 10^{80} optima are accessible from even the lowest ranked genotype.

This property means that any beginning sequence, no matter how poor, can climb to only a small range of local peaks via one-mutant neighbors and is limited to their fitnesses. As in New England, in rugged landscapes "you can't get there from here."

A Small Fraction of the Genotypes Can Climb to Any One Optimum.   It is useful to calculate an upper bound on this fraction. First, note that in general there is only a single genotype which is the global optimum in the landscape. (We ignore "ties.") Note next that the set of genotypes which can climb to that optimum via one-mutant fitter neighbors is identical to the set of genotypes that the global optimum could *descend to* via one-mutant *less fit neighbors.* Then the fittest genotype could reach D less fit one-mutant neighbors. Each of those on average has $D/2$ still less fit neighbors, and each of those has on average $D/4$ less fit neighbors. By the now familiar argument, this process continues for about $\log_2 D$ steps. Therefore, an upper bound on the total number of genotypes which can climb to the global optimum is

$$D + (D)\left(\frac{D}{2}\right) + (D)\left(\frac{D}{2}\right)\left(\frac{D}{4}\right) + (D)\left(\frac{D}{2}\right)\left(\frac{D}{4}\right)\left(\frac{D}{8}\right)$$
$$+ \cdots (D)\left(\frac{D}{2}\right)\left(\frac{D}{4}\right)\cdots\left(\frac{D}{D}\right)$$

This formula is exact only if D is a power of 2, and it is an upper bound because it ignores the possibility that a genotype might climb by two or more routes to the global optimum. It leads to an expression for the number of genotypes which can climb to the global optimum:

$$\sum_{i=1}^{\log_2 D} D^i 2^{-(i)(i-1)/2} \tag{2.10a}$$

Consider genotypes with $N = 256$ loci and $A = 2$ alleles. The number of genotypes is 2^{256}, or about 10^{77}, while the number of these which can climb to the globally optimal genotype is only about 2^{29}, or 10^9. Thus only a tiny faction of the genotypes can climb to even the global optimum if constrained to pass via fitter one-mutant variants. This constraint implies that the vast majority of adaptive walks via one-mutant fitter variants end on optima which are below the global optimum.

I am indebted to D. Lane (private communication) for an alternative simple expression for the number of genotypes which can climb to the global optimum. Consider a genotype on a random landscape able to descend to two less fit variants, each of which, on average, can descend to a single still less fit variant. That initial genotype is thus able to descend to four less fit variants. But its predecessor in descending from the global optimum had, on average, four less fit variants, and the next higher predecessor had eight less fit variants, and so on until the descent started at the global optimum. This leads to

$$\frac{4}{2} \times 2^1 \times 2^2 \times \cdots \times 2^M = \frac{4}{2} \times 2^{M(M+1)/2} \approx 2 \times 2^{M^2/2} \tag{2.10b}$$

where $2^M = D$ and M is the number of steps descended from the global optimum. In this notation, the total number of local optima $2^D/(N + 1)$ is $2^{2^M}/(2^M + 1)$, which is vast compared with $2 \times 2^{M^2/2}$ even for modest M.

A Complexity Catastrophe: As Complexity Increases, the Heights of Accessible Peaks Fall Toward the Mean Fitness. We now investigate the inexorable onset of a novel complexity catastrophe which limits selection. It is the consequence of attempting to optimize in systems with increasingly many conflicting constraints among the components: Accessible optima become ever poorer, and fitness peaks dwindle in height.

We have already noted that, as K increases, the number of conflicting constraints increases. Thus if $K = 0$, each gene can assume its most valuable allele independent of the choice at any other allele. If $K = 1$ and genes i and j influence each other, the optimal choice of allele for i in the context of all possible choices of alleles at j will typically not be identical to the optimal allele at j for all possible alleles at i. These conflicting constraints mean that the best mutual choices of alleles tend to be poorer overall. As K increases, the web of constraints becomes enormously complex. When K increases to $N - 1$, we might expect the conflicts among constraints to reach a maximum and hence tend to lead to poorer local optima than for smaller values of K. Indeed, this proves to be the case. We examine this complexity catastrophe first in the limit of fully random landscapes, when $K = N - 1$, as N grows larger. Below I show that the tendency for local optima to fall in fitness occurs for a range of NK landscapes in which K increases as a constant fraction of N.

To be concrete and make the general argument more transparent, I shall modify the NK model slightly and randomly assign fitness values not from the *uniform* distribution but from the extreme of a *U-shaped distribution* having only the extreme values in the range—0.0 and 1.0. Also for simplicity, I shall consider the case where each gene has two alleles. For each genetic locus, a fitness contribution of either 0.0 or 1.0 is assigned to each of the 2^N combinations of the two alleles of the $K + 1 = N$ loci bearing on the fitness contribution of that locus. As before, the fitness of any genotype is the average of the contributions from all N loci. Thus the fitness of any genotype is just the fraction of the N loci whose fitness contributions are 1.0. Therefore, the distribution of fitness values among the genotypes is now just the binomial distribution of the sum of N random variables x, $x = 0.0$ or 1.0. The mean of the distribution is 0.5, and the distribution approaches Gaussian rapidly as N increases.

Note that I am now considering not a landscape in which fitness is only rank-ordered but one in which the fitness values are real-valued and are drawn at random from a defined range, 0.0 to 1.0.

A simple way to think about the lengths of adaptive walks is that they continue until the expected number of fitter one-mutant neighbors drops below 1. Since each genotype has N neighbors, on average, walks stop when the expected fraction of fitter neighbors just falls below $1/N$.

The fitness of a genotype f is the fraction of its N loci which make fitness contributions of 1.0. Because the fitnesses of one-mutant neighbors are random, the expected proportion of one-mutant neighbors which have higher fitness than that of a given genotype is simply the *probability* in the "right tail" of the binomial distribution above the fitness of that genotype (Figure 2.3). As f increases, say from 0.6 to 0.7, the probability above this increased fitness decreases. Walks will continue to higher fitness values until the expected fraction of fitter neighbors falls to $1/N$ (Figure 2.3).

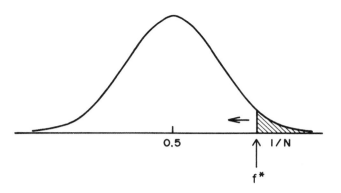

Figure 2.3 Gaussian approximation to the distribution of fitness values in an uncorrelated landscape with a true binomial distribution and large N. The fraction of fitter neighbors at any given fitness f is the probability in the right tail of the distribution above f (shaded area). Adaptive walks stop, on average, at the fitness f^* whenever the number of fitter neighbors drops below one; hence when the fraction of fitter neighbors drops below $1/N$, as shown by the shaded area. As N increases, f^* decreases toward 0.5.

I now show that as N increases, the fitness values of attainable local optima decrease toward 0.5. For N large, the central limit theorem shows that the fitness has approximately a normal distribution with mean 0.5 and variance $1/4N$. Using the above arguments, the expected fitness f of an attainable local optimum is found from

$$\Pr(x > f) = 1/N \tag{2.11}$$

where x has a normal distribution with mean 0.5 and variance $1/4N$. If $f^* = f - 0.5$ and the approximation

$$\int_y^\infty \frac{1}{\sqrt{2\pi}} e^{-u^2/2} \, du \tag{2.12a}$$

$$\approx \frac{e^{-y^2/2}}{y\sqrt{2\pi}} \tag{2.12b}$$

is used, then Equation 2.11 gives for f^*:

$$\left(\frac{0.199\sqrt{N}}{f^*}\right)(e^{-2(f^*)^2N}) = 1 \tag{2.13}$$

This expression is adequate for reasonably large values of N.

Equation 2.13 implies that, as the number of genes N increases, the accessible optima dwindle in height toward the average unselected fitness in the space of genotypes (Figure 2.4). Thus inexorably in these landscapes, adaptive walks terminate on poorer "solutions" as N increases.

I believe this to be a genuinely fundamental restraint facing adaptive evolution. As systems with many parts increase both the number of those parts and the richness of interactions among the parts, it is typical that the number of conflicting design constraints among the parts increases rapidly. Those conflicting constraints imply that optimization can attain only ever poorer compromises. No matter how strong

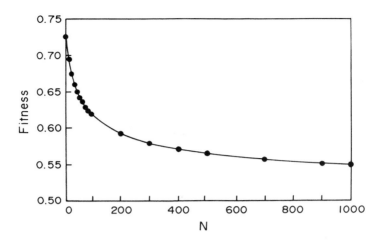

Figure 2.4 The complexity catastrophe. As N increases, the expected fitness of local optima f^* decreases toward the mean of the space, 0.5.

selection may be, adaptive processes cannot climb higher peaks than afforded by the fitness landscape. That is, this limitation cannot be overcome by stronger selection.

It might be objected that *normalization* is essential to this limitation on selection and that such normalization is an arbitrary assumption of the model. Yet if we include "costs per part," it is clear that some form of something like normalization is a natural and general consideration. Further, it is clear that conflicting constraints are a very general limit in adaptive evolution. Each part of a complex system costs something. For example, additional genes and proteins require metabolic energy.

As a concrete example, suppose we do not normalize fitness in the NK model and consider the total fitness of the system, ignoring any cost per part. Then as N increases, both the total complexity, which equals N, and the maximum possible fitness, which also equals N, increase *without bound*. Nevertheless, increasingly high total costs will typically bound the overall fitness which can be achieved. In the $K = N - 1$ case, total fitness increases as N increases, albeit ever more slowly. But suppose cost per part is constant, so that total cost rises linearly. At some point, total cost exceeds total fitness. Further increase in complexity—increasing N—is no longer profitable. This argument shows that there is again a limit on the complexity which can be attained. The marginal increase in fitness for the next part must be positive. The complexity catastrophe due to conflicting constraints captured in Equation 2.13 is therefore a general property of complex systems.

The Tunable NK *Family of Correlated Landscapes*

The NK model was invented not to explore the two extreme landscapes, but to have in hand a model which allowed construction of an ordered family of tunably correlated landscapes. In order to investigate the statistical properties of landscapes for different values of the fundamental parameters of the model, numerical simulations were carried out (Kauffman, Weinberger, and Perelson 1988; Weinberger 1989). The four main parameters of the model are N, K, the distribution of K among the N, and the number of alleles at each site A. A fifth parameter is the underlying distribution from which fitness values are assigned to each site for each combination of alleles at the $K + 1$ sites bearing on the site in question. The properties we investigated include

- The fitness of local optima
- The lengths of adaptive walks to optima
- The dwindling fraction of fitter neighbors at each step along a walk
- The mean waiting time to find a fitter variant
- The number of local optima
- The similarity between local optima
- The number of walks from different initial genotypes which climb to each local optimum, hence the attracting basin size of each optimum
- The autocorrelation of the fitnesses encountered along a random walk in the landscape as a measure of the correlation structure of the landscape

To investigate the first four properties, numerical simulations were carried out for different random examples of the NK landscapes for fixed N and K values, initiating first adaptive walks from a random initial genotype and then hill climbing via a randomly chosen fitter one-mutant variant of each successively fitter genotype to the nearest optimum. Walks were carried out on 100 different randomly chosen landscapes for the same values of the model parameters. The numbers reported are the means of those 100 simulations for each value of the parameters.

Perhaps the most surprising feature of the results is that most aspects of these landscapes are so nearly insensitive to any parameters but N and K.

Table 2.1 shows the average fitness of local optima in the case where the K epistatic inputs to each gene were chosen to be its flanking $K/2$ neighbors to either side. To avoid boundary effects, we consider "circular" genomes. Each gene was limited to $A = 2$ alleles, 0 and 1. The table shows first that, for $K = 0$, the fitness of optima are independent of N and equal to about 0.66. This is expected because the fitness values were drawn at random between 0.0 and 1.0 for allele 0 and allele 1. Order statistics shows that the average value of the less fit allele will be ⅓ and that of the more fit allele will be ⅔. Since each site contributes additively to the overall fitness, which is merely the mean fitness per site, the global optimum should be independent of N and should

TABLE 2.1 Mean Fitness of Local Optima (Nearest-Neighbor Interactions)

	N				
K	8	16	24	48	96
0	0.65(0.08)*	0.65(0.06)	0.66(0.04)	0.66(0.03)	0.66(0.02)
2	0.70(0.07)	0.70(0.04)	0.70(0.08)	0.70(0.02)	0.71(0.02)
4	0.70(0.06)	0.71(0.04)	0.70(0.04)	0.70(0.03)	0.70(0.02)
8	0.66(0.06)	0.68(0.04)	0.68(0.03)	0.69(0.02)	0.68(0.02)
16		0.65(0.04)	0.66(0.03)	0.66(0.02)	0.66(0.02
24			0.63(0.03)	0.64(0.02)	0.64(0.01)
48				0.60(0.02)	0.61(0.01)
96					0.58(0.01

*Numbers in parentheses are standard deviations. Along main diagonal of table, K = N, actual K value is K − 1.

be ⅔. By a similar argument, for an arbitrary but fixed number of alleles A, the average fitness of the fittest of these alleles if drawn at random from the uniform distributuion between 0.0 and 1.0 is $A/(A + 1)$.

Table 2.1 shows that the complexity catastrophe does certainly occur in the $K = N - 1$ limit. The fitnesses of accessible local optima begin high and dwindle toward 0.5 as N and K increase. In order to investigate whether the same complexity catastrophe would occur, as expected, regardless of whether the fitness contributions per site were drawn from the uniform interval between 0.0 and 1.0 or from some other distribution, we investigated a V-shaped distribution which favored values nearer 0.0 or 1.0 and an inverted V, or "humped," distribution which favored decimals near 0.5. In all cases, as expected, in the uncorrelated landscape limit with $K = N - 1$, the fitnesses of optima recede toward 0.5 as N and K increase (data not shown).

Three further features of Table 2.1 are important. First note that, if K is fixed while N increases, then the fitnesses of optima do not fall. Second, the fitness of optima for fixed small values of K ($K = 2$, $K = 4$, $K = 8$) is higher than for $K = 0$. Thus low levels of epistatic interaction appear to buckle the landscape like heaving-up mountain ranges and yield fitter optima than those available in the simplest additive Fujiyama landscape. Third, the table shows that, as K increases as a constant proportion of N, the fitness values may transiently increase but ultimately *fall toward 0.5*. This third point demonstrates that if K increases linearly with N, the complexity catastrophe sets in. Therefore, there is some very broad set of landscapes within the NK family subject to the limitation that optima recede to the mean of the space as N increases. In contrast, the complexity catastrophe as N increases appears to be averted for small fixed values of K.

These behaviors of optima fitness as a function of N and K suggest that two major regimes within the NK family of landscapes exist: (1) K remains small, of order 1, as N increases and (2) K grows with N, hence is of order N, as N increases. In the latter case, the complexity catastrophe arises. In the former case, optima remain high. Presumably these two regimes are different from each other in many basic respects which require investigation.

Table 2.2 shows similar results for the case where the K genes which epistatically affect each locus were chosen entirely at random for each locus. The constraint to circular genomes is removed, and no reciprocity in epistatic influence is assumed. The main feature of Table 2.2 is that it is nearly the same as Table 2.1. Therefore, within the NK family of correlated landscapes, optima fitness is largely insensitive to the distribution of the K among the N.

Table 2.3 shows the average number of steps via fitter one-mutant variants with

TABLE 2.2   Mean Fitness of Local Optima (Random Interactions)

K	\multicolumn{5}{c}{N}				
	8	**16**	**24**	**48**	**96**
2	0.70(0.06)	0.71(0.04)	0.71(0.03)	0.71(0.03)	0.71(0.02)
4	0.68(0.05)	0.71(0.04)	0.71(0.04)	0.72(0.03)	0.72(0.02)
8	0.66(0.06)	0.69(0.04)	0.69(0.04)	0.70(0.02)	0.71(0.02)
16		0.65(0.04)	0.65(0.03)	0.67(0.03)	0.68(0.02)
24			0.63(0.03)	0.65(0.02)	0.66(0.02)
48				0.60(0.02)	0.62(0.02)
96					0.58(0.01)

Along main diagonal of table, K = N, actual K value is K − 1.

TABLE 2.3   Mean Walk Lengths to Local Optima
(Nearest-Neighbor Interactions)

K	N				
	8	16	24	48	96
0	1.5(1.2)	8.6(1.9)	12.6(2.2)	24.3(3.4)	48.8(4.6)
2	4.1(1.9)	8.1(3.2)	11.2(3.1)	22.5(4.6)	45.2(6.6)
4	3.2(1.8)	6.6(2.5)	9.4(2.9)	19.3(3.9)	37.3(6.1)
8	2.7(1.5)	4.7(2.3)	7.7(3.0)	15.3(4.3)	27.7(5.3)
16		3.3(1.7)	4.8(2.1)	9.6(3.0)	19.3(4.2)
24			3.5(1.4)	7.4(3.0)	5.0(3.9)
48				3.9(1.9)	8.9(3.0)
96					5.1(2.4)

Along main diagonal of table, K = N, actual K value is K − 1.

K drawn from neighbors as in Table 2.1. For $K = 0$, walk lengths to optima are about $N/2$. For $K = N - 1$, walk lengths are close to ln N. The deviation may reflect sampling effects. Thus walk lengths vary from linear in N to logarithmic in N as K increases. Table 2.4 shows similar data for the randomly chosen K case of Table 2.2. Again, Tables 2.3 and 2.4 are very similar. Thus whether the K epistatic inputs to a gene are its neighbors or random among the N has almost no bearing on the lengths of walks to optima.

As walks proceed toward optima, the number of fitter one-mutant neighbors dwindles to zero. The reciprocal of the fraction of fitter one-mutant neighbors is the expected wait time to find such a fitter variant. Figure 2.5a shows the natural logarithm of the average number of fitter mutants; Figure 2.5b shows the natural logarithm of the wait time for $N = 96$ as K increases for the neighboring K cases. Data for K random are similar. Note that for $K = 0$ the fraction of fitter neighbors dwindles slowly, and as K increases the fraction dwindles ever more rapidly. Furthermore, for $K = 2$ or more, the fraction of fitter neighbors falls off by approximately a constant fraction at each improvement step. That is, the fall off in fraction of fitter neighbors is approximately exponential for $K \geqslant 2$. The slope increases as K increases, toward the expected limit (log$_2$) of a fully random landscape. That the slope is approximately log-linear even for $K = 2$ is interesting, for it suggests that this is a general feature of rugged landscapes even when those landscapes remain quite highly correlated.

Figures 2.6a–d examine the ruggedness of the fitness landscape in the one-mutant

TABLE 2.4   Mean Walk Lengths to Local Optima
(Random Interactions)

K	N				
	8	16	24	48	96
2	4.4(1.8)	8.1(2.8)	12.5(3.8)	26.5(5.1)	49.6(6.1)
4	3.6(1.8)	7.3(2.9)	10.9(3.3)	22.9(5.6)	44.5(7.9)
8	2.7(1.5)	5.3(2.5)	8.0(3.2)	17.0(4.3)	34.7(6.5)
16		3.3(1.7)	4.8(2.1)	10.1(3.4)	21.6(4.8)
24			3.5(1.4)	7.4(2.6)	16.0(4.3)
48				3.9(1.9)	9.3(2.6)
96					5.1(2.4)

Along main diagonal of table, K = N, actual K value is K − 1.

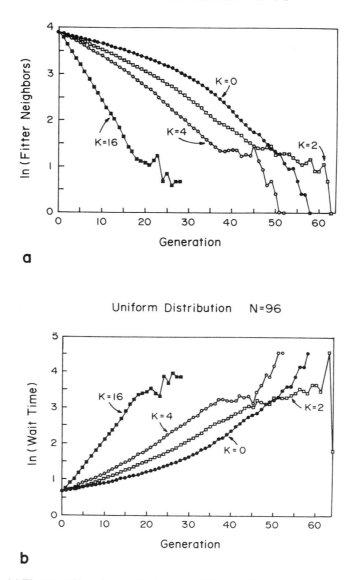

Figure 2.5 (*a*) The natural logarithm of the fraction of fitter neighbors at each step on walks to local optima for $N = 96$ and various values of K. Curves are each the mean of 100 walks on landscapes with the same values of N and K. (*b*) The natural logarithm of wait time to find a fitter neighbor at each adaptive step.

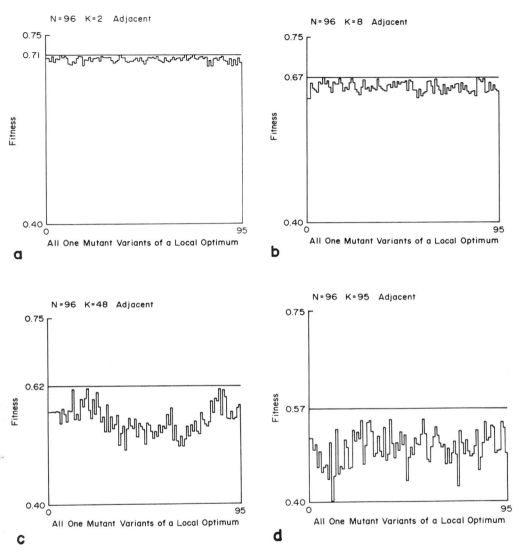

Figure 2.6   The ruggedness of *NK* landscapes for *N* = 96, showing the fitness of all 96 one-mutant variants of a random local optimum on one landscape for increasing values of *K*. (*a*) *K* = 2; (*b*) *K* = 8; (*c*) *K* = 48; (*d*) *K* = 95.

vicinity of local optima. More precisely, walks were carried to local optima, and then the fitness of all one-mutant variants of such optima were assessed. Our expectation was that, as *K* increases, the landscapes become progressively less correlated. Therefore, the fitness drops away from optima that occur for high values of *K* should tend to be more precipitous than the drops in the gentler landscapes having low values of *K*. This is clearly seen, implying that the selection gradients back to optima are steeper in rugged than in gentle landscapes. This variation in gradients will clearly

TABLE 2.5   Number of Optima

	N	
K	8	16
2	5*	26
4	15	184
7	34	
8		1109
15		4370

*All values are a mean of 20 landscapes.

affect how populations adapt on such landscapes, as we discuss below. Similar results are found for K random landscapes as K increases.

Properties 5 through 7 (p. 55) are *nonlocal* features of NK landscapes. To investigate these properties, we randomly chose a specific NK landscape and carried out many adaptive walks to local optima from randomly chosen initial genotypes in the space. To establish the number of local optima, searches were carried out either until no further optima were uncovered or until 10 000 optima were discovered. Any such algorithm has the difficulty that optima with very small basins of attraction may be missed. The sampling is therefore biased by the distribution of basin sizes. Therefore, current numerical values are estimates. Table 2.5 shows the pooled results for adjacent and random choices of K and different values of N and K.

A Massif Central in K = 2 Landscapes

Among the most surprising features of the NK family of landscapes is the fact that, for small values of K and two alleles, the local optima are not distributed randomly in genotype space but instead are near one another. Thus there is a *global structure to the fitness landscape.* More precisely, *the highest optima are nearest to one another.* The natural measure of the distance between two genotypes with only two alleles per locus, 1 or 0, is the *Hamming distance,* which is the number of positions at which the alleles differ. Thus (00000) and (10000) have a Hamming distance of 1. If local optima were distributed randomly in genotype space, the average Hamming distance between two local optima would be $N/2$. Furthermore, the Hamming distance from the highest local optimum (that is, the fittest found) to the second highest local optimum would on average be $N/2$. This is clearly not the case for small values of K, whether the K epistatic inputs per site are adjacent to that site or even randomly distributed among the N. Figure 2.7 shows the correlation between the fitness of local optima and their Hamming distance from the highest optimum found. Note the striking fact that for K small—for instance, $K = 2$—the highest optima are nearest one another. Further, optima at successively greater Hamming distances from the highest optimum are successively less fit. Thus, as stated, there is a global order to the landscape. LIke the Alps, our landscape here possesses a kind of Massif Central, or

Figure 2.7   The correlation between fitness of local optima and their Hamming distance from the fittest local optimum found. In all cases, $A = 2$ alleles at each site. (*a–c*) The K sites affecting each site are drawn at random from among the N. (*d–f*) The K sites affecting each site are that site's flanking neighbors.

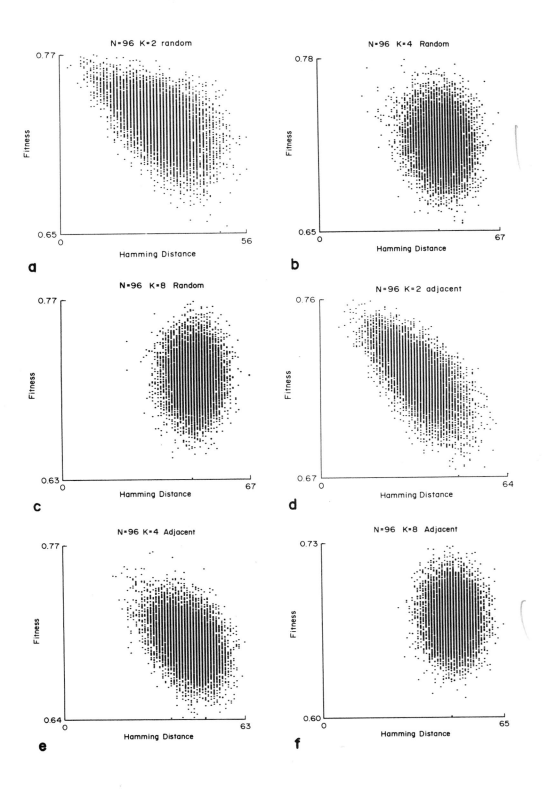

61

high region, of genotype space where all the good optima are located. As K increases, this correlation falls away, more rapidly for K random (Figures 2.7a–c) than for K adjacent (Figures 2.7d–f). Some insight into this phenomenon is offered in Chapter 6.

The NK family of landscapes offers a further surprise: The distribution of basin sizes climbing to specific optima can be very nonuniform. Some basins are enormous. Further, for K small, there is a tendency for the *highest optima to have the biggest basins.* These phenomena are shown in Figure 2.8. Thus, simultaneously, the global order of the Massif Central in these landscapes is expressed by the fact that the highest optima are nearest one another and have the largest drainage basins. In turn, this implies that one high local optimum has information about where other good local optima are. And further, the region *between* two high local optima is a good

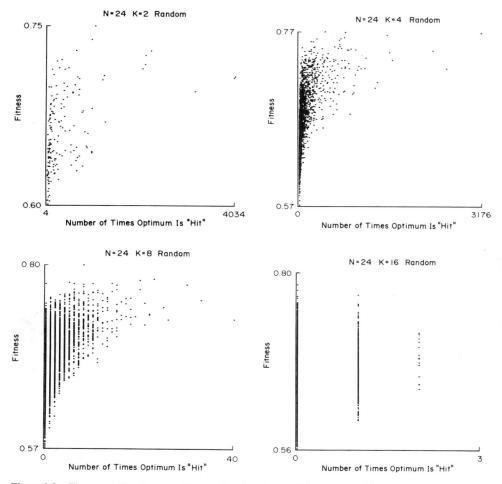

Figure 2.8 The correlation between fitness of local optima and the number of times each local optimum was attained on independent adaptive walks from random initial genotypes on the same landscape. In all cases, $A = 2$ alleles at each site.

area to search for still higher local optima. I show below that this mutual information carries marked implications about the usefulness of genetic recombination as a search strategy in rugged correlated fitness landscapes.

Note, finally, in Figure 2.8 that, as K increases relative to N, the tendency for the highest optima to have the largest drainage basins dwindles. But even for modest values of K, some very large basins persist. The grounds for this are not clear.

The results given apply to the case where each gene has two alleles. Less is known for larger numbers of alleles per site. It is clear, however, that the complexity catastrophe still occurs as K increases, and a tendency for high optima to have large drainage basins also persists for small values of K. However, with a large number of alternative alleles, high optima are not near one another for small values of K.

Measuring the Correlation Structure of Fitness Landscapes

The eighth statistical property of landscapes was studied with a procedure due to Weinberger (private communication). Earlier in the chapter, I defined a correlated fitness landscape as one in which the fitness values at one point were more or less similar to the fitness values at neighboring points. Weinberger (1991) has suggested use of the autocorrelation function to measure the correlation structure of fitness landscapes and has applied it to the NK model. His idea is to begin at an arbitrary genotype, then walk randomly via one-mutant neighbors across the landscape. At each step, however, the fitness value of the genotype encountered is recorded. If the landscape is correlated, then nearby steps along the walk should have similar fitness values. This is just what Weinberger found. The autocorrelation function relates the fitness of two genotypes along the walk which are s steps apart:

$$R(t,s) = \frac{E(f_t \times f_{t+s}) - E(f_t) \times E(f_{t+s})}{\text{variance } (f)} \qquad (2.14)$$

where E is the expected or mean value, f_t is the fitness of the genotype at the tth step, and f_{t+s} is the fitness of the genotype s steps farther along the walk. Weinberger estimated Equation 2.14 by considering random walks of 2048 steps, for all $2048 - s$ possible pairs of genotypes s steps apart along such a walk. His results are shown in Figure 2.9. The first feature to note is that the landscape is correlated for $K < 95$ but, as expected, uncorrelated for the fully random $K = N - 1$ landscape. Second, for each value of K the autocorrelation initially falls off exponentially with increasing distance between genotypes along the random walk in sequence space. Thus there is a natural correlation length for each landscape. Third, as K increases, the autocorrelation drops off faster. Hence as K increases, genotypes a fixed distance apart in the space have fitness values which are less correlated. Weinberger suggests the possibility that if these properties are generally insensitive to the details of the underlying distributions from which fitness values were assigned in the NK model or, more generally, are insensitive to a wide range of Gaussian statistical landscapes of which the NK model is an example, then the autocorrelation function may be a useful measure of many fitness landscapes.

Other Combinatorial Optimization Problems and Their Landscapes

The purpose of this chapter has been to discuss the structure of rugged fitness landscapes. I have done so in terms of the NK model. I bring the discussion to a close by

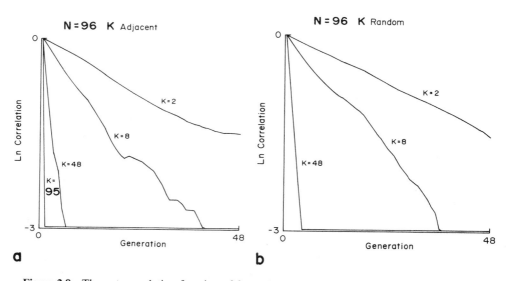

Figure 2.9 The autocorrelation function of fitness between pairs of configurations encountered along a random walk on a rugged landscape for different values of K relative to N. (a) The K sites are neighbors. (b) The K sites are randomly chosen.

mentioning a few other examples of complex combinatorial optimization problems. Each has a complex fitness or cost landscape.

I have already briefly mentioned spin-glasses. These disordered magnetic materials have afforded a rich source of models of complex potential surfaces (Edwards and Anderson 1975; Sherrington and Kirkpatrick 1975; P. W. Anderson 1985; Binder and Young 1986; Stein, Baskaran, et al. 1987). Analysis of these systems has explored topics such as the number of local minima, the spin overlaps between minima, the heights of potential barriers between minima, and slow relaxation times at finite temperatures as a spin-glass explores its potential surface.

The close relation between the NK model and spin-glasses again warrants comment. As noted, the NK model is a version of a spin-glass in which the fitness contribution of each site is written as a sum of terms depending upon the "allele" at that site, plus terms for pairwise interactions of that allele with each of the K others impinging upon it, plus more terms for all the triadic combinations among the $K + 1$ sites, plus terms for all combinations up to the "K-adic." This version is closely related to a spin-glass model investigated by Gross and Mezard (1985), which examines a Hamiltonian energy surface given by the sum of all possible K-adic interactions among the spins. The Gross–Mezard model appears to be almost identical to a model introduced by Amitrano, Peliti, and Saber (1991) to study molecular evolution on rugged landscapes. The limit of the Gross–Mezard spin-glass when K approaches N is Derrida's random-energy spin-glass model (1981). In fact, the random-energy model is identical to the NK model in the $K = N - 1$ limit. Thus results we have obtained for this limiting case also apply to the corresponding model in statistical physics.

The most famous combinatorial optimization problem is the traveling salesperson problem (TSP). The task is to begin at one of N cities, travel in turn to each city, and return to the initial city by the shortest total route. This problem, so remarkably simple to state, is extremely difficult (Lin and Kernighan 1973; Johnson and Papadimitriou 1985; Kirkpatrick and Toulouse 1985; Brady 1986; Kauffman and Levin 1987). At present, it is believed that the only way to find the globally optimal tour is to examine all possibilities. Any tour through the N cities is one of the possible $N!$ permutations if choice of the initial city is ignored. Therefore, exhaustive search explodes rapidly as N increases. The TSP is the paradigm case of an "NP" (nonpolynomial) complete problem (Lin and Kernighan 1973).

Tours can be located in a "tour space," each tour next to "one-mutant" neighbor tours, which differ by exchanging the links between two pairs of cities. An adaptive walk via fitter one-mutant tours begins at a tour and walks via fitter neighbors to a local optimum. The distribution of optima found in this way appears to be roughly Gaussian and far from the optimal tours (White, Solla, and Sorkin 1986).

Problems similar to the TSP arise in computer design in attempts to minimize wiring placement on silicon chips (Kirkpatrick, Gelatt, and Vecci 1983; Anderson 1985). Another such problem is that of partitioning a graph consisting of a set of points connected by a set of lines into two disjoint sets of points such that each set has connections only to members of the other set (Anderson 1985; Stadler and Happel 1992).

Despite the body of research on how to optimize in these problems and the very large number of well- and poorly studied complex combinatorial optimization problems, there is only the beginning of a literature on the structure of these landscapes (Stadler 1992; Stadler and Schnabl 1992).

The NK family of correlated landscapes is but one such family. It is an entirely open question whether there may be a few fundamental families of correlated landscapes or an extremely large number of such families. This is a critical question for further investigation. The grounds to hope that there may be only a few fundamental families is that in many areas of statistics a few fundamental distributions have proved important. Should we eventually discover that rather few families of correlated landscapes exist, then it might prove possible to measure a few parameters of a given rugged correlated landscape and discern both the family to which it belongs and the way adaptive strategies based on mutation/selection or some other process might best optimize on that particular landscape. Whatever the answer may prove to be, the utter simplicity of the NK family lends credence to my impression that it is likely to be a fundamentally important family of correlated landscapes.

There are, in fact, clues that many complex combinatorial optimization problems can be mapped by a few parameters onto the NK family of rugged landscapes. Weinberger (personal communication) recently applied the autocorrelation measure of fitness landscapes to an apparently very different problem (described in Chapter 3): RNA folding stability for model RNA sequences. He found that the autocorrelation structure of the model RNA stability landscape corresponded, via the NK model, to an effective K of 8 for RNA sequences of length $N = 70$. Utilizing these parameters and the NK model, he was able to predict the observed number of local optima in model RNA sequence space for folding stability. Finally, Weinberger (1991) has proposed that a wide class of landscapes are at least qualitatively described by NK landscapes having intermediate values of K. He shows that NK landscapes are generic members of a class of so-called AR(1) landscapes, in which all correlations are completely determined by correlations between neighboring points in the space. In this case, the sequence of fitness values obtained via an unbiased random walk on the landscape must, for large landcapes, be an AR(1) (Ornstein–Ulenbeck) process. Such processes are the most general stationary, Gaussian random process that are also first-order Markovian. Such processes are common among complex combinatorial optimization problems (Sorkin 1988), including the traveling salesman problem (Stadler and Schnabl 1991).

The success in mapping a model RNA folding problem, the traveling salesman problem, and others into AR(1) landscapes and onto the NK model, which is a generic member of the AR(1) family of landscapes, suggests that the statistical properties of the NK model can be used to predict features of a wide range of other complex cost surfaces.

SUMMARY

This chapter has introduced the concept of rugged fitness landscapes. Such landscapes undoubtedly underlie adaptive evolution at the molecular and morphological levels. In order to study the structure of such landscapes, I have carefully eschewed discussion of the fitness of whole organisms in their environment. I have instead defined fitness narrowly, as, for example, the capacity of peptides to bind to a specific antibody molecule or of proteins to catalyze a specific reaction under standard conditions.

The major discussion of the chapter has explored the expected structure of discrete fitness landscapes corresponding to sequence spaces, such as protein space. Here, evolution can be considered as an adaptive walk from proteins having low fitness for a specific function to or toward proteins having high fitness for that function. In the simplest case, walks proceed via fitter one-mutant variants to local optima in the space of possibilities. Our analysis has focused upon the statistical features of such walks as a function of the ruggedness of fitness landscapes. Important aspects of adaptive walks include the number of local optima, the number of steps on a walk to a local optimum, the average number of mutants tried on such a walk, the average number of mutants accepted on such a walk, the rate at which fitter one-mutant neighbors dwindle to zero along an adaptive walk as local optima are attained, the average number of local optima accessible by alternative adaptive walks from a given starting point, the number of points which can climb to the global optimum, the relative locations of local optima with respect to one another and with respect to the global optimum, and the basin sizes draining to local optima as a function of the height of those optima.

We have found that all these properties differ as a function of how rugged and multipeaked the fitness landscape is. In particular, I introduced the NK family of fitness landscapes to explore these issues. This family is among the first studied to explicitly invite us to explore the statistical structure of fitness landscapes. In the model, tuning the epistatic coupling parameter K relative to N increases the ruggedness of the landscape in a controlled manner from single-peaked and smooth for $K = 0$ to fully random for $K = N - 1$.

I have developed and presented the NK model in detail for the following purposes:

1. There can be no doubt whatsoever that the real adaptive evolution of proteins for specific catalytic tasks, ligand binding tasks, and other purposes confronts fitness landscapes with some statistical characteristics. If the NK model were to serve no other purpose than to tune our intuitions about what such landscapes might look like, that alone would warrant our attention.

2. In the interpretation of this model as proteins, the *NK* family of landscapes may or may not prove to be useful in predicting the structure of real fitness landscapes with respect to protein evolution.

3. The model is a serious sustained one of epistatic interactions in a haploid genetic system. It is open to direct generalization to diploid genetic models in which the fitness contribution of each allele on each chromosome now is a randomly assigned function of the alleles at the $K + 1$ sites on both chromosomes.

4. The *NK* family already has led us to a clear and largely unrecognized feature of many complex combinatorial optimization processes: As the number of conflicting constraints increases, not only do the corresponding landscapes tend to become more multipeaked and rugged, but for an unknown and very large family of landscapes the peaks recede toward the mean fitness in the space of possibilities. This is clearly important. Organisms are complex. Do they avoid this limitation? If so, how? If not, what are the implications?

5. The model suggests at least one means of mitigating the conflicting-constraints complexity crisis. If the number of epistatic interactions K remains small while N increases, landscapes retain high accessible local optima. This is a first hint of something like a construction requirement to make complex systems with many interacting parts which remain perfectible by mutation and selection. Each part should directly impinge on rather few other parts.

The complexity catastrophe is averted in the *NK* model for those landscapes which are sufficiently smooth to retain high optima as N increases.

CHAPTER 3

Biological Implications of Rugged Fitness Landscapes

We have now investigated in some detail the *structure* of multipeaked fitness landscapes. The present chapter begins to discuss the biological implications of such landscapes, for it is the fixed or deforming structures of fitness landscapes which are the nexus of adaptive evolution. The first sections of the chapter take up a number of phylogenetic implications of the fact that adaptive evolution occurs on rugged fitness landscapes. The second half of the chapter discusses the implications of rugged fitness landscapes for the flow of a population under the drives of mutation, recombination, and selection.

The first issues we must examine are the consequences of relaxing the assumption that adaptive walks occur via fitter one-mutant or few-mutant variants. In the limit, adaptive walks might pass via very distant mutants—in other words, via long jumps across genotype space. We shall find a "universal law" showing that, in such long-jump processes, the rate of finding fitter variants slows very rapidly. We shall also find that such long-jump adaptation confronts the complexity catastrophe already identified: Optima attained fall ever lower.

The characteristic structure of rugged landscapes and simple adaptive walks lead one to expect radiation and stasis as a generic feature of adaptation on relatively fixed fitness landscapes. Yet familiar theories of radiation and stasis place no emphasis on these features of fitness landscapes. The generic features of rugged landscapes also appear to bear on the famous Cambrian explosion, when many new phyla were created in short evolutionary order. Landscape ruggedness also influences the degree of divergent and convergent evolution, and hence is necessarily important in assessing morphological evolution and the systematists' problem of "homoplasy" due to convergent evolution. Homoplasy, usually viewed as "noise" by systematists, is in fact data about the structure of fitness landscapes. Finally, the simple points raised in this section may also have bearing on the famous embryological and evolutionary dicta known as von Baer's laws. We shall first find a puzzle in these laws and then attempt to rederive them.

PHYLOGENETIC IMPLICATIONS
OF RUGGED LANDSCAPES

A "Universal Law" for Long-Jump Adaptation

In Chapter 2, we limited our discussion to adaptive walks via fitter one-mutant variants. However, in real biological populations with a fixed mutation rate, members of the population exhibit a distribution of point mutations (Crow and Kimura 1965, 1970; Ewens 1979). In addition to point mutations, which lead to substitutions of single amino acids, *frame-shift mutations* due to deletion or addition of nucleotides can alter most or all of the amino acids incorporated into a protein downstream from the mutation site. Such mutations, in a single event, "jump" a long distance across protein space. More generally, recombination between two genes can, in a single event, lead to a hybrid protein made up of the left half of one gene and the right half of the other (Yourno, Kohno, and Roth 1970). Such recombined proteins—or, more generally, genes—represent long jumps across sequence space. Similar jumps can occur in the space of genomic regulatory systems controlling cellular differentiation and morphogenesis. Thus, real populations evolve on rugged landscapes by a mixture of mutations which search the immediate vicinity and some fraction of mutations which jump long distances in genotype space.

In keeping with the sensible intellectual strategy of gaining insight into complex behavior by proposing simple initial caricatures, it is of interest to find that a simple, near-universal law describes the behavior of an adapting population in the limit of search by long jumps across sequence space (Kauffman and Levin 1987). The "law" is equivalent to Feller's (1971) theory of records, which asks the waiting time for successive new athletic records to be set. The answer is that the waiting time *doubles* after each record is set. In our context, suppose adaptation is occurring on a highly rugged but correlated landscape. An example which suffices is the traveling salesman problem. Consider a search procedure with two individuals. One remains on the current best tour while the second jumps a long distance in tour space. If the jumper does not find a fitter tour on the first trial, she or he keeps jumping long distances from the current best tour until a better tour is encountered. When the jumper finds a fitter tour, both individuals gather at that new tour and the second again jumps a long distance in tour space. The question is, What is the expected waiting time to find each fitter tour?

The critical idea is that if the searcher jumps *beyond the correlation lengths of the space*, then whether or not the landscape is correlated, the searcher is encountering *a fully uncorrelated random landscape.* Therefore, at each successful jump, the *expected fitness* of the new tour is *halfway* between the relative rank-orders of the present tour and the best tour possible. Since the waiting time to find a still fitter tour is the reciprocal of the relative rank-order, *the waiting time doubles after each fitter variant is found.* This leads to the relation

$$s = \log_2 g \qquad (3.1a)$$

where s is the expected number of improvement steps which have occurred and g is the number of trials. In the population context, where we can think of each trial as constituting a generation, g stands for the number of generations.

The results expressed in Equation 3.1a do not require a single searcher but extend to a fixed-size, hypothetical population of searchers who leave a single individual on the current best tour while the rest all jump long distances. Then the whole population moves to the fittest tour found.

Consider an adaptive process begun with the entire population of N^* individuals located at a particular randomly chosen entity. The relative rank-order of that entity x/T is, on average, 0.5. Thus $0.5N^*$ of the long-jump variants sampled in the first generation are fitter than the initial entity. The best among these fitter variants has an expected relative rank-order of $(N^* - 1)/N^*$. At the second generation, all N^* long-jump variants are again sampled, and the expected number of fitter variants is 1. The best of these, on average, is ⅔ of the rank-order distance from the current best entity to the top-ranking entity. At the third generation, the expected number of fitter variants uncovered by the N^* long-jump variants sampled is 0.5. Consequently, the expected waiting time until a third fitter variant is found is the reciprocal, $1/0.66 = 1.5$ generations. On average, a single fitter variant is found, and on average it lies halfway between the last fittest variant and the top rank T. Therefore, after the third improvement step, the expected waiting time to find the next (fourth) improved variant doubles to three generations. The cumulative number of improved variants s uncovered in g generations is therefore approximately

$$s = \log_2 g + 1 \qquad (3.1b)$$

This simple expression is independent of population size as long as the size is fixed and large.

This is a remarkable simple result. It is also quite accurate. For example, we recently tested it on the traveling salesman problem (Kauffman and Levin 1987) and found the mean cumulative number of fitter tours at any generation to be $0.8 \log_2 g$.

The fact that, for N^* large, we considered a "greedy" process—that is, one in which the *fittest* variant found is chosen—makes a difference only initially because the relative rank-order of the population is soon so high that, at most, one member will find a fitter variant in each generation. Thus if one selects randomly from among all fitter variants found, it will take slightly longer than Equation 3.1b predicts to reach rank $(N^* - 1)/N^*$. An upper bound for the average is $\log 2(N^* - 1)$ improvements, but asymptotically, the rate of finding new improvements yields the same slope as that found in Equations 3.1a and 3.1b. For $N^* = 16$, for example, the greedy algorithm saves less than one step, on average, compared with a nongreedy algorithm.

Equation 3.1b is an overestimate of the fitness reached after a fixed number of generations. The argument leading to this equation ignores the fact that steps which reach more than halfway to the top are more likely to slow the rate of finding still-fitter variants than are steps which reach less than halfway to the top likely to hasten that rate. Weinberger (personal communication) recently rederived Equation 3.1b taking account of these facts and found

$$s = \ln g \qquad (3.1c)$$

Equations 3.1a–c can be thought of as very simple, universal aspects of long-jump adaptation on, presumably, any sufficiently rugged multipeaked landscape. The range of landscapes to which this "universal law" applies is not clear. Nor is it clear what length is needed in order for the long jump to jump beyond the correlation lengths in the landscape. For example, spin-glasses and the NK model appear to have self-similar landscapes (Stein 1987) in which small hills nestle into the sides of larger hills, which nestle into the sides of still larger hills. These landscapes thus have a variety of length scales. If an adaptive process jumps a moderate distance which remains shorter than the longest length scales in such a landscape, then that process is sampling an uncorrelated landscape with respect to shorter length scales but a correlated landscape with respect to longer length scales. With respect to the latter, the process is climbing a local hill. Nevertheless, we have explored this law both for the traveling salesman problem and for the very different problem of adaptation in model genetic regulatory networks (discussed in Chapter 5) and the *NK* model. Our results confirm the simple universal law (Kauffman and Levin 1987). Presumably its applicability is very wide.

This theory showing the long-jump behavior of an adapting population is important in two main respects. First, long jumps encounter the complexity catastrophe. Second, three time scales arise naturally for adaptive processes on rugged landscapes.

Long Jumps on Correlated Landscapes Suffer the Complexity Catastrophe

Since the landscape encountered in the long-jump limit is uncorrelated, this limit of an adaptive process suffers from the *complexity catastrophe*. In collaboration with my colleague Lloyd Clark, long-jump adaptation was carried out for $K = 2$ landscapes, shown to be rugged but highly correlated in Chapter 2, for values of N from 10 to 200. At each generation, a single copy of the current best genotype was left in place, while a second fully random mutant copy explored a single-long-jump new genotype. If no fitter variant was found, the process was iterated, leaving the current best marker genotype in place. Otherwise, the process moved to the fitter variant. Trials were carried out over 200 generations in 100 different landscapes for each value of N. Figure 3.1*a* shows the average results for each genome size. Clearly, as N increases, the fitness attained after any fixed number of generations declines toward

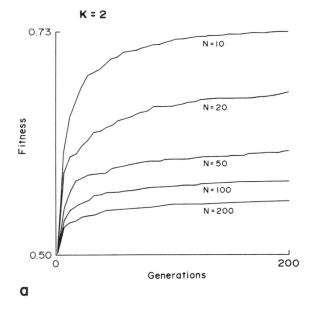

a

Figure 3.1 (*a*) The complexity catastrophe seen in long-jump adaptation on correlated $K = 2$ *NK* landscapes. "Generation" is the cumulative number of independent long-jump trials. Each curve is the mean of 100 walks. Fitness falls as N increases.

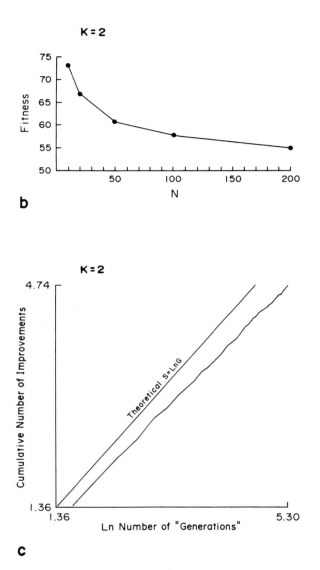

Figure 3.1 (*b*) The complexity catastrophe. Data from (*a*) at 200 generations used to plot fitness attained as a function of *N*. Note similarity to Figure 2.4. (*c*) Testing the universal law for long-jump adaptation. Vertical axis is cumulative number of improvements from each adaptive walk in (*a*).

the mean of the genotype space, 0.5. The rate of falloff as *N* increases (Figure 3.1*b*) closely parallels that found in Chapter 2 for the complexity catastrophe in searches via fitter one-mutant neighbors in fully random landscapes (Figure 2.4).

This simple observation means that the general applicability of the complexity catastrophe is far wider than one might have supposed. In examining the *NK* model, we found that the complexity catastrophe due to conflicting constraints sets in with respect to adaptive walks via one-mutant variants when *K* increases proportionally to N. This led us to ask for possible *construction requirements* for achieving epistatic systems which adapt on well-wrought, well-correlated landscapes which might maintain high optima as complexity (N) increases. A sufficient condition appears to be that the richness of epistatic interaction (K) remain low. If, however, even on these "good" landscapes, adaptive processes which proceed by reasonably long jumps confront the complexity catastrophe, selection is boxed in. Selection had best confine searches to the vicinity of good local hills and not take excessively long jumps. Blind long jumps become an ever more wasteful strategy as complexity increases, even on the best of landscapes.

Figure 3.1c tests again the hypothesis that the rate of finding fitter variants in the long-jump limit varies as given in Equation 3.1c. Clearly, the experimental slope is very close to that predicted by theory.

Three Time Scales in Adaptation on Rugged Landscapes

Adapting populations exhibit a range of mutations per individual, some due to point mutations substituting single amino acids and some due to frame shifts and recombination. This feature of real adaptation on rugged landscapes leads us to identify *three natural time scales* to adaptation.

1. Early in an adapting process, fitter variants in the vicinity of an initial poorly fit protein are easily found but are constrained by the correlation structure of the landscape to being only *slightly* fitter. In contrast, fitter *distant* mutants beyond the correlation lengths of the space are *as easily found* because the initial protein is poorly adapted. Because these distance proteins are not constrained by the local correlation structure, however, they can be *very much fitter*. Since the fittest variants sweep the population fastest (Gillespie 1983, 1984), the population will tend to flow to these distant points. After each such distant fitter variant is found, however, the waiting time to find another doubles. After relatively few such long jumps, the waiting time has become long compared with the rate at which fitter mutants are found in the *immediate vicinity* due to the existence of the correlations in the landscape.

2. Therefore, in the midterm, adaptation finds nearby fitter variants faster than it finds distant fitter variants and hence climbs a local hill. As the hill is climbed, however, the rate of finding fitter nearby variants first dwindles and then stops as a local optimum is reached.

3. On the longer time scale, the process, before it can proceed, must await a successful long jump to a better hillside some distance away.

I emphasize that this picture of complex diffusion in rugged sequence spaces is a caricature in assuming that the entire haploid population moves to a single fitter variant whenever one is found and in assuming that fitness landscapes are fixed in structure. But the outline frames in the behavior of an adapting population when the rate of finding fitter variants is low compared with fitness differentials.

Radiation and Stasis as Generic Properties of Adaptation on Rugged Landscapes

In our consideration of the rugged character of adaptive landscapes, we found a number of general features which, on a little thought, are virtually inevitable. Such landscapes have a large number of local optima. Walks toward optima in high-dimensional spaces which begin with poorly fit entities—proteins, DNA sequences, or organisms with some set of traits—initially can climb uphill in many alternative directions. As optima are approached, the number of ways uphill must decrease to zero. It follows from this that *if the adaptive process can send branches* upward along alternative adaptive walks, then initially many branches can emerge from a first poor genotype but, ultimately, single lineages wend their way uphill until trapped on local optima.

A simple conclusion follows from this trapped-lineage model, with the permission of the reader to think of such branching walks as *phylogenetic evolutionary lineages. In a fixed but rugged landscape, radiation and ultimate stasis are utterly generic.* We require no special mechanism to account for such phenomena. I return to this shortly, for many authors have felt constrained, in accounting for radiation and stasis, to offer special explanations, typically in terms of initially empty niche space which is progressively filled.

Rederiving von Baer's Laws

Von Baer's laws are familiar to all embryologists and are one of the foundations of comparative anatomy (de Beer 1958). They state that, in well-established lineages, notably the vertebrates, early embryos of the various lineages are more similar than late embryos. Thus early fish, frog, chick, and human embryos are remarkably similar (de Beer 1958, Gould 1977). The familiar explanation for these laws is that mutants affecting early ontogeny are more disruptive than mutants affecting late ontogeny. Thus mutants altering early development are less likely to accumulate, and early embryos remain more similar from one order of organisms to another than do late embryos.

Is this plausible argument actually so plausible? Even if the rate of finding early beneficial mutants were a constant low fraction of the rate of finding late beneficial mutants, hundreds of millions of years have elapsed in the vertebrate lineage. How can we be so assured that, over a time span of 600 million years, *no beneficial mutants affecting early embryos* should have arisen and altered early fish from chick from human?

The simple ideas about rugged landscapes may more adequately explain von Baer's laws. A first step in rederiving the laws makes use of Wimsatt's (1986a, b) idea of "generative entrenchment," which denotes the number of consequences a given mutant has in development. On average, Wimsatt argues, mutants affecting early development cause many alterations and thus have high generative entrenchment; those acting late in development have few consequences and hence low generative entrenchment. The immediate consequence is that early mutants are *adapting on a highly uncorrelated fitness landscape* and late mutants are *adapting on a well-correlated landscape.*

From the *NK* model, we know that adaptation on correlated landscapes typically implies that the rate of finding fitter variants decreases far more slowly than on uncorrelated landscapes. Thus early mutants, because they are on a nearly uncorrelated landscape, confront the consequences of Equations 3.1: The waiting time to find fitter early-acting variants doubles after each fitter variant is found.

This ever-doubling waiting time implies that the rate of finding fitter mutants altering early ontogeny decreases rapidly. Thus, to state the result in idealized form, given a fixed mutation rate in a fixed-size population with one generation a year, after

1000 years, ten early mutants would have accumulated; after 1 000 000 years, another ten would have accumulated; and after 1 000 000 000 years, only another ten would have accumulated. By contrast, late-acting mutants can continue to accumulate at a more constant rate, since their fitness landscapes are correlated. It follows that early development should soon become more conservative than late development: hence von Baer's laws.

The Cambrian Explosion and Permian Quiescence

The existence of three time scales for adaptive evolution on rugged landscapes coupled with the greater generative entrenchment of early mutants appears to offer a simple explanation for the vast explosion of biological diversity which took place at the onset of the Cambrian. It may also account for the relative quiet following the great Permian extinction.

The Burgess Shale, which was deposited about 550 million years ago in British Columbia, bears witness to the Cambrian explosion (Gould 1989), which established essentially all the major animal body forms, or baupläne, and hence all the major phyla which would exist thereafter. Lewin (1988) points out that, compared with the 30 or so phyla extant today, the Cambrian explosion may have generated as many as 100 phyla, many of which subsequently became extinct. Not only did a very large number of novel body forms originate rapidly, but the Cambrian explosion exhibited another novelty: Species which founded taxa appear to have built up the higher taxa from the top down. That is, exemplars of major phyla were present first, followed by progressive filling in at class, order, and lower taxonomic levels (Raup 1972, 1983; Valentine 1977, 1980; Erwin, Valentine, and Sepkoski 1987; Jablonski and Bottjer 1988). In contrast to this evolutionary burst of invention are the consequences of the great Permian extinction about 200 million years ago, when approximately 96 percent of all species became extinct. Following the Permian extinction, there is a rapid increase in species abundance and taxa up to the family level, but no new phyla or classes were created (Bambach 1985; Erwin, Valentine, and Sepkoski 1987; Jablonski and Bottjer 1988). Rather, the existing phyla replenished from the bottom up.

The puzzle is why the Cambrian and Permian are so different from each other. Valentine (1980) and Valentine and Walker (1986), echoed by Erwin, Valentine, and Sepkoski (1987), favor an argument based on early ecological opportunity followed by progressive filling of niche space. Early Cambrian organisms confronted an empty niche space. Valentine, for example, assumes that mutants of nearby morphologies occupying nearby niches are more likely to be viable than those which jump to distant morphologies and thus occupy distant niches. Further, Valentine suggests that invasion of a niche requires that it be unoccupied. If not, the invader is blocked by competitive exclusion. In the early Cambrian, such distant niches *were* unoccupied. This vacancy allowed at least some fraction of distant mutants to colonize distant niches, thereby founding new phyla. But, Valentine argues, as niche space filled, the Cambrian radiation slowed. By the time of the Permian extinction, the surviving 4 percent of species were still located across vast tracts of ecospace and hence left little opportunity for a new evolutionary explosion. New phyla were not founded, but innovation at the family and lower levels was abundant.

More recently, Valentine (1986) and Valentine and Erwin (1986) also suggest the alternative possibility of a change in the genomic system over time such that development was less constrained, or "canalized" (Rendel 1979), during the Cambrian explosion than it later became. On the genomic hypothesis, by the time of the Permian extinction, the genome had become more canalized, and so development was less able to vary. Hence new phyla were not founded, and taxa filled in from the bottom up.

Analysis of the generic properties of adaptation on rugged but correlated fitness landscapes suggests an alternative hypothesis: the Cambrian–Permian asymmetry may be a direct expression of adaptation on such landscapes. There are three basic ideas in such a framework.

1. Adaptive evolution occurs on rugged fitness landscapes.

2. The rate of finding successively fitter variants decreases very rapidly both on very rugged landscapes and on smoother landscapes explored by long jumps; in contrast, the rate of finding successively fitter *nearby* variants decreases gradually on smoother landscapes.

3. Mutants affecting early ontogeny typically have more profound consequences than mutants affecting late ontogeny (Wimsatt 1986a, 1986b). Thus the rate of finding fitter mutants altering early development and fundamental baupläne decreases more rapidly than the rate of finding fitter mutants affecting late ontogeny.

Together these ideas imply that early, poorly fit multicellular organisms could rapidly explore a large diversity of improved alternative basic morphologies, thereby establishing phyla. As the rate of finding fitter mutants altering early ontogeny decreased but fitter variants affecting later ontogeny were still readily found, variant species founding classes, orders, and the lower taxa became established. Taxa filled in from the top down. In contrast, by the Permian extinction, early ontogeny was largely frozen, and so fitter variants altering fundamental baupläne became very hard to find. But mutants altering late ontogeny remained easier to find. New genera and families arose. Taxa replenished from the bottom up.

As noted above, branching radiation which gradually slows and developmental locking-in are derived from the generic features of innovation on rugged fitness landscapes. There is no need in this model for filling of ecospace or competitive exclusion to slow radiation, as suggested by Valentine (1980), although such exclusion may also occur. Also, a locking-in of early development, and hence von Baer's laws, do not represent a special mechanism of developmental canalization, the usual sense of which is a buffering of the phenotype against genetic alterations (Waddington 1940, 1957, 1962, 1966; Rendel 1979). Instead, locking-in of early development is a direct reflection of the fact that the number of ways to improve organisms by altering early ontogeny has dwindled faster than the number of ways to improve by altering late development.

It might be objected that this argument rests on the assumption of fixed fitness landscapes and that, if fitness landscapes changed significantly during the Phanerozoic, some lineages could have again found themselves at the bottom, and long-jump mutations would again have a good chance of being adaptive. This concern is important. Landscapes cannot be fixed in general, if only because the physical environment changes. If landscapes heave rapidly in time and dramatically in extent, such that an organism's fitness in the old environment is nearly uncorrelated with its fitness in the new environment, then the insights derived from thinking in simplified terms about fixed fitness landscapes are surely mistaken. If landscapes change relatively slowly in time, however, and—equally important—if the fitness values in the new environment are strongly correlated with those in the old environment, then in the new environment organisms remain highly fit. Thus in the new environment adaptation will be dominated by *local search* from highly but not optimally adapted organisms, rather than by long-jump adaptation. If mutants affecting early development typically adapt on more rugged landscapes than those affecting late development, then as the fitness landscapes warp, adaptation should continue to occur via modification of late development. Once established, major baupläne remain locked in. Despite deformations of the fitness landscapes, formation of new phyla is difficult once locking-in of early development occurs. Formation of new families and genera is easy.

This account may also offer insight into rapid, bottom-heavy origination below the ordinal level in rebounds following mass extinctions (Jablonski and Bottjer 1988). If mass extinction is assumed to be related to environmental and ecological change, then following the change, fitness landscapes may be sharply altered. Such altered landscapes would be expected to cast many organisms back to lower fitness levels, followed by fairly dramatic alterations of late ontogeny in many directions. This would yield bottom-heavy radiations, with families tending to arise early in the history of their orders and genera tending to arise early in the history of their families. Jablonski and Bottjer (1988) note evidence for this in the early Paleozoic and early Mesozoic for marine organisms and in the early Cenozoic for mammals. Focus on landscape structure and on the number of improvement directions as adaptation occurs and as landscapes alter raises the possibility that post-extinction rebound radiations might be due not to lack of competitive exclusion in depauperate environments, but to the increased number of directions open for adaptive improvements in an altered environment. Gould, Gilinsky, and German (1987) argue that such bottom-heavy clades, which dominate in the fossil record from the Cambrian through the Ordovician, constitute an asymmetry which shows the direction of evolutionary time. Focus on the fact that bushy radiation dwindling to stasis is generic on rugged landscapes may help account for this fundamental asymmetry.

These general expectations can be assessed readily using a modified version of the *NK* model. The modifications incorporate (1) the hypothesis that some traits are more generatively entrenched and therefore cause more disruption if modified and (2) the possibility that the world changes, thereby changing fitness landscapes. In the *NK* model as presented, each site receives exactly K epistatic inputs and epistatically affects on average K other sites. In order to model differential generative entrenchment, my colleague Sonke Johnsen and I modified the model such that each site is still affected by K epistatic inputs, but some sites epistatically affect a large number of other sites, while others epistatically affect no other sites. To be concrete, we distribute the outputs hierarchially, falling off exponentially from the first site to the Nth site, such that the weighted proportion of epistatic outputs from each site i is $e^{-0.1i}$. This particular choice means that, for $N = 50$ sites, site 1 affects about 12 times as many sites as does site 50.

In order to consider the effects of changing the external world, the *NK* model can be modified to include an external world consisting in a list of N features, present or absent. Hence the world is a vector of 1 and 0 values. A convenient way to couple the *NK* landscape on which adaptation occurs with the world is to suppose that each site in the landscape is affected by K other sites in the adapting organism and by w sites in the world. Then each site in the organism is affected by $K + 1 + w$ sites, and for each of the 2^{K+1+w} combinations of organism traits and world features, a random fitness contribution by that site is assigned. In effect, this expands the random fitness table of each site to include the 2^w alternative combinations of world features which impinge upon that site. Given this coupling, the fitness landscape of the organism changes as the features of the world change. If w is large and if each site in the world is coupled to many sites in the organism, then a slight change in the world yields a drastic change in the fitness landscape. If, instead, w is small, then small changes in the world yield small changes in the landscape.

Simulations with this modified NK model were carried out under three "move" procedures. In the first, adaptive steps occurred via the fittest one-mutant variant. This corresponds to a greedy gradient ascent and to the situation in which the fittest variant sweeps the population. In the second, one of the fitter one-mutant variants, if any existed, was chosen at random at each iteration. In the third, a random one-mutant variant was sampled and, if fitter, the adaptive process stepped to that variant. The results are basically similar under all three move procedures and confirm the following:

1. During adaptation, sites of high generative entrenchment are more likely to change when the fitness of the adapting entity is low. As fitness increases, sites of lower entrenchment become more likely to change. This is shown in Figure 3.2. Site 1 has the highest generative entrenchment; site 50, the lowest. Thus when fitness is low, adaptation occurs initially on a very rugged fitness landscape via sites which influence the fitness contributions of many other sites. As fitness increases, successive adaptation occurs via progressively minor variants which affect fewer and fewer other sites.

2. If the world changes slightly but often, a characteristic distribution is found in which sites of high generative entrenchment change rarely or not at all, while sites of progressively lower entrenchment change progressively more often. That is, sites which influence many other sites become frozen into given states, 1 or 0, and are resistant to change, and sites which influence few other sites can continue to change rapidly. As the world changes more radically but equally often, all sites change more often in an attempt to climb hills on the rapidly deforming landscape. The most entrenched sites (that is, the lowest-numbered sites) change least often (Figure 3.3). Notice that as the world deforms more dramatically, highly entrenched sites change.

3. If the world changes dramatically but only occasionally, then fitness drops dramatically and sites of high generative entrenchment tend preferentially to change during the initial adaptation. As fitness increases, sites of progressively lower entrenchment change. This is summarized in Figure 3.4. The effects are clearest for the greedy move procedure (Figure 3.4a).

These results echo and support the arguments advanced above, when we looked at the Cambrian–Permian asymmetry. Early development, when associated with high generative entrenchment, locks in rapidly, while later development, due to lower entrenchment, continues to alter. This remains true if the world continues to change rapidly but only slightly at each change. Thus early development should remain conservative, as should major features which define higher taxa, such as the vertebrates. In contrast, progressively more minor features change more readily. Once early development locks in, continued rapid but slight changes in the world lead to continued proliferation of lower taxa. Following drastic changes in the world, adaptive variation of major features occurs as well, unleashing formation of higher taxa during the adaptive rebound.

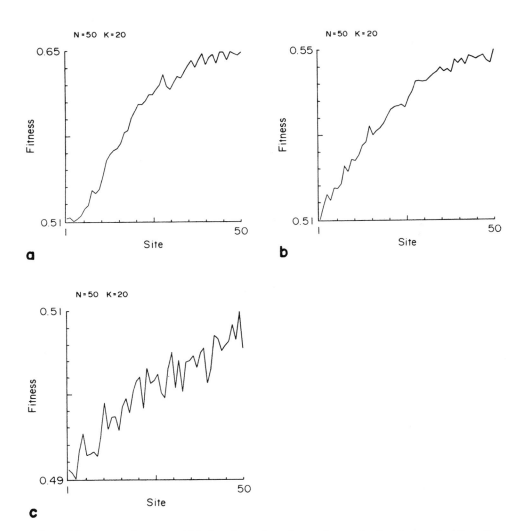

Figure 3.2 Average fitness at which sites change in order to take an adaptive step as a function of the number of the changing site, averaged over 200 runs. Hierarchical *NK* model, $N = 50$, $K = 20$, $w = 20$, world changes each five generations. Data are for the three move algorithms described in the text: (*a*) greedy, (*b*) fitter, (*c*) random. Low-numbered sites have epistatic outputs to many other sites and tend to change only at low fitness. High-numbered sites have few or no epistatic effects on other sites and tend to change at high fitness.

Figure 3.3 Average, over 100 runs, of the number of times a site changes as a function of site number: (*a*) greedy algorithm, (*b*) fitter algorithm, (*c*) random algorithm. Hierarchical *NK* model, $N = 50$, $K = 20$, $w = 5$, world changes at each generation. $wm = 1, 2, 4, 8$ sites in the world, a binary vector N long, change at each moment; thus $wm = 8$ causes more rapid deformation of fitness landscapes than $wm = 1$. Low-numbered sites affect many other sites; high-numbered sites affect few or no other sites. Note that, in (*a*) and (*b*), low-numbered sites almost never change. Progressively less highly entrenched sites (higher numbers) change more often and do so more often as the world deforms more dramatically.

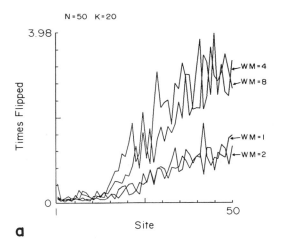

a

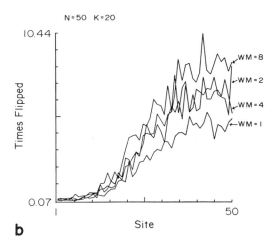

b

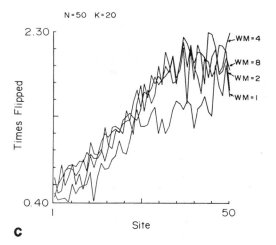

c

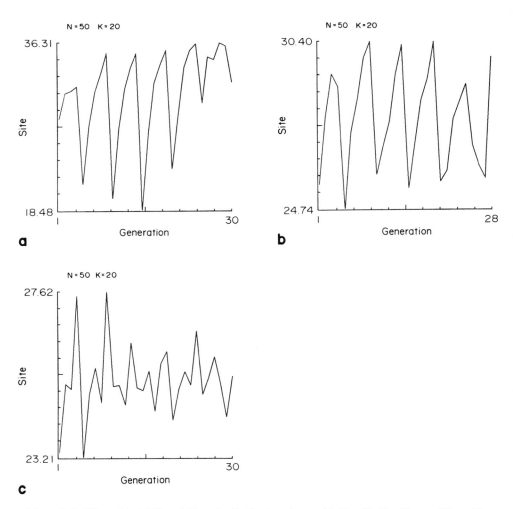

Figure 3.4   Hierarchical *NK* model in episodically changing world. $N = 50$, $K = 20$, $w = 20$, world changes every five generations, changing $wm = 20$ sites in the N sites of the world. Vertical axis is mean site number which changes at each generation: (*a*) greedy algorithm, (*b*) fitter algorithm, (*c*) random algorithm. After the world changes, fitness falls and low site numbers, with epistatic effects on many other sites, preferentially change. Thus the average number of the site which changes falls. As fitness increases, sites which influence fewer other sites (hence in the hierarchical model have higher site number) change preferentially. Sawtooth variation shows that, when the world changes dramatically, at first low-numbered sites with high generative entrenchment preferentially change, then sites with less influence on other sites change.

I note briefly that if these arguments and results are roughly correct, then von Baer's laws might apply to the evolution of complex forms on almost any "rerun" of the evolution of multicellular organisms. Von Baer may have enunciated a universal law.

Another point here is of considerable interest: The existence of a hierarchy of entrenchment, or of K values, implies that adapting organisms are *buffered* with respect to the variable extent to which their fitness landscapes deform. Thus when confronted with landscapes which typically deform slightly but occasionally deform dramatically, organisms can respond to the former by evolving on relatively smooth landscapes via changes in moderately entrenched traits and to the latter by evolving on very rugged landscapes via alterations in highly entrenched traits. In effect, such buffering affords a range of responses as a function of how drastically and how often the environment changes. This feature bears on the problem of homoplasy, discussed next.

The Problem of Homoplasy: Convergent Evolution Confounds Phylogenetic Tree Construction but Carries Information About the Structure of Adaptive Landscapes

Conceiving of adaptive evolution as occurring on more or less rugged fitness landscapes may afford a means for turning "noise" in the data of systematists into useful data. Convergent evolution, as we shall see, makes construction of phylogenetic trees difficult but harbors information about the structure of rugged fitness landscapes. Therefore, it is very much worth investigating how to extract this information.

Here is the problem in brief and informally. When systematists consider a collection of species or, more typically, a collection of higher taxa, one of the major aims is to attempt to reconstruct the historical branching phylogenies leading to the group under investigation (Patterson 1987; Felsenstein 1988; Levinton 1988). Alternatively, the aim of some members of the cladist school of systematists is to construct the minimal and natural branching pattern among the taxa, regardless of whether or not that pattern fits the evolutionary branching pattern. Analysis by cladists distinguishes "ancestral" traits and "derived" traits. The former are traits shared by all members of a group and some further outgroup. The latter are those which arise during evolution of the group under investigation. Derived traits shared by any two or more subgroups of the group count as evidence that the members of these subgroups are more closely related to one another than to members of the other subgroups (Hennig 1966). Shared derived traits are called synapomorphies. In either case, the aim is to construct the "proper" branching pattern, or cladogram, among the taxa.

With respect to morphological or molecular criteria, it is often the case that the systematist considers each taxon to be characterized by some number of traits N, each of which is coded as ancestral (0) or derived (1). Thus each terminal taxon is like a sequence in the NK model with $A = 2$ alleles. In reconstructing the branching lineage tree, which is derived from a cladogram by assigning taxa to intermediate nodes and assigning one taxon as a basal ancestral root, the aim is maximum parsimony and consistency (Patterson 1987; Felsenstein 1988; Levinton 1988). The difficulty with the consistency requirement is that two lineages which may in fact be distant in the true historical phylogeny may have *independently* evolved the same new derived trait in parallel. This is called parallel, or convergent, evolution. If many derived traits evolve in parallel independently in separate lineages, the task of discriminating the true tree is made much more difficult. Such lack of consistency, termed "homoplasy" by cladists, is therefore unwelcome noise in the effort to uncover the true branching patterns among taxa. The noise, however, is not without regularity. Sanderson and Donoghue (1989) have recently shown that, as the number of taxa in a tree increase, the consistency decreases in a characteristic way.

I describe next the measure of homoplasy, called the consistency index, used by Sanderson and Donoghue, and their results. Then we shall consider the implications that branching adaptive walks on rugged fixed and deforming landscapes hold for the consistency index and other measures of homoplasy. Finally, I shall return to the implications of Sanderson and Donoghue's results and point out a striking consideration: Viewed casually, their data appear consistent with a null model of neutral branching phylogenies in a high-dimensional trait space; viewed with more care, however, their results suggest that the observed homoplasy is consistent with adaptive evolution on landscapes which deform persistently and over a range from minor alterations to massive deformations.

Sanderson and Donoghue collected 60 recent cladistic analyses run on a wide variety of organisms—animals, plants, and fungi—and based on morphological and molecular data, including proteins and DNA. The independent variables of their analysis were (1) the number of terminal taxa in the study, which ranged up to 70; (2) the number of derived traits analyzed in each study; and (3) the taxonomic rank of the taxa analyzed, which ranged from species to classes/phyla. The dependent variable was a consistency index, widely used by systematists as an overall measure of homoplasy. For a given branching tree, the ancestral root can be thought of as defined by a list of N "0" values, representing the ancestral states of each of the N traits which will become derived in one or more of the terminal taxa. At each ascending branch, one or more traits have changed states. Thus traits 1 and 17 may change from ancestral to derived in the first left branch. Ascending higher in the tree, other traits may change from ancestral to derived. If there were no inconsistency in the deduced cladogram, then each trait would switch from ancestral to derived just once and then remain in the derived state in all terminal taxa ascending from that branch. Homoplasy, or inconsistency, occurs when trait 1, for example, which changed from ancestral to derived in the first left branch, must be assumed to switch from ancestral to derived in some *other* branch also. Homoplasy also occurs if trait 1 must be assumed to revert from derived back to ancestral in some branch.

The consistency index is a ratio whose numerator is the total number of derived traits which occur in one or more of the terminal taxa. Here the numerator is N. The denominator is the sum of the number of times each of the N traits switches from ancestral to derived or from derived to ancestral in the tree. Thus a great deal of parallel evolution leads to a large denominator and lowers the consistency index, whose maximum value is 1.0.

Sanderson and Donoghue carried out a regression analysis testing whether the consistency index changes systematically with number of taxa, number of derived traits, or taxonomic rank. The main results are interesting. The index shows only one clear trend. It decreases as the number of taxa in the study increases. The authors find that a linear regression analysis accounts for 50 percent of the variance if the natural logarithm of the index is plotted as a function of number of taxa. The slope observed is -0.0158 (Figure 3.5). In contrast to the clear trend with respect to number of taxa, no trend is found with respect to number of traits or with respect to taxonomic rank. The results are nearly the same for plants and animals considered separately and for morphological and molecular studies.

It is, of course, reasonable that, as the number of taxa in a study increases, the probability that traits will exhibit homoplasy increases. Thus the consistency index would be expected to decrease as the number of taxa increases. It is not obvious, however, that data for so many different systems, studied in such different ways, would be so well described by a single regression.

I turn next to ask whether study of branching adaptive processes on fixed or varying fitness landscapes can be expected to yield insight into divergent and convergent

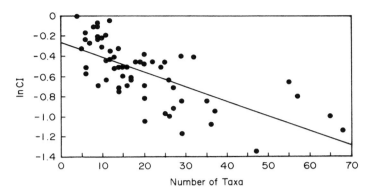

Figure 3.5 Natural logarithm of the consistency index as a function of number of taxa in each cladistic study. (From Sanderson and Donoghue 1989)

evolution, and, conversely, whether measures of divergence and convergence can be expected to yield insight into the structure of fitness landscapes. Then I return to ask what the results of Sanderson and Donoghue suggest about large-scale evolution.

The first important point to realize is that both divergent and convergent evolution carry information about the structure of the adaptive landscape. Convergence can represent adaptation climbing from different initial points to either the same or nearby peaks in the fitness landscape. Consider the NK landscapes. Let each sequence correspond to the list of the N traits, present or absent, in each taxon. Confine attention to the $K = 2$ landscapes for some large number of traits, say $N = 100$. In these $K = 2$ landscapes, the highest optima are nearest one another. Imagine branching adaptive walks on such a rugged multipeaked landscape, where branching reflects branching of taxa. Then those walks will in part *diverge* from one another in the high-dimensional space. That is, the Hamming distances between the diverging taxa will tend to increase. Conversely, if those walks climb to the *same* local optimum or even to nearby high optima clustered near one another in the Massif Central, then independent lineages will evolve the same traits in parallel, and the branching process will also tend to *converge* in Hamming distances. Quite clearly, both divergence and convergence will occur during adaptive walks on rugged multipeaked fitness landscapes. The extent of convergence will be governed by the structure of the landscape, by how close peaks are to one another, by the rate of branching in the adaptive walks, and by whether adaptation is constrained to pass only via fitter neighbors or is a more general flow. In addition, the relative rates of divergence and convergence will depend upon whether the landscape is fixed, as I have supposed it to be for most of this chapter, or whether it deforms as a result of environmental changes or coevolution, as discussed just above and in Chapter 6.

Figure 3.6 illustrates these ideas. To model branching phylogenies using the NK model, Sonke Johnsen and I assumed that an adaptive process might branch as alternate taxa clamber uphill. Specifically, we assumed that, at each time step, each current taxon examined all its one-mutant neighbors and chose a fitter neighbor at random. If two or more one-mutant neighbors were fitter, we supposed that the taxon branched to two derived taxa with a fixed probability P. Therefore, P fixes the rate of branching and forming new taxa as a function of the rate of change of traits. The set of trait values of the initial randomly chosen taxon was defined to be the ancestral

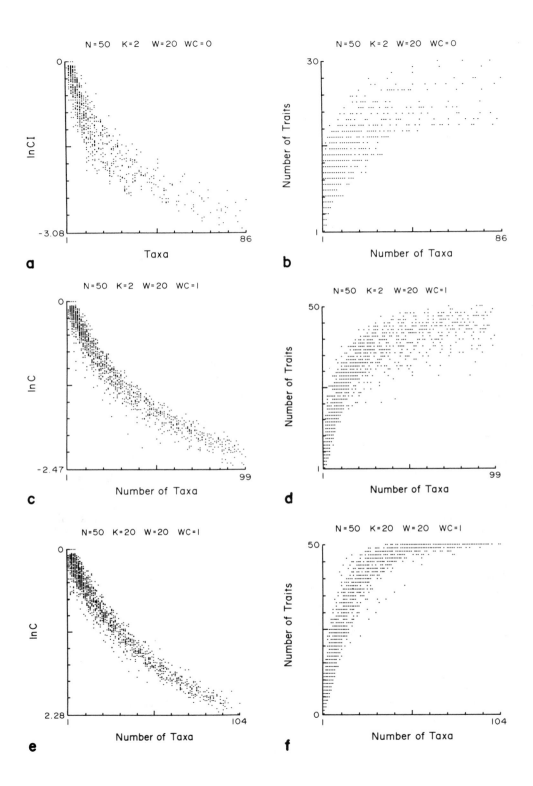

state values, and the subsequent branching taxa were therefore each characterized by their particular mixtures of ancestral and derived trait values. To model the effect of changing the external world through physical or coevolutionary processes, each site in each adapting taxon was coupled to w sites in an external world given by an N vector of 1 and 0 values, as described above. Changing the world then consists in altering one or more bits in the N vector. Note in Figures 3.6a and 3.6c that both when the world is fixed and when it is changing, the logarithm of the consistency index decreases in a concave-up fashion with respect to the number of taxa. Furthermore, the curve is remarkably similar whether the world is fixed or changing.

It is important that a neutral null model can account for the main relation between consistency index and number of taxa. Thus divergence and convergence do not logically require that the branching phylogenies be climbing hills on rugged fixed or deforming landscapes. Consider a null hypothesis according to which branching phylogenies form by ignoring fitness values and merely walk and branch randomly on the N-dimensional trait space. Figure 3.7a shows the results for $N = 50$, with the probability of branching at each moment set to .3. Thus at each step, each evolving taxon changes one trait at random either from 0 to 1 or from 1 to 0 and branches into two taxa with probability .3. If taxa branch, each flips a different randomly chosen trait. The results are nearly identical to those shown in Figures 3.6a, c, and e. Figure 3.7b shows the traits-versus-taxa curve for the neutral random-branching model.

These results show that, for a wide range of landscapes and a wide range of alteration rates, and even for random nonadaptive branching walks in an N-dimensional trait space, the relation between consistency index and number of taxa is nearly identical. Thus it appears that the consistency index is a relatively insensitive measure of the underlying structure of adaptive landscapes. This insensitivity suggests that the index is a poor guide for distinguishing alternative hypotheses about adaptive evolution on rugged fixed or deforming landscapes via random branching walks in high-dimensional trait spaces. Other, more sensitive measures of divergence and convergence which might prove useful are worth mentioning.

Two measures I now introduce, H/D and H/P, appear to be reasonably sensitive to landscape structure. Each measures how rapidly taxa spread apart in trait space relative to how far they have evolved from the ancestral root. Consider a branching lineage of taxa whose organisms are characterized by the presence or absence, 1 or 0, of N traits. Three natural distances relate the terminal taxa, or twigs, in the phylogenetic tree. The first is the Hamming distance between each pair of terminal taxa and gives the mean Hamming distance between the two members of each pair. This distance measures how similar the phenotypes of the terminal taxa are to one another. Call this measure H. A second natural distance measures the number of

---

Figure 3.6 (a) Natural logarithm of the consistency index as a function of number of taxa, derived from analysis of branching phylogenies in the NK model. The $wc = 0$ means that the external world does not change and hence the landscape is fixed. The probability of branching into two taxa at each generation is .3. (b) Number of derived traits which have switched from ancestral to derived one or more times in the branching lineage, plotted as a function of number of taxa in the lineage. Data averaged over 100 trees for each condition. (c) As in (a), except that here one site in the world changes at each generation. Hence the landscape deforms while the taxa branch upon it. (d) As in (b), except that here one site in the world changes at each generation. (e) As in (a), except that here $K = 20$ and one site in the world changes at each moment. Thus the landscape is more rugged and deforms as taxa branch upon it. (f) As in (b), except that here $K = 20$ and the world changes one site at each generation.

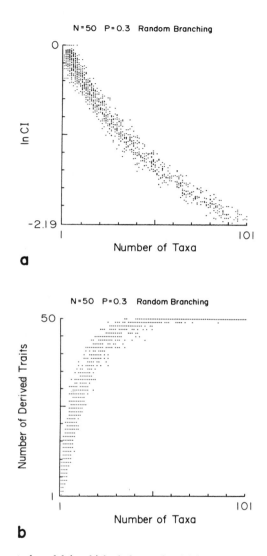

Figure 3.7 (*a*) The neutral model, in which phylogeny branch in an $N = 50$ trait space. Branching probability per generation is .3. (*b*) Trait-versus-taxa scatterplot for the neutral model with $N = 50$ and branching probability per generation = .3.

traits by which each terminal taxon differs from the presumed ancestral taxon. This measures the distance D each terminal taxon has diverged from the ancestral taxon. The mean of D over the tree is a reasonable measure of the mean distance all taxa have come. (D is called the degree of advancement by systematists.) The third natural measure traces the total number of trait changes from one terminal taxon to another via their most recent common ancestor. Summed over the entire branching tree, this measure shows how many trait changes, on average, separate the two members of each terminal pair from their common ancestor. Call this measure P. (P is sometimes called a patristic distance.) The ratios H/D and H/P for any pair of terminal taxa averaged over the total tree then characterize how rapidly the phenotypes of the terminal taxa diverge from one another as a function of how far the taxa have diverged from their most ancestral or most recent common ancestors. If convergence due to the adaptive landscape is high, then H/D and H/P should be low. If convergence is rare either because the landscape does not harbor convergence or because the taxa are ignoring fitness values and diverging in trait space by random-branching walks or because the landscape is deforming rapidly as a result of changes in environment due to physical or coevolutionary processes, then H/D and H/P should be large. (In the H/D ratio I describe below, H is the mean over all pairs of terminal taxa and D is the mean of the distance from each terminal taxon to the presumed root. In contrast, the H/P ratio is calculated for each pair of terminal taxa and averaged over the entire tree.)

Preliminary studies using the NK model confirm these intuitions, as Figures 3.8a–c and 3.9a–e show. Unlike the relatively insensitive consistency index, H/D and H/P are clearly strongly sensitive to the structure of the adaptive landscape and to its rate of deformation. Furthermore, the distributions on fixed or deforming landscapes can clearly be discriminated from those of the neutral random-branching model. The results shown in these figures are for 100 independent branching walks on landscapes of the same structure and same rate of deformation. The broad spread of H/D and H/P under each landscape condition suggests that assessing which model landscape best fits a specific observed branching walk must be undertaken with care and with sufficient data.

The H/D and H/P ratios depend not only upon the structure of the fitness landscape but also upon how often branching occurs on that landscape and upon how rapidly the landscape deforms. It would be useful to have independent evidence bearing on how often and how dramatically landscapes deform.

Both the patterns of homoplasy within trees and the rates of branching within trees afford evidence about how dramatically and how often landscapes deform. If the landscape remains fixed for long periods and then alters dramatically, currently fit taxa will be cast back to lower fitness and should rebound with a burst of radiation because the number of directions allowing improvement should increase suddenly. This radiation burst should show up in two ways. First, the rate of branching should be high right after the dramatic deformation and then decrease gradually until the next upheaval. Second, after a marked deformation, highly entrenched traits should change but then become locked in thereafter. This locking-in should be seen as a *bias in the locations of homoplastic traits;* these traits should be found in nodes in a tree where the trait changes but then remains consistent in higher nodes closer to the terminal taxa.

If landscapes deform somewhat more frequently and less dramatically, smaller bursts of radiation would be expected. If the landscape deforms often but only slightly, then taxa should continue to branch at a more or less uniform rate. Such continuous radiation should show up, again, in two ways: The rate of branching should be uniform, and homoplastic traits should be distributed uniformly over trees.

These features of rugged landscapes suggest that analysis of the locations of homoplastic traits in trees and analysis of *the variance* in the rate of branching per unit trait change in the D measure give information on how often and how dramatically landscapes change. In addition, the mean rate of branching per trait change in the D measure gives information about how often, on average, taxa branch. The H/D and $H/$

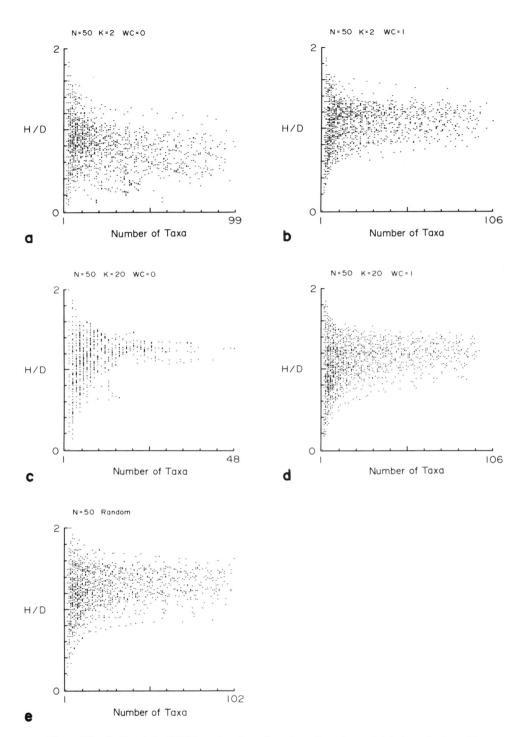

Figure 3.8 Scatterplots of *H/D* as a function of number of taxa in model phylogenies branching on an *NK* landscape. In all cases, *N* = 50. Branching probability at each generation is .3. (*a*) *K* = 2, *wc* = 0 (fixed landscape). (*b*) *K* = 2, *wc* = 1; hence at each moment, one site in the world changes. Each site in each organism is coupled to *w* = 20 sites in the world. (*c*) *K* = 20, *wc* = 0. (*d*) *K* = 20, *wc* = 1. (*e*) *K* = 2, *wc* = 0, data reflect the neutral model with random branching in an *N*-dimensional trait space.

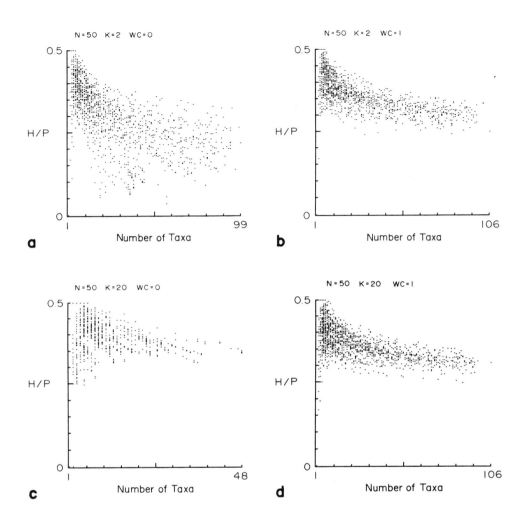

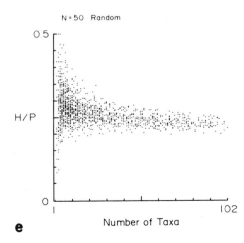

Figure 3.9 Scatterplots of *H/P* as a function of number of taxa in model phylogenies branching on an *NK* landscape. In all cases, *N* = 50 and branching probability at each generation is .3. (*a*) *K* = 2, *wc* = 0 (fixed landscape). (*b*) *K* = 2, *wc* = 1. (*c*) *K* = 20, *wc* = 0. (*d*) *K* = 20, *wc* = 1. (*e*) *K* = 2, *wc* = 0, data reflect branching in the neutral model in an *N*-dimensional trait space.

P measures give information about the extent of convergence among terminal taxa relative to their distance from either root or common ancestors.

Attempts to utilize measures of homoplasy to assess the structure and deformation of fitness landscapes must also discriminate whether convergence and homoplasy reflect adaptive evolution or merely constraints and biases in developmental mechanisms. Consider an organism with a set of N hypothetical traits which, for developmental reasons, have only two possible states, 0 and 1, and have the property that mutations are four times more likely to change each trait from 0 to 1 than from 1 to 0. Numerical simulations confirm that, if the anscestral root species starts with all N traits either in the unstable 0 state or in the stable 1 state, the biased 0-to-1 mutation of each trait will lead to a phylogeny in which H/D and H/P are biased low or high compared with the H/D and H/P values for a random branching phylogeny generated with an equal conversion rate between the 0 and 1 states. However, if the ancestral species already has an equilibrium distribution of 1 and 0 states over its N traits—80 percent 1 and 20 percent 0 in the current example—then H/D and H/P are not distorted by those biases. In any reconstructed phylogeny, it is possible to assess for each trait whether its conversion between ancestral and derived states is biased in favor of the former or the latter. Thus it is possible to assess whether the presumed root taxon begins with an equilibrium distribution with respect to those biases over the N traits. In short, it may be possible to discriminate convergence on rugged landscapes due to adaptive evolution toward fitness peaks from selectively neutral branching phylogenies in trait space biased by the directional mutability of developmental mechanisms.

In summary, adaptive branching evolution on deforming landscapes is undergoing both divergence and convergence. The process is, if you will, surfing the wavetops of a tossing fitness sea. The aim is to extract information about the structure and rate of deformation of fitness surfaces underlying phylogenetic evolution. Doing so requires discrimination of adaptive convergence and homoplasy based on adaptation from biased mutations among a restricted number of morphologies or molecular traits and neutral random branching phylogenies. Using sensitive measures such as H/D and H/P plus careful analysis of the distribution of homoplastic traits in phylogenies, such an assessment should be possible. Then homoplasy, currently noise in the data, can assume its proper role as data.

I return now to the implications of the Sanderson–Donoghue results. At first glance, it appears that a null model can account for the data. Consideration of the data at different taxon levels, however, suggests a selection-based account.

Within the null model, the first major factor tuning the rate of decrease of the consistency index with the number of taxa is N, the number of traits which might alter from ancestral to derived. As N increases, the probability that any trait will be chosen to change in two independent lineages decreases. Hence, as N increases, the consistency index is higher for each fixed number of taxa. A second major factor is the branching probability P per time step. If the branching rate is very low, then a large number of traits will have changed states during the formation of very few taxa. Hence the chances of multiple switches for any single trait are large, and the consistency index will be lower for that number of taxa than were the branching probability higher. In short, at any fixed number of taxa, a high branching rate leads to a higher consistency index.

Using these two parameters, the null model can account for the Sanderson–Donoghue results. Figure 3.10a shows the predicted relation between the logarithm of the consistency index and number of taxa for the random branching model with $N =$

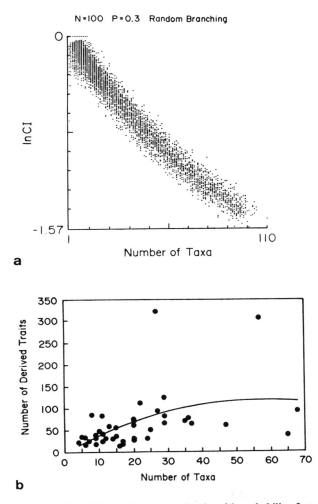

Figure 3.10 Neutral model in which taxa branch randomly, with probability .3 at each generation, in a 100-dimensional trait space: (*a*) natural logarithm of consistency index as a function of number of taxa in the model phylogeny; (*b*) number of derived traits as a function of number of taxa in studies accumulated by Sanderson and Donoghue (1989), together with their rough quadratic model regression fit to data.

100 and $P = .3$. The mean slope, crudely -0.015, is very close to the -0.0158 observed by Sanderson and Donoghue (Figure 3.5). The scatterplot of Figure 3.10*a* is slightly concave upward. So, too, are the data in Figure 3.5 when fitted with a quadratic estimate.

The capacity of the null model to fit these data is, in fact, better than mere curve fitting. Although much the same curve as that in Figure 3.10*a* can be achieved either by increasing N and decreasing P or by decreasing N and increasing P, any specific pair of N and P values makes specific predictions about the relation between the *number of traits and the number of taxa* observed in these studies. The choice of N

= 100 and $P = .3$ implies that the initial slope of traits versus taxa is about 3. Further, since the total number of traits is N, the curve must saturate at 100. Figure 3.10b shows Sanderson and Donoghue's data and their quadratic regression line. The initial slope is, convincingly, roughly 3. Less convincingly, because of two major outlying taxa with more than 300 traits each, the asymptote is about 100. It is not obvious that a single choice of N and P would allow the null model to fit both Figure 3.10a and 3.10b. This success suggests that some predictive power lies in the null model.

What inferences should be drawn from the Sanderson–Donoghue results? First, caution in two guises. The number of studies, and hence the number of data points, remain very limited. Further, the "data" are consistency indexes drawn from reconstructed trees inferred from taxonomic morphological and molecular data. The results may tell us more about how systematists classify, however, and about how reasonable the tree-construction algorithms are than about the world of biological diversity. Given these caveats, Figures 3.6a, c, e, and f and Figure 3.7 show that very different adaptive processes give much the same curves as that shown in Figure 3.10a. The slight differences in slopes in these graphs can be altered at will by increasing or decreasing N in the different models. In short, both the NK model of adaptive evaluation and a null model of random branching are compatible with the consistency-index data analyzed by Sanderson and Donoghue. Thus these data do not appear to discriminate readily among alternative models of forces, adaptation, or drift which may underlie branching phylogenies.

While the null model of random branching cannot be easily excluded by the bulk data, one feature of the Sanderson–Donoghue data is hard to reconcile with a random branching model. The authors find that the same statistical features arise at each taxonomic level, from species to class/phyla. While the data are too scant for this conclusion to be firm, assume it holds. The curious problem, on a random model, is to account for the very different *time scales* on which trait changes sufficient to accord species versus class/phyla rank occur. Changes that accord species rank are garden variety, those that accord class/phyla rank very rare.

One attractive hypothesis to account for these phenomena derives from the hierarchial version of the NK model. If some sites are highly generatively entrenched and thus epistatically affect many other sites, while other sites have few or no epistatic downstream consequences, the highly entrenched sites will tend to change only when the world alters dramatically and then will become locked in. Since alterations in more highly entrenched traits cause more downstream consequences, it is natural to associate those changes with changes accorded higher taxonomic rank. In turn, given a distribution of changes in the world, with many slight changes and few massive changes, adaptive alterations by changing highly entrenched traits will be rare, while those associated with minor features will be common. Thus hierarchial epistatic couplings plus a world which often changes slightly and rarely changes dramatically yield the observed time-scale results as a consequence of adaptive evolution struggling to climb deforming landscapes.

To test the compatibility of the hypothesis of hierarchical epistatic coupling with the Sanderson–Donoghue results, Johnsen and I used the hierarchial NK model and defined alterations in the most entrenched ten sites as phylum/class-level changes, while those in the ten lowest entrenched sites were defined as species-level changes. The resulting cloud of data on the graph of consistency index plotted as a function of number of taxa lies on a common curve matching the data observed by Sanderson and Donoghue. Thus a hierarchical epistatic model appears consistent with the present data.

The problem of homoplasy emerges as an opportunity. Without doubt, the patterns of divergence and convergence in evolution bear witness to complex optimization processes coupled with random drift in high-dimensional trait spaces at the morphological and molecular levels. The human genome project promises vast new data in the future. The abundant efforts of systematists and molecular biologists to utilize these data for the construction of phylogenetic trees has borne substantial fruit. An effort to discern in the same data evidence for the structure and deformation of the fitness landscapes foreseen by Wright (1931, 1932) and governing evolution has just begun. It is too early to be confident that the results will be worthwhile. It is not too early to be confident that the effort is worthwhile.

POPULATION FLOW
ON RUGGED FITNESS LANDSCAPES

Adaptive evolution in real populations is necessarily a search process driven by mutation, recombination, drift, and selection over either fixed or deforming fitness landscapes. Our task in this section is to explore how the multipeaked rugged structure of fitness landscapes governs both the *evolvability* of adapting populations and the *sustained fitness* of those populations. We must begin with the plausible prejudice that properties of organisms which govern the structure and rate of landscape deformation, as well as such parameters of the adaptive search process, as mutation and recombination frequencies, are subject to natural selection. If so, then our basic questions must be, which properties of landscapes and search parameters optimize adaptive evolution? and, Can selection attain such optimization?

In the present section, I discuss adaptive evolution on smooth and rugged landscapes in asexual haploid organisms under the drives of mutation, selection, and drift. We shall find that, when landscapes are very smooth, an error complexity catastrophe sets in and adapting populations slide down fitness peaks to the valleys below. In contrast, when landscapes are very rugged, a second complexity catastrophe sets in: Adaptation becomes trapped in small local regions of the space and hence cannot search effectively. Evolvability, the capacity to search a reasonable fraction of the space, may be optimized when landscape structure, mutation rate, and population size are adjusted so that populations just begin to "melt" from local regions of the space. Sustained fitness may be optimized when landscape structure is tuned so that the sides of fitness peaks are steep enough to offset the mutation rate and the rate at which the landscape is deformed by abiotic or coevolutionary forces. Whether both evolvability and sustained fitness can be jointly optimized is unclear.

In the last part of this section, I discuss the role of recombination as a search strategy in rugged landscapes. The capacity of recombination to aid adaptive search depends upon the structure of the landscape. Recombination is useless on uncorrelated landscapes but useful under two conditions: (1) when the high peaks are near one another and hence carry mutual information about their joint locations in genotype space and (2) when parts of the evolving system are quasi-independent of one another and hence can be interchanged with modest chances that the recombined system has the advantages of both parents.

The Mutation/Selection Error Catastrophe
and the Weak Maxwell's Demon

Chapter 2 deduced general features of adaptive evolution under the assumption that selection was always strong enough to pull an adapting population to the fittest variant found. Thus hill climbing was limited only by the topology of the adaptive land-scape, the existence of local optima for one-mutant adaptive walks, and the metasta-ble durations of fitter variants in long-jump evolution. In the present section, we discuss a very general limitation to selective adaptation on reasonably smooth land-scapes alluded to above:

> For a fixed mutation rate, the number of mutants per individual increases as the complexity of entities under selection increases. As this occurs, a threshold is passed beyond which selection cannot hold a population at the locally fittest variant, errors accumulate, and the population falls from rare optima toward less fit but more typical members of the ensemble.

The general idea of a mutation/selection error catastrophe is well known in clas-sical population genetics. I shall discuss first the case of the simplest possible diploid model in population genetics, which corresponds very closely to the haploid *NK* model in the $K = 0$ limit. The landscape has a single optimal genotype and is as highly correlated as possible. Thereafter, we shall examine the more complex and realistic problem of populations adapting on multipeaked rugged landscapes. I should emphasize that this field of inquiry is still largely uncharted, and so we shall accordingly raise more questions than answers. Therefore, one major point of the following section is to view it as an indication of areas which require further effort.

Classical Genetics and the Error Threshold

Classical single-locus theory considers two alleles, $A1$ and $A2$, in diploids with genotypes $A1A1$, $A1A2$, and $A2A2$; a mutation rate v from the more favored $A2$ to $A1$; and a mutation rate u from the less favored $A1$ to $A2$. If both alleles are equally fit, the equilibrium frequency P_x of $A1$ in the population is $v/(u + v)$. In the presence of selection against $A1$, where $A1A1$ has fitness $1 - s$, $A1A2$ has fitness $1 - s/2$, and $A2A2$ has fitness 1, s, $(0 \leq s \leq 1)$, is the difference in the fitness of the two homo-zygotes), the frequency distribution of $A1$ in the population is given by the confluent hypergeometric series expression

$$P_x = \frac{u}{u + v} \times \frac{m(4Nu + 1, 4Nu + 4Nv + 1, 2Ns)}{m(4Nv, 4Nu + 4Nv, 2Ns)} \tag{3.2}$$

where N is the number of individuals in the population and m refers to terms in the series (Ewens 1979). If the term $2Ns$ is large and negative, requiring s to be negative and not too close to 0, then Equation 3.2 simplifies to

$$P_x = 2v/|s| \tag{3.3}$$

where $|s|$ is the absolute value of s. The frequency of the less favorable allele becomes

$$f_x = \frac{N2v}{N|s|} = \frac{2v}{|s|} \tag{3.4}$$

Single-locus theory can be directly extended to consider a total of t independent genetic loci, each with two alleles, one less and one more favorable, and each con-tributing *additively and independently* to the total fitness of the system.

To ensure that maximal fitness remains 1.0 as t increases, the total fitness of each genotype is normalized through division by t. Thus it is necessary to define the fitness contribution per locus s' as $s' = s/t$. This implies that the proportional contribution of each locus to fitness *decreases* as the number of genetic loci in the system, t, *increases*. Therefore, as t increases, the fitness contribution of each favored allele over its less favored version decreases, but the mutation rate converting the more favored allele to the less favored one remains constant. One would expect that, as t increases, mutation would eventually become stronger than selection. This happens. Eventually, less favorable alleles begin to accumulate in the population.

Assume for simplicity that the forward and back mutation rates are equal. Then at the equilibrium distribution of mutations per individual in the population, the *expected fraction of less favorable alleles* per individual due to the balance between mutation and selection is

$$f_x = \frac{2vt}{|s|} \qquad (3.5)$$

Consequently the expected number of less favorable alleles c is

$$c = \frac{2vt^2}{|s|} \qquad (3.6)$$

This simplifies to

$$v = \frac{c|s|}{2t^2} \qquad (3.7)$$

Equations 3.6 and 3.7 imply that, to hold the total number of unfavorable alleles at some constant number c per individual as the total number of loci t increases, the mutation rate v must *decrease inversely to t^2*. Alternatively, for a fixed mutation rate, as the complexity of the entities under selection t increases, *the number of "bad" alleles which accumulate increases as t^2*.

Let us recapitulate these implications of classical population genetics. The argument assumes that each locus contributes additively to fitness, independently of all other loci, as in the *NK* model for $K = 0$. Further, fitness is normalized, as in the *NK* model for $K = 0$. Third, there is a *single global optimum,* and *connected adaptive pathways via fitter mutants* to that optimum exist from any suboptimal genotype. The single optimal genotype is that with the favorable allele at each of the t loci. Any less favorable genotype can always be improved by mutating any less favorable allele to its more favorable counterpart at any locus.

Despite the fact that connected pathways to the single global optimum exist, selection fails to hold a population at that optimum as t and hence the complexity increase. Each locus contributes proportionally to total fitness. Thus as t increases, the contribution of each favorable allele versus its unfavorable allele to overall fitness decreases, but the mutation rate from favorable to unfavorable allele remains constant. In short, as t increases, the "gradient" of selective force pulling the population uphill to the global optimum decreases, while the mutational forces tending to disperse the population away from that optimum remain constant. Below a critical complexity t_c, the selective force is stronger than the mutational force, and selection can either hold the population at the global optimum or pull it there from any suboptimal genotype. In this condition, the population exists as a *tight cloud centered* at the global optimum. Above the critical complexity t_c, the dispersing mutational pressure becomes stronger than the selective force and the population disperses away from the global optimum. In that case, the population falls from the global optimum to a *stationary state distribution,* a fixed average "distance" from the global optimum, where the number of unfavorable alleles is steady and proportional to t^2.

How does such a population displaced a fixed mean distance from the global optimum behave? Note that in the present model, where all loci make equal contribution to fitness, a very large number of genetic states are all at the same fitness distance from the global optimum. Concretely, suppose 10 percent, plus or minus a bit, of the t loci are in their less favorable allele. A very large number of alternative genotypes have about 10 percent of the loci in the unfavorable allele. These many genetic states are all almost equally fit and form a kind of thin shell around the global optimum at a constant distance from it. *Movement in and on that shell is almost selectively neutral.* Further, each genotype on that shell is a one-mutant or a two-mutant neighbor of many other genotypes on the shell. *Thus within each shell, there are connected walks which are selectively neutral.*

Will the population diffuse to a uniform distribution on that shell, remain at a single point, or wander as a coherent cloud across the shell surface? The answer tends to be the last. The reason is straightforward. Consider any fixed population. At each generation, some members may fail to leave offspring, not due to fitness differences but merely by chance. In a haploid nonsexual population, this failure implies that after a number of generations, *all* members of the last generation will have descended from a single founder in the first generation. In small populations, this founder effect tends strongly to trim the dispersing cloud on the shell and lead to a coherent random walk of a clustered cloud across the shell surface. The size of the cloud increases with population size, of course. Similar arguments apply to sexual populations.

These results are easily demonstrated using the *NK* model in the $K = 0$ limit, as Figure 3.11 shows. In this numerical simulation, $N = 50$, we have assumed two alleles per locus, a mutation rate from each to the other of 0.001, and 50 genotypes evolving. At each generation, each genotype is sampled at random and produces an offspring at each sampling in proportion to its fitness. Whether or not that potential parent produces an offspring, it is placed back in the parental pool and so may be sampled as a potential parent again. This process normalizes fitness. After 50 offspring genotypes constituting the next generation are created, each site in each genotype is mutated with probability 0.001, and the simulation iterates. A mutation rate of 0.001 is high enough that selection is here unable to hold the population at the optimum. It falls and wanders at a level with about seven to nine genes in their less favorable allele. If released instead from a random genotype, the population climbs to the same steady state level with seven to nine genes in their less favorable allele. Thus from both the global optimum and a random initial starting genotype, the system falls or climbs to the same intermediate level of fitness. This represents the neutral shell within which the population thereafter wanders.

In summary, classical population genetic models demonstrate a tendency, as the complexity of entities under selection grows, for the population to fall from optima due to an accumulation of "errors." This tendency has been rediscovered in different contexts which generalize the classical result.

More on Mutation/Selection Error Catastrophes

Eigen and Schuster (1977, 1978a, 1978b, 1979) were the first to stress these kinds of results. These authors constructed detailed models of evolution among replicating

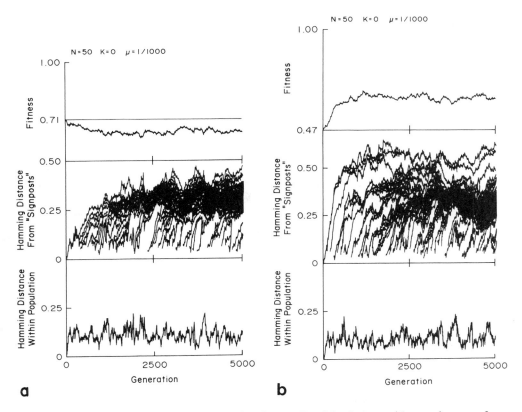

Figure 3.11   The error catastrophe. Population flow on $K = 0$ landscapes, with mutation rate of 0.001. *Top panels:* 50 genotypes released (*a*) at a local optimum and (*b*) from a random initial genotype. Note that in (*a*) selection cannot hold the population at the global optimum, while in (*b*) the population climbs to the same stationary intermediate fitness level seen in (*a*). *Middle panels:* mean Hamming distances (measured in fractions of N) of the adapting populations, measured from consensus-sequence signposts created every 100 generations along the adaptive flow. Distances are measured to each signpost. Many signpost curves lie on top of one another. *Lower panels:* mean Hamming distance within each population.

molecules, such as DNA or RNA, competing with one another and subject to fixed mutation rates per nucleotide. In a population of competing replicating sequences which initiates from some single sequence, the dynamics of the population depend on these factors:

1. The replication rate of each sequence with respect to competing sequences.

2. The forward mutation rates by which each sequence mutates into its many one-mutant, two-mutant, . . . neighbor sequences, together with the back mutation rates by which those neighbors mutate to each specific sequence. These mutations generate a cloud of related sequences, called a quasispecies, each member of which subsequently becomes a competitor of all the rest.

3. The boundary conditions specifying the competition. Typically, this is given by a constant rate of influx of monomers to a "chemostat" and a constant rate of efflux

of material from the chemostat. Under these conditions, the fastest replicating molecular sequence, called the "master sequence," plus its quasi-species cloud wins the competition and dominates the chemostat until, at rare intervals, a truly improved variant occurs and dominates the chemostat with its own new quasi-species cloud. More precisely, the quasispecies is approximately the fastest growing eigenvector (Chapter 14), or fixed concentration ratio of the master sequence and its one-mutant, two-mutant, . . . variants.

Eigen and Schuster prove a basic result showing that, as the sequence lengths of replicating molecules under selection increase, a threshold is reached beyond which selection cannot maintain the fittest variants. Errors accumulate in the replicating molecules. The error threshold these authors find depends on the ratio of the selective superiority of the wild-type sequence with respect to its competitors:

$$E_T \propto \frac{s_m}{u} \tag{3.8}$$

where u is the mutant rate per symbol (for example, nucleotide within a sequence) and s_m is the relative selective superiority of the sequence. When this threshold is surpassed, the predicted result is that the most highly fit variant will become a minor component of the spectrum of replicating sequences. In the absence of a further theory about the distribution of fitness values in sequence space, the Eigen–Schuster error catastrophe states merely that the population will flow away from the current master sequence and that the information encoded in this sequence will be lost. Where the population may flow thereafter is not definable without some further hypotheses about the structure of the fitness landscape.

Elegant experiments have confirmed these predictions (Biebricher 1987; Eigen 1987; Biebricher and Eigen 1988). Moreover, Eigen and Schuster have pointed out that viruses in fact live within and close to the error thresholds given by the known rates of nucleotide mutations (Eigen and Schuster 1979; Eigen 1987). For example, single-stranded RNA viruses such as Q_B replicate via specific replicases with an error rate per nucleotide on the order of 5×10^{-4}. For fitness advantages of the wild type with respect to its mutant neighbors ranging from 2 to 200, this error rate leads to the prediction that the maximum number of nucleotides can range from 1386 to 10 597. In fact, Q_B has 4500 base pairs.

The classical genetics result and Eigen and Schuster's error threshold are clearly intimately related. They differ in whether fitness is independent and additive across the symbols within a sequence. For the additive model, fitness is independent. For Eigen and Schuster's general result, the assumption is merely that the current master sequence has replication superiority over its competitors. No further specification of the fitness landscape is made. Thereafter, the general result of these models is that, even with connected pathways to a single global optimum, scaling laws exist and show that, when the system is sufficiently complex, selection is unable to hold a population near that optimum in the face of mutation. In terms of the image from an earlier chapter, in which Maxwell's demon plays the role of selection, as the system becomes complex, the demon becomes weaker than the mutational back pressure driving the system toward the average properties of the ensemble. Thus both models point toward another critical limitation on selection. This same limitation will arise in a different form in Chapters 11 and 13, when we consider the capacity of selection to achieve arbitrary genomic cybernetic regulatory systems.

Population Adaptation on Multipeaked Rugged Landscapes

The classical genetic model concerns very simple fitness landscapes with single optima. But real landscapes in genotype space—or in protein space or sequence space or phenotype space—are almost certainly multipeaked and rugged. While population geneticists have implicitly and explicitly studied population dynamics in complex fitness spaces (Crow and Kimura 1970; Ewens 1979), we lack a general theory which relates the *known statistical ruggedness of landscapes with population flow upon them* under conditions of mutation, recombination, and selection. We now examine recent provocative numerical simulations by Fontana and Schuster (1987) on a model of adaptive evolution of RNA-like polymers which replicate and mutate in a chemostat. (See also Schuster 1986, 1987; Eigen 1987; Eigen, McCaskill, and Schuster 1988; Fontana, Schnabl, and Schuster 1989; Fontana, Griesmacher, et al. 1991; and Fontana, Stadler, et al. 1992 for discussions of landscapes in sequence spaces.) In addition, I shall briefly describe work using the *NK* model.

Fontana and Schuster (1987) take, as their primitive model of a single-stranded RNA sequence, a sequence N long of 0 and 1 bits. These strings, like true single-stranded RNA molecules, are subjected to a folding rule in which 0 bits bond to 1 bits. This rule yields model RNA sequences, which I hereafter will call RNA sequences, with secondary structure consisting in hairpin loops and bonded stems, plus open loops whose bits cannot bond to one another (Figure 3.12). The folding rule utilized is complex and seeks to maximize the number of 1–0 bonded pairs possible in any RNA molecule, depending upon the number and arrangement of the 0 and 1 bits along the RNA string.

The authors' purpose is to investigate the flow of an adapting population of self-replicating RNA molecules across RNA space. The space, of course, is just the 2^N possible strings of 1 and 0 bits. Each string is a one-mutant neighbor of N other strings, which differ from one another by changing a single bit to the opposite value. Throughout the simulations, N is 70. To model the evolution of replicating RNA

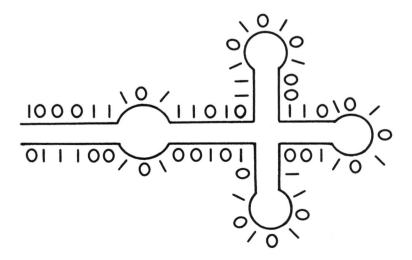

Figure 3.12 Model RNA molecule with 1 and 0 "bases." The molecule folds under the constraint that 1 and 0 are template complements which bind. (From Fontana and Schuster 1987)

molecules, the authors need to assign a fitness landscape over RNA space. They invent two such landscapes, a *thermodynamic landscape* and a *kinetic landscape.* Each is very rugged. Each helps tune our intuitions about the kinds of phenomena found and about sensible ways to try to build toward a general theory.

In the thermodynamic landscape, the fitness of each RNA molecule depends upon the stability of the folded strand and so is proportional to the total fraction of the N bits which are bonded in the folded strand. Therefore, it is straightforward to deduce what the optimal sequences look like: Each is a single hairpin with a stem of bonded bits about $N/2$ long, with about four bits at the loop end of the hairpin reflecting steric hindrance in real single-stranded RNA hairpins. As the authors point out parenthetically, substitutions of 1 and 0 values in this terminal loop are selectively neutral.

In the kinetic landscape, the authors imagine a more complex rule to ascribe fitness to each string. Specifically, they suppose that the rate of replication of a string is increased in open, unbonded regions but that the loss of stability in these regions leads to easier degradation of the molecule. Thus the fitness of any folded molecule is a complex mixture of ease of replication and resistance to degradation. Here the optimal folded forms are not at all clear.

Fontana and Schuster carefully emphasize, and I echo, that the point of constructing these landscapes is not their utter realism from the molecular point of view. Rather, the point is to have in hand some explicit and complex landscapes upon which evolution of a population may be studied.

Selection is carried out in a model chemostat. Each molecule replicates at a rate proportional to its fitness. Influx of monomers or bits 1 and 0 into the chemostat and efflux of all molecules at a constant flow rate supply the selection conditions. Under such conditions, the total number of bits, either free or bound into polymers, falls to a constant number. The remaining selection dynamics occur subject to this constraint. Therefore, selection is reflected in the *increasing proportion* of the total number of bits which are in the form of one specific RNA molecule. In other words, selection is just an increase in the relative abundance of one specific RNA molecular sequence in the adapting population of replicating RNA sequences.

In addition to the selection dynamics, the authors impose a mutation rate at which 1 bits mutate to 0 bits and vice versa. The mutation rate is given as $1 - Q$, where Q is the probability that any bit is copied accurately. Therefore, $1 - Q$ is the probability of a replication mistake per bit. This means that each replicating molecule produces mutant strings at some frequency. Since each bit in a string may mutate, the number of mutations introduced into any string after it replicates is a Poisson process with a mean of $(1 - Q)N$. Typical values of Q were on the order of 0.995. Thus most strings have no mutations, some have one mutant bit, and a few have a fairly large number of mutants.

As mentioned above, both landscapes are very rugged. In this respect, they are reminiscent of the rugged one-mutant fitness spectrum around local optima in the NK model for modestly large values of K relative to N (Figure 2.6a–d). Indeed, the thermodynamic landscape shows a Gaussian distribution of fitness values and, as noted in Chapter 2, apparently can be mapped onto an NK landscape with $N = 70$, $K = 8$ to 10.

Three aspects of the simulations carried out by Fontana and Schuster are particularly interesting:

1. They studied how well the population of 2000 replicating molecules climbed hills as a function of the mutation rate. For small enough mutation rates, hill climbing was successful and faster with a slightly higher mutation rate. As in the example of adaptation in the traveling salesman problem (Kauffman and Levin 1987), too low a mutation rate limits the rate of finding fitter variants on the current hill. A slightly higher rate explores not only one-mutant but two-mutant, three-mutant, and higher-mutant variants more extensively, and the population as a whole adapts more effectively. Typical values of Q are 0.999 to 0.997.

2. At a still higher mutation rate, selection cannot hold the population at a local optimum. When Q is reduced to 0.996, the population rapidly falls from the global optimum to very modest fitness values. In addition to falling from the optimal RNA sequence, the population rapidly encounters a very large number of different sequences, all of modest fitness. The population as a whole is wandering through some reasonably large volume of sequence space. Fontana and Schuster carry over the concept of a master sequence into the current context. In a simple fitness landscape with a single global optimum, the master sequence is that optimal sequence. For low mutation rates, the master sequence is also the most abundant in the population. On a rugged landscape, the master sequence becomes simply the most abundant sequence present at any moment, regardless of its fitness. The wandering population therefore encounters a very large number of master sequences, many of which have very low fitness.

3. The authors utilize a Hamming-distance description of how the population flows through the multipeaked landscape over time. Recall that the Hamming distance between two strings of 1 and 0 values N long is just the number of sites where the two strings differ. Figure 3.13 shows the Hamming distance, fitness, and abundance of strings over the course of adaptation. Hamming distance of all strings was measured with respect to the final master sequence with the highest fitness. The fascinating feature of Figure 3.13 is that the population is initially localized as a clump far from the ultimate master sequence. In the midterm of the adaptive process, a small number of sequences appear halfway to the ultimate master sequence. This small number grows into a substantial cluster of sequences. Then a new cluster appears in the vicinity of the ultimate master sequence. This implies that the mutant cloud surrounding the initial master sequence included some members with a modest number of mutations which were only slightly less fit than the initial master sequence and hence could gain purchase on a new hill some distance from the initial hill. The population can therefore reach across valleys of low fitness, sending scouting parties far enough away to encounter good terrain. Yet from Figure 3.11 we know that, if the mutation rate is too high, the population will simply flow down from peaks and wander the lowlands of the fitness landscape.

These simulation results make it clear that we need a real theory relating the structure of rugged multipeaked landscapes to the flow of a population upon those landscapes. We do not yet have such a theory.

Population Flow on NK Landscapes

The advantage of the NK family of landscapes in studying population flow is that the structure of the landscape can be tuned from smooth to rugged. This ability allows us to investigate both the error catastrophe on smooth landscapes and *trapping or freezing* into small regions of the space at a fixed mutation rate as landscapes become progressively more rugged.

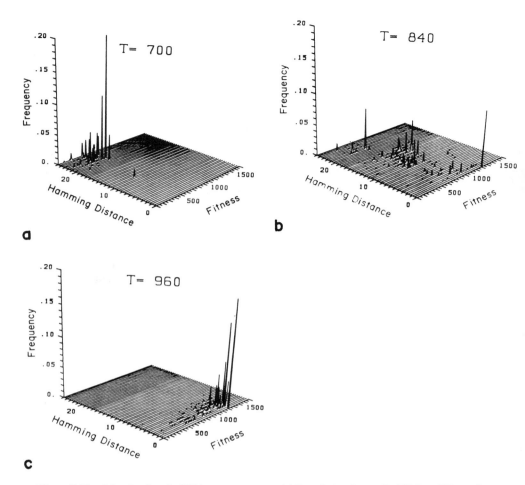

Figure 3.13 Adaptive flow in RNA sequence space. (*a*) Population fitness (0–1500) and Hamming distance (1–25) from ultimate master sequence as a histogram in two dimensions. Simulation time step $t = 700$. (*b*) As in (*a*) but at a later time step, $t = 840$, when the adapting molecules have populated an "island" partway between the initial region of sequence space and the terminal region of higher fitness. (*c*) As in (*a*) but at $t = 960$. The population is now clustered around the ultimate master sequence. (From Fontana and Schuster 1987)

Figure 3.14 All panels as in 3.11*a*. (*a*) Population flow in $K = 2$ landscapes. The mutation rate is 0.001. Population is released from an average local optimum and shows no overall increase in fitness over the walk, and population walks over large reaches of genotype space *(middle panel)*. (*b*) The same landscape as in (*a*), but the mutation rate has been reduced to 0.0001. The population appears to show slow but steady increase in fitness during the last 5000 generations and does not walk far across genotype space. (*c*) Population flow on a more rugged ($K = 25$) landscape. The mutation rate is 0.001. Population released on an average local optimum. Note fitness initially decreases but then jumps to a new higher level. Population is frozen into a small region of genotype space *(middle panel)* over the last 8000 generations. Freezing appears to coincide with the jump in fitness and hence with flow to a good small region of the space. Note also that, at the same mutation rate, the population in (*a*) on the smoother $K = 2$ landscape is not frozen into a small region of genotype space.

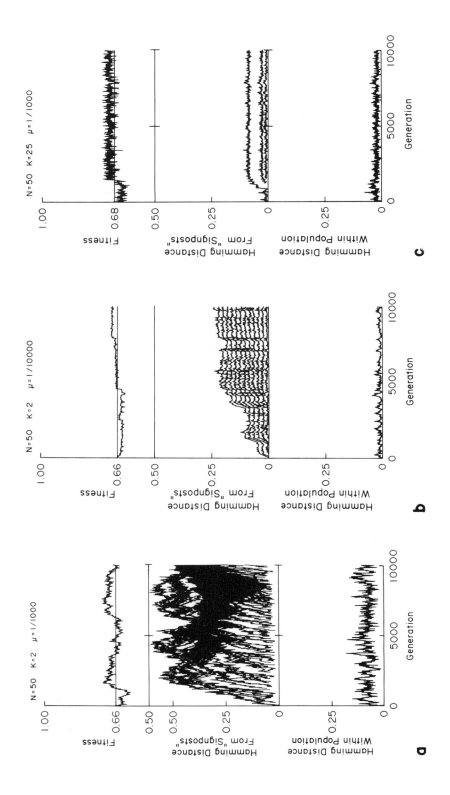

We have carried out initial simulations (Kauffman 1989a) for $N = 50$; $K = 0, 2$, and 25; and mutation rates ($1 \leftrightarrow 0$) of 0.001 and 0.0001 per allele per genotype per generation. Populations were fixed at 50 haploid "organisms." The initial population was released either at a single local optimum or at a single random "genotype." As described above, in order to normalize fitness, at each generation, each organism was sampled randomly and then, whether or not it produced an offspring, was placed back into the pool. Sampling continued until 50 offspring were generated. Thereafter the process was iterated for 5000 or 10 000 generations.

The questions of importance concern the flow of the population over the landscape as a function of the evolutionary parameters concerning landscape structure, population size, and mutation rate. To monitor how spread out the population was over genotype space during its wandering, we measured the mean Hamming distance within the population at each generation. To measure how far the population moved across the space, we defined consensus-sequence signposts every 100 generations. At each generation, the mean distance of each of the 50 genotypes to each of the previously planned signposts was measured. Thus flow across the space shows up as increasing distance from previously planted signposts.

The error catastrophe on smooth landscapes is shown in Figure 3.11. Lack of trapping or freezing at a high mutation rate and freezing into a local region at a lower rate on modestly rugged landscapes are shown in Figures 3.14a and b. Finally, as landscapes become more rugged, trapping into local regions occurs at a mutation rate which would not cause trapping on a smoother landscape, as shown in Figure 3.14c.

Landscape ruggedness and mutation rate also govern sustained fitness, which can be higher on more rugged landscapes with lower peaks. Recall that, as K increases, the *heights* of local optima fall ever lower, while the *steepness* of fitness peaks increases. The increasing steepness implies that the selective gradient back toward the peaks or high ridges is *greater* in more rugged than in less rugged landscapes. Thus in the face of a modestly high mutation rate, an adapting population might fall farther from the high peaks on a smooth, gentle landscape than from the somewhat lower peaks on a more rugged landscape. Figure 3.15 compares the mean fitness maintained by adapting populations on $K = 0$, $K = 2$, and $K = 25$ landscapes in the face of a mutation rate of 0.001. The striking feature is that mean fitness is *highest* on the most rugged, $K = 25$, landscape. Yet the local optima on this landscape are clearly lower than on the smoother $K = 2$ and smoothest $K = 0$ landscapes.

The implications of these results with respect to evolvability and sustained fitness include the following:

1. In rugged landscapes, many local optima exist, perhaps in a self-similar form.

2. At sufficiently low mutation rates, the population will climb the nearest fitness peaks and remain clustered about one or another of those peaks. The process is then limited merely by the heights of those peaks. On rugged landscapes, this behavior implies that the population will typically be trapped on a relatively poor local optimum.

3. As the mutation rate increases for a fixed population size and landscape, the population may "melt" with respect to a local small region of genotype space and flow within it. However, fitness valleys between this region and other, perhaps better regions offer barriers to traversing the space for greater distances. Thus a hierarchy of mutation rates coupled with a hierarchy of sized regions which can be explored is expected. At any mutation rate, a local region can be explored relatively quickly

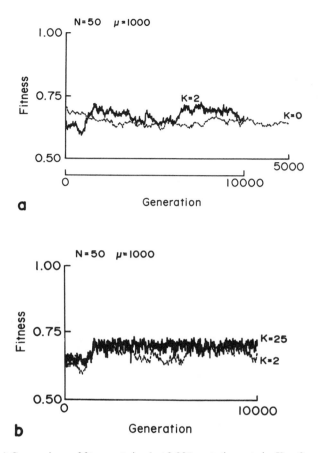

Figure 3.15   (*a*) Comparison of fitness attained at 0.001 mutation rate in $K = 0$ and $K = 2$ landscapes. Data redrawn from Figures 3.11 and 3.14*a*. (*b*) As in (*a*), except comparison is between fitness attained in $K = 2$ and $K = 25$ landscapes. Data redrawn from Figures 3.14*a* and 3.14*c*.

while escaping to other regions past the fitness-loss barriers will be a very slow process.

4. At sufficiently high mutation rates, the population can flow across vast tracts of genotype space. It will have entirely melted. At these high mutation rates, no accumulation of heritable information beyond the coherence of walks due to founder effects is to be expected.

5. An optimum mutation rate at which populations just begin to melt seems likely to optimize both evolvability and sustained fitness. At a low mutation rate, populations become trapped on poor local peaks. At a high mutation rate, populations are driven far below the peaks and drift in fitness lowlands. Thus sustained fitness should be optimized at some intermediate mutation rate. That rate is likely to occur when populations are just beginning to melt from peaks. First, in pop-

ulations which are just melting, evolvability should be optimized. This intermediate mutation rate seems to be a good compromise between maintenance of heritable information and search across the space. But sustained fitness should also
be optimized. Analysis of the NK family of landscapes also shows that, typically,
the highest optima have the largest drainage basins. Thus if the population just
melts off local peaks, it has a reasonable chance to flow into the basins climbing
to high peaks. If so, it will tend to hug the higher peaks and therefore optimize
sustained fitness.

6. The mean fitness maintained by a population can be higher on a more rugged
landscape, which harbors lower optima, than on smoother landscapes. Thus again
we see that sustained fitness reflects both landscape structure and the properties
of the stochastic adaptive process on that landscape. The highest average maintained fitness need not occur on those landscapes with the highest peaks but will
occur on those landscapes whose ruggedness best suits the character of the adaptive process. In the present case, for any fixed size of genotype N, population size,
and mutation rate, some particular value of K will yield a landscape whose ruggedness is optimal for maintaining high average fitness. Thus some presently
unknown curve in the NK plane corresponds to the optimal landscape to match
a given stochastic flow exploring that landscape in order to achieve the highest
average fitness. Grounds to suppose that this curve is related to the mutation rate
at which populations just begin to melt from local regions, hence that both evolvability and sustained fitness can be jointly optimized, were noted above.

The freezing and melting alluded to above bear analogies to thermal physics and
bear on the famous selectionist–neutralist controversy: Once off the peaks, populations can often wander vast distances across genotype space among a percolating
cluster of near-neutral mutants. We consider this briefly next.

Neutral Percolation Domains and the Selectionist–Neutralist Argument

When examining the classical additive population genetic model, we found that the
error catastrophe led an adapting population to fall below the single optimum.
Thereafter, the population wandered neutrally as a coherent cloud within a thin shell
at a fixed mean fitness distance below the global optimum.

In a rugged multipeaked landscape, if a population is displaced from an optimal
point in genotype space and falls to lower fitnesses, there may well be a very large
number of different genotypes with very nearly the same fitness, each of which is a
one-mutant neighbor of many other genotypes within the same narrow fitness band.
Whether there is a single peak in the landscape or many peaks, suppose mutation is
strong enough to force the population off the peak or peaks down to such a fitness
band and *hold it there;* then the population *must wander neutrally through this band.*
Thus the distance across the space over which the population wanders depends on
how far the narrow band of fitness values extends across the landscape. The intuitive
image is of a cloud layer hanging below the peaks of a mountain range. As we note
below, this image oversimplifies the flow of an adapting population, for such a population will typically not be held in a constant narrow range of fitnesses but may more
or less hug the landscape, moving higher when below the higher peaks, falling lower
when below lower peaks. Nevertheless, the simplified image helps tune intuition.

Mathematically, such a band is a *connected cluster of one-mutant near-neutral genotypes.* More precisely, restrict attention to any defined narrow band of fitness values and imagine coloring all genotypes within that range red. Then connect any red genotype to those other red genotypes which can be reached via one-mutant neighbors. This process forms connected clusters of near-neutral genotypes ranging in size from isolated red genotypes to perhaps much larger clusters. If the size of the largest cluster scales with the number of genotypes in the space, the cluster can be thought of as percolating, or extending, across the space (Stauffer 1985). Recently, Flesselles et al. (1988) investigated the statistics of clusters on the N-dimensional Boolean hypercube, corresponding to a genotype space in the NK model with two alleles, 1 and 0. These authors examined the relationship between the fraction x of red sites and cluster size, under the assumption that the locations of the red sites are uncorrelated. Thus their example corresponds to the $K = N - 1$ limit of the NK model or to any other fully random, uncorrelated fitness landscape. As noted by Amitrano, Peliti, and Saber (1988), a good estimate for the critical value of x is

$$x_c \approx \frac{1}{z} + \frac{3}{2z^2} + \frac{15}{3z^3} + \frac{83}{4z^4} + \cdots \tag{3.9}$$

where $z = N - 1$. Intuitively, when x, the fraction of red genotype, is slightly more than $1/N$ of the 2^N genotypes, each red genotype will have slightly more than one red one-mutant neighbor, and therefore a very large cluster will form. Were the population to wander randomly on such a cluster, arbitrarily different genotypes would be encountered. By contrast, if x is less than x_c, then small isolated red clusters will form, and the population will be confined to wander neutrally in one or another of these clusters. Thus a small increase in x from below to above x_c yields a *phase transition:* The population will melt from a well-localized distribution to one which is delocalized and can wander neutrally across vast tracts of genotype space.

Three features of this simple image should be emphasized. First, the red-connected clusters of near-neutral genotypes may be few in number, very large, and scale in size with the number of genotypes in the space, or they may be many and small. Second, small changes in x can lead to dramatic changes in population behavior. Third, any such cluster may correspond to an interconnected web of fitness ridges of high fitness, just below the fitness peaks of the landscape. The population will then tend to flow along this system of interconnected ridges. Although the dimensionality of the entire genotype space may be high, the connectivity patterns among the genotypes on the ridges may be very low, each being connected to only a few others. As noted by Eigen (1987), the adapting population will tend to be guided along such ridges, with selection operating on the entire mutant distribution in genotype space at once.

The analysis of Flesselles et al. (1988) is valid only for fully random fitness landscapes. On correlated landscapes—for example, in the NK model with $K < N - 1$ neighboring genotypes have more or less similar fitness values; thus the locations of genotypes within any narrow fitness range is not random. Among the important issues in understanding population flow on rugged multipeaked landscapes are the sizes, numbers, and relative locations of such connected neutral clusters. A question of fundamental interest is the rate at which such clusters increase in size as fitness levels are lowered. The increase may be gradual, as in a $K = 0$ landscape, or a single, sudden, dramatic increase in size may be encountered, as in the case of fully random landscapes, $K = N - 1$. In the latter case, a slight increase in mutation rate, pushing the population slightly farther down from the optima, may suddenly leave the population on a very much larger cluster. Then a very slight increase in mutation rate will yield a sudden, dramatic increase in the range over the fitness landscape across which the population wanders as a coherent cloud. In landscapes with $0 < K < N$ or in other complex landscapes, a variety of rates of increase in cluster sizes as successively lower fitness levels are examined may arise. Slight increases in mutation rate would then sometimes lead to jumps in the range of genotype space accessible by diffusion among near-neutral mutants.

Another question of importance is how far below the fitness peaks neutral percolating clusters spanning large tracts of genotype space lie. On rugged landscapes, such neutral clusters should hover close to the peaks. On smooth landscapes, such clusters must be farther below peaks. It is easy to see that, on fully random landscapes, such spanning clusters need be only slightly below fitness peaks. Consider a point one step from a local optimum. Typically, only a single direction uphill is available: that which leads to the local optimum. If one considers a point about two or three steps from a local optimum, however, then typically one or two directions do not lead directly to local optima. Therefore, a connected web of ridges only a few steps below local peaks, hence typically of rank order about $2/(N + 1)$ or $4/(N + 1)$, may participate in such neutrally connected clusters spanning much of genotype space. In contrast, consider the single-peaked $K = 0$ landscape. To gain access to large tracts of genotype space, mutation would have to be strong enough to force the population far below the peak. For example, access to half the genotype space would require that the population be driven down to a level of fitness at which half the N genes, on average, are in their less favorable state. Since landscapes are undoubtedly quite rugged, neutral percolating domains are typically not far below fitness peaks.

As remarked above, these neutral percolating domains oversimplify the problem of understanding population flow, for an adapting population is not held in a fixed fitness band. Instead, it is expected to follow the contours of the fitness landscape in a complex distribution reflecting fitness gradients, sizes of genotype basins able to climb to each local optimum, distribution of fitness-loss barriers between alternative basins, mutation rate, and population size. Nevertheless, the sizes of near-neutral percolating domains is one estimate of the range of the landscape accessible and of how that range increases as mutation rate increases.

The behavior of an adapting population on a rugged landscape obviously should have high significance in the selectionist–neutralist debate. The neutralists assert that the bulk of evolution is selectively neutral drift. In particular, this claim has been held to apply to the molecular level. But it seems likely that most protein fitness landscapes are rugged, not neutral. For example, in the next chapter we shall find that we can use the NK model to fit many features of adaptive evolution of antibody molecules by assuming that each amino acid in the 110 to 120 amino acid "variable" region of the antibody molecule is influenced by about $K = 40$ other amino acids. The corresponding landscape is very rugged and corresponds to what we know about how many amino acid substitutions significantly lower a protein's function. Then does such ruggedness assure us that the neutralist claims are wrong? Not necessarily. Consider instead that, in much of molecular evolution, the mutation rate relative to population size and landscape structure may be high enough to drive the population below the peaks; the population can then diffuse vast distances among a percolating cluster of near-neutral mutants. Then, although the landscape is rugged, the population experiences a largely neutral drift across sequence space. Therefore, to the extent that molecular evolution occurs on *fixed* fitness landscapes, the neutral theory would seem to require that very large near-neutral clusters occur in genotype space and guide evolutionary flow. Since fitness landscapes undoubtedly are *not* fixed, the plausibility of the neutral theory is even greater. As a landscape deforms, the population may drift across a large connected cluster of near-neutral genotypes whose components alter as the landscape does.

The flow of an adapting population on rugged landscapes under the drives of mutation and selection is closely analogous to a physical system, such as a spin-glass at a fixed finite temperature. These analogies are important because ideas from statistical physics are likely to prove generally useful in population-flow problems. For example, in spin-glass models at $0°K$, the spins can flip only if the new configuration is of lower energy than the old configuration. In the typical Glauber dynamics used to model such a system, only one spin is allowed to flip at a time (Sherrington and Kirkpatrick 1975). Thus at $0°K$, a spin-glass walks to a local energy minimum via fitter (lower-energy) one-mutant neighbors. Higher temperature is analogous to a higher mutation rate. At a finite temperature, a spin is allowed to flip to a higher-energy, unfavorable state with a probability which increases as the temperature increases. Thus at higher temperature the system can jump to one-mutant neighbors which are much less favored and can move in a zone of quite high energy across configuration space (Stein, Palmer, et al. 1984; Stein and Ogielski 1985).

The consequences of spin-glass behavior on rugged energy landscapes at any finite temperature are freezing into local regions and slow relaxation times. At a fixed temperature, the system remains frozen into some region of configuration space, through which it wanders on some typical time scale but eventually can escape to other regions (Stein, Palmer, et al. 1984; Stein and Ogielski 1985). This phenomenon leads to nonexponential, or slow, relaxation times in the exploration of the space. At a fixed temperature, a system with two quadratic potential wells separated by an energy barrier typically shows an exponential escape from the higher-energy to the lower-energy well. The exponential rate gives the time scale of the escape. In spin-glasses with complex potential energy landscapes, slow relaxation is manifest by a multiplicity of time scales of exploration of ever larger regions of configuration space. The parallels to population flow on rugged landscapes are clear.

The parallel between statistical physics and population biology is welcome. The same tools apply in both areas. Ebling, Engel, et al. (1984) have begun such studies for fully random landscapes. Their analysis is based on the eigenvalues and eigenvectors of the genotype system under the influences of mutation and selection. As described in Chapter 14, the eigenvalues and eigenvectors give the behaviors of the independent modes of the system. The largest positive eigenvalue is associated with a specific eigenvector, or linear combination of underlying genotypes, whose relative concentrations are increased over time. Localized population flow occurs if the largest positive eigenvalues amplify the concentrations of only a small fraction of the genotypes. Nonlocalized melting corresponds to amplification of the concentrations of a large fraction of the genotypes which are well dispersed across genotype space. While these analyses have been carried out for random fitness landscapes with some success (Ebling, Engel, et al. 1984), it is necessary to extended them to correlated landscapes (for example, $K \ll N$) for large N. Eigen (1987) has made useful contributions to this general subject.

A combinatorial optimization procedure called *simulated annealing,* based on concepts from statistical physics, offers insights for population biology. Simulated annealing (Kirpatrick, Gelatt, and Vecci 1983; Aarts and van Laarhoven 1985) derives from the fact that the range of states explored by a system such as a spin-glass is a function of temperature given by the Boltzmann distribution. At high temperature, the system is not trapped by potential barriers. At low temperature, the system spends most of its time confined to deep potential wells. The idea behind simulated annealing is to gradually lower an analogue of temperature appropriate to the optimization problem. If temperature is lowered gradually enough, the system tends to become trapped in deep, wide energy wells. Ultimately, when the temperature is near 0°K, the system ends up in a very good energy minimum. In contrast, if the process were carried out at very low temperature, the system would rapidly fall into a poor local energy minimum and become trapped there.

Simulated annealing is a powerful optimization strategy. In effect, the use of high temperature smoothes out the free-energy landscape. More precisely, at high temperature the landscape is smooth. As temperature is lowered, the landscape becomes more rugged and the system eventually becomes trapped in an energy valley.

The analogue of simulated annealing in population flow on rugged landscapes would be to begin with a high mutation rate and gradually lower it. In effect, a high mutation rate smoothes out the analogue of the free-energy landscape of a physical system. The potential power of this procedure is great, and in Chapter 4 I shall describe Eigen's efforts to utilize it in applied molecular evolution.

Although it seems unlikely that the precise analog of simulated annealing plays a role in population search on rugged landscapes, the search behavior of nearly melted populations may mimic important aspects of simulated annealing. A precise analog would require adapting populations to gradually tune their mutation rates from high to low. I find it highly implausible that a natural biological population could control its mutation rate to decrease gradually enough for effective annealing. On the other hand, annealing works well only in landscapes in which deep energy wells also drain wide basins. It does not work well on either a random landscape or a "golf course" potential, which is flat everywhere save for a unique "hole." In the latter cases, the landscape offers no clues to guide search. But we have good grounds to think that very many complex fitness landscapes have the property that high peaks drain large basins. Thus, as noted above, if mutation rate is tuned relative to landscape structure so that populations are just melting, then two features mimic simulated annealing. First, at the melting mutation rate, *very small* changes in mutation rate or population size can dramatically alter the extent of genotype space explored. Second, the population has a good chance of flowing into a basin climbing to a high peak and hence of optimizing sustained fitness. These possibilities lie behind the intuitive hope that proper evolutionary tuning of mutation rate, population size, and landscape structure might simultaneously optimize both evolvability and sustained fitness.

Recombination as a Global Search Strategy on Rugged Landscapes

In this final portion of the chapter, I want briefly to discuss the following issue: *When is recombination an effective adaptive strategy on rugged landscapes?* As we shall see, there appear to be some surprising answers to this question, for the effectiveness of recombination depends in subtle ways upon the ruggedness of the landscape.

So far we have focused on adaptation by accumulation of advantageous point mutations and considered largely the structure of fitness landscapes. But organisms make use of recombination in evolution. Most generally, recombination occurs between two strands of genetic material, DNA or RNA, within or between genes. For example, recombination between parental chromosomes with alleles $A1$ and $B1$ on the maternal chromosome and alleles $A2$ and $B2$ on the paternal chromosome can lead to two new chromosomes with $A1$ and $B2$ on one and $A2$ and $B1$ on the second. This event occurs if the recombination break points lie between the A and B loci on both chromosomes. Recombination occurs not only between genes but within genes as well. In the latter case, the left half of the gene coding for protein $A1$ is recombined with the right half of the gene coding for protein $A2$, forming a recombined gene coding for a recombined protein. The question we shall address shortly is whether such recombination, even within a protein whose amino acids cooperate in complex ways, is a useful hill-climbing strategy.

From the geneticist's point of view, recombination is often considered the reason for sex. The problem is this: Why bother with diploid organisms and the genetic cost of two parents, rather than blissful haploid fission? Why do even bacteria, normally content with the haploid state and mitotic division, engage in occasional bouts of passion?

Why sex has evolved remains a matter of debate which I shall not discuss beyond a brief note. The classical argument for the existence of sexual recombination, due to Fisher (1930) and Muller (1932), is that recombination favors the incorporation into the population of favorable new alleles arising at different loci, since recombination is more efficient in allowing such favored genes to occur in the same individual. Ewens (1979) discusses this in an informal argument:

> Suppose a favorable mutation $A1$ arises at a locus A and begins to spread throughout a population. If a favorable mutation $B1$ subsequently arises at a locus B, then without recombination $A1$ and $B1$ cannot both become simultaneously fixed (in the population) unless the initial $B1$ happens to arise on an $A1$ chromosome. This is unlikely to occur until the frequency of $A1$ is substantial, and thus either the evolution at other loci is slowed down by the evolution of the A locus or the favorable mutation $A1$ is lost through the increase in frequency of $B1$ (and hence the linked allele $A2$) at the B locus. With recombination, both $A1$ and $B1$ genes can eventually arise on the same chromosome so that evolution, under this argument, proceeds more rapidly than with no recombination.

Crow and Kimura (1965) later attempted to quantitate this argument for diploid populations, attempting to show the rate of incorporation of useful new mutations into a population as a function of population size, the selective advantage of each mutation, the overall constant rate of finding favorable new mutations, and the recombination rate.

As Ewens (1979) points out, many such arguments concern long-term optimization and ultimately rely on intergroup competition rather than selection acting solely on individuals. More recent models which attempt to quantitate the advantages of sexual over nonsexual existence have looked in the direction of rapidly changing environments, parasite–host coevolution, and other mechanisms to explain the virtues of recombination. The focus on rapidly changing environments reflects the fact that recombination not only builds up useful combinations of genes but destroys them as well. Thus once a population has attained a fitness peak, recombination will tend to disrupt useful combinations of genes. If one considers recombination as a search for peaks not yet found, this symmetry can be ignored. I shall make this assumption for the remainder of the discussion. In the context of the current debates about the usefulness of sex in a changing environment, we see next that the usefulness of recombination demands something of the underlying fitness landscape.

Fisher's and Muller's arguments, as well as that of Crow and Kimura, appear to require something like additivity. If mutation $A1$ is favorable compared with $A2$ and

$B1$ is favorable compared with $B2$, then the joint presence of $A1$ and $B1$ on the same chromosome is still more favorable. But our experience with the NK model of random epistatic interactions has by now shown that, in general, this additive feature is false.

Recombination is a useless search strategy in random landscapes. Consider again the $K = N - 1$ limit of an entirely uncorrelated landscape. Let two peptides exist. Let the left half of the first happen to be fitter than the left half of the second, while the right half of the second is fitter than the right half of the first. Break and join the better halves. What is the result? The new peptide has a fitness which is *entirely random* with respect to the fitness values of the two parental peptides. The recombinational move has not helped at all other than in the sense of constituting a long jump across peptide space. And with respect to that process, we know that the rate of improvement slows as improved variants are found according to $s = \log_2 g$ (Equation 3.1a). Evidently, whether or not recombination is a useful strategy depends upon the character of the landscape.

Any account of the origin of sex presumably must show that recombining a variety of potentially interacting genes, or parts of genes, on one chromosome is typically useful. And beyond theory, such recombination is a fact of life. Does it typically help adaptive evolution, and why? These issues have also been stressed and studied in some detail by Holland and colleagues in the genetic algorithm (Holland 1981; Goldberg 1989).

Conditions for Useful Recombination in NK *Rugged Landscapes*

It is obvious that recombination is useful in the NK model in the $K = 0$ limit. Here the genes in the model genome make independent contributions to fitness. This simple case carries over to the argument that recombination is useful whenever the adapting system comprises functionally independent parts.

Consider next the NK model as a model of either genotypes or amino acid sequences. Focus attention first on the general class of models in which each gene or amino acid is affected only by its neighbors along the sequence. In the genetic context, this amounts to assuming that epistatic interactions are local. It is intuitively plausible that adaptive hill climbing will achieve local optima in which neighboring genes are coadapted. Nevertheless, depending upon the value of K, the correlation length along the sequence for such coadaptation is bounded. Then the fitness contributions of sites at the left end of the sequence are expected to be little affected by the alleles at the right end.

The local-epistasis model leads to the intuition that, when K is small, recombination between long genotypes can be useful, since distant regions are functionally independent. Thus if the left half of genotype 1 is fitter than the left half of 2 while the right half of 2 is fitter than the right half of 1, the breaking and joining will yield two recombined strands, one of which has the two better halves. This recombination event will tend to disrupt any coadapted complexes only at the break point but leave intact coadaptation among alleles far from the break point. If the two fitter halves are sufficiently fitter, the fitness of the recombinant may already be higher than that of either initial genotype. In any case, further adaptive hill climbing may ameliorate the disruption at the break point.

The idea that recombination is generally useful because local coadapted gene complexes exist faces a serious objection: Knowledge of the cybernetic structure of genomic regulatory systems, discussed in Chapters 12 and 13, makes it clear that genes at a variety of sites distributed over the entire set of chromosomes may directly or indirectly regulate the activity of genes anywhere else in the genome. In other words, the genomic regulatory network is rich in widespread epistatic interactions. Worse than the image due to simple recombination, mating of two parents to form a zygote typically unites genomic regulatory systems which harbor detailed differences in distributed circuitry and control logic.

In short, we must ask whether recombination might be a useful strategy when the epistatic interactions are not localized but instead are spread out over the entire genome.

To gain qualitative insight into the role of recombination, we must change our perspective from the genotype sequences to the structure of the entire fitness landscape. That space of 2^N genotypes harbors some very large number of local fitness optima. The question we must ask is this: If we know where one such optimum is, does that give us any information about where other optima may be? As we saw above, the answer can be "yes." As described earlier, in the NK family of landscapes, when K is small relative to N, the *highest optima are near one another and also have the largest drainage basins climbing to them* (Figures 2.7 and 2.8). Clearly in such a landscape the location of one high optimum carries information about where other good optima are located.

Consider an analogy with the Alps. Like many other mountain ranges, the Alps have a Massif Central of very high peaks clustered on one another's shoulders. About these lie lower peaks, and yet farther away lie the foothills ebbing to the plains beyond. It now becomes intuitively plausible that locating one peak gives some information about where other peaks may be. But even more important, if two peaks are located, it is reasonable to think that high peaks lie between those two peaks. More generally, quartering the region between many peaks is an organized way of quartering the region which might well contain high peaks. Thus *pairs of peaks contain mutual information about good regions of the space.*

Recombination is just a means of looking between the two recombined genotypes in high-dimensional space. Thus recombination between local adaptive peaks is a means of quartering and searching the prospectively good region of the fitness landscape, where more high peaks are likely to be found. More precisely, in the present case, any two local optima are identical at some fraction of the N sites and differ at some others. Let the Hamming distance H stand for the sites which differ. Recombination between the two genotypes cannot alter the alleles, 1 or 0, at the sites which are identical but can alter the combination of H alleles which occur on each of the recombined genotypes. There are 2^H such possible combinations, each of which lies *between* the two initial genotypes in the Boolean hypercube.

In order to test this numerically, my colleague Lloyd Clark and I have used the NK model and schematized the effects of recombination in a simple way. We released a fixed number (100) of randomly chosen genotypes upon an NK fitness landscape and allowed each to walk via randomly chosen one-mutant fitter variants to a local optimum. In general, 100 or fewer independent local optima were found. Thereafter, we mated and recombined randomly chosen pairs of local optima at randomly chosen positions within each genotype, to form 100 recombined genotypes. These 100 recombinants were then allowed to walk via randomly chosen one-mutant fitter variants to local optima. Thereafter, the cycle of recombination followed by hill climbing to optima was repeated. This numerical procedure clearly asks whether the regions between local optima help direct the adaptive process to yet higher local optima.

The control experiment consisted in choosing 100 random initial genotypes and allowing them to walk to local optima. Thereafter, 100 entirely new fully random initial genotypes were chosen and allowed to walk to local optima. This cycle was repeated. This control procedure merely samples local optima repeatedly but never uses information about the locations of local optima.

Figure 3.16*a* shows that, over successive generations of recombination and hill climbing, the mean fitness of optima gradually increases. In contrast, the control

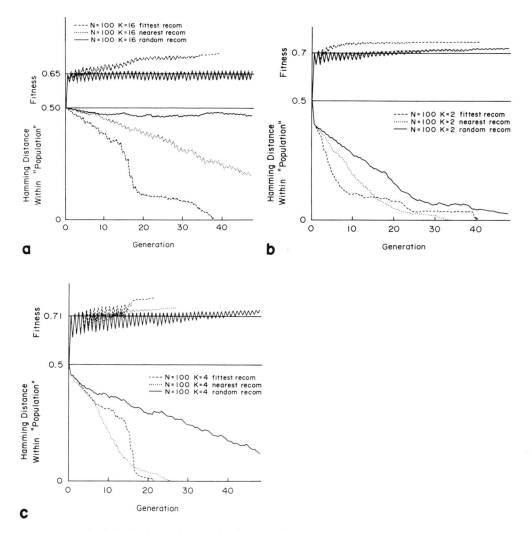

Figure 3.16   Usefulness of recombination tested on an *NK* landscape with $N = 100$, $K = 16$, and coupling to adjacent sites. *Lower panel:* mean Hamming distance between 100 "walkers" at the beginning and end of each cycle of walks to local optima. Walkers at local optima are recombined and walk from consequent genotypes to local optima. *Top panel:* random recombination between walkers at optima, biased recombination with nearest neighbors, and recombination restricted to the fittest half of the walkers at each cycle. (*b*) Same as (*a*), except that $K = 2$ and the epistatic couplings among the genes are random. (*c*) Same as (*b*), but $K = 4$.

shows no gradual improvement; rather, the mean fitness of optima attained fluctuates narrowly around the level found on the first trial. Thus recombination has helped the adaptive search process.

Happily, recombination also aids search on rugged landscapes when the K sites are randomly assigned. Figures 3.16b and 3.16c show results when the epistatic inputs to each site are chosen entirely randomly. Again, random pairs of local optima are recombined. The same results are found.

1. The mean fitness of local optima increases. Thus adaptation is helped more by recombination than by repeated jumps into genotype space.
2. The local optima after one step have a mean Hamming distance of about 40 rather than 50.0. Thus for K small, the optima are in a *local region in hyperspace,* even though the inputs to each site are random.
3. After recombination but before hill climbing again, the mean fitness of the genotypes rises. This means that much of the region between the local optima is well above the average of the space as a whole. The peaks occur in an entire region of high fitness.
4. The Hamming distance between the optima continues to decrease over recombination cycles. Since recombination alone cannot reduce the mean Hamming distance in the population of genotypes, this decrease means that the net effect of each cycle of hill climbing is to pull the genotypes closer together. We see why in a moment.

Two features of the landscape structure noted above suffice to account for the observations.

1. There is a Massif Central where most of the good optima lie and where the higher optima are closer together than the average Hamming distance among all local optima.
2. The higher optima drain larger basins; that is, the higher optima can be climbed to via adaptive walks from a greater volume of the space than can lower optima.

If these two properties hold, then, on average, each new set of walks from recombined walkers to local optima will climb to the higher local optima via the larger drainage basins; hence mean fitness will increase. And because the higher optima are closer together than the average distance between optima, the mean Hamming distance will decrease.

These numerical results demonstrate a fundamental feature of adaptation on rugged correlated landscapes: Recombination can be an effective search strategy in rugged landscapes without requiring functional isolation of the epistatically interacting parts of the system.

SUMMARY

This chapter investigates some of the biological implications of the structure of fitness landscapes. These include a universal law for long-jump adaptation, where genetic alterations jump beyond the correlation length of the landscape. The rate of finding fitter variants slows exponentially in the long-jump limit. Furthermore, the complexity catastrophe sets in; as the complexity of the entities under selection increases, long-jump adaptation becomes an ever poorer search strategy. These general results suggest three natural time scales in adaptive evolution. Early in an adaptive process, distant variants of a poorly fit entity can be much fitter than nearby variants; hence distant variants sweep through a population faster than do nearby variants. Therefore, the population adapts by long jumps. As this occurs, however, the rate of finding fitter distant variants slows exponentially. Thus in the mid-term of an adaptive process, fitter variants are more readily found nearby. Adaptation climbs local peaks. On a longer term, long jumps to distant promising hillsides occurs, or landscapes deform.

Radiation and stasis are fundamental features of phylogenies in the evolutionary record. Radiation is often explained in terms of free ecospace. Stasis has been explained as a consequence of filled niches in a crowded ecospace and normalizing selection, or as a consequence of special genetic mechanisms which canalize, or buffer, development against phenotypic alterations. In contrast, the very structure of rugged fitness landscapes, where the number of directions uphill dwindles as adaptive walks climb toward peaks, implies that radiation and stasis are inherent features of adaptation on rugged landscapes. Branching radiation should typically be bushy at the base, dwindling to single lineages which become trapped on alternative local optima. Bushy-based radiation is, in fact, typical in the evolutionary record and exhibits an asymmetry in time which has puzzled many thinkers. All may well be simple consequences of adaptation on rugged landscapes.

The wonderful burst of diversity of the Cambrian explosion, the quiescence of the rebound in diversity following the Permian extinction, and von Baer's laws may also reflect the same deep properties of adaptive evolution on rugged landscapes. Mutants altering early ontogeny typically affect development more than do mutants affecting late development. Thus mutants affecting early ontogeny adapt on more rugged landscapes than do mutants affecting late development. It follows that the rate of finding fitter variants altering early ontogeny slows exponentially, and does so far faster than the rate of finding fitter variants altering late development. Early development locks in, not necessarily due to the filling of ecospace or to special genetic canalization mechanisms, but simply because discovery of fitter variants affecting early development and making massive changes to the pathways of development becomes exceedingly rare. It follows that early vertebrate embryos should be more similar than late-stage embryos; hence von Baer's laws are expected. And both the Cambrian explosion and the Permian quiescence also are expected. In the former, the higher taxa filled in with founder species from the top down. Naturally: Early in the evolution of multicelled organisms, mutants altering early development, and hence long-jump adaptations to distant fitter variants, were easily found. Species founding phyla were established. Later, nearby variants were more readily found. Species founding orders, classes, and the lower taxa arose. In the post-Permian extinction rebound, early development had locked in; hence higher taxa were replenished from the bottom up. Finally, branching phylogenies climbing fixed or deforming landscapes will show patterns of convergent or parallel evolution which reflect that landscape structure. Thus homoplasy in systematists' phylogenies should be viewed as data, not noise.

Adaptive evolution is a search process—driven by mutation, recombination, and selection—on fixed or deforming fitness landscapes. An adapting population flows over the landscape under these forces. The structure of such landscapes, smooth or rugged, governs both the evolvability of populations and the sustained fitness of their members. The structure of fitness landscapes inevitably imposes limitations on adaptive search. On smooth landscapes and a fixed population size and mutation rate, as the complexity of the entities under selection increases, an error threshold is reached. Beyond that complexity, selection is a weaker force than mutation. The weak Maxwell's demon sets in. The population "melts" from the adaptive peaks and flows across vast reaches of genotype space among near-neutral mutants. Conversely, on sufficiently rugged landscapes, the evolutionary search process becomes trapped in very small regions of genotype space. Evolvability is compromised. On a fixed rugged landscape, as mutation increases in frequency, one or a series of phase transitions occurs at which the population "melts" out to ever larger regions of the space of possibilities. Reciprocally, for a fixed mutation rate and population size, adaptive search will remain frozen in small regions on sufficiently rugged landscapes, but melt over ever larger tracts of smoother landscapes.

An attractive hypothesis is that evolvability may be optimized if landscape structure and search parameters yield a population which is *nearly melted.* Such a population has the advantageous property that small changes in mutation rate can dramatically alter the extent to which the fitness landscape is explored and hence can best balance preservation of heritable information with exploration. That is, the population can optimize evolvability. Furthermore, such a poised mutation rate should optimize sustained fitness. If the mutation rate is too low, the population will be trapped on poor local peaks. If the mutation rate is too high, the population will be driven far from the peaks and drift in fitness lowlands. An intermediate mutation rate and population size relative to landscape structure must optimize sustained fitness. Further, very many rugged landscapes have the property that the highest peaks drain the largest basins. In such landscapes, a search process which is not too frozen into small regions has a reasonable chance of reaching basins that climb to high peaks. Thus optimizing evolvability may often optimize sustained fitness.

Fitness landscapes are not fixed, but deform as a result of alterations in the physical environment and coevolutionary effects. Organisms may be buffered against a range of alterations in their fitness landscapes by harboring a hierarchy of epistatic couplings among genes or traits. Alterations of highly epistatic traits alter phenotype drastically and hence adapt on very rugged landscapes. Alterations of traits with few epistatic consequences cause minimal alterations to phenotype and hence adapt on relatively smooth landscapes. Possession of a hierarchy of epistatic interactions among genes or traits automatically permits organisms to tune their response to the deformation rate of the fitness landscape being explored.

Recombination is presumed to be the reason that sex and diploidy have evolved. The general supposition is that recombination is a useful search process which hastens the assembly of useful combinations of genes or parts of genes. However, the usefulness of recombination itself is governed by the structure of fitness landscapes. For example, on fully random landscapes recombination suffers all the defects of long-jump adaptation. Conversely, recombination is a useful search procedure on rugged landscapes in which high peaks nestle close together. In such cases, the locations of high peaks carry information about the locations of other high peaks. The region between those high peaks is then likely to harbor even higher peaks. Recombination is precisely a process which quarters the region between such peaks. Clustering of high peaks can arise due to functional modularity, as in the NK model with the K epistatic inputs to each site chosen among its nearest neighbors, or in additive genetic models. But clustering can also arise as a feature of landscape structure without obvious modularity, as shown by NK landscapes with the K epistatic inputs to each site chosen at random among the sites.

A major purpose of Chapters 2 and 3 has been to lay the foundations to examine the relation between self-organization and selection, the two overarching themes of this book. For concreteness, I introduced an example of such self-organization in Chapter 2. The number of cell types in an organism is about a square-root function of the number of genes in that organism. This property holds over a large number of phyla. But it is also a *typical property* of a very large class of genetic regulatory systems which we will discuss in detail in Chapter 12. Unless we are utterly stubborn, we must wonder at this correspondence and ask whether this beautiful and simple ordered property of many organisms over many phyla which diverged 600 million years ago reflects selection or not. Perhaps it is a self-ordered property so deep within genetic regulatory systems that selection cannot avoid it.

Within the framework introduced in Chapters 2 and 3, this ordered property is widely spread in a large ensemble, or space of genomic regulatory systems. The ordered property is not randomly spread in the space of systems, but distributed in some broad way. On the other hand, isolated zones in the space which do not exhibit the property exist.

The results of this chapter suffice to say that selection can be unable to avoid spontaneous order. The limitations on selection arise because of two inexorable complexity catastrophes. These two "catastrophes" conspire to limit selective adaptation. Each arises where the other does not, one on rugged landscapes trapping adaptive walks, the other on smooth landscapes where selection becomes too weak to hold adapting populations in small untypical regions of the space of possibilities. Between them, selection is sorely pressed to escape the typical features of the systems on which it operates. Order for free should shine through.

CHAPTER 4

The Structure of Adaptive Landscapes Underlying Protein Evolution

We turn now to the evolution of proteins in "protein space." There can be no doubt that such evolution occurs on more or less rugged fitness landscapes. I emphasize again that by the "fitness" of a protein in protein space I mean the measurable capacity of the protein to carry out some defined function. The distribution of such a fitness measure over the space of possible proteins is the fitness landscape with respect to that specific function.

The most important theme of this chapter is the need for experimental work to establish the structure of protein fitness landscapes. Given such information and a companion theory about how populations may adapt on such landscapes under the simultaneous action of mutation, recombination, and selection, we shall be better able to understand major features of biological evolution, at least at the molecular level for protein function.

If our most basic task is to discover the statistical character of fitness landscapes with respect to specific protein functions, the next most important task is to build reasonable mathematical models of such landscapes. The theory outlined in the previous chapters was meant to provide us with a mathematical framework for thinking about adaptive evolution of genotypes or proteins in sequence spaces in which the notions of neighbor sequences and fitness can be defined. It led us to focus upon the characteristics of fitness spaces, such as number of local optima and lengths of adaptive walks to optima. Since real fitness landscapes are correlated in some way, useful mathematical models will help us characterize their statistical structure. I introduced the *NK* model in Chapter 2 as a first attempt to build a theory of such landscapes. In the present chapter, I ask whether this model is useful in characterizing protein fitness landscapes. The tentative answer is "yes." For example, the model does quite well in predicting a number of features of maturation of the immune response.

Our concept of a fitness landscape now needs to be broadened. As defined, a fitness landscape for proteins is merely the distribution of the capacity to perform some function over protein space. Yet protein evolution often is not optimization of an initial catalytic action but evolution of an entirely *new* catalytic action. The fitness landscape with respect to this new function may be quite different from the landscape with respect to the initial catalytic action. Thus we need a conceptual framework for analyzing the evolution of new catalytic activities. To build this framework, I introduce the concept of a *catalytic task space.*

Catalytic task space is an abstract representation, or mapping, of all chemical reactions which can be catalyzed onto a space of tasks. Here any point represents one catalytic task, and an enzyme covers a "ball" of similar tasks. This abstract representation is useful and carries surprising and potentially important implications. First, the idea of a task space carries with it the idea of neighboring tasks; hence the evolution of an enzyme to carry out "nearby" reactions maps into the idea of evolving an enzyme to cover a neighboring ball. A second important implication rests on the fact that apparently very different reactions may constitute the *same catalytic task.* A third is the striking possibility that, since any enzyme covers a ball in task space, a finite number of roughed-in enzymes might cover all of catalytic task space and be capable of catalyzing virtually any reaction. Such a set would be a set of universal protoenzymes for all catalytic tasks. As we shall see, recent work which shows that antibodies binding to the transition state of a reaction can catalyze that reaction strongly supports both the idea of a catalytic task space and the possibility that a finite number of enzymes might catalyze all reactions.

This chapter discusses four major topics: the structure of specific fitness landscapes in protein space, the adequacy of the *NK* model for such landscapes, the concept of catalytic task space and evidence supporting it, and the use of genetic engineering to generate large numbers of random genes, RNA molecules, and proteins to explore protein fitness landscapes on a massive scale. These conceptual and experimental questions are now open to investigation in a powerful new way. In the final section of this chapter, I shall describe a new experimental program that explores the ligand-binding and catalytic properties of proteins. This program uses recombinant DNA technology to generate billions, indeed trillions, of novel genes, followed by either selection or screening procedures to identify those with specific catalytic or ligand-binding properties. Current methodologies, in short, for the first time allow us to explore the character of adaptive landscapes in protein space. These technologies promise new drugs, new vaccines, new enzymes catalyzing single reactions and connected sequences of reactions. I believe we are on the verge of a tremendous new field: applied molecular evolution. This study of landscapes is no mere academic quest, then. Theories of adaptive walks in rugged landscapes are likely to find practical application in the near future.

ADAPTIVE MATURATION OF THE IMMUNE RESPONSE

During the immune response, a remarkable and rapid adaptive evolution of antibody molecules occurs (Kauffman, Weinberger, and Perelson 1988). During this adaptive evolution, the genes coding for the antibody secreted in response to a specific antigen accumulate successive mutations which progressively increase the affinity of antibody for antigen. Such adaptive improvements constitute one of the most rapid and easily studied adaptive walks in sequence space (here, the space of antibody sequences). Therefore, maturation of the immune response is a testing bed for ideas about the structure of mountainous fitness landscapes. In thinking about these issues, we are really considering three bodies of questions:

1. What does the affinity landscape, or fitness landscape, in antibody space look like, and how does its structure bear on the behavior of the immune system?

2. Is this landscape a good harbinger of other adaptive landscapes with respect to catalytic function?

3. Is the *NK* or any other mathematical model a good one for representing the structure of such adaptive landscapes, and if so, why?

When an organism is exposed to an antigen and mounts an immune response, the complex sequence of events which ensues includes binding of the antigen to immature antibody-secreting B cells. Those B cells whose antigen receptors, each with the same specificity as the antibody molecule it will later secrete, best match the incoming antigen proliferate most rapidly. This process is called *clonal selection* (Burnet 1959) and leads to an abundance in the serum of antibodies which match the antigen.

The antigen specificity of an antibody immunoglobulin is determined by the amino acid sequences in the heavy-chain (H) and light-chain (L) variable (V) regions of the immunoglobulin. The diversity in these two regions is generated by the combinatorial assembly of five different V-gene segments during the formation of the V genes. A complete heavy-chain V domain results from the joining of V_H, diversity (D), and H-chain-joining (J_H) gene segments in the genomic rearrangements in each stem cell. Similarly, the light-chain V domain is created by the joining of V_L and J_L gene segments. Each of these segments is chosen from a repertoire of several to hundreds of alternatives, to build up combinatorially a very large number of alternative heavy and light variable regions (reviewed in Honjo 1983; Yancopoulos and Alt 1986). Estimates of the minimal diversity in the mouse generated by these mechanisms range from 5.1×10^7 (Honjo 1983) to 10^9 (Berek, Griffiths, and Milstein 1985).

In addition to this combinatorial diversity, a further source of diversity is generated by variability in the locations of joining at the junctions of V-gene segments during assembly with insertion of random nucleotides (Tonegawa 1983). In addition, another source of diversity results in nucleotide replacement and is termed *somatic mutation*. In principle, somatic mutation allows almost limitless V-region diversity. From analysis of clonally related cells, it now appears that there exists a special hypermutation system which specifically alters bases in the V region at a rate of 10^{-3} per base pair per generation, a rate approximately six orders of magnitude higher than the spontaneous mutation rate (McKean, Huppi, et al. 1984; Clark, Huppi, et al. 1985; Manser, Wysocki, et al. 1985; Sablitzky, Weisbaum, and Rajewsky 1985; Sablitzky, Wildner, and Rajewsky 1985; Wabl, Burrows, et al. 1985; Wysocki, Manser, and Gefter 1986).

Antigen Selection Theories

The cellular and molecular mechanisms by which the immune response matures are still being uncovered. Classical theories suggest that competition for limited amounts of antigen may drive a selection process (Siskind and Benaceraf 1969). The argument goes as follows: The amount of antigen bound to cell-surface immunoglobulin depends upon the product of the antigen concentration and the affinity of the receptor for antigen. During an immune response, the antigen concentration should decrease. If there is a critical amount of bound antigen required to stimulate a B cell into antibody production, then as antigen concentration falls, only those B cells with increasing affinities for the antigen will remain stimulated and continue to secrete antibody. Because the antibody secreted by a cell has the same affinity for the antigen as when that antibody acts as the cell's receptor, the average affinity of serum antibody should increase during an immune response. According to this theory, based on clonal selection, smaller and smaller subsets of preexisting B cells are selected by antigen during the immune response. Mathematical models based on this theory were developed by Bell (1970, 1971).

Somatic Mutation Theories

Doria (1982) pointed out that certain observed patterns of affinity changes are not consistent with the classical theory. For example, antigen selection theories predict that antibody affinity should be higher for low doses of antigen than for higher doses. Instead, lower doses lead to antibodies with *lower* affinity (Siskind, Dunn, and Walker 1968).

Recent evidence has been obtained by studying the messenger RNA (mRNA) sequences for different specific antibody molecules called *monoclonal antibodies.* These monoclonal antibodies are created experimentally by fusing B cells with cancerous immune-system cells to create hybridomas. These hybridoma cells divide and secrete the specific antibody initially secreted by the B cell. Sequencing of mRNAs from different stages of the immune response to a single antigen indicate a more complex process (Kohler and Milstein 1976; Kennett, Denis, et al. 1978). In particular, it now appears that somatic mutation plays a major role in maturation of the immune response such that, over time, the affinities of the antibodies secreted increase.

In response to a specific antigen, clonal proliferation of those germ-line genes whose variable regions most precisely match the antigen leads to amplification in the serum of an initial set of roughed-in antibodies from a restricted number of cells, which contain V regions. The initial fraction of B cells which responds to an antigen is on the order of 10^{-5} (Press and Klinman 1974; Cancro, Gerhard, and Klinman 1978). These germ-line genes have little or no somatic mutation evident (Kaartinen, Griffiths, et al. 1983; Tonegawa 1983; Manser, Wysocki, et al. 1985; Wysocki, Manser, and Gefter 1986). Later in the primary or secondary response, the majority of antibodies no longer directly correspond to germ-line varieties but instead show extensive somatic point mutations. The accumulation of these mutations is correlated with an increase in the affinity of the antibody for the antigen (Urbain, Van Acker, et al. 1972). According to present somatic mutation theories, the increased affinity is a direct consequence of further clonal selection. Those somatic mutations which result in an alteration of the protein sequence of the V region may alter the binding affinity of the antibody. Then those mutated B cells whose antibodies bind the antigen with higher affinity proliferate more rapidly and come to dominate the immune response by clonal selection. Over a succession of somatic mutations in the V region of the initial roughed-in B cells, the mean affinity of the antibodies increases sharply. Typical increases over the course of maturation are from $5 \times 10^4 \, M^{-1}$ to $5 \times 10^7 \, M^{-1}$ (Kaartinen, Griffiths, et al. 1983; Fish and Manser 1987).

Maturation of the immune response is an adaptive walk in antibody space from the initial V region, through a succession of higher affinity variants, toward some locally optimum antibody which has higher affinity for the antigen than do its one-mutant neighbors. All the questions we posed in the previous chapter about the character of adaptive walks come to the fore and point to a central experimental question: How correlated is the landscape?

I again define precisely the fitness landscape in question. Consider the incoming antigen and a single *epitope,* or molecular feature, on that antigen. Then consider measuring the affinity of all possible antibody molecules for that epitope. The distribution of affinity values across antibody sequence space constitutes a well-defined affinity landscape with respect to that epitope. Presumably it is the statistical character of that landscape which largely determines the character of adaptive walks in antibody space. Therefore, we would hope that studies of a model like the NK model might provide this kind of statistical information. I discuss next a method of applying the NK model to these issues, based on work with my colleagues (Kauffman, Weinberger, et al. 1988; Kauffman and Weinberger 1989). The first attempt to apply the concept of an affinity landscape to maturation of the immune response was based on the idealization that such landscapes are entirely uncorrelated (Kauffman and Levin 1987). Further analysis using this limiting idealization has been carried out by Macken and Perelson (1989).

Application of the NK *Model to Maturation*
of the Immune Response

The fundamental assumptions in applying the NK model to immune-response maturation are that a representative member of the population of maturing antibodies can be identified at any time and that the affinity of these antibodies for the antigen steadily increases, due to fortuitous point mutations, until a locally optimal antibody is obtained. We define a locally optimal antibody as one having higher affinity for the antigen than any of its one-mutant neighbors. The experimental results in the preceding section confirm that this is, in general, a plausible scenario. However, it is not known whether mature antibodies are in fact local optima. It is known that the V regions continue to mutate without substantial changes in affinity even after they have attained maximum affinity for the antigen. This continued mutating may reflect mutational dispersal among near-neutral mutants in the immediate vicinity of the local optimum.

In addition to these two main assumptions, there are a number of others, which I describe now.

Choice of **N.**   We identify the parameter N with the number of amino acid sites in the V region which has between 110 and 120 amino acids. In applying the NK model to immune-response maturation, $N = 112$ was used because it was slightly easier to with K amino acids drawn randomly from the chain.

Choice of Starting Place.   We assume that the fact that one in 100 000 B cells responds to a given antigen implies that those that do respond secrete antibodies that are in the 99.999th percentile in ability to bind to the antigen. In other words, walks start well up on adaptive hillsides. From the point of view of numerical simulations, the fitness contribution of each amino acid in the model antibody is a random number. Therefore, finding model antibodies in the appropriate percentile reduced to the problem of finding random-number seeds that give a sequence of $N = 112$ random numbers whose average is in the same percentile. Although use of this procedure implies that there will be some fluctuation in the starting fitness of the model antibody, departure from the bottom boundary of this top percentile was insignificant.

Choice of Neighborhoods.   The NK model suggests that lengths of walks to optima and fitnesses of the optima achieved do not depend strongly on the details of which sites interact with each other. However, preliminary simulations assumed that there could be only two amino acids per site and that the walks started from randomly selected initial proteins rather than from proteins already in the 99.999th percentile in fitness. In modeling V regions, we considered both extremes: Each amino acid interacts directly only with its K neighbors; to avoid boundary effects, we therefore idealized the V region as a circular protein. Alternatively, each amino acid interacts with K amino acids drawn randomly from the chain.

Choice of 19N Neighbors or Neighbors via the Genetic Code.   A V-region length N can be thought of as having $19N$ one-mutant neighbors. However, at the DNA level, many single amino acid substitutions require two or three base pair changes. Restriction to single base changes at the DNA level implies a reduction on the number of one-mutant neighbors at the protein level. Both cases were studied. "Coding" was incorporated into the model by explicitly including translation. In particular, coding was modeled by assuming that the evolving entity was a pair of polymers: a "protein molecule" consisting of the 112 "amino acids," as before, and a (single-stranded) "DNA molecule" consisting of $112 \times 3 = 336$ sites, each with one of the four "bases." The initial DNA molecule was back-translated from the starting-model V region to a DNA sequence coding for that model V region. The codon assigned to each position in the back-translation was chosen randomly from the synonymous codons for that amino acid. A step consisted in a point mutation of one of the DNA sites, using the genetic code to translate the new DNA sequence into the corresponding protein, and then computing the fitness of the protein. Since model adaptive walks pass only to *fitter* neighbors, the adaptive walk in this procedure did not pass to a one-mutant neighbor which was a silent mutation to a synonymous codon. A DNA mutation which resulted in an internal stop codon in the model V region was scored as a lethal mutation with fitness 0.

Use of the genetic code reduced the number of one-mutant neighbors. Each DNA sequence had only 1008 one-mutant neighbors, obtained by substituting any of the three other bases in each of 336 sites. In addition, due to synonymous codons, only about 75 percent of these substitutions resulted in substitution of a new amino acid. Thus in the versions of the model based on coding, each model V region had about 756 one-mutant neighbors rather than the $19 \times 112 = 2128$ based on mutation of any amino acid to one of the 19 remaining possibilities.

Complement-Determining Regions or Not.   The NK model in its general form is isotropic. It assumes that all sites make a direct contribution to fitness of the overall string, whether that string is interpreted as a genotype or as a protein. Proteins, however, may be more hierarchically constructed, with some sites—for example, amino acids at the active site of an enzyme or the binding site of an antibody molecule—having direct bearing on function and other sites playing a support role.

In the V region of antibody molecules, special hypervariable regions called *complement-determining regions* (CDRs) are known to play a critical role in antibody diversity and in antigen binding. The surrounding parts of the V region are thought to be a supporting framework for the binding site (Kabat 1976; Kabat, Wu, and Bilofsky 1979). A simple way to begin to model the distinction between CDRs and framework is to assume that only the amino acids in the CDRs have a direct impact on the fitness of the V region, while those in the framework influence the CDR amino acids. Thus, as a first effort, the existence of CDRs was modeled by assigning three contiguous regions of amino acid positions in our model V regions, matching those in V regions, and measuring the fitness contributions only of the CDR amino acids. Because the framework amino acids interact with the CDR amino acids, the former still have an indirect bearing on fitness.

Choice of **K.**   The experimental data described below show that walk lengths in affinity landscapes average between six and eight steps, but with considerable variance. Walks start well up on adaptive hillsides, where the starting germ-line V region initially amplified by clonal selection is in the highest 99.999th percentile. Thus we seek a value of K such that walks to local optima from that starting percentile average six to eight steps. This is the central parameter-matching step in applying the NK model. We use two features of the immunological data: (1) the fraction of B cells which respond to an antigen sets the starting percentile in affinity space and (2) the number of mutations substituting amino acids in the V region during maturation sets the mean walk length to optima. Given these conditions, it is possible to find the value of K which yields walks with the appropriate length by carrying out numerical simulations at various trial values of K (Kauffman and Weinberger 1989).

Affinity Landscapes Are Correlated

An immediately interesting point arising from framing these questions is that the appropriate value of K must be less than the maximum, $K = N - 1$, and therefore that antibody affinity landscapes must be *correlated*. This conclusion follows from examining walk lengths at the upper extreme value $K = N - 1$, which corresponds to a fully random landscape. Here the probability that a model V region with fitness x is fitter than its $19N$ fitter neighbors is x^{19N}. Thus any starting protein that is in the top 99.999th percentile in fitness has roughly a 98 percent chance of already being fitter than its 2128 one-mutant neighbors. That is, *if affinity landscapes were entirely uncorrelated, initially selected germ-line variants would already be local optima.* Since antibody molecules mature, we can conclude both that affinity landscapes are correlated and that K must be less than $N - 1$.

Experimental Features of Affinity Landscapes

The immune response matures on a rugged affinity landscape whose structure is only partially known. In general, all the questions raised previously regarding abstract landscapes are *a fortiori* of interest with regard to the immune system.

1. How many improvement steps must be taken from any initial antibody molecule to a local optimum; that is, how many somatic mutations accumulate in the V region of an initial roughed-in germ-line variant antibody molecule during maturation? The answer, as mentioned above, appears to be a range, with a mean of 6 to 8 (Bothwell, Paskind, et al. 1982; Tonegawa 1983; Heinrich, Traunecker, and Tonegawa 1984; Berek, Griffiths, and Milstein 1985). For example, Crews, Griffin, et al. (1981), studying the V_H gene responding to phosphorylcholine, found between one and eight residues changed; Bothwell, Paskind, et al. (1982) found three mutations in a lambda light-chain V region and six in a lambda$_2$ light-chain V region; McKean, Huppi, et al. (1984) studying the V_K region of antibodies against a determinant on influenza found seven or eight replacements; Clark, Huppi, et al. (1985), studying the secondary response to influenza, found 20, 12, and 19 V_H coding mutations and 9, 8, and 15 V_K coding mutations.

2. What fraction of one-mutant variants of the initial roughed-in germ-line antibody have higher affinity for the antigen? How does that fraction change, presumably dwindling to zero, as successively higher affinity antibody molecules are selected as the immune response matures? Here it is known that a large fraction of the one-

mutant variants have lower affinity, but the exact fraction with higher affinity at any step is unknown.

3. How rugged is the affinity landscape in the one-mutant vicinity of local optima? The question of whether affinity falls off dramatically in some directions and slowly in others translates directly to whether mutations at some positions in the V region cause dramatic loss of affinity, while those at other positions cause little loss of affinity. Restated, the distribution of the number of amino acids which can be substituted at a site with retention of function is a direct picture of the local ruggedness of the affinity landscape.

4. How many alternative local optima can be reached from any initial roughed-in germ-line antibody amplified by initial clonal selection? Further, what is the probability of climbing to each of those alternative optima and hence the density of their occupancy? Here work with inbred mice (Perlmutter 1984; Slaughter and Capra 1984) has demonstrated that multiple local optima are accessible. In many cases, initial clonal selection opts for the same initial V region, which then climbs to different mature forms by accumulating different somatic mutations. It appears from these and similar experiments that the number of alternative optima accessible from the initial antibody may be at least modestly large. Typically, comparison of five to ten monoclonal antibodies deriving from the same V gene shows that all differ from one another. Because only small numbers of sequences have been compared in this way, it is unknown whether a much larger number of local optima are accessible. These experiments are ambiguous. As remarked above, we have assumed that mature antibodies are local optima, and one of the predictions of the NK model will be that many local optima should be accessible. However, the fact that different mature antibodies emerge from the same V gene is insufficient to confirm this conclusion. From the work of Eigen and Schuster (1979), from classical population genetic analyses (Ewens 1979), and from our analysis of adaptation on rugged landscapes as a function of mutation rate as discussed in Chapter 3, a distribution of antibody sequences around an optimum can be expected, and the rate of hypermutation is known to be high. Thus the diversity seen in mature antibodies derived from one V gene may reflect the incapacity of clonal selection to eliminate near-neutral variants.

5. How similar are the local optima? Maturation climbs to alternative local optima from an initial roughed-in V region. The typical observations when several different monoclonal antibodies derived by maturation are compared is that many amino acids are "conserved," while a smaller fraction are repeatedly mutated. Furthermore, some sites repeatedly have mutated in parallel to the same alternative amino acid (Perlmutter 1984; Slaughter and Capra 1984). This topic is equivalent to that of the systematist's problem of homoplasy at the molecular level, as discussed in Chapter 3.

Predictions of the Model and Comparison with Experimental Results

The Appropriate Value of* K *Is Near 40.   Numerical simulations were carried out for all versions of V-region models (Table 4.1; Kauffman and Weinberger 1989). Of these, presumably the most realistic combination includes both the CDR and the genetic code. But as we shall see, all possibilities predict the same landscape features

TABLE 4.1 Results of Adaptive Walks from Best Initial Model V Region in 100 000, Based on Protein V Regions with $19N$ One-Mutant Neighbors or on the Genetic Code with About 775 One-Mutant Neighbors at the Protein Level

	K	Average walk length to optima	Average number of fitter neighbors on first step	Mean number of allowed substitutions per site
Protein (no CDR)				
	30	13.6 (3.2)*	65.9 (20.4)	15.4 (3.6)
Adjacent	40	8.6 (4.5)	24.7 (9.9)	4.1 (3.9)
	50	4.9 (2.3)	10.3 (4.9)	0.8 (1.4)
	30	17.7 (4.4)	83.7 (13.0)	17.9 (2.4)
Random	40	11.5 (5.3)	42.1 (6.9)	8.5 (5.8)
	50	6.6 (2.2)	17.5 (3.5)	2.1 (3.0)
Protein (with CDR)				
	30	26.1 (7.1)	89.7 (27.8)	12.5 (6.6)
Adjacent	40	9.8 (3.7)	27.7 (11.3)	5.0 (5.5)
	50	7.0 (3.3)	12.7 (6.0)	1.5 (2.9)
	30	16.1 (5.3)	81.8 (16.2)	15.5 (4.5)
Random	40	12.0 (3.9)	36.9 (9.3)	7.4 (5.8)
	50	7.3 (3.2)	17.0 (5.8)	1.9 (3.0)
Genetic code (no CDR)				
	30	8.4 (2.3)	20.1 (12.7)	10.5 (5.1)
Adjacent	40	5.2 (2.7)	7.3 (4.3)	3.4 (3.8)
	50	2.9 (2.1)	3.9 (2.5)	0.3 (0.7)
	30	11.7 (4.5)	24.6 (8.2)	14.1 (5.1)
Random	40	6.9 (3.3)	11.2 (4.9)	5.7 (5.1)
	50	3.1 (1.9)	4.5 (2.8)	1.0 (2.0)
Genetic code (with CDR)				
	30	15.1 (5.6)	41.8 (17.5)	9.7 (7.5)
Adjacent	40	6.7 (3.1)	17.1 (12.1)	3.2 (4.6)
	50	3.5 (2.1)	7.0 (7.0)	0.6 (1.7)
	30	11.2 (4.1)	27.4 (9.3)	11.9 (6.1)
Random	40	7.6 (2.7)	10.6 (5.5)	3.8 (4.5)
	50	4.1 (2.2)	5.6 (3.3)	0.7 (1.7)

*The values in parentheses represent standard deviations.

and qualitatively agree with the available experimental data. The results are remarkably robust. As Table 4.1 shows, whether CDRs were included or not, whether all $19N$ neighbors of the V gene were used or translation via the genetic code was used, and whether the K sites were constrained to be flanking adjacent sites or chosen at random, a value of K around 40 gives rise to walks of between six and 12 steps. A value of $K = 30$ typically yields walks which are too long; $K = 50$ typically yields walks which are too short. Since walk lengths are largely insensitive to the remaining parameters to a very good first approximation, the dominant parameters are N and K.

There is considerable dispersion about this mean value of K. As shown in Figure 4.1, for a given value of N and K and under defined conditions for the rest of the model conditions, walk lengths might range from two or three steps to 15 to 20. This dispersion reflects again the ruggedness of the landscape and is encouraging, given the fact that there is a similar dispersion in the experimental data.

Ultimately, the NK model predicts some specific distribution of walk lengths to optima, not just a mean and a standard deviation. Thus accumulation of adequate

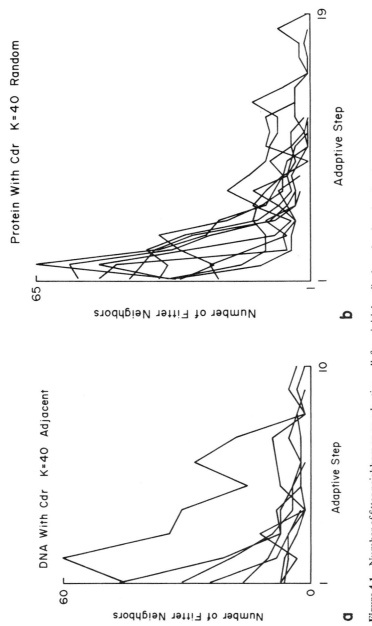

Figure 4.1 Number of fitter neighbors on an adaptive walk from initial antibody molecule to local optimum plotted as a function of current adaptive step on that walk. (*a*) Variable region modeled as a protein with 19*N* one-mutant neighbors and *K* = 40 randomly chosen amino acids bearing on each amino acid's fitness contribution. (*b*) As in (*a*), except walk carried out with respect to model DNA coding regions and hence there are about 775 one-mutant neighbors at the protein level. Ends of lines indicate the step at which walks end.

data can ultimately establish the distribution of walk lengths for comparison with either the *NK* model or some improved version of it.

Finding a specific value for *K* is in itself interesting. If the model is taken literally, *K* stands for the number of amino acids which bear on the fitness contribution of each amino acid. Then, if *K* is roughly 40, alteration in a single amino acid could affect the behavior and function of about 40 amino acids in the V region. Is this plausible and is there any evidence bearing on the issue? In a well-folded protein, an amino acid is open to influence not only by its neighbors in the primary sequence but also by those amino acids which are near it in the folded form even though they are distant in the primary sequence. One approach to studying how many amino acids in a protein can affect any amino acid is to use hydrogen exchange data, which can detect subtle motions in a folded protein as it twists and unfolds slightly in different ways. Wand, Roder, and Englander (1986) and Roder, Wand, and Englander (1989) have studied such hydrogen exchange in numbers of proteins. While they have not yet analyzed differences between a protein and a one-mutant variant, they have looked at the oxidized and reduced forms of cytochrome *c*, a protein containing 106 amino acids. Oxidization and reduction, due to the presence or absence of a charge on the heme group, correspond very roughly to substitution of a charged for an uncharged amino acid in that vicinity. These authors have examined 50 hydrogen-bonded hydrogens and found that at least 30 of them alter their exchange behavior when they go from the oxidized to the reduced form. The very crude conclusion to be drawn is that a charge alteration at one point in a protein can affect at least 30 amino acids. Since these authors studied only half the hydrogen-bonded atoms, the number of amino acids affected by altering one amino acid may be greater than 30. This point has obvious caveats. The study is not of an amino acid substitution but of an altered heme group. Further, to have found a statistically significant alteration in hydrogen exchange by an amino acid does not yet say that such alterations are in any way relevant to protein function. Third, the cytochrome *c* molecule may be well evolved to undergo alterations when the heme group is charged. Many fewer alterations in hydrogen exchange behavior might be found by randomly substituting amino acids in proteins. Nevertheless, the data suggest that any amino acid might be affected by, and affect, as many as 30 amino acids in a protein region of about 106 amino acids. Direct testing in antibody molecules would require study of hydrogen exchange in the V region of a mature antibody and its one-mutant variants.

The NK *Model Makes Plausible Predictions About the Fraction of Fitter One-Mutant Variants.*   Given a value of $K = 40$, the *NK* model makes clear predictions about the fraction of fitter one-mutant variants of the first roughed-in V region and about the fraction of fitter one-mutant variants of each improved variant on the adaptive walk. The expected number of fitter variants to the first V region is on the order of 1 or 2 percent in all the combinations of conditions mentioned in Table 4.1. In those runs that used all $19N$ one-mutant neighbors, 24 to 42 among the 2128 one-mutant variants are typically fitter. When translation via the genetic code and the implicit constraints in the one-mutant neighbors were added, typically there were seven to 17 fitter one-mutant variants among the 1008 one-mutant nucleotide substitutions and about 756 one-mutant V regions at the protein level (Table 4.1).

On individual walks, there is moderate variance in the fraction of fitter one-mutant variants of the initial V region. The minimum found on the initial step is one, and the maximum is 70, or over 3 percent. The fraction of fitter variants dwindles, but not smoothly on any specific walk, to zero over the steps to the local optimum, as shown in Figure 4.1.

The fraction of fitter variants in maturing antibody molecules is not yet known in detail, but the experimental procedure for finding this fraction is clear: Monoclonal antibodies at different stages during an adaptive walk must be obtained, the genes cloned, and the one-mutant spectrum examined for the affinities of the one-mutant variants.

Studies of the *lac repressor* in the lactose operon of *Escherichia coli* provide an indirect estimate of the number and nature of fitter one-mutant variants of a roughed-in V region. Here the observed fraction of fitter one-mutant variants is less than 1 percent (Jacob and Monod 1961, 1963; Muller-Hill, Rickenberg, and Wallenfels 1964; Burstein, Cohn, et al. 1965; Zubay and Chambers 1971).

The repressor monomeric unit has 360 amino acids. Miller and his colleagues (1979) studied a collection of over 300 altered proteins, each by a single substitution, with respect to two ligand-binding activities. Mutant phenotypes in which the repressor could no longer bind to either operator DNA or allo-lactose (or a synthetic inducer called IPTG) and mutants showing an increased affinity for the DNA or IPTG allowed Miller's group to study the consequences of such mutations conveniently. The results from 323 single amino acid replacements is that 42 percent result in a detectable change in the capacity to bind either IPTG or the operator DNA. The remaining 58 percent appear to be silent mutations. About 33 percent of the replacements decrease capacity to bind to the operator DNA, although only 15 percent of the substitutions destroyed 25 percent or more of the capacity, and only 8 percent became fully inactive. About 11 percent of the replacements reduced affinity for the IPTG. On the other hand, 1 percent of all one-step mutants increased affinity for the operator DNA, in some cases by as much as 100-fold. No one-step mutant was found which increased affinity for IPTG.

From the study by Miller and his colleagues (1979) we can draw the following conclusions. First, even well-tuned proteins may have rare variants which improve a given function. Here, 1 percent of the one-mutant neighbors at a restricted number of sites showed increased affinity for the operator DNA, and no mutant showed increased affinity for the inducer IPTG. Second, 58 percent of the single amino acid substitutions had no obvious effect. Because the assays employed are rough measures, the reported fraction of silent mutants is probably an overestimate. However, since 42 percent of the mutants clearly do reduce affinity for the operator or inducer or both, it is very unlikely that more than a small fraction of the one-step neighbors subtly increase affinity.

Our simulation results suggest that an initial germ-line V region with $K = 40$ would be open to improvement by about 1 to 2 percent of the one-mutant variants (Table 4.1). This value is very close to the observed data for the lac repressor. Thus tuning K to fit observed walk lengths yields a value which, having tuned the ruggedness of the fitness landscape, predicts a plausible value for the expected fraction of fitter one-mutant variants of the initial germ-line V region amplified by clonal selection.

These predictions of the NK model are fairly sensitive to K. When $K = 30$, roughly 3 to 5 percent of the one-mutant variants of the first clonally selected V region have higher affinity, while walks to optima average about 13 steps. For $K = 20$, the average walk length is 22 steps, and about 7 percent of the one-mutant variants of the initial antibody are fitter.

The NK *Model Predicts Conserved and Variable Sites in the V Region.*   In real proteins, some amino acids cannot be substituted without drastic loss of function, while others can be substituted with relative impunity. It is therefore of interest to ask whether the NK model predicts this phenomenon without further assumptions.

To answer this question, the fitnesses of all one-mutant neighbors of the local maxima obtained in the adaptive walk simulations were examined. In order to make valid comparisons between simulations with and without the genetic code, all 19 other amino acids were substituted in each site of the V region, including those amino acids that required several mutations of the corresponding DNA sequence. The results, for different values of K and for different assumptions about whether the entire V region or just the CDRs are used in computing fitness, are as expected (data not shown). The first main result is that, as K increases, the jaggedness of the landscape increases. The second result is that, as K increases, the *mean loss in fitness* and the *variance* in loss of fitness among the different possible one-mutant neighbors both increase. Thus as K increases, some mutants persist in causing relatively minor loss in fitness, while others now cause a dramatic loss in fitness. The third major result, for $K = 40$, is particularly surprising: At some sites, any model amino acid substitution causes a dramatic loss of fitness, while at others all substitutions cause almost no loss. At still other sites, some substitutions cause almost no loss of fitness, while other substitutions in the same site cause drastic loss. Thus, without further assumptions, the NK model for these parameters gives a highly rugged landscape in which amino acids at some sites in the locally optimal V region must be entirely conserved to preserve function, while amino acids at other sites can be substituted indiscriminately.

Note that, in constructing the general NK model, no site is *a priori* more important than others. It is instead the fact that K is high, resulting in a rugged landscape, which predicts that some sites are conserved, while others are broadly substitutable.

A particularly interesting view of these results is the following. We have no direct scale relating fitness in the NK model with real affinities of antibody molecules. However, real antibody walks start with those antibody molecules already the best in 100 000, and such molecules typically have an affinity of 10^4 M^{-1} for the antigen. In contrast, matured antibodies have affinities around 10^7 M^{-1}. Then it is sensible to define the fitness of the first member of the model walks (0.618) as corresponding to a modest affinity of 10^4 M^{-1} and let this fitness serve as a threshold separating model V regions which do and do not bind antigen. Given this threshold, one can test the number of substituted amino acids at each site in the 112-unit-long optimal V region which preserve at least above-threshold function. Table 4.2 shows the results for four values of K. Similar results are found for the different versions of the model, with and without CDR, coding, or choice of adjacent or random epistatic connections (Table 4.1). Again, it is K which determines the qualitative features of the landscape. When K is small, one-mutant neighbors of a local optimum have only slightly lower fitness and the variance in those fitness levels is small; hence none of the one-mutant neighbors fall below the threshold starting fitness, demarking modest affinity. Thus, for $K = 20$, each of the 112 sites can be substituted by all 19 other amino acids, and the affinity of the mutated model V region remains above threshold. As K increases, the loss of fitness and the variance in that loss increase. For $K = 30$, most sites are substitutable by 19 amino acids, but some sites can be substituted by only 15, 16, 17, or 18 amino acids without falling below a fitness threshold of 0.618. For $K = 50$, the mean loss in fitness among the one-mutant variants of a local optimum is dramatic, and almost no sites can be substituted with any amino acids and preserve above-threshold affinity. For K about 40, however, a wide distribution is found. Some sites can be substituted by 19 other amino acids, some by 15, some by ten, some by five,

TABLE 4.2   Number of Allowed Substitutions of a Locally Optimal Model V Region

Allowed substitutions per site	Distribution of allowed substitutions									
	$K = 20$	$K = 30$	$K = 30$	$K = 30$	$K = 40$	$K = 40$	$K = 40$	$K = 50$	$K = 50$	$K = 50$
0	0	0	0	0	7	3	7	105	63	85
1	0	0	0	0	4	4	14	6	31	20
2	0	1	0	0	11	5	6	1	13	6
3	0	1	0	0	10	14	8	0	4	1
4	0	2	0	0	15	9	3	0	1	0
5	0	2	0	0	15	9	5	0	0	0
6	0	3	0	0	12	13	5	0	0	0
7	0	5	0	0	11	9	3	0	0	0
8	0	4	0	0	7	6	5	0	0	0
9	0	20	0	1	6	8	6	0	0	0
10	0	7	0	0	3	5	6	0	0	0
11	0	18	1	5	5	6	5	0	0	0
12	0	10	5	6	2	4	7	0	0	0
13	0	11	6	3	2	5	6	0	0	0
14	0	8	4	7	1	5	4	0	0	0
15	0	10	7	12	0	2	9	0	0	0
16	0	5	12	16	1	4	8	0	0	0
17	0	4	12	14	0	0	3	0	0	0
18	0	1	21	25	0	0	2	0	0	0
19	112	0	44	23	0	1	0	0	0	0
Mean	19.00	11.13	17.14	16.37	5.49	7.23	8.17	0.07	0.65	0.31
SD	0.00	3.27	2.17	2.39	3.36	4.29	5.62	0.29	0.88	0.61

and some by zero. Thus a value of K near 40 yields the broadest distribution. I emphasize that this broad distribution is a *prediction* of the NK model.

This broad distribution reflects the fact that, for $K = 40$, the mean loss in fitness among the one-mutant variants of an optimal antibody roughly equals the difference in fitness between the beginning antibody on the adaptive walk, 0.618, and the fitness attained at the local optimum. Because the one-mutant variants span across this threshold, the distribution of substitutability per site is broad. As we are about to see, the observed distribution does appear to be broad; thus we are also led to predict that the mean affinity of the one-mutant variants of a locally optimal, mature antibody should hover near the affinity of the initial antibody from which adaptive walks began.

The experimental data to test the prediction that the distribution of allowed substitutions per site is broad would consist in a high-affinity, mature monoclonal antibody against a defined epitope and its entire one-mutant spectrum with respect to V-region mutants. The affinities of that mutant spectrum constitute the data set. It is not available, but the experiment is obviously feasible using cloned antibody molecules. Nevertheless, a rough approximation to this experiment is available. Geysen, Barteling, and Meloen (1985), Geysen, Rodda, and Mason (1986, 1987), Fieser, Tainer, et al. (1987), and Getsoff, Geysen, et al. (1987) have studied how all possible one-mutant variations in an antigen change the antigen's affinity for antibody. More precisely, these authors raised, on a protein antigen, polyclonal sera or monoclonal antibodies against a defined epitope six amino acid long and then made synthetic hexamers identical to that epitope and demonstrated that the sera and antibodies had high affinity for the hexamers. Then the authors looked at all 19 variants at each of the six positions, one position at a time. The results for nine such epitopes are summarized in Figure 4.2. The striking feature is that the distribution of allowed substitutions is again very broad.

Four comments are warranted. First, it is clearly encouraging that the broad distribution for $K = 40$ predicted by the NK model is found. Second, we have defined a threshold affinity as the fitness of the first model V region in the walk, the best in 100 000; we do not know how this threshold bears on the measured affinities. Third, the data need to be used cautiously in this context, since they concern free hexamers bound to polyclonal sera or to monoclonal antibodies, not the number of substitutions at each position within the V region of a mature monoclonal antibody; the constraints within a V region may or may not dramatically alter the observed distribution. Fourth, taking data and model at face value for the moment, the same value of K which fits walk lengths to optima also predicts a reasonable fraction of fitter one-mutant variants and genuinely predicts that some sites allow no substitutions while others are more permissive. Were K much smaller, say 20, almost all sites would be open to substitution by most model amino acids; the prediction is thus sensitive to K.

The Number of Alternative Local Optima Found from an Initial V Region. The NK model for these parameters allows us to examine the number of alternative optima accessible from the initial model V region and also to test whether alternative accessible optima are typically attained equally often on independent walks or with biased preferences. The experimental data on repeated walks from the same V region remain scant, but clearly suggest that multiple optima are accessible from the same initial region. The true number of such local optima is not known experimentally but presumably is greater than the five to ten alternatives often observed.

Numerical simulations with $K = 40$ were carried out from initial model V regions and were stopped by limitations of computer storage. In two simulations making 797 and 315 walks from the same initial region, 150 and 235 optima were found. Because many of these optima were found only once, it is difficult to know how many more remain to be accessed from the same initial V region. However, it is a clear prediction

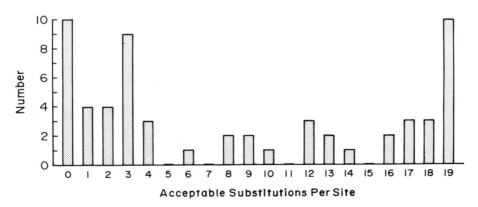

Figure 4.2 Observed number of amino acids which can be substituted per site in hexapeptides and preserve discriminable affinity of the monoclonal antibodies or polyclonal sera to the initial peptide. An arbitrary threshold of about 10 percent affinity for the correct hexamer was used.

of the *NK* model that a given initial germ-line V region can give rise to hundreds if not thousands of mature antibodies, each a local optimum in affinity space.

A second feature of these studies is shown in Figure 4.3, which is the histogram of the numbers of times each local optimum was encountered on independent random walks from the same initial model V region. As can be seen, three optima are each encountered many times. Analysis of these three showed that each is a one-mutant variant of the initial V region from which walks started. Ultimately, the model predicts a distribution which is open to experimental testing. The density distribution with which nearby local optima are reached is another expression of the ruggedness of the fitness landscape.

Similarity of Alternative Local Optima: Conserved Sites and Parallel Mutations.
Comparison of alternative mature V regions obtained experimentally reveals that not all sites in the region accumulate somatic mutations equally. In particular, some sites are rarely mutated, and among the sites which are preferentially mutated, sometimes the *same* amino acid is substituted on two or more independent walks. These are called *parallel mutations.* To see whether these phenomena are observed in *NK* landscapes, we compared five to ten alternative optima accessed from the same initial model V region. Similarly, experimental data often compare five to seven V regions obtained by independent walks from the same initial region (Perlmutter 1984; Slaughter and Capra 1984). Table 4.3 shows that, for experimental and model V

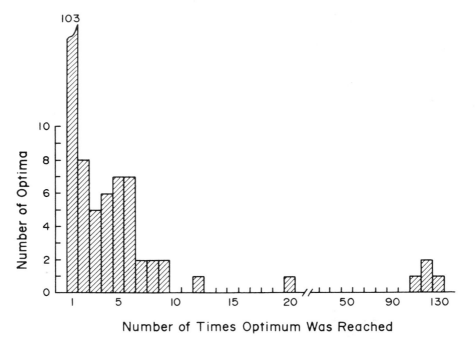

Figure 4.3 Number of times each locally optimal model V region was reached on adaptive walks from the same initial V region of modest affinity. Note that some optima are encountered many times; typically these optima are only one or two mutant steps from the initial V region from which walks begin.

Table 4.3 Number of Observed Sites in Model and Real V Regions Which Have Mutated One or More Times with Number of Expected Mutations.

K	Number of observations	Number of experiments	Number of sequences*
A† 20	79	< 91	9
A 25	68	< 83	9
A 25	55	< 68	8
A 30	48	− 48	7
A 30	52	< 62	10
A 40	41	< 54	5
A 40	36	< 43	9
A 40	35	< 39	4
A 40	27	< 33	10
A 40	41	< 54	7
A 40	34	< 37	7
A 50	20	< 20	4
A 50	18	< 27	11
A 50	15	< 17	4
A 50	21	< 23	6
R† 40	47	< 53	5
Antiarsonate vs. germ line	13	< 23	6
Antiarsonate vs. prototype antiars	28	< 42	6
Antiphosphocholine vs. germ line	21	< 26	8

*Number of alternative optima from the same model V region or number of observed alternative mature V sequences to the same antigen compared.

†"A" means the K sites bearing on each site were adjacent, "R" means those sites were chosen randomly.

regions, the number of sites which accumulate mutations in one or more local optima is *less than expected by chance.* This means that some sites are preferentially not mutated and others are mutated more often than expected in real and model V regions. Figures 4.4a and 4.4b show experimental data for two clusters of V regions, one for the arsonate system (Slaughter and Capra 1984), the other for phosphocholine (Perlmutter 1984). In addition, Figure 4.4c shows five local optima and the initial model V region for an example with $K = 40$. Note that, in the real and experimental sets, some sites have similar parallel mutations.

How seriously should we take the NK model as an account of the structure of affinity landscapes? With considerable but not unbridled enthusiasm. It is the first effort at a statistical model for predicting the structure of fitness landscapes in sequence space. A single choice of parameter values—$N = 112$ as set by the known length of the V region, and K about 40 as tuned to fit known walk lengths to mature antibodies—predicts a number of features of antibody affinity landscapes well. It is premature to say that the model predicts these features accurately. All we can say now is that the predictions are very plausible. Investigations with cloned V regions at different stages of maturation are needed to test the predictions. Even more directly, one might imagine carrying out an entire adaptive walk by fitter one-mutant variants *in vitro* beginning with a cloned V gene from the initial B cells which respond.

Although broadly successful, the NK model as tested does exhibit certain failures. During maturation of the immune response, there appears to be a tendency for mutation causing amino acid substitutions to accumulate preferentially in the CDRs. Fur-

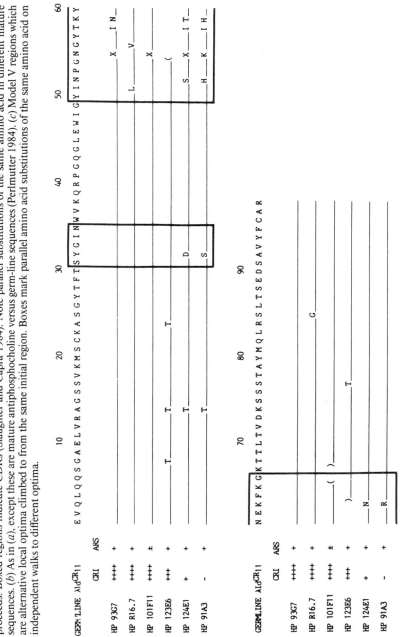

Figure 4.4 (*a*) Antiarsonate versus germ-line V region in matured sequences compared with initial canonical V region from which maturation proceeds. Boxed regions indicate CDRs (Slaughter and Capra 1984). Note parallel substitutions of the same amino acid in different mature sequences. (*b*) As in (*a*), except these are mature antiphosphocholine versus germ-line sequences (Perlmutter 1984). (*c*) Model V regions which are alternative local optima climbed to from the same initial region. Boxes mark parallel amino acid substitutions of the same amino acid on independent walks to different optima.

138

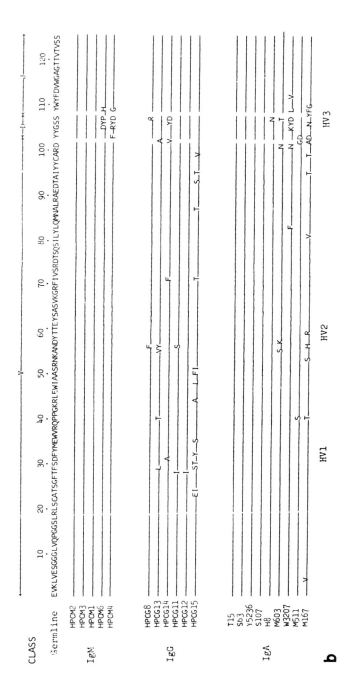

Figure 4.4 *(continued)*

K = 40   Adjacent   Number of Mutant Sites = 27

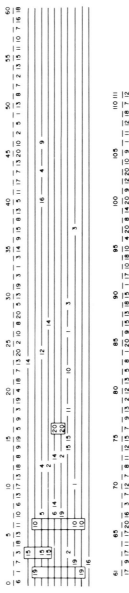

C

ther, there may be a tendency for silent synonymous codon mutations to accumulate in the framework regions outside the CDRs. If true, these biases are not captured in the current application of the NK model to V regions. Such biases might reflect evolutionary specialization of the framework to create the fundamental structure of an antibody binding site, while the CDRs specialize for antigen binding. In this view, the framework is highly adapted and easily disrupted, leading to overall loss of binding by the entire V region. Modeling such a highly adapted character of the framework is ignored in this work. Instead, we tested the case in which only CDRs contribute directly to fitness, while the framework acts indirectly via the CDRs. An alternative approach would be to allow the framework amino acids, on average, to affect primarily other framwork amino acids, which CDR amino acids would affect only other CDR amino acids. Such functional uncoupling of framework and CDR regions would allow the former to be optimized independently in evolution.

The fact that the NK model appears to succeed as well as it does is encouraging in at least three respects. First, it suggests that a statistical model may well capture the structure of fitness landscapes. Second, if either the NK model or an improved one can predict the statistical structure of antibody affinity landscapes, it may also be able to predict the structure of fitness landscapes with respect to the evolution of improved enzymatic function. Both landscape types involve the evolution of a structure with a "business end," the binding site of the antibody and the active site of the protein. Third, if the NK model is close to right, it may be telling us something fundamental about how proteins work. In solid state physics, spin-glass mdoels (Edwards and Anderson 1975; Sherrington and Kirkpatrick 1975; Binder and Young 1986) capture the real behavior of physical spin-glasses by assuming that interactions are so complex that the statistical distribution of their effects can be captured only by random assignments of coupling energies. The same may be true for proteins. I now comment briefly on these issues.

Certainly the most important implication of the rough success of the NK model is the hint that some statistical theory may someday fit well-established data on the structure of the affinity landscape. Obvious refinements of the model would include more details of protein chemistry. Thus some sites should interact more than others, reflecting the fact that some amino acids have more hydrogen bonds, hydrophobic bonding, and salt bonds than others. One would therefore like K to be chosen from a distribution of possible values. In this simplest application of the NK model, the identity of the amino acid at each site bears no relation to its identity at another site. "Amino acid 7," say, is merely a name specific to one site. In reality, alanine at any site is the same amino acid. Thus the nature of the interactions should reflect the fact that alanine at each site is the same amino acid and that different amino acids have different chemical properties.

The NK model is but one theoretical approach to the structure of affinity landscapes. Shenkin, Yarmush, et al. (1987) have carried out detailed simulations predicting antibody hypervariable loop conformations using a realistic model of molecular dynamics. Among the intersting ways of relating realistic molecular dynamic models of protein conformations of the NK model will be efforts to establish whether or not the two models can predict similar statistical features for real antibody molecules. Thus both models should predict similar numbers of fitter one-mutant variants at each adaptive step in the maturation process. Ultimately, it must be possible to predict the observed structure of affinity landscapes for real antibody molecules.

Having focused on the structure of affinity landscapes in maturation of the immune response, I should emphasize that it is but one piece of the immune system puzzles. Even maturation of the immune response requires an understanding of the relation between the affinity of antibody molecules on B-cell surfaces and the selective advantage increased affinity may confer. Since increased clonal selective advantage is probably mediated by faster cell division times and since there is some minimal time required for cell division, we can expect selective advantage to saturate as affinity increases. Further, as noted, I have idealized adaptive maturation as if the mature form is a local affinity optimum, whereas our analysis in Chapter 3 makes it clear that a clonally expanding population of antibody molecules will flow in complex ways across the affinity landscape under the drives of somatic mutation and clonal selection. Finally, I have ignored interactions between B cells and the rest of the immune system.

Are Protein Adaptive Landscapes and Folding Landscapes Related?   Adaptive landscapes appear to be very rugged and may be captured by something like the *NK* model. The clear relation between the model and spin-glasses was noted in Chapter 2. Spin-glass models are currently proving useful as models for protein folding, a complex process in which the protein binds to itself rather than to another molecule (Richardson 1981; Debrunner and Frauenfelder 1982; Karplus and Cuschick 1983; Ansari, Berendzen, et al. 1985; Stein 1985; Bryngelson and Wolynes 1987; Bryngelson 1988). Spin-glass models stress the idea that the potential surface guiding protein folding is likely to be very complex with many local minima, as do initial attempts to apply the *NK* model to protein folding (Kauffman and Stein 1989). Indeed, one way of stating the deep puzzle in understanding protein folding is to ask how the potential surface is constructed such that proteins reliably fold, or refold, to essentially the same shape without becoming trapped on the many incorrect local minima expected in the potential surface. However such reproducible folding is accomplished, proteins, once folded, presumably "breathe" by undergoing transitions between neighboring low-energy minima. Clothia and Lesk (1987) and Karplus, Brunger, et al. (1987) comment that families of evolutionarily related proteins undergo shape deformations on the same scale as the breathing deformations of a single protein. This suggests that the range of readily available protein shape deformations, guided by intramolecular forces, is closely related to the range of shape and function deformations in protein evolution. In turn, the function of proteins in binding ligands and catalyzing reactions is primarily due to the similar shape and force properties. Might it be that the statistical character of the potential surface underlying protein folding is intimately related to the statistical character of adaptive landscapes in protein evolution? If so, then spin-glass models, the *NK* model, or a similar but improved model may capture the right statistical features of both.

EVOLUTION OF NOVEL CATALYTIC FUNCTIONS

In the present section, we discuss the evidence that evolution of novel catalytic functions has occurred repeatedly. Such evolution, the fundamental stuff of novelty, is not mere improvement of an initial catalytic action but instead represents a new direction toward a new end. As we shall soon see, this requires us to broaden the idea of a fitness landscape.

   The fundamental fact underlying the capacity of selection to achieve enzymes which catalyze "new" reactions is that enzymes exhibit a range of activities, such that any enzyme typically catalyzes one reaction with maximal velocity but accepts related substrates and catalyzes related reactions with lower velocities. Selection for increased catalytic activity with respect to such peripheral reactions can accomplish two critical alterations. First, it can successfully increase catalytic activity for the initially peripheral reaction. Second, in so doing, the altered enzyme becomes capable of catalyzing still further peripheral reactions which the initial enzyme did not catalyze at all. Thus, in a third transition, a novel catalytic function can be culled from an initial enzyme.

Such protein *differentiation* (Dickerson 1977) is both well established in evolution and, more recently, the subject of exciting experimental work. The first examples to be worked out, by du Vigneaud's group in 1953, were the peptide hormones vasotocin, oxytocin, bovine vasopression, and porcine vasopressin. Numerous studies of protein evolution have been carried out (for example, Neurath 1984; Carrell, Pemberton, and Boswell 1987). Indeed, comparison of sequences and functions has been widely used to construct a phylogenetic tree (Schulz and Schirmer 1979). Lactalbumin and lysozome are a well-known case of two proteins with similar sequences but different functions. Perhaps the most famous family of proteins with related but diverged functions are the trypsin-like serine proteases. Trypsin cleaves lysyl and arginyl bonds quite promiscuously. Relatives of trypsin act in a similar manner but are more specific in their action, cleaving only one or a few peptide bonds in one specific protein (Schulz and Schirmer 1979). A rather more surprising group is the immunoglobulins, the transplantation antigens, and superoxide dismutase. In the first two, the relation is based on sequence homology and chain-fold homology; the grouping with the dismutase is based on chain-fold homology alone.

Protein differentiation has often been accompanied by (1) gene duplication or multiplcation (Rigby, Burleigh, and Hartley 1974; Markert, Shaklee, and Whitt 1975), allowing a duplicated form to evolve away from the initial task, and (2) genomic rearrangements which fuse different protein domains carrying out different catalytic or ligand-binding tasks (Yourno, Kohno, and Roth 1970). Yet the root form of protein differentiation occurs within a single protein which can evolve to carry out a different catalytic task. Evidence for protein evolution comes not only from the evolutionary record, based on sequence and fold homologies between proteins which now have distinct functions, but also from the rapid evolution of soil organisms able to survive on—and even become obligate to—formerly toxic environments, and the repeated acquisition of antibiotic resistance (Clarke 1976). Direct experimental analysis of protein differentiation includes evolution of an acetamidase to a phenylacetaminidase (Betz, Brown, et al. 1974), a ribitol dehydrogenase to a xylitol dehydrogenase (Rigby, Burleigh, and Hartley 1974; Hartley, Altosaar, et al. 1976), and betagalactosidase activities (Hall 1976, 1978; Hall and Zuzel 1980).

Neighboring Catalytic Tasks Deform the Fitness Landscape and Suggest a Catalytic Task Space

Protein differentiation for a new catalytic function requires that we broaden our ideas about fitness landscapes. A fitness landscape is the distribution over protein space of the capacity to carry out one specific reaction. Protein differentiation means that proteins evolve to carry out new but "neighboring" reactions. Change of the function, or task, measured in protein space will change the fitness landscape. Therefore, we would ultimately like a theory which not only specifies the actual or at least statistical structure of fitness landscapes, but also allows us to say how a given alteration in a task *deformed* the landscape.

We are quite clearly far short of such a theory. In the present section, I discuss experimental work evolving novel catalytic functions in proteins. Among the most interesting features of this work is the hint that some new reaction tasks are "between" two other tasks. The evidence is simple: selection from one of the two "outside" tasks for the other increases catalytic efficiency with respect to the "middle" task.

De Novo Experimental Evolution of Beta-Galactosidase Activity

The experimental evolution of beta-galactosidase function is a particularly good example of protein differentiation. *Escherichia coli* deleted for the lactose operon has a second gene, EBG (evolved beta-galactosidase), whose wild-type function is unknown but which has been shown repeatedly to code for a protein capable of evolving beta-galactosidase activity (Hall 1976, 1978; Hall and Zuzel 1980). The wild-type protein does have activity toward an analogue of lactose, the synthetic substrate ONPG, but very little activity toward lactose.

Mutations which result in the capacity to catalyze lactose breakdown fall into two classes. Class I strains grow rapidly on lactose but not on the disaccharide lactulose. Class II strains grow slowly on lactose but somewhat more rapidly on lactulose and have a broader range of activity than class I strains, being active toward lactose, lactulose, ONPG, and another substrate, ONPF. Both classes arise as single mutations at each of two distinct sites in the EBG gene, about 1000 base pairs apart.

In addition to lactose and lactulose utilization, selection for lactose or lactulose catalysis is associated with an increase of another activity: the capacity to cleave a related metabolite, galactose-arabinoside. The wild-type EBG gene does not allow galactose-arabinoside to be used as a sole carbon source, while both class I and class II mutants, selected for lactose utilization and due to mutations at two different sites, permit minimal ability to use the new sugar source. On the other hand, all attempts to select for galactose-arabinoside growth directly from wild-type EBG have failed. This means that, in order to achieve the capacity to grow on galactose-arabinoside, bacteria had first to be selected to grow on lactose or lactulose. Then the bacteria could be selected a further step away to the peripheral capacity to cleave glactose-arabinoside.

Selection on either class I or class II strains for yet further improved capacity to utilize lactose repeatedly led to formation of the double intracistronic mutant having both class I and class II mutations at the two sites with the EBG gene. These double mutants were more capable of growing on galactose-arabinoside than either single mutant.

A further class of mutants was found—this one capable of utilizing lactobionate. The evidence strongly suggests that here three point mutants are needed and that an *evolutionary pathway* via one-step mutants, each selected for a specific carbon source, is needed to step successively to this altered form. In particular, this mutant class is formed via a sequence that starts with either phenylgalactoside or lactose selection first and the other second, followed by lactulose selection and finally lactobionate selection. Each transition sets the stage for the next.

A number of features of these studies are of basic interest:

1. We can calculate that all or almost all one-mutant variants have been sampled. For typical genes in *E. coli,* the mutation rate is reckoned to be about 10^{-10} per

nucleotide per generation. The EBG gene is large (almost 1000 amino acids long), and so the rate of mutation to it is on the order of 10^{-7} per generation. Typical experiments examined platings of 10^9 cells, and so about 100 variants are expected in each trial. The gene admits about 10 000 one-step neighbors at the DNA level and about 20 000 at the protein level, some of which require two base changes (Hall 1976, 1978; Hall and Zuzel 1980). On the order of 100 trials explored most one-step variants. Thus the one-mutant spectrum has been saturated.

2. One striking feature of the results is that *so few mutants improved each given function.* Each enzymatic function appears to saturate after one or two mutations. Thus selection for increased lactose utilization yields either a class I or a class II mutant in a single step. Each mutation occurs at a distinct site. While more than one variant (up to three or four) may have been recovered in the class I and class II groups based on heterogeneity of the dissociation constant K_m and of the maximum velocity of the catalyzed reaction V_{max} for the enzymes, this conclusion is not firm. No evidence supports the possibility that further selection on class I mutants yields further improvements at the class I site; rather, the second mutation which improves lactose utilization is at the class II site. Therefore, under these conditions, two mutants appear to saturate the capacity to improve lactose utilization due to mutations in the EBG gene.

It must be stressed that the observations underestimate the numbers of mutations which may marginally increase each catalytic activity, since the selection conditions were sufficiently crude that a substantial increase in activity (on the order of twofold) was probably necessary for recovery of an improved variant. With that caveat, it nevertheless appears that rather few neighbors are (much) better than the initial form at a given peripheral catalytic task, and that rather few steps are needed to achieve a local maximum for that peripheral function when it is selected. Even if these are underestimates, they leave the strong sense that, among contemporary well-evolved proteins, selection for a neighboring catalytic task achieves a local maximum after very few steps. In comparison, in the immune system maturation, it appears that a larger number of mutations (typically six to eight) intervene between the roughed-in canonical V segment selected initially and the highest affinity variants subsequently selected.

3. Some reactions are "between" other reactions. Another feature of the experimental evolution of EBG which deserves attention is the complex interplay of selection for one catalytic function and increases/decreases in catalytic activities for other tasks. All cases of selection from class I and class II *lactose*-utilizing strains to class III or class IV *lactulose*-utilizing strains simultaneously *increased* activity for galactose-arabinoside; at the same time, most strains showed a decrease in capacity to utilize lactose. Conversely, a class III strain which utilized lactose poorly but lactulose well was selected to increase its efficiency of lactose utilization and simultaneously showed decreased activity toward lactulose and *increased* activity toward galactose-arabinoside. Overall, the data suggest not only that increase for one function may often cause correlated increases or decreases in other functions but also that, with three functions, a notion of "between" makes sense. Galactose-arabinoside utilization appears to be *between* lactose and lactulose utilization. Selection *from* either of these outside functions *toward* the other increases galactose-arabinoside utilization.

Measuring and Covering Catalytic Task Space

In this section, I examine the possibility that the notion of between-ness in catalytic tasks can be formalized as a catalytic task space. This abstract representation has important implications:

1. Because catalytic task space contains a notion of nearby catalytic tasks, it provides a framework for thinking about the evolution of divergent enzymatic activities.

2. The concept of catalytic task space implies that different reactions can represent the same catalytic task.

3. Because any enzyme covers a ball of similar tasks in task space, a finite number of enzymes might catalyze all possible reactions. To estimate the number of crude enzymes required, I shall derive measures of the number of effectively different catalytic tasks in catalytic task space. The number appears to be between 10^6 and 10^8. Thus about 1 000 000 to 100 000 000 enzymes might suffice as a universal enzymatic tool kit.

The hypothesis of a catalytic task space has experimental support in the recent discovery that *antibody molecules* which bind the *transition state of a reaction can catalyze that reaction.* Therefore, provisionally, we might think of a catalytic task as the binding of the transition state of a reaction.

A useful way to think about catalytic task space is to consider first the immune system and the concept of *shape space* introduced by Perelson and Oster (1979). (See also Perelson 1988; Segel and Perelson 1988.) As noted above, the immune system supports on the order of 10^8 antibody variants due to combinatorial diversity. Consider each such antibody molecule with its antigen-combining site as a specific "shape" able to bind a complementary antigenic epitope shape. The familiar image is lock and key.

The recognized features of an antigenic epitope are spatial extent—on the order of 1 to 5 nanometers in length, width, and depth—and some number of chemical features such as charge and dipole moment. Consider, then, some general "shape space," each of whose axes corresponds to one such feature. Since the physical lengths involved in an epitope are bounded between zero and some modest number of nanometers and since the other physical quantities are also bounded between some minimum and maximum values, shape space itself is bounded. Then any epitope shape can be represented as a single point in this shape space, and epitopes with similar shapes are a neighboring cluster in shape space. Similarly, a given antibody-combining site is a specific point in shape space. Then any antibody molecule will bind an epitope with the appropriate complementary shape. Further, since specificity is not infinite, any antibody molecule with bind a ball of epitopes with sufficiently similar shapes.

Perelson and Oster (1979) use this picture to ask the question, How many antibody molecules does it take to cover shape space? That is, how many antibody molecules are needed to make sure that almost all possible epitope shapes are matched by at least one antibody molecule? To think about this problem, they imagine that each antibody molecule has a combining site with a shape that is drawn at random and hence able to recognize some complementary ball of epitope shapes. By hypothesis, the location of each such epitope ball is random. Then the question of how many antibody molecules are needed to cover shape space becomes the question of how many balls of a given average volume must be placed at random to ensure that almost the entire volume is covered by at least one ball.

The fraction of shape space covered by placing a fixed number of balls at random locations by a Poisson process depends only on the ratio of the volume of each ball to the total volume. Perelson and Oster then note that the simplest immune systems have an estimated antibody diversity of about 10^4 and argue that, to be minimally effective, such a system must cover a sizable fraction of the space. They pick $1/e = 0.37$ as a convenient threshold. Therefore, they are led to say that 10^4 random balls cover at least 0.37 of the volume. In turn, this threshold value allows them to calculate the average volume of one ball, allowing for possible overlaps when many random balls are thrown into shape space. Then it is easy to calculate that an immune system with 10^8 antibody molecules would cover the space very well indeed. Virtually any antigen shape would have at least one antibody molecule able to bind it.

In addition to this inventive line of thought, Perelson and Oster advance a further argument to estimate that the *number of features, or dimensions, of shape space must be small*. The idea is disarmingly simple. A finite number of antibody molecules can cover all of shape space because each recognizes a finite ball of shapes. This quantizes shape space into a finite number of balls which just suffices to cover the space. The size of each ball reflects the range in antigenic shape which an antibody can tolerate. Alternatively stated, ball size represents imprecision in specificity. Suppose the range in each dimension of such a ball were imagined to be 0.5 of the possible range. Then along each dimension, an average of two nonoverlapping balls could be placed adjacent to each other. From this supposition, we can calculate the number of dimensions which must exist in shape space. If the immune system requires 10^8 antibodies to cover shape space and if two nonoverlapping balls occupy each of L dimensions, then $2^L = 10^8$ and $L = 26$, where L is the number of dimensions in shape space.

But suppose the permissible fraction of the range in each dimension is only 0.1. Then ten adjacent balls can be placed along each dimension; hence $10^L = 10^8$ and $L = 8$. If the imprecision tolerated by an antibody in each dimension is 0.05, then $20^L = 10^8$ and L is about 6.

This is a powerful line of reasoning, for it is undeniably plausible that imprecision in shape with respect to spatial or other "shape" features is on the order of 0.05 to 0.1 with respect to each feature. Hence only about six to eight spatial and chemical features matter! Without the concept of shape space, it would be hard indeed to derive such a general conclusion from such simple premises.

A more precise statement of Perelson and Oster's idea is that many different possible contributions to "shape," including spatial extent in three dimensions and an unknown variety of chemical and physical properties, project onto about six axes. "Projection" merely means that each of these six axes may itself be made up of some combination of a large number of spatial, physical, and chemical aspects of antigens and antibodies. These spatial and chemical features can be projected to an independent set of six to eight features which suffice to characterize shape space.

Different Molecules Can Have the Same Local Shape

Among the most critical features of the concept of a shape space is that very different molecules may have essentially identical epitopes. That is, the *same point in shape space* corresponds to *indefinitely many different molecules,* all of which have essentially the *same local shape.* This idea is quite different from the familiar notion that *similar molecules have similar shapes.* That concept is captured in the idea of a ball in shape space. Rather, the newer concept is that quite different molecules can, on the relevant scale of an epitope, have the same shape.

This idea has become familiar as the concept of *internal images* in the immune system (Jerne 1974, 1984; Sege and Peterson 1983; Greene and Nisonoff 1984; Gaulton and Greene 1986; Greenspan and Roux 1988). When exposed to an antigen, an organism mounts an immune response by secreting a set of antibody molecules which are shape complements of the antigen determinant(s). But these *rank 1* antibody molecules have V regions which are themselves antigenic determinants which elicit a second rank of antibodies. In turn, the rank 2 antibodies stimulate rank 3 antibodies against rank 2 idiotypes. But rank 1 antibodies are *shape complements of the external antigen,* and rank 2 antibodies are *shape complements of rank 1 antibodies.* Hence rank 2 antibodies might be expected to include some molecules with shapes which *mimic* the shape of the external antigen. Shape, shape complement, shape. Think of the rank 1 antibody as a lock; if two keys, antigen and a rank 2 antibody, fit the same lock, the two keys must have similar shapes. Such rank 2 antibodies are then *internal images of external antigens.*

Indeed, such internal images exist. And, as we shall see, their existence carries a host of implications. It is known that a rank 2 antibody can mimic the shape of an external antigenic determinant even when the antibody site is a *sequence of amino acids* but the antigenic determinant is a *carbohydrate* (Sacks, Kirchhoff, et al. 1985). That is, very different molecules can have the "same" shape on a local part of the molecule. Many examples exist beyond the immune system. The action of opiates derives from the fact that they are close mimics of peptide endorphins and bind the same receptors, despite marked chemical differences.

From Shape Space to Catalytic Task Space

In shape space, a point represents an antigen shape. Implicit in this concept is the idea of a corresponding task. That is, the task is the requirement for some antibody to have a complementary shape and hence be able to *bind the antigen.* Shape space is then a space of binding tasks.

We are used to conceiving of chemically similar reactions. Indeed, all of biochemistry is based on classifications of reactions into similarity groups based on which functional groups participate in the reaction and by which reaction mechanism. Any enzyme carries out a catalytic task. Nearby ractions are nearby tasks. A catalytic task space is then a space in which each point represents a different catalytic task, while neighbors in the space represent neighboring catalytic tasks. Just as we do not know exactly what spatial, chemical, and physical features constitute shape but can still make substantial use of the concept of shape space, so too can we make use of the concept of a catalytic task space without knowing in detail the spatial, chemical, and physical features which constitute a task. To be concrete, however, I suggest that, to a first approximation, a *catalytic task is the high-affinity binding of the transition state of a reaction.*

According to transition state theory (Lienhrd 1973), catalysis occurs because the enzyme has a catalytic site capable of binding the transition state of a reaction with high affinity. The binding energy of the enzyme for this intermediate state deforms bonds in the substrate(s) and thus lowers the activation energy of the reaction leading to the products. This lowering of the activation energy catalyzes the reaction.

The suggestion that a catalytic task is as simple as binding the transition state with high affinity is an oversimplification (Fersht 1974, 1985; Fierke, Kuchta, et al. 1987). Indeed, as Jencks (1987) points out, the statement that enzymes catalyze reactions by stabilizing the transition state is not a hypothesis but is required by the definitions of "catalysis" and "transition state." However, mere stabilization of the transition state is only necessary for catalysis, not sufficient, for the critical issue is the energy barriers from the enzyme–substrate complex and from the enzyme–product complex to the transition state. The necessary and sufficient condition for catalysis is that the enzyme lowers these energy barriers. Typically, this is achieved when the enzyme binds to the transition state with sufficient affinity to stabilize the state while increasing the energy of the enzyme–substrate and enzyme–product complexes. The increase in free energy of the ES and EP complexes can be due to physical strain, desolvation, or other mechanisms and is relieved in the transition state. Thus the enzyme utilizes the specific binding energy of the ES or EP complex to overcome the strain and entropy loss which destabilize the complex and make the transition state more available (Jencks 1987).

Albery and Knowles (1976), Hermes, Blacklow, et al. (1987), and Knowles (1987a, 1987b) discuss the same issue in considering the evolution of catalytic perfection. These authors conceive of three stages in the evolution of catalysis. In the first, a primordial catalyst binds the ES complex, the transition state, and the EP complex with equal affinities, lowering the energies of all three. In a second step, differential binding occurs with higher affinity for the transition state than for the two complexes, thus meeting Jenck's proposed minimum requirements for catalysis. In a third stage, the energy minima of S alone, the ES complex, the EP complex, and P alone are tuned to be monotonically increasing to maximize catalytic efficiency in conversion of S to P or P to S, without unduly high concentrations of ES or EP complexes unable to release S or P.

These considerations make it clear that picturing a catalytic task as merely binding of the transition state with high affinity is too simple. At best, the concept of task conforms to Albery and Knowles's second stage (differential affinity for the transition state over the ES and EP complexes). However, the simpler image will suffice for a start.

Conceive, then, of a catalytic task space whose axes represent some number of shape and chemical features relevant to catalysis. These features obviously include those relevant to shape, since fitting of substrates and enzyme are critical features of effective catalysis. If we accept either the simpler or the more complex understanding of transition state theory, then we are saying that the generalized shape of the transition state compared with the generalized shapes of the ES and EP complexes corresponds to a point in catalytic task space. The task is accomplished by an enzyme *binding* the transition state with high affinity and the ES and EP complexes with lower affinity.

To be concrete in discussing catalytic task space, I shall assume a simple transition state theory. Each catalytic task is to bind a shape afforded by the transition state of a reaction. Using this concrete idea, however, should not obscure the fact that the same principles carry over to more complex theories of what might constitute a catalytic task.

Different Reactions Can Represent the Same Catalytic Task

Corresponding to the idea of a ball of shapes recognized by a given antibody is the idea that any highly evolved enzyme catalyzes a set of similar reactions. That is, a shape space ball is a set of similar shapes; a catalytic task ball is a set of similar reactions. Further, just as more than one molecule can have the "same" shape, so too can *more than one reaction represent,* in some sense, the "same" catalytic task.

As above, this is not the idea that similar substrates and products constitute similar reactions (this idea is incorporated in the concept of a ball in catalytic task space) but that very different substrates and products might constitute roughly the same catalytic task, by virtue of having locally identical transition states. If so, then the same enzymatic site and machinery would be expected to catalyze two nonhomologous reactions. In the logical limit, two quite different substrates and reactions might constitute such highly similar catalytic tasks that any enzyme capable of catalyzing one would catalyze the second as well.

How Many Enzymes Might Cover Catalytic Task Space?

The concept of catalytic task space and the idea that quite different reactions can constitute the "same" reaction really mean the following. We have assumed that an indefinitely large number of different reactions can each be identified with a single point in catalytic task space. And we have assumed that all reactions can be mapped to some point in task space. Further, to capture the idea of similar reactions, we have assumed that nearby points in task space can represent nearby reactions. But since quite different reactions can map to the same point, nearby points really represent similar tasks, not merely similar reactions.

We can then ask, How many enzymes are required to be certain that at least one enzyme can catalyze each task in task space? A highly evolved enzyme covers some volume in catalytic task space. Just as Perelson and Oster can reason that each antibody molecule covers a fraction of shape space and can estimate how many molecules are needed to cover shape space completely, we can use the idea that an enzyme covers a ball in task space and try to estimate the number of balls needed to cover the whole space. I discuss next two crude estimates of the number of enzymes required to cover catalytic task space. Both estimates are based on asking the probability that two apparently different chemical reactions constitute nearly the same task. If so, an enzyme catalyzing one might cover a ball in task space which covers the second. From that, we can estimate the number of balls which might cover task space.

Suppose a square dart board is divided into N small square sections and m darts are thrown at it at random. At first, as more and more darts land on the board, all sections have zero or one dart. Eventually, some sections begin to have two darts. The number of darts which must be thrown in order that some sections have two darts is on the order of the $\sqrt{N}$. Thus finding the number of darts which must be thrown such that some sections have two darts is a means of measuring the number of sections N.

We can apply this measure to estimate how many enzymes cover task space. Any well-evolved enzyme covers some small ball of catalytic tasks. Imagine throwing random "reaction darts" at catalytic task space, where each section corresponds to one ball, and ask how many must be thrown to find two reactions which land in the same ball. Stated otherwise, one enzyme covers one ball. That such an enzyme catalyzes one reaction corresponds to being hit by one dart. Excluding similar or cognate reactions present in the same ball by simple virtue of similarity, are there any enzymes known which catalyze reactions whose substrates and products are seemingly very dissimilar? Such cases correspond to two darts in the same section.

Indeed, there are some such cases. For example, glucose oxidase can catalyze the production of hydroquinone from benzoquinone, and galactose oxidase can catalyze the stereospecific oxidation of aliphatic alcohols (Klibanov, Berman, and Alberti 1981; Klibanov, Alberti, and Marletta 1982). An immobilized sulfatase can separate alpha and beta napthols (Cambou and Klibanov 1984). Cambou and Klibanov report these and other examples of unusual catalytic activity, many of which are transesterifications catalyzed by carboxyl esterase. In addition, other workers have recently succeeded in constructing novel substrates for specific enzymes by considering the structure of the active site. Here the substrate is chemically distant from the normal substrates of the reaction catalyzed by the enzyme; hence this work shows that one enzyme can act on quite diverse substrates but of course cannot be taken as evidence for the probability of finding such distinct reactions, since the novel substrate was constructed with a chemist's insight.

The fact that such cases have been found allows a rough estimate of the number of effectively different catalytic tasks. The number of reactions which have been studied is more than 1000 but probably less than 10 000. At least some of these reactions are catalyzed by the same enzyme and hence fall in the same ball. Then the number of catalytic tasks is on the order of the square of these numbers, or between 1 000 000 and 100 000 000.

These numbers are striking. A maximum estimate of the number of catalytic tasks is on the order of 10^8, but this is also just about the number of human antibody molecules needed to cover shape space. The estimates are based on entirely different approaches, yet *a posteriori* it is encouraging that they are similar. One would expect that shape matching is critical to effective catalysis; hence the two estimates might well be about the same order of magnitude. This correspondence is all the more encouraging since, as we shall see in a moment, antibodies which are the shape complement of the transition state of a reaction can catalyze that reaction.

An independent argument leads to a similar conclusion. Suppose that 10 000 enzymes have been studied and that, at a maximum, each has been tested for the capacity to catalyze 100 apparently unrelated reactions. Assume further that any possible reaction must lie in one of a total number x of balls in task space. Assume that there are ten to twenty known cases of enzymes which catalyze two quite different reactions. Then among the 10 000 $\times$ 100 reactions tried, ten to twenty have landed in the same catalytic ball. From this, x is trivially estimated: $100 \times 10\ 000/x = 10$; $x = 10^5$. Were we to assume that fewer than 100 apparently unrelated reactions had been tried, then x would be even smaller.

In either estimate, however, 10^5 or 10^8, the major issues to focus on are (1) that catalytic task space should be covered by a finite number of balls representing more or less well-tuned, highly specific enzymes and (2) that that number might be on the order of 100 000 to 100 000 000.

There are a number of reasons to consider these estimates cautiously. First, estimates from the first method depend on the idea that random reactions are thrown at catalytic task space. But reactions that take place in an organism which appear to be sufficiently different to count as independent darts may have evolved because of hidden similarities in substrates and products. Equally obvious, the choices of reactions studied have not been random, and estimates of the number of reactions studied are crude.

Without the concept of a catalytic task space, the conclusions we have reached would be hard to state. Task space allows us to map indefinitely many diverse reactions to the same point and each possible catalyzable reaction to some point. Thereafter, it becomes obvious that any enzyme covers a finite volume ball, thus that some finite number of enzymes might cover the space. While the estimates of the number of enzymes required to cover catalytic task space may be wrong, the basic idea seems likely to be right. Just as 10^8 antibody molecules may suffice to cover shape space and recognize any epitope on the relevant size scale, 10^8 enzymes might cover catalytic task space and catalyze any reaction on the relevant molecular scale. Such a set of enzymes would be a universal chemical catalyst set.

Testing the Hypothesis of a Coverable Catalytic Task Space: Catalytic Antibodies

The recent discovery of antibodies which catalyze specific reactions (Pollack, Jacobs, and Schultz 1986; Tramontano, Janda, and Lerner 1986a, 1986b) affords the first independent tests of the concept of a catalytic task space and of the hypothesis that a finite number of catalysts might be able to perform all catalytic tasks. Indeed, the repertoire of antibodies in humans may already be one such universal toolbox.

According to transition state theory, as noted above, an enzyme lowers the activation energy of a reaction by having high affinity for the transition state. By the enzyme's binding of that deformed state with high affinity, the binding energy is thought to be harnessed to catalysis. Based on this idea, Pollock, Jacobs, and Schultz (1986) and Tramontano, Janda, and Lerner (1986a, 1986b) reported preparation of monoclonal antibodies raised against stable *molecular analogues of the transition state* of a reaction. Such antibodies might therefore serve as catalysts for the reaction. In fact, several such catalytic antibodies, termed *abzymes,* have been recovered. They show high specificity for the substrate and Michaelis–Menten kinetics and increase reaction rates by 770-fold or more (Tramontano, Janda, and Lerner 1986a, 1986b; Tramontano, Janda, et al. 1987). Since the turnover rate of an enzyme is related not only to the enzyme's affinity for the transition state but also to its capacity to release substrates and products K_m, achieving truly efficient catalytic antibody molecules will require tuning the affinity for the transition state so that it is much higher than the affinity for the substrate or product conformations. Pollock, Jacobs, and Schultz (1986) and Pollock and Schultz (1987) demonstrated that a phosphorylcholine-binding antibody which binds the transition state analog 4-nitrophenylphosphorylcholine selectively catalyzes the hydrolysis of the corresponding choline carbonate. Tramontano, Janda, et al. (1987) report the capacity of catalytic antibodies to catalyze a stereospecific reaction with Michaelis–Menten kinetics and a 167-fold rate acceleration. Jacobs, Schultz, et al. (1987) report that antibodies elicited to an aryl-phosoponate transition state analog selectively catalyze the hydrolysis of the corresponding aryl-carbonate.

These are very exciting results. First, they tend to support transition state theory. They show that catalytic activity can be driven by affinity for the transition state. Second, since we know that quite different V regions can bind to the same epitope, it is legitimate to surmise that quite different monoclonal antibodies might bind to the same transition state analog and catalyze the same reaction. Third, since the abzymes were selected not by action on the target reaction but by *affinity for an analogue of the transition state,* it follows that, if there are *other reactions* with nearly the same transition state, the same analogue should allow recovery of an abzyme which catalyzes these *different nonhomologous reactions* as well. In short, the success strongly suggests that different reactions with nearly the same transition state, even if the substrates are highly dissimilar, will be catalyzed by the same abzyme. These reactions represent the same catalytic task. Fourth, the immune system is thought to be capable of responding to nearly any epitope on the proper size scale and of mounting a response locating antibodies with high affinity for that epitope. If an analogue for the transition state is an epitope, then on the order of 10^8 antibodies ought to suffice to allow recovery of a few which catalyze the corresponding ball of reactions. If antibodies can cover shape space, they may also cover catalytic task space. Then 10^8 will suffice. Finally, from the analysis above concerning maturation of the immune response, we would expect on the order of ten improvement steps to occur between a roughed-in antibody binding to the analogue of a transition state and a well-honed, highly specific antibody with high affinity for that state. Thus the number of improvement steps needed to achieve a quite good enzyme from a rough starting point may be limited.

These results suggest that a well-developed theory of adaptive walks in correlated landscapes may ultimately allow a modest number of protoenzymes, say 100 000 to 100 000 000, with low specificity to be kept in a library as starting material for *in vitro* or *in vivo* selection procedures allowing attainment of highly efficient enzymes for almost any catalytic task.

Evolving Novel Catalytic Tasks Projected onto Catalytic Task Space

The evolution of proteins to carry out new neighboring reactions, so fundamental to evolution at the molecular level, can be interpreted within the image of catalytic task space. If an enzyme covers a ball in task space, then evolution to a new neighboring function moves the ball to a nearby point in task space. Thus the concept of a task space allows us to ask how enzyme specificity is related to protein differentiation. This raises three issues:

1. How many *crude* enzymes might cover task space?
2. Is protein differentiation easier with crude enzymes?
3. How many improvement steps separate an enzyme specialized for one task from one specialized for a task at the first enzyme's periphery?

Let us consider question 1 first. Presumably a low-specificity enzyme, by virtue of its lower specificity, covers a larger volume of catalytic task space than a high-specificity enzyme. Borrowing from Perelson and Oster, we can say that roughly a doubling in range of imprecision—from 0.05 to 0.1 or from 0.1 to 0.2 of the range of each of six to eight dimensions in catalytic task space—suffices to increase the volume covered by an imprecise enzyme by a factor of 100 to 1000 over the volume covered by a high-specificity enzyme. Therefore, if we crudely estimate that 10^5 to 10^8 specific enzymes might cover task space, it follows that 10^2 to 10^6 low-specificity enzymes might cover task space. I confess that 10^2 appears quixotically low, but 10^6 might be reasonable.

In regard to question 2, differences in ball size imply a difference in how easy it is for evolving proteins to carry out novel enzymatic tasks. Like other adaptive walks in protein space, presumably walks which evolve novel function must always maintain or improve some other function. We saw in the experimental evolution of EBG that obligate sequences of selection transitions were required, from an enzyme performing a task A well and a peripheral task B poorly to one performing B well but a new peripheral task C poorly. Ultimately, via a succession of selection steps, enzymes for C were obtained. This argument implies that walks to novel functions by less-specific enzymes should be far easier than walks by highly specific enzymes. The ball in task space covered by each crude enzyme is larger than the ball covered by a specific enzyme. If each ball is bigger and if a fixed number of crude enzymes N are considered, each covering its own ball, then any one crude enzyme overlaps more neighboring balls than were the same number of smaller balls considered. It follows that, with crude enzymes, a greater diversity of peripheral intermediate tasks can be found which lie on one or another pathway leading from an initial task to a distant target task. Therefore, the number of *alternative routes via possible neighboring tasks* which lead to a distant task is greater for crude enzymes than for highly specific ones. We must conclude that the variety of selection conditions that can lead from task A to task W is wider for crude enzymes than for highly specific ones. These ideas suggest that evolution of novel function was easier when life first appeared on earth than it is now. We shall return to this point in Chapter 7 in considering the origin of life.

To answer question 3, consider that, since high-specificity enzymes cover smaller balls, the number of mutation steps to an optimal enzyme for a peripheral task should be smaller with high-specificity enzymes than with crude ones. This expectation seems to be borne out experimentally. The data on selection for peripheral catalytic tasks—for example, in EBG (Hall 1976, 1978; Hall and Zuzel 1980)—suggest that only one or two single-neighbor mutant steps separate a current well-tuned enzyme from a local maximum for peripheral tasks. In contrast, a roughed-in V region with low affinity for an antigenic determinant takes eight to ten adaptive steps before reaching an optimum for that binding task. Buried in this, however, is the assumption that EBG is already of high specificity, while an initial antibody for an antigen is of low specificity.

The general implication of catalytic task space and of data on evolution of novel neighboring catalytic functions is that such adaptation might be expected to proceed via a small burst of substitutions until a local optimum is reached. Evidence of such bursts is available in the rates of alteration of protein and nucleotide sequences in evolution (Gillespie 1983, 1984). In fact, Gillespie has proposed a model of deforming fitness landscapes to explain just such phenomena, which we discuss next.

Bursts of Substitutions in Molecular Evolution

In phylogenetic comparisons of the amino acid sequences of a particular protein, the observation that evolution rate is nearly constant over extended periods of time led Zuckerkandl and Pauling (1965) to call the amino acid substitution process a "molecular evolutionary clock." Detailed studies of both a large number of proteins over a few species and particular proteins over a large number of species have substantiated this general idea. They have also substantiated the observation of Ohta and Kimura (1971) that the variance in evolution rate is higher than what would be expected if substitution were a Poisson process.

There are serious problems estimating variance in substitution rates, both because estimating the number of substitutions rests on procedures such as maximum-parsimony procedures, which assume the minimum number of mutational changes needed to account for the observed transitions, and because variance in the evolution rate is of interest only in comparison to the mean, as the ratio $K = \text{var/mean}$. For a simple Poisson process, $K = 1$. As Gillespie points out, however, the data are strong enough to establish that the observed value of K is approximately 2.5. This higher value means that substitutions tend to occur in small bursts. The molecular clock stutters.

Hudson (1983) asked whether the pure neutral-allele model could account for a value of K as high as 2.5 and concluded that, in order to do so, we have to assume population sizes and mutation rates which conflicted with other data. If the neutral model is hard pressed to account for such bursts, can selection?

To address this, Gillespie proposed a general model of molecular evolution similar to that presented in Chapter 2. It is based on fitness landscapes in sequence space. He supposes that any sequence has D neighbors and that adaptive steps proceed via one-mutant neighbors. Gillespie's concern is to model the adaptive behavior of a genetic system as the environment changes. He supposes that the current fittest allele or protein carrying out a given task is fixed in the population. When the environment changes, he assumes that it *deforms to a similar fitness landscape* such that the current fittest allele remains one of the two to five best among the $D + 1$ neighboring sequences and itself. He then examines the expected number of evolutionary steps to a local maximum in the new environment. At each adaptive step, Gillespie draws fitness values at random from a predefined, fixed probability distribution—Gaussian or exponential, for example—and assign the randomly chosen fitness values to the new $D - 1$ neighbors which become accessible at that step. The adaptive process moves to a fitter neighbor not equiprobably but weighted in favor of the fitness differences with respect to the currently fixed allele. The process stops when a sequence which is fitter than all its neighbors is found.

The main results of Gillespie's efforts are that, when the rank-order of the previously fixed protein in the new environment is two to five among the D, evolution to a local maximum occurs in a small number of steps which constitute a burst, yielding a value of 2.5 to 3 for K. Further, the results are very insensitive to the underlying probability distribution, whether it is Gaussian or exponential.

Gillespie concludes that selective evolution is able to account well for the observed bursting character of molecular evolution but that such correspondence alone cannot rule out either a neutral-allele model or the still more plausible hypothesis that both selective and neutral drift changes occur in protein evolution.

The close relation between Gillespie's ideas and the development of Chapter 2 and this chapter is obvious. His assumption that, in a new environment, the rank-order of the previously fixed allele becomes the second to fifth among the $D - 1$ neighbors is a reasonable approach to modeling the statistical deformation of fitness landscapes as the catalytic task changes. This approach is underpinned by the concept of a catalytic task space, where selection for a neighboring task begins with a protein already weakly able to carry out that task and moderately high up in the global rank-ordering for the task. Within the context of the NK model of fitness landscapes, Gillespie's ideas amount to modeling expected landscape deformations when an external task requirement, corresponding to a catalytic task, changes. Development of such a model seems an important direction for further work. If either the NK model or better successors capture the statistical structure of fixed fitness landscapes underlying adaptive evolution in protein space, then extension to landscape variation as the task varies should follow.

We have now discussed the possible structure of fitness landscapes in sequence space, developed the NK model of such landscapes, and applied the model to maturation of the immune response. We have considered the concept of catalytic task space. These ideas invite a massive experimental program to discover the structure of fitness landscapes for specific protein functions in protein space. It is therefore remarkably fortunate that we have passed the threshold of just such a program. I turn to this in the final section of this chapter.

APPLIED MOLECULAR EVOLUTION: DIRECT EXPLORATION OF DNA, RNA, AND PROTEIN SEQUENCE SPACES

I believe we have crossed the threshold of a terribly important new era. Applied molecular evolution of useful biopolymers has become feasible. It has become clear in the past several years that, for the first time in history, molecular cloning techniques afford us the possibility of truly exploring sequence spaces. In particular, we can for the first time generate billions, trillions, indeed many orders of magnitude more, of totally or partially stochastic novel DNA sequences, RNA sequences, and protein sequences, either introduce these sequences into viruses, bacteria, or higher cells or amplify them *in vitro,* and then utilize selection or screening procedures to find biopolymers of interest. Those interests range from basic science—the distribution of structural, catalytic, and ligand-binding capacities in sequence space—to practical—the generation of novel drugs, vaccines, catalysts, ligands, DNA regulatory elements, ribozymes, biosensors, and even self-reproducing biochemical systems.

The first expressions of interest in this field, to my knowledge, occurred in 1985. In March of that year, my colleague Marc Ballivet and I applied for international patents bearing on many of these purposes. Patents have now been issued in France (Ballivet and Kauffman 1987), Great Britain (Ballivet and Kauffman 1989), Germany (Ballivet and Kauffman 1991), and India (Ballivet and Kauffman 1991). Smith (1985) and Childs, Villanueva, et al. (1985) subsequently proposed some aspects of the field. Since then, the area has continued to grow dramatically (Cull, Miller, and Schatz 1992).

The basic science interests in DNA, RNA, and protein sequence spaces are by now obvious (Eigen 1985; Kauffman and Levin 1987; Kauffman, Weinberger, and Perelson 1988) and range from the issues mentioned above to the fundamental question of the requirements of protein folding. The discovery several years ago that RNA molecules can act catalytically (Kruger, Grabowski, et al. 1982; Zaug and Cech 1985; Cech 1986a, 1986b; Been, Barfod, et al. 1987), discussed in more detail in Chapter 7, has only intensified the importance of exploring these spaces. We wish to know not only how easily current well-evolved sequences can either evolve to improve a given function or differentiate to carry out a neighboring function, but also how difficult it was in evolution to find DNA, RNA, or protein sequences with any function at all. Therefore, we are hardly well advised to limit ourselves to the biopolymers present in contemporary organisms. Even if we examine one-mutant, two-mutant, or higher variants of enzymes and utilize selection regimes to seek fitter variants, as Hall (1978) and others have done, we limit ourselves to exploring the vicinities of proteins which have had the benefit of 3.8 billion years of evolution. One cannot map a continent by walking about a small subset of its highest peaks. On two counts, our understanding of biological molecular evolution implies that, at best, only a tiny region of sequence space can have been searched since life began. First, the total number of DNA, RNA, or protein polymers which might plausibly have formed during the history of the earth is about 10^{65}, as discussed in Chapter 1. But the total number of DNA sequences of length 300, which is the minimum length needed to code for an average-size protein, is $4^{300} \approx 10^{180}$. Second, within the local reaches of sequence space which historically might have been explored by evolution, trapping in local peak subregions will have markedly constrained further search. The number of alternative solutions to functional design problems in sequence space is undoubtedly far larger than we have supposed. In fact, the concepts of shape space and catalytic task space clearly point in that direction. There are far fewer effectively different shapes and chemical reactions than there are sequences. Function must be highly redundant in sequence space. If so, life itself is far more readily attained than we have thought heretofore.

We need to explore DNA, RNA, and protein spaces in earnest, but doing so confronts obvious combinatorial and practical problems. The number of potential polymers of interest is larger than the number of hydrogen atoms in the universe. If our questions concern the distribution of catalytic or liganding properties, then we must pose questions which we can answer by sampling locally at many points in DNA, RNA, or protein space, exploring the immediate vicinities of those points, and then building up a unified picture of the statistical structure of fitness landscapes with respect to *single* structural, catalytic, or liganding properties. Further, we need the mapping from the set of organic chemical reactions onto catalytic task space and from the set of biopolymers or other organic molecules onto shape space. That is, we want to know not only how many local optima there may be for a given catalytic task in sequence space but also how, as we move in sequence space, we move from enzymes for one catalytic task to enzymes for neighboring tasks. Similarly, we wish to know not only the distributions of local optima to exhibit a given shape and hence a given capacity to bind but also how, as we move in polymer space, we move in the corresponding shape space such that the liganding capacity covers a nearby ball in shape space. Toward these ends, we need to find a way to generate a very large number of random DNA, RNA, and protein sequences; find those which exhibit some structural, catalytic, or liganding property of interest; and then mutate, explore, and hill climb in the vicinity of the local optima and utilize recombination to search sequence space between local optima.

The basic idea of applied molecular evolution is simple. Use any one of a variety of techniques to generate extremely large numbers of fully or partially stochastic DNA or RNA sequences, which may be of interest either in themselves or in terms of the vast array of different proteins for which they code. Select or screen such sequences for properties of interest and then amplify those sequences of interest either *in vitro* or *in vivo* (or else amplify the sequences first—for example, by cloning into viral or plasmid expression vectors, with passage through bacterial or host cells—to create a vast library expressing novel sequences and then select or screen the library for properties of interest). In short, rather than attempting to guess the design of useful biopolymers from first principles, we can hope to utilize either selection or screening techniques to find even very rare sequences with specified properties. Alternatively, we can use large numbers of variants of best-guess-designed DNA or RNA sequences for further selection. Since it remains improbable that we shall soon be able to design DNA, RNA, or proteins optimally for desired functions, any biotechnology based on finding novel sequences with desired functions seems almost certain to use best-guess sequences followed by adaptive evolution for improved function among those showing a modicum of the desired activity.

Among the immediate obvious advantages of *in vitro* or *in vivo* cloning procedures over the examining of mixtures of nonamplifiable sequences is that the DNA, RNA, or protein sequence of interest can be amplified and selection carried out on it. For example, if the novel gene is cloned into a plasmid introduced into a host cell, the gene coding for a useful protein is identified simultaneously with the protein having an interesting function. Therefore, any such protein can be isolated from a mixture of novel proteins and produced in abundance. Furthermore, it becomes straightforward to utilize mutation and recombination in adaptive hill climbing to seek improved function.

Applied molecular evolution requires large-scale efforts and sophisticated use of population genetics theory. All the rugged landscape theory of the previous two chapters becomes immediately applicable. Simultaneously, the enormity of sequence space suggests that practical applied molecular evolution cannot be a timid undertaking. We must explore trillions, not hundreds, of sequences with high efficiency. I turn next to discuss this opportunity in more detail.

Four approaches are now being taken in exploring protein space. In the first, mixtures of random proteins or peptides are generated and analyzed for function. In the second, random fragments of genomic DNA are used to search for a specific function. In the third, defined mutant spectra around a current well-evolved protein sequence are introduced as replacements into that protein, and function is sought. In the fourth, entirely random, partially random, or pseudorandom DNA coding sequences are constructed and used to create libraries of random or partially stochastic proteins in the search for function. All approaches are promising.

Interest in the properties of random peptides or proteins is longstanding. There is wide evidence that mixtures of synthetic peptides exhibit a range of weak catalytic activities, ligand-binding activities, and so forth (Cavadore 1971). Further, examination of a small collection of protein fragments by Orgel (1987, personal communication) abets this impression. Very small peptides can exhibit catalytic activity. For example, tripeptides with tryptophane can recognize and cleave DNA at apurinic sites (Behmoaras, Toulme, and Helene 1981a, 1981b). In addition, as I shall discuss in more detail in Chapter 7, Fox and his colleagues (Rohlfing and Fox 1969, reviewed in Fox and Dose 1977 and Fox 1980) have for years studied the catalytic properties of quasi-random thermal proteinoid polymers made abiogenically by thermal-driven condensation reactions. Like many mixtures of random peptides, such mixtures of thermal proteinoids catalyze a number of reactions weakly. Yet the fundamental problem with mixtures of large numbers of synthetic peptides is that isolation of individual useful ones is impractical. Thus even if weak catalytic activity with respect to a given reaction is found in the mixture, it is difficult to know if a modest fraction of the members of the mixture exhibit weak activity or if only one or two exhibit strong activity. Thus ascertaining the probability that a given peptide or protein catalyzes a given reaction is difficult. Conversely, analysis of small collections of peptides to find one which may catalyze a given reaction is hampered by the fact that only a few candidates can be studied.

Random Genomic DNA Codes for Sequences with Function

Several workers have utilized restriction enzymes to cleave genomic DNA from bacteria, yeast, and even human cells, and then cloned such fragments into one or another expression vector and sought a specific function. The first surprise is that it has proved relatively easy to find the desired function. The major limitation of such studies should be mentioned immediately, however. Random genomic fragments code not for random peptides but for more or less random fragments of highly evolved proteins. Any distribution of function among such fragments, therefore, may bear little relation to the distribution of function in peptide–protein space. An obvious reason for this concern is the possibility that function requires stable folding, which might be rare in random peptides but already selected for in any fragment of a well-evolved protein.

The first reported effort along these lines, by Kaiser, Preuss, et al. (1987), concerned the search for DNA fragments which confer signal-peptide function. In more detail, many proteins are secreted from the cell which synthesizes them. Secretion is mediated by specific signal peptides which are at one end of the protein to be secreted. The cell recognizes the signal peptide and transports the entire protein molecule from the cell, cleaving off the signal peptide on extrusion. Kaiser, Preuss, et al. cleaved human DNA into small fragments with restriction enzymes, replaced the normal signal-peptide sequence of a protein with the DNA fragments, cloned the construct into an expression vector which ensured transcription and translation of the new chimeric protein + random peptide, and introduced the vector into host yeast cells. Fully 20 percent of the DNA fragments tried coded for peptides able to function as signal peptides. Sequence analysis of these peptides revealed essentially no homologies but demonstrated that most sequences had a high frequency of hydrophobic amino acids. The authors speculate that such hydrophobic residues help intrude the signal peptide into lipid membranes and hence aid transport from the cell. Whatever the underlying cause, a principle implication is that finding signal-peptide function in small genomic fragments is almost trivially easy.

In a similar series of experiments, Baker and Schatz (1987) showed that random digests of the *Escherichia coli* genome can functionally replace, at a similar frequency, the signal sequence involved in the mitochondrial targeting of yeast cytochrome *c* oxidase subunit IV.

A second exciting body of work has been carried out by Ptashne and his co-workers (Ma and Ptashne 1987) utilizing the Gal4 gene in yeast. This gene encodes a protein which binds to yeast DNA at a specific site and activates transcription of an adjacent yeast gene, presumably by "touching" a transcription factor related to transcription of that adjacent gene. Fragment analysis of the Gal4 protein demonstrated that it has two binding domains, one specific for the DNA site, the other for transcription activation. The latter site is at the carboxyl end of the protein. Ptashne's group replaced the coding region for the carboxyl part of the protein with random restriction fragments of *E. coli* DNA which were enclosed in an expression vector which was then transformed back into yeast cells. Analysis of 1500 yeast colonies revealed that 151 activated transcription of the adjacent gene. Thus an astonishing 1 percent of random genomic fragments from a bacterium code for a peptide which confers the capacity to activate a specific *yeast gene* when used as the carboxyl terminus of a "fusion protein," next to the DNA-binding part of the Gal4 protein. Since it is reasonable to suppose that such activation requires some kind of touching of a transcription factor and some kind of reasonably specific molecular recognition, we are left with the impression that finding such function in random fragments of evolved bacterial genes is easy. Analysis of the peptides, ranging from 14 to 80 amino acids in length, revealed a propensity to form amphipathic helices with positive charges on one face and negative charges on the other. Ginger and Ptashne (1987) then synthesized a completely synthetic sequence predicted to be amphipathic and demonstrated that it, too, activated transcription of the adjacent gene when used to make a fusion with the remainder of the Gal4 protein. Neither Kaiser, Preuss, et al.,

Baker and Schatz, nor Ptashne and co-workers have yet carried out mutational or recombinational hill climbing toward improved function in their experimental sequences.

Defined Mutant Spectra

A third approach to searching sequence space is to replace the active site of an enzyme with a mutant spectrum. This work follows more limited mutagenesis and analysis of catalytic activity (Raines, Straus, et al. 1986; Raines, Sutton, et al. 1986). In the spectrum approach, cloning techniques which replace ten to twenty amino acids with sequences containing an average of one, two, three, or more mutants are utilized. Very large numbers of such alternatives are generated and transformed into host cells which are deletion mutants for the enzyme under study, and then selection experiments are carried out to seek sequences which have the enzymatic function (Hermes, Blacklow, and Knowles 1987; Hermes, Blacklow, et al. 1987; Hermes and Knowles 1988). The first results (Hermes, Blacklow, and Knowles 1990) were obtained by screening 150 000 mutants of a sluggish triose–phosphate isomerase mutant, thereby covering more than 99 percent of the single-base changes and about 75 percent of the two-base changes. With a single-base change, on average only 5.7 of the 19 other amino acids can be encoded. Among two-base changes, on average 15.7 of the 19 other amino acids can be encoded. Thus these workers examined almost the entire one-mutant spectrum and parts of the two- and higher-mutant spectra. Among these 150 000, six second-site suppressor mutants which partially restored catalytic function were found. All map within the first or second shell of amino acids near the active site in the folded enzyme.

Interestingly, none of the six second-site suppressors is itself a local optimum. These workers have now examined all possible pairs of the six second-site suppressors, and all double-suppressor mutants have higher enzyme activity than any of the single-suppressor mutants. In some cases, the effect in the double-suppressor mutant is more than additive with respect to either single suppressor (Knowles, personal communication). One important implication of this, which precisely parallels our expectations derived from application of the NK model to maturation of the immune response, is that these suppressor mutants are on pathways that climb to *a variety of local optima that are different from* the optima found with the initial wild-type triose–phosphate isomerase. It should be possible to use the NK model, or an improved variant, to predict the number of local optima accessible via such adaptive walks from sluggish mutants.

Search for enzymatic function need not be confined to one- or two-mutant spectra. Dube and Loeb (1989) replaced the active site in a gene coding for beta lactamase, the enzyme that cleaves the beta-lactam ring of penicillin antibiotics, with sequences which preserve the codon for the active serine 70 but also contain 15 base pairs of entirely random sequences coding for 3.2×10^6 amino acid substitutions. From *E. coli* harboring these substitutions, the authors identified seven new active-site mutants, several mutants at more than one site compared with wild type. Oliphant and Struhl (1989) subsequently performed similar experiments in which they substituted the region from Arg[61] to Cys[77] with an average of three mutant codons per molecule, with results similar to those of Dube and Loeb. Gill, Sadowski, and Ptashne (1990) recently identified, scattered over a 65-residue region of Gal4, multiple amino acid substitutions that increase transcription rates over their wild-type levels. Again, these mutants may be on pathways to many alternative local optima.

The use of defined and biased mutant spectra is obviously a powerful way to explore adaptive walks to original or other local optima for a given catalytic function from defined mutant distances in sequence space (Straus, Raines, et al. 1985). Not only can the initial enzymatic activity be sought, but experiments evolving new catalytic activities can be undertaken. The strength of this approach is that it explores the vicinity of known mountains with known, more or less random peptide replacement sequences rather than with random fragments of evolved proteins, as described above. The weakness is that, in exploring near the tops of known mountains represented by well-evolved enzymes, this approach may give little insight into how widely spread weak enzymatic activity may be with respect to any specific reaction in sequence space. This issue requires full-bodied courage. We must use fully nonevolved and perhaps fully random peptides. I turn to this next.

Generating Fully or Partially Stochastic DNA: The Stop-Codon Problem, Optimizing Biological-like Sequences, and Optimizing Evolvability

Exploration of sequence space with respect to protein function can be based on either fully or partially stochastic sequences. Optimizing search involves confronting a number of choices. Among these, it may often be useful to minimize the frequency at which stop codons are encountered in random RNA sequences and truncate translation. It may typically be useful to ensure that the stochastic sequences code for all amino acids and, perhaps more critically, all pairs, triplets, quadruplets, and higher ordered sequences of amino acids. It may be useful to bias DNA or RNA sequences in favor of codons which are highly utilized by a host cell. It may be useful to bias sequences toward those yielding random proteins with size, charge, and hydrophobicity indices typical of evolved proteins. Finally, it may often be useful to construct sequence libraries which permit not only ready excision but also easy fragmentation and recombination both to increase the diversity of sequences under selection and also to ease the use of recombination in adaptive searches over rugged molecular landscapes. Alternative techniques to address these issues are available.

Three of the 64 triplet codons are stop codons. Consequently, if random polymers of the four nucleotides A, T, C, and G are formed, the mean interval between stop codons is on the order of 21 amino acids. One possible solution to this problem (Ballivet and Kauffman 1987, 1989) uses small sets of double-stranded oligomers which, when ligated in tandem, code for all 20 amino acids in both orientations, in all reading frames, without stop codons. A simple example is based on the use of six specific double-stranded octamer restriction enzyme linkers: EcoR1, Pst1, HindIII, Nde, ClaI, and SalI. Restriction enzyme linkers are double-standed DNA oligomers made up of eight to 12 nucleotides and having an internal dyad of symmetry; they are useful because each is also the site of cleavage of a specific endonuclease.

A practical procedure for cloning such copolymers of linkers is shown in Figure 4.5. It suffices to blunt-end-ligate a random mixture of the linkers, digest briefly with a restriction enzyme corresponding to one of the linkers to produce overhanging cut ends, separate the different size classes on a standard agarose gel, elute the size fragments desired, and then ligate them into a host vector which has already been cut with the same restriction enzyme, thereby creating complementary overhanging ends. The resulting plasmids, or viral DNA molecules—each containing a different random copolymer of linkers and hence an open reading frame coding for a novel peptide inserted into the cloning site—are then available for further analysis. The

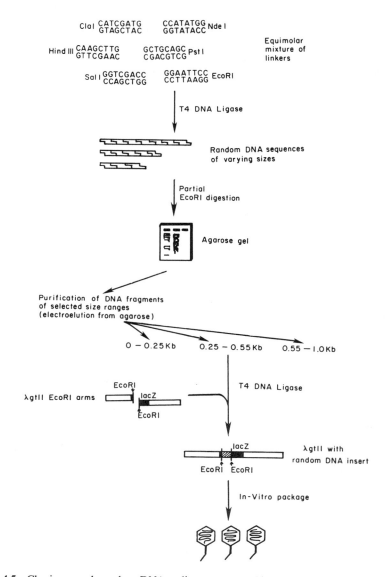

Figure 4.5 Cloning pseudorandom DNA coding sequences. Blunt-end-ligate a mixture of restriction enzyme linkers; cut with one of the restriction enzymes; size-separate the sequences on an agarose gel; ligate the cut sequences into a cloning site prepared in a vector such as lambda-gt11 by cutting with the same restriction enzyme; package the vector *in vitro* into infectious viral particles each containing a pseudorandom coding sequence; infect *E. coli* host cells to express the novel coding sequences in lytic or lysogenic phases.

procedure has several practical advantages. By using a unique linker at the 3′ and 5′ extremities of previously generated linker copolymers and then cloning into a site corresponding to the same linker, one generates open-reading-frame novel genes which can later be excised intact. Further, the linker copolymers, built of a smallish number of linker species, have internally repetitive motifs. This small number provides a simple and convenient method for *in vivo* mutagenesis of the novel gene sequences via recombination. When two viruses coinfect a host cell, recombination commonly occurs between homologous DNA sequences. Homologous recombination requires that a stretch of 12 to 20 nucleotides be identical. Because each linker is eight nucleotides long, a sequence of two in common between two different random copolymers in two different viruses allows recombination. Indeed, two linkers in tandem at two positions within one random novel gene allow recombination within that sequence.

These procedures have proved practical in our laboratory for generating a library with on the order of 10^{10} novel genes. The sequences were cloned into the EcoR1 site in beta-galactosidase in the lambda-gt11 vector. Using this vector, we confirmed that over 80 percent of the vectors contained inserted DNA sequences of the expected length distributions. Using "western blots" probed with antibody against beta-galactosidase, we confirmed that the experimental sequences code for beta-galactosidase proteins containing added amino acids. Insertion of novel sequences into the wild-type beta-galactosidase protein to create a fusion protein would be expected to decrease or destroy enzymatic function. Perhaps surprisingly, about 20 to 30 percent of our experimental sequences exhibited low levels of enzymatic activity on color indicator plates. Since only about 33 percent of the inserted DNA sequences would leave the downstream portion of the beta-galactosidase gene in the proper reading frame, these results suggest that most in-frame insertions permit at least a low level of enzymatic activity to persist in this fusion enzyme system. Detailed examination of western blots supports this. Foreign proteins synthesized in *E. coli* are subject to degradation by host cell enzymes. Fusion proteins which exhibited weak beta-galactosidase activity were substantially less degraded than those exhibiting no activity. This is consistent with the hypothesis that those with activity are in-frame insertions, hence less degraded, while those without activity are out-of-frame insertions and hence more readily degraded.

Linker-generated libraries are practical. Furthermore, because of their capacity for *in vivo* or *in vitro* recombination, it is straightforward to amplify the diversity still further. The advantages of this linker technique are clear. These libraries are now under investigation for catalytic and ligand-binding functions.

An alternative approach to the stop-codon problem would generate random coding sequences by use of specific triplets for each amino acid or for all 400 pairs of amino acids or for 8000 triplets of amino acids strung together in random order. For example, the 8000 codons for all ordered triplets, chosen among the redundant codons for each amino acid to favor those utilized by the host cell, can be synthesized and cloned. Such sequences would have open reading frames in one orientation and frame but stop codons in the remaining orientation and frames. Still another approach to the stop-codon problem (Scott and Smith 1990) uses random nucleotides in the first two positions of a triplet and only C or A in the third position. Based on a related idea, Mandecki (1990) has generated a library of random proteins having an average molecular mass of 30 kDa and capable of being expressed in *E. coli*.

Evolved proteins are unlike fully random sequences of amino acids with respect to average charge, hydrophobicity, and other features. T. LaBean, a graduate student in my laboratory, has developed algorithms which bias DNA sequences to optimize with respect to minimizing the stop-codon frequency and these other features (LaBean, private communication). A library of such random proteins is currently under investigation for folding, catalytic, ligand-binding, and other structural properties (LaBean, Butt, and Kauffman 1992; LaBean, Kauffman, and Butt 1992).

Other approaches to the generation of random DNA sequences have been proposed by Oliphant, Nussbaum, and Stuhl (1986). These authors used random DNA sequences to search for consensus sequences for specific regulatory functions as DNA sequences (Oliphant and Struhl 1987) and so were not concerned with the stop-codon problem.

Selection for Proteins Which Catalyze an Arbitrary Reaction

If a fundamental question is the probability that an arbitrary protein possesses a specific catalytic activity, then a direct approach is to use cloned novel genes in a very large number of host cells to select or screen for such catalytic activity. A straighforward example explores this (Ballivet and Kauffman 1987, 1989). Mutants of *E. coli* which lack beta-galactosidase and hence are unable to grow with lactose as sole carbon source are readily available. Transform a large number of such host cells so that each expresses a novel gene and synthesizes the corresponding novel protein. Attempt to select for capacity to grow either on lactose alone or on lactose-dominated medium. Any cells which survive and grow might do so either because the novel protein has the desired catalytic activity or because the host cell's genome has mutated to produce an altered enzyme with beta-galactosidase activity. These two conditions can be discriminated by reinfecting naive beta-galactosidase minus mutant bacteria with the novel gene. If survival is due to the novel gene, all recipients will fare well. If not, only rare mutants of the bacteria will survive. Thus repeated cycles either reinfecting or retransforming naive bacteria with the population of novel genes derived from survivors at each cycle will amplify the useful novel genes. Finally, it is necessary to demonstrate that the catalytic activity is due to the novel gene and not to another mutation which may have accumulated in the vector. This in general can be demonstrated by excising the novel gene, transferring it to a naive vector, and demonstrating that it confers the expected activity. Thus this process allows recovery of very rare useful novel genes.

Screening for a Protein Catalyzing Any One of Many Reactions

Demanding that any peptide catalyze one specific reaction places needlessly constricted requirements on any analysis of the distribution of catalytic capacities in protein space. It is obviously advantageous to have, as the target of screens or selection, as many different reactions as possible. For example, rather than demand that a novel protein catalyze lactose cleavage, we might demand that the protein act as an endonuclease and cleave double-stranded DNA at some internal site. Any long, fixed, double-stranded DNA has a very large number of sites. For example, if one considers eight nucleotide sites, there are $4^8 = 64\,000$ potential sites for cleavage. Thus in a long, specific random sequence of DNA, all such sites will be present, each a potential cleavage target. By radioactively labeling the sequence, exposing it to one or a mixture of potential protein catalysts, and then analyzing the DNA for cleavage fragments on agarose gels, cleavage of all such site can be screened in parallel. Similarly, in asking for ligand-binding capacity as well as catalytic activity, it might be sensible to select or screen in parallel for binding or catalysis at a wide variety of points in the system. For example, penicillin kills only dividing bacteria. Then, by exposing to penicillin a large library of bacteria, each expressing a novel protein, all those bacteria whose novel proteins interfere with any of the many processes needed for growth will be spared, the rest killed. That the sparing was due to the novel protein can be determined by transferring the novel gene to a new host cell.

Note that such procedures amount to a new kind of *additive mutagenesis* analysis. By inserting novel genes into a virus or cell and interfering with any specific function, one has a new way of probing for the components performing that function. The novel protein presumably binds to, cleaves, or interacts with some component in the underlying normal control system within the virus or cell.

A Toolbox of Rough Enzymes for All Catalytic Tasks?

In a previous section, I introduced the idea of a catalytic task space. Just as the immune system may have enough antibody diversity covering balls in shape space to saturate the space, so too enough enzymes may cover and saturate catalytic task space. Maturation of the immume response suggests that adaptive hill climbing can begin with a roughed-in antibody, which is then refined to match a given epitope shape. Catalytic antibodies complementary to the transition state of reactions have been discovered. All this leads to the idea that a finite number of roughed-in enzymes might be generated and selection procedures used to sharpen candidates for any catalytic task needed. This generation-selection process might proceed via cloning random DNA into the variable regions of antibody molecules, yielding an *in vitro* means of generating antibody diversity and taking advantage of the evolved capacity of antibodies to bind. Alternatively, a universal enzymatic toolbox might take the form of entirely novel proteins. In either case, we might guess that on the order of 10^6 to 10^9 such sequences would suffice. Those numbers are no longer at all beyond the possibility of discovery, analysis, and exploitation.

Just this approach has recently been taken by Huse, Sastry, et al. (1989). These workers generated a combinatorial V_L and V_H library in phage lambda. They utilized mRNA from a mouse immunized with KLH-coupled *p*-nitrophenyl phosphonamidate antigen and coding for Fab fragments. They amplified the mRNA derived from spleen cells or hybridomas with the polymerase chain reaction (PCR), cloned the V_L sequences into one library and the V_H sequences into a second library, and then united these combinatorially in a master library with a diversity of at least 10^8. The diversity of this phage library can be increased by reisolating the heavy- and light-chain sequences, recombining them into new combinations, and recloning. Thus the approach parallels the use of linker recombination in this regard. Diversity can also be increased by massive mutagenesis of the CDRs, which control detailed binding affinity. The diversity may also be increased as a result of copy errors made during the initial amplification via the PCR reaction; such errors therefore create stochastic diversity prior to cloning. These authors propose to use this system to find catalytic antibodies. If the concept of a saturable catalytic task space is correct, their system is already a selectable universal toolbox.

Requirements for Shape

What fraction of random peptides or polypeptides will fold into a stable shape? How does that fraction change as a function of peptide length? If we begin with a peptide which folds imprecisely and then carry out adaptive evolution by mutating sites in the peptide to produce variants which fold more precisely, can well-folded peptides—or larger polypeptides or proteins—be obtained? No one knows. Yet it would be hard to find more fundamental questions. These questions are directly accessible by making large numbers of random peptides of defined lengths, genetically or chemically, and then using a variety of means to look for shape. For example, Richardson (personal communication) noted with respect to this problem that the simplest approach would be to run a mixture of such random peptides on electrophoresis gels under nondenaturing and denaturing conditions. If a fraction of the polymers are folded, then their migration patterns under denaturing conditions should be different from the patterns under nondenaturing conditions. The density distribution shift would indicate the fraction with folding. Alternative bulk methods suitable for use on complex mixtures of random peptides or proteins include circular dichroism and hydrogen exchange. For any specific peptide, other physical techniques—such as nuclear magnetic resonance studies and, ultimately, attempts to crystallize the protein—are available. Adaptive evolution to achieve well-folded peptides or polypeptides would begin with a specific sequence showing some signs of folding and then carry out a mutation-selection adaptive walk (LaBean, Kauffman, and Butt 1992).

Mapping Shape Space and Its Practical Implications for Mimicking Arbitrary Shapes: Novel Drugs and Vaccines

As noted earlier, the distribution of shape features across the set of all organic molecules constitutes an obvious and fundamental problem. Conversely stated, how do organic molecules map to shape space? Since proteins are able to mimic carbohydrates and other epitopes, use of random proteins to establish their mapping to shape space is a step toward mapping organic molecules in general to shape space.

A sensible approach to answering this question is to use a specific monoclonal antibody molecule to probe a very large number of entirely novel small peptides to find those which are bound by that antibody molecule and then mutate the positive peptides and examine how binding affinity for the monoclonal antibody varies in the vicinity of each. Thus it becomes possible to investigate very large numbers of entirely different novel peptides, sampling wide regions of protein space, to test whether quite different primary sequences give rise to peptides which are all bound by the same monoclonal antibody. That is, we have a reasonably efficient way of beginning to map clouds scattered across protein space, all mapping to the same ball in shape space. A concrete example would consider the set of all $20^6 = 64 \times 10^6$ hexamers organized into a sequence space with each peptide next to those differing in a single amino acid. Clouds bound by the monoclonal antibody become one ball in this space.

Then, by using different monoclonal antibodies, we can begin to map the overlaps between clouds in hexamer space bound by one antibody and those bound by another. Since the set of peptides bound by each antibody molecule maps to a single ball in shape space, these overlaps should given an ordering relation among antibody molecules. The overlaps define which antibody molecules are complements to neighboring shapes in the space. Thus this same information can be used to assess which shapes are neighbors of one another. Concretely, the overlaps in binding by antibody molecules define which clouds in hexamer space, each corresponding to one ball in shape space, are near one another in shape space. It should ultimately be possible to use such balls to map shape space. Equally important, it should be possible, using this space, to see where the shape complement to each shape lies in the space. Using this information, one might know how to walk across shape space from an antibody binding to one shape to an antibody binding to another shape, via intermediate shapes in the space. Since those intermediates would be represented by a consecutive sequence of protein ligands for any sequence of antibodies or other liganding molecules, it would be clear how to choose a succession of epitope targets to move across the space. Such information would help guide adaptive evolution of catalytic antibodies—or other sequences—to novel reactions of interest.

Such experiments are also potential sources for new drugs and vaccines. The clear implication of the concept of complementary shapes in a shape space and of the fact that very different molecules can have the same local shape is that it is possible to find proteins which mimic the shape of other organic molecules. A direct approach to this is the following. Consider an arbitrary molecule of interest, say a hormone. Immunize and obtain polyclonal or monoclonal antibodies against the hormone. Each antibody has a shape complementary to some epitope on the hormone. Utilize the antibody molecules to screen a large library of novel peptides. Any peptides bound by the antibody molecules must have shape features similar to that of the initial hormone. Thus any such peptide is a plausible candidate to mimic those actions of the hormone which depend on that shared shape feature. Thus if the antigen is a hormone, the peptides may bind to its receptor and antagonize, agonize, modulate, or simulate the hormone. Such a peptide is a candidate drug. If the initial antigen happens to be a protective antigen on a pathogen, then the novel peptides mimicking the shape features of that initial antigen are candidates for a vaccine (Ballivet and Kauffman 1987, 1989; Parmley and Smith 1988; Scott and Smith 1990).

The immune system readily accomplishes the same feat of finding and improving mimetic proteins. The initial round of rank 1 antibody molecules which arise to bind the antigen are its shape complements. Those antibodies serve as antigens which engender rank 2 antibodies binding the rank 1 ones. Among the rank 2 antibodies, some have shapes which mimic the shape of the initial antigen: key-lock-key (Jerne 1974, 1984; Bruck, Co, et al. 1986). That the rank 2 antibody mimics the antigen shape is demonstrated by showing that the rank 1 antibody binds both the antigen and the rank 2 antibody by the same binding site.

Not only can rank 2 antibodies mimic the shape of an initial antigen, but examples are now known where these antibodies mimic antigen function. Examples include mimicry of biological receptors (Gaulton and Greene 1986). In other cases, the initial antigen is the hormone insulin, and the rank 2 antibody is able to induce insulin responses. Further, there is now an exciting series of results demonstrating that rank 2 antibodies can serve as vaccines against primary protein antigens (Bluestone, Sharrow, et al. 1981) and even against primary carbohydrate antigens (McNamara, Ward, and Kohler 1984; Stein and Sonderstrom 1984; Sacks, Kirchhoff, et al. 1985). This result is critical, for it unambiguously demonstrates that chemically different kinds of polymers—proteins and carbohydrates—can have the same local shape.

Since the immune system, with a diversity of 10^8, can, via the essentially random DNA sequences in the V region, utilize complementary shapes in a shape space spanning carbohydrates and proteins, entirely novel mimetic proteins can be found in the same fashion. The probability that an arbitrary epitope binds with modest affinity to an arbitrary antibody is about 10^{-5} (Press and Klinman 1974). Thus if each novel protein presents a single epitope, fewer than a million will reveal some novel proteins which cross-react with a monoclonal antibody to an arbitrary antigen.

A second estimate yields a similar number of proteins which need to be screened to find a protein mimic. Studies of peptide hexamers which are parts of a protein epitope (Geysen, Barteling, and Meloen 1985; Geysen, Rodda, and Mason 1986, 1987; Getsoff, Geysen, et al. 1987) show that about 100 variants bind the antibody. Since the number of possible hexamers is $20^6 \approx 64\,000\,000$, these results suggest that about 640 000 random hexamers, slightly fewer than 10^6, would have to be screened to find a mimetic peptide able to bind a given antibody.

Scott and Smith (1990) and Cwirla et al. (1990) have recently found just such mimetic peptides. Both groups cloned sequences coding for random hexapeptides into the protein coat of the filamentous fUSE5 factor. Phage which displayed the desired determinant bind antibody. They were selected from a background of non-binding phage by affinity purification. Scott and Smith screened, in theory, 21 percent of the $64 \times 10^6 \approx 64\,000\,000$ hexamers with two monoclonal antibodies specific for hexapeptide DFLEKI and found 19 fusion proteins with high affinity for the antibodies. Thus slightly more than one in a million random fusion proteins must be screened to find one which can cross-react. Importantly, none of the 19 is identical to DFLEKI. Indeed, on average, the novel sequences differ in three positions. Presumably, as in other such cases, all the fusion protein lie on pathways to different local optima with respect to mimicking DFLEKI. Cwirla et al. (1990) used the same phage system and selected for fusion proteins able to bind an antibody raised against the N terminus of beta-endorphin. They screened about 3×10^9 and found 51 cross-reacting proteins. None is like any of the known ligands of beta-endorphin. All are dissimilar to one another, except that all share a tyrosine at the N terminus and 48 share a glycine adjacent to it.

As a practical matter, it is now possible to mimic the shape—and hence the effects—of an unknown variety of biologically active molecules. This can only be useful in finding novel drugs and vaccines. For example, contemporary molecular biological techniques to create vaccines require cloning the DNA coding region of the pathogen which codes for part of a protective antigen. The cloned protein fragment is then used to immunize. This can work brilliantly but requires that the antigen, at a minimum, be a protein rather than a carbohydrate or another molecular species. Further, the pathogen, the protective antigen, and the gene coding for it must all be identified. In contrast, an effort to find novel proteins which mimic the shape of the protective antigen does not even require, in principle, that the pathogen be known, or that any gene coding for it be identified, or that the protective antigens be protein. All that is required is that high-titer antibody against the protective antigen be available for screening against a large battery of novel proteins. Given such antibodies, it should be possible to find mimetic proteins. Thus, using antibodies from an infected individual, it becomes possible, in principle, to find vaccines for diseases where the pathogen is not yet even known! The implications with respect to many orphan infectious diseases for which it is not now practical to manufacture vaccines looms large. Further, consider fashioning individualized treatment for autoimmune disease: Use a patient's own autoimmune serum to obtain mimetic proteins, which then become candidate drugs to bind to and titrate out the autoimmune antibodies. Scott and Smith (1990) have independently made similar points.

Applied molecular evolution includes procedures to evolve novel proteins binding specific cellular or biochemical targets, such as DNA regulatory sites (Ballivet and Kauffman 1987, 1989). Devlin, Panganiban, and Devlin (1990), using the same phage system as Scott and Smith but coding for 15 random amino acids, screened 2 $\times$ 10^7 phage for those binding to streptavidin, which had no previously known protein-binding activity. Nine binding fusion proteins were found, all shaping a four-amino consensus sequence but otherwise differing widely. The capacity to find and perfect novel proteins binding to arbitrary molecular targets is likely to be of great medical and applied interest in the near future. For example, it must be anticipated that proteins capable of regulating the transcription or translation of specific viral or cellular genes can be found and perfected by mutation and selection, that proteins capable of binding to components of the immune network and modifying their activities can be evolved, and that such liganding properties will help develop biosensors.

Selecting for Mimetic Proteins by Closure of an Autocrine Feedback Loop

Screening procedures to find mimetic proteins work, but *selective* procedures would be preferable. A selective procedure would provide a selective advantage to a cell which synthesized a protein mimicking some arbitrary hormone or pathogenic antigen. A general approach to this problem can be based on closure of autocrine feedback loops (Kauffman, patents pending). Many cells have specific receptors for specific growth factors. For example, many cells have such receptors for epidermal growth factor (EGF) and will multiply only in its presence. But EGF is a small peptide hormone. Suppose a cell with EGF receptors happens to synthesize and *secrete* EGF. Then that EGF will bind to the receptors on the same cell and stimulate that cell to divide continuously. This is called closure of an autocrine feedback loop. Such loop closure may often underlie the onset of cancer (Sporn and Todaro 1980).

Suppose we wished to find a random peptide which mimics EGF. Use a library of random DNA coding fragments to construct a library of expression vectors, each containing a random coding sequence adjacent to a signal-peptide coding sequence, and transform the library into cells containing surface EGF receptors. Incubate the cells under conditions such that only cells containing EGF proliferate. Then any EGF-mimicking peptide secreted by a transformed cell will bind to the receptors on that cell and trigger its proliferation. This offers a selective procedure to search for peptides mimicking EGF. Any cells which grow have a gene which encodes a random peptide which mimics EGF. Practical demonstration of this procedure has been achieved. Stern, Hare, et al. (1987) have constructed a vector which secretes EGF, transformed the vector into cells with EFG receptors, and showed that such cells secrete EGF, close the autocrine loop, and proliferate.

Clearly, closure of autocrine feedback loops generalizes to any protein mimic of a specific growth factor which stimulates cell division. Therefore, the most exciting potential use of this procedure is based on the realization that, for B lymphocytes of the immune system, the antibody molecule which any B cell secretes is present on the cell membrane at an early stage in the cell's history and acts as a specific receptor for a specific growth factor, and this factor is just the antigen which binds to the antibody molecule. That is, the antigen acts exactly as EGF does; each binds a specific receptor and triggers cell division. Therefore, a general procedure to select for arbitrary shape mimics is this: To mimic arbitrary epitope X, immunize with X and obtain immature B lymphocytes with anti-X antibodies as surface receptors. Transform this population of anti-X lymphocytes with a library of vectors, each causing secretion of a different random protein. Any protein which mimics X will bind to the antibody receptor on the cell which secreted it and stimulate that cell to divide. Thus the proliferating B cells contain genes coding for proteins which mimic X. More generally, stimulation of some B cells requires that antigen be presented to them by antigen-presenting cells, such as macrophages. Transform the macrophages such that they each process different random proteins for presentation, co-culture such macrophages with the immature B cells bearing anti-X antibodies, observe which B cells proliferate, and select macrophages in the vicinity of those B cells for further analysis.

Screening for Useful DNA and RNA Sequences

Applied molecular evolution also includes procedures for evolving useful DNA or RNA sequences (Ballivet and Kauffman 1987, 1989). For example, Horwitz and Loeb (1986; see review by Horwitz and Loeb 1988) substituted 16 or 19 base pairs centered at the -35 promoter element of the plasmid pBR322 tetracycline resistance gene with chemically and enzymatically synthesized random DNA sequences to produce a population of plasmid DNA in which each molecule contained a unique sequence. After transfection into *E. coli* and selection on tetracycline, the authors selected from among the 3×10^{11} possible sequences those which had promoter activity. Oliphant and Struhl (1987) subsequently carried out similar experiments. Among the 170 active sequences recovered, the resulting consensus sequence differed from the known 263 natural variants of the promoter. Interestingly, the strongest promotors departed from the consensus sequence.

Another procedure selects and amplifies single-stranded RNA or DNA sequences able to bind a protein target. The process rests on incubation of a complex mixture of single-stranded RNA or DNA with the protein target, then nitrocellulose-affinity selection of such polynucleotide-protein complexes from the mixture, followed by amplification of the selected RNA or DNA sequences (Ballivet and Kauffman 1987, 1989). For example, in an elegant use of this procedure, Tuerk and Gold (1990) affinity-purified, from the pool of 65 536 possible single-stranded RNA sequences containing a random eight-base-pair sequence, those able to bind T4 DNA polymerase. These authors generated the stochastic sequences, bound them to the polymerase, and selected nitrocellulose-bound complexes. DNA copies of the selected RNA sequences were made by reverse transcription, the copy DNA was amplified via the PCR reaction, and further RNA was obtained by transcription of the amplified DNA. The RNA sequences were then subjected to further rounds of affinity-purification and amplification. Two RNA sequences were found. One is the wild-type sequence; the other varies from the wild type in four positions and has nearly the same affinity for the polymerase.

In another recent experiment, Ellington and Sjostack (1990) generated stochastic single-stranded RNA sequences 100 nucleotides long, passed them over affinity columns with a diversity of dyes similar to metabolic cofactors, and found a number of sequences able to bind several dyes. The authors estimate that they have screened on the order of 10^{13} random RNA sequences and that the complexity of the pool of sequences which bind the dyes is on the order of 100 to 1000 sequences, although it may be as high as 100 000 sequences. Assuming they have recovered 1000 distinct binding sequences, the probability of success is about 1 in 10^{10}. At a minimal estimate, if their pool has 10^5 binding sequences, the probability of finding an RNA sequence which binds a specific dye is on the order of 10^8. These important results show that single-stranded RNA sequences capable of binding arbitrary organic molecules can be found. By screening for those binding to a stable analogue of the transition state of a reaction, these authors hope to find novel ribozymes. Presumably, like the antibody repertoire, RNA sequence space is enzymatically universal.

The experimental evolution of novel ribozyme activities has already taken place. Joyce (1989) has developed a procedure for selectively amplifying RNA sequences carrying out specific catalytic functions and, with Robertson (Robertson and Joyce 1990), selected among variants of a ribozyme those capable of catalyzing cleavage of a DNA sequence rather than an RNA sequence. Finally, Eigen (1985) has for several years been carrying out a variety of experiments evolving RNA sequences. He has been an early, influential, and highly imaginative proponent of the power of applied molecular evolution.

All these experiments are exciting in a variety of respects. Not the least, it is clearly important that a figure on the order of 10^8 arises in so many contexts. We find it in our estimate of the number of enzymes required to cover catalytic task space, as well as an estimate of the complexity of the immune repertoire thought to be able to cover shape space. Catalytic antibodies from a set of 10^8 may be a universal enzymatic toolbox covering task space. The probability of finding cross-reacting proteins to a single monoclonal antibody, and hence the probability of mimicking a single shape, are on the order of 10^{-6}; the probability of finding RNA sequences binding a dye is perhaps as high as 10^{-8} or as low as 10^{-10}. All these probabilities are small. The corresponding number of sequences to be searched is large but not at all beyond our power of manipulation—or the power of evolution to find and mold. The number of effectively different shapes, or of effectively different catalytic tasks, is vastly smaller than sequence space. Hence, function is highly redundant and far more readily available in sequence space than we have thought.

I emphasize again that development in this area toward large-scale applied molecular evolution is critical even in the more restricted domain of modification of known proteins to improve or alter the tasks they perform. Whether one is interested in catalytic antibodies or in *in vitro* evolution of novel enzymatic functions via mutant spectra, it will be necessary to generate very large numbers of mutants, search for improved variants, and carry out adaptive hill climbing on rugged landscapes by mutation, recombination, and selection. Thus we must soon develop insight into the structure of such fitness landscapes. The quest is not merely academic.

SUMMARY

The present chapter has looked at adaptive evolution in proteins with respect to the ligand-binding properties of antibody molecules, which improve during maturation of the immune response, and with respect to the improvement of ligand-binding properties of the lac repressor. We haved applied the NK model to maturation of the immune response with substantial success. The model captures major statistical features of adaptive landscapes with respect to affinities for ligand. As such, it supports the idea that the statistical features of such landscapes can be captured with relatively simple models. Equally, we must suspect that the NK model will account for statistical features of landscapes concerning the adaptive evolution of enzymatic function. Hence we can look forward to a new kind of statistical theory about the adaptive evolution of proteins that may be brought to bear on the comparative molecular data now being assembled. In addition, the NK model, or variants of it, can be expected to be useful in applied molecular evolution.

Biological evolution has required the emergence and improvement of biopolymers capable of performing either catalytic or ligand-binding functions. We do not yet know how improbable such functions are. However, recent results evolving mimetic peptides, novel ribozymes, and single-stranded RNA sequences which bind organic dyes indicate that function is far more probable in sequence space than we have suspected. This chapter has discussed a theoretical and experimental framework for exploring this broad question.

We have looked at data on improved catalytic function by examining the evolution of novel catalytic capacities by hill climbing from an enzyme for one task to a better enzyme for a neighboring task.

We have been led to formulate the ideas of shape space and catalytic task space, allowing definition of the ideas of neighboring shapes and neighboring catalytic tasks. The concept of a catalytic task space allows us to pose new questions. Different reactions with very different substrates and products may constitute the same catalytic task. Thus we must map the indefinitely large set of reactions to catalytic task space. Given such a task space, an enzyme catalyzes a ball of neighboring tasks. Protein differentiation is, then, evolution to cover a neighboring ball. Most striking, a finite number of enzymes, perhaps on the order of 100 000 000, might saturate catalytic task space and constitute a universal enzymatic toolbox.

The recent discovery of catalytic antibodies, obtained by selecting for those which bind a stable analogue of the transition state of a reaction, strongly supports the hypothesis of a catalytic task space. Roughly, a catalytic task corresponds to binding the transition state with high affinity and binding substrates and products with lower affinity. The immune repertoire in humans, with a diversity of about 10^8 antibody molecules, is thought to be capable of recognizing any antigenic epitope. The human antibody repertoire may therefore already be one universal enzymatic toolbox. RNA sequences may be another.

These considerations are not merely academic. The emerging field of applied molecular evolution has now made fully feasible the search for novel DNA, RNA, or proteins which perform a variety of functions. These possibilities are offered by our capacity to generate, screen, or select, and amplify wholly or partially stochastic DNA or RNA sequences of interest either in themselves or in terms of the proteins for which they code. This allows us to begin to explore protein space and make practical use of adaptive hill climbing and recombination in rugged fitness landscapes to evolve new drugs, vaccines, enzymes, ribozymes, biosensors, and DNA or RNA regulatory sites.

CHAPTER 5

Self-Organization and Adaptation in Complex Systems

Eighteenth-century science, following the Newtonian revolution, has been characterized as developing the sciences of organized simplicity, nineteenth-century science, via statistical mechanics, as focusing on disorganized complexity, and twentieth- and twenty-first-century science as confronting organized complexity. Nowhere is this confrontation so stark as in biology. Nowhere are new conceptual tools so deeply needed. In this chapter, I describe some of those tools.

Living systems—organisms, communities, coevolving ecosystems—are the paramount examples of organized complexity. The genomic system of any higher metazoan cell encodes on the order of 10 000 to 100 000 structural and regulatory genes whose joint orchestrated activity constitutes the developmental program underlying ontogeny from the fertilized egg. The human immune system harbors a repertoire capable of deploying on the order of 100 000 000 distinct antibody molecules in harmonized patterns. Neural systems, even in relatively simple organisms, enlist the joint parallel activities of perhaps billions of neurons to assess, categorize, and respond to the exterior and interior milieu. Each of these systems is the consequence of evolution. How, we must ask, can such wonderful systems emerge merely through random mutation and selection? For if Darwin told us that adaptation occurs through the gradual accumulation of useful variations, he has not yet told us what kinds of systems are capable of accumulating useful mutations. Nor has he addressed the issue of whether selection may play a role in attaining systems capable of adaptive evolution.

In this chapter, I seek the adaptation principles of such complex and beautiful ordered systems. I explore two components of those principles: self-organization and selection. First of all, contrary to our deepest intuitions, massively disordered systems can spontaneously "crystallize" a very high degree of order. Much of the order we see in organisms may be the direct result not of natural selection but of the natural order selection was privileged to act on. Second, selection *achieves* complex systems capable of adaptation. Moreover, I shall suggest that there are general principles characterizing complex systems able to adapt: They achieve a "poised" state near the boundary between order and chaos, a state which optimizes the complexity of tasks the systems can perform and simultaneously optimizes evolvability.

In this chapter, I first outline the mathematical theory of dynamical systems, for such theory provides the natural language needed to describe the "integrated behavior" of systems coordinating the actions of many elements. Here I introduce the concept of dynamical attractors, which may be either simple or chaotic. Dynamical attractors "box" the behavior of a system into small parts of its state space, or space of possibilities. Hence attractors literally are most of what a system does. It is the boxing of behavior into small parts of state space which constitutes much of the self-organization we shall encounter.

The second section of the chapter introduces random NK Boolean networks. This ensemble of networks permits us to study the emergence of order in systems coordinating the activities of thousands or even billions of elements. In NK Boolean networks, each element has two possible states of activity: active or inactive; a network links the activity of each of its N elements to the prior activities of K other elements. Random Boolean networks are a vast family of disordered systems. Yet we shall find that they exhibit three broad regimes of behavior: ordered, complex, and chaotic. In the *ordered* regime, many elements in the system freeze in fixed states of activity. These frozen elements form a large connected cluster, or *frozen component,* which spans, or *percolates,* across the system and leaves behind isolated islands of unfrozen elements whose activities fluctuate in complex ways. In the *chaotic* regime, there is no frozen component. Instead, a connected cluster of unfrozen elements, free to fluctuate in activities, percolates across the system, leaving behind isolated frozen islands. In this chaotic regime, small changes in initial conditions unleash avalanches of changes which propagate to many other unfrozen elements. These avalanches demonstrate that, in the chaotic regime, the dynamics are very sensitive to initial conditions. The transition from the ordered regime to the chaotic regime constitutes a phase transition, which occurs as a variety of parameters are changed. The transition region, on the edge between order and chaos, is the *complex* regime. Here the frozen component is just percolating and the unfrozen component just ceasing to percolate, hence breaking up into isolated islands. In this transition region, altering the activity of single unfrozen elements unleashes avalanches of change with a characteristic size distribution having many small and few large avalanches.

The second section of the chapter also examines self-organization. The transition from chaos to order in random Boolean networks occurs either as K decreases to 2 or as other parameters are altered in simple ways. Thus systems with millions of elements can crystallize order if each element is affected by only a few others. The emergence of order does not require that all details of structure and "logic" be controlled precisely. Hence a rich vein of order lies available for selection's further sifting. Indeed, in Chapter 12 we shall see that the spontaneous order seen in these networks accounts for many aspects of ontogeny in higher organisms.

Complex systems, just like simple peptides, must evolve by accumulating useful variations. In the chapter's third section, I examine the structure of fitness landscapes of networks in the ordered and chaotic regimes. The former correspond to smooth landscapes; the latter, to rugged landscapes. The implications of rugged landscapes for evolvability, limitations on selection, and adaptation seen in the *NK landscape* model of Chapters 2 and 3 remerge in these more complex systems. Moreover, it is a plausible hypothesis that Boolean networks near the edge of chaos have fitness landscapes whose ruggedness optimizes the capacity of the networks to evolve by accumulating useful variations.

In the final, fourth section, I take up the new and difficult question of the capacities of such large parallel-processing networks for performing complex tasks. We examine the attractive hypothesis that networks poised at the edge of chaos can perform the most complex tasks. Furthermore, we consider whether selection can achieve such poised systems. If both answers are yes, as they begin to appear to be, then we may have succeeded in discovering the characteristic kind of complex system which selection achieves in order to optimize both evolvability and fitness.

In order to focus in this chapter on self-organization and adaptation in complex systems, I defer to later chapters discussion of many of the biological implications. In particular, the topics addressed here will permit us, in the next chapter, to extend our thinking to coevolution in interacting systems. The analogues of percolating frozen components and avalanches of changes in Boolean networks reemerge as connected clusters of species with fixed optimal genotypes and avalanches of coevolutionary change and extinction events.

DYNAMICAL SYSTEMS AND THEIR ATTRACTORS

State Space and Simple Attractors

The most natural language for describing the behavior of an integrated system is dynamical systems theory. The most familiar form derives from Newton and consists in a system of differential equations in which the rate of change of each variable is written down in terms of the present values of all the variables which influence the variable in question. To be concrete, suppose there are three chemicals reacting in a vessel. The rate of formation and disappearance of each chemical depends on the concentrations of (1) those chemicals either forming it or influencing its formation and (2) those influencing its conversion to other chemicals. In addition, each of the three chemicals may be added to or removed from the vessel, or their concentrations may be held constant or caused to change in arbitrary ways by outside forces. The most natural representation of such a system is a three-dimensional *state space,* where each axis corresponds to the concentration of one of the three chemicals (Figure 5.1). The concentration of all three chemicals at any instant then corresponds to a single point in this three-dimensional space.

Over each small interval of time, the concentration of each chemical may increase or decrease slightly as a result of the various reactions possible. Therefore, after one small time interval, the representative point will have moved to some (in general) new point in state space. Over a succession of such small intervals, the representative point will move to a succession of (generally) new points. Thus the succession of points may be connected by a smooth line which shows the *trajectory* of the system across state space as time passes (Figure 5.1). Notice in the figure that the trajectory no longer explicitly shows time increments. (This qualitative description is slightly inaccurate mathematically. More precisely, at each point in the state space, the trajectory is tangent to a small arrow, or vector which shows the direction and rate of movement of the system at that point over the next infinitesimal time increment.)

If the system had been initiated at a different initial combination of concentrations, or *state,* it will follow some different trajectory. A general theorem for continuous systems of differential equations shows that only one trajectory passes through each point in state space (Hirsh and Smale 1974). This at a minimum implies that each state flows *deterministically* to a well-defined succession of other states. Further, the theorem implies that two states which lie on different trajectories will always, in

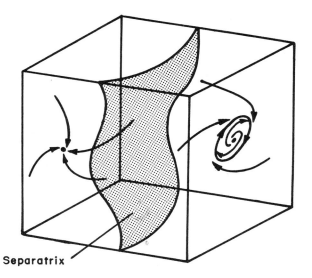

Separatrix

Figure 5.1   Three-dimensional state space with two basins of attraction, separated by a wall, or separatrix. Within each basin of attraction, trajectories flow deterministically toward an attractor and remain on the attractor thereafter unless disturbed by an outside perturbation. The attractor in the basin is a single point to the left of the wall corresponding to a steady state. The attractor in the basin on the right corresponds to an oscillatory limit cycle around which the sytem flows repeatedly.

the finite future, lie on different trajectories. Two trajectories do not merge into one in continuous dynamical systems. However, two states which were initially quite far apart and on different trajectories may come to be arbitrarily close to each other, for their respective trajectories may *converge*. In particular, different trajectories may converge toward a single state which does not change in time—that is, a steady state—reaching it in the limit of infinite time. Then the steady state is a zero-dimensional, or point, *attractor*, and an entire volume of states which lie on trajectories flowing to that attractor is its *basin of attraction* (Figure 5.1).

Many but not all dynamical systems have attractors. Among those which do not are the classical Hamiltonian systems of physics, exemplified by the frictionless pendulum. If released at any defined position and initial velocity, the pendulum swings on a periodic, closed orbit in its state space without loss of energy. If displaced to a slightly larger or smaller orbit by a perturbation, the pendulum follows a different closed, periodic orbit in its state space, with a slightly different energy. Each orbit is neutrally stable, for the system will remain in any orbit once placed there. No orbit drains a basin of attraction. Below we consider limit cycles, or periodic trajectories which do drain a basin of attraction. The existence of attractors in physical systems often requires some form of driving and friction which prevents conservation of energy within the system itself.

The idea of basins of attraction and steady-state point attractors is essentially the same as the idea of a mountainous region with hills, ridges, valleys, lakes, and a water-drainage system. Lakes correspond to point attractors; drainage basins, to the basin of attraction. Just as a mountainous region may have many lakes and drainage basins, so may a dynamical system have many attractors, each draining its own basin. Therefore, it is natural to conceive of the state space as being partitioned into *disjoint basins of attraction*. When released from an initial state, the dynamical system is on a trajectory lying in *only one* basin, and the system flows to that basin's attractor. This restriction means that each disjoint basin leads to only one attractor and thus that *the different attractors constitute the total number of alternative long-term behaviors of the system*. In due course, the system winds up on one or another of its finite number of attractors. Note also that, since attractors are typically much smaller than the volume of states in their basins, the system becomes boxed into an attractor unless perturbed by an outside force.

Attractors may or may not be stable to small outside perturbations. The image of a mountainous region helps immediately characterize the stability of the different kinds of attractors. A *stable* steady state is represented by a lake at the bottom of a valley. A drop of water displaced slightly in any direction will return to the lake as soon as the drop is free to move. An *unstable* steady state is represented by mountain peaks and by saddles between valleys. A drop of water will remain on a peak if left undisturbed but will flow away from the peak if perturbed in any direction. (Strictly speaking, an unstable state is not an attractor, since the only state within its basin of attraction is the steady state itself.) A drop of water in a saddle will return if displaced up the hills on either side of the saddle but will flow away if displaced toward either valley below. In other words, saddle steady states are stable with respect to perturbations in some directions and unstable with respect to perturbations in other directions. The existence of saddles reflects the fact that two stable basins of attraction must abut, and the ridge where they do is a basin of attraction having one fewer dimension than the number of variables in the system and is called a *separatrix* (Figure 5.1). In two dimensions, the ridge is a line separatrix. If the system is released on that ridge line, it flows down the ridge to the saddle steady state. If perturbed off the ridge, the system flows to the corresponding valley. Beyond these fundamental properties of stability, the trajectories may also approach a stable steady state either without spiraling around it or by spiraling into it. Similarly, the system may depart from an unstable steady state either without spiraling around it or in ever-widening spirals.

These qualitative descriptions have precise mathematical formulations (Hirsh and Smale 1974). The stability of any steady-state point attractor is analyzed by linearizing the differential equations about the steady-state point and then assessing the characteristic way small displacements away from the steady state change in time. If small displacements in one or more directions increase, the steady state is unstable in the corresponding directions. The behavior of the displacement is given by the eigen values of the linearized equations: stable if those eigen values are all negative, unstable if any is positive, and spiraling if any is complex. In Chapter 14, I shall examine this issue in more detail.

More Complex Attractors

The general definition of an *attractor* is a set of points or states in state space to which trajectories within some volume of state space converge asymptotically over time. Thus, in addition to simple steady states, continuous dynamical systems may admit of more complex attractors. The simplest of these is a *limit cycle,* or hoop of states. If released on the hoop, the system flows around the hoop repeatedly. Over time, the variables exhibit a repetitive oscillation. The hoop is called a limit cycle because points not on it lie on trajectories which spiral either in or out and ultimately converge on it in the limit of infinite time (Figure 5.1). Thus a stable limit cycle drains some basin of attraction. Just as a steady state is a zero-dimensional attractor in an N-dimensional state space, a limit cycle is a closed one-dimensional attractor in higher-dimensional state space.

Antilimit cycles exist, having the property that a point released exactly on an antilimit cycle will remain on it but, if displaced off the limit cycle, will be on a trajectory which spirals away from the limit cycle. Thus antilimit cycles are unstable with respect to small perturbations off the antilimit cycle.

It is easy to see that donutlike, or toroidal, attractors can exist. Suppose the dynamical system has a total of ten chemicals. Suppose four chemicals set up one oscillation and six set up another oscillation; then in the ten-dimensional state space, the representative point moves on two hoops at once. This two-hoop movement can be represented by motion on a donut, with flow in one direction passing through the hole and flow in the other direction being around the donut but not passing through the hole. This simultaneous flow produces a spiraling trajectory which winds both through the hole and around the donut. If the periods of the two oscillations are a ratio of whole numbers, say 3 to 1, then the trajectory comes back exactly to its initial point after three of the faster oscillations. If the relative periods are irrational, then the point representing the whole system winds through the hole and around the donut at an irrational angle to the donut's natural axes and eventually covers an arbitrary amount of an entire two-dimensional surface of the donut. Thus this is a two-dimensional attractor in a ten-dimensional state space. In such a flow, motion is quasi-periodic: After sufficient time, the trajectory will pass arbitrarily close to any initial point. Higher-dimensional toroidal attractors exist.

In addition to these classes of attractors, *strange, or "chaotic," attractors* exist (Lorenz 1963; Ruelle 1979; Grassberger and Procaccia 1983; Mayer-Kress 1986). In such a dynamical system, which might be ten-dimensional, the flow might, for example, bring all trajectories onto a two-dimensional attractor a bit like a Moebius strip with a pleat or some other folded form (Figure 5.2). The interesting property of such attractors is that, if the system is released from two points on the attractor which are *arbitrarily close* to each other, the subsequent trajectories remain on the attractor surface but *diverge away* from each other. After a sufficient time flowing on the attractor, the two trajectories can be arbitrarily far apart on it.

The first critical novel feature found in strange attractors but not in steady states, limit cycles, and so forth is due to this divergence of trajectories on the attractor. It is a *sensitivity to initial conditions* (Ruelle 1979). Tiny differences in initial conditions make vast differences in the subsequent behavior of the system. In contrast, a system with a stable limit cycle squeezes all flows onto the same hoop of states; hence nearby initial points are still nearby later on. A good example of this feature in a strange attractor, and historically the first, is provided by the weather or, more certainly, by a mathematical model of the weather introduced by Lorenz (1963), where small alterations in initial conditions lead to vast differences in due course (Figure 5.2). this sensitivity to initial conditions is amusingly called the butterfly effect. A butterfly in the Amazon might, in principle, ultimately alter the weather in Kansas. Dorothy's trip to Oz rode on tiny wings. This effect is the basic reason long-term forecasting is notoriously weak.

The second feature to notice about strange attractors is that they may be of very low dimensionality even in a high-dimensional state space. Thus a system may have 100 variables, but flow may be restricted to a strange attractor of two dimensions, a folded surface closing back on itself in that 100-dimensional space. From the point of view of the entire state space, the attractor is a very small object indeed. The system is boxed into a tiny volume of state space even though its behavior within that small

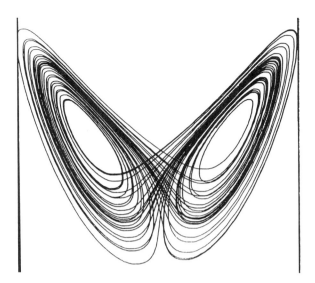

Figure 5.2   The Lorenz strange attractor which models aspects of weather. States not on the attractor flow to it and then both diverge and converge on the attractor. The local divergence leads to sensitivity to initial conditions. (From Holden 1986)

volume is chaotic in the precise sense of high sensitivity to initial conditions. This point is very important to stress, for the behavior of such a system which exhibits *low-dimensional chaos* is much more orderly than the behavior of a system which wanders through vast tracts of state space on *high-dimensional attractors.*

The dimensionality of a strange attractor is often not an integer. Rather, it is natural to define a *fractal dimension* (Mandelbrot 1977) for the attractor, which might be 2.3 for an attractor which occupies more than two but fewer than three dimensions in the 100-dimensional space (Packard, Crutchfield, et al. 1980; Farmer 1982; Farmer, Ott, and Yorke 1983; Mayer-Kress 1986). The definition of dimension depends on how the density of points on the attractor changes with radius in all 100 dimensions away from any arbitrary point on the attractor. The intuitive idea is simple. Consider a hoop limit cycle in a three-dimensional state space. The attractor is one-dimensional. Pick a point on the attractor, and examine a sphere of radius r around this point. The volume of the sphere increases as r^3, but the number of points on the limit cycle and within the sphere increases only as r. The ratio of rates of increase of attractor points within the sphere allows one to determine that the attractor is a one-dimensional object at the point measured. The generalization of this idea provides a measure of the dimensionality of an attractor at each point on the attractor and hence an average dimensionality for many points on it. Such fractal attractors are already being found in biological systems—for example, in cardiac and neural electrical activity patterns (Holden 1986; Mackey and Glass 1988).

Parameter Space and Dynamical Bifurcations as Parameters Change Smoothly

We need another fundamental idea from dynamical systems theory: parameter space. The behavior of a chemical system depends on how rapidly chemicals are added or removed, on temperature, on solvent, and so forth. These values are typically held constant in time and are the *parameters* of the dynamical system. If a system has some number of parameters, then we must conceive of a *parameter space* where each axis corresponds to one parameter. A point in parameter space then reflects a specific combination of all parameter values. Any point in parameter space corresponds to a fixed set of parameters and thus to a fixed set of basins of attraction and attractors in the corresponding state space of the dynamical system. The set of basins of attraction is often called the *basin portrait* of the dynamical system.

The next question to address is this: What happens if the parameters are changed gradually? The key idea is that, as parameters change slowly, the trajectories and attractors typically change slowly also. That is, the basin portrait of the system alters smoothly. For particular changes of the parameters, however, sudden dramatic changes in trajectories and attractors can occur. Such sudden changes are called *bifurcations* in the behavior of the dynamical system (Hirsh and Smale 1974).

A familiar example of bifurcation is the onset of turbulence in a fluid flow. A gentle stream flowing past a rock shows smooth laminar flow. As stream velocity increases, eddies begin to form. An eddy either is or is not present in the fluid flow: not present when velocity is slow but present above some critical velocity (Nicholis and Prigogine 1977). Thus the character of the flow changes suddenly.

A second famous but less familiar example is the onset of Benard cells (Nicolis and Prigogine 1977). If a pan of liquid is heated gently from below, water warmed at the bottom of the pan is less dense than water higher up and tends to move upward gently. But if the heating is more vigorous so that the temperature difference through the fluid from bottom to top becomes great, then there are formed rising columns of hot water surrounded by descending columns of cooler water. Together, these columns form convective cells. Viewed from the top of the pan, beautiful roll or hexagonal patterns can be seen. This is a bifurcation at a critical value of the temperature difference from bottom to top surfaces. Below that parameter value, convective cells do not form; above it, they do. Incidentally, this is the first example we have mentioned in which energy flux through an initially homogenous system sets up a spatially ordered pattern. This idea has received enormous attention, with implications in pattern formation in embryos, to which we shall return in Chapter 14.

Bifurcation Surfaces Partition Parameter Space into Disjoint Volumes

The immediate implication of bifurcations is this: *Each* point in parameter space specifies the dynamical system defined in the differential equations and hence also specifies its state space and all its basins of attraction. As the parameters change slowly, some or all of the trajectories and basins of attraction also change slowly. For specific values of the parameters, dramatic changes occur. For example, a basin might contract to nothing, or a new basin might appear. The values of parameters at which bifurcations occur therefore divide parameter space into disjoint volumes. Crossing from one volume to another causes a bifurcation. If there are three parameters, then the three-dimensional parameter space is, in general, divided by two-dimensional surfaces into three-dimensional subvolumes. Crossing these two-dimensional surfaces causes bifurcations in the dynamical behavior. In general, an N-dimensional parameter space is partitioned by $N - 1$ dimensional surfaces.

Bifurcations are profound features of dynamical systems with profound implications in biology. For example, in Chapter 14, where we discuss morphogenesis, we shall find that bifurcations in the behavior of developmental mechanisms readily account for the evolutionary generation of families of related forms. Such families of "normal" forms go some real distance toward fulfilling the rational morphologist's dream of laws of form. In the present context, where we are interested in ways of characterizing the integrated behaviors of dynamical systems, the existence of parameter space and bifurcations in dynamical behavior carries implications about the ruggedness of adaptive landscapes, as described next.

The Concept of Structural Stability

The concept of *structural stability* (Thom 1970, 1972; Hirsh and Smale 1974) concerns the idea that, typically, volumes in parameter space defined by bifurcation surfaces are like soap bubbles. The volumes are reasonably large relative to the bifurcation surfaces which divide them. Thus for *most changes* in the parameters, the system remains within one volume in parameter space and the dynamical behavior does not change dramatically. Dynamical systems having this property are said to be structurally stable. Their dynamics typically changes only slightly as parameters change but does jump crossing bifurcation surfaces. For some time it was thought that almost all dynamical systems exhibit this property. However, recent work on strange attractors indicates that a large class of systems does not exhibit structural stability. In many systems with strange attractors, tuning the parameters leads to a succession of bifurcations at successively smaller intervals in parameter space (Mackey and Glass 1988). Such is seen in the famous period-doubling bifurcations studied by Feigenbaum (1978), which may underlie the onset of turbulence in fluid flow. In these cases, the volumes separating qualitatively different behaviors become sinuous, intertwined labyrinths in parameter space. Here, tiny changes in parameters in almost any direction can lead to successive dramatic changes in the dynamical behavior of the correlated dynamical system.

One implication of the occurrence or nonoccurrence of structural stability is that, in structurally stable systems, smooth walks in parameter space lead to mostly smooth changes in dynamical behavior. In the third section of this chapter, I discuss the idea that adaptive evolution, or learning in dynamical systems, is achieved by adaptive walks through parameter space to find "good" dynamical behavior. Thus structurally stable systems adapt on correlated landscapes. By contrast, chaotic systems, which are not structurally stable, adapt on uncorrelated landscapes. Very small changes in the parameters pass through many interlaced bifurcation surfaces and so change the behavior of the system dramatically. The link between order, chaos, and landscape structure will loom below as a large issue. In particular, it immediately becomes obvious that alterations in systems that cause the systems to pass from ordered to chaotic behavior will inevitably alter the statistical structure of their fitness landscapes from smooth to rugged. We might imagine, in prospect, that an intermediate ruggedness optimizes evolvability. It now appears that that optimum may lie near the boundary between order and chaos.

SPONTANEOUS ORDER AND CHAOS IN COMPLEX DYNAMICAL SYSTEMS

I turn in this section to a new body of theory about the behavior of dynamical systems containing extremely large numbers of coupled elements. In setting the stage for our analysis, I might first make comparative allusion to statistical mechanics, which until recently provided the best examples of systems in which large numbers of variables interact. I described in Chapter 2 the outlines of statistical mechanics with respect to perfect gases. Each of N particles of gas in a box interacts with all the other particles via the Newtonian force laws. The theory develops by considering the phase space of all possible $6N$ position and momentum variables describing the current state of the N particles in a box containing them. For a system released at a point in phase space, the flow under the drive of collisions is an ergodic wandering over phase space. Because the system conserves energy, the phase volume occupied by a flowing cloud of points, representing the system started at neighboring points in phase space, remains constant as it flows. From this constancy and ergodicity, it becomes possible to calculate the probability that the system is in any specific volume of phase corresponding to some macroscopic state, such as all particles being in a corner of the box.

The contrast with biological systems exhibiting ordered complexity is enormous. First, in genomic, immune, neural, or other evolved systems, the law governing the behavior of one element is not the same as the laws governing the behaviors of other elements. Biochemistry admits of a variety of molecular couplings which control and alter catalytic or ligand-binding actions in complex and diverse ways. The behavior of one element, enzyme, antibody molecule, or neuron as its regulatory inputs change in general differs from the behavior of another element. Second, unlike closed physical systems, which conserve energy, biological systems are open thermodynamically, typically dissipate energy, and have attractors. Third, the regular unfolding of ontogeny alone suffices to say that biological systems cannot wander randomly and ergodically over their space of possibilities. The essence of development from the fertilized egg is its astounding combination of complexity and utter regularity. Fourth, the most profoundly random aspect of biological systems is random mutation in the *space of possible systems*. That is, evolution is an adaptive, or drifting, process which searches across the space of biological systems. Thus unlike statistical mechanics, which can be characterized as a more or less ergodic flow within the state space or phase space of a single system, evolution is a more or less adaptive flow across a space of systems. What we need, therefore, is a new kind of statistical mechanics, one which analyzes the properties of ensembles of complex systems with very many coupled elements. By understanding the characteristic structure and behaviors of the members of such ensembles, we may be able to understand both the emergence of order in organisms and its adaptive evolution. This section develops aspects of such a new statistical mechanics.

Discrete Dynamical Systems: Introducing Boolean Dynamical Networks

Boolean networks are systems of binary variables, each with two possible states of activity (on and off), coupled to one another such that the activity of each element is governed by the prior activity of some elements according to a Boolean switching function.

Switching Boolean networks are of central importance to the construction of a statistical mechanics over ensembles of systems and to an adequate theory of complex but ordered systems. I collect the reasons for this:

1. We are concerned with dynamical systems containing thousands or millions of coupled variables. These variables might be active or inactive genes coupled in a genetic regulatory cybernetic network (Kauffman 1969, 1971a, 1971b, 1971c,

1974, 1984a), the linked cellular and molecular components of the immune system and idiotype network (Jerne 1984; Kaufman 1988), the interacting polymers in an autocatalytic polymer system (Farmer, Kauffman, and Packard 1986; Farmer, Packard, and Perelson 1986; Kauffman 1986b), or the interacting neurons in a neural network (McCulloch and Pitts 1943; Hopfield 1982a, 1982b; Rummelhart and McClelland 1986). The idealization to on–off switching elements allows us to study such enormously complex systems. The corresponding problems are often intractable using continuous nonlinear differential equations.

2. For many systems, the on–off Boolean idealization is either accurate or the best idealization of the nonlinear behavior of the components in the system.

3. Use of the Boolean idealization implies that a well-defined ensemble of all possible networks exists. Given well-defined ensembles, the averages of structural and behavioral properties over each defined ensemble can be assessed. Like temperature or pressure in familiar statistical mechanics, the statistically expected averages of such structural and behavioral properties are the macroscopic observables explained by the new statistical mechanics.

4. Disordered complex Boolean networks, it has turned out, exhibit three major regimes of behavior: ordered, complex, and chaotic. Thus analysis of these extremely complex systems reveals unexpected simplicity with important biological implications for development and evolution.

5. Use of the Boolean idealization allows us to pose and answer the question, What are the requirements for collective order in very complex switching networks?

6. The same properties which ensure orderly dynamics and hence spontaneous order simultaneously yield systems which adapt on highly correlated fitness landscapes.

7. Because disordered Boolean networks exhibit three broad regimes of behavior and can pass from one regime to another by relatively simple alterations in a few parameters characterizing each network, selection can readily tune networks into each regime, thereby controlling spontaneous order, complexity of behavior, and evolvability.

Positive Cooperativity, Sigmoidal Response Functions, and the On–Off Idealization

A short example demonstrates why a Boolean, or on–off, idealization captures the major features of many continuous dynamical systems. Many cellular and biochemical processes exhibit a response which changes an S-shaped, or sigmoidal, curve as a function of altered levels of some molecular input (Monod, Changeux, and Jacob 1963). For example, hemoglobin is a tetrameric protein, with each monomer binding oxygen. But the binding behavior of the four monomers exhibits *positive cooperativity:* Binding of oxygen by one monomer increases the affinity of the remaining three monomers for oxygen. This cooperativity implies that as oxygen levels increase from a base level, the amount of oxygen bound by hemoglobin increases faster than linearly at first. At sufficiently high oxygen concentrations, however, all four monomers have almost always bound an oxygen, and so further increases in oxygen concentration do not increase the amount bound per hemoglobin molecule. In short, the response saturates. This means that a graph of bound oxygen concentration (output) as a function of free oxygen concentration (input) is sigmoidal (Figure 5.3).

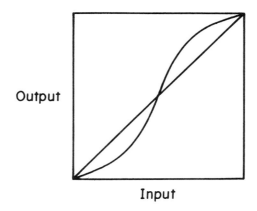

Output

Input

Figure 5.3 A sigmoidal curve and a 1:1 proportional response curve. Note that, for low values of input, output on the sigmoidal curve is lower than input. For input levels above the point where the sigmoidal curve intersects the 1:1 response curve, output is higher than input. Crossing of the two curves separates these two response regions.

The vital issue is to realize that even with a soft sigmoidal function—that is, one whose maximum slope is less than vertical—*coupled systems* governed by such sigmoidal functions are often properly idealized by on–off systems. It is easy to see intuitively why this might be so. In Figure 5.3, I graph both a sigmoidal function and a constant, or proportional, response curve, where output is equal to input—in other words, the slope is 1.0. The sigmoidal function is initially *below* the proportional response. Here a given input leads to an output that is less than the input. Were that reduced output fed back as the next input, then the subsequent response would be even less. Over iterations, the response would dwindle to zero. The sigmoidal response becomes steep in its midrange, however, and crosses above the proportional response. An input above this critical crossing point leads to an output that is greater than the proportional-response output. In turn, were that output fed back as a next input, the output would be still greater than that input. Over iterations, the response would climb to a maximum. That is, feedback of signals through a sigmoidal function tends to *sharpen to an all-or-none response* (Walter, Parker, and Ycas 1967; Glass and Kauffman 1972; Hopfield and Tank 1986a, 1986b). This is the basic reason the on–off idealization of a flipflop in a computer captures the essence of its behavior. We show this next for a small system of two variables each required to activate or synthesize the other.

Among the common mathematical expressions used in biochemistry to capture sigmoidal responses are Hill functions of the form

$$Y = \frac{X^N}{\theta^N + X^N} \tag{5.1}$$

where N is called the Hill coefficient (Mackey and Glass 1988) and θ is a threshold and constant. The curve showing Y as a function of X is sigmoidal. As N increases, the midregion slope of the curve becomes steeper, approaching the all-or-none Heavyside step function in the limit as N goes to infinity.

Consider a small dynamical system with two hypothetical chemical variables, X

and Y. Suppose the synthesis of each depends positively on the concentration of the other by a Hill function law and that each is lost at a rate (K_1 or K_3) proportional to its concentration. Then the differential equations for this system are

$$\frac{dX}{dt} = \frac{Y^2}{\theta^2 + Y^2} - K_1 X \tag{5.2a}$$

$$\frac{dY}{dt} = \frac{K_2 X^2}{\theta^2 + X^2} - K_3 Y \tag{5.2b}$$

Intuitive insight into the behavior of this system is simple: Neither X nor Y can be synthesized if the other is not present. Thus the state $X = 0$, $Y = 0$ must be a steady state. At the other extreme, if both X and Y are present in high concentration, each induces the synthesis of the other, and we expect the pair to remain in abundant presence. (I assume some kind of external supply of precursors to X and Y, of course.) Thus there should be a steady state in which both concentrations are high. For intermediate levels of X and Y, then, it seems plausible that the system will go either to the low steady state or to the high steady state. This is essentially correct. A one-dimensional line separatrix divides the XY state space into two alternative basins of attraction flowing to these two stable steady states. The separatrix is a locus of points flowing in to an unstable saddle steady state on the separatrix. We see this next.

A convenient means of analyzing the system represented by Equations 5.2a and 5.2b is to set both equations equal to zero. Doing so corresponds to seeking the concentrations of X and Y such that, for Equation 5.2a, the concentration of X is not changing and, for Equation 5.2b, the concentration of Y is not changing. Such a locus of points is called a *nullcline*. The X and Y nullclines are

$$\frac{1}{K_1} \left(\frac{Y^2}{\theta^2 + Y^2} \right) = X \tag{5.3a}$$

$$\frac{1}{K_3} \left(\frac{K_2 X^2}{\theta^2 + X^2} \right) = Y \tag{5.3b}$$

Graphing these two equations yields two S-shaped curves in the XY state space (Figure 5.4a). Since each curve represents the locus of states where one of the two variables is unchanging, places where the two curves cross are steady states of the entire system. The curves cross in three locations: a low steady state, $X = Y = 0$; a high steady state; and an intermediate steady state. The lower and upper steady states are stable to perturbation. The middle steady state lies on a separatrix. If it is on the separatrix, the system flows to the middle steady state. That state is a saddle, however. If displaced toward either stable steady state, the system flows to the corresponding attractor. Thus this small system of coupled sigmoidal functions has two extremal stable steady states and one unstable interior steady state.

Next, consider two on–off Boolean logical functions $X = Y$ and $Y = X$. The equality $X = Y$ asserts that, if Y is 1 at a given moment, X will be 1 at the next moment; if Y is 0 at a given moment, X will be 0 at the next moment. Similarly, $Y = X$ asserts that if X is 1 at a given moment, Y will be 1 at the next moment; if X is 0 at a given moment, Y will be 0 at the next moment. This small logical system has two steady states: $X = Y = 0$ and $X = Y = 1$. These two states correspond to the two extremal steady states of the continuous system.

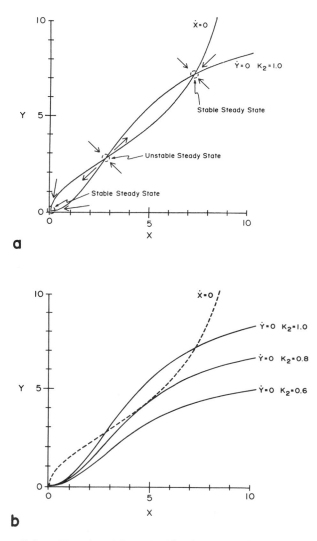

Figure 5.4 The nullclines (Equations 5.3*a* and 5.3*b*) of the dynamical system with two coupled variables, each activating the othe according to a Hill function. The $\dot{Y}$ and $\dot{X}$ represent dy/dt and dx/dt, respectively. Parameter values are $K_1 = K_3 = 0.1$, $K_2 = 1.0$, $\theta = 20$. For $K_2 = 1.0$, the two nullclines cross at three points, each a steady state. The middle point corresponds to an unstable saddle point lying on a separatrix between two basins of attraction. Within each basin, trajectories flow to the enclosed stable steady state. In one extreme stable state, both X and Y are 0; in the other, both are high. This reflects the fact that the two variables jointly activate one another. The separatrix is therefore a threshold line in the state space. Below it, X and Y go to 0; above it, X and Y go to high steady levels. (*b*) As in (*a*). As K_2 decreases to 0.8 and 0.6, the Y nullcline (solid lines) shifts to become tangent to the X nullcline (dashed line) and no longer crosses the X nullcline in three places. Thus as K_2 decreases past 0.8, a bifurcation occurs and eliminates the upper and middle steady states.

The small continuous system in Equation 5.2 can show bifurcations in terms of the parameters K_1, K_2, K_3. Tuning K_2 from 1.0 to 0.8 to 0.6 (Figure 5.4b) causes the nullclines of Equation 5.3 to shift position relative to each other such that the two curves cross only at a single point. Then for those values of the parameters, the system has only a single stable steady state. The bifurcation occurs as parameters change when the two nullclines shift so that, rather than crossing, they are tangent at a point. Then the two adjacent steady states have merged to one and will vanish when the two nullclines shift slightly farther so that they do not touch at all in that vicinity. Boolean functions can capture the consequences of bifurcations by altering the logical rule computed by an element.

Figure 5.5 shows a slightly more complex dynamical system with three variables—X, Y, and Z—in which X is activated by Y and Y is activated by either X or Z via cooperative sigmoidal functions. The X and Y nullclines cross, yielding three steady states, when $Z = 0$. The lower and upper steady states are stable, as in Figure 5.4a. Increasing Z, however, activates Y and shifts the Y nullcline, causing a bifurcation which eliminates the lower steady state. If Z remains high, the XY system flows to and remains at the remaining (high) steady state. Thereafter, if Z is lowered to 0, the system will remain in the upper stable state. In short, transient activation of Z can irreversibly switch the XY system from the low steady state to the high one.

These behaviors of a continuous nonlinear dynamical system are captured by the corresponding Boolean system. The variables X and Y activate each other, and Y is also switched on by Z. If $Z = 0$, the XY system has two steady states: $X = Y = 0$ and $X = Y = 1$. If $Z = 1$, then the low XY steady state disappears and the XY system remains in the high steady state. If Z is transiently 1, then the XY system irreversibly switches to the high steady state, $X = Y = 1$. The Boolean XY system remains active, $X = Y = 1$. Note that, as in Figure 5.4, the unstable intermediate steady state of the continuous XYZ system does not show up in the Boolean idealization.

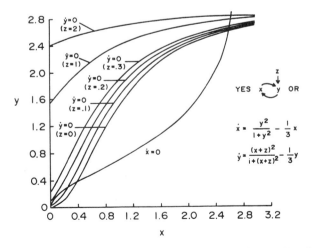

Figure 5.5 Nullclines of a three-variable system in which two variables, X and Y, activate one another but one of the two is controlled by a third, independent variable Z. Note that, as the third variable changes from low to high to low, the bistable XY system can be switched irreversibly from one to the other steady state.

Dynamical behaviors such as these occur frequently in cellular and developmental biology. For example, the activities of many receptor proteins and enzymes are controlled by phosphorylation and dephosphorylation cascades. Like the system in Figure 5.5, these cascades often have feedback loops with alternative stable maintained patterns of phosphorylation which can be irreversibly switched into one of the patterns by an exogenous input.

In summary, *logical switching systems* capture major features of a homologous class of nonlinear dynamical systems governed by sigmoidal functions because such systems tend to *sharpen* their responses to *extremal values of the variables.* The logical networks can then capture the logical skeleton of such continuous systems. However, the logical networks miss detailed features and in particular typically *cannot represent the internal unstable steady states* of the continuous system. Thus Boolean networks are a caricature, but a good one, a powerful idealization with which to think about a broad class of continuous nonlinear systems as well as switching systems in their own right. I stress that it is now well established that switching systems are good idealizations of many nonlinear systems (see, for example, Glass and Kauffman 1972; Glass and Pasternack 1978a, 1978b; Glass and Perez 1982; Hopfield and Tank 1986a, 1986b). But characterizing the class of nonlinear systems which are homologous to switching networks remains a large mathematical problem.

The State-Space Dynamics
of Autonomous Boolean Networks

Boolean networks are made up of binary, on–off variables. A network has N such variables. Each variable is regulated by some of the variables in the network, which serve as its inputs. The *dynamical behavior* of each variable, whether it will be active (1) or inactive (0) at the next moment, is governed by a *logical switching rule, or Boolean function.* The Boolean function specifies, for each possible combination of current activities of the input variables, the activity of the regulated variable at the next moment. For example, an element with two inputs might be active at the next moment if either one or the other or both inputs are active at the current moment; this is the Boolean "Or" function. Alternatively, the element might be active at the next moment only if both inputs are active at the present moment; this is the Boolean "And" function.

Let K stand for the number of input variables regulating a given binary element. Since each element can be active or inactive, the number of combinations of states of the K inputs is just 2^K. For each of these combinations, a specific Boolean function must specify whether the regulated element is active or inactive. Since there are two choices for each combination of states of the K inputs, the total number of Boolean functions F of K inputs is

$$F = 2^{2^K}$$

The number of possible Boolean functions increases rapidly as K increases. As we shall see, special subclasses of the possible Boolean functions are important for the emergence of orderly collective dynamics in large Boolean networks.

If a network has no inputs from outside the system, it is considered to be autonomous and its behavior depends on itself alone. Such a network is specified by choosing, for each binary element, which K elements will serve as its regulatory inputs and assigning to each binary element one of the possible Boolean functions of K inputs.

Figure 5.6a shows an autonomous Boolean network made up of three elements. Each receives inputs from the other two. Element 1 is governed by the "And" function; 2 and 3 are governed by the "Or" function. The simplest class of Boolean networks is *synchronous,* which means that all elements update their activities at the same moment. To do so, each element examines the activities of its K inputs, consults its Boolean function, and assumes the prescribed next state of activity. This is summarized in Figure 5.6b. Here I have rewritten the Boolean rules. Each of the 2^3 possible combinations of activities of the three elements corresponds to one state of the network. Each state at one moment causes all the elements to assess the values

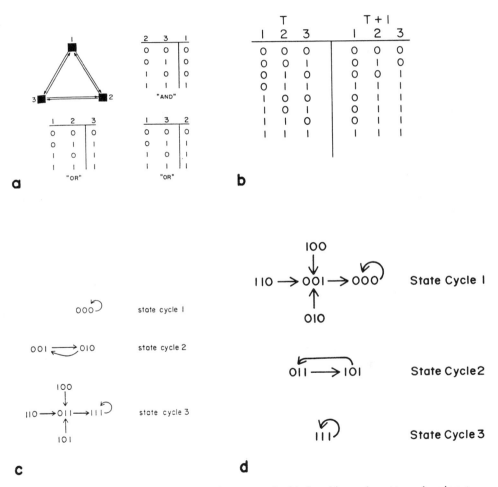

Figure 5.6 (a) The wiring diagram in a Boolean network with three binary elements, each an input to the other two. (b) The Boolean rules of (a) rewritten to show, for all $2^3 = 8$ states at time T, the activity assumed by each element at the next time moment $T + 1$. Read from left to right this figure shows the successor state for each state. (c) The state transition graph, or behavior field, of the autonomous Boolean network of (a) and (b), obtained by showing state transitions to successor states connected by arrows. (d) Effects of mutating the rule of element 2 from "Or" to "And."

of their regulatory inputs and, at a clocked moment, to assume the proper next activity. Thus, at each moment, the system passes from a state to a unique successor state. Over a succession of moments, the system passes through a succession of states called a *trajectory*. Figure 5.6c shows these trajectories.

The first critical feature of autonomous Boolean networks is this: Since there is a finite number of states, the system must eventually reenter a state previously encountered; thereafter, since the system is deterministic and must always pass from a state to the same successor state, the system will cycle repeatedly around this recurrent *state cycle*. These state cycles are the *dynamical attractors* of the Boolean network. The set of states flowing into one state cycle or lying on it constitutes the *basin of attraction* of that state cycle. The *length* of a state cycle is the number of states on the cycle and can range from 1 for a steady state to 2^N.

Any such network must have at least one state cycle attractor but may have more than one, each draining its own basis of attraction. Further, since each state drains into only one state cycle, the set of state cycles are the dynamical attractor of the system, and the cycles' basins *partition the* 2^N state space of the system.

The simple Boolean network in Figure 5.6a has three state cycle attractors (Figure 5.6c). Each is a discrete *alternative recurrent asymptotic pattern* of activities of the N elements in the network. Left to its own, the system eventually settles down to one of its state cycle attractors and remains there.

The stability of attractors to minimal perturbation may differ. A minimal perturbation in a Boolean network consists in transiently flipping of the activity of an element to the opposite state. Consider Figure 5.6c. The first state cycle is a steady state, or state cycle of length 1 (000), which remains the same over time. Transient flipping of any element to the active state—for instance, to (100), (010), or (001)—causes the system to move to one of the remaining two basins of attraction. Thus the (000) state cycle attractor is unstable to any perturbations. In contrast, the third state cycle is also a steady state (111), but it remains in the same basin of attraction for any single perturbation (011), (101), or (110). Thus this attractor is stable to all possible minimal perturbations.

A structural perturbation is a permanent "mutation" in the connections, or Boolean rules, in the Boolean network. In Figure 5.6d, I show the result of mutating the rule governing element 2 from the "Or" function to the "And" function. As you can see, this alteration has not changed state cycle (000) or state cycle (111) but has altered state cycle (101). In addition, state cycle (000), which was an isolated state, now drains a basin of attraction and is stable to all minimal perturbation, while (111) has become an isolated state and is now unstable to all minimal perturbations.

To summarize, the following properties of autonomous Boolean networks are of immediate interest:

1. The number of states around a state cycle (called the *length* of the cycle). The length can range from one state for a steady state to 2^N states.

2. The number of alternative state cycles. At least one must exist, but a maximum of 2^N might occur. These are the permanent asymptotic alternative behaviors of the entire system.

3. The sizes of the basins of attraction drained by the state cycle attractors.

4. The stability of attractors to minimal perturbation, flipping any single element to the opposite activity value.

5. The cascade of changes in dynamical behavior (called *damage*) caused by transiently altering the activity of a single binary variable.

6. The changes in dynamical attractors and basins of attraction due to mutations in the connections and the Boolean rules. These changes will underlie the character of the adaptive landscape on which such Boolean networks evolve by mutation to the structure and rules of the system.

Boolean networks are discrete dynamical systems. The elements are either active or inactive. The major difference between a continuous and a discrete deterministic dynamical system is that, in a discrete system, two trajectories can merge. To be concrete, Figure 5.6c shows several instances where more than one state converge on the same successor state.

Attractors carry strong biological implications and hence warrant our attention. Because many complex systems harbor attractors to which the systems settle down, the attractors literally are most of what the systems do. The remaining states in state space are visited only along transients leading to such attractors. Dynamical systems ranging from genomic cybernetic systems to immune systems, neural networks, organ systems, communities, and ecosystems all exhibit attractors. The alternative attractors in a genomic regulatory network, for example, can be interpreted as the alternative cell types in the organism (Kauffman 1969, 1974, 1986a, 1986c, 1986d; Chapter 12); the alternative attractors in immune networks correspond to different immune states (Jerne 1974, 1984; Kaufman, Urbain, and Thomas 1985; Farmer, Packard, and Perelson, 1986; Kaufman and Thomas 1987; deBoer 1988; Hoffman, Kion, et al. 1988; Kaufman 1988; Perelson 1988; Sieburg 1988); the alternative attractors in neural networks have been interpreted as alternative memories or categories by which the network "knows" its world (McCulloch and Pitts 1943; Little 1974; Hopfield and Tank 1986a, 1986b; Rummelhart and McClelland, 1986; Derrida, Gardner, and Zippelius 1987; Kurten 1988a, 1988b); the alternative attractors in cardiac systems correspond to normal and abnormal rhythms (Mackey and Glass 1988); the alternative attractors in ecosystems correspond to alternative stable patterns of species abundances (May 1976). Thus the characteristics of attractors in complex systems with hundreds, thousands or millions of interacting elements are inevitably of basic importance in both development and evolution. It is just the expected characteristics of such attractors in complex systems which we can evaluate in ensembles of Boolean networks.

Chaos and a Phase Transition to Order in Random Boolean Networks

In the first chapter, I discussed the role of the second law of thermodynamics, statistical mechanics, and the gas model in tuning our intuitions about the need for outside work if systems are to achieve order. This model is conceptually connected with the idea that selection is the sole source of order in evolution. However, the second law really states that any system will tend to the maximum disorder possible, *within* the *constraints* due to the dynamics of the system. The dynamical laws of gases with perfect elasticity ensure ergodicity. That is, the trajectory in state space wanders randomly all over state space without collecting onto small attractors. An enormous amount of outside work must be done to box such a system into small subvolumes of its state space. If, by contrast, many other dynamical systems can have basins of attraction draining to small attractors, then such systems *spontaneously* box themselves into small volumes of their state spaces. Such systems spontaneously exhibit order in the absence of outside work. Thus the possibility of the emergence of small attractors is of deep importance for biology. The order which might characterize cell types, immune responses, categorization in neural networks, and stability in coevolving systems will lie at hand for further employment by natural selection.

Boolean networks coordinating the mutual activities of thousands of elements can yield vastly ordered dynamics. Our intuitions have been wrong. Evolution has had unexpected order as the very groundstuff it might utilize from the outset. Here I summarize the behaviors of Boolean networks as a function of N, the number of elements in the net; K, the average number of inputs to each element in the net; and P, which measures particular biases on the subset of the possible $(2^2)^K$ Boolean functions used in the net. We find that, as K decreases from N to 1, a phase transition occurs at $K = 2$. For $K > 2$, behavior is chaotic. At $K = 2$, order crystallizes. The same phase transition from chaos to order occurs as P is changed smoothly. Later we shall focus on the transition region between order and chaos. In this regime, complex dynamics arises. We shall find grounds for thinking that the ordered regime near the transition to chaos is favored by, attained by, and sustained by natural selection. Then the generic properties of complex parallel-processing networks in this regime will emerge as possible biological universals.

In order to assess the expected influence of these parameters, I have analyzed the typical behavior of members of the ensemble of Boolean networks specified by any fixed values of N, K, and P. The first results I describe allow no bias in the choice of Boolean functions; hence N and K are the only parameters. I further simplify and require that each binary element have exactly K inputs.

To analyze the typical behavior of Boolean networks with N elements each receiving K inputs, it is necessary to sample at random from the ensemble of all such networks, examine their behaviors, and accumulate statistics. Numerical simulations to accomplish this therefore construct exemplars of the ensemble entirely at random. Thus the K inputs to each element are first chosen at random and then fixed, and the Boolean function assigned to each element is also chosen at random and then fixed. The resulting network is a specific member of the ensemble of NK networks.

I therefore stress that random NK Boolean networks are examples of strongly disordered systems (Kauffman 1969, 1984a, 1986a; Fogelman-Soulie 1984, 1985a, 1985b; Derrida and Pommeau 1986; Derrida 1987a, 1987b; Stauffer 1987a, 1987b; Flyvberg and Kjaer 1988). Both the connections and the Boolean functions are assigned at random. Were any such network examined, its structure would be a complex tangle of interactions, or "input wires," between the N components, and the logical rule characterizing the behavior of one element would typically differ from its neighbors in the network. Such Boolean networks are spiritually similar to spin-glasses and to the NK family of landscapes introduced in Chapters 2 and 3. Here, however, we generate networks with random wiring diagrams and random logic, and ask whether orderly behavior emerges nevertheless. Note that such behavior is occurring in a *parallel-processing network*. All elements compute their next activities at the same moment. If we find order in random networks, then random parallel networks with random logic have order despite an apparent cacophony of structure and logic.

Table 5.1 summarizes the salient features for the following cases: $K = N$, $K > 5$, $K = 2$, $K = 1$.

$K = N$: *The Grand Ensemble.*

This ensemble of networks is maximally disordered. Since each element receives an input from all other elements, there is only one possible wiring diagram. The Boolean function assigned to each element is chosen at random from among the 2^{2^N} possible logical rules. Therefore, this ensemble corresponds to the largest possible ensemble of Boolean networks of N variables. Any

TABLE 5.1   Properties of Random Boolean Nets for Different Values of K

	State cycle length*	Number of state cycle attractors	Homeostatic stability	Reachability among cycles after perturbation
$K = N$	$0.5 \times 2^{N/2}$	N/e	Low	High
$K > 5$	0.5×2^{BN} $(B > 1)$	$\sim N \left[\dfrac{\log \left(\dfrac{1}{\frac{1}{2} \pm \alpha} \right)}{2} \right]$; $\alpha = p_{(K)} - \frac{1}{2}$	Low	High
$K = 1$	$(\pi/2)^{1/2}(N^{1/2})$	Exponential in N	Low	High
$K = 2$	$\sqrt{N}$	$\sqrt{N}$	High	Low

*Column 1: State cycle length is median number of states on a state cycle. Column 2: Number of state cycle attractors in behavior of one net. ($\alpha = P_K - 1/2$, where P_K is mean internal homogeneity of all Boolean functions on K inputs; see text). Column 3: Homeostatic stability refers to tendency to return to same state cycle after transient reversal of activity of any one element. Column 4: Reachability is number of other state cycles to which net flows from each state cycle after all possible minimal perturbations, due to reversing activity of one element.

ensemble with $K < N$ or with constraints, for example, P, on the subset of Boolean functions is a subset of this grand ensemble.

Since each element is assigned at random a Boolean function on N elements, the successor to each state in these maximally disordered sysetms is a completely random choice among the 2^N possible states. For this reason, this ensemble is sometimes called the *random-map model.* Networks in which $K = N$ have received increasing attention (Kauffman 1969, 1971a, 1971b, 1974, 1984a, 1986a, 1986c, 1986d; Wolfram 1983, 1984; Gelfand and Walker 1984; Coste and Henon 1986; Derrida and Flyvberg 1987b; Derrida and Bessis 1988). In part, the interest reflects the analytic simplicity of this extremely disordered case, but in part it reflects the fact that even this maximally disordered system exhibits unexpected order. While attractors are very long and unstable to most perturbations, these systems have few alternative attractors.

Chaotic behavior in these Boolean networks shows up in two major ways: the lengths of state cycles and sensitivity to initial conditions. Table 5.1 shows that the median expected length of state cycles is $0.5 \times 2^{N/2}$.

Showing that cycle lengths scale as $\sqrt{2^N}$, where 2^N is the total number of states, is quite easy and instructive. Let $2^N = M$. Start with an initial state and pick its successor at random. The chance that the state picks itself as a successor and forms a loop is $1/M$. Thus the chance that the chosen successor state is some other state is $1 - 1/M$. The chance that the second state chooses either the initial state or itself as *its* successor state and forms a state cycle is $2/M$. Hence the chance that the second state fails to form a state cycle and that the sequence continues to a third state is $1 - 2/M$. Therefore, over a succession of L steps, the probability P_L that the sequence continues without forming a state cycle is

$$P_L = \Pi \left(1 - \frac{1}{M} \right) \left(1 - \frac{2}{M} \right) \left(1 - \frac{3}{M} \right) \cdots \left(1 - \frac{L}{M} \right) \qquad (5.4)$$

By an argument familiar from Chapter 2, as L increases, this product decreases and eventually falls below 0.5. At that value of L, half of such sequences will have closed upon themselves to form state cycle loops, and half will not have. The value of L is a median estimate of when cycles will form. It is easy to see that, when a cycle does form, the feedback loop is as likely to end on any one of the L states as on any other. Hence, on average the cycle will be about $L/2$. Some simple algebra leads to the result

$$\text{Median cycle length} = 0.5\sqrt{M} = 0.5\sqrt{2^N} \qquad (5.5)$$

State cycle lengths of $0.5\sqrt{2^N} = 2^{N/2}$ are vast as N increases. For $N = 200$, for instance, lengths average $2^{100} \approx 10^{30}$. At a microsecond per state transition, it would require about a billion times the age of the universe to traverse the attractor. Here is surely a "big" attractor wandering through state space before finally returning. I shall call such attractors, whose length increases *exponentially* as N increases, *chaotic*. This does *not mean* that flow on the attractor is divergent, as it is in the low-dimensional chaos discussed earlier. A state cycle is the analogue of a one-dimensional limit cycle.

Networks in which $K = N$ also exhibit maximum sensitivity to initial conditions. Because the successor to each state is randomly chosen, perturbing a state by flipping the activity of a single element will lead, a moment later, to a successor state which is totally random with respect to the successor state which would have followed the unperturbed state. Thus minimal changes yield enormous differences in activity patterns virtually immediately.

Despite these signs of chaotic behavior, the random-map model exhibits one startling sign of order: The number of cycles, and hence the number of basins of attraction, are small, N/e. Thus a system containing 200 elements would have only about 74 alternative asymptotic patterns of behavior. More strikingly, a system containing 10 000 elements and chaotic attractors with median lengths on the order of $2^{5\,000}$ would harbor only about 3700 alternative attractors. This is already an interesting intimation of order even in extremely complex disordered systems.

Two other features of $K = N$ random-map models are important. First, the sizes of basins of attraction are not uniform. Rather, a few enormous basins drain to a few large state cycle, and many small basins drain to smaller state cycles. Second, because successor states to each state are completely random, the stability of any state cycle to perturbations is just equal to the size of its basin of attraction. This equality implies that most attractors are unstable to many perturbations and that each attractor is *next to* all other attractors in the sense that some minimal perturbation, transiently reversing the activity of one binary element, will move the network from the first attractor into the basin of any of the other attractors. Thus the network can reach any attractor by minimal perturbations to each attractor.

K $\geqslant$ 5: *Internal Homogeneity in Boolean Functions, the* P *Parameter, and a Mean-Field Approach.* The main feature of $K = N$ networks, a modest number of chaotic attractors, persists as the number of inputs per binary variable decreases to five and perhaps to three. Analytic results are available for networks with about five or more inputs per element. Numerical results are available for $K = 4$ and $K = 3$. The exponential rate at which attractors lengthen is small for small values of K and increases to $N/2$ as K approaches N. The analytic results depend on what physicists call a mean-field approximation.

The expected lengths of state cycle attractors depend on the extent to which trajectories converge in state space. If two trajectories only rarely converge, cycles will tend to be very long. If there is high convergence in state space, attractors will tend to be shorter. This intuition can easily be made useful by defining the *internal homogeneity P* of a Boolean function.

Any Boolean function of K input variables has 2^K possible combinations of values for these variables. For each of these 2^K possibilities, the function assigns a 1 or a 0 value as the next activity of the regulated element. A Boolean function might have all 1 values (tautology), all 0 values (contradiction), or, more typically, about 50 percent 1 values and 50 percent 0 values. The internal homogeneity P of a Boolean function is defined as the fraction of the 2^K positions with 1 values or with 0 values, whichever is greater than 50 percent (Gelfand and Walker 1984). For example, a Boolean function with an internal homogeneity of .8 has either 80 percent 1 values or 80 percent 0 values in the 2^K positions specifying the function.

Without loss of generality, consider a Boolean network with $K = N$ but *biased to* $P = .8$ such that all functions have the 1 value in 80 percent of the 2^N positions. Then clearly a specific central state $(11111 \ldots 11111)$ is a preferred state. This state is the successor to very many states which therefore converge onto it. Indeed, the number of states which converge onto the central state is enormous:

$$\text{Convergence} = .8^N 2^N = 1.6^N \qquad (5.6)$$

These $K = N$ networks are *biased random mappings,* biased by P. Given this fixed value of $P = .8$, we can obtain a lower bound estimate of median state cycle lengths as a function of N and P by assuming that all states have convergence upon themselves as high as the central state (Kauffman 1984a). Then an argument like that leading to Equations 5.4 and 5.5 yields

$$\text{Expected median cycle length} = (0.5)\left(\frac{1}{\sqrt{P}}\right)^N \qquad (5.7)$$

Let $B = 1/\sqrt{P}$. Then, since $B > 1$, cycle lengths increase exponentially as N increases:

$$\text{Median cycle length} = 0.5B^N \qquad (5.8)$$

A critical implication of Equation 5.8 is that no *fixed* internal homogeneity P, and hence no corresponding convergence in state space alone, will suffice to ensure that state cycles remain small as N grows large. (Here I mean by "small" that state cycles do not grow faster than, say, linearly as N increases.)

The insights just gained can be extended from $K = N$ networks biased by internal homogeneity P to *unbiased* networks with $K \geq 5$. The main idea is simple. If we consider Boolean functions of $K = 1, K = 2, K = 3, K = 4, \ldots$ inputs, each defines a set of 2^{2^K} functions. For each function, we can define its internal homogeneity P. Hence we can define the average internal homogeneity of all Boolean functions with $K = 1, K = 2, \ldots$ inputs. It can be shown (Kauffman 1986d) that, as a function of K, the mean internal homogeneity P_K is

$$P_K = \frac{1}{2} + 2^{-(2^K+1)} \cdot \binom{2^K}{2^{K-1}} \qquad (5.9)$$

Internal homogeneity reaches a peak at $K = 2$ and declines thereafter as K increases (Table 5.2). The reason is straightforward. In $K = 2$ networks, eight of the 16 Boolean functions have either three 0 values and one 1 value or three 1 values and one 0 value. These functions have $P = .75$. As K increases, functions with such high values of P become rare, and so the mean value of P declines.

Mean internal homogeneity P_K as a function of K can be used to obtain a lower

TABLE 5.2 Internal Homogeneity P as a Function of Inputs K

K	P
1	.5
2	.6875
3	.6367
4	.5982
5	.5699
6	.5497
7	.5352

bound on expected median state cycle lengths in networks with random choices for Boolean functions of K variables. We know from Equation 5.8, where $K = N$, that a fixed P allows prediction of a lower bound on state cycle lengths. When $K = N$, the particular positions in any function where 1 or 0 values occur are completely random, biased only by P. For $K = 5$ and $N = 50$, to take a concrete case, the positions where 1 and 0 values occur in each function depend on the particular five input elements to each element and hence are *not* random. But let us accept a rough argument: When $K \geqslant 5$, the positions in the function, determined by random choices of the five inputs to each element, are nearly randomly scrambled. Then, using this rough mean-field approximation, we can use Equation 5.9 to calculate values of B (Table 5.3). Then, using these values of B in Equation 5.6, we can compute expected cycle lengths for values of K which are moderately large, up to $K = N$. The results, of course, show that, for any fixed K, $K = 5$ to $K = N$, *state cycle lengths* still increase *exponentially* as N increases.

This argument is a mean-field approximation. It is therefore important that numerical simulations fairly strongly tend to confirm it (Figure 5.7) (Gelfand and Walker 1984). Note that, for $K \geqslant 5$, observed cycle lengths increase exponentially as N increases. Further, the expected lengths of state cycles predicted by the mean-field theory for $N = 20$ are reasonably close to observed results, shown in Figure 5.7 (Table 5.4). When we use the mean-field approximation, the theory should be more accurate as K increases. Clearly the estimate is too high for $K = 3$; by $K = 6$ and $K = 7$, observed results are close to theory.

The important result is this: Random Boolean networks with a modestly high number of inputs per element, say $K = 5$ or $K = 10$, and randomly chosen Boolean functions have state cycle attractors whose lengths increase exponentially as N

TABLE 5.3 Exponential Growth Rate B of Attractors as a Function of K and N

K	B
3	1.2565
4	1.2929
5	1.3246
6	1.3487
7	1.3669

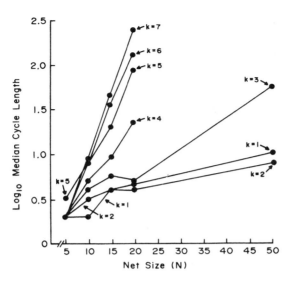

Figure 5.7 $\log_{10}$ median cycle lengths in random networks with K ranging from 1 to 7. Cycle lengths grow less than linearly for $K = 1$ and $K = 2$ but exponentially for $K > 3$. The slopes are predicted moderately well by the mean-field theory. (From Gelfand and Walker 1984)

increases. The exponential rate of growth increases as K increases toward a limit of $N/2$.

While state cycles are long for $K \geqslant 5$, disordered networks have few attractors. Fully random genomic systems with $K = N$ have only N/e distinct attractors. Recent analytic results strongly indicate that for $N \geqslant K \geqslant 5$, the number of attractors is at most a linear function of N and increases to N/e as K approaches N. Letting $a = P_K - \frac{1}{2}$ (that is, a is the deviation of the mean internal homogeneity above 0.5), Coste (reported in Kauffman 1986d) was able to obtain bounds on the number of attractor cycles:

$$ N \frac{\log\left(\dfrac{1}{\frac{1}{2} + a}\right)}{2} \leqslant \text{number of cycles} \leqslant N \frac{\log\left(\dfrac{1}{\frac{1}{2} - a}\right)}{2} \qquad (5.10) $$

For fixed K, the mean internal homogeneity is fixed and so cycle numbers increase roughly linearly within the bounds. For $K = 2$, Coste's approximation is wrong. As

TABLE 5.4 Mean Field Predicted $\log_{10}$ Cycle Lengths as a Function of K

K	$\log_{10}$ cycle length for $N = 20$
3	1.680
4	1.928
5	2.141
6	2.297
7	2.414

we shall see, for $K = 2$, the number of attractors increases less than linearly as N increases.

In summary, random Boolean networks with $K \geqslant 5$ exhibit chaotic behavior. State cycles lengthen exponentially as N increases, the systems show sensitivity to initial conditions in the exact sense that initial minor differences in activity patterns increase over time, the number of attractors is roughly linear in N, and their stability to perfurbations is at best modest. These characteristics of the chaotic regime change drastically when K decreases to 2.

$K = 2$: A Phase Transition to Order in Random Boolean Networks.   It has now been known for over 20 years that Boolean networks which are *entirely random* but subject to the simple constraint that each element is directly controlled by $K = 2$ elements spontaneously exhibit very high order (Kauffman, 1969, 1971a, 1971b, 1971c, 1974). A number of workers have attempted to understand the order observed (Kauffman 1969, 1971a, 1971c, 1974, 1984a, 1986a, 1986c, 1986d; Cull 1971; Aleksander 1973; Babcock 1976; Cavender 1977; Sherlock 1979a, 1979b; Thomas 1979, 1984; Walker and Gelfand 1979; Atlan, Fogelman-Soulie, et al. 1981; Fogelman-Soulie, Goles-Chacc, and Weisbuch 1982; Gelfand and Walker 1982, 1984; Fogelman-Soulie 1984, 1985a, 1985b; Derrida and Flyvbjerg 1986; Derrida and Pomeau 1986; Derrida and Stauffer 1986a, 1986b; Derrida and Weisbuch 1986, 1987; Hartman and Vichniac 1986; DeArcangelis 1987; Hilhorst and Nijmeijer 1987; Stauffer 1987a, 1987b, 1989; DeArcangelis and Coniglio 1988; Weisbuch 1989a, 1989b).

Three approaches to understanding that order—two analytic, one numerical—have been taken. I shall begin with rather remarkable recent analytic work by the solid-state physicist B. Derrida and his colleagues (Derrida and Pomeau 1986).

Derrida's approach is called the *annealed approximation.* The problem is to understand the behavior of deterministic Boolean dynamical systems sampled at random from defined ensembles specified by N and K. Once the connections and Boolean functions are chosen for a particular net, they remain constant and define that network. Call such a fixed structure *quenched,* meaning fixed in place. Derrida and his co-workers have gained substantial insight into the emergence of orderly dynamics in quenched Boolean networks by making the drastic assumption that the entire structure and logic of the network change at each moment. That is, suppose that at each moment, after each state transition, the newly computed state is left intact and the identity of each of the N elements is left intact but *the connections and the Boolean functions among the element are chosen at random again,* specifying an entirely new network at each successive moment. This is the *annealed model.* The dynamics of the state transitions in this model bears a close relationship to that of the quenched model.

In the quenched model, of course, after at most 2^N time moments, the system must hit a state previously encountered and thereafter will cycle around a recurrent set of states on an attractor state cycle. In the annealed model, no periodic behavior can be expected, since the connections and functions change at each moment. On the other hand, there is a deep connection between the quenched and annealed models. Consider a quenched network with very large N, say 100 000, and $K = 2$. Focus on the behavior of a given element i. Start the network in some arbitrary state at time 0. In order to calculate the state of element i at the next time moment $T = 1$, the values of its $K = 2$ inputs at $T = 0$ must be assessed. In order to calculate the state of i at time $T = 2$ as a function of the state of the network at $T = 0$, the $T = 0$ values of the total of four inputs must be examined. Specifically, the values of the four inputs to the two direct inputs to i must be examined. Thus, over successive moments, the value of i as a function of the initial state at $T = 0$ depends on K^1, K^2, K^3, K^4 ... values of different element according to different Boolean functions. If N were infinite, then after any finite number of state transitions, the behavior of the ith element would depend on exponentially increasing numbers of values evaluated at time $T = 0$, but always by a random new set of Boolean functions which changes as time increases by each moment. This suggests that the annealed and quenched models are closely related in a number of respects.

In particular, let us define the *Hamming distance* H between two different states of N elements as the number of positions in which the two states are different. Thus (1111111111) and (0111111110) overlap in eight of the ten positions and differ in two, so that the Hamming distance is 2. The *overlap* between two states is just the number of elements which have the same activity in both states. Overlap is just $N - H$.

The question Derrida addresses is this: Do annealed Boolean nets tend to cause initially different patterns of activities among the N elements to *converge* to the same pattern of activity over time? Remarkably, the answer is "yes" for $K = 2$ annealed networks but not for $K > 2$. More precisely, Derrida shows that, for large N, the quenched and annealed models predict nearly the same time evolution in the *overlap between two different initial states of the network* over a sequence of states corresponding roughly to the *transients* flowing toward the state cycle attractors of the quenched model.

The approach is this: Start with two randomly chosen states 1 and 2. Choose a random network in the NK ensemble. Place the network in state 1, and allow it to undergo a transition to a successor state 1'. Place the network in state 2, and allow it to undergo a transition to a successor state 2'. Measure the overlap between the two initial states and that between the two successor states. Is the latter overlap greater or less than the former? Next, randomly rewire the network, and randomly assign to each element a new Boolean function on its K inputs. Use this new network to calculate, for states 1' and 2', their transitions to their own successors 1" and 2". Calculate the overlap between 1" and 2". Keep repeating this procedure in order to determine whether the overlap between pairs of successor states increases to complete overlap in all the elements or approaches some fixed fraction of the N elements.

It is simple to show that the answer to Derrida's question depends on K. If the overlap between two states at time T is $N(a_{12})$, then, after we divide by N, the fractional overlap at the next moment is

$$a_{12}(T + 1) = \frac{1 + a_{12}^K}{2} \tag{5.11}$$

This may be derived as follows. If the overlap at time T is $Na_{12}(T)$, then some fraction of the N elements have all K of their inputs in *the same value* in both states 1 and 2. The expected number of these is just $(Na_{12})^K$. Each such element whose K inputs are in identical values in states 1 and 2 at time T will certainly be in the identical value in the two successor states 1' and 2' at time $T + 1$. The number of sites at time T which have at least one of their K inputs different is just $N\{1 - [(a_{12}(T)]^K\}$. Each of these elements which have at least one input in different values in states 1 and 2 have a probability of ½ of becoming the same at time $T + 1$. Therefore,

$$a_{12}(T + 1) = [a_{12}(T)]^K + \frac{1}{2}\{1 - [a_{12}(T)]^K\}$$

which is Equation 5.11.

Equation 5.11 demonstrates a dramatic difference between $K = 2$ networks and $K > 2$ networks. For $K = 2$, over successive time moments, the difference H between pairs of successor states *decreases* and vanishes at infinite time. That is, in the annealed model, for $K = 2$, the overlap between any two initial states increases between successors until it becomes essentially complete. For $K > 2$, the difference between successor state does *not* vanish. Rather, the overlap approaches a fixed fraction which depends on K. This difference between $K = 2$ and $K > 2$ networks is most readily seen in Figure 5.8, which shows the iteration of the mapping expressed in Equation 5.11. The 45° line represents those cases where the difference does not change over one time step. Therefore, the positions where the curves for the different values of K cross the 45° line represent the steady-state fixed points. For $K > 2$, there is a fixed point at a difference H_s, which is greater than 0; each of these curves crosses the 45° line at H_s from above for lower values of H at time T to below for higher values of H at time T. Thus if H is less than the crossing point H_s at time T, it increases at time $T + 1$; if H is more than the crossing point at time T, it decreases at time $T + 1$. Thus the point H_s where each curve crosses the 45° line, is a stable steady state which H approaches as time increases. This means that, for $K > 2$, any two neighboring initial states, no matter how close initially, will be expected to diverge to a mean Hamming distance H given by the steady-state value in Figure 5.8 and to fluctuate around that value thereafter. This divergence of initially neighboring states for $K > 2$ demonstrates in another form sensitivity to initial conditions. It is somewhat similar to the sensitivity discussed earlier for flow on chaotic attractors.

The behavior for $K = 2$ is radically different, for in this case there is no crossing

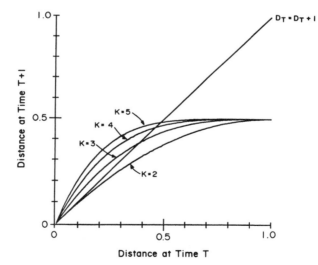

Figure 5.8   Recurrence relation showing the expected distance D_{T+1} between two states at time $T + 1$ after each is acted upon by the network at time T, as a function of the distance D_T between the two states at time T. Distance is normalized to the fraction of elements in different activity values in the two states being compared. That is, Hamming distance at time T is $H_T = N[1 - a_{12}(T)]$, $D_T = H_T/N$. For $K = 2$, the recurrence curve is below the 45° line, and hence distance between arbitrary initial states decreases toward zero over iterations. For $K > 2$, states which are initially very close diverge to an asymptotic distance given by the crossing of the corresponding K curve at the 45° line. Thus $K > 2$ networks exhibit sensitivity to initial conditions and chaos, not order. Based on annealed approximation. (From Derrida and Pommeau 1986)

of the 45° line (Figure 5.8); the curve is beneath that line everywhere. Thus, remarkably, over time, the differences between successor states vanish! (More precisely, the fraction of sites which differ approaches zero.) This dramatic phenomenon is a *phase transition*. As K decreases from $K = N$ to $K = 2$, in the annealed model, networks are initially in the chaotic regime but undergo a sudden transition to ordered behavior when $K = 2$. Numerical simulations comparing the quenched model and the annealed model, for different values of K and reasonably large values of N, confirm Derrida's theory (Derrida and Weisbuch 1986). For the annealed model, the infinite N predictions are close to the quenched model even for $N = 20$ to 30. In the quenched case, of course, the Boolean network is fixed and settles to a state cycle. Unless that cycle is a steady state and there is only one attractor, it cannot be the case for any value of K that overlap between two arbitrary initial states increases to 100 percent. Nevertheless, the annealed theory fits remarkably well the observed evolution of overlaps as a function of K along pairs of transient sequences of states flowing toward state cycles. Therefore, it seems clear that Derrida has found evidence for a phase transition between the behavior of $K = 2$ Boolean networks and $K > 2$ networks. The evidence points in the right direction. There is indeed a phase transition when K decreases to $K = 2$.

Numerical simulations of randomly chosen members of the ensemble of $K = 2$ input networks in which N ranges up to 10 000 have been carried out (Kauffman 1969, 1971a, 1971b, 1971c, 1974, 1984a, 1986d). The following are the dominant results:

1. The expected median state cycle length is about $\sqrt{N}$. That is, the number of states on an attractor scales as the square root of the number of elements. A Boolean network with 10 000 elements which was utterly random within the constraint that each element is regulated by only two elements would therefore have a state space of $2^{10\,000} = 10^{3000}$ but would settle down and cycle recurrently among a mere $\sqrt{10\,000} = 100$ states. Thus, in contrast to random Boolean networks with $K \geqslant 5$, where cycle lengths increase exponentially in N and rapidly become hyperastronomical, randomly assembled Boolean systems with $K = 2$ spontaneously confine their dynamical behavior to truly tiny subvolumes of their state space. A system of 10 000 elements which localizes its dynamical behavior to 100 states has restricted itself to 10^{-2998} parts of its entire state space. Here is spontaneous order indeed.

2. The distribution about this median is skewed. Most networks have short state cycles, while a few have very long ones. If plotted as the logarithm of cycle length, the distribution remains skewed rather than becoming a familiar bell-shaped Gaussian distribution.

3. The number of state cycle attractors is also about $\sqrt{N}$. Therefore, a random Boolean network with 10 000 elements would be expected to have on the order of 100 alternative attractors. A system with 100 000 elements, comparable to the human genome, would have about 317 alternative asymptotic attractors.

4. If the stability of each state cycle attractor is probed by transient reversing of the activity of each element in each state of the state cycle, then for about 80 to 90 percent of all such perturbations, the system flows back to the same state cycle. Thus state cycles are inherently stable to most minimal transient perturbations.

5. When the perturbed system flows back to the same state cycle, it typically reaches that state on the state cycle which it would have reached in the same time interval

had the system not been perturbed. Thus if one thinks of position around a state cycle as carrying "phase" information, a perturbed system tends to return to the same cycle and maintain phase after the perturbation has ended.

6. For perhaps 5 to 15 percent of the minimal perturbations transiently reversing the activity of a single element, the system leaves the state cycle from which it was perturbed and flows to another state cycle.

7. A perturbed state cycle can directly change only to a small number of other state cycle attractors in the system. It follows that many of the 5 to 15 percent of the perturbations which cause the system to change from attractor *A* induce change to the *same neighboring attractor B.* In short, a variety of different stimuli acting on different elements in the system induce the same specific response.

8. A large fraction of the *N* elements, typically 70 percent or more, fall to a fixed active or fixed inactive state which is *identical* on all the alternative attractors of the Boolean network. (I return to this shortly, for it tokens a new principle of collective order in massively parallel networks.)

9. The mean difference in patterns of activity on different attractors is a few percent.

10. Altering the activity of a single element transiently typically propagates and causes alterations in activity of a small fraction of the total number of elements in the system. In other words, "damage" is slight.

11. Deleting any single element or altering its Boolean function typically causes only modest changes in attractors and transients.

The results discussed here provide the conceptual framework of Chapter 12, where I shall interpret the binary elements as genes switching one another on and off and the Boolean network as the cybernetic genetic regulatory network. There I shall interpret a state cycle attractor of recurrent patterns of gene activity as a cell type in the behavioral repertoire of the genomic regulatory system. Then

• The size of attractors maps to how confined a pattern of gene expression corresponds to one cell type.

• The number of attractors maps to the number of cell types in an organism.

• The stability of attractors maps to homeostatic stability of cell types.

• The number of attractors accessible by perturbing the states of activities of single genes maps to the number of cell types which any cell type can differentiate into. Since this number is small compared with the total number of cell types in the organism, then ontogeny must be, and is, organized around branching pathways of differentiation.

• The overlap in gene activity patterns on attractors maps to the similarity of cell types in one organism.

• The alteration of attractors by mutations maps to evolution of novel cell types.

The spontaneous order we have just uncovered in $K = 2$ networks and their generalizations described below underlie a serious hope to account for much of the order seen in the orderly, coordinate behavior of genetic regulatory systems underlying ontogeny in the absence of selection. "Random" genetic programs can behave with order.

K = 1: *Networks in the Ordered Regime.* In these networks, each element has only a single input. The structure of the network falls apart into separate loops with descendant tails. If the network connections are assigned at random, then most elements lie on the tails and do not control dynamical behavior, since their influence propagates off the ends of the tails. On the order of $\ln N \sqrt{N}$ of the elements lie on loops. The number of loops scales roughly as $(\ln N)/e$. Each loop has its own dynamical behavior and cannot influence the other, structurally isolated loops. Thus such a system is *structurally modular.* It comprises separate, isolated subsystems. The overall behavior of such a system is the product of the behaviors of the isolated subsystems. As Table 5.1 shows, the median lengths of state cycles increase rather slowly as N increases ($\sqrt{\pi/2} \sqrt{N}$) (Jaffee 1988); the number of attractors increases exponentially as N increases, and their stability is moderate.

There are four Boolean functions of $K = 1$ input: "Yes," "Not," "True," and "False." The last two are constantly either active or inactive. The values in Table 5.1 assume that only "Yes" and "Not" are utilized in $K = 1$ networks. When all four functions are allowed, most isolated loops fall to fixed states and the dynamical behavior is dominated by those loops with no "True" or "False" functions assigned to elements of the loop. Flyvberg and Kjaer (1988) and Jaffee (1988) have derived detailed results for this analytically tractable case.

Percolation of Frozen Clusters: A New Principle of Order in Massively Parallel Boolean Systems

What principles allow $K = 2$ networks to exhibit such profound order? The basic answer appears to be that such networks develop a connected mesh, or *frozen core*, of elements, each frozen in either the 1 or the 0 state. The frozen core creates spanning, or *percolating,* walls of constancy which break the system into functionally isolated islands of unfrozen elements cut off from influencing one another by the walls of frozen elements. The formation of such functionally isolated islands by a percolating frozen core appears to be a sufficient condition for order in Boolean networks; conversely, failure of a frozen core to percolate and leave functionally isolated unfrozen islands is a sufficient condition for chaos. The boundary regime where a frozen core is just percolating and, more important, the unfrozen region is just breaking into unfrozen islands is the phase transition between order and chaos.

Two related means of forming such percolating walls are now established. The first is called *forcing structures* (Kauffman 1971a, 1971b, 1974, 1984a, 1986a; Fogelman-Soulie 1984, 1985a, 1985b; Gelfand and Walker 1984). The second has as yet no specific name. I propose to call this mechanism *internal homogeneity clusters.*

Forcing structures will be introduced here and discussed again in Chapter 12, for they appear to arise in genomic regulatory systems and crystallize orderly dynamics. Consider the Boolean "Or" function. This function asserts that, if either one or the other of the two regulating inputs is active at a given moment, then the regulated element will be active at the next moment. Notice that this function has the property that, if the *first input is currently active,* that alone guarantees that the regulated element will be active at the next moment, regardless of the current activity of the second input. That is, this Boolean function has the property that, if the first input is active, the regulated element can be *fully insensitive* to variation in the activity of the second input.

I define as a *canalyzing Boolean function* any Boolean function having the property that it has at least one input having at least one value (1 or 0) which suffices to guarantee that the regulated element assumes a specific value (1 or 0). "Or" is such a function. So is "And" since, if either the first or the second input is 0, the regulated locus is guaranteed to be 0 at the next moment. By contrast, the "Exclusive Or" function, in which the regulated locus is active at the next moment if one or the other input—but not both—is active at the present moment, is not a canalyzing Boolean function. No single state of either input guarantees the behavior of the regulated element. A striking feature of known regulated genes is that virtually all are regulated by canalyzing functions in the Boolean idealization. Just this feature will allow us to deduce large-scale order in genomic regulatory systems underlying ontogeny. The order arises as a result of the formation of forcing structures.

Consider a system of several binary variables, each receiving inputs from two or three of the other variables and each active at the next moment if any one of its inputs is active at the current moment (Figure 5.9). That is, each element is governed by the "Or" function on its inputs. As shown in Figure 5.9, this small network has a feedback loop. Now, the consequence of the fact that all elements are governed by the "Or" function on their inputs is that, if a specific element is currently in the 1 state, at the next moment all the elements it regulates are *forced* to be in the 1 state. Thus the 1 value is guaranteed to propagate iteratively from any initially active element in the net to all descendents. But the net has loops, and so the forced 1 value cycles around such a loop. Once the loop has filled up with 1 values at each element, it remains in a fixed state with 1 at each element and cannot be perturbed by outside influences of other inputs into the loop. Further, the "fixed" 1 values propagate to all descendents of the feedback loop, fixing them in the 1 value as well. Such circuits are called *forcing loops* and *descendant forcing structures* (Kauffman 1971, 1974a, 1974b, 1974c, 1984a, 1986a; Fogelman-Soulie 1984, 1985a, 1985b; Hartman and Vichniac 1986). Note that the fixed behavior of such a part of the network provides

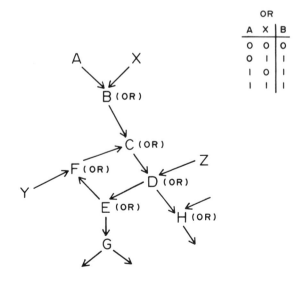

Figure 5.9 Forcing structure among binary elements governed by the Boolean "Or" function. The forcing 1 value propagates down the structure and around the forcing loop, which eventually is frozen into the forced state with 1 values at all elements around the loop. The loop then radiates fixed forced 1 values downstream.

walls of constancy. No signal can pass through elements once they are frozen in their forced values.

The limitation to the "Or" function is made here only to make the picture clear. There are $(2^2)^2 = 16$ Boolean functions of switches with $K = 2$ inputs. In Figure 12.16, I show a network with a forcing structure in which a 1 state at some specific elements forces a descendant element to be in the 0 state, which in turn forces its descendant element to be in the 1 state. The defining feature of a forcing structure in a Boolean network is that, at each point, a single element has a single value which can force a descendant element to a specific value regardless of the activities of other inputs. Propagation of such forced values occurs via the forcing connections in the network. In order for a connection between two regulated elements to be classed as forcing, the second element must be governed by a canalyzing Boolean function and the first element, which is an input to the second, must directly or indirectly (that is, via $K = 1$ input connections) be governed by a canalyzing Boolean function. Finally, the value of the first element which can be forced must be the value of the first element which forces the activity of the second element. Thus an "And"-governed element does not force an "On"-governed element. Clearly a network of elements governed by the "Or" function does meet these requirements. More generally, forcing is a transitive relation such that if A forces B and B forces C, then A indirectly forces C via B. Forced values must propagate down a connected forcing structure.

Formation of large forcing structures suffices to account for the order seen in large random $K = 2$ networks. The forcing structures form a large interconnected web of elements which percolates across the entire network (Kauffman 1971, 1984a, 1986a, 1986c, 1986d; Fogleman-Soulie 1985a, 1985b; Hartman and Vichniac 1986). This web falls to a fixed state, each element frozen in its forced value, and leaves behind *functionally isolated islands of elements* which are not part of the forcing structure.

The occurrence of walls of constancy depends on the character of the switching network and in particular on the number of inputs to each variable—that is, on the connectivity of the dynamical system. Large, connected forcing structures percolate spontaneously in $K = 2$ networks because a high proportion of the 16 possible Boolean functions of $K = 2$ inputs are canalyzing functions. If two elements regulated by canalyzing Boolean functions are coupled, one as the input to the second, then the probability that the connection is a forcing connection is .5. This means that, in a large network with all elements regulated by canalyzing Boolean functions, on average half of the connections are forcing connections.

The expected size and structure of the resulting forcing structures are a mathematical problem in random-graph theory (Erdos and Renyi 1959, 1960; Harary 1969; Kauffman 1974, 1983, 1986). A mathematical graph is a collection of points, called vertices, connected either by lines or by arrows. If the vertices are connected by unoriented lines, the system is an *undirected graph.* If the vertices are connected by directed arrows, the system is a *directed graph.* In a random graph, the lines or arrows connecting vertices are assigned in some random way. The fundamental questions of interest in graphs concern the connectivity properties among the vertices. In particular, percolation thresholds occur in random graphs and determine when large, connected webs of elements will form. Below the threshold, such webs do not form; above the threshold, they do. The percolation threshold for extended forcing structures in a random Boolean network requires that the ratio of forcing connections to elements be 1.0 or greater (Kauffman 1971b, 1983, 1986), as discussed more fully in Chapter 12. Thus in large networks using elements regulated by canalyzing functions on two inputs, half of the $2N$ connections are forcing. Therefore, the ratio of forcing connections to elements, $N/N = 1$, is high enough for formation of extended large forcing structures. More generally, both for $K = 2$ random networks and for networks with $K > 2$ restricted to canalyzing functions, such forcing structures form and crystallize a frozen state which induces orderly dynamics in the entire network.

Because percolation of a frozen component also accounts for the emergence of order due to *homogeneity clusters* (discussed just below), I defer for a moment describing how the frozen component arising from either forcing structures or homogeneity clusters induces orderly dynamics.

Percolation of Homogeneity Clusters: P > P_c

Random Boolean networks exhibit three regimes of behavior: chaotic, ordered, and complex. Passage from chaotic to ordered behavior is a phase transition driven by a variety of control parameters. One such parameter is the connectivity K of the elements of the network. Low connectivity is a sufficient condition for orderly behavior in disordered switching systems, but other parameters can drive the transition from chaos to order. These parameters tune the biases on the class of Boolean switching rules utilized in the network.

The new control parameter is P, mentioned above, which is a measure of the internal homogeneity of a Boolean function. Consider a Boolean function of four input variables. Each input can be on or off, and hence the function must specify the response of the regulated switching element for each of the 2^4 combinations of values of the four inputs. Among the 16 responses, either 1 or 0 might occur equiprobably, or else one of them may occur far more often than the other. Let P be the fraction of the 2^K positions in the function with either a 1 response or a 0 response, whichever is the larger fraction. Thus P ranges from .5 to 1.0. If P approaches 1.0, then most combinations of activities of the four variables lead to a 1 or a 0 response. The deviation of P above .5 measures the internal homogeneity of the Boolean function.

In Figure 5.10, I show a two-dimensional lattice of points, each of which is an on–off variable and each of which is regulated by its four neighboring points. Each is assigned at random one of the possible Boolean functions on four inputs, subject to the constraint that the fraction of 1 values in that Boolean function is a specified percentage P, $P > .5$.

Derrida and Stauffer (1986a, 1986b), Weisbuch and Stauffer (1987), and de Arcangelis (1987), summarized in Stauffer (1987a, 1987b) and in Weisbuch (1989a), studied two- and three-dimensional lattices with nearest-neighbor coupling and found that, if P is larger than a critical value P_c, then the dynamical behavior of the network breaks up into a connected frozen web of points fixed in the 1 value, and isolated islands of connected points which are free to oscillate from 0 to 1 to 0 but are functionally cut off from other islands by the frozen web. In contrast, if P is closer to .5 than to P_c, then such a percolating web of points fixed in 1 values does not form. Instead, small isolated islands of frozen elements form, and the remaining lattice is a single connected, percolating web of elements which oscillate between 1 and 0 in complex temporal cycles. In this case, transiently altering the value of one point can propagate via neighboring points and influence the behavior of most of the oscillating elements in the lattice. Thus the critical value P_c demarks a phase transition in the behavior of such a dynamical system.

The arguments for the percolation of a frozen component for $P > P_c$ do not require that the favored value of each on–off variable in the lattice be 1. The argu-

```
     8   8   1   1228228228228228228228228   1   1   1   1   1   1   1   1   1   1   1   1   1
     8   8   1   1   1   1   1228228228228   1   1   1   1   1   1   1   1   1   1   1   1   1
     8   8   845645645622822822822822822822   8   1   1   1   1   1  10  10  10   1   1   1   1
     1   8   1   1228228228228228   1   1   1   1   1   1   1   1  10  10  10   1   1   1   1
     1   1   1228228228228228228228   1   1   1   1   1   1   1   1   1   1   1   1   1   1
     1   1   1   1   1228228228228228228228   1   1   1   1   1   1   1   1   1   1   1   4   4
     1   1   1   1   1   1   1   1228228228228   1   1   1   1   1   1   1   1   1   1   1   1
     1   1   1   1   6   1   1228228228228   1   1   1   4   1   1   1   1   1   1   1   1   1
     1   4   1   6   6   6   1   1228228228228228228   4   1   4   1   1   1   1   1   1   1
     1   4   1   6   6   6   622822822822   1   1   1   4   1   4   1   1   1   1   1   1   1
     4   4   1   6   6   6   6   6228228228   1   1   1   1   1   1   1   1   1   1   4   4
     1   4  12   6   6   6   1228228228   1   1   1   1   1   1   8   8   8   1   1   1   4
   220   1   1   1   1   1   1   1   1228228228   1   1   1   1   1   8   8   8   8   1  1220
 220220   1   1   1   1   1   1   1228228228228   1   1   1   1   8   8   4   8   1   1   1
 220220   1   1   1   1   1   1   1228228   1   1   1   1   1   1   1   1  1220110   1
 1220110110   1   1   1   1   1228228   1   1   1   1   1   1   1  20  20110110
 1110110110   1   1   4   1   1228   1   1   2   4   1   1   1   1   1  20  20110110
110110110110110   1   4   1   1   1   1   1   2   4   1   1   1  20  20  20  20  1110
110110110  22   1   1   1   1   1   1   4  4228   1   1   1  20  20  20  20  20  20  20110
110110   1   1   1   1   1   1   1   1   1228   1   4   1  20  20  20  20  20  20  20110
110  22  22  22  22   1   1228228   1   1228228   1   4   4   1   1   1   1   4  20   2  22
 22  88  22  22   1   1   1   1228   1228228228   1   1   1   1   1   1   1   1  20   2   1
  1  88   1   1   1228228228228228228228228228   1   1   1   1   1   1   1   4   4   4   1
  1   8   1   1228228228228228228228228228   1   1   1   1   1   1   1   1   1   1   1
```

Figure 5.10 Two-dimensional lattice of sites, each a binary state spin which may point up or down. Number at each point in lattice is period of spin on the lattice state cycle. Hence, sites with 1 are frozen active or inactive. Each variable is coupled to its four neighbors and is governed by a Boolean function on those four inputs. When P is increased, the bias in favor of a 1 or a 0 response by any single spin leads, above a critical value P_c, to percolation of a frozen component of spins which spans the lattice and leaves isolated islands of spins free to vary between 0 and 1. (From Weisbuch and Stauffer 1987)

ments carry over perfectly if half the on–off variables respond with high probability, $P > P_c$, by assuming the 1 value and the other half respond with $P > P_c$ with the 0 value. In this generalized case, in the frozen web in the lattice, each frozen value is frozen in its more probable value. Thus for arbitrary Boolean lattices, $P > P_c$ provides a criterion which separates two drastically different behaviors: ordered versus chaotic.

The value of P for which this percolation and freezing out occur depends on the kind of lattice and increases as the number of neighbors of each point in the lattice increases. On a square lattice for $K = 4$, P_c is .72 (Stauffer 1987a, 1987b, 1989). On a cubic lattice, each point has six neighbors and P_c is closer to 1 than on square lattices.

Let me call such percolating frozen components for $P > P_c$ *homogeneity clusters* to distinguish them from extended forcing structures. I choose this name because freezing in this case depends on the internal homogeneity of the Boolean functions used in the network. That the two classes of objects are different in general is clear. In forcing structures, the characteristic feature is that, at each point, a single value of an element suffices to force one or more descendent elements to their own forced values. Homogeneity clusters are more general: Elements might each be held in a frozen state by the joint activity of several elements. Thus consider two pairs of elements—$A1$, $A2$, $B1$, and $B2$. Elements $A1$ and $A2$ might receive inputs from both $B1$ and $B2$ as well as from other elements, while $B1$ and $B2$ receive inputs from $A1$ and $A2$ as well as from other elements. But, because of the high internal homogeneity of the Boolean functions assigned to each ($P > P_c$), *simultaneous 1 values* by both $A1$ and $A2$ might jointly guarantee that both $B1$ and $B2$ be active regardless of the activities of other inputs to $B1$ and $B2$. At the same time, simultaneous 1 values by both $B1$ and $B2$ might jointly guarantee that both $A1$ and $A2$ be active regardless of the activities of other inputs to $A1$ and $A2$. Once the four elements are jointly active, they mutually guarantee their continued activity regardless of the behavior of other inputs to them. They form a frozen component. Yet this homogeneity cluster is not a forcing component, since the activity of two elements, either $A1$ and $A2$ or $B1$ and $B2$, must be jointly ensured to guarantee the activity of any single element.

While there appear to be certain differences between forcing structures and homogeneity clusters, those differences are far less important than the fact that, at present, the two are the only established means of obtaining orderly dynamics in large, disordered Boolean networks.

Whether percolation of a frozen phase is due to an extended forcing structure or to a homogeneity cluster arising from $P > P_c$, there are two main implications. First, if a frozen phase does not form and a percolating unfrozen phase does form,

- The attractors in such a system are very large and grow exponentially as the number of points in the lattice N increases. Indeed, the attractors are so large that the system can be said to behave chaotically.

- A minor alteration in the state of the lattice propagates alterations in behavior throughout the system. More precisely, consider two identical lattices which differ only in the value of one element at a moment T. Let the two lattices behave dynamically according to their identical Boolean rules. Define the damage caused by the initial alternation to be the total number of sites in the lattices which, at the succession of time moments, are induced to be in *different* states, 1 or 0. Then for P closer to .5 than P_c, such damage propagates across the lattice with a finite speed, and a large fraction of the sites are damaged (Derrida and Stauffer 1986a, 1986b; Weisbuch and Stauffer 1987; Stanley, Stauffer, et al. 1987; Stauffer 1987a; Lam 1988). Propagation of damage from a single site difference implies that dynamical behavior is highly sensitive to small changes in initial conditions. Thus chaotic behavior in Boolean networks, like that in continuous dynamical systems, shows this typical talisman of chaotic dynamics.

- Consequently, many perturbations by single alterations drive the system to an entirely different attractor.

- Damage by deletion of an element or alteration of its Boolean function tends strongly to alter many attractors. Thus such systems adapt on very rugged landscapes.

Second, if the ratio of forcing connections to elements is greater than 1.0 or if P is closer to 1.0 than to P_c

- A large frozen component and percolating walls of constancy do form, leaving behind functionally isolated islands which cannot communicate with one another.

- The result is that *attractors are small,* typically increasing as the number of nodes to some fractional power (Kauffman 1969, 1974, 1984a, 1986a, 1986c, 1986d; Stauffer 1987a; Weisbuch and Stauffer 1987; Kurten 1988a, 1988b). This means that the sizes of attractors increase less than linearly as the number of points in the lattice N increases. Such attractors are small indeed, for the entire state space is the 2^N possible combinations of the on–off values in the lattice. An attractor comprising fewer than N states is tiny compared with 2^N. Thus the existence of a frozen component—due either to forcing structures or to homogeneity clusters for $P > P_c$—implies that such systems spontaneously box themselves into very small volumes of their space and exhibit high order.

• Further, damage does not spread. Transient reversal of the state of an element propagates alterations in behavior only locally if at all (Kauffman 1969, 1974, 1986; Stauffer 1987). This means that attractors tend strongly to exhibit *homeostatic return after perturbation.*

• For frozen components resulting from either forcing structures or homogeneity clusters, the system is typically not much altered by mutations deleting single elements or altering their Boolean rules. Any element buried in the frozen component cannot propagate alterations to the remainder of the network. A mutated element within one functionally isolated island communicates only within that island. Damage does not spread. Thus such systems adapt on highly correlated landscapes.

To summarize: The percolation of a frozen component yields disordered Boolean systems which nevertheless exhibit order. They have small attractors precisely because a large fraction of the variables remain in fixed states. Furthermore, because there is this frozen component in the lattice, minor modifications of connections in a Boolean function, or substitution of one for another Boolean function at one point, or alterations in other parameters lead to only minor modifications of dynamical behavior. Thus such networks have attractors which adapt on highly correlated landscapes. This is not surprising; the properties of the system which give it small attractors and hence homeostasis tend to make it insensitive to small alterations in the parameters affecting any one part of the system. Selection for one of these connected set of properties is selection for all. Self-organization for one, bootstraps for all.

ADAPTATION IN DYNAMICAL SYSTEMS

In this section we turn from a description of behavior in dynamical systems in general and of Boolean networks in particular to the fundamental topic of the capacities of such systems to adapt by mutation and selection. This is the stuff of evolution. Whether we consider genomic regulatory networks, morphologies, immune systems, or organ systems, gene mutations directly or indirectly affect the dynamical systems engendering organic forms. Selection acts on the resulting phenotypes. We must ask that which Darwin did not broach: What kinds of integrated dynamical systems harbor the ability to adapt? I begin the section by introducing the ideas of Ross Ashby, who first examined the concept of adaptation as an adaptive walk in the parameter space of a dynamical system toward parameter values corresponding to a dynamical system with "good" attractors. I then examine the adaptive capacities of random Boolean networks. The two central points to note are (1) that the multipeaked rugged landscapes seen in Chapters 2 and 3 remerge here and constrain adaptive evolution and (2) that landscape structure varies from correlated to very uncorrelated as Boolean systems pass from ordered to chaotic behavior. These twin features suggest that evolution may control landscape structure in part by controlling location on the order–chaos axis.

At the beginning of this chapter, I introduced continuous dynamical systems and their parameter spaces. In particular, I discussed a hypothetical chemical system containing ten chemical species. If we choose to think of an *ensemble of dynamical sys-*

tems as the set associated with all the different values of the parameters of the system, then we have the notion of a system and its neighbors within a small volume of parameter space. Then we can construct the idea of an adaptive walk through parameter space, a walk undertaken in order to achieve a dynamical system having desired properties. For example, it might be of interest to find parameter values such that the underlying dynamical system had at least one attractor which kept a subset of the ten chemical variables within some prescribed concentration bounds. This is precisely Ashby's fundamental image for the basic idea of adaptation in a system which exhibits coordinated, integrated behavior.

Ashby's *Design for a Brain* is delightful, elegant, and extremely clear and simple, reflecting very great care on the author's part. He sets himself the task of attempting to capture the central problem of adaptation in a system with many interacting parts. He is thinking of physiological and neural adaptation, not evolutionary adaptation. He conceives of a system (an organism) and an environment which are mutually coupled. He supposes that the system-cum-environment, hereafter written System, is deterministic. If released from any initial state, the System will flow to an attractor and remain there. The critical idea is to suppose that a *subset* of the organism's internal variables constitutes *essential variables,* which must be maintained within certain bounds. For example, the critical variables might be physiological ones, such as blood glucose level or body temperature. For an autopilot-airplane-air System, the essential variables might be those characterizing straight and level flight.

From the idea of essential variables in an organism which must be kept in bounds, it follows that the System, after reaching an attractor, either does or does not keep the essential variables within bounds. Ashby's next essential idea is simple: If the System on the attractor keeps the essential variables in bounds, change nothing. If the essential variables are not kept in bounds, however, then make a *jump change* in one of the parameters in the organism. Making this change may alter the state transitions in the System and hence alter the basins of attraction. Consequently, the state which the System was in prior to the jump change may now find itself in a newly formed basin of attraction, and the System may flow to a new or modified attractor which keeps the essential variables within bounds. If the new attractor is successful in boxing the essential variables, stop changing the parameters. If not, make another jump change, at random, in some other parameter. Keep making random changes in parameters until a parameter setting is found such that the System keeps the essential variables in bounds.

This is the core of Ashby's idea. It works. In effect, Ashby was able to build a crude autopilot, called a homeostat, which learned to hold an airplane in straight and level flight prior to crashing despite being wired at random to the controls. This capacity was termed *ultrastability.* At the time Ashby wrote, the concepts of bifurcation theory were not familiar. Consequently, he focused on jump changes in parameters to obtain changes in the attractor structure of the corresponding dynamical system. In terms of bifurcation theory, however, his paradigm consists in walks in parameter space. Smooth changes in attractors occur within any single bifurcation volume in parameter space; sharp changes in attractors occur when bifurcation surfaces are crossed. Obviously, the ruggedness of fitness landscapes is governed in part by how closely intertwined bifurcation surfaces are. If they are closely interwoven, landscapes are very rugged.

Not only did Ashby first introduce the idea of adaptation as a walk in parameter space seeking good attractors, but in his efforts to understand the conditions which would allow parallel-processing networks of elements to adapt, he also introduced the idea of percolating walls of constancy. The percolating frozen components which arise in Boolean networks in the ordered regime are precisely the kind of walls of constancy Ashby envisioned. We see next that the presence or absence of such walls alters the structure of fitness landscapes from smooth to rugged.

Selective Adaptation of Integrated Behavior in Boolean Networks

Ashby's paradigm is central to adaptive evolution. In the context of Boolean networks, the essential variables correspond to any subset of the N variables. Keeping those variables in bounds corresponds most simply to holding them fixed active or fixed inactive. In the limiting case where all the variables are essential, holding them in bounds corresponds to a *fixed steady state of the network as a whole,* at which each element is fixed in the proper active or inactive value. Thus the parallel to Ashby's idea applied to Boolean networks is to mutate the connections and logic of networks and tune their attractors until a good attractor with the desired steady state is found.

Let me be clear about the question I want to ask. Boolean networks exhibit a wide range of properties. Among these, the attractors commend themselves to our attention. A central question therefore is whether an adaptive process which is constrained to altering the input connections between elements in a network and the logic governing individual elements can hill climb to networks with desired attractors.

Notice that, as in protein space, we again confront a *space of systems.* Here the space is the space of Boolean NK networks. Each network is a one-mutant neighbor of all those networks which differ from it by alteration of a single connection or by alteration of a single Boolean function. More precisely, each network is a one-mutant neighbor of all those which alter the beginning or end of a single input connection or which alter a single bit in a single Boolean function.

In considering protein space, I defined a fitness landscape as the distribution over the space of proteins of any specific measurable property of those proteins. This definition led us to examine the statistical features of fitness landscapes, including their correlation structure, the number of local optima, the lengths of walks to optima, and the number of accessible optima. Similarly, in considering adaptation in Boolean network space, any measurable property of such networks yields a fitness landscape over the space of networks. Again we can ask what the structure of such landscapes looks like.

For our current purposes, I shall define the fitness of a Boolean network in terms of a steady target pattern of activity and inactivity among the N elements. This target is the (arbitrary) goal of adaptation. Any network has a finite number of state cycle attractors. I shall define the fitness of any network by the *match* of the target pattern to the closest state on any of the net's state cycles. A perfect match yields a normalized fitness of 1.0. More generally, the fitness is the fraction of the N elements which match the target pattern.

In Chapters 2 and 3, I introduced and discussed the NK family of rugged landscapes. There we encountered a number of general properties of rugged landscapes, including a universal law for long-jump adaptation. The waiting time to find fitter variants doubles after each fitter variant is found; hence the mean number of improvement steps S grows as the logarithm base 2 of the number of generations. We derived a complexity catastrophe during adaptation via fitter one-mutant variants, which leads, on sufficiently rugged landscapes, to an inexorable decrease in the fitness of attainable optima as the complexity of the entities increases. We found that in the long-jump limit, a similar complexity catastrophe applies to an even wider class of rugged landscapes. There the fitness attained after a fixed number of generations dwindles as complexity increases. Finally, we found that some landscapes—those in which the number of epistatic interactions K remained small—retained high optima as N increases. These landscapes have good correlation structures. Together, these properties identified limits to selection as complexity increases. In smooth landscapes, fitness differentials between one-mutant neighbors dwindle below critical values and selection cannot overcome mutation. The mutation error catastrophe sets in. In rugged landscapes, adaptive search is trapped in local regions and the complexity catastrophe sets in.

Here we are dealing with adaptation in the *coordinated dynamical behavior of Boolean networks*. It is not obvious that the same generic features and limitations exist, but they do.

Long-Jump Adaptation in K = 2 Networks Confirms the Universal Law

Figures 5.11a and 5.11b show numerical results of long-jump adaptation in Boolean networks with $N = 100$ binary elements, each receiving $K = 2$ inputs. The population consists in 20 networks, located "at" the current fittest network found in the adaptive search process. At each generation, each of 20 networks mutated 25 percent of the bits in its N Boolean functions (Figure 5.11a) or 50 percent of the connections (Figure 5.11b). The attractors of the grossly mutated nets were tested for their match against a predefined target pattern. If a fitter net was found, the entire population of 20 hopped to that fitter net and searched via long-jump mutations from that new site in network space on the next generation. If no fitter net was found on that generation, the search repeated with 20 new long jumps from the current best-fit network.

Figures 5.11a and 5.11b compare numerical simulation data with the expectation from Chapter 3 that the cumulative number of improved variants should increase as $\log_2$ of the number of generations. The agreement is extremely close.

The range of applicability of the universal law for long-jump adaptation, closely related to the theory of records (Feller 1971), is not yet clear but seems to be broad.

The Complexity Catastrophe Occurs in Long-Jump Adaptation

The complexity catastrophe occurs not only in NK landscapes but also in long-jump adaptation in Boolean networks. That is, as N increases, *long-jump adaptation achieves substantially less-fit* networks at any fixed generation. To test this, adaptation was carried out in the long-jump limit in which half the connections among binary variables in $K = 2$ input nets were mutated in all members of a population except one left at the current best network. As N increased from 20 to 100, the fitness achieved after 100 generations declined from .88 to .67. The difference is statistically significant. Thus as N increases, fitness after a fixed number of generations clearly is falling. In principle, it falls toward .5, the mean fitness of networks in the space.

This result is of considerable interest. As in the NK landscape family and the traveling salesman problem, the rate of finding improved variants in Boolean nets depends on the mutant search range and how well it matches the correlation structure of the landscape. Often, search via fitter two-mutant variants is better than via one-mutant variants. In the limit of long jumps on rugged landscapes, however, the

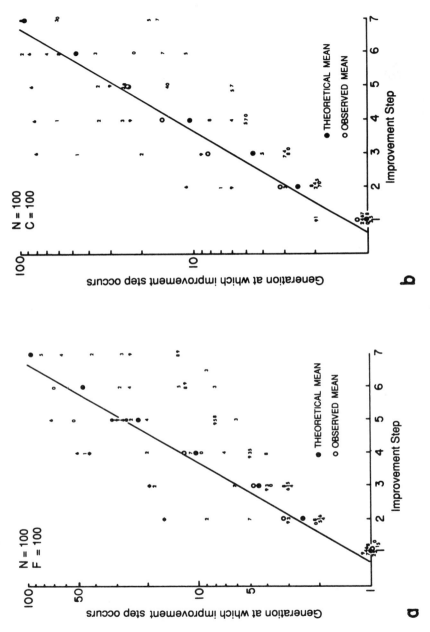

Figure 5.11 Tests of the universal law for long-jump adaptation in $K = 2$, $N = 100$ Boolean networks. $\log_2$ of generation at which improvement step occurred plotted as a function of number of improvement steps in each member of the population (except for a current-best placeholder). (a) Following mutation in 25 percent of the bits in the $F = 100$ Boolean function; (b) following mutation in 50 percent of the $C = 100$ connections.

213

rate of improvement slows to the $\log_2$ law and the *complexity catastrophe sets in.* Thus long-jump adaptation is a progressively worse adaptive strategy as the complexity of Boolean networks increases.

Boolean Network Space Is Full of Local Optima Which Trap Adaptive Walks

In assessing the *NK* family of landscapes, we found that rugged landscapes have many local optima, which trap adaptive walks. In the current case, we are asking whether Boolean networks can adapt via mutation and selection to have as a steady-state attractor, a specific pattern of activities among the *N* binary elements. Note first that no mathematical constraint foredooms such an effort. Any network in which each element which is to be active in the target pattern is active for all input patterns while each element which is inactive in the target pattern is inactive for all input patterns, fills the bill. The constantly active rule is the Tautology Boolean function; the constantly inactive rule is the Contradiction Boolean function. It follows that adaptation by alteration of single bits in Boolean functions can in principle achieve such a network.

In Figure 5.12*a* I show the results of adaptive walks via fitter one-, two-, and five-mutant variants of Boolean networks. The number of mutants refers to the number of bits altered in the net's Boolean functions. As before, a population of 20 nets is adapting from the best net found in the current generation. Figure 5.12*b* shows similar phenomena when one, two, or five connections in the networks are mutated. Note the following general features:

1. Improvement is rapid at first and then slows and typically appears to stop. Walks have arrested on *local optima.* The fact that improvement slows shows that the fraction of fitter mutant neighbors dwindles as optima are approached.

2. Walks always stop at local optima *well below the global optimum.* Trapping is rife in network space. This has a critical consequence: *adaptive walks typically cannot achieve arbitrary patterns of activities on attractors!* Adaptation via fitter variants in network space becomes grossly hindered by the rugged structure of the landscape. Walks become frozen into small regions of the space. Any intuition we may have harbored that mutation and selection alone could tune attractors to arbitrary patterns of behavior appears to be wrong. Such problems are very complex combinatorial optimization tasks, and selection confronts enormous problems moving successfully in such spaces. These limitations appear to be very important. Earlier in this chapter I described a number of examples in which the attractors of parallel-processing networks are interpreted in biological contexts ranging from memories in neural networks to cell types in genetic networks. In general, learning or adaptation is imagined to occur by alteration of couplings among the network elements in order to achieve desired attractors. These results suggest that this mechanism may typically be either extremely difficult or impossible. If so, then either alternative means of searching rugged adaptive landscapes in network spaces must exist or adaptation and learning do not achieve arbitrary attractors.

3. Adaptation via two- and five-mutant variants is more rapid and reaches higher optima than adaptation via fitter one-mutant variants. Thus the correlation structure favors search at slightly longer distances. Figures 5.13*a* and 5.13*b* show adaptive walks for $K = 10$ networks. Because cycles are long, we studied small net-

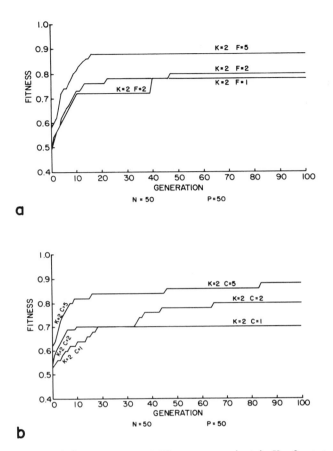

Figure 5.12 Adaptation via fitter one-, two-, and five-mutant variants in $K = 2$ networks. Mutation alters the bits within the Boolean functions in the networks of the adapting population. N = number of genes per network; P = population size; F = number of mutations in Boolean functions. (b) Same as (a), except that one, two, and five of the network connections were mutated. N = network size; P = population size; C = number of mutations in connections.

works. The same basic features of trapping on local optima were found as occurred in $K = 2$ networks.

Networks for which $K = 2$ adapt on a more correlated landscape than do $K = 10$ networks. Figure 5.14 compares the ruggedness of fitness landscapes in $K = 2$ and $K = 10$ networks. The salient feature is that, in $K = 2$ networks, the fitness of the one-mutant neighbors of the best network is nearly the same as the fitness of the best network. In other words, the landscape is highly correlated. This impression is confirmed by looking at the two- and five-mutant variants. The spread in fitness increases only slightly. In contrast, for $K = 10$ networks, the spread in fitness is wider and increases rapidly as one- to five-mutant variants are examined. Thus $K = 10$ networks adapt on a very much more rugged landscape than do $K = 2$ networks.

Also, $K = 2$ networks exhibit the complexity catastrophe, but only slowly. In

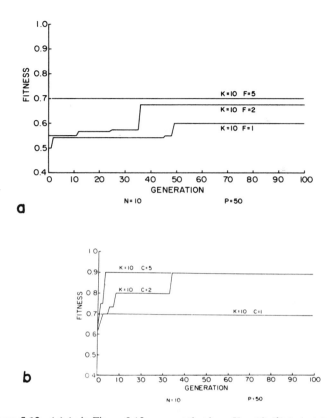

Figure 5.13   (*a*) As in Figure 5.12*a*, except that here $K = 10$. (*b*) As in (*a*), except that here connections rather than Boolean functions were mutated.

Table 5.5 I examine the fitness of optima attained after 100 and 200 generations of adaptation in $K = 2$ networks having $N = 20$ and $N = 100$ elements. The important result is that, as N increases, fitness appears to reach a local optimum after 200 generations. Nevertheless, fitness *decreases* as N increases. This means that, even though $K = 2$ networks adapt on well-correlated, good landscapes, they cannot avoid the complexity catastrophe. Presumably, as N increases the fitness attained will ultimately be hardly better than chance, .5. On the other hand, comparison with long-jump adaptation for the same class of $K = 2$ networks suggests that the rate of decrease of fitness as N increases is faster in the long-jump limit. Thus adaptation via near neighbors on the correlated $K = 2$ fitness landscape does not fall prey to the complexity catastrophe as rapidly as would occur were the landscape fully uncorrelated.

A general summary of our results is that the features of adaptive landscapes found for sequence spaces and in the *NK* family of landscapes extend to adaptation in the integrated dynamical behavior of Boolean networks. It was not obvious that the same features would be found, for sequence space and landscapes over proteins might be very different from fitness landscapes over spaces of dynamical systems with respect

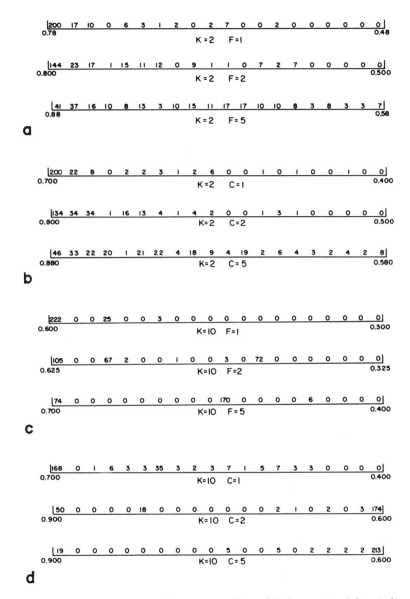

Figure 5.14 Fitnesses of one-, two-, and five-mutant variants of the fittest network found after adaptive hill climbing in (*a*) and (*b*) $K = 2$ networks and (*c*) and (*d*) $K = 10$ networks. F = mutations in Boolean function bits; C = mutations in connections.

TABLE 5.5 Mean Fitness in $100K = 2$ Networks
Attained After 100 and 200 Generations for Networks
Containing $N = 20$ and $N = 100$ Elements

N	Mean fitness	
	After 100 generations	After 200 generations
20	.90	.91
100	.78	.79

to their attractors. Nevertheless, similar features are found. Landscapes are rugged and multipeaked. Adaptive processes typically become trapped on such optima. The long-jump law obtains. Importantly, as the complexity of the entities under selection increases (here the number of binary switching variables in a disordered Boolean network), the attainable optima again fall toward the mean of the space. We do not know at this stage just how general this complexity catastrophe limiting the power of selection when operating on complex systems may be, but it appears likely to be a powerful factor in evolution. Finally, Boolean networks in the ordered regime, $K = 2$, and in the chaotic regime, $K = 10$, clearly adapt on radically different landscapes. Ordered systems encounter relatively correlated landscapes; chaotic systems adapt on very rugged landscapes.

Landscape structure governs both evolvability and sustained fitness. Having investigated the basic behavior of parallel-processing networks with respect to dynamical behavior and capacities to adapt, I turn next to ask whether there may be a characteristic "poised" state of parallel-processing networks—located in the ordered regime but near the edge of chaos—which may simultaneously optimize the complexity of tasks such a network can perform and also optimize the capacity of the network to evolve.

Selective Adaptation Toward the Liquid Region at the Edge of Chaos

All the results of this chapter indicate that a *phase boundary* separates networks that exhibit frozen, orderly dynamics from those that exhibit chaotic dynamics. The existence of this boundary leads us to a very general and potentially very important hypothesis: Parallel-processing systems lying in this interface region between order and chaos may be those best able to adapt and evolve. Further, *natural selection* may be the force which pulls complex adaptive systems into this boundary region. If so, we begin to have a powerful tool with which to examine the collaborative interaction between self-organization and selection.

The analysis of Boolean networks indicates that several *control parameters* underlie this transition. Boolean networks can be characterized by N, the number of binary variables; K, the number of inputs per variable; and P, one choice of a parameter which reflects biases in the choice of Boolean functions utilized in the network. Holding N fixed, we showed that K is a control parameter. For $K = 2$, networks crystallize order with percolating frozen components, state cycles are short and few, damage does not propagate, and such systems adapt on highly correlated fitness landscapes. For $K \geq 5$, networks are clearly in the chaotic regime. No frozen component forms, attractor sizes scale exponentially in N, damage propagates, and such systems adapt on highly uncorrelated landscapes. Thus in passing from $K = 2$ to $K \geqslant 5$, a transition from order to chaos is passed. Theoretical results due to Derrida and Pomeau (1986), described above, powerfully indicate that $K = 3$ networks with randomly chosen Boolean functions are already chaotic.

The P parameter introduced by Derrida and his colleagues, denoting the internal homogeneity of Boolean functions, is also a control parameter, as we have seen. Tuning P from near 1.0 toward 0.5 causes a system to pass from a regime having ordered dynamics due to frozen components to the chaotic regime. Similarly, biases in Boolean functions with respect to the use of canalyzing Boolean functions allow networks to pass from ordered dynamics to chaotic dynamics as a result of percolation of forcing structures. Thus the fraction of functions which are canalyzing is also a control parameter.

Other features, such as the structure of the wiring diagram among the Boolean variables (whether a regular lattice, random, or distributed in other ways), are also control parameters. With P constant, for example, systems pass from ordered to chaotic if the number of inputs per variable increases. Undoubtedly, further control parameters will be discovered.

While the zoo of control parameters awaits full disclosure, a critical fact remains: Boolean networks, among the most general class of massively parallel-processing systems, exhibit three broad regimes of behavior. Systems may lie in the ordered regime with frozen components, in the chaotic regime with no frozen components, or in the boundary region between order and chaos where frozen components just melt.

The existence of this phase transition suggests that the boundary region might be a particularly interesting region for useful behavior in complex parallel-processing networks. This suggestion has been made by myself (1985c), by Packard (1988), by Langton (1986, 1990), and most recently by Crutchfield (private communication). The central idea is that, if a network is deep in the frozen phase, then little computation can occur within it. At best, each small unfrozen, isolated island engages in its own internal dynamics functionally uncoupled from the rest of the system by the frozen component. In the chaotic phase, dynamics is too disordered to be useful. Small changes at any point propagate damage to most other elements in the system. Coordination of ordered change is excessively difficult. At the boundary between order and chaos, the frozen regime is melting and the functionally isolated unfrozen islands are in tenuous shifting contact with one another. It seems plausible that the most complex, most integrated, and most evolvable behavior might occur in this boundary region. It is not yet unambiguously clear that this hypothesis is correct. I describe next supporting reasons and, more important, an approach now under way to investigate whether complex adaptive systems attain the edge of chaos. The preliminary results are encouraging.

Langton (1986, 1990) has considered simplifications of Boolean networks—specifically, two-dimensional lattices in which each site has eight possible states 0, 1, . . . 7. Each site is assigned a function determining its next state as a function of the current states of its four neighbors and itself. The same rule is assigned to each site. Systems of this kind are called two-dimensional cellular automata (Wolfram 1983, 1984). Langton characterized his cellular automata rules by a lambda parameter which is the analogue of the internal homogeneity parameter for Boolean functions. Lambda measures the fraction of the input states which result in a 0 output. Specifically, λ is 1 minus the fraction of 0 output values; hence $\lambda = 0$ means that all output values are 0. As λ increases, the fraction of 0 outputs, or internal homogeneity, decreases.

Langton finds three main ranges of λ. When λ is near 0.0, large frozen components, all of whose sites have value 0, arise and percolate across the system. At a critical value λ = .28—apparently the same percolation threshold as that found by Stauffer (1985, 1987a, 1987b, 1989)—the frozen component melts. As λ increases further, the dynamical behavior becomes ever more chaotic in the senses described above. In analogy with the solid, liquid, and gas phases of ordinary matter, Langton calls the low-λ frozen state "solid." The critical value of λ where the frozen component just melts corresponds to the "liquid" phase. As λ increases further and behavior becomes massively chaotic, the system passes to the "gas" phase. The analogy with three phases should not be taken too literally, however. Matter exhibits a true liquid phase. The boundary between order and chaos does not appear to be a distinct phase.

Langton analyzes the dynamics in his two-dimensional cellular automation with a familiar measure of *mutual information* between two elements in a network. Mutual information is the entropy of site 1 plus the entropy of site 2 minus the entropy of the pair of sites jointly. Here the entropy of a site is the familiar $\Sigma_{i=1}^{n}(P_i \log_2 P_i)$ for the frequencies (P_i) of each site's values 0 . . . 7. The entropy of the pair of sites is over the sum over the $8 \times 8 = 64$ possible pairs of site values. According to this measure of mutual information, if the behavior of the two sites is uncorrelated, the mutual information is 0. Conversely, if the behavior of each site is very simple, its own entropy is near 0 and hence mutual information is again near 0. In the gas phase, behavior is so chaotic that mutual information decreases to 0. Knowing what occurs at one site carries no information about what happens at the other site. In the solid phase, where lattice sites are largely frozen, the behavior of each site is so simple that its entropy is low; hence the mutual information between the two sites is also low. Intuitively, one learns little new about site 2 by knowing the state of site 1. In the liquid interface, however, where the frozen component is just melting, mutual information is maximized. The behavior is complex but sufficiently ordered that mutual information is propagated throughout large regions of the system by confined cascades of damage. Thus information processing within a parallel-processing network is maximized at the edge of chaos. Such systems should be able to carry out the most complex computations.

Packard (1988) has found evidence that adaptation for the capacity to carry out complex computations may result in adaptation to the edge of chaos. He selected on one-dimensional cellular automata, which like other Boolean networks, follow state trajectories to state cycles. Packard carried out adaptation by mutation and selection. Each member of the population was a specific cellular automation. Its mutant neighbors were those which altered single bits in the Boolean function governing all sites in the automation. The *fitness* of the automation was defined as the capacity to discriminate initial states with .5 of the sites or fewer in the 0 value from those with more than .5 of the sites in the 0 value. Discrimination was achieved by flowing from the initial state to a terminal attractor either of 0 values at all sites or of "1" values at all sites. At each generation, a population of cellular automata was evaluated for the fitness of each automation, and the best were kept together with mutants to seed the next generation.

Packard found that selection to accomplish this computational task induced an adaptive walk in the space of Boolean rules toward rules which lie on the boundary of chaotic dynamics. He surmises that this may be a very general feature of complex parallel-processing systems which have adapted to perform complex computation.

Crutchfield and Young (1990) have considered *symbolic dynamics.* Here the idea is to characterize the behavior of a continuous dynamical system by dividing its state space into arbitrary regions. For example, if the state space is broken into two regions labeled 1 and 0, then any continuous trajectory of the continuous dynamical system results in a sequence of 1 and 0 symbols reflecting the way the trajectory passes through regions 1 and 0 over time. If one chooses the two regions with care, the sequences of 1 and 0 values carry maximum information about the underlying continuous dynamical system. But then comes the pretty addition. Since one has characterized the continuous dynamics by a symbol sequence of 1 and 0 values, the continuous system can be modeled by a binary Boolean network of some sort. Crutchfield and Young argue that this procedure allows one to define the optimal (minimal) binary model for the continuous process. Furthermore, they find the same three categories of behavior—solid, liquid, gas—we find in analyzing Boolean networks. The most complex and interesting dynamics again corresponds to the liquid interface. In short, there may be a generalization from these three classes of behavior in binary and other finite state systems to continuous dynamical systems. In all cases, the most complex computations may occur in the liquid region.

These ideas are attractive. Orderly dynamics is due to the percolation of a frozen phase containing functionally isolated islands. Conversely, the chaotic phase has no percolating frozen clusters. Deep in the frozen, orderly phase, each functionally isolated island can perform its own computations but is unable to communicate with other islands. Conversely, in the chaotic phase, orderly computation seems improbable since any slight perturbation will cause damage to spread exponentially. At the margin, normally isolated islands might be in tenuous contact with one another. Thus it is more than plausible that adaptation for complex behaviors in parallel-processing networks requires that the networks be close to the boundary of chaotic behavior.

How Networks Play Games

I consider next how we can investigate more fully whether natural selection generically seeks and attains systems at the edge of chaos in order to perform complex computations. There are four subissues:

1. The ability of an optimal parallel-processing network to perform some task or computation is governed *by the task.* If the task is very simple, a simple system will suffice. If the task is more complex, a more complex network will be required. Thus we need a means of generating tasks of differing complexity which we can ask parallel-processing networks to carry out.

2. For any given task, we must characterize the network properties—in particular, the position on the order–chaos axis that optimizes the capacity of the network to carry out the task. Thus for simple tasks, networks either deep in the ordered regime or deep in the chaotic regime might perform as well as those near the edge of chaos. For complex tasks, networks near the complex regime might be clearly superior to both solid and gas networks. If for most tasks networks near the liquid region are superior, we have broad grounds for claiming that such networks are in general the most useful class.

3. We must assess whether natural selection can achieve and maintain networks which are those best able to carry out a given task. If this is also typically true, then we have grounds for believing that selection has achieved such systems in the biological world.

4. We must assess whether organisms are characterized by the putative optimal class of parallel-processing networks.

A coherent approach to this set of issues makes use of game theory to set a range of tunably complex tasks for networks to carry out. The application of game theory to coevolution, introduced into evolutionary theory by Maynard Smith and Price (1973) and Maynard Smith (1974, 1982), is the central focus of the next chapter. Here I describe its outlines briefly. In game theory, each player has a set of possible actions which it might take. When two players play a game, the payoff of each player depends on its own action and upon that of its opponent. Thus the payoff of each player is given by a *payoff matrix* showing the results the player obtains for each combination of its own action and that of its opponent. Each player's objective is to maximize its payoff. Players may play only once or repeatedly. The main point to emphasize here is that a payoff matrix often has a payoff landscape like those we have already encountered in considering adaptive evolution. The game-theory analogue of a fitness landscape arises if each player has a set of actions in which there is a notion of *distance* between actions, such that some actions are near a given action, while others are far from it. If there is a distance between each player's actions, then if one player plays a *constant* action, the second player confronts a payoff landscape as that player moves to nearby or distant actions. Further, by moving to a neighboring action, the first player deforms the structure of the second player's payoff landscape to some extent. Therefore, given the concept of neighboring actions, payoff matrices can be characterized both by how multipeaked and rugged they are for a fixed action by the opponent and by how much the landscape deforms for each neighboring move by the opponent. Clearly, these two features tune how difficult a task each player confronts.

The next step asks Boolean networks to carry out increasingly complex tasks by playing increasingly complex games. I describe next work with my colleague Sonke Johnsen which has just begun and has not yet yielded conclusive results. We are analyzing the adaptive behavior of a population of 20 Boolean networks which play defined games with one another. Each network begins with $N = 20$ binary variables. The first six are defined as match sites. Networks play games with one another by matching their six match sites and obtain a score which depends on both the current activity states, 1 or 0, of their own six match sites and the activity of the six match sites of the opponent. Thus each player has $2^6 = 64$ actions, each action is one move away from six other actions which differ in the activity of a single match site, and the payoff matrix is 64×64. The concrete game we are currently analyzing is a mismatch game. Each player receives a score given by the fraction of its match sites which *differ* in activity from its opponent's match sites. Thus (000000) versus (000111) yields a score of ⅚ for each player. The mismatch game generates a simple payoff matrix landscape, with peaks for maximum mismatch and payoffs which fall off smoothly as mismatches decrease.

This game is subtle if there are more than two players. With two players, each can optimize payoff by settling to a fixed pattern exactly opposite that of its opponent. With many players, a trivial constant strategy yields to each player a score of about .5. Achieving higher payoff requires that each network sense what the other network is currently doing at its match sites, such that each can alter behavior to increase mismatches. To enable sensing, we let each match site also be an "eye" which looks at the current activity of the corresponding match site in its opponent via an input from that match site. To sense and react to its opponent, each player must play each opponent more than once. In our simulations, opponents play one another ten times in a row. Between each of the ten plays, each network takes ten state transitions to "think" about its next response. Each network plays all the other networks in the population. The fitness of each network is given by its mean score after playing all the other networks.

To study the adaptive behavior of game-playing Boolean networks, we utilize the standard population genetics paradigm. At each generation, a new generation is created by biased sampling such that the chance of having an offspring is proportional to fitness. After a new generation of networks has been created, different classes of mutations occur with differing probabilities in each network. In order of frequency, the Boolean function governing a site may be altered bit by bit, thereby changing the logic of the network; an input to or output from a binary element to another element in the network may be altered, thereby changing the wiring diagram of the system; or a binary element may be entirely deleted from or added to the network. This general flexibility allows networks to change the values of P, K, and N. In addition, subtle sculpting of the basins of attraction and attractors can occur.

Our early results show that networks can and do solve the mismatch game by adapting toward the ordered regime near the edge of chaos. How often this solution is sought is still not clear. We have studied systems begun deep in the frozen regime ($K = 1$), deep in the chaotic regime ($K = 7$), and for K values crossing the boundary region into the chaotic regime ($K = 1, 2, 3, 4, 5, 7$). We have analyzed the typical changes in K, P, and N and in a measure of convergence in state space defined below.

Several results appear reliable. Networks do improve their mean scores well above chance, K changes toward the boundary region, P changes toward the boundary region, and N increases.

The first observation is important. Boolean networks do adapt and partially solve the mismatch game. Mean scores typically increase from .5 to about .6 to .7, well above chance. Thus changes in structure and logic during adaptation are in fact abetting improvement.

Systems deep in the frozen regime ($K = 1$) typically *increase* the mean number of inputs per element to values between $K = 2$ and $K = 3$. Systems deep in the chaotic regime ($K = 7$) typically decrease the mean number of inputs per element to at least $K = 4$.

Simulations show that, in networks begun in the chaotic regime, P increases toward the boundary values. Similarly, in networks begun with $K = 1$, P moves toward the boundary region.

The increase in N is interesting, for it suggests that, typically, larger, more complex networks can solve the mismatch task better than smaller networks.

Although these results suggest that networks are adapting toward the boundary region between order and chaos, these measures do not adequately characterize whether Boolean networks are in the solid, liquid, or gas regime. Two improved measures concern mean convergence in state space and the complexity of state transition behavior while playing a game. Only the first measure, described next, has begun to be employed. The second measure, based on Crutchfield's measure of the complexity of a stream of 1 and 0 symbols in terms of the simplest finite automaton which can generate that sequence, might prove useful.

In the chaotic regime, nearby states diverge from one another. In the ordered regime, nearby states converge. This suggests that we should measure average convergence in the state space of a Boolean network as a function of how far apart initial pairs of states are. The measure is analogous to the recurrence mapping in Derrida's annealed approximation, described above. The convergence measure is obtained by considering pairs of states which differ in one site value, in 10 percent of their site values, in 20 percent of their site values, and so forth. For each pair, the normalized

Hamming distance D_T between the pair is calculated, the successor to each state in the pair generated by the Boolean network is determined, and the normalized Hamming distance between the successor states D_{T+1} is determined. If the two initial states are on diverging trajectories, $D_{T+1} > D_T$. If the trajectories are converging, $D_{T+1} < D_T$. To obtain reliable statistics, 100 randomly chosen pairs of states at each initial distance were sampled. The recurrence mapping, with D_{T+1} plotted as a function of D_T, is essentially identical to that derived by Derrida for the annealed model (Figure 5.8). For $K = 3$ networks, the recurrence curve is above the 45° slope for small values of D_T; hence $D_{T+1} > D_T$ and nearby states diverge. The system is chaotic. In contrast, for $K = 2$ networks, the recurrence relation is below the 45° line; hence states at all initial distances lie on trajectories which converge. The system is ordered.

If adaptation in Boolean networks is associated with a resculpting of trajectories toward the ordered regime near the boundary of chaos, then we might expect to see initially chaotic networks alter their recurrence curve from *above* the 45° line to *below* it. Visualization of changes in the recurrence relation is aided by a slightly altered graphing procedure, which plots, at each time moment during the adaptive process, for each initial normalized Hamming distance 1, 10 percent, 20 percent, . . . 80 percent, the ratio D_{T+1}/D_T. If this ratio is greater than 1.0, states are on diverging trajectories; if the ratio is less than 1.0, states are on converging trajectories.

Figure 5.15 shows results for populations of networks initiated in the ordered or in the chaotic regime playing the mismatch game. Each curve in each figure corresponds to a unique value of D_T. The topmost curve at the zeroth generation represents the minimal, or one-neighbor, distance. Lower curves correspond to $D_T = 10$, 20, . . . 80 percent. The results suggest that networks adapt toward the edge of chaos.

These results are preliminary. They require extension to obtain more data both for Boolean networks playing the mismatch game and for Boolean networks playing a variety of other harder and easier games. The preliminary results support the hypothesis that adaptation seeks the edge of chaos. The avenue of investigation is obvious, and further supporting evidence has been found (Kauffman 1993).

Solid, Liquid, and Gas: The Implications for Landscape Structure and Evolvability

If there are grounds to suppose that parallel-processing networks near the boundary of chaos can perform the most complex computations, there are also beginning to emerge reasons to think that the fitness landscapes of such systems may optimize the capacity to evolve. The analysis of Chapters 2 and 3 demonstrates that the structure

---

Figure 5.15 Results of evolution experiments in which 20 Boolean networks play the mismatch game against one another. All networks begin with $N = 20$ binary variables. Different populations begin with different numbers of inputs per variable: (a) $K = 1$; (b) $K = 5$; (c) $K = 7$. The $K = 1$ networks begin deep in the ordered regime, $K = 2$ networks are near the boundary of order and chaos, and $K \geq 3$ networks begin in the chaotic regime. Figures plot a measure of convergence or divergence in state space D_{T+1}/D_T against generations elapsed, for different initial distances D_T. Top curve in each figure at generation 0 corresponds to $D_T = 1$ site different; second curve from top at generation 0 corresponds to $D_T = 10$ percent; third curve corresponds to $D_T = 20$ percent, and so on. The value $D_{T+1}/D_T = 1$ is the boundary between divergence and convergence. As expected, states which are initially distant, $D_T > 50$ percent, always converge in all networks, ordered or chaotic. States which are initially close (top two curves) may diverge or may converge. Over adaptation time, initially chaotic systems become more ordered. Deeply ordered ($K = 1$) and deeply chaotic ($K = 5$, $K = 7$) systems appear to approach the boundary between order and chaos.

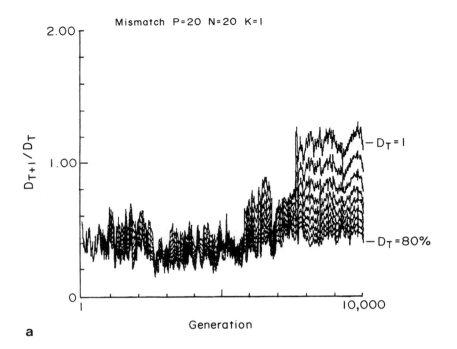

a

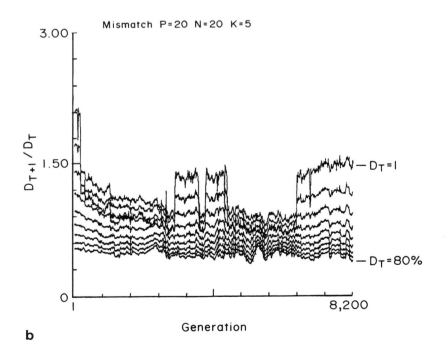

b

225

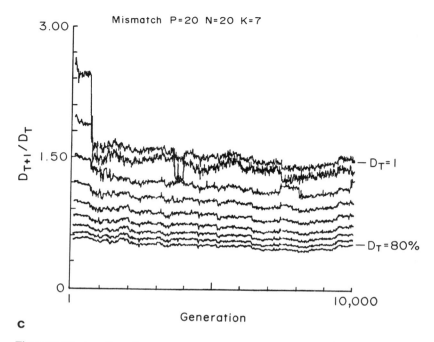

c

Figure 15.15   (*continued*).

of fitness landscapes depends critically on whether or not the adapting system is modular. For example, in the *NK* landscape model, for $K = 0$, each site makes a fitness contribution which is independent of the contributions of all other sites. That is, the fitness contributions are made by independent modules. Consequently, the fitness landscape is very smooth and often single-peaked. Modular construction in parallel-processing Boolean networks can be attained in two very different ways. First, the network may be constructed of structurally independent modules. The $K = 1$ Boolean networks exemplify such construction. These networks fall apart into structurally isolated feedback loops, each with a descendant set of elements controlled by the loop. Each loop plus its tails is structurally independent of all other loops. Mutations altering one loop have no affect on the behaviors of other loops. Adaptation occurs on relatively smooth fitness landscapes.

The second way to achieve modularity is radically different. It rests on the percolation of a frozen component, leaving behind functionally isolated islands unable to influence one another. Thus Boolean networks in the solid ordered regime are inherently *functionally modular*. Behavior within any one island may be simple or chaotic, but such systems adapt on relatively smooth fitness landscapes. Mutations altering the behavior of one island in useful ways can accumulate without altering the behavior of other islands.

In contrast, landscapes are very rugged in the chaotic regime. This ruggedness is a direct consequence of the fact that damage spreads widely in networks in the chaotic regime. Almost any single mutation will dramatically alter landscape structure. But the changing sensitivities of landscapes as networks pass from the ordered to the

chaotic regime again focuses attention on the boundary regime, where single mutations cause a wide variety of changes in dynamical behavior, from minor to major.
This distribution parallels the power-law distribution of damage which spreads after
transient alteration in the activity of any single element for networks at the phase
transition. Recall from Chapter 3 our discussion of *NK* landscape models with a hierarchy of *K* values. Alteration of sites which influence the fitness contributions of
many sites leads to adaptation on very rugged landscapes. Alteration of sites which
epistatically affect few sites results in adaptation on smooth landscapes. Recall that
this hierarchy yields a deep buffering in the adapting system. If the landscape deforms
dramatically, highly entrenched sites affecting many other sites tend to alter. If landscapes deform only slightly, slightly entrenched sites preferentially change. But precisely the same features hold true for Boolean networks near the edge of chaos. Many
mutations cause minor changes in behavior, and some cause massive changes.
Although we have concentrated on adaptation on fixed fitness landscapes, we shall
see in the next chapter that fixed landscapes are a fiction. Real landscapes deform as
a result of coevolutionary effects and alterations in the abiotic world. Based on this
reality, it is eminently plausible that Boolean networks in the ordered regime but near
the boundary of chaos may harbor both the capacity to perform the most complex
tasks and the capacity to evolve most adequately in a changing world.

Are parallel-processing networks performing complex tasks poised at the edge of
chaos? The answer is unknown. However, the simple overview summary of Chapter
12, which discusses the emergence of ordered behavior in genomic regulatory systems, is that genomic systems are in the solid phase, poised near the edge of chaos. If
true, this is remarkable. First, it provides a kind of statistical mechanics of genomic
regulatory systems and hence a theory for the structure, organization, behavior, and
capacity to evolve in such systems. But in addition, *if* genomic systems of plants to
animals, separated for the past 600 million years, are all poised near the edge of chaos,
then we would virtually have to conclude that selection has achieved such a poised
state. If true, this finding would provide striking evidence that parallel systems with
nearly melted frozen components possess the construction requirements which permit complex systems to adapt. Hence such features might be quasi-universals in
complex adaptive systems.

Implications for Neural-Network Models

The results we have discussed on the behavior of disordered Boolean networks may
have an important bearing on neural networks: Attractors in neural networks will be
chaotic in the absence of some further ordering principle. Learning may be that principle.

Since McCulloch and Pitts (1943) introduced the "formal neuron" as an on–off
idealization of real neurons, it has been clear that neural networks can be modeled
as complex networks of binary switching elements. Neurons receive excitatory and
inhibitory inputs from other neurons. McCulloch and Pitts realized that this mechanism could be captured by utilizing a special subclass of Boolean functions called
threshold functions. A formal neuron with many excitatory and inhibitory inputs
fires at any moment if the sum of the excitatory activity from its active excitatory
inputs minus the sum of the inhibitory activity from its active inhibitory inputs
exceeds a threshold. The Boolean "Or" function is such a threshold function. If the
synaptic weight on each of two inputs is 1.0 and the element has a threshold of .5, it

will fire if either or both inputs were active at the previous moment. If the threshold is 1.5, the element will be active only if both inputs were active and hence compute the "And" function.

Current models of neural nets, which may use a sigmoidal output response from the neuron to the input activity level rather than an all-or-none output from the neuron (Hinton, Sejnowski, and Ackley 1984; Hinton and Sejnowski 1986; Hopfield and Tank 1986a, 1986b; Grossberg 1987), often seek to model pattern-recognition capacities and associative memories by these parallel-processing networks in terms of the attractors of such networks (Hopfield 1982a, 1982b; Toulouse, Dehaene, and Changeux 1986). For example, the attractors might be thought of either as memories held by the neural network or as concepts. Then such networks are naturally content-addressable. This phrase means that, if released in the basin of attraction of a specific memory or concept, the system will flow under the dynamics of the network to that attractor. Hence partially correct information, releasing the system in the correct basin of attraction but not yet at the attractor, can lead to recovery of the proper memory or concept. Note that a system with attractors naturally *classifies* and *generalizes.* All initial states in the same basin converge to the same attractor and hence are classified as identical. In order that this classification be "natural," however, it is reasonable to suppose that *similar* things are typically classified as the same. But this property is *not* an inevitable property of systems with attractors. In chaotic Boolean or neural networks, nearby initial states typically flow to entirely different attractors. There is no natural sense in which similar states are classified as similar in such networks. Thus the capacity to classify similar things together via flow to the same attractor requires that the state-space flow of the classifying dynamical system be quite ordered. Nearby states must typically flow to the same attractor. In short, sensible classification requires that the dynamical behavior of the network not be chaotic. This requirement suggests that natural classification requires neural networks to be either in the solid phase of orderly dynamics or at the edge of chaos. But can this be expected? And if so, on what basis?

The central problem raised by the results we have reported on disordered Boolean networks is that networks with a large number of inputs per element and randomly chosen Boolean functions tend overwhelmingly to have chaotic attractors. Thus it is important to ask whether the same chaotic behavior occurs in networks with randomly chosen *threshold* functions on many inputs. The answer is "yes." Consider binary networks with equally many excitatory and inhibitory inputs per element, and let each element have a threshold of 0.0. Numerical simulations show that the lengths of attractors increase exponentially as the number of model neurons increases (Kauffman 1984a; Kurten 1988a, 1988b). Based either on the theory of percolating forcing structures or on the theory of homogeneity clusters, this is what we should expect. When the threshold value is near 0.0, the corresponding Boolean threshold functions have nearly equal numbers of 0 and 1 outputs on their 2^K input patterns, and K is large. Therefore, a frozen component does not percolate. Conversely, if the threshold is much higher or lower than 0.0, then the output of the neuron is biased sharply toward 0 or 1 over the set of input states, and one might expect internal homogeneity clusters to percolate and thus create a frozen component. Results from Derrida (1987c) and Weisbuch (1989a) show that, in order for such a frozen component to percolate in networks in which the K inputs to each binary neuron are chosen randomly among the N neurons, the absolute value of the threshold must be greater than $\sqrt{K} \ln K$ when the synaptic weights are randomly chosen to be

+ 1 or − 1 for each of the K inputs to each neuron. When the number of input neurons is large, this critical threshold is a reasonably small fraction of K. Thus alteration of the threshold levels of elements in model neural network can convert a parallel-processing system with chaotic attractors to one with orderly attractors.

Learning in neural-net models occurs by tuning of the synaptic weights which govern how strongly one input neuron affects the postsynaptic neuron (Rummelhart and McClelland 1986). Insofar as it is imagined to achieve useful attractors, learning is a *walk in synaptic weight space seeking good attractors.* Presumably, such a walk would be extremely difficult for chaotic attractors. Some general mechanism must temper such chaos and corral it into order.

Learning itself may be the fundamental mechanism which converts chaotic attractors to orderly ones. The postulated "Hebbian synapse" increases the synaptic weight between pre- and postsynaptic neurons if they fire at closely correlated times. This alteration in synaptic weight *increases the probability that the presynaptic neuron fires the postsynaptic neuron.* Thus it increases P toward P_c and correspondingly increases the probability that *homogeneity clusters will percolate.* Indeed, Kurten (1988a, 1988b) has shown, both for this learning rule and for a variety of similar rules, that these alterations in synaptic weight convert networks from chaotic to ordered attractors. He analyzed model neural networks having three inputs per neuron and allowed the networks to tune synaptic weights as a function of their own spontaneous firing patterns. His results are clearly related to those of Weisbuch (1989a), for altering synaptic weights relative to a fixed threshold is related to holding weights fixed and altering the threshold. In either case, transition from a chaotic to an ordered regime occurs. Indeed, a difficulty with most versions of the Hebbian synapse is that the system tends to dig itself into a Hebbian hole deep in the solid regime. Attractors tend to become too deeply grooved into the system. This embedding may inhibit flexible learning by trapping the system too readily in suboptimal responses. It remains to be seen whether or not, under modified learning rules that couple anti-Hebbian and Hebbian rules in order to suppress Hebbian holes, complex learning is better achieved near the edge of chaos.

Another implication of our general analysis of networks and landscapes is important. Models of neural networks utilize the attractors of such networks as either internal memories or categories. In the context of neural networks one seeks training procedures such that a network may come to have the desired attractors. The attractors in turn, depend on the synaptic weights in the network. The problem is whether there are procedures tuning the synaptic weights, hence moving in synaptic weight space, which can move a neural network with an initial set of attractors to one with the desired attractors. With the exception of the Hopfield model (1982a, 1982b) where weights can be calculated beforehand to ensure the structure of a potential surface leading to desired point attractors, learning is thought to occur by some form of updating of synaptic weights to mold attractors to match more closely those which are desired. Back propagation, a procedure which corrects synaptic weights in different layers of a network by calculating backward from the difference between output and desired output (Rummelhart and McClelland 1986) works well in feedforward layered networks is far more difficult in networks whose neurons are coupled in rich cyclic webs (Pineda 1987).

Recall our results from a previous section attempting to mold state cycle attractors in Boolean networks by mutating connections and/or Boolean functions. The landscapes are very rugged. We were unable to begin with an arbitrary network and hill climb to a network with an arbitrary steady-state pattern of activities. That is, we were unable to hill climb to an arbitrary attractor. All the phenomena discovered for the *NK* family of rugged landscapes—trapping on local optima, freezing, the complexity catastrophe, and so forth—were found again. There is no reason to believe that the same problems will not typically afflict the attempt to achieve *arbitrary attractors* of binary neurons from initial neural networks by successive modification and hill climbing in synaptic weight space.

Similar trapping problems evolving in synaptic weight space can be expected to arise in model neural networks which use continuous sigmoidal kinetics rather than binary elements. Rugged landscapes in synaptic weight space should exist. Rummelhart (private communication) has pointed out a particularly interesting way in which learning in continuous neural networks is related to landscape ruggedness. Learning in such networks begins with all synaptic weights near zero. This level corresponds to firing activity levels at the midpoint of the sigmoidal response curve, far from the upper or lower asymptote. By construction, the sigmoidal response function is nearly linear in this middle region. As learning begins, synaptic weights begin to change slowly from zero and so move away from the midpoint of the response curve. In this linear region, however, a gradient search method minimizing the difference between the desired attractor and the actual attractor is solving a linear problem. The corresponding landscape in synaptic weight space has only a single peak with smooth sides, just as in the *NK* landscapes for $K = 0$. In short, the landscape is very smooth for small synaptic weights, and hence the optimum is easily sought. As synaptic weights increase, the system is driven farther from the midpoints of the sigmoidal response functions into the nonlinear regions near the asymptotes, and landscapes become progressively more rugged. Trapping on suboptimal peaks occurs. The importance of Rummelhart's observation is that continuous sigmoidal systems may afford a means of exploring a relatively smooth landscape early in the adaptation process, thereby locating a good region of weight space. Thereafter, as synaptic weights diverge from 0, the landscape becomes increasingly rugged and the system becomes trapped.

A final feature of networks and adaptation on rugged landscapes may bear on characteristic learning curves in neural networks. The number of directions uphill dwindles in characteristic ways as local optima are approached. If learning in neural networks is a search process in synaptic weight space, then we might expect similar generic features to apply and have implications for the rate of learning as a function of the number of learning trials. We saw in Chapter 2 that, for a wide range of K values in the *NK* landscape model, the resulting landscapes had the property that the number of ways uphill dwindled by a constant fraction after each step uphill. For example, on completely random landscapes, the number of directions uphill drops by half after each improvement step. In contrast, for very smooth landscapes—for example, the $K = 0$ additive landscape—the number of directions uphill dwindles by a constant number (here by one) at each adaptive step. Thus on reasonably rugged landscapes, the cumulative number of improvement steps taken increases—rapidly at first, then more gradually—as the logarithm of the number of trials. On smoother landscapes, the cumulative number of improvement steps increases even faster than this initially, then slows.

Remarkably, many learning curves in psychology and even economics follow just such curves (Herrnstein, private communication; Newell and Rosenbloom 1981). Perhaps all these phenomena are near-universal consequences of the structure of complex combinatorial optimization landscapes.

Means of Tuning the Correlation Structure of Fitness Landscapes

I conclude these considerations by summarizing a variety of means of tuning the structure of fitness landscapes.

We have already focused on modularity. Familiar examples include the domain structure of proteins, which allows new proteins to evolve by recombination between different structural genes, assembling new combinations of functional domains. The same principle is found in the combinatorial building of antibody molecules, in operon behavior, in the construction of tissues from cellular modules, and perhaps in the metameric segemented character of many organisms, where each segment can be modified independently of the others.

In richly coupled parallel-processing networks, such as Boolean or neural networks in the ordered regime, functionally isolated, unfrozen islands resulting from percolating frozen components are a related second fundamental means of achieving correlated fitness landscapes. Indeed, the isolated islands provide functional modularity.

A third means lies in structural stability, due to the general fact that, for most familiar continuous dynamical systems, small changes in parameters typically do *not* cause bifurcations but only small changes in the behavior of associated dynamical systems. That is, most neighboring dynamical systems behave similarly and hence exist in a correlated landscape (Thom 1972; Hirsh and Smale 1974).

A fourth means lies in the diversity of time scales operating in real complex systems (Goodwin 1976). System variables which change on a very fast time scale exhibit only their averages to variables which operate on a slower time scale. An enzyme population "sees" the average concentration of its substrates; conversely, the current pattern of gene transcription making new enzymes may change so slowly on the time scale of a catalyzed reaction that those transcription rates are slowly altering parameters for the enzyme system. Thus, in general, time scales break up a system into functionally independent subsystems having similar response times and damp out small fluctuations in faster variables, while slower ones continue to change smoothly.

A particularly important means of tuning landscape structure lies in how *compressed* a system's description of itself is. For example, consider an algorithm to generate a particular computer output. One measure of the complexity of an algorithm is the minimal program length which yields the desired output. The more minimal an algorithm is, the less redundancy it contains. Consequently, minor modifications of minimal programs grossly alter the output of the algorithm. In contrast, minor alterations of highly redundant algorithms modify output only slightly. In short, compression of an algorithm increases the ruggedness of the landscape it adapts upon. The minimal program would adapt on a fully random landscape. The analogue of compression arises in parallel-processing networks. Consider specifying a subset of the variables as input and output variables, while the remainder are "hidden units," or internal control variables. In this context, compression lies in reducing the number of hidden units. As this number is reduced, the input–output mapping adapts on a more rugged fitness landscape. I suspect that this form of compression, favored because it reduces redundancy and construction costs but increasingly harmful as landscape ruggedness approaches a fully uncorrelated limit, plays a major role in evolution in tuning landscape structure and evolvability.

A Bold Hypothesis: Knower and Known, Life Adapts to the Edge of Chaos

We have now seen that the transition region between order and chaos gives rise to the most complex dynamics. In addition, tentative evidence supports the hypothesis that parallel-processing systems coevolving to carry out complex tasks, such as the mismatch game, do in fact evolve both from the ordered regime and from the chaotic regime toward the edge of chaos. Thus we are led to a bold hypothesis:

> Living systems exist in the solid regime near the edge of chaos, and natural selection achieves and sustains such a poised state.

Such a conjecture may well be overstrong, but the analysis in this chapter renders it more than slightly plausible. Boolean systems, and by extension some large family of homologous nonlinear dynamical systems, with nearly melted frozen components can carry out the most complex, yet controllable behavior. Such poised systems are also highly evolvable. They can adapt by accumulation of successive useful variations precisely because damage does not propagate widely. Useful alterations in the behavior of one functionally isolated island can accumulate with useful alterations in another island. Furthermore, evolvability is high in networks near the order–chaos boundary because here many mutations cause minor changes and some mutations cause major changes. In a changing environment, this range of responses provides adaptive buffering: If the abiotic or coevolutionary world changes dramatically, large useful changes due to single mutations can be found rapidly; if the world changes only slightly, minor useful changes in behavior lie to hand.

In contrast, systems deep in either the ordered regime or in the chaotic regime are probably neither capable of complex behavior nor highly evolvable. Deep in the ordered regime, islands are small and functionally isolated. Complex behavior cannot propagate across the system. Here, virtually all mutations cause only minor changes in behavior. While landscapes are smooth, achieving large alterations in behavior is cumbersome. Further, in the absence of a range of effects of single mutations, the system can respond to large deformations of its fitness landscape only with small changes in behavior. Adaptive buffering is not present. Deep in the chaotic regime, slight changes in structure almost always cause vast changes in behavior. Complex controllable behavior seems precluded. Such deeply chaotic systems must also adapt on very rugged landscapes. Thus it is plausible that systems in the solid regime near the edge of chaos possess the internal organization which permits, and may be required for, subtle and successful computation and adaptation.

But it is also plausible that systems poised at the boundary of chaos have the proper structure to *interact* with and internally *represent* other entities of their environment. In other words, complex living systems must "know" their worlds. Whether we consider *E. coli* swimming upstream in a glucose gradient, a tree manufacturing a toxin against a herbivore insect, or a hawk diving to catch a chick, organisms sense, classify, and act upon their worlds. In a phrase, organisms have internal models of their worlds which compress information and allow action. An old philosophical skeptic puzzle due to Hume is "Why is the future like the past?" It is not an answer to Hume, but neither is it beside the point to notice that all organisms live in worlds which require practical action. Such action requires that the world be sufficiently stable that the organism is able to adapt to it. Were worlds chaotic on the time scale of practical action, organisms would be hard pressed to cope. But how do organisms know their worlds? Again, no adequate theory exists, but our examination of Boolean systems may provide part of the answer.

Consider a system—neural, genetic, molecular, even logical—characterizable as a Boolean network. For any such system, its information about the other entities in its world and about other aspects of its environment can be thought of as connections from the other entities or the environment to the binary variables within the system. Then the system is driven through its own dynamical behavior in part by its internal states and in part by the barrage of data arriving from its world. What does it mean for such entities to know one another, and how might that knowing depend on whether the system is in the solid, liquid, or gas regime? My best guess is that systems in the solid regime near the boundary of chaos can best know one another.

A central part of this problem considers the criteria for natural classification. The capacity to know a world requires that sufficiently similar states of that world be able to be classified as "the same." Consider *E. coli* and a receptor for glucose. Because of the way receptors bind ligands, a family of similar ligands is typically bound by any given receptor. All members of the set of ligands bound by the receptor are, roughly, classified as "the same." (I permit myself the word "classified" because we may imagine that the bacterium responds more or less identically to any ligand binding the receptor, be it glucose or some other molecule.) Of course, further molecular evolution allows modification of the receptor and hence permits discriminating among ligands formerly classified as the same.

But Boolean networks classify as well. As described earlier, the current enthusiasm for connectionism and for parallel-processing neural networks is precisely that any such systems almost inevitably classifies. Any such network has internal dynamics whose attractors represent alternative asymptotic states of the network. The alternative attractors in a fixed environment from which the network receives inputs can be thought of as alternative classifications of the same environment. Similarly, alternative environments which map to the same attractor can be seen as classifications of those different environments as "the same" by the network. Thus such networks inevitably classify and have internal models of their worlds.

Nevertheless, the classes formed by arbitrary Boolean networks are not necessarily natural, useful, or evolvable. The sensible image underlying connectionism is that *similar states* of the world *are classified as the same.* Yet just that property is *absent* in networks whose behavior is chaotic. In such networks, arbitrarily nearby points in state space map to arbitrarily different attractors. Attractors cannot readily be paradigmatic cases of a class of similar objects. Nor can such classes be achieved readily by altering the structure and dynamics of a network in the chaotic regime. Conversely, just this property is *present* in networks in the solid regime. Indeed, three important properties are present in solid-regime networks. First, similar states typically flow to the same attractor and hence are classified as the same. Second, minor alterations in network structure and logic can cause nearby states which formerly flowed to the same attractor to flow to two different attractors; hence discrimination can be achieved. Third, states along trajectories flowing to the same attractor converge on one another. Organisms undoubtedly know and respond to their worlds not only by reaching attractors and then responding, but also by responding while on transients. Yet the same issues of control arise. In chaotic systems, nearby initial states diverge from one another even before reaching attractors. Minor alterations in initial conditions inevitably imply drastic change in response. Conversely, for ordered systems, nearby initial states converge along transients. Similar conditions elicit similar responses.

These points are obvious. They suggest that organisms and other entities which interact with their worlds are likely to couple to those worlds in such a way that smooth classification occurs, and the world is seen as relatively stable. Then the "knower" should not be chaotic, nor should its classification, the "known," be. It is a reasonable guess that both the knowing system and the known world are in the solid regime, perhaps near the edge of chaos.

The bold hypothesis will undoubtedly prove oversimple. Yet it does point in a direction. We may ultimately form a theory of evolving entities which optimize how they know their worlds by adapting under the aegis of natural selection to the edge of chaos. Strikingly, we see in the next chapter that coevolving entities may also achieve the edge of chaos.

SUMMARY

This chapter's overarching purpose is to seek the construction requirements which permit adaptation in complex systems. Our broad question is whether there may be, in such complex systems, nearly universal principles of construction. If so, we might suppose that selection achieves such systems.

In pursuing this broad objective, we have introduced the concepts of dynamical systems theory, state spaces, trajectories, simple and complex attractors, parameter spaces, and structural stability. Most biological systems confront us with vast numbers of connected elements. Typically we are ignorant of the details of structure and "logic" by which the elements of such systems are coupled. And in the evolution of genomic regulatory networks, molecular adaptation during immune responses, or learning in neural networks, the very structure and logic of the system are being modified, indeed scrambled, due to either mutations or their analogues. To study the behavior and evolution of these systems, we need a new kind of statistical mechanics, one which defines ensembles of dynamical systems and seeks generic behaviors in distinct ensembles. The generic properties then become the macroscopic observables expected from the theory.

It is just such a new statistical mechanics we have broached. I have done so in terms of defined ensembles of random Boolean networks. These are massively parallel-processing systems linking the activities of thousands or millions of variables. Boolean networks exhibit three regimes: ordered, complex, and chaotic. The ordered regime arises because a large fraction of the elements fall to fixed active or inactive states. These fixed elements form a connected, percolating, *frozen* component which spans the system. The frozen component leaves behind functionally isolated islands of unfrozen elements, free to fluctuate in activities in complex ways. In the chaotic regime, the unfrozen elements form a connected component which percolates across the system, leaving isolated frozen islands. The border between order and chaos—the complex regime—is a phase transition when the frozen component is just percolating and the unfrozen region is just breaking into islands.

In the chaotic regime, attractors are very large and scale as an exponential function of the number of binary variables in the system. Transient alterations in the activity of any single variable unleash a cascade of alterations which propagate to many or most other elements in the network. Thus such systems exhibit sensitivity to initial conditions.

In the ordered regime, attractors are small, boxing behavior into localized regions of state space. Further, because the frozen component percolates across the system, transient or structural alterations in the network do not cause cascades of damage to propagate throughout the system. Systems in the ordered regime therefore adapt on correlated fitness landscapes. In short, such systems literally crystallize order. We have found that tuning such simple construction parameters as the mean number of elements which act as inputs to each variable suffices to drive massive dynamical systems into the ordered regime. Random NK Boolean networks with $K = 2$ inputs to each of 100 000 binary elements yield systems which typically localize behavior to attractors with about 317 states among the $2^{100\,000}$ possible alternative states of activity. Whatever else you may mark to note and remember in this book, note and remember that *our intuitions about the requirements for order in very complex systems have been wrong.* Vast order abounds for selection's further use.

Having marked to note that complex systems exhibit spontaneous order, mark a second, bold and fundamental possibility: *Adaptive evolution achieves the kind of complex systems which are able to adapt.* The lawful property of such systems may well be that they abide on the edge of chaos. This possibility appears to me to be terribly important. The thought that selection achieves systems able to adapt leads ultimately to the question of whether there may be attractors of that selective dynamics. If it is the case that systems poised between order and chaos are indeed the natural culmination of selective evolution, we shall have found deep laws indeed. But caution obviously is required about so large an issue. We do not yet know the range of tasks for which systems in the complex regime are the optimal solution. Nor most important, do we know with conviction what biological systems, if any, abide at the edge of chaos. Adaptation to the edge of chaos may ultimately become a general principle in biology. At present, it must be held as a working hypothesis.

CHAPTER 6

The Dynamics of Coevolving Systems

The true and stunning success of biology reflects the fact that organisms do not merely evolve, they *coevolve* both with other organisms and with a changing abiotic environment. If in the preceding chapter we began to explore the requirements for order and the capacity to evolve in parallel-processing networks, in the present chapter we must take up the very deep issues of how organisms coevolve with one another. It is not at all obvious that such a process should often permit its participants even the most rudimentary success. Thus we seek overarching principles which may permit and govern the emergence of entities capable of coevolving with one another. Since, we may suppose, selection is critical in this emergence, we must seek to understand the ways in which selection attains systems able to coevolve.

In the preceding chapter we found evidence that parallel-processing, nonlinear, dynamical systems—in particular, Boolean networks—crystallize order. Notably, we found evidence that a phase transition occurs between frozen "solid" and chaotic "gas" behaviors. Between these two extremes lies a "liquid" region with nearly melted frozen components, poised at the edge of chaos. Such systems appear able to carry out the most complex computations and yet may harbor sufficiently ordered fitness landscapes that the systems are able to evolve well. Quite strikingly, we shall uncover evidence that natural selection, in a selective metadynamics, may drive coevolutionary systems to a liquid state poised on the edge of chaos. At present, it is an attractive hypothesis that complex coevolving systems ultimately tend to a state in which each system internally is poised at the edge of chaos and that all such systems may coevolve to the edge of chaos as an "ecosystem."

We can make a rough distinction between evolving complex systems and coevolving complex systems. In the former, the components of the system do not replicate, and hence selection cannot act directly on them. Instead, selection acts only on the system as a whole. In the latter, the components of the system replicate, and so selection may act on the level of the parts of the system as well as on the system as a whole.

The simplest example of coevolution may be the hypercycle model for the origin of life, which we address in detail in Chapter 9. In this model, putative template replicating RNA molecules are coupled cyclically such that each molecule helps its neighbor around the cycle to replicate better. The system is symbiotic. Selection may act on individual template-replicating RNA sequences or at a higher level: on the cyclically coupled system as a whole. Multicellular organisms also exhibit coevolutionary processes, for the cells of the organism must mesh their behaviors for the good of the whole organism. As in the hypercycle model, multicellular organisms exemplify the fact that selection works at many levels. When the initial multicellular organisms formed, individual cells, which might have had eternal futures on their own, gave up the propensity for direct cellular descendants in order to take part in the body, or soma, of the organism. Only the gametes pass their genes to the future. Buss (1988) in particular has argued that the emergence of individual multicellular organisms having distinct cell lineages giving rise to soma and to gametes was a major step in evolution. The most familiar examples of coevolutionary processes occur in ecosystems. Flowers and insects have coevolved. The flower evolves to attract the insect, which acts to pollinate; the insect evolves to harvest food from the flower.

There is a fundamental difference between simple adaptive evolution and coevolution. Evolution on a fixed fitness landscape, as discussed in Chapters 2 to 5, is similar to the behavior of a physical system on a well-defined potential energy landscape. In both cases, the attractors of the "adaptive" process are local optima which are single points. In a coevolutionary process, however, the adaptive landscape of one actor heaves and deforms as the other actors make their own adaptive moves. Such coevolving systems may not in general have a potential function. Thus coevolving behavior is in no way limited to attaining point attractors which are local optima, nor is it even clear that coevolving systems must be optimizing anything whatsoever. This mathematical issue parallels long-standing debates in evolutionary theory about what, if anything, evolution is optimizing.

The topic of coevolution is far too large for this book. In the present chapter I summarize some current views of coadaptation at the ecological level. In particular, I describe the game theoretic framework introduced into evolutionary theory by Maynard Smith (1982). I then explore what the *NK* landscape model implies about the success of such coevolutionary processes. This analysis is carried out by thinking of coevolution as a process which *couples* the *NK* landscapes of different "species" such that adaptive moves by one species deform the landscapes of its partners. We seek conditions on landscape structure and coupling such that coevolution can lead to improvements rather than to chaotic fluctuations in fitness as landscapes twitch spasmodically. Discovering such conditions will lead us to identify possible selective processes which tune the structure of a coevolving system such that its members typically are able to adapt successfully. I shall also extend these results to preliminary work applying the *NK* model to ecosystems which exhibit an analogue of the frozen components seen in Boolean networks: Some coevolving species may exhibit stasis, while their partners continue to change. Finally, our results will hint at the possibility that selective tuning of landscape structures and the couplings between landscapes underlying coevolution may achieve a kind of poised state which optimizes sustained fitness of the partners and is characterized by nearly melted frozen components. Instabilities causing local breakdown of the frozen components of ecosystems may underlie bursts of speciation and extinction events which propagate through the system to various extents. The size distribution of these avalanches of change may also account for the size distribution of extinction events in the fossil record.

NK *Boolean Games: A First Hint of Frozen Components in Coevolving Systems*

In this chapter we consider coevolution cast in the framework of game theory. This framework was introduced by Maynard Smith and Price (1973) and Maynard Smith (1974, 1982). The core image of such a coevolutionary process envisions a set of players, which might be members of a single species coevolving with other members of the same species. In addition, the game theoretic framework also extends to interspecific coevolution, where the players are members of different species in an ecosystem. Each encounter between each pair of phenotypes results in a payoff to each of the players. The payoff is interpreted in terms of Darwinian fitness. The process is governed by a *payoff matrix* showing, for each phenotype of each player, its fitness in an encounter with each of the other possible phenotypes.

An oversimple model of coevolution, developed in discussion with economist John Miller, will help introduce the central ideas of game theory. The model consists in a slight reinterpretation of the NK landscape model introduced in Chapter 2. Recall that the NK model consists of N sites, each with A alternative states, or alleles. Each site receives epistatic inputs from K other sites chosen among the N. The fitness contribution of each site for each of the A^{K+1} combinations of alleles of itself and the K other sites is assigned at random from the uniform interval between 0.0 and 1.0. Thus given any genotype for the N sites, the fitness contribution of each site in the context of its K inputs can be calculated. The fitness of the entire genotype is the mean of the fitness contributions of its N sites.

In our simplified version of a coevolutionary game, let each site now be interpreted as an independent "agent." Let each of the N agents be coupled, exactly as in the NK landscape model, to K other agents. Let each have two possible actions: 1 or 0. As in the NK landscape model and the discussion just above, assign a random payoff between 0.0 and 1.0 to each of the 2^{K+1} combinations of actions of the $K + 1$ agents whose action, 1 or 0 bears on a given agent's payoff. The payoff shows how well the agent does with that specific action when played in the context of the 0 or 1 actions of the K agents who affect it.

To complete a simple NK coevolutionary game, let each agent at each moment assess the current action of itself and its K inputs, assess its payoff matrix, and change to whichever action maximizes its payoff under the assumption that the K other agents *do not* change action. This is an extreme myopic strategy. From the evolutionary perspective, it amounts to an adaptive step to a currently fitter phenotype in the context of the phenotypes of the other species in the ecosystem under the assumption that the other species are not changing. From a game theoretic perspective, it is an action exhibiting no foresight with respect to possible moves by other agents.

Note that an action by one player to improve its own fitness may simultaneously lower the fitness of the players with which it is coupled. Successive moves by the players, each attempting to improve its own fitness, may drive the coupled system to ever-lower fitness. In other words, the overall process need not proceed uphill. Thus such a coevolutionary process is not occurring on a potential surface.

Under these assumptions, an NK Boolean game (Boolean because each agent has only two actions) is exactly equivalent to a random NK Boolean network, as described in Chapter 5. The equivalence can be seen trivially. Let the K connections among the N agents be set, say at random. The payoff matrix for each agent for each of the 2^{K+1} combinations of actions of itself and the K agents which bear on it is randomly assigned a value between, say, 0.0 and 1.0. Thus for each fixed combination of actions of the K other agents, the preferred myopic action which the agent should take to maximize its payoff is, as randomly specified by the payoff matrix, either 0 or 1. Since this is true for all combinations of actions of the other K agents, the NK Boolean game, played myopically, specifies at random a given Boolean function for each agent with respect to its K inputs. At each step of the game, each agent acts to maximize its payoff while assuming the other agents do not change. All agents act simultaneously at each clocked moment. Players are restricted to pure strategies; that is, for each combination of actions of the K agents bearing on it, any agent must play either a 1 or a 0 with probability 1.0. (In contrast, games with mixed strategies allow a player to play 0 with probabiliy P and 1 with probability $1 - P$.) Over time, the trajectory of this myopic coevolutionary process describes a flow through the state space, or strategy space, of the N agents. In this limit of *pure strategies,* the dynamics of the myopic coevolutionary game is identical to that of a random Boolean network.

With the restriction to pure strategies, this identification between an NK Boolean game and a random NK Boolean network helps yield intuition about such a simple coevolutionary process. Suppose the NK Boolean game has a fixed steady state where all agents do not change over time—in other words, a steady-state attractor. Then, the action of each agent currently maximizes that agent's payoff under the assumption that no other agent deviates from its current action. Thus a steady state in the game corresponds to a *pure-strategy Nash equilibrium* (Nash 1951). A Nash equilibrium is a combination of actions by a set of agents such that, for each agent, granted that the other agents do not alter their own actions, its action is optimal. As we see shortly, Maynard Smith (1982) extended the idea of pure or mixed Nash equilibria to an evolutionary stable strategy, or ESS.

Analysis for NK Boolean networks with $K = 2$, or for larger numbers of input, for the canalyzing ensemble or for those with high internal homogeneity shows that such networks typically form a frozen component. This frozen component is a percolating, partial pure-strategy Nash equilibrium in the NK Boolean game interpretation of NK Boolean networks. A percolating cluster of agents is at a local mutual optimum and is myopically frozen into that state, even though other agents continue to oscillate through their own myopic actions. This frozen component is a first hint of the possibility that in an ecosystem some species may be frozen, while others continue to evolve. I return to this possibility below, for we shall find the same phenomenon with more complex models of coevolution.

In contrast to Boolean games having a frozen component, analysis of NK Boolean networks shows that, for N large and $K > 2$ and with no other constraints on the Boolean functions, such networks generically have no steady-state attractors. Thus with very high probability, NK Boolean games with $K > 2$ couplings among agents have no pure-strategy Nash equilibria. If the agents are limited to pure strategies, the coevolving system will oscillate. Such oscillations are similar to the children's game of "rock, paper, and scissors," in which players select one of the three actions and then rock breaks scissors, scissors cuts paper, and paper covers rock in a cycle with no stable pure equilibrium.

For readers not familar with game theory, and hence with its implications in coevolution, it is important to stress that a Nash equilibrium is merely a *local combination* of actions or strategies such that each agent is locally happier as long as other agents do not deviate from their own fixed strategy. But there is no general claim that such an equilibrium is particularly good for the players caught in its local sway, either

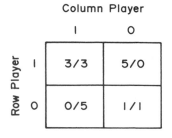

Figure 6.1 Payoff matrix for the prisoner's dilemma (1 = collaborate, 0 = defect). Right number in each cell is row player's payoff; left number is column player's payoff.

relative to the set of possible alternative equilibria or, more dramatically, relative to possible chaotic flow in the space of actions. That is, it is easy to construct cases in which the Nash equilibria are, for all players, of lower fitness than alternative behaviors. A simple example is the "prisoner's dilemma" (Axelrod 1984, 1987; J. H. Miller 1988), whose payoff matrix is shown in Figure 6.1. Here 1 means "collaborate," 0 means "defect" for each of two prisoners, and the payoff matrix shows the malicious consequences. If both players collaborate, each is paid 3. If one collaborates and the other defects, the defector is paid 5 while the collaborator is paid 0. If both defect, both receive 1. The only Nash equilibrium is when both players defect. This equilibrium is stable, since either player is better off defecting rather than collaborating, given that the other player is defecting. But clearly the best choice, both collaborate, is not an equilibrium because it pays either player to defect assuming the remaining player continues to collaborate. The prisoner's dilemma has been subjected to considerable analysis to determine the conditions under which collaboration can emerge (Axelrod 1984, 1987; J. H. Miller 1988).

In short, it is a general and interesting question to study the conditions under which Nash equilibria are both better than most alternative actions and attainable. I return to this question below in analyzing coevolution in terms of coupled *NK* landscapes.

Order in Boolean Networks Predicts Order in NK *Landscapes*

Let me return for a moment to the puzzle seen in Chapter 2, where for the *NK* landscape model with $K = 2$ we found that the highest optima were nearest one another. We have just seen that, for $A = 2$ actions, *NK* models are equivalent to random *NK* Boolean networks. For $K = 2$ Boolean networks, the percolating frozen component reflects a set of mutually nonfrustrated agents in the network. Each agent is myopically maximizing its payoff in a way which is held in place by the other agents in the percolating web. This same frozen component implies that, in the *NK* landscape with $K = 2$, the high optima will share an equivalent frozen component. The high optima will therefore be identical to one another with respect to that component and hence near one another. Thus the emergent order seen in random Boolean networks plays back upon the order seen in *NK* landscapes.

COEVOLUTION IN ECOSYSTEMS

Consider an ecosystem. Each kind of organism has, as parts of its environment, other organisms of the same kind and of different kinds. As is now clear, adaptation by one kind of organism alters both the fitness and the fitness landscape of the other organisms. Two alternative pictures have emerged. In the first, coevolution is viewed as resulting in an unceasing evolutionary process in which all species continue to change. Both the fox and hare must run faster and faster in order to maintain the same relative fitness. This model has variously been called the Rat Race (Rosenzweig 1973) and the Red Queen Hypothesis (Van Valen 1973). The alternative class of models has led to the idea of evolutionary stable strategies (ESS) (Maynard Smith and Price 1973), in which the phenotypes of the coevolving species stop changing at a balance which is stable in the sense that any other mixture of phenotypes attempted by any species in the system would be less fit. That is, an ESS is a stable, optimum mixture of phenotypes for the coevolving species and has the property that it is not invadable by any other mixture of phenotypes.

There has been considerable effort among evolutionary ecologists to distinguish the conditions under which the Red Queen behavior obtains from the conditions under which settling down to an ESS occurs. For example, Rosenzweig, Brown, and Vincent (1987) suggest a broader formulation of adaptive evolutionary games in which one considers standard kinds of ecological models, such as predator–prey interaction systems, where predators encounter prey at some frequency, eat them, and convert them to more predators, while prey run from predators and produce more prey when unmolested. To these familiar models, these authors add the idea the each species can adapt—for example, by running faster—and include the critical idea that, in such adaptive processes, *constraints within the organism exist.* Thus the prey's capacity to run faster means that its basal metabolic rate may increase; thus a price is paid elsewhere for running faster. For the predator, running faster may mean less capacity to search thoroughly, and so on. By assuming that some traits are constrained in this way while others can increase without bound, the authors show that the constrained traits in each species coupled in the ecosystem tend to fall to ESS values, while those which can increase or decrease indefinitely tend to become trapped in Red Queen races wherein the coevolving species keep relative position while their properties evolve continuously.

The NK Model Again:
Generalized Epistatic Fitness Landscapes

That traits which are subject to constraints and tradeoff within each organism reach stable values, while traits which can increase without bonds evolve continuously, is a curious conclusion to reach. In real organisms, no feature can increase without bounds; all are constrained.

Part of our difficulty in thinking about fitness in organisms is that the ways different traits in one organism combine to contribute to fitness, and the constraints and tradeoffs among those traits with respect to fitness, are very complex. Thus manufacture of a given insect-repelling substance may aid the fitness of a plant but cost in drain of a substrate needed in metabolism. The simultaneous presence of the insect-repelling substance and rough bark may help repel insects and more than offset the energetic cost of producing both the substance and the rough bark. In general, any trait might be involved in tradeoffs with some unknown number of other traits in arbitrarily complex ways. We encountered this problem in Chapter 2. The problem is that of *epistatic linkages either among traits or among the genes controlling them.* In general, those linkages may be extremely complex and reflect an elaborate diversity of conflicting constraints.

I introduced the *NK* family of rugged fitness landscapes in Chapter 2 to describe landscapes engendered by arbitrarily complex epistatic constraints, recalled this family of landscapes immediately above in defining *NK* Boolean games, and recall it now to think about rugged fitness landscapes where each organism has *N* traits, present or absent.

In the context of a whole organism with *N* traits, the *NK* model assigns, to each of the *N* traits, the *K* traits among the *N* which are epistatic inputs to that trait, assigns random fitness contributions to each trait in the context of all the combinations of states of itself and its *K* inputs, and defines the fitness of any "genotype," or combination of the states of the *N* traits, as the mean of the fitness contributions of its *N* sites. Recall that the fitness values defined over the *N*-dimensional Boolean hypercube yield a fitness landscape. For $K = 0$, each trait makes an independent fitness contribution and the fitness landscape is highly correlated about a single global peak; for $K = N - 1$, the landscape is fully uncorrelated and has many local peaks. More generally, as K increases, the number of local peaks increases, the steepness of the sides of the peaks increases, and the mean fitness of local optima decreases. These features reflect the increasing number of conflicting constraints as K increases relative to N. Bear in mind these three features as K increases relative to N. In the ensuing model of coevolution, they appear to be fundamental to the attainment of a state poised on the edge of chaos.

Coevolving Species and Coupled Dancing Landscapes

Consider an ecosystem with *S* species. For simplicity, let us imagine that each species is homogeneous—that is, all organisms in the species are identical; hence the species currently occupies a particular combination of its *N* traits. Then an *NK* landscape represents the fitness landscape of one homogeneous species. This assumption corresponds to the limit discussed in Chapters 2 and 3, due to Gillespie (1983, 1984), in which mutations in an adapting population are infrequent relative to the fitness differences between initial and mutant forms. Recall that, under these conditions, the population encounters advantageous mutants on a long time scale and moves as a whole on a short time scale to the new fitter variant. Thus in these limiting cases, the population can be approximated as homogeneous. With a higher mutation rate or under other conditions, a species is not homogeneous. It is possible to extend the model to allow the population representing one species to be a cloud distributed over its landscape, in which frequency-dependent and density-dependent coevolution within the species occurs.

In a coevolutionary system, we need to represent the fact that both the fitness and the fitness landscape of each species are a function of the other species. Thus, in general, it is necessary to couple the rugged fitness landscapes for each species, such that an adaptive move by one species projects onto the fitness landscapes of the other species and alters those fitness landscapes more or less profoundly. Over time, each species jockeys uphill on its own landscape and thereby deforms the landscapes of its ecological neighbors. Any such move by one species may increase or decrease the fitness of each neighbor on the latter's landscape and alter the uphill adaptive walks accessible to that neighbor.

In the context of the *NK* model, the natural way to couple landscapes is to assume that each trait in species 1 depends epistatically on *K* other traits internally and on *C* traits in species 2. More generally, in any ecosystem with *S* species, each trait in species 1 will depend on *K* traits internally and on *C* traits in each of the S_i among the *S* species with which it interacts. It is also natural to assume symmetry. If species 1 is in the niche of species 2, then species 2 is in the niche of species 1.

To represent the effect of the *C* traits from species 1 which are coupled to each trait in species 2, we expand the fitness tables defining the landscape of species 2 to incorporate these *C* traits. Hence, for each of the *N* traits in species 2, the model will assign a random fitness between 0.0 and 1.0 for each combination of the *K* traits internal to species 2, together with all combinations of *C* traits in species 1. In short, we expand the random fitness table for each trait in species 2 such that the trait looks at its *K* internal epistatic inputs and also at the *C* external epistatic inputs from species 1. Given these expanded tables, the fitness landscape of species 2 is a function of the current location of species 1 on species 1's fitness landscape. Therefore, as species 1 adapts, it both changes the fitness of species 2 and deforms 2's fitness landscape. In turn, each trait in species 1 must be coupled to *C* traits in species 2, and the fitness values for the *N* traits in species 1 must be expanded similarly. This procedure couples the two landscapes; each species is in the niche of the other.

In a system of *S* species, the interactions can be represented by a web of such projections. By representing the interactions in this way, we have at our disposal a choice of (1) the number and identity of the traits *C* which couple from one species to each single trait in another species, (2) the number and identity of other species among the *S* which project onto each species, and (3) the number of species *S* in the ecosystem.

In the first part of the ensuing discussion, I shall assume that each species in an *S*-species system is coupled directly to all other species. This richest coupling is undoubtedly unrealistic. A vast literature studies the hierarchical structure of food webs (Pimm 1982). In a sense, the completely coupled ecosystem is probably the worst possible case for achieving coevolutionary progress. In a more limited web, the links are fewer and each partner's "dancing" has a direct effect on only a few neighbors. I return to this limited-web situation below.

I shall also assume a second worst-case condition—that, in an *S*-species ecosystem, there is no similarity between the species and hence the effects of species 1 on 2 and of 3 on 2 are randomly assigned. In reality, rabbits and hares probably look much the same to a fox. Similarity of species presumably can be thought of as reducing the number of *effectively different species* with which each species interacts.

I shall also initially consider a case that is naive from an ecological standpoint—that each coadapting partner interacts at each moment with all other partners. In the next section, I shall extend this *NK* coevolutionary model to a more formal ecological context in which species are represented by populations which are coupled to one another by the familiar Lotka–Volterra class of equations, which model both population growth reflecting density-dependent inhibition within a species and mutualistic or competitive interactions between species. The results of the present section extend to these population models.

Landscape Ruggedness and Couplings Between Landscapes Tune Coevolution

The game theoretic models which have been explored to study coevolution have not as yet been built to take account of three issues: the statistical ruggedness of the landscapes of the coevolving partners, the richness of coupling of those landscapes, and the implications of those features on coevolution. But surely, these three issues are major aspects of the problem. The *NK* model affords tunably rugged landscapes whose richness of coupling can also be tuned; hence we can study the influences of these factors on coevolution.

Simulations of coevolving systems were carried out under the assumption that each species acts *in turn,* in the context of the current state of the other species. On its turn, each species tries a random mutation and moves to that mutant variant if the variant is fitter; if the variant is not fitter, the species does not move. Thus any movement by a species at least transiently increases the fitness of that species but may increase or decrease the fitness of its coevolving partners. In addition to this random dynamics, we also examine two alternative cases. In the fitter dynamics, each species in turn examines all its single-mutant variants and chooses at random one of the fitter variants, if any exist. In the greedy dynamics, each species in turn chooses the fittest single-mutant variant.

Nash Equilibria

In general, such a coevolutionary process admits of two behaviors. Either the partners keep dancing or the coupled system attains a steady state at which the local optimum of each partner is consistent with the local optimum of all the other partners via the *C* couplings. Such a steady state is the analogue of a pure-strategy Nash equilibrium in the current context. I use the word "analogue" for the following reason. A true Nash equilibrium assumes that each agent can, at each moment, choose any one of its possible actions. In the present context, this wide choice range corresponds to each species, in a single moment, altering its current genotype to any of the 2^N other possible genotypes. In assuming that each species is able, at each moment, to mutate a single gene or trait, I am constraining the range of alternative genotypes, or actions, locally accessible to the species. Thus the steady states we shall find are similarly constrained. In the remainder of this chapter, I shall use the term "Nash equilibria" with the understanding that such equilibria are with respect to the mutant search range.

A second caution is required. The concept of an ESS is a further refinement of the concept of a Nash equilibrium, in which the condition of noninvadability by a mutant at an *initial low frequency* in the population is analyzed. In the simplified dynamics used here, where the whole population moves in an instant to a fitter variant, I have not analyzed invadability. Initial studies with fuller population dynamics (see below) which do include analysis of invadability confirm the simpler dynamics.

Simulations were carried out between pairs of coevolving species, each modeled as a single organism on an independent *NK* landscape. In addition, simulations were carried out for larger numbers of species *S*. The first major result is that Nash equilibria are encountered. It is not obvious that this should occur, for each species has 2^N genotypes among which it is evolving. An *S*-species system has the product of these genotypes in its joint strategy space.

Figure 6.2 shows eight species coevolving over 2500 generations. Here, then, each site within one species is epistatically affected by 13 other sites within that species and by one site in each of the seven other species. Over eight generations, each species in turn tries a random mutation and moves to that new genotype only if it is fitter than the current genotype in the context of the current genotypes of the other seven species. At each generation, the fitnesses of all species are recalculated in the context of the genotype of each and their couplings. As can be seen, for the first few hundred

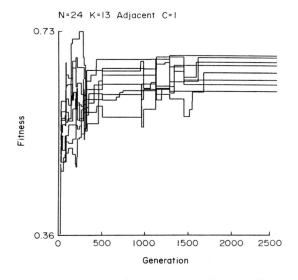

Figure 6.2   Coevolution among eight species, each governed by an *NK* landscape. Each of the *N* traits in each species is affected by *C* = 1 trait in each of the seven other species. System reaches a steady state about generation 1600. Note that mean fitness in the absence of selection is .5.

generations, the mean fitness of the whole set of species increases, rapidly at first, and then more slowly. Increasingly long intervals with no change occur, reflecting the fact that, as fitness increases, the waiting time to fitter variants increases for each partner. Sudden bursts of change by many species, however, are instigated by occasional changes by a single partner. By about 1600 generations, a Nash equilibrium has been found such that each species is locally fitter than all one-mutant variants, granted that the other species do not change.

Waiting Time to Encounter Nash Equilibria

In order to examine how *N*, *K*, and *C* bear on the waiting time to encounter a Nash equilibrium, simulations were carried out between two species. For each value of *N*, *K*, and *C* tried, 100 coevolving pairs were released. Over generations, a successively larger fraction of the coevolving pairs will have encountered Nash equilibria and hence stopped evolving (Figure 6.3). The main point to note in Figure 6.3 is that, as *K* increases relative to *C*, the waiting time to hit a Nash equilibrium *decreases*. Thus as the ruggedness of landscapes increases (*K* increasing), the expected waiting time to find Nash equilibria is decreasing. Presumably, this decrease in waiting time reflects the increased number of local optima in *NK* landscapes as *K* increases for fixed *N*. When *K* > *C*, Nash equilibria are found rapidly. When *K* < *C*, Nash equilibria are still found, but the mean waiting time becomes very long. In short, for a pair of species which are coevolving, *K* = *C* is a crude dividing line separating these two regimes.

   Similar studies as *N* increases for fixed *K* and *C* show that the waiting time to hit Nash equilibria increases, presumably because the density of local optima decreases. Since *N* is a crude measure of the complexity of the coevolving entities these results

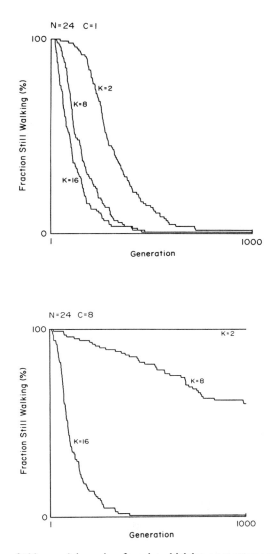

Figure 6.3 Fraction of 100 coevolving pairs of species which have not yet encountered a Nash equilibrium and hence are still walking, as a function of generations elapsed.

suggest a general tradeoff in complex adapting systems: Higher complexity may have advantages but be paid for in part by delays in reaching Nash equilibria.

Coevolution When Two Interacting Species Have Different K Values

It is of considerable interest to study the outcome of coevolution in which partners are on landscapes of different ruggedness. Figure 6.4 reports the results of simulations in which pairs of species have K values of 2, 4, 8, 12, and 16 for C values of 1, 8, and

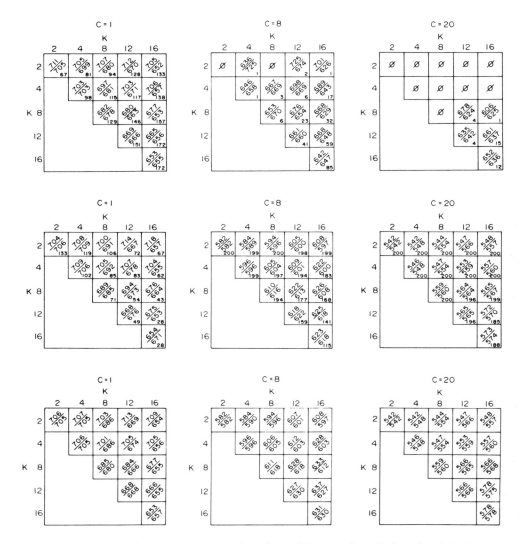

Figure 6.4 Coevolution among 200 pairs of species for 250 generations. Various C and K values. In all cases, $N = 24$ and the dynamics is random. Upper three grids correspond to cases where Nash equilibria were encountered. Each cell contains three numbers. The numerator of the fraction is the mean fitness of the row player. The denominator is the mean fitness of the column player. The number in the lower right corner shows how many of the 200 pairs encountered a Nash equilibrium. The middle three grids show cases where no Nash equilibrium was encountered in 250 generations. The three numbers in each cell are the same as in the upper grids except that here the fitness values correspond to mean fitness over the last 85 generations. At each generation a random mutant was tried; the species moved to that variant only if the variant was fitter. The lower three grids show the total mean sustained fitness obtained for each cell by averaging the cell values from the upper and middle grids.

20. These three values of C were chosen to lie below, in the middle, and above the range of K values. For each set of parameter values, 200 pairs of coevolving species were released and evolved for 250 generations. By that time, a Nash equilibrum either had or had not been encountered. Thus after 250 generations, some fraction of the 200 pairs had encountered Nash equilibrium and some were still coevolving.

Figure 6.4 shows a number of interesting features:

1. For all values of K, the fraction of coevolving pairs which encounter Nash equilibria in 250 generations decreases as C increases. Conversely, for any fixed value of C, the fraction of pairs encountering Nash equilibria increases as K increases. This reexhibits the phenomena of Figure 6.3: High K leads to more rugged landscapes, and Nash equilibria are encountered more rapidly.

2. When $C > 1$, the fitness at Nash equilibria is higher than the corresponding fitness when the partners are still oscillating.

3. As C increases, the fitness of both coevolving partners during the oscillatory phase before encountering Nash equilibria decreases for all pairs of coevolving K values. Presumably, this decrease in pre-Nash fitness reflects the fact that, for high values of C, a single move by one partner sharply lowers the expected fitness of the remaining partner.

4. When C is high ($C = 20$), high-K players have higher mean fitness during the oscillatory period before hitting Nash equilibria than do low-K players. More strikingly, in playing against a second player with a fixed K value, the first player would increase mean fitness during the oscillatory period by increasing its own value of K. That is, a $K = 4$ player does better against a $K = 2$ player than would a $K = 2$ player, while a $K = 8$ player does even better, and so forth.

5. Equally remarkably, when C is high ($C = 20$), a low-K player achieves higher mean fitness during the pre-Nash oscillatory period if it plays against a species of high K. That is, a $K = 2$ player has higher fitness against a $K = 4$ player than against another $K = 2$ player. A $K = 2$ player fares even better against a $K = 16$ player. Thus when C is high, increasing the K value of one partner helps *both* coevolving partners.

6. This tendency seems to be reversed when $C = 1$. Hence, during the oscillatory period, low-K players seem to fare better than high-K players.

7. At the Nash equilibria encountered, the fitnesses of low-K players are clearly higher than those of high-K players for each value of C, and indeed seem roughly independent of C.

8. Finally, when C is high, overall average fitness is highest when K is high. When C is low, overall average fitness is highest when K is low. Thus fitness in coevolving systems would be enhanced were K able to adjust to match C or, more broadly, were K and C themselves evolvable.

Similar studies were carried out using the fitter and greedy dynamics. The main results are the same. A major difference arises in the greedy dynamics. Here, for any genotype, there is generically a unique best-fit one-mutant variant in the context of the other species. Thus if each species plays in a deterministic order, each changes to a unique next genotype and the set of coevolving species can enter a recurrent cycle in the total space of S genotypes. When $K < C$, such periodic attractors arise rather frequently.

The analysis above is based on mutating one site, gene, or trait at a time in each coevolving species. Figure 6.5 shows the results of coevolution as the mutation rate or, more accurately, the number of traits randomly mutated in each species increases to 2, 4, 8, 16, and 24 for coevolving pairs of species on increasingly rugged landscapes. As the number of genes mutated simultaneously increases, the number of local optima on a fixed landscape dwindles; hence the probability that the coevolving pair reaches a local Nash equilibrium falls. Thus the fitness seen in the coevolving pair reflects fitness during the pre-Nash period. Figure 6.5a shows that, for all values of K, as the number of mutations increases, the maintained fitness reaches a maximum for two or four simultaneous mutations and falls thereafter. The decrease in maintained fitness is greatest for small K values and less marked for large K values. Figures 6.5b and 6.5c show that, in general but not uniformly, the optimal mutation rate decreases as C increases. These results suggest that there may typically be an optimal—and *typically low*—mutation rate for maintaining fitness in coevolutionary processes.

These results are also interesting from the general game theory perspective. Lower fitness with increasing search range (here number of sites mutated) presumably reflects search beyond the natural correlation length of a fixed landscape, plus the fact that each landscape deforms more if the other players change more dramatically. Thus if each species makes a large change by changing many sites in an attempt to climb uphill, that change dramatically deforms the landscapes of the species' partners and can lead to lower sustained fitness for all partners. In the general context of game theory, if it is assumed that all players can change at each instant to any action, these ideas do not apply. These ideas do arise, however, in any context where each player can search only its local strategy space—for example, because there is some cost to searching—and where the distance moved in strategy space by one player is related to the deformation in the payoff landscape of the other players. In such contexts, myopic play, as is typical of biological evolution, may typically lead to higher sustained fitness—or payoff—than nonmyopic play.

Possible Coevolutionary Dynamics of the Coevolutionary Parameters

The results we have seen make a number of intriguing suggestions. All point to the possibility that a coevolving system of species may collectively tune the parameters governing its own coevolution. The fundamental parameters—N, K, C, and S—may conspire under selection to lead to coevolving systems which couple landscapes in such a manner that the partners coevolve well. If so, the point is far from trivial, for it is clear that coevolution can cause coupled landscapes to heave so erratically that no partner can long maintain a fitness much above that which is merely average in the space of systems and attainable even without selection. I discuss these suggestions first with respect to completely coupled ecosystems. Then I extend them to structured ecosystems in which each species interacts with a few neighboring species in an ecosystem web.

Figure 6.5   Effects of increasing mutation rate on mean fitness maintained in coevolving species. Both coevolving species randomly mutate 1, 2, 4, 8, 16, or 24 of the $N = 24$ genes at each moment. Figure plots average fitness over last third of 250 coevolutionary steps as a function of mutation rate. Note that, for most K values, maintained fitness increases and reaches a maximum for two- or four-mutant search ranges. For most values of K, maintained fitness falls for greater mutation rates.

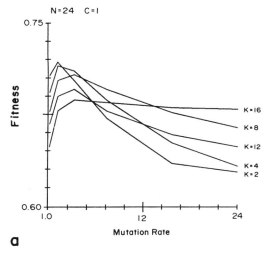

N = 24   C = 1

a

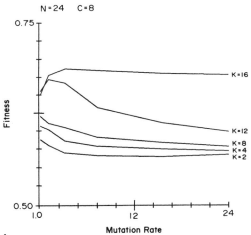

N = 24   C = 8

b

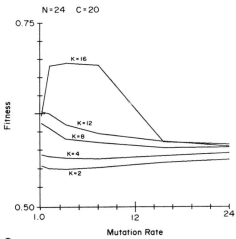

N = 24   C = 20

c

First, there may well be selective processes which match K to C in order to optimize the coevolutionary capacities of the coevolving partners. For instance, K "should" increase when it is low relative to C and decrease when it is high relative to C. As shown in Figure 6.4 for the biologically most plausible case in which each species tries random mutations and moves to a variant if that variant is fitter, if C is high relative to K, any player increases its fitness by increasing its own K value. When C is high, increasing K has two beneficial effects. First, Nash equilibria are encountered more rapidly at higher K values and are fitter than the prior oscillatory period. Second, fitness during that prior oscillatory period is higher than it is at lower K values. Thus it is advantageous to any player to increase K in a high-C environment. Perhaps equally remarkably, in the biologically reasonable case of random mutations (Figure 6.4), such a move by one species *also helps the second species.* Each has higher pre-Nash fitness and finds Nash equilibria sooner. Conversely, suppose K is high relative to C. Then, as is clear from Figures 6.3 and 6.4, Nash equilibria are encountered very rapidly. Thus the fitness in the pre-Nash oscillatory period is of less importance when K is high, and the fitness of Nash equilibria is more important. But local optima at Nash equilibria are higher for low-K players than for high-K players. Thus if K is too high relative to C, it should be advantageous to decrease K. In short, at this group level of coevolving species, it seems clear that there are reasonable selective advantages to a species as a whole to tune K to match C. At that match, given a fixed C, Nash equilibria will be encountered rapidly and will be highly fit, a condition that optimizes mean fitness during any pre-Nash periods, minimizes the mean duration of those periods, and maximizes the fitness of the Nash optima attained.

The analysis above is based on the idea that it is advantageous to the species *as a group* to increase K. In order to avoid positing group election mechanisms, we should seek selective conditions acting on *individual members* of a species which might increase K in members of that species and hence in the coupled ecosystem. Within the framework of the NK model, a change in K would naturally be envisioned as a mutation which altered the epistatic coupling between traits or genes such that a trait or gene now depended on either one more or one fewer epistatic inputs. That is, we must let K evolve. In this framework, the natural way to express the consequences of such a mutation which increases K is to expand the fitness table for that trait such that the table looks at the new trait as well as at the K traits it initially looked at. That new epistatic connection, in the context of the current genotype in which the new connection is formed, might increase or might decrease the fitness of the current genotype. Thus we can envision three ways in which selection on an individual level may allow an increased K value at one genetic locus to spread throughout a population:

1. The new epistatic link, when it forms, causes the genotype to be fitter and is selected and hence spreads.

2. The new epistatic link is near neutral and spreads through the population by random drift.

3. The new link not only has a direct effect on the fitness of the current genotype, but also increases the *inclusive fitness of the individual and its progeny.* This long-term effect is due to the increased fitness of those progeny in the coevolutionary process, which in turn is due to increased rapidity of finding Nash equilibria and higher fitness during the pre-Nash oscillatory period.

These considerations suggest the possibility of a coevolutionary dynamic which optimizes K relative to C in an ecosystem, such that partners maintain high mean fitness.

We turn next to consider the coevolution of S in the coupled ecosystem. A clear process will tend to limit the number of species (but see below). Consider coupled landscapes with fixed K and C couplings. Let S increase, under the assumption that each species is C-coupled to all other species. When $K > S \times C$, all the coevolving partners encounter a Nash equilibrium rapidly. When $K < S \times C$, the coevolving partners do not encounter a Nash equilibrium for a long time. I stress that the exact relation between mean waiting time to encounter Nash equilibrium and K, S, and C is unknown. As a rough guide, we can use $K = S \times C$ as a rough division between the two regimes.

Numerical results are shown in Figure 6.6 for three values of S. In Figure 6.6a, $S = 4$ species are coevolving. Thus each species senses $C = 2$ inputs to each of its genes from each of the four species. Each species in turn randomly mutates and tries to find a fitter variant. Over 2000 generations, mean fitness increases and the four species find a Nash equilibrium. In Figures 6.6b and 6.6c, the number of species increases to 8, and then 16. Data to 2000 generations are shown for the 8- and 16-species cases. No Nash equilibrium was found for 8000 generations in these cases.

Note the following features in Figure 6.6:

1. As the number of species S increases, the waiting time to encounter a Nash equilibrium increases.

2. As S increases, the mean fitness of the coevolving partners decreases.

3. As S increases, the fluctuations in fitness of the coevolving partners increase dramatically.

Thus for the four-species case, mean fitness increases rapidly and achieves fairly high Nash equilibria. For the 16-species case, the entire system fluctuates about a mean fitness slightly above average, .5, with dramatic excursions below .4.

These results show that, as the number of mutually coupled species in the system increases, the mean fitness falls and fluctuations to very low fitness increase. Therefore, if we may assume that fluctuations to low fitness are associated with an increased chance of extinction of the unlucky species, these results suggest that, if S is too high, the coupled ecosystem will fluctuate dramatically and lead to the extinction of species, thereby lowering S. In turn, as S is lowered, the ecosystem behaves less chaotically and mean fitness of all partners improves both during the pre-Nash period and because Nash equilibria are encountered more rapidly. The remaining system coevolves well despite the fact that landscapes are coupled and deform as each actor moves.

Finally, it is interesting to consider that C may evolve. For example, prey may try to reduce C, while predators try to increase it. Further C, may not be freely variable with respect to K but instead may be some more or less fixed fraction of K. That is, it is not implausible that the epistatic couplings of a trait within a member of one species to other traits within the organism will be greater than the couplings to traits in organisms of other species. Even without ecological considerations, K should be larger than C.

In short, with clear hesitations and heralded caveats, this framework begins to sug-

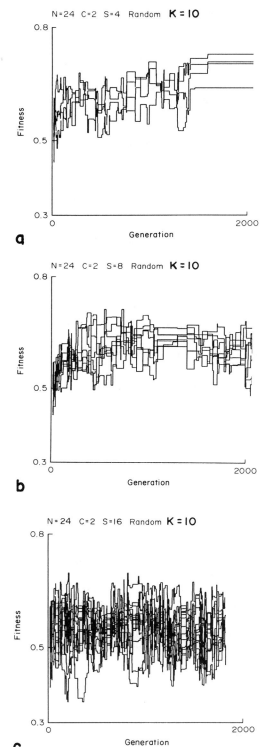

gest the possibility that the coevolutionary parameters governing the ruggedness of landscapes, the couplings among landscapes, and the number of coevolving partners might themselves coevolve without group selection, to continuously re-create well-formed ecosystems which are able to coadapt successfully. No mean feat, this, for as Figure 6.6c makes clear, coevolution among coupled species can lead to chaotic fluctuations with no accumulation of improvement. If one wishes a Red Queen, here is one to reckon with.

STRUCTURED ECOSYSTEMS AND SELF-ORGANIZED CRITICALITY: COEVOLUTION TO THE EDGE OF CHAOS

Real ecosystems are not totally connected. Typically, each species interacts with a subset of the total number of other species; hence the system has some extended web structure. We now extend our own results to such ecosystems. The supposition that selection can act on a coevolutionary system to control its connectivity, and therefore its dynamics, points in a very interesting direction. It might be the case that coevolving ecosystems tend toward a state of "self-organized criticality" in which parts of the ecosystem are frozen for long periods, such that the species in the frozen component do not change, while other species continue to coevolve. Avalanches of changes initiated at local points in the ecosystem web may propagate to various extents throughout the ecosystem. Such avalanches may trigger speciation and extinction events. Furthermore, the endogenous dynamics of the coevolving system acted on by selection may tend toward this poised state in which such avalanches can propagate on a variety of size scales with a power-law distribution between sizes of avalanches and their frequencies. Indeed, the theory comes close to predicting the size distribution of extinction events in the evolutionary record.

The term "self-organized criticality" was coined by physicist Per Bak (Bak, Tang, and Wiesenfeld 1988) to refer to a quite generic pattern of self-organization. Bak asks us to consider a tabletop onto which sand is dropped at a uniform rate. As the sand piles up, it begins to slide off the edges of the table. Eventually, the system reaches a steady state at which the mean rate of dropping sand onto the pile equals the mean rate at which sand falls over the edges. At this stage, the slopes from the peak to the edges of the table are near the rest angle for sand. Bak asks the following question: If one adds a single grain of sand to the pile at a random location and thereby starts an avalanche, what will the distribution of avalanche sizes be? He finds a characteristic power-law distribution relating the frequencies and sizes of avalanches, with many tiny avalanches and few large ones. He argues that this distribution is characteristic of a wide range of phenomena, including distribution of earthquake sizes and other examples. The argument requires that the system under investigation attain and maintain a kind of poised state able to propagate perturbations—avalanches—on all possible length or size scales.

There are at least two ideas derived from Bak's theory which seem interesting in the coevolutionary context. First, avalanches of perturbations, made up of "packets" of coevolutionary change and having a characteristic relation between size scale and

---

Figure 6.6 Coevolution among (a) four, (b) eight, and (c) 16 species. Note that, as the number of species increases, the mean fitness decreases and the variance in fitness increases. For (b) and (c), no Nash equilibrium was found in 8000 generations.

frequency, may propagate through an ecosystem. This possibility requires some part of the ecosystem to be at rest while other parts change. Second, the propagating changes are likely to be associated with fluctuations to low fitness, which may engender both extinction and speciation events. Extinction events would be expected because of low fitness. Speciation events might be expected at low fitness because the number of directions of improvement is increased and because isolated subpopulations would form. If the probability of branching speciation is proportional to the number of directions of improved fitness, then low fitness episodes should trigger speciation events. Thus the propagation of avalanches through the system would be linked to speciation and extinction phenomena. We shall see that these properties are likely to be associated with the existence of nearly melted frozen components in ecosystems. I discuss these ideas next.

Fluctuating Frozen Components: Nash Equilibria Extended to Lattice Ecosystems

A first hint that the ideas derived from Bak's theory may apply to ecosystems was found in the simplified *NK* Boolean games described above. Recall that these models are formally identical to *NK* Boolean networks. A frozen component of players, each a single site which has optimized its 0 or 1 choice in the context of its *K* inputs, can form across the network. Thus the frozen components seen in Boolean networks in the solid or liquid regions, as described in Chapter 5, are already versions of simple ecosystems with many players frozen at a Nash equilibrium while other players continue to change.

The second hint that Bak's ideas may apply to ecosystems arises in extending the coevolving *NK* model to structured ecosystems in which each species interacts with only a few other species: Here, too, *parts of the system may be fixed at Nash equilibria while other parts continue to coevolve.*

To investigate the behavior of structured ecosystems, my colleague Sonke Johnsen and I have carried out simulations on "square" ecosystems, in which each interior species interacts with its four neighbors. Corner and edge species interact with two and three neighbors, respectively. Model ecosystems have varied from $3 \times 3 = 9$ species to $10 \times 10 = 100$ species. In addition to square ecosystems, which have corners and edges, we have investigated toroidal ecosystems, in which the square is first folded into a cylinder by joining left and right edge species and then bent into a torus by joining top and bottom species. We have also investigated randomly connected ecosystems with similar general results.

Figure 6.7 shows 12 successive times in the temporal coevolution of a 10×10 ecosystem. At each time moment, one of the 100 species plays and greedily chooses the fittest one-mutant variant if any is fitter than its current genotype. Each species plays in turn, and thus 100 plays constitute an ecosystem generation. After each ecosystem generation (hereafter "generation"), any species may have changed its genotype or remained the same. If the species changed, we color it white; if it remains unchanged, we color it black. The simulation was run over 200 generations, and Figure 6.7 represents generations from a period in the middle of the 200 generations.

The first question to ask is, Can frozen regions and changing regions coexist in the ecosystem? The salient features to note are these:

1. A large fraction of the species are frozen over single generations.
2. Some regions remain frozen over very many generations. In Figure 6.7, species in the upper left and lower left corners remain frozen over about 48 generations.

N = 24   K = 10   C = 1   S = 100   Greedy   ■ = Frozen   ☐ = Unfrozen

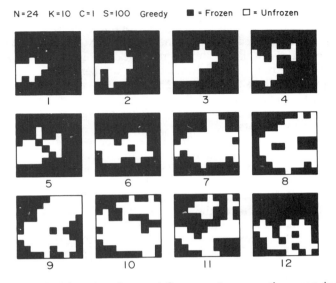

Figure 6.7   Twelve successive generations, each four ecosystem generations apart, in a 10 × 10 = 100 species ecosystem wherein each species plays with its immediate neighbors. Thus corner species coevolve with two immediate partners, edge species with three, and interior species with four. As time passes, frozen regions (black), where species are not changing genotype at that generation, emerge, expand, and contract over the distributed ecosystem. If all species stop changing, the entire ecosystem is frozen, each species at a local Nash equilibrium.

3. One or more unfrozen regions may exist.

4. Over time, the location of the frozen region changes and the size waxes and wanes. That is, *a fluctuating frozen component* can exist and extend throughout some or much of the ecosystem.

5. In the simulation carried out here, ultimately, the frozen region encompasses the entire ecosystem (not shown in Figure 6.7). That is, the ecosystem comes to rest at a combination of genotypes which are local Nash equilibria for all 100 species. In the absence of exogenous perturbations, the system will remain in this frozen state thereafter.

6. In many simulations, particularly those using the greedy algorithm, the coevolving ecosystem encounters a limit cycle. Typically in these cases, a fraction of the system remains permanently frozen, while the remainder oscillate through a recurrent set of genotypes.

These results show that one region of an ecosystem can be frozen, while other regions continue to coevolve. One region persists in something like an ESS, while adjacent regions in the same ecosystem persist in Red Queen antics. These results echo the frozen components discussed in an earlier section concerning Boolean networks and partial percolating Nash equilibria in the *NK* Boolean games. This behavior in itself is very interesting, since the existence of a frozen component in Boolean networks appears to be a requirement for orderly dynamics and the capacity to evolve. Therefore, it seems important to ask, What are the general principles under which frozen components emerge on higher scales, such as these complex, gamelike contexts? Note that the *NK* Boolean games and the *NK* models of coupled ecosystems are remarkably similar. In the former, each site is an independent myopic agent. In the latter, defined collections of sites correspond to each agent. In all cases, sites within one agent are coupled to sites in other agents such that the payoff to each agent depends on the actions of all the other agents. The only difference between NK Boolean games and NK coevolutionary models is the number of sites which collectively correspond to each independent agent. The emergence of frozen components suggests that some kind of scaling laws in N, K, C, and S, in ecosystem structure, and in the number of sites corresponding to each independent agent will govern the emergence of frozen components.

One approach to scaling behavior for the emergence of frozen components has been taken by W. Fontana and Alan Kaufman and deserves brief mention. We consider a square lattice of sites, each coupled to its four neighbors, according to the familiar NK rules. We can define each site to be an independent species adapting on a landscape which deforms as the four neighbors of the species change. This re-creates the NK Boolean games discussed above. Such systems will, of course, be chaotic, with no frozen components, since random $K = 4$ Boolean lattices are in the chaotic regime. Alternatively, we can define the entire square lattice to be a single species and recover the fixed NK fitness landscapes discussed in Chapters 2 and 3. Here the system has local peaks as point attractors. While the system is climbing to such a peak, more and more of its traits will become frozen into either the 1 or the 0 state, and finally the system will stop changing entirely at a local peak.

Since chaos emerges when each site is a separate species but frozen components emerge when the entire lattice is one species, it must be the case that, as the size of the sublattice which counts as each species grows from one site to the entire lattice, a phase transition occurs at which frozen components begin to percolate. Thus let the lattice be broken into nonoverlapping sublattices, each constituting a single species. Each species senses the species adjacent to it across the common boundary separating them. The adaptive moves by each species are now constrained by the boundary-site values of its neighbors, which are now like the C couplings between species discussed above. Hence each species coevolves on a fitness landscape which is deformed as the neighbors of that species move on their own landscapes.

Our preliminary results show that, as the size of the sublattice which counts as a single species increases, there is a critical size at which frozen components begin to emerge both within and between species. The critical size appears to be closely related to the local range in which each site is coupled to nearby sites on the lattice. Once each sublattice is large enough to include one site and all its immediate neighbors via the couplings between sites, frozen components emerge. This critical state presumably corresponds to Bak's critical state with a power-law distribution of avalanches. A study of the critical sublattice size, and of how it changes with lattice size and with K couplings, should yield the scaling laws we seek.

In addition, three related ideas are appealing:

1. Subregions of a lattice which count as species, and hence coevolve with one another, create an interesting analogue of temperature in a physical system with a complex potential surface. A sufficiently high fixed temperature keeps a physical system from settling down into a local minimum. Coevolutionary dynamics, as a result of deforming landscapes, tend to keep each coevolving species from settling down at a local fitness optimum. The persistent deformations of landscapes in a coevolutionary setting are thus at least crudely analogous to temperature in a physical system.

2. A selective coevolutionary dynamics might optimize the size of sublattices which count as individuals, a process that is analogous to finding, for a physical system, a fixed temperature which optimizes energy over a given time scale. For any rugged potential surface and *fixed time* that the physical system is allowed to explore the landscape, there is an optimal *fixed intermediate temperature* at which the system minimizes its energy over that time scale. If the temperature is too high, the system does not stay in good minima long enough. If the temperature is too low, the system because trapped in poor local minima for too long. As the time scale permitted for exploration increases, the optimal fixed temperature decreases. Therefore, the second intriguing possibility is that the fitness achieved over a fixed time scale may *first increase and then decrease* as the size of subregions which count as an organism increases. Thus there may be an optimal size grouping of sites which count as a single organism in a species for sustained fitness over a given time scale. The time scale would be set by the rate at which abiotic or biotic influences alter the coevolutionary system.

3. As the size of the grouping changes, the surface-to-volume ratio of the entity changes. Couplings to other entities are across the boundary surface. The surface-to-volume ratio is a crude measure of the extent to which adaptive moves by one partner deform the fitness landscapes of other partners—deformation is high if the ratio is high and low if the ratio is low. If there is an optimum grouping size which corresponds to an organism, we might imagine an evolutionary process whereby an organism evolves the ratio of its surface boundaries with other organisms versus its volume, to optimize sensitivity to its coevolving partners for sustained fitness. I return to these rough ideas below.

Might There Be a Coevolutionary Adaptive Process Leading Toward a Self-Organized Critical State with a Percolating Frozen Component?

The results in the previous section concerning unstructured ecosystems suggest that a coevolutionary dynamics might tune the parameters of coevolving species such that the species coevolved well. Here I discuss results suggesting that these ideas may extend to structured ecosystems. Species may selfishly tune both the K values and the number of species with which they interact, such that the coupled system as whole coadapts well. At the optimal state, a frozen component percolates across and covers the ecosystem rapidly.

Figures 6.8a and 6.8b show the results of simulations of 5×5 square ecosystems in which the average fitnesses of corner, edge, and interior species are accumulated. For each value of K, 50 ecosystems were analyzed over 200 generations each. The figures reveal the following features:

1. For all interspecies connections—two for the corner species, three for the edge species, and four for the middle species—there is an optimum value of K at $K = 8$ to 10 which optimizes sustained mean fitness. For lower or higher values of K, the average fitness declines.

2. Mean fitness increases as connections to other species decreases.

If there is an optimal value of K for sustained fitness in coevolving systems, then selective effects might pull K values of coevolving partners toward this optimum.

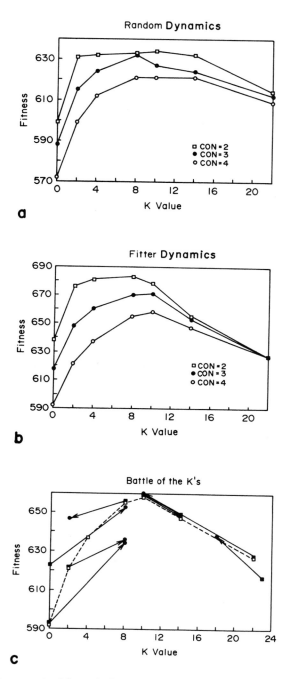

Figure 6.8 (*a*) Mean sustained fitness in 5 × 5 ecosystems as *K* varies from 0 to 22. In all cases, *N* = 24, *C* = 1. Corner species are connected to two others (top curve), edge species to three others (middle curve), interior species to four others (bottom curve). Note that sustained fitness first increases and then decreases as *K* increases. Random dynamics was used. (*b*) As in (*a*), except that here fitter dynamics was used. (*c*) Selection force toward K_{opt} value of *K* = 8 to 10. Two experimental species located adjacent to the central species in the 5 × 5 ecosystem were constructed with a *K* value that was different from that of the remainder of the ecosystem. In all cases, deviation of the *K* values of the experimental pair (●) toward K_{opt} increased the sustained fitness of the experimental species relative to the unperturbed ecosystem (□) and relative to the control species (■) in the perturbed ecosystem.

This intuition is confirmed in Figure 6.8c. Here we investigate how increasing and decreasing K affects the fitness of two test species, located adjacent to the center of the ecosystem. To sample the rest of the ecosystem fairly, we monitored the fitness of two control species, also adjacent to the center of the ecosystem. Figure 6.8c shows that the presence of two test species had little effect on the fitness of the two control species. More critically, if the rest of the ecosystem had suboptimal K values, 0 or 2, or above-optimal K values, 14 or 23, then deviations of the K values of the test species *to* or *toward* the optimal values of $K = 8$ to 10 *increased* the fitness of the test species. Conversely, if the ecosystem is at the optimal values of $K = 8$ to 10, then deviations of the K values of the test species *away* from the optimal values *decreases* the fitness of the test species. In short, there is a *selective force toward the optimal value of K =* 8 to 10 which can act on single species, presumably via individual members of that species, and pull each toward the jointly optimal K value.

These results support and extend those for completely connected ecosystems, discussed above. There we found evidence that it was advantageous for a single species to increase or decrease K toward an optimal value relative to C. The results in Figure 6.8c are a powerful indication that a previously unexpected kind of selective metadynamics may very well tune K, the ruggedness of landscapes among coevolving species, toward a *joint optimum* where all partners coevolve well. The efficacy of such a metadynamics, however, remains to be established.

Optimization of Sustained Fitness by Optimization of Landscape Ruggedness Yields Ecosystems with Nearly Melted Frozen Components

The optimal ruggedness of fitness landscapes $K = 8$ to 10 achieved by selective tuning of the ruggedness by adapting species, as shown in Figure 6.8c, corresponds to coevolving systems which have achieved the poised transition regime between order and chaos. As shown in Figure 6.9, the rapidity with which the lattice ecosystem becomes frozen at a Nash equilibrium increases as K increases. For values of K less than and including $K = 8$, no freezing of the ecosystem occurs over 200 generations. Such systems are in the chaotic regime. For $K = 10$, entire ecosystems freeze at Nash equilibria gradually over the 200 generations. For $K > 10$, ecosystems freeze rapidly. Such ecosystems are well into the ordered regime. The optimal value of ruggedness of landscapes $K = 8$ to 10 occurs just at that value where freezing begins. Thus model ecosystems optimize coevolutionary fitness when frozen components are tenuously extending across the ecosystem, when the system is in the transition region between order and chaos.

A New Hypothesis: Ecosystems Coevolve to the Edge of Chaos

The results discussed above lead to a new and perhaps startling hypothesis: In coevolution, organisms adapt under natural selection via a *metadynamics* where each organism myopically alters the structure of its fitness landscape and the extent to which that landscape is deformed by the adaptive moves of other organisms, such that, as if by an invisible hand, the entire ecosystem coevolves to a poised state at the edge of chaos. I state this hypothesis in bold form. If true and general, it would constitute a powerful new framework to understand evolutionary biology. In the remain-

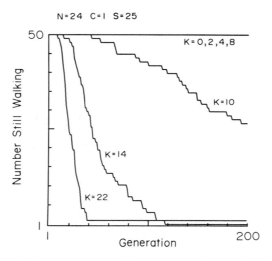

Figure 6.9 Fraction of 5 × 5 ecosystems which have not yet become frozen in an overall Nash equilibrium plotted against generation. Note that for $K \leqslant 8$, none of the ecosystems attained a frozen Nash equilibrium in the time available. For $K \geqslant 10$, some or most systems freeze at Nash equilibria and do so more rapidly as K increases.

der of this chapter, I discuss further current evidence bearing on this hypothesis. The data currently supporting such an hypothesis are these:

1. In the current coupled NK lattice ecosystem model, optimization of K, governing the ruggedness of landscapes with respect to a fixed value of the couplings between landscapes C, does optimize mean fitness of the coevolving partners.

2. More surprisingly, Figure 6.8c shows that selection can act on each species independently to pull it toward and hold it at the optimal value of K. Thus each species, via an increase in the inclusive fitness of members of that species, can adapt myopically to alter the structure of its fitness landscape; yet the consequent achievement of the poised state benefits all.

3. Control of the number of species with which any species interacts helps control whether a model ecosystem is in the chaotic regime or the ordered regime, or is poised between them. As discussed below, species in ecosystems do regulate the number of species with which they interact. Thus the means to tune this aspect of coevolution actually occurs.

4. We will see that at the poised state, coevolutionary avalanches of change propagate through the ecosystem with a power-law distribution of avalanche sizes. Such avalanches might be expected to lead to extinction events. The actual distribution of sizes of extinction events in the evolutionary record appears close to a power-law distribution. This suggests that ecosystems across the past 600 million years may be in the ordered regime near the edge of chaos.

5. Finally, an independent model of coevolving entities, self-reproducing programs coevolving in a computer core, also creates an ecosystem which exhibits a power-law distribution of extinction events. Thus this model ecosystem of myopic adaptive agents appears to coevolve to the edge of chaos.

In presummary, the present evidence is suggestive, not conclusive. However, I believe that the hypothesis, if true, is very important. Indeed, if it is generally true that myopic adaptive agents that can modify the games they play with one another, each to its own myopic advantage, universally coevolve to the edge of chaos, then the hypothesis has implications ranging from biology to economics and beyond. Adam Smith's "invisible hand," slightly reworked, may span many domains.

Coevolving Species Control Their Mutual Connections

Clearly, species can tune coevolution not only by tuning the ruggedness of their own landscapes, but also by tuning how many other species impinge on them. For example, in the present model, as shown in Figures 6.8a and 6.8b, fitness is in general increased by decreasing the number of other species to which a species is coupled. Similar results arise in completely connected ecosystems, where fitness increases as the number of interacting species decreases (Figure 6.6). Thus these results suggest that coevolutionary fitness may be optimized if the couplings among species are sparse. In reality, such decoupling cannot be complete, since predators need their prey to survive. These food-web requirements are not explicitly part of the current model but are readily incorporated into it. The question becomes, Do organisms regulate the species connectance of ecosystems?

Connectance is, in fact, controlled in food webs. Good evidence suggests that the number of connections in food webs is adjusted such that each species maintains roughly a constant number of connections to other species, regardless of the number of species in the web. Sugihara, Schoenly, and Trombla (1989) summarize recent data on more than 100 food webs—terrestrial, freshwater, and marine. A number of properties—such as lengths of food chains; connectance; ratios of top, intermediate, and bottom species; and ratios of predators to prey—appear to be *stable and scale-invariant,* both with respect to the number of species in the web and with respect to the aggregation of "guilds" of similar species into single "trophic species" or the aggregation of similar species into higher taxonomic units. The observations suggest that each species remains coupled to three or four other species. In short, real food webs do evolve couplings among species to maintain connectance as the number of species in the system increases or decreases.

Avalanches of Coevolutionary Changes in Coupled Ecosystems and the Distribution of Extinction Events in the Fossil Record

The edge of chaos also corresponds to a poised self-organized critical state with respect to coevolutionary avalanches. As seen above, as K decreases from above K_{opt}, the frozen component of the species at a Nash equilibrium melts. We investigate next the implications for coevolutionary avalanches and find that, at the optimized state for sustained fitness, avalanches propagate on all length scales in a power-law distribution.

Alterations in the species at one site in an ecosystem can often cause neighboring species to undergo coadaptive changes. Thus changes at one point may propagate to various extents throughout the ecosystem. Such avalanches of changes are the analog of Bak's sand-pile avalanches and also of the propagation of damage in Boolean networks.

To investigate such avalanches, Johnsen and I modified the ecosystem model to allow each species to be affected not only by its neighbors in the ecosystem but also by its external world. As described in Chapter 3, the external world of each species consists in a binary vector of length N. Each site in the species is coupled to W sites in its world. The fitness table of each site is augmented to look at the K internal sites, the C sites in each of the species impinging on the species being examined, and the W sites in that species' world. Thus alteration in the world of one species deforms the fitness landscape on which that species is coevolving. Typically, such alterations lower the fitness of the species enough to make its current genotype less fit than one or more of the one-mutant neighbor genotypes. If so, the species changes and may then unleash an avalanche of coevolutionary changes which propagates through the ecosystem.

The simplest avalanches of changes to visualize are those which are perturbations from the frozen state in which all species are at local Nash equilibria. In our modified model, we first used simulations with fitter-move dynamics to find the frozen state and then changed the world of a random member of the ecosystem. At the end of each generation, each species may have remained the same or may have altered its genotype. After the onset of such a change, coevolutionary changes continue until the system returns to a (perhaps new) frozen state with all species at local Nash equilibria. At that point, the avalanche has died out.

We used two measures of the size of such an avalanche: (1) total number of species caused to alter genotype, which measures the total number of species in the avalanche which have changed genotype at least once, and (2) total number of species which have changed at each generation from the start of the avalanche until the avalanche stops. Thus the second measure includes both the number of species affected and the number of generations in which each is affected. We denote this measure as species × time). The results are shown in Figures 6.10, 6.11, and 6.12.

We might expect a relationship between K and avalanche size. Intuitively, the frozen state is readily attained and solid when K is sufficiently large but becomes more tenuous as K decreases. When the frozen state is very solid, avalanches are not likely to propagate far. When the frozen state is nearly melted, any perturbation is likely to propagate farther. This is entirely in accord with the fact that, as K decreases from 18 to 12, the mean avalanche size increases and so does the variance. It also appears, at least crudely, that, as K decreases, the distribution of avalanche sizes approaches a power law. Such a power-law distribution will appear as a straight line in the ln-ln plots of Figures 6.10 through 6.12. The data for the 5 × 5 ecosystems of Figure 6.10 are convex for $K = 20$ and $K = 14$ but approaching linear for $K = 10$, the optimal value of K for sustained fitness in these ecosystems. The results for the 10 × 10 ecosystems (Figure 6.11) are similar. The linear slope in Figure 6.11c is an artifact arising because only two sizes of avalanches, one-species and two-species, occurred. Figure 6.12 shows that the ln-ln plots for $K = 18$ and $K = 14$ are clearly convex (hence not a power law) but may be approaching a power-law linear relationship by $K = 12$. Note that, as K decreases to 10 in the 5 × 5 ecosystems or to 12 in the 10 × 10 ecosystems, the spectrum of avalanches spreads out to an ever-greater range of length scales.

Figure 6.10 Avalanche size distribution in 5 × 5 ecosystems: natural logarithm of number of avalanches plotted as a function of natural logarithm of avalanche size (species × time) for (a) $K = 10$, (b) $K = 14$, and (c) $K = 20$. Note that, as K approaches the optimal value of $K = 10$, the avalanche size distribution appears to approach a power law. In all cases, $N = 24$, $C = 1$, $W = 20$, $S = 25$.

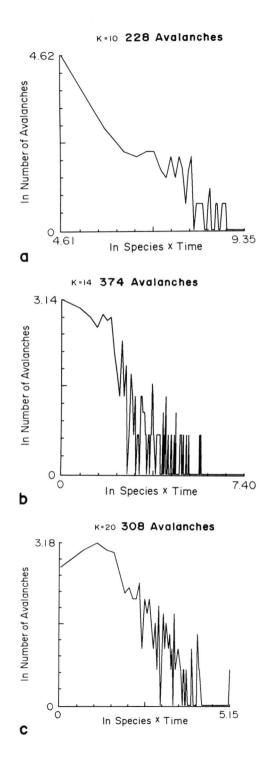

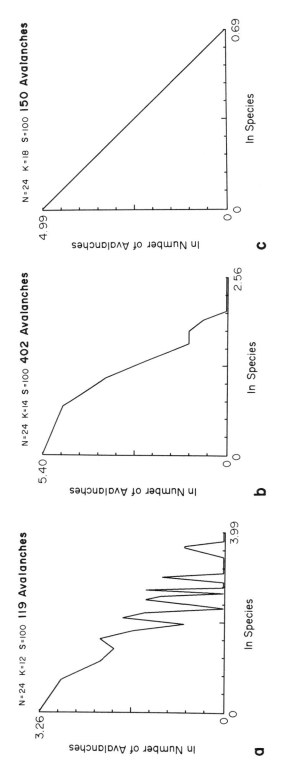

Figure 6.11 Propagation of coevolutionary avalances in 10×10 ecosystems: natural logarithm of number of avalanches of a given size as a function of natural logarithm of size, where size corresponds to the number of species affected by the avalanche, for (*a*) $K = 12$, (*b*) $K = 14$, and (*c*) $K = 18$. In all cases, $N = 24, C = 1, W = 20, S = 100$.

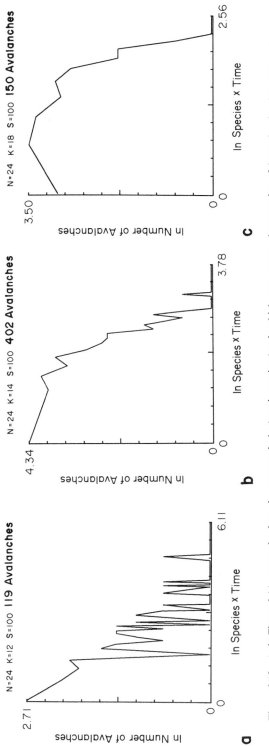

Figure 6.12 As in Figure 6.11, except that here size measure is the (species × time) unit, which sums, over each generation of the avalanche, the total number of species affected at that moment. (*a*) K = 12, (*b*) K = 14, (*c*) K = 18.

In sum, the following features are of interest:

1. On both measures of avalanche size, there are more small than large avalanches.
2. On both measures, as K decreases from above the optimal value of K, K_{opt}, the mean and median size of avalanches increases, the variance increases.
3. The distribution of avalanche sizes is clearly not a power law when K is much larger than K_{opt} but appears to be approaching a power law on both measures of avalanche size as K decreases toward a critical value at which waiting times to encounter Nash equilibria diverge.

During coevolutionary avalanches, species fall to lower fitness and hence are more likely to become extinct. Thus the distribution of avalanche sizes may bear on the distribution of extinction events in the evolutionary record. Raup (1986) has analyzed the intensity of extinction events at the family level during each of the 79 stages of the Phanerozoic. On average, each stage lasted about 7 million years. Figure 6.13a shows Raup's histogram of the number of extinctions per stage (intensity) graphed against the number of stages exhibiting that intensity. Clearly, there are many more small extinction events than large events. Raup makes the point that the distribution is also clearly continuous. Figures 6.13b replots Raup's data in ln-ln form. Although the data are obviously too weak to place much weight upon, the ln-ln plot is clearly convex, suggesting that the observed distribution is not quite a power law.

What conclusions are warranted by these results? A first general conclusion is the insight that *coevolutionary avalanches propagate* through ecosystems, that such avalanches have characteristic frequency-versus-size distributions which *change* depending on the parameters of the system. In particular, the distribution of avalanche sizes depends on how solid the frozen state is. If we tentatively accept Raup's data as weak evidence, the frozen state is modestly firm. Using Raup's data and improved evidence, we may ultimately be able to build a theory linking both ecosystem structure and extent of external perturbations to the size distribution of coevolutionary avalanches and to such phenomena as the distribution of extinction events.

A second and critical result is this: *Perturbations of the same initial size can unleash avalanches on a large variety of length scales.* This conclusion is clear and important. In these simulations, the perturbation in each case is a change in the external world of a single randomly chosen species in the ecosystem. If we may tentatively assume that avalanches can be linked to extinction events, then these results strongly suggest that *uniform* alterations in the external world during evolution can cause a *diversity* of sizes in extinction events. This possibility stands in contrast to the generally held hypothesis that small and large extinction events are associated with small and cataclysmic changes in the external world. Since the external environment has almost certainly undergone changes on a variety of scales, I do not wish to assert that high variability in extinction sizes does not in part reflect a heterogeneity in intensity of causes. But these results place part of the responsibility for extinction-size diversity on the dynamics of coupled ecosystems and on the ways in which damage propagates.

But the third conclusion is the most important: Raup's data suggest that ecosystems *do* coevolve to the edge of chaos. More precisely, the data suggest that ecosystems are slightly within the frozen regime. Thus we ourselves hover on the edge of a new view of coordinated coevolutionary processes among interacting adaptive entities. Here is a candidate universal law which warrants investigation.

Coevolutionary Intermittence and Punctuated Equilibrium

An important general feature of these results, whether on fully connected or on extended ecosystems, is that coevolutionary changes tend to occur in bursts followed by periods of quiescence. This unevenness is more notable in the random model than in the fitter and greedy models, but it occurs for all three. It is very tempting to associate these patterns with the general pattern of morphological stasis and bursts of punctuated change which appear to occur in the evolutionary record. One of the puzzles about such stasis is that it is normally accounted for as stabilizing selection which

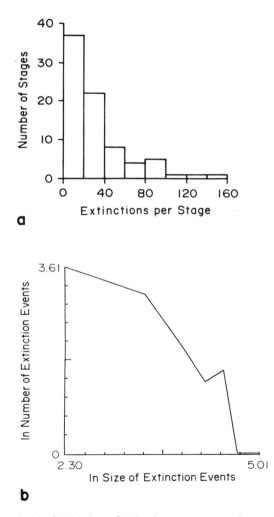

Figure 6.13 (*a*) Raup's data for the sizes of extinction events versus the number of events at that size. (*b*) Replot of Raup's data in natural logarithm form.

holds a phenotype at the optimum, in a stable environment. But as noted in Chapter 1, one of the problems is that the niche of an organism includes other organisms. Change by one alters the niche of others. How, then, can stabilizing selection opt for the same maximally adapted phenotype in a changing niche? Our results suggest an answer. Frozen components imply that species within or on the boundaries of such components have an *unchanging* optimal genotype and phenotype, despite changes in some of their coevolutionary partners. Thus familiar stabilizing selection *can* sustain the same phenotype in stasis, despite changes in the niche.

Inclusion of Density-Dependent Population Dynamics

Our analysis has relied on the use of a simplified population dynamics. Each species, at each generation, tries a mutant variant and moves there as a whole if the mutant is better. Each species interacts at each moment with all its immediate neighbors in the ecosystem. The natural extension of this class of models considers a population for each species, subject to population growth characterized by familiar ecological models. Such models are based on R and K factors, where R reflects the intrinsic growth rate of each population by itself and K reflects density effects resulting from the carrying capacity of the environment, which limits each population's increase to a standing abundance and includes possible competitive, mutualistic, and predator–prey interactions among species (May 1976; Roughgarden 1979; Pimm 1982).

To generalize our NK coevolving model, Johnsen and I defined the fitness of each genotype of each species in isolation from all other species in order to obtain its R value and then assessed, for each species i, the effect of a pairwise interaction with a member of each of the other species j connected to i. We defined R_i as the fitness of the species in isolation minus .5, the mean fitness in the space of genotypes. This definition allows a species to be either an autotroph, $R_i > 0$, or an obligate heterotroph, $R_i < 0$. We defined a_{ij} to be the fitness of species i when it interacts with species j minus .5. This definition allows interactions to be either mutualistic, $a_{ij} > 0$, or competitive, $a_{ij} < 0$. We assumed that each species had the same carrying capacity K. Then, for each species, we utilized the familiar logistic equation

$$\frac{dX_i}{dt} = X_i \left(R_i - \frac{X_i}{K} + \frac{\Sigma a_{ij} X_j}{K} \right)$$

where X_i represents the population number of the ith species. In the absence of interactions with other species, each species will attain the stationary population $X_i = R_i K$. Positive and negative interactions with other species, regarded as mutualism and competition, will in general alter the population attained by each species. In studies which allowed species to mutate, we first confirmed that a mutant form which was initially present at low frequency relative to the wild type and which increased in abundance more rapidly than the wild type continued to do so until the wild-type population was replaced. Having confirmed this capacity to invade and take over the initial population under a number of conditions, we substituted a simpler evolutionary dynamics: If a mutant form initially increased in abundance more rapidly in the current coevolutionary context than did the wild type, then the whole population moved to the mutant form.

Analysis of these population dynamics models appears to confirm the results based on the simpler dynamics discussed above but allows us to extend our results to look at the dynamics of population abundances as well as changes of genotypes. We analyzed 5×5 ecosystems in which interior, edge, and corner species interact with four, three, or two nearest neighbors, as above, and also modified 5×5 ecosystems in which each interior species interacted with its eight nearest neighbors (adjacent and diagonal), each edge species interacted with its five nearest neighbors, and each corner species interacted with its three nearest neighbors. The major results are similar for the two types of systems. The major differences are that ecosystems tend to be more volatile as the number of species connected to a given species increases, that easy attainment of Nash equilibria requires higher K values, and that consequently the optimal K value for maximal population size increases. Table 6.1 summarizes the data. Note that $K = 4$ yields the optimal sustained population size, the highest mean a_{ij} coupling, the highest fraction of mutualistic couplings, and the highest mean R_i, and, in mutations accepted, corresponds to a sharp drop from the value at $K = 2$. We see next that $K = 4$ is near the phase transition to order.

Figure 6.14 examines the onset of frozen components in 5×5 ecosystems in which each species interacts with eight, five, or three neighbors, by examining the number of mutations which are accepted among all 25 species at each generation. Note that when K is minimal relative to $C(a)$, the species continue to mutate persistently throughout the run. Thus the system never attains a Nash equilibrium. In contrast, when $K = 4(b)$ or K is much larger than $C(c)$, most species stop finding fitter mutants after a few hundred generations and hence the ecosystem is largely frozen and unchanging. Since here we use random rather than fitter or greedy dynamics, it is not entirely certain that the systems have attained true Nash equilibria.

Figure 6.15 examines the emergence of mutualism in these models. Each species grows on its own and hence can be thought of as an autotroph. In addition, however, each species might be helped or harmed by its interactions with other species. The NK model, via the C couplings, permits the possibility that each species can change its genotype such that it is helped by its coevolutionary partners. Table 6.1 shows that mean a_{ij} values are greater than 0 in all cases. Thus mutualism emerges in these systems. Note also in Table 6.1 that the a_{ij} values first increase and then decrease as K increases. Figure 6.15 shows the details. For $K = 0$, massive fluctuations occur in a_{ij} values as coevolutionary changes propagate through the unfrozen ecosystem. The fluctuations keep the a_{ij} values low. For $K = 4$, fluctuations are sporadic and cease

TABLE 6.1 5×5 Square Ecosystems

K	Population	a_{ij}	Fraction mutualistic	R_i	Mutations per species
0	1593	.009	.506	.158	47.4
2	2103	.077	.600	.205	38.2
4	2264	.138	.657	.210	14.0
8	2064	.090	.599	.196	10.2
16	1276	.027	.522	.158	5.3

Note: Interior species are each connected to eight neighbors; edge species, to five neighbors; and corner species, to three neighbors. Population is mean sustained population density of a species in the ecosystem, a_{ij} is mean value of this measure of mutualism among the 25 species of the ecosystem, fraction mutualistic is fraction of the couplings into each species for which $a_{ij} > 0$ (hence the fraction of the couplings between species which are mutualistic to some degree), R_i is mean self-growth rate (autotrophic if positive) of the 25 species, mutations per species is the number of mutations accepted per species over the simulated coevolutionary period.

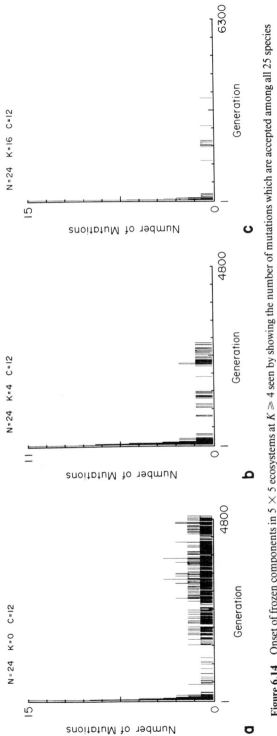

Figure 6.14 Onset of frozen components in 5 × 5 ecosystems at $K \geq 4$ seen by showing the number of mutations which are accepted among all 25 species at each generation. (a) $K = 0$, (b) $K = 4$, (c) $K = 16$. In all cases, $N = 24$, $C = 12$.

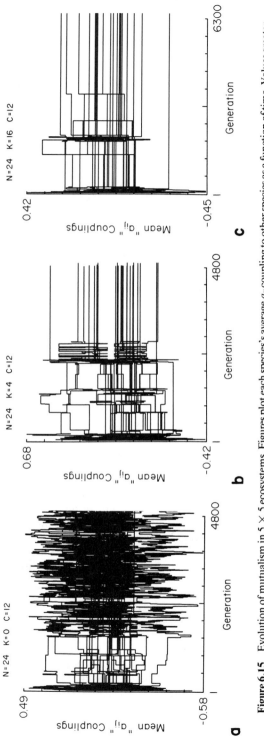

Figure 6.15 Evolution of mutualism in 5 × 5 ecosystems. Figures plot each species's average a_{ij} coupling to other species as a function of time. Values greater than 0 are mutualistic; those less than 0 are competitive interactions. (a) $K = 0$, (b) $K = 4$, (c) $K = 16$. In all cases, $N = 24$, $C = 12$.

after about 2400 generations, and the fraction of connections which are mutualistic is maximal. For $K = 16$, ecosystems freeze easily but the fraction of connections which are mutualistic has fallen and mean a_{ij} values have decreased from the optimum. Mutualistic interactions appear harder to attain when K is large, perhaps because each system harbors more conflicting constraints. Thus an intermediate value of K, $K = 4$, appears to optimize the ease of forming strong mutualistic interactions.

Figure 6.15 also gives (1) a dramatic view of the periods of quiescence and bursts of change which propagate through these model ecosystems and (2) for $K = 4$ and $K = 16$, clear evidence of the attainment of a Nash or near-Nash equilibrium.

Figure 6.16 shows the population dynamics for these conditions. Note that for the highest value of K, $K = 16$, population abundances increase smoothly to their carrying capacity. As K decreases, the population behavior becomes more erratic. Genotypic changes show up as discontinuities in the rate of population change. For $K = 0$, it appears that several species *decrease* monotonically in abundance during the run. As emphasized in Table 6.1, the total abundances of all species first increases and then decreases as K increases, reaching a maximum at $K = 4$. Thus there is an optimal intermediate of K that optimizes mean sustained fitness, as in the simpler dynamics studied above. Here, however, fitness is explicitly expressed as population abundance.

The results both for the coevolving ecosystems without population dynamics and for those with population dynamics are a first serious hint that selection may achieve coevolutionary systems with jointly optimized structures that optimize the capacity to evolve successfully and that the optimum structure is at the edge of chaos. It is quite amazing that such model systems approach a poised self-organized critical state. The deep questions concern how general such phenomena might be. The answers, of course, are unknown. Nevertheless, the general idea is plausible. In overview, the reasons for such a process in the NK landscapes appear to be that, as K increases relative to N, the fitness of local optima decreases, while the numbers of local optima and the steepness of sides of peaks increase. It is the balance of these opposing forces which dictates that an optimal value of K relative to C and S should exist and be attainable by coevolutionary processes tuning landscape structure and coupling. These properties may be very widely found in rugged landscapes, for increasing K relative to N merely captures the idea of increasing the conflicting constraints in complex systems. If in general fitness landscapes become more rugged and multipeaked as a result of such conflicting constraints, then coevolution to a poised state on the edge of chaos may turn out to be a near-universal feature of coevolution. Such a general result would be exciting indeed, with potential implications in many areas of the biological, social, and perhaps physical sciences.

Artificial Life at the Edge of Chaos: A Power-Law Distribution of Extinction Events in Tierra

An entirely independent investigation, carried out by ecologist Tom Ray, also suggests that coevolving systems create ecosystems which attain the poised edge of chaos. Briefly, Ray has begun studying model ecosystems in which the organisms are self-reproducing computer programs that live and spawn within the confined world of a computer's memory core. Ray seeds his world, Tierra, with a single type of self-reproducing computer program which literally copies itself into a neighboring region of

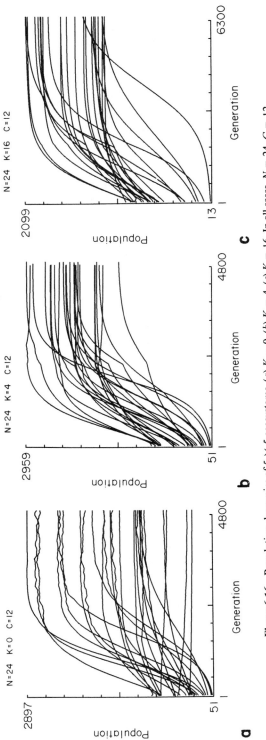

Figure 6.16 Population dynamics of 5 × 5 ecosystems. (*a*) K = 0, (*b*) K = 4, (*c*) K = 16. In all cases, *N* = 24, *C* = 12.

core. During this process, mutations to the code can occur. The resulting variety of model organisms compete with one another for access to computer time to carry out instructions. A "reaper" eliminates organisms which reproduce slowly. Ray's major interest in his intriguing system is to study the emergence of functional interactions among these model organisms. Indeed, many such functional roles emerge. Many are surprising. Many are even humorous. Inevitably, Ray's critters jointly create an ecosystem in which each makes its living—often by cooperating with, often by stealing from, its coevolving partners. But Ray's organisms also go extinct. Is there evidence for the edge of chaos?

I asked Ray to investigate the distribution of sizes of extinction events which occur in Tierra. His results are shown in Figure 6.17. The data are a power law except for very large extinction events. This fall-off presumably reflects finite size effects in the model which place an upper bound on the size of avalanches. While these data suggest that Ray's system is at the poised state, it shall be necessary to study his system in more detail to confirm this impression. If the results withstand further analysis, the importance seems great. Ray's model organisms are explicitly coevolving myopic adaptive agents that mutually construct the world they inhabit. Is the poised edge the general attractor of such a dynamics? It is hard not to notice that the slightly convex curve in Figure 6.17 is hauntingly like Raup's data for extinction events in the evolutionary record. Both hint, but do not yet prove, that complex ecosystems universally approach the edge of chaos.

I close this discussion of coevolution by raising the issue of the linking of speciation and extinction events into the evolution of ecosystem webs.

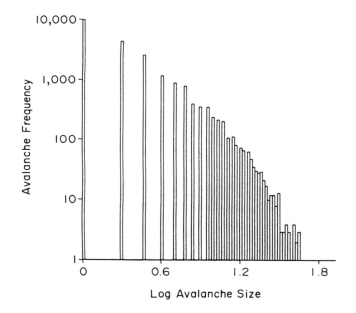

Figure 6.17   Logarithmic plot of the distribution of sizes of extinction events in Ray's Tierra simulations. The size of an extinction event corresponds to the number of genotypes that go extinct over defined short intervals. (Courtesy of Tom Ray)

Endogenous Processing Linking Speciation, Extinction, and Web Structure?

An endogenous coevolutionary process may trigger bursts of speciation and extinction as well as the integration of new species into the ecosystem. The issue is simple to state: Speciation events and extinction events, quite as much as changes in the external world, can cause alteration in the structure of the web and unleash avalanches of fluctuations in fitness which lead to further speciation and extinction events. Thus not only external shocks due to changing physical environments but also an *endogenous* dynamics may regulate speciation and extinction as a function of the current structure of the ecosystem. For example, suppose a speciation event at some point in the web leads to an incipient species connected to all the same other species as the parental species. As we saw in Figure 6.6, as the number of interacting species increases, the waiting time to encounter Nash equilibria increases, mean fitness during the pre-Nash period decreases, and the amplitudes of fluctuation in fitness increase. Thus emergence of an incipient species at a point in an ecosystem web locally increases the connectance of the web and should locally tend to melt the frozen component. This melting will lead to a lapse from the frozen Nash equilibrium for those few adjacent species. In turn, this lapse should be followed by increased fluctuation in fitness of all the species in that part of the web and lowered fitness (since fitness is typically lower during the pre-Nash fluctuations than at Nash equilibria). We have already supposed that fluctuations to lower fitness may increase the probability of extinction events, but during such fluctuations to lower fitness, the number of directions which are improved variants increases. Thus fluctuations to lower fitness should increase the probability of speciation events as well as the probability of extinction events. (The relationship between fitness and speciation is undoubtedly complex. Reduced population size leaves fewer explorers of new genotypes. Conversely, in sexual populations, reduced population size increases the number of peripheral isolates and hence increase speciation rates. And, on more familiar arguments, speciation rebounds are likely to follow extinction as a result of reduced competition.) In short, avalanches of coevolutionary changes are likely to be associated with both speciation and extinction events propagating through the system. In turn, the speciation and extinction events alter the structure of the ecosystem web and thereby may ease propagation of the avalanches. In addition to speciation and extinction events, migration of new species into or out of an ecosystem changes the structure of the system. The problem is to develop insight into the behavior of such a coevolving web as its members, their connectance to one another, and the structure of their adaptive landscapes change. At present, obviously, we have no such theory.

At least the following components seem requisite:

1. New species, resulting from either speciation or immigration, may adapt to a local increase in connectance either by increasing K or by reducing, through niche differentiation, the degree to which they are coupled to neighbors. Species may also lower effective coupling by interacting with an intersubstitutable group of species which are similar to one another, as perhaps the fox interacts with rabbits, hares, and squirrels. Such diffuse interaction lowers the sensitivity to alterations in any single coevolving partner.

2. The balance between speciation and extinction probabilities seems critical. If, as fluctuations to lower fitness deepen, speciation were always more probable than extinction, then local melting of the ecosystem web would increase fluctuations

and could lead to runaway speciation in which increasing connectance in the eco-system might cause the entire system to melt. This extensive melting could cause the entire system to fluctuate wildly, driving yet further speciation. Indefinite run-away speciation, however, is implausible, if only because total biomass is limited by solar energy intake, and each of very many species with few members is likely to be subject to rapid extinction. Thus, ultimately, extinction must be more rapid than speciation. For this or other reasons, speciation and extinction rates should eventually come into rough balance. Such a balance is observed in the evolution-ary record, where speciation exceeds extinction for the Cambrian and early Pha-nerozoic but the two rates are nearly the same for most of the rest of the Pla-nerozoic (Raup 1986).

3. Relative time scales are important. If speciation or immigration is slow relative to the speed of altering connectance among species, or K values, then a small excess of speciation plus immigration over out-migration and extinction might not melt an ecosystem. If these processes come into balance, the stability of the frozen state will depend on how rapidly old members leave and new members are accom-modated into the structure versus how rapidly that structure can shift.

These rough considerations make it clear that we need to develop one theory that relates all four phenomena: the balance between speciation and extinction, the bal-ance between immigration and out-migration, the integration of members into eco-system, and the propagation of avalanches through the ecosystems. Stenseth (1985) has considered some of these issues but without the concept of frozen components. The perspective discussed here, noting the role of frozen components, suggests that ecosystems may generally coevolve to a self-organized critical state having a frozen component that is firm enough to withstand the rate of integration of new members into the system, the loss of old members from the system, and external shocks from the physical environment.

Note finally that these themes lead again to a theme from Chapter 3. There we encountered the problems of homoplasy and of the rates of branching divergence and convergence on fixed or deforming fitness landscapes. The richness of landscape cou-pling, the number of coevolving partners, and the structure of the ecosystem tune the rate of coevolution. During such coevolution, each partner may speciate and diverge and converge on the deforming landscapes. Therefore, the parameters of coevolu-tion, here, N, K, C, and S, and the structure of the ecosystem must tune the diver-gence and convergence which will occur on such landscapes. Thus if we can see our way to a theory which relates the stationary structure of coevolving systems to the numbers of species, to speciation and extinction rates, and to ruggedness and cou-plings of landscapes, we might ultimately have a framework for linking divergence and convergence to the endogenous dynamics of coupled coevolving systems.

COEVOLUTIONARY CONCLUSIONS

We have examined coevolution in the limited context of coupled NK landscapes and without explicit consideration either of population flow on each landscape or of the behavior of coevolving ecosystems subject to invasions by new species (Roughgarden 1979; Pimm 1982). Further sophistication in these directions is certainly needed. Nevertheless, the approach we have used has the virtue of focusing attention on the relationships among landscape structure, landscape coupling, the number of coevolving partners, and the structure of the ecosystem in the overall dynamics. The

NK family appears to be the first family of landscapes which readily allows these questions to be posed. While the *NK* family of landscapes is just one family of correlated landscapes, it is critical to emphasize that the image of coupled landscapes is very general. It applies to many systems of coevolving components where each agent has many alternative actions, each with a fitness which depends on some analogue of a payoff matrix. If each agent must explore its own action space by more or less *local moves,* then that agent is attempting to hill climb on its own more or less rugged landscape. As it does so, it will cause its partner's landscapes to buckle and deform. While the structure and couplings of such coevolving systems are unknown, they surely exist. We have sought here some of the conditions necessary for coevolving systems to be able to accumulate successes and increase fitness to some maintained and fairly high level.

We may have found some of those necessary conditions. Landscapes must be tuned such that the richness of epistatic coupling within each species is roughly equal to the product of the coupling between species times the number of other species each species interacts with. This statement is rough but open to further investigation. It affords a jumping-off point for thinking about how landscape structure and coupling influence coevolution. Further, we may have identified selective forces acting at the individual level which can mold coevolution such that those remaining in the ecosystem fare well.

We have found evidence, hardly surprising, that the structure of an ecosystem governs coevolution. More surprising, we have found evidence for frozen components in ecosystems, and have raised the possibility that instabilities in the ecosystem web may propagate speciation and extinction events. Most important, the structure of fitness landscapes on which organisms adapt and the couplings among organisms' landscapes are themselves open to selective modifications which increase the inclusive fitness of each organism. Organisms jointly evolve the games they mutually play. The surprising result may be that each modifies its game so that the whole coevolving ecosystem attains the edge of chaos.

It bears repeating here that we have now encountered two quite different cases in which high K values may be better than low K values. In Chapter 3 we saw that high-K landscapes can lead to higher sustained fitness in the face of modest mutation rates, despite harboring lower local optima, because the sides of fitness peaks are steep enough to offset mutational dispersion. Here we have seen that, in fitness landscapes that are deforming rapidly as a result of strong coupling between landscapes, species with K high relative to C can coevolve to higher sustained fitness and attain Nash equilibria more rapidly than those with low K, despite the fact that optima are lower on high-K landscapes. Again, this is likely to reflect the fact that, as K increases, the sides of fitness peaks become steeper, allowing the species to respond more rapidly to deformations in its landscape as its partners move.

SUMMARY

The previous chapter discussed the conditions necessary for the emergence of order and for the capacity to adapt in complex parallel-processing systems; the present chapter has discussed coevolution. Both point in the direction of a new working hypothesis. Organisms which are internally constructed such that they are in the solid regime but near the edge of chaos appear to be best able to perform complex tasks and to adapt. Coevolving systems whose members have tuned the structure of their fitness landscapes and couplings to other members such that the entire ecosystem is poised at the edge of chaos appear to sustain the highest fitness. Thus we may adopt the hypothesis that selection attains systems which are poised both internally and collectively. If so, we have an important tentative result: Selection optimizes the capacity to evolve thanks to an attractor of the selective dyanmics, a generalized poised state. Let us look at this in more detail.

The capacity to coevolve successfully is not trivial, for mere chaotic twitchings of the angry Red Queen may occur. The results we have obtained suggest some tentative conclusions about the requirements for successful coevolution. Fitness landscapes need to be of sufficient ruggedness to offset the richness of couplings between landscapes and the number of partners whose moves impinge on each landscape. Otherwise viewed, to avoid the Red Queen, epistatic couplings within each member of a species need to be large enough to counterbalance epistatic couplings to the coevolving partners.

We have identified possible selective forces which may tune these parameters such that coevolution is typically successful. Selection, in a kind of selective metadynamics and as if by an invisible hand, may act on individual members of a species to alter the statistical structure of their fitness landscapes and the richness of their couplings to other partners so as to attain ecosystems poised at the phase transition between order and chaos. These structured ecosystems harbor nearly melted frozen components, optimize sustained fitness, and permit propagation of avalanches of coevolutionary change, ringing out the old species and ringing in the new. The resulting characteristic power-law distribution between the size and frequency of such avalanches offers one hypothesis for explaining the size distribution of extinction events in the fossil record. Caution, however, is necessary. The range of coupled landscapes that exhibit a phase transition from an ordered to a chaotic regime is unknown. The generality of the claim that the phase transition affords the highest fitness is unknown. The efficacy of selection to achieve and sustain the edge of chaos against drift is unestablished. The applicability to real coevolving systems is untested.

These results on the requirements for successful coevolution must bear on the internal structure and logic of adapting systems. If successful ecosystems coevolve to the edge of chaos, selection will have achieved this state by acting on the organisms within the system to tune (1) the ruggedness of their fitness landscapes, (2) the richness of couplings between organisms, (3) the sensitivity of one organism's fitness landscape to changes in the abundance of phenotypes or other organisms, and (4) the number and variety of other organisms in each species' niche. Our study of parallel-processing networks of Boolean elements revealed several major ways in which selection can alter the ruggedness of fitness landscapes. First, networks in the ordered regime adapt on very smooth landscapes, those in the chaotic regime adapt on very rugged landscapes, and those in the complex regime near the edge of chaos adapt on mixed rugged landscapes. In the complex regime, some mutations cause massive alterations in behavior, while most cause minor alterations in behavior. This mixture provides buffering against large and small deformations in fitness landscapes. A ready response lies near to hand for most such changes. Thus evolutionary tuning of an organism's position on the order–chaos axis tunes ruggedness of its landscape. In addition, we noted that compression of algorithms, via reduction in the number of variables which are "hidden units" lying between inputs and outputs in parallel-processing networks, increases ruggedness of landscape structure. If organisms typically adapt to the complex regime in order to be capable of complex behaviors, then within the complex regime, ruggedness of landscapes can be further tuned by controlling compression. Since these factors control landscape ruggedness, they are candidates for being optimized to achieve coevolving ecosystems which are collectively poised near the edge of chaos.

In short, frozen components, the requirements for order, the capacity to adapt, and adaptation to the edge of chaos through natural selection may reappear at many levels in complex adapting and coadapting systems. A generalized poised state may prove to be the ultimate attractor of an evolutionary dynamics.

This chapter is the last of five (2 to 6) which have attempted to sketch some of the major features of adaptive evolution in simple and complex evolving and coevolving systems. In the ensuing three chapters, we turn to the wonderful question of the origin of life.

PART II

The Crystallization of Life

The second part of this book, Chapters 7 to 10, explores a heretical possibility: The origin of life, rather than having been vastly improbable, is instead an expected collective property of complex systems of catalytic polymers and the molecules on which they act. Life, in a deep sense, crystallized as a collective self-reproducing metabolism in a space of possible organic reactions. If this is true, then the routes to life are many and its origin is profound yet simple.

This view is indeed heretical. Most students of the origin of life hold that life must be based on the self-templating character of RNA or RNA-like molecules. Because of such self-templating, any RNA molecule would specify its base pair complement; hence a "nude gene" might reproduce itself. After that, according to most thinkers, these simplest replicating molecules built up around themselves the complex set of RNA, DNA, and protein molecules which constituted a self-reproducing system coordinating a metabolic flow and capable of evolving.

Chapter 7 unfolds this new view, which is based on the discovery of an expected phase transition from a collection of polymers which do not reproduce themselves to a slightly more complex collection of polymers which do jointly catalyze their own reproduction. In this theory of the origin of life, it is not necessary that any molecule reproduce itself. Rather, a collection of molecules has the property that the last step in the formation of each molecule is catalyzed by some molecule in the system. The phase transition occurs when some critical complexity level of molecular diversity is surpassed. At that critical level, the ratio of reactions among the polymers to the number of polymers in the system passes a critical value, and a connected web of catalyzed reactions linking the polymers arises and spans the molecular species in the system. This web constitutes the crystallization of catalytic closure such that the system of polymers becomes collectively self-reproducing.

While heretical, this new body of theory is robust in the sense that the conclusions hold for a wide variety of assumptions about prebiotic chemistry, about the kinds of polymers involved, and about the capacities of those polymers to catalyze reactions transforming either themselves or other, very similar polymers. It is also robust in leading to a fundamental new conclusion: Molecular systems, in principle, can both reproduce and evolve without having a genome in the familiar sense of a template-replicating molecular species. It is no small conclusion that heritable variation, and hence adaptive evolution, can occur in a self-reproducing molecular system lacking a genome. Since Darwin's theory of evolution, Mendel's discovery of the "atoms" of heredity, and Weismann's theory of the germ plasm, biologists have argued that evolution requires a genome. False, I claim.

Also, this new body of theory is fully testable. If correct, sufficiently complex systems of RNA or protein polymers should be collectively autocatalytic.

In Chapter 8 these new concepts are extended to the crystallization of a connected metabolism. I strongly suspect that, rather than having formed piecemeal, a connected metabolism, like a self-reproducing set of catalytic polymers, emerged spontaneously as a phase transition when a sufficient number of potentially catalytic polymers were mixed with a sufficiently complex set of organic molecules. In this condition, a critical ratio of number of catalyzed reactions to number of molecular species present is surpassed, and a connected web of catalyzed transformations arises. Life began whole and integrated, not disconnected and disorganized.

Chapter 9 compares my new view with more standard hypotheses and attempts to place all into a coherent picture. In this integrated view, a self-reproducing metabolism involving catalytic RNA and perhaps peptides arose first, and then evolved an RNA polymerase, itself most readily supposed to have been a catalytic RNA sequence, *or ribozyme. The presence of such a ribozyme polymerase frees an autocatalytic polymer system of the requirement for collective catalytic closure and hastens further molecular evolution, leading ultimately to a selective transition to a code translating between polynucleotides and polypeptides.*

Finally, Chapter 10 extends the new view of the origin of life to what I shall call random-grammar models. Here, instead of polymers acting on polymers to form new polymers under the laws of chemistry, we have symbol strings acting on symbol strings to form new symbol strings under the "laws" specified by a particular random grammar. Chapter 10 is, in a sense, a conceptual culmination of both Part I and Part II of this book. Such grammar models afford a new framework for thinking about the emergence of functionally integrated systems which interact with, represent, and know their worlds. These systems provide models for the emergence of competition and collaboration between self-reproducing molecular systems and perhaps even functional integration and transformation in economic and cultural systems. Indeed, grammar models are natural testbeds for understanding the interweaving of historical contingency, with its avalanches of consequences cascading from frozen accidents, and the dominion of law, in biology and other deeply historical sciences.

Part I seeks the principles of organization in complex systems which permit the systems to adapt. It ends with the hypothesis that such systems attain the edge of chaos, both internally and in their coevolution. Knower and known, life is at the boundary of disorder. Yet provocative though such an hypothesis may be, it derives from mathematical theory which is silent on the deepest issues: What does it mean for Escherichia coli *to "know" its world? How is such knowing naturally represented in the dynamical models explored in Part I? The disturbing answer is that such knowing is not naturally represented in the models. Even in models of neural networks whose attractors are taken to be memories of external events, it is we the inventors of the networks who mandate that the attractor represent the input pattern. The network represents nothing to itself. But consider autocatalytic polymer systems, perhaps comprising sets of RNA sequences mutually reproducing themselves. Let two such systems interact by exchanging RNA sequences. A sequence injected by the first system into the second might poison that second system such that it no longer is able to reproduce itself. Heritable variation and natural selection—Darwinian evolution—might lead to the development in each system of defenses for warding off harmful sequences emitted by the other. Indeed, even the evolution of exchange of those sequences which are mutually helpful might occur. In short, autocatalytic polymer systems are primitive examples of a kind of agency, the locus of survival and death, the locus of integrated response to the environment. Given proliferation and Darwinian selection, we have a clear notion of what is "good" for such a system, and hence also of the functional import of any process to the system's proliferation. Such systems, in the same sense that holds for* E. coli, *come to know their worlds. If so, the principles may be very general. Perhaps IBM and* E. coli *know their worlds in much the same way.*

CHAPTER 7

The Origins of Life:
A New View

In the next three chapters we turn our attention to the wonderful problem of the origin of life. What more awesome problem could invite our attention? To many recent scholars—Wald (1954), Hoyle and Wickramasinghe (1981), and others— improbable features of current organisms imply improbable origins. If the probability that a protein catalyzes a given reaction is 10^{-20} and if a minimal contemporary organism such as a pleuromona-like organism has on the order of 1000 or 2000 enzymes, then the probability of their joint occurrence by chance is, say, $10^{-40\,000}$. More likely that, as Hoyle says, the whirlwind assemble a 747 from scraps in a junkyard. Yet here we are, in quite clear contravention to Hoyle's unhappy conclusion. We the lucky, or we the expected?

In the present chapter I shall develop a body of theory, tied to current experimental data, which argues that life is not improbable. On the contrary, I believe it to be an expected, emergent, collective property of complex systems of polymer catalysts. Life, I suggest, "crystallizes" in a phase transition leading to connected sequences of biochemical transformations by which polymers and simpler building blocks mutually catalyze their collective reproduction. This theory is heterodox, as we shall see, but fully testable.

In the first section of the chapter, I describe the historical background and current status of experimental work in the field. I review data concerning (1) the abiotic origin of the small organic molecules which are the building blocks of life, (2) the familiar hypothesis that DNA or RNA was the first "living" molecule, (3) the current status of research on RNA catalysts called ribozymes, including efforts to construct a ribozyme polymerase able to replicate itself, and (4) the alternative familiar hypothesis that the first living molecules were proteins. In my view, all contemporary accounts are inadequate. All have failed to note the phase transition to collectively autocatalytic sets of polymers which lies at the base of the theory I discuss. Not only is such a collectively reproducing polymer system relatively probable, I argue, but it is likely to have been the nexus from which polymers able to copy themselves, such as a ribozyme polymerase, later evolved. Further, I argue, collective self-reproduction in sufficiently complex polymer systems implies that life at the outset required a minimal complexity and has always been graced by an innate holism.

In the second section, I develop the mathematical theory showing that autocatalytic polymer sets, made up of RNA sequences, polypeptides, or both, are expected to form spontaneously. In the third section, I consider sets of polymers that are potentially infinite both in terms of polymer lengths and in terms of kinds of polymers. The phase transition to autocatalytic sets implies, in such potentially infinite sets of polymers, a phase transition between mathematically finite, or *subcritical,* catalytic sets and mathematically infinite, or *supracritical,* sets. We find critical values of the underlying parameters of the theory characterizing this phase transition. Further, among other implications, there is a characteristic distribution relating numbers of kinds of polymers in the set and the lengths of those polymers. As we shall see in Chapter 8, this size distribution comes close to fitting the size distribution of organic molecules. It is a candidate ahistorical universal in biology.

In the fourth section, we examine the capacity of collectively autocatalytic polymer systems to evolve without a genome and find that such evolution is to be expected. The final section discusses experimental consequences. Most notable, the successes of applied molecular evolution, discussed in Chapter 4, imply that attainably complex mixtures of single-stranded RNA sequences and/or peptides should be collectively autocatalytic. Experiments to test for the *in vitro* creation of self-reproducing biochemical systems are described.

BACKGROUND OF THE ORIGIN OF LIFE PROBLEM

In the present section, I briefly review the three standard theories of the origin of life: (1) self-reproducing polymers which replicate by virtue of template complementarity, typically with advocacy of RNA or RNA-like polymers; (2) collectively self-reproducing *sets* of catalytic polymers, typically with advocacy of proteins; and (3) self-reproduction by clays or other minerals.

Any account of life's origin necessarily includes answers to the question of the origin of the simple organic molecules whose traffic became metabolism. The standard answers, not disputed here although now doubted by some workers, involve appeal to an early reducing atmosphere on the primitive earth coupled with a variety of energy sources, a combination which rendered formation of covalent bonds energetically favorable. The famous Miller experiments sending electrical discharges through a mock primitive atmosphere containing hydrogen, nitrogen, and simple carbon compounds and producing a mixture of organic molecules with a number of amino acids are the progenitor of all later work in this direction. As is well known, Miller obtained modest yields of glycine and alanine and lower yields of many other amino acids—some among the standard 20 in present coded proteins, other not (Miller 1953, 1955, 1957). The results have been repeated and extended by many workers (Lawless and Boynton 1973; Miller, Urey, and Oro 1976; see Orgel and Miller 1974). Experiments in a similar spirit have shown that the fatty acid building blocks of lipids (Lowe, Rees, and Markham 1963) and the building blocks of RNA and DNA can be formed under presumptive primitive abiogenic conditions (Oro and Kimball 1961, 1962; Lowe, Rees, and Markham 1963; Ferris, Sanchez, and Orgel 1968; Schwartz and Chittenden 1977; summarized in Ferris 1987 and in Miller 1987). It is important to stress, however, that conditions allowing formation of the sugars, nucleosides, and nucleotide bases that are the building blocks of RNA and

DNA are less thermodynamically and kinetically favorable than those allowing synthesis of amino acids.

It should also be stressed, as Shapiro (1986) has done, that the argument for the abiogenic synthesis of the building blocks of life has a certain air of theater about it, with rapid scene changes as needed to advance the plot: Conditions A yield a modest amount of product X; then X plus Y, the latter obtained, in low yield, under quite different conditions B, can be joined in high concentrations under conditions C to obtain product Z. As Shapiro points out, the abiogenic conditions used for modest yields of many organic molecules differ from one type of molecule to another. Thus it behooves our caution when thinking that all transformations might take place under common conditions abiogenically. If not, then the congregation of these precursor compounds may be problematical. Be that as it may. Let us accept the sympathetic view: Let there be an abundance of small organic molecules in the promised dilute, or perhaps not dilute, early ocean, tidal pool, geyser basin, atmospheric droplet. This is not the problem I want to attack.

The problem I do want to attack is this: How hard is it to obtain self-reproducing systems of complex organic molecules capable of a metabolism coordinating the flow of small molecules and energy needed for reproduction and also capable of further evolution? Contrary to our expectations, the answer, I think, is that it may be surprisingly easy. In particular, I shall try to show that any sufficiently complex system of peptide or RNA polymers capable of catalyzing the formation and cleavage of new polymers either from other peptides or RNA polymers or from their monomeric building blocks can be expected to contain self-reproducing subsystems capable of metabolism and evolution.

The core of the theory is this: As the complexity of a collection of polymer catalysts increases, a critical complexity threshold is reached. Beyond this threshold, the probability that a subsystem of polymers exists in which formation of each member is catalyzed by other members of the subsystem becomes very high. Such sets of polymers are autocatalytic and reproduce collectively. Thus the new view I shall propose is disarmingly simple. Life is an expected, collectively self-organized property of catalytic polymers.

DNA or RNA First

There are two dominant views of the origin of life, once beyond the origin of small organic molecules. The overwhelming majority of workers favor the view that either DNA or, more probably, RNA is the primordial molecule of life. The loyal opposition holds that proteins were first, and sufficient. The remaining party to the discord is championed by Cairns-Smith (1982), who supposes that clay surfaces were able to replicate, by templating, specific "imperfections," or electronic arrangements, on successive layers and used such arrangements to catalyze subsequent reactions among organic molecules to invent a metabolism, proteins, and eventually genes.

Any DNA-or-RNA-first theory has two central and powerful arguments in its favor. First, all life is now based on DNA or RNA as the stable storage form for genetic information. Second, double-stranded DNA and RNA are beautiful examples of Schrödinger's aperiodic crystals, which are also point-point local templating complements. Since Watson and Crick remarked, with uncertain modesty, that the base pair rule associating A with T and C with G on complementary strands of the double helix carried obvious suggestions about how the genetic material might replicate, any sensible person considering the problem of self-replicating molecules must be drawn to the beauty of nature's apparent choice. The base pair sequence along each strand specifies the precise complementary sequence along the remaining strand. That one strand specifies the other is not an accident; it reflects the remarkable shape identity of the hydrogen-bonded AT and CG pairs which allows an undistorted double helix to form. Thus the two strands are, as just intimated, *local* shape complements. The complementarity holds at each base along the nucleotide chain.

Contemplation of the precision of double-stranded DNA or RNA engenders one of two opposing responses: either "how fortunate that such self-complementary molecular forms are occasioned by the laws of chemistry and physics" or "how unfortunately *special and untypical* is this almost miraculous self-complementary structure; are we really to think that life based on carbon chemistry requires such symmetry?"

While the obvious self-complementarity of DNA or RNA is a persuasive argument in its favor as the preeminent candidate for the first replicating molecule, two equally persuasive, perhaps overwhelming difficulties assail the hypothesis. First, compared with the abiogenic synthesis of the amino acid precursors of peptides and proteins, the abiogenic synthesis of RNA or DNA precursors is, respectively, hard and harder (see, for example, Schwartz and Chittenden 1977). Orgel (1987) points out that there is only one plausible prebiotic synthesis of ribose: the polymerization of formaldehyde. In this reaction pathway, ribose is formed as one of a large number of sugars and never as the major product. Similarly, the condensation of adenine or guanine with ribose leads to complex isomeric mixtures containing relatively small amounts of the natural nucleoside. Prebiotic synthesis of pyrimidine nucleosides is, Orgel suggests, even more difficult. And the possibility that the accurate replication of an oligoribonucleotide could occur directly in an unfractionated, racemic prebiotic soup seems remote (Joyce, Schwartz, et al. 1987; Orgel 1987). Second, no uncatalyzed replication of arbitrary sequences of single-stranded RNA or DNA has yet been attained (Joyce 1987; Orgel 1987). The experimental facts surrounding this present failure require attention.

Orgel and his co-workers have carried out extensive experiments aiming to achieve nonenzymatic template replication of arbitrary RNA or RNA-like sequences. Typical experiments utilize either single-stranded homopolymers of a single base—poly(U), poly(C), or poly(A)—or else copolymers of two or more of the four bases as a preformed "template" (Joyce, Inoue, and Orgel 1984; Chen, Inoue, and Orgel 1985; Joyce and Orgel 1986). The specificity of Watson–Crick pairing is used to bind monomers at complementary sites along the preformed template. Then, with the help of a suitable condensing agent, the adjacent template-bound monomers can be joined to form oligomeric products. The fundamental difficulty is that only preformed templates which are preferentially rich in C work well. Thus even when nonenzymatic synthesis leads to a complementary single-stranded RNA molecule of reasonable size, that molecule is preferentially rich in G rather than C. Hence the new molecule is unsuitable to act as a template to re-create the initial C-enriched RNA polymer (Joyce 1987). Templates rich in A, U, or G are unsuitable for different reasons, each of which appears difficult to overcome. For example, G-rich polymers tend to fold back upon themselves in stable secondary structures with hairpin loops. Additional problems concern formation of the proper $3'-5'$ bonds between adjacent residues rather than the more common $2'-5'$ bond. Thus using poly(U) as a template and activated A (adenosine $5'$-phosphorimidazolide, ImpA), the overall yield of poly(A) is $18:1$ in favor of $2'-5'$ bonds rather than $3'-5'$ bonds (Lohrmann and Orgel

1978). At present, the best results have been achieved with C-rich templates and another activated form of the monomers (2-MeImpG, A) in the presence of Zn^{2+} (Inoue and Orgel 1983).

As Joyce (1987) points out, the difficulty is severe and rests on the following central ideas: (1) RNA is to act as the carrier of genetic information; (2) such information requires *arbitrary* sequences of bases in the replicating RNA strands; and (3) replication of RNA strands is to be carried out, without an enzyme, by a *polymerase-like* function. Polymerases act to add successive monomers to a growing RNA chain, where the position of each new monomer is specified by the (arbitrary) base in the corresponding position in the preformed template. As Joyce notes, "In a nonenzymatic system . . . nucleotide polymerization should be regarded as consisting of 16 different reactions, involving the addition of any one of the four activated monomers to an oligomer terminated by any one of the four nucleotides. It will not be easy to find a reaction system that allows all 16 of these reactions to proceed at an acceptable rate." Note also that the proper 3'–5' bonds, rather than 2'–5' bonds, must be formed. Obviously, current difficulties do not mean that polymerase-like replication of arbitrary single-stranded RNA sequences without enzymes is impossible. Current work is beginning to focus on RNA-like polymers which are chemically simpler than RNA and yet may allow template-directed replication by successive addition of monomers complementary to an arbitrary sequence (Joyce, Schwartz, et al. 1987; Orgel 1987). If successful, the transition from simple polymers to RNA and DNA will require an account.

The recent discovery of highly efficient catalysis by RNA (Kruger, Grabowski, et al. 1982; Zaug and Cech 1985; Cech 1986a, 1986b, 1987; Szostak 1986; Been, Barfod, et al. 1987; Kay and Inoue 1987; Orgel 1987) has increased enthusiasm for the hypothesis of an RNA world (Buzayan, Gerlach, and Bruening 1986; Darnell and Doolittle 1986; Gilbert 1986, 1987; Gilbert, Marchionni, and McKnight 1986; Hutchins, Rathjen, et al. 1986; Orgel 1986) in which RNA molecules both carry out template replication and act as catalysts, including acting as polymerases for template replication of arbitrary RNA molecules. Specifically, eukaryotic genes are organized into two types of fragmented segments: exons, which code for parts of a specific protein, and intervening introns, which do not and are spliced out during maturation of messenger RNA (mRNA) molecules prior to their transport from the nucleus to the cytoplasm. It has now been shown that certain introns can catalyze their own exicison from single-stranded RNA. More exciting, the same RNA sequences can catalyze any one of three possible reactions: transesterification reactions which lead to elongation of one of the two RNA polymer substrates by one monomer, ligation of two independent single-stranded RNA molecules, or cleavage of one independent RNA sequence into two smaller sequences.

The ribozyme-catalyzed dismutation reaction described by Zaug and Cech (1986) is an example of the first kind of reaction. Here two five-carbon polymers are transformed to a four-carbon polymer and a six-carbon polymer. The reaction produces no new internucleotide bonds, however. It is encouraging because it leads to polymers that are longer than those present in the substrate molecules. Hence it may be a prototype for a polymerase-like activity in a ribozyme. But as Orgel (1987) points out, the ribozyme differs from a true polymerase in that the former utilizes an internal template specific for the substrates to be manipulated rather than an arbitrary external template. That is, the ribozyme acts by binding specific substrates and mediating a specific transesterification reaction. In this respect, the ribozyme is like pro-

Hexanucleotide template

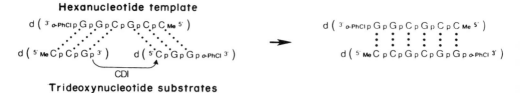

Figure 7.1   Autocatalytic replication of an RNA hexamer which specifically aligns with and ligates its component trimers. (From Joyce 1987)

teolytic enzymes (to be discussed below), which break proteins at specific sites and religate the fragments into new combinations in specific ways without creating new peptide bonds.

Given the difficulty of attaining template replication of *arbitrary* single-stranded sequences by a progressive addition of monomers to an arbitrary single-stranded template, it is very interesting that the autocatalytic replication of *specific* small oligonucleotides by a template mechanism has been demonstrated. At present, there are two examples. Both involve a self-complementary template. The first utilizes two trideoxynucleotide substrates and a hexanucleotide template (von Kiedrowski 1986) (Figure 7.1). The substrates are bound to the template by base pairing and, in the presence of a condensing agent, condense to form a new template molecule. The template–template complex can dissociate to yield two free template molecules, each able to bind additional substrate and begin a new round of synthesis. Thus this system is autocatalytic. The second example involves modification of the 3′ –OH of ribose to increase reactivity and use of a tetranucleotide template and two dinucleoside substrates (Zielinski and Orgel 1985) (Figure 7.2). Again, this specific template replicates autocatalytically given activated dinucleoside substrates.

I wish to stress an important feature of these two reactions. In each case, the template can be considered to be acting as a *specific ligase.* That is, the template molecule acts as a specific "enzymatic" site which holds and orients two substrates in proximity such that a condensing agent can create a bond between them. The template is *not arbitrary;* rather, it is specific for the pair of substrates which are joined. In this regard, the template functions as a *specific ligase* in the same sense as do proteolytic protein enzymes which recognize specific amino acid sequences within proteins and cleave or religate the protein substrate at specific sites. In the major part of this chapter, I shall attempt to show that complex systems of polymers mediating such ligation and cleavage reactions can be collectively autocatlytic. Thus I believe it is important to focus on the fact that even small RNA sequences can act as specific ligases.

Tetranucleotide template

Figure 7.2   Autocatalytic replication of an RNA tetramer which specifically aligns with and ligates its component dimers. (From Joyce 1987)

I should make a terminological point. I shall use the term "ligase" in a general sense to refer both to ligation mediated by a hexamer, as described above and based on simple base pairing, and to ligation mediated by ribozymes, where base pairing is part of the catalytic mechanism in aligning the substrate sequences but other mechanisms may also play a catalytic role in the reaction.

The ligase activities of small RNA templates would be expected to be less efficient than those of larger RNA enzymes, on the same general basis that one expects small peptides to be less efficient enzymes than larger proteins. In both cases, the larger molecule can bring higher specificity and more binding energy to the reaction. In general, the known ribozymes, derived from type II and type I introns, are reasonably large, consisting of several hundred nucleotides. That ribozymes can catalyze cleavage and ligation reactions on *exogenous RNA* has now been established for the self-cleaved L-19 IVS RNA from *Tetrahymena* pre-rRNA (Been, Barfod, et al. 1987). The fragment acts as a sequence-specific endoribonuclease, cutting other RNA molecules at specific sites and capable of religating the two cut fragments. Furthermore, mutation of the binding site on the ribozyme alters the substrates upon which it acts in predicted ways (Zaug et al. 1986). Doudna and Szostak (1989) have shown that the *Tetrahymena* ribozyme can ligate multiple oligonucleotide sequences lined up on a complementary RNA sequence. At present, attempts are under way to generate a ribozyme able to act as a polymerase and efficiently replicate any arbitrary single-stranded RNA sequence, including, of course, itself. Such an effort may succeed and would be a *tour de force* but faces the difficulties noted by Joyce above and by others enumerated below.

In summary, to date no means have been found to achieve template-directed replication of arbitrary RNA sequences by sequential addition of monomers to an arbitrary external template. Such mechanisms may be found, of course, and may obtain without ribozymes or with ribozyme polymerases. In the meantime, it is clear that specific RNA sequences can act as template ligases or as true ribozyme ligases to ligate small RNA sequences, or to cleave and ligate exogenous RNA sequences, as well as to mediate transesterification reactions. These established facts suggest that complex mixtures of RNA sequences are strong candidates to achieve collective self-reproduction. Ribozyme polymerases, rather than arising *de novo* and sustaining themselves in evolution against mutational degradation from Eigen and Schuster's (1977) error catastrophe, might then plausibly evolve from such stable, collectively autocatalytic RNA polymer systems.

Beyond the deep problems of achieving self-replication of arbitrary RNA molecules by template mechanisms lie other fundamental issues. Suppose a replicating double-stranded RNA molecule existed. Such a molecule, a *nude gene,* has as yet no capacity to gather about itself the rich, complex, interwoven web of chemical transformation we call a metabolism. This brief statement parses into two quite different sets of problems, one commonly commented upon, the other not.

The catalytic machinery of the cell is overwhelmingly due to proteins, which are the primary enzymatic effector mechanisms facilitating reactions and funneling metabolic flow down specific corridors of transformations. DNA carries out no direct catalytic activities; it is remarkably inert. While the discovery of the first examples of catalytic capacity in single-stranded RNA is obviously very intriguing, cells nevertheless do use proteins. Thus any theory of the origin of life based on self-replicating DNA or RNA must include a description of a mechanism by which DNA or RNA molecules can be coupled to the synthesis of proteins capable of catalyzing not only metabolic reactions but also the replication of the DNA or RNA. In contemporary cells, of course, the relationship between DNA and protein is mediated by the enormously sophisticated mechanisms of coding and translation. The key element in translation of a gene to the corresponding protein is, in fact, carried out by specific proteins which catalyze the attachment of the proper activated amino acid residues to each transfer RNA (tRNA). More precisely, DNA is transcribed to complementary single-stranded RNA, perhaps processed to excise introns and so on, and then translated. Translation is carried out by tRNA molecules, each with (1) an anticodon site which is complementary to and thus specific for a given base pair triplet and (2) a site to bind a specific corresponding activated amino acid. Binding of that amino acid is carried out by a specific protein synthetase enzyme. Thus the effective translation of the genetic sequence of base triplets constituting the coded information for the linear sequence of amino acids in the protein rests on the synthetase. Proteins translate genes into proteins. Any DNA-or-RNA-first theory confronts the origin of the code. The conceptual obviousness of replication by self-templating RNA or DNA here gives way to confusion about how such a system might emerge: Its emergence seems to require its prior existence. One elegant body of ideas about the onset of coding as a self-organized process, due to Bedian (1982), will be discussed in Chapter 9.

The coding problem, intense as it is, is really a fragment of a larger problem: How would a nude gene gather a coupled metabolism about itself? This, too, I believe, is a very deep problem. It can be examined in a different form, a form which has received very little attention. The simplest free-living entities, the pleuromona-like organisms, are on the order of 0.1 the size of a bacterium, have a very simplified cell boundary with a simple bilipid layer, and have a genome which encodes perhaps 1000 to 2000 proteins; these proteins play the usual structural and enzymatic roles in a coupled metabolism. Viruses can be as simple as RNA or DNA strands encoding perhaps a dozen proteins. But viruses are *not* free-living entities; they are obligate parasites forced to usurp the machinery of their host cells in order to carry out their own replicative life cycle. Viruses are highly sophisticated parasites which have virtually certainly managed to simplify their metabolic system because of that present in the host. Thus it is an observed fact that all free-living organisms exhibit *a minimal and substantial level of complexity.* The deep question is, Why?

It is not trivial that the DNA-or-RNA-first theory—in which we envision a single RNA molecule capable of self-replication—gives no obvious answer to this question. We have no conceptual problem in supposing populations of replicating RNA or DNA molecules, each competing with the other, such that the most rapidly replicating molecule plus its family of related sequences wins the Darwinian selection race. Indeed, this is exactly the picture developed by Eigen and Schuster and described in Chapter 3.

Using this fundamental theory, how might we account for the observed minimal complexity in free-living entities? We cannot assert that complexity is requisite for life. By hypothesis, simple complementary strands of DNA or RNA suffice for attaining evolving families of similar replicating molecules. The very simplicity of self-templating is just the conceptual virtue needed to help us see that life might have started without unduly improbable combinations of rare events. But such a "pseudospecies" family of sequences has, as yet, no collaborative metabolism, nor need it have a greater complexity.

The remaining alternative, starting with nude genes—for example, a self-reproducing ribozyme polymerase—is to assert that the original living molecules gradually gathered about themselves the increasingly complex machinery of catalysis and metabolism, thereby increasing the amount of information able to be stored in the

DNA or RNA genetic material without mutational degradation, all this a result of increased precision of replication under enzymatic control. In turn, this increased complexity gradually ensured the competitive advantage of the more sophisticated over the less sophisticated. Undoubtedly, this is the dominant view, and it does have merit. Yet its only answer to why we now observe a given minimal level of complexity is historical accident. Entities having the complexity of pleuromona-like organisms happen to be the simplest free-living survivors. Why do we not observe systems as simple as viruses, or even simpler, living in Darwin's shallow pond? Because, we respond, such hapless forms would soon be outcompeted by present-day organisms.

In short, the replicating-RNA theory, ribozyme polymerase or otherwise, offers no *theory* on why free-living organisms "must" exhibit a minimal complexity. We have another evolutionary just-so story. Like other just-so stories in evolution, of course, it may be true.

In summary, the nude gene theory holds fast to the idea that life started simple, with simple replicating molecules, and gradually learned to be complex. I hold just the opposite view. I shall argue that life started as a minimally complex collection of peptide or RNA catalysts capable of achieving simultaneously collective reflexive catalysis of the set of polymers (hence replication of an autocatalytic set of polymer catalysts) and a coordinated web of metabolism. Thus one of the virtues of the body of theory I shall develop is that it inherently accounts for the fact that free-living entities exhibit a minimal complexity.

Protein First

The alternative dominant strand in considering the origin of life proposes that proteins were the first living molecules. Any protein-first theory has several immediate advantages but a number of stellar problems. The advantages include the following. First, prebiotic experiments reliably show that it is quite easy to obtain many amino acids in moderate to low yields in a variety of plausible circumstances. Further, the spectrum of yield abundances, highest for alanine and glycine, closely mimics the spectrum of abundances in known proteins. Second, formation of peptides or of proteinlike polymers of amino acids, called proteinoids, under plausible prebiotic conditions has been demonstrated (Fox, Jungck, and Nakashima 1974; Fox and Dose 1977; Fox 1980, 1981; Fox and Nakashima 1980; Fox, Nakashima, et al. 1982). Perhaps the most critical advantage of the protein-first hypothesis is the fact that even small abiogenic peptides, as well as Fox's large proteinoid material, readily exhibit a wide variety of catalytic activities. While we are used to thinking of enzymes as highly adapted, well-perfected catalysts, which indeed they are, it appears to be the case that attaining at least weak catalytic activity is "easy" with peptides. Thus a number of simple tripeptides with amino acids having a ring structure, such as tryptophan, as the middle member appear rather readily to catalyze cleavage of depurinated DNA (Behmoaras, Touline, and Helene 1981a, 1981b; Pierre and Laval 1981). The tripeptide leu-lys-leu catalyzes its own self-condensation to form leu-lys-leu-leu-lys-leu.

Recognition that abiogenic peptides typically exhibit a spectrum of weak catalytic activities is not new. Cavadore (1971) synthesized large populations of random peptides having a mean of 50 amino acids per peptide. Such complex mixtures catalyzed numbers of reactions. Given the complexity of such mixtures, it has been impossible to discern whether many members catalyze the target reactions poorly or a few catalyze the reactions very well. The same observations were made repeatedly by Fox and his co-workers. When proteinoid material is placed in water, it forms hollow vesicles called microspheres, made up of a proteinoid shell surface and an aqueous interior. Suspensions of microspheres can carry out a variety of catalytic activities, albeit weakly. It is particularly important that proteinoid material can catalyze the *formation of additional peptide bonds,* often coupled with the degradation of ATP to ADP. Knowing this fact might lead to the expectation that formation of proteinoid material might exhibit an autocatalytic component. As the material forms, it might help catalyze further peptide bond synthesis to generate yet more proteinoid. If so, the rate of formation of proteinoid should be an accelerating function of time. This has been observed (Fox and Dose 1977; Fox 1980). In fact, proteinoid material also catalyzes the formation of nucleotide bonds, coupled with the degradation of ATP to ADP (summarized in Fox and Dose 1977).

Fox's microspheres exhibit a number of other remarkable properties. Because microspheres are closed vesicles, Fox has long suggested them as protocells, able to enclose an internal environment. In a mother liquor rich in proteinoid material, microspheres exhibit "budding" by an accretion mechanism, followed by "division" to form two independent microspheres. They exhibit selective diffusion of some small molecules across the peptide "shell" and even appear to support a trans-shell electrical potential reminiscent of that seen across lipid bilayer membranes of contemporary cells.

Protein-first theories become substantially more plausible if there are grounds to believe that the peptides and polypeptides which might have been utilized in early organisms were quite short. Very interesting evidence supporting this has been raised by Longberg and Gilbert (1985). These authors argue that contemporary RNA molecules contain clues to the earliest catalytic polypeptides, which might have typically been made of between 30 and 40 amino acids. Later proteins would evolve by recombination and divergence of relatively few molecular themes. A length of 30 to 40 amino acids is in the same size range as that explored by Cavadore. Naturally, evidence that primitive catalytic polypeptides were 30 to 40 units long neither proves nor disproves a protein-first theory. An RNA-world theory would want short peptides to be catalytic as well; otherwise, evolution of coding and the specificity to synthesize long proteins would be even more difficult to understand.

Protein-first theories confront significant difficulties. Foremost among these is the difficulty in envisioning how a protein might replicate itself. Large proteins are often globular, with hydrophobic groups on the interior and hydrophilic groups on the surface. Unlike RNA, where an intellectually simple and satisfying local point–point homology links complementary positive and negative strands by AU and GC base pair rules, no such local point–point complementarity is known for a protein. Thus there is no obvious way for a peptide or protein to specify its linear structure by a template-like mechanism. This conclusion is so transparent that it has stood as the dominant objection to a protein-first theory. There are other problems, however.

Suppose we conceive, even demonstrate, some mechanisms such that one or a collection of peptides or proteins is able to replicate autocatalytically. It is not obvious that such a single molecule or such a collection would be capable of further evolution. We are used to considering the fruits of selection as "stored" in the DNA. Where would these fruits be stored in an autocatalytic protein or set of proteins? In the structure of some single replicating protein? In the collective dynamics of an autocatalytic collection of proteins? Might such systems undergo selective adaptation, or even random drift, to form "new" autocatalytic systems? And if these and still further questions could be answered, the following arises: Just as a DNA-first theory must invent a code and proteins, any protein-first theory must invent translation and the genes.

A Double Origin?

Before continuing, it is very much worth a pause to ask the following. Suppose autocatalytic sets of polymers made of catalytic peptides *and* catalytic RNA sequences, coupled with the subsequent evolution of peptide or ribozyme polymerases and hence template-replicating RNA or DNA, are possible? Then we are entitled to imagine the coevolution and symbiosis of two primitive life forms: collectively autocatalytic systems and template-replicating systems. It is hard to avoid the thought that this dual existence would make life, forgive the craven, a lot easier. I shall return to this possibility in Chapter 9, but it is intuitively obvious that if autocatalytic sets of peptides and catalytic RNA can coexist and then couple to arbitrary template-replicating RNA or DNA sequences, the way is open for the evolution of protein coding. It is no longer necessary to solve the chicken–egg problem, which arises with the nude template-replicating RNA molecule, that useful proteins must first exist in order to code for themselves.

Protein-First Colleagues

The central conceptual problem with a protein-first view, as just stressed, is to understand whether and how either a single protein or a set of proteins might be able to reproduce. The first person to tackle this problem, to my knowledge, was Calvin (1969). He proceeds by first pointing out the well-established, but less well-known, fact that a number of peptides in contemporary bacteria and higher organisms are synthesized entirely by enzymatic means, rather than by coding. Examples include glutathione, which is a tripeptide in mammals, and several antibiotics, such as gramacidin, which is a peptide ring structure with L and D amino acids synthesized by bacteria. He then suggests two alternative pictures of replication in proteins. He imagines a single peptide which autocatalytically adds a succession of specific amino acids to one end. Thereafter, he imagines that the initial peptide cleaves the newly grown tail from the initial peptide seed. If the tail is the same as the seed, the polymer has replicated. This image is at least conceivable. It is hard to picture, however, how such a remarkable molecule could *evolve* to mutant forms which retained that remarkable property. Further, such a molecule replaces the nude RNA gene with a nude protein. That is, it is hard to imagine such a peptide gathering around itself a connected metabolism.

In his second alternative, Calvin supposes a set of peptides having the property that each member had its formation catalyzed by one or more members. Thus it is the *set* of peptides which is collectively autocatalytic by virtue of reflexive catalysis among its members. Calvin briefly postulates this possibility but carries the analysis no further. What is interesting is that Watson (1965), in attempting to demonstrate that replication *necessarily* is based on the point–point complementarity of RNA or DNA, also considers reflexively autocatalytic sets of peptides but dismisses the possibility as too complex. On the other hand, serious attempts have been made in articles by Kauffman (1971b, 1986b), Rossler (1971, 1974, 1983), and Cohen (1988), and in a delightful book by Dyson (1985) modeled after Schrödinger's *What Is Life?* (1944). My own view of the best formulation of an autocatalytic polymer set theory is presented next.

AUTOCATALYTIC SETS OF CATALYTIC POLYMERS

Building a Theory of Emergent Collective Autocatalysis Among Catalytic Peptides or Catalytic RNA

The remainder of this chapter discusses the theory of autocatalytic sets of polymers. It is important to emphasize that the theory applies equally to *peptides* with catalytic properties and to *catalytic RNA*. This is no small point. We noted above that template replication of arbitrary RNA sequences has not been attained without a polymerase. In contrast, ligase reactions by RNA catalysts are well known and even occur with small, specific template RNA sequences. Ribozyme polymerases may prove chemically possible, but highly unlikely to form *de novo* and evolutionarily unstable to Eigen and Schuster's error catastrophe. Thus, as noted above, a body of theory which shows that the formation of collectively autocatalytic sets of RNA sequences that catalyze ligation and cleavage reactions may be better able to account for replication among RNA catalytic sequences than does ribozyme-polymerase–mediated template replication of arbitrary RNA sequences. Indeed, if both peptides and RNA can catalyze reactions involving the other kind of polymer, there is no reason not to envision mixed polymer systems. Some indication that ribozymes can act on other kinds of polymers has been reported by Shvedova, Korneeva, et al. (1987), who find that a specific RNA molecule containing many modified bases is the catalytic component of a polyglucan branching enzyme.

The theory we develop here, coupled with the discussion of ordered dynamics in Boolean networks in Chapter 5, is a step toward a deep theory of homeostasis. Homeostasis may be an expected dynamical property of complex systems of reflexively catalytic organic molecules. It is a direct consequence of molecular specificity: Dynamical systems in which each variable is affected by few other variables spontaneously are homeostatic. But low connectivity in a dynamical system, such that each variable is affected by few others, is just a renaming of *high molecular specificity* of catalysis and ligand binding in chemical systems. In other words, I hope to show that self-reproduction and homeostasis, basic features of organisms, are *natural collective expressions of polymer chemistry.*

Some Requirements for Autocatalysis in Sets of Peptides or Ribozymes

In order for autocatalytic sets of peptides or RNA molecules to exist, at least the following are necessary:

1. Peptides and polypeptides must be capable of catalyzing the formation and cleavage of peptide bonds; ribozymes must be capable of catalyzing the formation and cleavage of internucleotide bonds.

2. In order for autocatalytic sets of peptides or RNA molecules to have played a role in prebiotic evolution, it is necessary that abiogenic formation of such polymers capable of catalyzing formation and cleavage of peptide and phosphodiester bonds be feasible.

3. In order for reactions to occur effectively, the reactants must be confined to a sufficiently small volume.

4. An anabolic flux synthesizing larger peptides from some maintained "food set" of amino acid monomers, small peptides, or other molecules must be thermo-

dynamically feasible; an anabolic flux synthesizing larger RNA polymers from a food set of nucleotides, small oligonucleotides, or other molecules must be thermodynamically feasible.

5. Catalytic "closure" must be achieved and maintained. That is, it must be the case that every member of the autocatalytic set has at least one of the possible last steps in its formation catalyzed by some member of the set, and that connected sequences of catalyzed reactions lead from the maintained food set to all members of the autocatalytic set.

The main focus of the model I shall present is on this last item, the conditions required for catalytic closure. Nevertheless, the first four issues in the list above are of basic importance and require at least brief discussion here.

1. It is perfectly familiar that proteins can catalyze the cleavage and synthesis of peptide bonds. Trypsin, the well-known gut enzyme, is merely the best-studied case. An enormous variety of proteases have been analyzed, from trypsinlike enzymes of relative low specificity to members of the clotting cascade which cleave essentially single target polypeptides. As noted, Fox's proteinoid material also catalyzes formation of peptide bonds. Our discussion of ribozymes above makes it clear that these catalyze the formation of specific $3'-5'$ phosphodiester bonds between nucleotides.

2. As noted above, a variety of abiogenic origins of peptides and proteinoids are plausible (Fox and Dose 1977; Fox, Nakashima, et al. 1982), and such peptides and proteinoids carry out a variety of catalytic activities (Cavadore 1971; Behmoras, Toulme, and Helene 1981a, 1981b; Pierre and Laval 1981). The recent discovery of catalytic antibodies (Pollack, Jacobs, and Schultz 1986; Pollack and Schultz 1987) and of mimetic and liganding peptides among random epitopes (Cwirla et al. 1990; Devlin, Panganiban, and Devlin 1990; Scott and Smith 1990) strongly supports the conclusion that sufficiently complex mixtures of peptides having chemical function will abound. Abiogenic formation of single-stranded RNA seems rather more difficult, as noted by Orgel (1987). However, achieving RNA sequences with *catalytic* activity need not require that all bonds be $3'-5'$, or that proper optical activity be maintained at each residue. The results of Tuerk and Gold (1990) and those of Ellington and Szostak (1990) discussed in Chapter 4 suggest that the probability of finding RNA sequences able to bind arbitrary substrates is not excessively low. Presumably, achieving ribozymes is on the same order of difficulty. In short, it is reasonable to think that abiogenically derived peptides, RNA sequences, and related polymers are feasible and that a complex collection of sequences would be blessed with an abundance of chemical liganding and catalytic functions.

3. I shall return below to the problem of confinement of polymers to a small enough volume to permit reaction. A number of suggestions are familiar, including Oparin's coacervates, Fox's proteinoid microspheres, and even aerosol droplets. A hypothesis which I find congenial and discuss in the final section of this chapter is enclosure of amino acid monomers and polymers in hollow lipid membrane spheres called liposomes. Although synthesis of peptide bonds and of phosphodiester bonds is thermodynamically unfavored in aqueous media, we shall see that a kind of "osmotic drive" in liposomes may push the synthesis progressively as the reaction proceeds. A particularly interesting discussion of the possible routes to the origin of confining environments and initial cells may be found in Cavalier-Smith (1987).

4. Maintenance of an anabolic flux leading to the synthesis of larger polymers from smaller peptides or larger RNA polymers from smaller ones forces us to focus on the thermodynamical problems associated with the possible existence of peptide or RNA autocatalytic sets. Because an enzyme merely speeds up forward and reverse reactions proportionally, it does not shift the equilibrium ratio of reactants and products.

Focus for a moment on peptides. The free energy associated with a single peptide bond is on the order of 1400 calories. Peptide bond formation is associated with abstraction of a water molecule; thus the equilibrium is shifted in favor of cleavage by the presence of water. Simplifying with the assumption that the energy of any one peptide bond is essentially the same as that of any other, regardless of surrounding primary sequence structure, it follows that the rate of dissociation of a dipeptide to its constituent amino acids is higher than the reassociation rate by a factor of about 10 when both reactants and products are present in 1 M concentration. The consequence is that the equilibrium concentration of a specific peptide with N amino acids falls off by a factor of 10 as N increases by 1:

$$[P_N] = C^N K^{-(N-1)} \tag{7.1}$$

where K is the dissociation constant for a peptide bond, C is the molar concentration of a species of amino acid, and N is the length of the peptide.

Equation 7.1 shows two things: (1) for polymers composed of a single type of amino acid, at thermodynamic equilibrium, most of the amino acids are present as monomers or very short polymers, while a very few are present as high-molecular-weight polymers and (2) any specific polymer of high weight is present at infinitesimal concentration. Dehydration of the reaction mixture such that the concentration of monomers increases shifts the equilibrium toward the synthesis of larger polymers. Thus in the absence of other metabolic energy sources—such as ATP or pyrophosphate, which can "activate" the monomers to a higher energy state which then favors formation of peptide bonds as a downhill step—dehydration is one major means of shifting the equilibrium mixture toward larger polymers. Adsorption onto a surface is a second means of shifting the equilibrium mixture toward large polymers.

In their interesting discussion of the origin of enzymes, Dixon and Webb (1960) formulate Equation 7.1 and then note that if, in contrast to a system using a single kind of amino acid, one considers a system with A species of monomers at concentration C, then the equilibrium distribution becomes

$$\Sigma[P_N] = A^N[C^N]K^{-(N-1)} = \left(\frac{AC}{K}\right)^N K \tag{7.2}$$

(See also Flory 1953.) Dixon and Webb point out that, as A increases, the ratio AC/K becomes larger than 1.0 and hence $(AC/K)^N$ *increases* as N increases. This implies that, for a sufficiently high number of amino acid species, at equilibrium, the vast bulk of the amino acids are bound up in polymers of *high molecular weight*. Of course, the concentration of any *specific* high-molecular-weight polymer remains very small.

The simple thermodynamic fact that dehydration shifts the equilibrium distribution toward larger polymers is well illustrated by the plastein reaction (Wasteneys and Borsook 1930; Levin, Berger, and Katchalski 1956; Neumann, Levin, et al. 1959; Silver and James 1980, 1981a, 1981b; Fruton 1982). If a complex mixture of large proteins is incubated with trypsin, the latter cleaves the proteins to small peptides. If the mixture is then evaporated slightly, the equilibrium shifts in favor of synthesis of peptide bonds, and trypsin catalyzes the formation of large polymers. If the large polymers are removed from the reaction mixture and the remainder is again concentrated, the reaction again runs in the direction of synthesis of large polymers. Among the interesting facts about the plastein reaction is that the synthesis of peptide bonds by trypsin requires *no* high-energy compounds such as ATP. It simply reflects a shift in equilibrium.

This thermodynamic description and the plastein reaction argue that, in principle, thermodynamic considerations *allow* an anabolic flux of amino acid monomers to form a web of high-molecular-weight peptide and polypeptide polymers without the exogenous aid of a coupled metabolism linking degradation of high-energy phosphate or other compounds. Obviously, to maintain an anabolic flux, the system must be open via the influx of a food set of amino acids or small peptides and the efflux of waste products.

These considerations with respect to the formation of peptide bonds carry over directly to the formation of phosphodiester bonds between nucleotides to generate polynucleotide RNA sequences. In the absence of other energy sources, in principle, dehydration or adsorption to a surface shifts the equilibrium toward larger polymer species. Note that in practical experiments carried out by Orgel and his co-workers, however, activated forms of nucleotides plus a condensing agent are used (Joyce 1987).

I shall base initial discussion of the model of autocatalysis in sets of peptide and/or RNA polymers, *for simplicity only,* on the severe hypothesis that an autocatalytic polymer system be required to be composed purely of such peptide or RNA polymers and monomers. In fact, refusal to include couplings to additional energy sources which can abet the formation of larger polymers markedly limits the maximum polymer size which can occur in modestly high concentrations in such autocatalytic systems and hence limits their potential complexity and capacity to evolve. As we shall see in the next chapter, this restriction is *unnecessary,* since the same considerations which lead us to expect the formation of autocatalytic sets of peptides and RNA sequences also lead us to expect that such systems will automatically "crystallize" about themselves a connected metabolism whereby energy released from catabolic processes helps drive anabolic processes, such as polymer synthesis, which are thermodynamically uphill. Thus the general body of theory will suggest that a self-reproducing coupled metabolism including RNA and/or peptide polymers plus a web of other organic molecules collectively crystallizes as a whole.

Catalytic Closure: Autocatalytic Sets of Peptides or RNA Sequences

The central problem to address is that of achieving catalytic closure in a set of catalytic polymers, either peptides or RNA sequences. We proceed in four steps:

1. Consider the set of all possible polymers up to some maximum length M.
2. Consider the set of all possible legitimate reactions by which these polymers can be formed from one another.

3. Consider simple models of the distribution, in the space of polymers up to length *M*, of the capacities to catalyze the different reactions among the polymers.

4. Consider the resulting probability that the set of polymers contains a subset which is reflexively autocatalytic and show that, as the complexity of the polymer set increases past a sharp threshold, the probability that an autocatalytic set exists jumps sharply to 1.

Consider first a set consisting of two monomer species—say, alanine and glycine or C and U—and polymers up to length *M*. Then the number of polymers of length *M* is just 2^M, and the size of the set up to length *M* is $2^{M+1} - 2 \approx 2^{M+1}$. Thus as *M* increases, the number of polymers increases exponentially. If *B* = 20 amino acids are considered, the number of polymers of length *M* is 20^M and the size of the set is slightly larger than 20^M but not as large as 20^{M+1}. Similar expressions hold for the set of *B* = 4 nucleotides.

Peptides and RNA sequences are oriented polymers: Left and right ends differ. The simplest possible reactions among such polymers are (1) cleavage to two smaller fragments and (2) the reverse synthesis reaction. Consider a polymer of length *M*. It contains *M* − 1 bonds and so can be synthesized by *M* − 1 different condensation reactions from smaller polymers making up its "left" and "right" termini. A polymer which is smaller than the maximum length *M*, say one of length *L* < *M*, can be formed by condensation of still smaller polymers in *L* − 1 ways. In addition, however, the polymer of length *L* can itself be a component of larger polymers of lengths $L + 1, L + 2, \ldots, M - 1, M$. If the polymer of length *L* is either the left terminal residue or the right terminal residue, then it can be cleaved free from the larger polymer in a single reaction. A specific polymer L* can be cleaved from 2×2^K larger polymers of length $L + K$, where $L + K$ is less than or equal to *M*. Thus, the total number of ways to form L* is

$$R^M_{L^*} = \sum_{i=L+1}^{M} (2 \times 2^{i-L}) + (L - 1) \tag{7.3}$$

Summing over all polymers, the total number of reactions R_T by which they can interconvert is

$$R_T = 2^M(M - 1) + 2^{M-1}(M - 2) + \cdots + 2^{M-(M-2)}(M - (M - 1)) \tag{7.4}$$

(This equation involves minor double counting of mirror symmetric polymers.)

The critical issue to examine is the *ratio* of the number of reactions by which polymers can interconvert to the number of kinds of polymers, as the maximum length of polymers increases from *M* to *M* + 1. Since the maximum-length polymers, those of length *M*, can be formed in *M* − 1 ways, it is obvious that, as *M* increases, more new kinds of reactions arise than new kinds of polymers, at a marginal ratio of *M* − 1. Dividing Equation 7.4 by the total number of polymers up to length *M* (which is approximately 2^{M+1}), it is simple to show that the ratio of reactions to polymers is just

$$\sum_{i=1}^{M} \frac{M - i}{2^i} \approx M - 2 \tag{7.5}$$

The critical implication of Equation 7.5 is that, as *M* increases, the number of polymer species increases exponentially, but the number of condensation and cleavage reactions by which these species interconvert increases still *faster,* such that the ratio of reactions to polymers increases linearly in M. Simply put, as M increases, there comes to exist vastly more legitimate reactions by which polymers can interconvert than there are polymers. Clearly, this imbalance reflects the simple combinatorics of polymer strings made up of two monomer units. If we consider instead $B > 2$ monomer units, the same general results are found and the ratio of reactions to polymers tends to $2(M - 2)$ as B increases.

Reaction Graphs and Centripedal Specific Reaction Density

We encountered the concept of a graph earlier: A graph is a set of points, called vertices, and a set of either edges (for undirected graphs) or arrows (for directed graphs) which connect pairs of points. The general concept of graphs can be applied to chemical reactions. Specifically, we can represent the forward synthesis reaction ligating two small peptides to form a single larger peptide, or two small RNA sequences to form a single larger RNA sequence, by representing the polymers as points and drawing arrows from the two smaller polymers to the single larger polymer, and representing the reverse cleavage reaction by drawing arrows from the larger to the two smaller polymers. Alternatively, since the reactions are reversible, pairs of arrows in opposite directions can be replaced with a simple unoriented edge. Then one reaction is represented by a pair of edges from the two smaller to the single larger polymer. The set of all such edges among all polymers up to length M constitutes the *reaction graph* for this polymer system.

Figure 7.3 shows reactions forming a polymer of length $M = 5$ and one of length $L = 2$. Clearly, the dimer (Figure 7.3b) is formed by more reactions than the pentamer (Figure 7.3a), since the dimer can be cleaved off the ends of many larger polymers. Specifically, from Equation 7.3, as L increases to $L + 1$, the ratio of the number of reactions by which a specific polymer of that length can be formed decreases by about ½. Therefore, the reaction graph among polymers up to length M is strongly *nonisotropic,* with the numbers of ways to form polymers increasing centripedally as L become smaller.

Models of the Distribution of Catalytic Capacities in Peptide and RNA Space

Insofar as our central question is whether or not, and why, a set of polymers might become reflexively autocatalyic, it is natural to ignore the fact that condensation reactions forming peptide or phosphodiester bonds can occur spontaneously at a slow rate in the absence of catalysis. Such spontaneous reactions can only help our argument, but they are beside the point at issue; therefore, let us for the moment suppose that reactions do not occur unless catalyzed. (More realistically, we might suppose that spontaneous reactions occur on too long a time scale to matter. Later, consideration of spontaneous reactions will be needed.) This then leads us to our major experimental and theoretical question: What is the distribution of catalytic capacities in the space of peptides or RNA sequences? Enough has been said in previous chapters about adaptive evolution of improved catalytic capacities, about catalytic task space, and about adaptive searches in statistically rugged landscapes to indicate how little is known about this fundamental problem. For the experimentalist, this ignorance is a call to work creating or designing ways to explore sequence space. For the theoretician interested in whether autocatalytic sets of peptides or

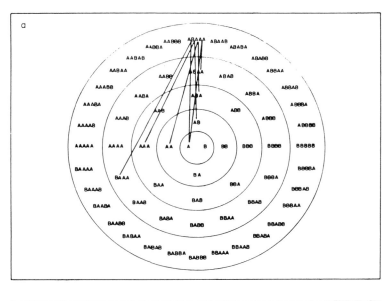

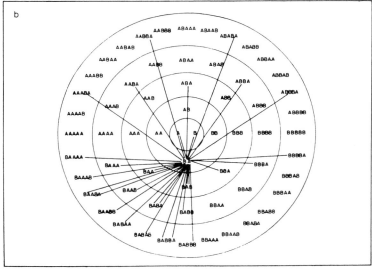

Figure 7.3 (*a*) Pairs of edges meet at the juncture of successive monomers in the pentapeptide ABAAA and lead to those smaller pairs of polymers which form ABAAA by condensation. (*b*) Edges connect dimer BA with A and B, which form BA by condensation, and with each larger polymer which yields BA by cleavage.

RNA sequences are possible, the sensible approach in the face of our present ignorance is to consider alternative simple models of the distribution of catalytic capacities in peptide and RNA space, assess whether these models lead to results allowing one to understand the conditions necessary to achieve reflexive autocatalysis, and then assess how robust the theoretical conclusions are with respect to the idealizations used in building the initial models of the distribution of catalytic capacities. If the conclusions are robust, then it is at least plausible that the detailed distribution of catalytic capacities does not matter to the general conclusions.

Begin, then, with the familiar. Proteases catalyzing cleavage and formation of peptide bonds range widely in specificities. If we are concerned with a set of $B = 20$ amino acids, we might imagine 400 low-specificity, trypsinlike protoenzymes able to cleave a bond between any one of the 400 possible pairs of amino acids, regardless of the position of that bond within a primary sequence. Because reactions are reversible, each such protoenzyme would catalyze the reverse condensation reaction, ligating any two smaller peptides ending with the appropriate carboxy and amino terminal amino acids. Such a hypothetical set of protoenzymes would be a kind of *universal constructor set,* able to catalyze all possible cleavage and condensation reactions among peptides made up of the 20 amino acids.

If we consider RNA polymers and ribozymes able to catalyze cleavage and ligation at specific sites, exactly the same arguments apply. Were there a set of 16 ribozymes which recognized each of the 16 possible dinucleotide pairs regardless of remaining sequence context and either cleaved such sequences or ligated sequences ending in the given nucleotides, that set of 16 ribozymes would also be a universal constructor set able to catalyze all possible cleavage and condensation reactions among RNA sequences. Alternatively, 256 ribozymes recognizing all 16×16 possible pairs of dinucleotides terminating sequences and ligating them would be a universal constructor set. I shall focus on peptides, but the general ideas carry over directly to RNA.

Several comments are in order:

1. If, for simplicity, the 400 protoenzymes all catalyze their respective reactions with the same velocities, this universal constructor set is merely a fancy means of flowing quickly to the thermodynamic equilibrium distribution of polymers in a closed system.

2. The universal constructor set is *reflexively and collectively autocatalytic* in a trivial sense. The 400 protoenzymes catalyze the formation of all polymers; hence they catalyze their own formation as well. The concentration of each polymer will depend on details of flux through such a system when displaced from thermodynamic equilibrium.

3. The idealization of 400 protoenzymes each specific for a given pair of amino acids utterly regardless of the context of that pair in the primary sequence is obviously false. Even trypsin exhibits significant sequence sensitivity. Consider what, roughly, occurs as sequence sensitivity increases. Then in the very much higher dimensional space of all cleavage and condensation reactions among discriminated polymers up to length M, many but not all reactions are catalyzed. Holes begin to occur in the space of reactions, reflecting those reactions which are either not catalyzed or catalyzed only very inefficiently. The first immediate consequence of such holes is that the open system need not approach a distribution of polymer species at all like either the equilibrium distribution for a closed system or that for the equivalent open system when all reactions are catalyzed equally

well. Second, it may or may not remain the case that the specific 400 protoenzymes catalyze their own formation in the presence of uncatalyzed holes in reaction space.

4. Let the specificity increase to the limit, such that any protoenzyme catalyzes cleavage or ligation of a specific pair of amino acids in a way which is so dependent on primary sequence that discrimination is infinite. Only one among the possible reactions is catalyzed by one protoenzyme. This extremely overstates the case in two ways. Most obvious, infinite discrimination is never achieved. An enzyme recognizes its substrates with finite precision; hence it always can catalyze reactions with a smaller or larger range of similar substrates. That is, any enzyme catalyzes a "ball" in catalytic task space. Less obvious, but as discussed in Chapter 4, an enzyme can catalyze quite distinct reactions using the same active site because quite different organic molecules can have the same local chemical features. In other words, apparently different reactions can constitute the same catalytic task. Thus a realistic picture of an enzyme is that it catalyzes a local ball in catalytic task space; the size of the ball reflects the obvious fact that discrimination is imprecise and obviously similar substrates are acted upon. The subtler notion is that different sets of substrates are members of the same equivalence class, each set the cloud of similar substrates centered about a different best substance(s) for the "same" catalytic task.

As just noted, in the limit of 400 protoenzymes catalyzing formation of peptide bonds between all possible pairs of amino acids, the trivial existence of reflexively autocatalytic sets of peptides is just that, trivial. An equivalent statement is that 16 or 256 ribozyme ligases catalyzing all possible 3′–5′ condensation reactions among RNA molecules are trivially autocatalytic. As specificity increases and uncatalyzed holes appear in the space of reactions, it is far from obvious that reflexively autocatalytic sets of peptides or ribozymes are to be expected. In particular, the limit of infinite discrimination offers the more difficult conditions for the emergence of reflexively autocatalytic sets. Thus for the purpose of developing an initial model of the conditions for the emergence of autocatalytic sets, I shall adopt this limiting idealization. I shall, however, preserve the idea that such a protoenzyme can catalyze quite distinct reactions among different sets of substrates because they constitute the same task. I make a further assumption here for convenience but shall relax it later. It is realistic to suppose that catalytic capacities of peptides or RNA sequences are correlated in some way with length. Small peptides or RNA sequences cannot fold to form efficient sites. On the other hand, small peptides and RNA sequences are known to exhibit catalytic activities, albeit weak ones, and weak suffices at the start. Thus in the initial development of the model, I shall assume that catalytic capacity is not correlated with polymer length.

These idealizations lead to an initial simple model of the distribution of catalytic capacities in the space of polymers. I shall suppose that any polymer has a constant probability P of catalyzing any reaction. This rule, in effect, represents a cloud of similar reactions by a single central reaction and a cloud of similar polymers able to catalyze that reaction by a single central polymer. It preserves the idea that the same polymer may catalyze quite different reactions constituting the same task.

The Catalyzed Reaction Subgraph

Consider again the set of polymers up to length M and the conjugate space of condensation and cleavage reactions induced by that set of polymers. By the simple hypothesis of a fixed probability P, we may now randomly assign to each polymer those reactions, if any, it catalyzes. That is, each polymer has a chance P of catalyzing the first reaction, the second reaction, and so forth. If a given reaction is catalyzed, we color the pair of edges representing it red. In addition, we can keep track of which reactions each polymer catalyzes by a pair of blue arrows emanating from that polymer and ending on the pair of red edges. When this simple game has been played for all the polymers in the space, some number of reaction edges may have been colored red. These red edges constitute the catalyzed reaction subgraph of the initial graph. The entire system of polymers, red catalyzed reactions, and blue arrows is called a hypergraph. The question to which we now turn is whether the polymers of the red catalyzed subgraph are expected to be *connected* to one another in such a manner that a reflexively autocatalytic set of polymers in the hypergraph is formed. Since we have utilized a simple probability P independently for each assignment of who catalyzes which reaction, the subgraph is a *random* subgraph of the initial graph. Whether that random subgraph is connected, and in what manner, turn out to be percolation problems in graph theory.

Connectivity Properties of Random Graphs Exhibit "Phase Transitions"

The central intuition I want next to communicate, already mentioned in Chapter 5, is that the connectivity properties of random graphs exhibit very sudden transitions—in effect, phase transitions—as the ratio of the number of edges to the number of points increases. The graphs of Figure 7.4 are *isotropic random* graphs in the sense that the probability an edge joins any pair of points is equal. A *connected component* of a graph is a connected set of points such that it is possible to walk from each point to every other point in the connected set via one or more edges. A *tree* is a branched acyclic component. *Cycles* of various lengths are obviously possible. The sequence of panels tunes our intuitions quickly. When the ratio of edges to points is low, say 0.1, most points are isolated, a few pairs are connected, and a few small trees may form; hence there are many small components disconnected from one another. As the ratio increases toward 0.5, the sizes of connected components grow larger and their numbers decrease as small components become connected into larger ones. As E/N increases past 0.5, Erdos and Renyi (1959, 1960) showed that a rapid transition occurs in which a single gigantic connected component containing most of the points emerges, leaving only a few isolated trees and points behind. Remarkably, cycles are not expected to occur until E/N passes 1.0, when cycles of all lengths have a finite and equal chance of occurring (Erdos and Renyi 1959, 1960; Cohen 1988).

Two intuitions must be borne away from this example. First as E/N increases, an isotropic random graph crosses a threshold when the system passes from nearly unconnected to nearly connected. This is called the percolation threshold and is related to theories of pore density describing how water percolates through a connected set of pores in a solid (Stauffer 1985). Second, this kind of threshold property is rampant in graph theory. Component size, the existence of cycles, and so on, all show sharp transitions as discrete values of E/N.

Erdos and Renyi initiated studies on isotropic random graphs. Their results do not directly apply to the connectivity properties of random catalyzed reaction subgraphs among peptide or RNA polymers, since those subgraphs are markedly nonisotropic, there being more reactions creating small polymers than reactions creating large polymers. Nevertheless, the fundamental results on isotropic random graphs carry an obvious implication: as E/N increases, sudden changes in connectiv-

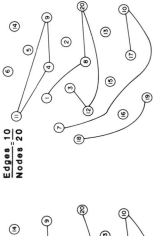

Edges = 5
Nodes 20

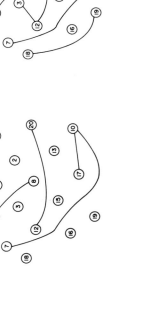

Edges = 10
Nodes 20

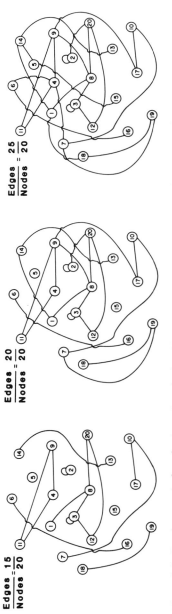

Edges = 15
Nodes 20

Edges = 20
Nodes 20

Edges = 25
Nodes 20

Figure 7.4 Random graphs for a fixed number of points N connected at random by an increasing number of edges E. For large values of N, as E/N increases past a threshold of 0.5, most points become connected in one gigantic component. As E/N passes 1.0, cycles of all lengths begin to emerge.

ity properties are to be expected. Thus, in our present case, when the fraction of reactions which are catalyzed is high enough, connected sequences of catalyzed reactions are to be expected.

The Main Idea

The pieces are now in place to state the main idea, which in fact is simple. As the maximum length of polymer M increases, the number of polymers increases exponentially but the number of reactions by which these polymers might interconvert increases yet faster, such that the ratio of reactions to polymers grows linearly, as $M - 2$. Let each polymer be a catalyst for each reaction with a fixed probability P. Then as the ratio of reactions to polymers increases, it must eventually become large enough that the number of red catalyzed reactions is larger than the number of polymers whose formation requires catalysis. Thus, eventually, almost all polymers will have at least one last step in their formation catalyzed by some polymer in the system. More formally, as the ratio of *catalyzed* reactions to polymers increases, some threshold will be reached when a connected reflexively autocatalytic set of transformations will "crystallize." In short, almost any sufficiently complex set of catalytic polymers will be expected to be collectively autocatalytic. I turn now to a slightly more formal statement of the central idea and, in the subsequent section, attempt to characterize this threshold more precisely.

The connectivity requirements allowing an autocatalytic set of polymers to exist are simply stated. Each member of the set must have its formation catalyzed by at least one member of the set. Furthermore, there must be connected catalysis pathways leading from a maintained exogenous food set to all members of the autocatalytic set.

These requirements allow us a simple sufficient condition for a set of polymers to be reflexively autocatalytic. Consider the longest polymers of interest, those of length M. Any specific polymer M* can be formed in $M - 1$ ways by condensation of smaller polymers. The chance $\bar{P}$ that none of the 2^{M+1} polymers in the set catalyzes any of these $M - 1$ reactions is just

$$\bar{P} = (1 - P)^{(M-1)(2^{M+1})} \approx e^{-P(M-1)(2^{M+1})} \tag{7.6}$$

where P is the *a priori* probably of one polymer species catalyzing any specific reaction. If we require that $\bar{P}$ be low, say 0.001, then we have stated a condition such that, with a probability of 0.999, the formation of M* *will* be catalyzed by at least one member of the set. Equivalently, 0.999 of the 2^M polymers of length M will have their formation catalyzed by some member of the set. But consider smaller polymers, those of length $M - 1$. According to Equation 7.3, these smaller polymers can each be formed in about twice as many ways as polymers of length M. Thus if 0.999 of the longest polymers will have their formation catalyzed, then 0.9995 of the 2^{M-1} polymers of length $M - 1$ will have their formation catalyzed, 0.99975 of the 2^{M-2} polymers of length $M - 2$, and so on. Therefore, the nonisotropic character of the underlying graph ensures that essentially all smaller polymers will have a last step in their formation catalyzed by at least one member of the set, thereby satisfying the requirement for connected catalyzed transformations leading outward to larger polymers from the monomers and small polymers in the food set.

The central conclusion is now straightforward. The combinatorics of polymers and their reactions, coupled with a simple model of the distribution of catalytic capacities in peptide or RNA space, leads to the expected existence of collective

TABLE 7.1   Stringent Criterion for an
Autocatalytic Set Using Only Ligation and
Cleavage Reactions

$$\bar{P} \approx e^{-P(2^{M+1})(M-1)} = \frac{1}{e^8} < 0.001$$

P	M	2^{M+1}
10^{-4}	11.85	7382
10^{-5}	14.83	58 251
10^{-6}	17.86	475 801
10^{-7}	20.94	4.02×10^6
10^{-8}	24.05	34.73×10^6
10^{-9}	27.19	306.22×10^6

Note: $\bar{P}$ is the probability that no last step in the synthesis of M*, a specific polymer of length M, is catalyzed by any other member of the set of polymers. P is the *a priori* probability of catalysis of any specific reaction by one polymer species. Noninteger values of M were used to obtain more precise estimates; M is the radius of the autocatalytic system, that is, the critical polymer length. 2^{M+1} is the number of polymer species in the set.

reflexive autocatalysis due to the percolation properties of random graphs. We have the beginning of a theory for the minimum complexity required to achieve autocatalytic closure. Below that minimum, disconnected subsystems exist. Above it, a connected whole emerges.

Equation 7.6 allows us to calculate, as a function of the probability of catalysis, the expected number of polymers and their length distribution such that the set is expected to contain an autocatalytic system. Table 7.1 gives the results for $P = 10^{-4}$ to $P = 10^{-9}$.

We reach a new and fundamental conclusion: For any fixed probability of catalysis P, *autocatalytic sets must become possible at some fixed complexity level* of numbers of kinds of polymers. The achievement of the catalytic closure required for self-reproduction is an *emergent collective property in any sufficiently complex set of catalytic polymers.*

Exchange Reactions Make Achievement of Catalytic Closure Easier

The size of polymer set required to achieve reflexive autocatalysis obviously depends on the ratio of reactions to polymers. The simplest reactions among polymers are ligation and cleavage. However, trypsin can catalyze more complex *exchange reactions,* cleaving a terminal fragment from one peptide and an internal bond on a second peptide, and then ligating the first fragment to either fragment of the second cleavage (Figure 7.5). Such a compound reaction is made up of at least three elementary reactions. The critical point is that a single enzyme can catalyze this entire transformation. If the numerator of the reactions/polymers ratio is expanded to include exchange reactions as well as ligations and cleavages, the result is $R = M^2 2^M$. This ratio grows extremely fast, very much faster than $M - 2$, which is the ratio when we have ligation and cleavage alone. In addition, the reaction graph remains non-

A ABA + BBBB $\rightleftharpoons$ A ABB + A + BBB

Figure 7.5 An exchange reaction involving cleavage of two old bonds and formation of one new one.

isotropic when exchange reactions are included, with more reactions forming smaller than larger polymers. Thus, as M increases, the ratio of uncatalyzed reactions by which those polymers can interconvert to polymers grows explosively. Maintaining the hypothesis that any polymer has a fixed probability P of catalyzing any reaction, one would expect that the size of the set needed to achieve autocatalysis is very much smaller when exchange reactions are included. This is indeed true, as Table 7.2 shows. Note that now, for a probability of catalysis of only 10^{-9}, a mere 18 000 to 19 000 polymers should achieve the critical minimum complexity necessary for collective autocatalysis!

The exchange reaction described above has three elementary steps. If we consider simpler transpeptidation exchange reactions in which only a single peptide bond is broken and a new one formed, the ratio of reactions to polymers grows as 2^{M+1}; hence these exchange reactions alone with ligation and cleavage reactions require somewhat larger sets of polymers to achieve autocatalysis.

Exactly the same considerations apply to ribozymes. Recall the ribozyme-catalyzed transesterification of two five-carbon units to yield a four-carbon unit and a six-carbon unit. This is of the same form as simple transpeptidation reactions creating no new peptide bonds. If transesterification and ligation and cleavage by RNA polymer catalysts are considered, the ratio of reactions to polymers grows almost as 2^{M+1}. Thus, using the logic of Equation 7.6, it is easy to calculate the size of the system of RNA monomers and polymers which would be expected to be collectively autocatalytic.

We are led to the following conclusion. If random peptides or RNA molecules, or their less ordered analogues, can catalyze only cleavage and ligation reactions, then Table 7.1 gives an estimate of the complexity of a polymer system needed to achieve catalytic closure. Estimates of the probability of binding or catalysis by random pep-

TABLE 7.2 Stringent Criterion for Autocatalytic Set Using Exchange Plus Cleavage and Ligation Reactions, Such that Almost All 2^{M+1} Members of the Set Have a Last Step in Their Formation Catalyzed by at Least One Other Member of the Set

$$\bar{P} \approx e^{-P(M-1)(1+2^{M}+2)(2^{M+1})} = \frac{1}{e^{8}} < 0.001$$

P	M	2^{M+1}
10^{-4}	5.55	94
10^{-5}	7.01	258
10^{-6}	8.54	944
10^{-7}	10.04	2105
10^{-8}	11.59	6165
10^{-9}	13.15	18 179

Note: $\bar{P}$, P, and M as in Table 7.1.

tides and RNA sequences, based on data discussed in Chapter 4, range from about 10^{-5} to 10^{-10}. While this range is wide, even 10^{10} sequences is a very small fraction of the possible peptide sequences of length 20 or even of RNA sequences of length 20. Further, many copies of each of 10^{10} kinds of peptide or RNA polymers can be enclosed in very small volumes and hence interact. Given a midrange estimate that the probability that a polymer catalyzes a specific reaction is 10^{-8}, Table 7.1 suggests that about 34×10^6 random polymers would achieve catalytic closure. On the more optimistic hypothesis that such polymers might equally well catalyze complex exchange reactions, the same estimated midrange chance of catalysis suggests that about 6200 polymers would achieve catalytic closure (Table 7.2). If this range of probabilities is correct, life may be far more probable than we have supposed.

GROWTH ON THE INFINITE GRAPH OF POLYMERS AND THERMODYNAMIC BEHAVIOR

Growth of the Catalyzed Reaction Graph: Supracritical Versus Subcritical Behavior and a Critical Size for the Food Set

The emergence of catalytic closure in sufficiently complex sets of catalytic polymers is a phase transition which arises in the hypergraph of polymers and the catalyzed reactions among them. Connected webs of catalyzed transformations percolate across the set of polymers and are mediated by the polymers. But the set of potential polymers is infinite both in length and in kind. The phase transition seen in the crystallization of autocatalytic sets in the hypergraph is reexpressed in this infinite set of polymers by a phase transition between the two regimes. One is a finite, hence *subcritical,* regime in which polymer catalysis of reactions that form polymers from some initial food set of monomers and small polymers can increase to produce only a *finite* number of kinds of polymers. In contrast, in the infinite, or *supracritical,* regime, polymer catalysis of the formation of polymers from an initial food set can, ignoring thermodynamic limitations, increase to produce an infinite number of kinds of polymers. We shall characterize this phase transition in terms of two major parameters of the underlying model: the probability of catalysis P and the complexity of the food set. We shall find that, in this two-dimensional parameter space, a critical line separates the plane into the supracritical and subcritical regions.

It is important to stress that, for the moment, I shall entirely ignore considerations of thermodynamic forces and chemical concentrations. The questions we are posing concern the potential extension of the catalytic subgraph of the hypergraph among the infinite set of possible polymers. This subgraph may be either finite or infinite. It is a further fundamental question, in a given physical situation with defined concentrations of the polymers, how many of the polymers which participate in the catalyzed hypergraph are formed in the system. In short, given that such a catalyzed structure exists, our current questions concern the mathematical properties of the catalyzed subgraph and not the physical behavior of the polymers. Of course, I must and shall return to these thermodynamic considerations below and show that these potentially collectively autocatalytic systems are in fact *expected* to be autocatalytic.

I turn now to a more detailed investigation of the catalyzed reaction graph and the minimal conditions such that a phase transition occurs and catalytic closure is attained. Let us recast the problem slightly. Suppose that we can maintain, at a constant high concentration, a food set comprising all monomer amino acid species as well as all small polymers out to some length L_c. This set of molecular species can simultaneously serve as substrates for cleavage or ligation reactions and as potential catalysts for those reactions. Let us remove the restriction that the longest polymers under consideration are bound to be length L_c or less, thus letting ligation reactions produce polymers of lengths ranging up to $2L_c$. If some of the possible reactions are catalyzed, then new polymers, larger than those in the food set and ranging in length from $L_c + 1$ to $2L_c$, may be formed.

If ligation reactions among the polymers in the initial food set can create new polymers longer than those in the food set, the new polymers become available as new substrates for new cleavage and ligation reactions. These new reactions may be catalyzed by the members of the initial food set. In addition, the new polymers are available as potential new catalysts for both the original reactions among polymers of length L_c or less and all the new reactions involving the new polymers either alone or together with the original polymers and monomers of the food set. Thus, consider what can occur on a second round. Both the old and the new first-round polymers may catalyze yet more reactions, leading to the formation of molecules ranging up to $4L_c$ in length. These new second-round molecules may become substrates in new reactions which may be catalyzed by all existing polymers from the food set and the first round, and the new polymers from the second round may catalyze all previous and new reactions. Thus, over a succession of rounds, the catalyzed reaction graph may grow larger in two senses: The number of catalyzed reactions increases, and the number of kinds of polymers larger than those in the maintained food set up to length L_c may increase.

I note again that we are ignoring the polymer concentrations, which may in fact be infinitesimal. Therefore, we need to define the concept of a *virtual* catalyzed reaction graph, which shows all reactions catalyzed by monomers or polymers in the total system. Since the actual concentrations of such polymers are a function of the flux of material through the system, the boundary conditions, and thermodynamic issues, and since those concentrations may be infinitesimal, the virtual catalyzed reaction graph asserts that, were the substrates and catalysts present in sufficient concentration, the reactions would proceed. In short, the virtual graph gives the catalytic structure of the polymer system. As stressed above, the dynamical behavior of flux and concentrations are a further consideration. For the remainder of this section, I shall use the term "reaction graph" to refer to the virtual catalyzed reaction graph.

The issue we are now concerned with is the *growth* of the number of polymers in the reaction graph over successive rounds. That growth is perfectly well specified algorithmically; it can exhibit only one of two alternative ultimate behaviors. Either over successive rounds, the number of newly added polymer species at each round dwindles to zero so that the graph stops growing and the number of different polymer species in the system remains finite, or at each round, a finite positive number of new polymer species are added to the system, in which case the reaction graph grows without bound and the number of kinds of polymers in the system is infinite. The former case is *subcritical,* the latter is *supracritical.*

I emphasize the obvious: The existence of an infinite number of polymer species in the supracritical reaction graph applies to the *virtual* reaction graph as defined above. No real system of interacting polymers can be infinite. However, the behavior of a real polymer system may depend fundamentally on whether the underlying virtual graph is supracritical or subcritical. I return to this shortly.

Because the growth properties of reaction graphs are either subcritical or supracritical, it is to be expected that there is some relationship among the parameters defining the way in which the graphs grow; this relationship constitutes a critical threshold condition. On one side of the threshold, graph growth will die out; on the other, it will blow up. The parameters which enter into the problem are the number of kinds of amino acids B, the maximum length of polymers in the maintained food set L_c, and the probability that any polymer catalyzes each reaction P. From these, it is possible to derive critical sufficient conditions such that graph growth is supracritical (Farmer, Kauffman, and Packard 1986; Kauffman 1986b). The result for $B = 2$ monomers is

$$2^{L_c^*+1} = \sqrt{\frac{1}{2P}} \tag{7.7}$$

where L_c^* is the critical value of L_c dividing subcritical from supracritical behavior. Since the number of molecular species in the maintained food set is then 2^{L_c+1}, Equation 7.7 asserts that the number of molecules in the food set must be on the order of $\sqrt{1/2P}$. If the probability of catalysis is 10^{-6}, then about 1000 species must be maintained in the food set if supracritical growth of the reaction graph is to occur. As B increases, the requisite number of molecules in the food set simplifies to

$$B^{L_c^*} = \sqrt{\frac{1}{2P}} \tag{7.8a}$$

$$P \approx B^{-2L_c^*} \tag{7.8b}$$

The phase transition from supracritical to subcritical behavior is given quite accurately by Equations 7.7 and 7.8. Figure 7.6 shows numerical data testing Equation 7.7 and its generalization as B increases. A computer program implementing reaction graph growth was written by my colleagues D. Farmer and N. Packard. The critical threshold values of parameters separating subcritical and supracritical behavior were ascertained by fixing a value of L_c and B, and then tuning the probability of catalysis P. If a specific value of P gave supracritical growth and a second value gave subcritical growth, then a new trial value which was the mean of those two values of P was used. Successive iterations converge on the critical value of P for the specified values of L_c and B. As can be seen, the observed and predicted curves are close and parallel. The offset of the theoretical curve reflects the fact that Equation 7.7 is a slightly-more-than-sufficient condition for supracritical behavior. The fact that the slopes are parallel reflects the correct analytic result that the size of the mantained food set necessary for supracritical graph growth scales with the square root of $1/P$. Figure 7.6b shows the correspondence between food set size and supracritical graph growth for $B > 2$, according to Equation 7.8b.

Figure 7.7 shows the number of new polymer species added to the reaction graph at each round of growth. For subcritical values of the parameters, this number typically dwindles to zero. For slightly supracritical values of the parameters, this number may behave erratically. For clearly supracritical values, the graph grows faster than exponentially.

One intriguing feature of this new body of theory is that it predicts a relationship between polymer size and number of polymer kinds. The size distribution of polymers in the system is *unimodal,* with most polymer kinds having an intermediate length and fewer kinds being shorter or longer than this intermediate length. The underlying noncatalyzed reaction graph for polymers is inherently nonisotropic,

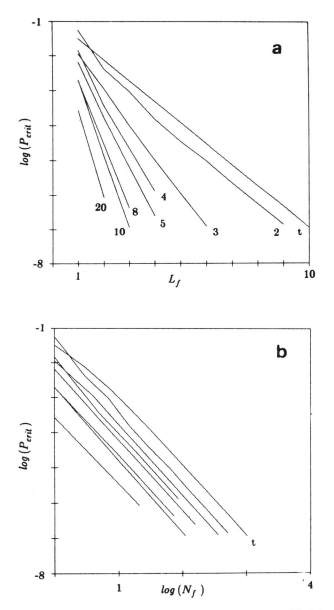

Figure 7.6 (*a*) Dependency of P_{crit}, the probability of catalysis at the supracritical threshold, on L_c, the maximum length of the molecular species in the initial food set. The food sets contain different numbers B of monomer species, as indicated; the line labeled T is from theoretical results for $B = 2$ amino acids. (*b*) As in (*a*), except that L_c is replaced by log N, where N is the total number of species in the food set. (From Farmer, Kauffman, and Packard 1986)

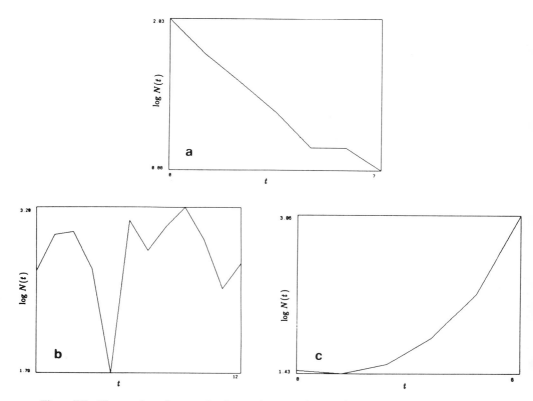

Figure 7.7 The number of new molecular species created at each iteration during the generation of a random catalyzed reaction graph. In all cases, $B = 2$ amino acids; $L_c = 6$ is the maximum length of polymers maintained in the food set, and t is the time, or iterations of graph growth. (*a*) Below the supracritical point, graph growth decays and stops. (*b*) Near the supracritical point graph growth behaves erratically. (*c*) Above the supracritical point, the graph increases without bound. Note the semilog scale. (From Farmer, Kauffman, and Packard 1986)

always having many more reactions forming smaller than larger polymers. Indeed, for cleavage and ligation reactions, the number of polymer species formed falls by about ½ as length increases by one amino acid. Thus, as the graph grows, a larger *fraction* of the small polymers will have their formation catalyzed. At the same time, the number of possible kinds of polymers of length L is B^L and hence increases with L. When these two conflicting effects are considered together, the number of distinct polymers of length L whose formation is catalyzed will be unimodal, starting low for small L, increasing to a maximum for intermediate values of L, and then dwindling as L continues to increase.

Figures 7.8*a* and 7.8*b* show such curves of the distribution of numbers of polymer kinds as a function of polymer length. The distributions are indeed unimodal, increasing to a peak and then declining approximately exponentially as polymer length increases. Figures 7.8*c* and 7.8*d* plot the logarithm of the number of polymer species of each length as a function of length. The declining tail is nearly linear, indicating a nearly exponential tail in the linear plots of Figures 7.8*a* and 7.8*b*.

Unimodal distributions of organic molecules may be very general in biology. The fundamental point is that the centripedal, nonisotropic character of the underlying

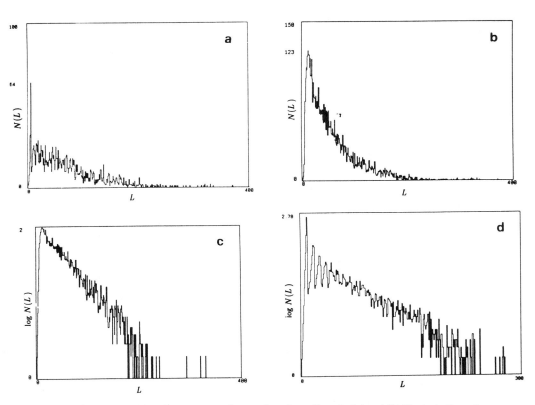

Figure 7.8 Distribution of polymer species as a function of length. (*a*) and (*b*) illustrate the action of one iteration of the graph update procedure. (*c*) Same as (*b*), but with a semilog plot indicating the exponential tail. (*d*) Length distribution for a system with only two amino acids used to form the monomers and dimers of the foot set, which causes a cycling, or increased abundance, at small multiples of the maximum polymer length maintained in the food set. (From Farmer, Kauffman, and Packard 1986)

uncatalyzed reaction graph leads to a quite characteristic unimodal distribution. This distribution is almost certainly insensitive to the detailed assumptions about the distribution of catalytic capacities in sequence space. In the next chapter, I shall suggest that life originated with a connected metabolism which crystallized as a phase transition occurred in the space of organic molecules. Thus very similar considerations apply to the *distribution of numbers of kinds of organic molecules in a metabolism as a function of atoms per molecule.* Therefore, it is interesting that curves remarkably like those of Figure 7.8 are known to relate the number of kinds of organic molecules to the number of carbon atoms per molecule (Morowitz 1968). Indeed, I shall suggest that a distribution like that in Figure 7.8 may prove to be an ahistorical universal characterizing any connected metabolism which might evolve.

The parameter values which correspond to a phase transition from subcritical to supracritical behavior depend on the ratio of number of reactions to number of polymers present. Thus the phase transition will shift position if different classes of reactions are considered. Equation 7.7 reflects the constraint to consider only cleavage and ligation reactions. It yields the result that the critical number of maintained food species scales inversely as the square root of the probability of catalysis. If exchange reactions are also used, the ratio of reactions to polymers grows much faster, and consequently a still *smaller* food set would be supracritical. No analytic results are yet available. Obviously, the critical values expressed in Equation 7.7 reflect the assumption that the probability of catalysis is independent of polymer length. However, the general results are very robust to this assumption, as I shall comment on below.

Thermodynamic Considerations

We have now seen that a phase transition occurs from subcritical to supracritical behavior of graph growth and that this transition is correlated with the emergence of catalytic closure in atuocatalytic sets of polymers. It is now time to consider whether physically realizable autocatalytic sets will form when account is taken of the thermodynamic and kinetic properties of real polymer systems. The fundamental point is that, under appropriate—and apparently realistic—conditions, the answer is "yes." I focus first on consideration of peptide systems; parallel considerations apply to RNA systems. Indeed, I suspect that the latter are more likely candidates to achieve collective reproduction in experiments carried out in the near future. Thus it should soon be possible to construct real collectively self-reproducing sets of polymers.

The forward and reverse reactions between two monomers A and B ligating to form the dimer C plus water can be represented as

$$A + B \rightarrow C + H_2O; \quad C + H_2O \rightarrow A + B$$

In an aqueous environment, where the water concentration is about 55 M, the energy of a peptide bond is 2 to 3 kcal/mol. Therefore, hydrolysis of the peptide bond is spontaneous, while polymerization is thermodynamically unfavored. These conflicting tendencies occur in a ratio determined by Boltzmann's formula $e^{-\Delta G/RT}$, where ΔG measures the free energy change from reactions to products. For peptide bonds with $\Delta G = 3$ kcal/mol, the formula yields $e^{-5} = 0.007$ as the ratio of dimer to monomer. For peptide bonds with $\Delta G = 2$ kcal/mol, the ratio is $e^{-3.33} = 0.036$ (Fox 1988).

The energy of a peptide bond depends on which amino acids are joined. Dixon and Webb (1960) suggest that an overall useful approximate expression for the "typical" peptide bond yields a dissociation constant of about 10 favoring smaller peptides produced by cleavage; hence in a closed thermodynamic system at equilibrium, the ratio of dimer concentration to each monomer concentration is about 10^{-1} to 1. Similarly, the ratio of a trimer to a monomer is 10^{-2} to 1. Similar considerations apply to ligation and cleavage of single-stranded RNA molecules. The energy of a phosphodiester bond is about 5 kcal/mol (Fox 1988). The corresponding ratio of dimers to monomers is $e^{-8.33} = 0.0002$. Thus, as noted above, in an aqueous environment, formation of peptide bonds is substantially easier than formation of phosphodiester bonds.

The exponential character of the equilibrium distribution implies, as noted above, that the concentration of large polymers at equilibrium decreases exponentially with length. For N linkages at 3 kcal/mol, the ratio is e^{-5N}. Thus, assuming monomers in 1 M concentration, for $N = 20$, the expected concentration of a specific polymer of length N is on the order of $e^{-100} = 3.7 \times 10^{-44}$ M. As Fox (1988) notes, organisms must use complex strategies to circumvent this fundamental difficulty.

Autocatalytic behavior requires that the system be thermodynamically open to the flux of monomers and small polymers. In order for autocatalytic behavior to be physically realizable, a collection of polymers at some finite concentration in a rea-sonably small volume must be able to be formed and sustained by addition of matter and energy from the outside. The concentrations of those formed polymers must be high enough to allow a sustained metabolic flux from the exogenous food set to all the members of the autocatalytic set, and the concentrations of the members of the latter must be persistently sustained at finite concentrations. From the dynamic sys-tems perspective, a plausible form for such an autocatalytic set to take is some kind of dynamical attractor, ranging from steady state or periodic orbit to chaotic attrac-tor.

We consider next the simplest open system, driven only by influx of monomers and small polymers. In order that large polymers be formed, thermodynamic equi-librium must favor their synthesis. In contemporary organisms, this is achieved by the coupling of exergonic and endergonic reactions. In a first, simplest system, the only mechanism we shall consider for shifting equilibrium in favor of peptide syn-thesis is dehydration. In the simulations I report, we "tuned" the dehydration of the system to favor the synthesis of peptide bonds, drove the system at a fixed rate by adding monomers and small polymers, and assumed that only catalyzed reactions occur. The last assumption, consistent with those utilized in stating the model, cor-responds to assuming that the rate of catalyzed reactions is much faster than the rate of spontaneous reactions. Since the rate of spontaneous hydrolysis of a peptide bond in water is on the order of one per month (Fox 1988), a 1000-fold increase in reaction rate due to catalysis amounts to about one reaction turnover event per hour.

We may use a chemostat as a concrete model of an open thermodynamic system. Monomers and small polymers are exogenously added to the chemostat at a fixed rate. The volume of the chemostat is kept constant by allowing overflow of excess material; such overflow removes monomers and all polymers at a rate proportional to their concentrations. Implicitly, such overflow leads to competition since poly-mers formed more slowly than their mean residence time in the system are diluted out. It is a simple calculation to show that, since the rate of food addition is constant but loss of food molecules and of more complex polymers is proportional to concen-tration, the chemostat will fall to a steady state in which the total number of mono-mers, either free or bound into polymers, is constant in time. Because the system is open to influx of food, however, the distribution of monomers into diverse polymers is not bound by the equilibrium distribution of a closed thermodynamic system and will reflect the kinds of catalytic activities present among the peptides in the che-mostat. These catalytic activities will channel the flux of material preferentially down particular pathways. Note again that—no matter whether the underlying virtual cat-alyzed reaction graph is subcritical/finite or supracritical/infinite—the mass of monomers plus polymers is finite and thus the number of kinds of polymers present in the system is finite. Restated, if an autocatalytic set is formed, it remains a *finite* entity which may nevertheless "live" on an *infinite* virtual catalyzed reaction graph.

Dynamics of Autocatalytic Sets in an Expandable State Space

In order to show that autocatalytic polymer systems are expected to be physically realizable after taking account of thermodynamics, D. Farmer, N. Packard, and I implemented a computer program (Farmer, Kauffman, and Packard 1986). We defined a maintained food set, a chemostat, an input rate for food, an overflow rate, a probability of catalysis, and the number of monomer species. The following addi-tional features were required.

1. We had to specify a minimal concentration needed in order for a species to be considered present in the chemostat. For example, a 10^{-9} M solution corresponds to one molecule in the volume of a bacterium, while 10^{-23} M is about one molecule per liter. Therefore, choosing a minimal concentration amounts to scaling the volume of the chemostat. In the limit where this minimal concentration goes to zero, the system approaches the virtual graph.

2. It was necessary to monitor the concentration of each reactant and product species and of the various bound forms of intermediates. We utilized a technical means to circumvent having to keep track of all the diverse intermediates without loss of rigor.

3. We were forced to confront a novel feature of this class of models: The set of variables in the system, and the dynamical equations coupling them, are open and evolving. This is new. Conventionally, the set of variables in a dynamical system, and the dynamical laws coupling the change of each variable as a function of the variables, are fixed at the outset, thereby specifying a fixed *state space* of the system. Recall from Chapter 5 that a state space has coordinates that are the different variables of the system. The current state of the set of variables is represented by a point in this space. The system's dynamical trajectory over time is shown as a trajectory through the state space. As described in Chapter 5, typically, such a system either settles down to an attractor such as a steady state (in which case the representative point does not move) or enters a sustained oscillation in which the representative point travels either around a closed orbit in the state space or around a more complex attractor in which the point falls onto the surface of a torus or of a more complexly folded object called a strange attractor.

In the present case the set of polymers which will constitute the variables of the system is *not* specified ahead of time. Rather, we specify an algorithm which decides, at each moment in the temporal evolution of the system, whether the reactions catalyzed among the existing monomers and polymers are forming new species of polymers and whether such new polymers have increased in concentration above our defined threshold and so count as present. Any new polymers that do count as present are added to the system. These new polymers may participate in new reactions either as catalysts or as substrates, while all monomers and old polymers may participate in the new reactions either as catalysts or as substrates. Therefore, the algorithm decides by chance, according to a fixed probability of catalysis P for each polymer with respect to each possible reaction, which new reactions and old reactions are catalyzed by the new and old species present at above-threshold concentrations. These newly catalyzed reactions are installed in the catalyzed reaction graph of the system. Subsequently, the reaction dynamics of this new, altered system is followed over an interval of time. Still more new polymers may increase above the concentration threshold and so count as present. If so, they are added to the system and any newly catalyzed reactions are installed in the expanding graph. This process continues until the system settles into some attractor in which no new polymer species are being added.

In short, the set of variables is itself open and variable in the potentially infinite space of possible polymers.

Using this system, we confirmed that autocatalytic polymer systems are perfectly plausible after taking account of thermodynamic and kinetic questions. In particular, we showed that the behavior of the system changes when the underlying virtual reaction graph passes from subcritical to supracritical, and, most important, that stable autocatalytic sets with sustained concentrations of small and large polymers do form.

Figure 7.9 plots total number of polymer species formed in the chemostat at steady state, as a function of probability of catalysis and of concentration threshold (Farmer, Kauffman, and Packard 1986). Given the values of the parameters—two amino acid species to form a food set of monomers and small polymers—the expected transition from finite to infinite virtual graph occurs when P is about 0.015 and the threshold concentration is 0. As expected, for low threshold values, as P crosses this critical value and the graph becomes infinite, the actual number of species of polymers explodes rapidly. Thus when the phase transition occurs and the virtual graph has become infinite, the diversity of the set of polymers present above the threshold concentration increases because the set is able to expand on the infinite graph. Restated, the fixed total mass of monomers and polymers spreads out on the infinite supracritical virtual catalyzed reaction graph to populate a larger variety of polymer species.

Figure 7.9 also shows that, as the threshold concentration increases, the number of polymers in the chemostat at the terminal steady state declines. This is as expected. If the threshold were chosen to be larger than the maintained concentrations of the food set, then obviously no polymers would be allowed to catalyze any reactions at all, and the system would remain the food set alone.

Both dehydration and an increase in the rate of food addition can drive the physically realized system from nonautocatalytic to autocatalytic. By construction, this polymer system is innocent of any metabolism coupling breakdown of high-energy

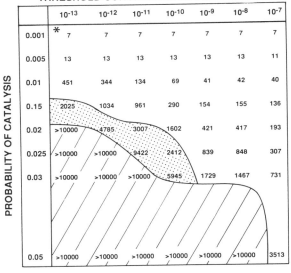

THRESHOLD CONCENTRATION FOR CATALYSIS

PROBABILITY OF CATALYSIS	10^{-13}	10^{-12}	10^{-11}	10^{-10}	10^{-9}	10^{-8}	10^{-7}
0.001	* 7	7	7	7	7	7	7
0.005	13	13	13	13	13	13	11
0.01	451	344	134	69	41	42	40
0.15	2025	1034	961	290	154	155	136
0.02	>10000	4785	3007	1602	421	417	193
0.025	>10000	>10000	9422	2412	839	848	307
0.03	>10000	>10000	>10000	5945	1729	1467	731
0.05	>10000	>10000	>10000	>10000	>10000	>10000	3513

✱ Total number of polypeptides in autocatalytic set

Figure 7.9   Plot of the asymptotic number of polymers formed in an autocatalytic set as a function of P, the probability of catalysis, and of $\log T$, the logarithm of the concentration threshold. All simulations used two amino acids combined in various ways to form a food set consisting of monomers and dimers.

metabolites to the synthesis of peptide bonds. Consequently, the only way to shift the equilibrium to favor synthesis of large polymers is to dehydrate the system and increase the concentrations of all species. Then, like the plastein reaction and Dixon and Webb's argument, the bulk of the mass shifts from being in monomers and small polymers to being bound in high-molecular-weight species. Since a polymer must be present above a minimal concentration in order to count as present and have a chance to be a new substrate or catalyst, it is to be expected that shifting the hydration state of the chemostat will shift the number of kinds and sizes of polymers formed. Therefore, the qualitative effect of tuning the dehydration parameter is clear. When water is present in high concentration, only small polymers will be found in appreciable concentrations; larger polymers, if formed, will be in too low a concentration to count as present in a defined volume. Therefore, even if the virtual catalyzed reaction graph is supracritical, the thermodynamically attainable polymer system may have too few components to achieve catalytic closure; no autocatalytic set will exist. As dehydration increases, more and more of the polymers longer than those in the food set will increase in concentration over the threshold. Eventually a sufficient number will be present to achieve catalytic closure, and an autocatalytic set will be present.

For a similar reason, as the rate of food addition increases from a low level relative to a *constant* dilutional overflow rate from the chemostat, the total number of monomers and small polymers at steady state increases. Therefore, the attainable diversity of the set of longer polymers present above threshold increases. As that diversity increases and spreads out over the supracritical graph, eventually catalytic closure will be attained and an autocatalytic polymer system will be present.

The most important conclusion derived from these studies is this: The model engenders sustained autocatalytic polymer systems. Figure 7.10 shows one example. This is a small system, allowing ready graphical presentation. Variation in thermodynamic terms, food set size, and probability of catalysis yields much larger systems, with hundreds or thousands of polymers present.

More recent work by my colleagues Bagley, Farmer, and Fontana (1992) and Bagley (1991) shows that, by tuning the catalytic efficiency of polymers acting at different points in the autocatalytic system, it is possible to achieve a nearly uniform distribution of polymer concentrations as sizes range from monomers and dimers up to sequences with 15 to 30 residues. Thus even restricting ourselves to autocatalytic systems lacking a coupling of exergonic and endergonic reactions, it is still possible, merely by dehydration, to achieve dynamically stable autocatalytic systems with high concentrations of some large polymers. Another simple potential mechanism for driving synthesis reactions beyond dehydration is coupling of high-energy pyrophosphate to nonspecific phosphorylation and dephosphorylation cycles (Fox 1988).

The Emergence of Autocatalytic Sets
of Sequence-Specific RNA Ligases

I now extend our results to show that sufficiently complex mixtures of single-stranded RNA, acting on one another by mediating template-specific ligation and cleavage reactions, are very likely to be collectively self-reproducing. If this supposition is correct, direct demonstration of collective autocatalysis by catalytic polymers may soon be possible.

In the first section of this chapter, I discussed the evidence showing that replication of arbitrary RNA molecules copying an arbitrary template without a polymerase

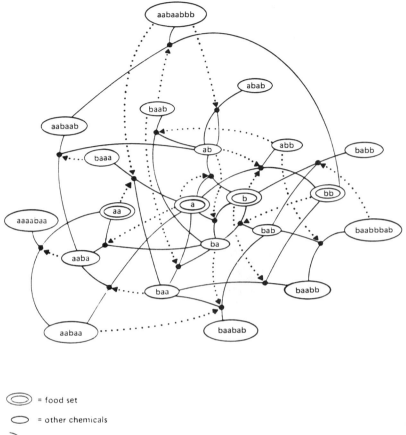

Figure 7.10 A typical example of a small autocatalytic set. The reactions are represented by points connecting cleavage products with the corresponding larger ligated polymer. Dotted lines indicate catalysis and point from the catalyst to the reaction being catalyzed. Monomers and dimers of A and B constitute the maintained food set (double ellipses).

appears to be very hard to achieve. In contrast, I noted evidence that some RNA sequences could act as ribozymes which cleaved or ligated other single-stranded RNA sequences. Such catalytic action is sequence specific and depends on the particular base pairs at the active site in the ribozyme and upon template recognition for binding of the RNA substrates. In addition, we saw examples of autocatalytic replication of a specific tetramer from two dinucleotides and a specific hexamer from two trinucleotides. In each case the larger molecule acts via base pairing as a specific ligase to orient and ligate the smaller molecules into a replica of itself. These results already demonstrate that autocatalytic sets, via a specific ligase, can occur.

In addition, three exciting lines of evidence were noted in Chapter 4, all indicating that novel single-stranded RNA sequences having new catalytic properties might be found. Recall that Joyce (1989) was able to evolve a ribozyme such that it came to catalyze a reaction involving a DNA rather than an RNA substrate. Tuerk and Gold (1990) screened about 65 000 stochastic RNA sequences for those able to bind a polymerase and found the wild type and a novel sequence. But most important, Ellington and Szostak (1990) passed a library of about 10^{13} stochastic 100-base-pair single-stranded RNA sequences over affinity columns and found about 100 to 1000 able to bind specific organic dyes with fairly high affinity. Since the capacity to bind a dye is closely related to the capacity to bind to the transition state of a reaction and catalyze the reaction, these results suggest that the probability of catalyzing an arbitrary reaction is about 10^{-10}. Since it is very much easier for an RNA molecule to bind its template complement than to bind an arbitrary organic molecule, the probability that an RNA sequence can catalyze the ligation and cleavage of two other RNA strands ending in appropriate complementary base pairs may be much higher than 10^{-10}.

We now ask whether the concepts underlying the emergence of autocatalytic sets in which each polymer catalyzes *any* possible reaction with a fixed probability P generalizes to the case in which polymers must be the *template shape complement* of their substrates in order to have a chance of catalyzing a reaction between those substrates. The clear answer is "yes." I shall base my discussion on the concept that shape complementarity is afforded by template complementarity, as is familiar in RNA and DNA molecules. I confine discussion to model RNA sequences comprising two bases: "one" and "zero." Template complementarity then requires only one and zero bases to bond. The idea immediately generalizes to four bases. In fact, the idea generalizes to peptides whose generalized shapes must be complements in order for one peptide to have a chance of catalyzing a reaction involving the other. For the moment, the string models of polymers we consider stand for RNA sequences.

The specificity of observed cases of hexamer- and ribozyme-catalyzed ligation rests on normal Watson–Crick base pairing. Sequence-specific recognition and binding among RNA or DNA molecules therefore can require as few as four or, more plausibly, six base matches. Hybridization experiments with DNA and RNA have long demonstrated that over lengths of six to ten base pairs, binding may occur in the presence of occasional mismatches. Such mismatches lower the affinity of the binding.

My colleague R. Bagley and I have utilized these facts to implement a model to study the emergence of autocatalytic sets of specific "ribozyme" ligases. In reality, single-stranded RNA molecules fold to form secondary structures with stems, hairpin loops, and so forth. Presumably the probability that any short region on such an RNA molecule is physically able to act as a prospective ribozyme catalytic site is correlated with the length of the molecule. Larger molecules may form better, and more, potential catalytic sites. For example, such sites might be loops with free nucleotides or with unpaired tail regions. As RNA sequences become longer, and depending on the particular ratios of bases, the expected distribution of numbers of loops, sizes of loops, and numbers of unpaired tails within each folded RNA sequence shifts. In our simplified model, we assumed that the number of potential catalytic sites in a model RNA molecule was proportional to its length and that RNA molecules of length six or more had at least one such prospective catalytic site.

To model template complementarity, Bagley and I implemented a template-match rule. We assumed that any model ribozyme site had a defined length in nucleotides drawn from some distribution. In the results I report, we assumed that each site was of length four or six. The match strength of a prospective catalytic site was given by the number of nucleotides among the four or six which were the match complement to the substrates in question. Thus match strength might vary from 0 to 6. Since a ribozyme or another enzyme must recognize sites on both substrates to be ligated, we required that our model ribozymatic sites be the template complements of the left and right terminal two or three nucleotides of the two substrates to be joined. This template-match rule gives a strength, or affinity, to each possible match between each potential ribozyme site and any pair of substrates.

The remaining fundamental parameter we utilized was a probability of catalysis P, considered as a function of match strength. Concretely, we assumed that a perfect match corresponded to some maximum probability that the site was able to act as a catalyst of the reaction. Further, we assumed that, as match strength declined, the probability that the site catalyzed the reaction declined to zero. The assumption that even perfect matching corresponds only to a *maximum probability* that the site catalyzes the reaction models the fact that many features of prospective ribozymes and their potential substrates beyond template complementarity must be present before catalysis can occur. The assumption that the probability of catalysis drops to zero as or before match strength declines to zero models the idea that a ribozyme site must be a shape complement as a precondition to act as a catalyst. The existence of a finite range of match strengths with a positive probability of catalysis models the possibility that sites might catalyze reactions whose substrates have occasional mismatches to the catalytic site.

Two features of this assignment of catalysts to reactions generate clouds in reaction space. First, consider a model ribozyme site of length 4, required to match perfectly two nucleotides at the ends of each of its two substrates. Consider a set of potential substrates each having ten nucleotides. To match a specific potential ribozyme, two terminal nucleotides must be specified. But the remaining eight are free. Thus there are $2^8 = 256$ "left" substrates and an equal number of "right" substrates which are the shape complement of the potential ribozyme site. Each possible pair of these left and right substrates might be joined in a reaction; thus there are $2^8 \times 2^8 = 2^{16}$, or about 64 000, potential reactions which this potential ribozyme site might catalyze. The 64 000 reactions constitute a cloud in reaction space. The reactions are highly similar, of course, in that all rely on the same pair of terminal dinucleotides. The situation is identical to restriction endonucleases which recognize specific four-base sequences.

The second feature which plays on the cloud in reaction space is the allowance of mismatches. Such mismatches diffuse the cloud further.

It is trivially obvious that this system can form autocatalytic sets under trivial assumptions. Thus if the probability of catalysis granted a perfect match is set to 1.0, then many model RNA sequences which are entirely a catalytic site (such as those shown in Figures 7.1 and 7.2) will ligate two substrates corresponding to their left and right parts into a replica. Such ligation corresponds to the observed autocatalysis of the tetrameric and hexameric sequence noted above. Further, it is obvious that, if the probability of catalysis is independent of match strength, this model reverts to the basic model already studied, in which each polymer has a fixed probability of catalyzing any reaction.

The important fact is that the model yields autocatalytic sets of model ribozyme ligases under the plausible assumption that the probability of catalysis as a function of match strength is much less than 1.0. Specifically, Bagley and I have modeled four- and six-nucleotide ribozyme sites and required either perfect matching or only single mismatches between the ribozyme site and the terminal nucleotides on the two substrates it ligates. We considered a maintained food set of all possible polymers of length 6, 8, or 10. For each we found an approximate minimum probability of catalysis such that the resulting virtual catalyzed reaction graph grew supracritically. The fundamental result, therefore, is that this model definitely supports the growth of an infinite reaction graph. Despite the use of template-matching rules rather than a simple probability of catalysis independent of shape matching, the two versions of the model belong to the same universality class, a term physicists use to describe a family of models all of which share some fundamental assumptions leading them to share a set of consequences. That the two versions are so similar implies that autocatalytic sets of specific ribozyme ligases are, in principle, possible.

The minimum probability of catalysis for supracritical graph growth, given that a prospective catalytic site first matches its substrates, can be very low. A crude overestimate of that probability can be seen from our consideration of the cloud of reactions which might be catalyzed by one potential tetrameric site acting on substrates of length 10. The number of potential reactions, about 64 000, grows as the square of the number of substrates which match one end of the site. Thus substrates of length 10 offer $2^8 = 256$ left or right ends, and the product of these, $2^{16} = 64\,000$, is the number of reactions the site might catalyze. Suppose the probability of catalysis were $1/64\,000$. Then each potential site would catalyze, on average, at least one reaction among the set of all possible substrates of length 10. In this case, a large number of novel polymers would be formed and might act as new substrates and potential catalysts for yet more reactions. Over rounds of graph growth, the catalyzed reaction graph will grow supracritically.

The square of the number of substrates matching a site leads to an overestimate of the minimum probability of catalysis which will support supracritical graph growth, since the condition that all potential sites catalyze at least one reaction is overstrong. All that is required is that some new reactions proceed and some new polymers be formed at each round. In fact, numerical simulations for tetrameric sites which allow a single mismatch, operating on a food set of all possible polymers of length 10, show that a probability of catalysis of only 5×10^{-6} suffices for supracritical graph growth.

Pause to focus on this result. It says that, if a collection of all 1024 possible model RNA molecules of length 10 was assembled in a vat, if potential ribozymatic sites were four units long, if any ten-unit polymer had at least one such potential site, if terminal dinucleotides in pairs of substrates must each match the ribozymatic site at least at one nucleotide, then if the probability that the site catalyzes the reaction is only 2 in a million, the system would generate supracritical graph growth. If thermodynamic conditions were also satisfied, the system would become genuinely autocatalytic. The vat in the computer contains a self-organized, collectively self-reproducing system of specific ribozyme ligases.

The results generalize immediately to four nucleotides, since a pair of adjacent model nucleotides, 1 and 0, can be thought of as one of four possible true nucleotides. These simulations slightly strain the Cray system at Los Alamos; thus tracking the minimum probability of catalysis for supracritical graph growth is difficult. For the four-nucleotide case and four-base ribozyme sites, our overestimate utilizing the square of the number of substrates matching one-half of the ribozyme sites implies that a set of all $4^7 \approx 16\,000$ possible heptamers will become autocatalytic when the probability of catalysis granted a match is about 1 in a million. Clearly, if the probability of catalysis is lower, then a correspondingly larger set of small RNA oligomers would be required to achieve supracritical graph growth.

We noted above the data from Ellington and Szostak (1990) yielding, even without the advantages of template matching, a plausible estimate of 10^{-10} for the probability of catalysis. Thus an RNA polymer system having a sequence complexity of perhaps 10^5 to 10^6 and lengths of perhaps 20 to 40 nucleotides to allow folding and formation of stable active sites might well be collectively self-reproducing. I suggest below that such experiments are now fully feasible. Indeed, a sequence diversity of 10^{12} or more can be utilized.

I should remark that we have here considered only supracritical graph growth in the case of ligation and cleavage reactions, without yet considering the possibility that model ribozymes might also catalyze more complex exchange reactions. Precisely those more complex transesterification reactions are catalyzed by known ribozymes. As in the basic model discussed above, inclusion of exchange reactions will rapidly increase the ratio of reactions to polymers, hence markedly increase the ease of achieving supracritical graph growth.

In summary, extension of the basic model of autocatalytic self-organization to the template-matching case strongly suggests that self-replication by sets of specific ribozyme ligases or by sets of specific peptide ligases is possible.

Relaxing the Idealizations of the Model

For ease of analysis, a number of idealizations were introduced in the formulation of these models for the self-organization of reflexive autocatalysis. In addition, some of the idealizations reflect our ignorance. It therefore becomes essential to investigate whether and where the main conclusions hold when the idealizations are relaxed. Is the model robust? We now see that it is.

The basic model, with a fixed probability of catalysis P, assumes perfect recognition. Does autocatalysis still arise if each polymer might catalyze a cloud of similar reactions? A fixed probability properly captures the possibility that one polymer might catalyze more than one quite distinct reaction because both reactions represent the same catalytic task. But a fixed P erroneously requires infinite discriminability such that the polymer may catalyze a given reaction but not catalyze any of the cloud of similar reactions whose substrates are almost identical to the substrate of the catalyzed reaction. More realistically, suppose any polymer catalyzes such a cloud of similar reactions. Then more of the reactions in the underlying reaction graph are catalyzed and hence are part of the catalyzed reaction subgraph. From the point of view of attaining a critical catalyzed connectivity, such additional catalyzed reactions only help. Thus inclusion of clouds abets the formation of an autocatalytic set. However, it is equally obvious that such extraneous reactions may often induce flow down pathways to useless polymer products not needed in the autocatalytic set. But this recognition of nonproductive pathways is simply another way of seeing that autocatalytic sets may live on supracritical virtual catalyzed reaction graphs. In any such system, as the mass of the open thermodynamic system increases, the system will spread out on the underlying connected virtual catalyzed reaction graph. Extension of the basic model to template matching as a requisite for potential enzymatic sites, coupled with allowed mismatches, captures one model of such clouds in a defined shape space.

In the basic version of the model but not in the RNA version with minimum-length template match sites, I have assumed that even monomers might catalyze reactions. In fact, even the bifunctional histidine can play catalytic roles. However, it is plausible that, with increasing polymer length, it becomes easier to build a reasonable catalytic site. An extreme version of this hypothesis might suppose that all polymers less than some length have no catalytic activity, while all those above this threshold length have a constant probability of catalyzing any reaction. It is easy to show that this supposition makes little difference to the fundamental conclusion (Kauffman 1986b). The ratio of number of reactions to number of polymers present increases so rapidly that reasonable-size sets of polymers which are potential catalysts still form autocatalytic sets. Two important differences emerge, however. By this hypothesis, all smaller polymers are catalytically unactive. Thus the metabolic chains leading out to large catalytically active polymers are longer than otherwise; hence if too many extraneous reactions are catalyzed, the concentrations of catalytically active polymers may be too low to be sustained above the threshold value. No analytic or simulation work has been done on this aspect of the problem. A related difference is that the supracritical explosion of a critically complex food set of small polymers cannot occur since, by hypothesis, such small polymers are not catalytically active. In the template-match version of the model, we have assumed that a minimum length is required to form a catalytic site and that the number of sites is proportional to polymer length. Since autocatalytic sets arise, the model is robust with respect to this assumption. In the RNA template model, the probability of catalysis is a function of total polymer length, such that long polymers, because they are likely to have more active sites, are more likely to be catalytically active than short polymers.

The basic model, with a fixed probability of catalysis, does not yet make use of the implications of shape complementarity. The implications of shape matching, however, suggest that achieving collective autocatalysis is easier than in the basic model. The basic model (without shape matching) assumes that the distribution of catalytic capacities is isotropic and unbiased, as generated by an independent probability of catalysis P. Thus the fact that peptide X catalyzes the cleavage of peptide Y carries no implication that Y may catalyze a reaction involving X as a substrate. Yet X acts on Y in part by virtue of shape complementarity; hence there should be an increased probability that Y also acts on X. We encountered this idea earlier, in the concept of a shape space. If Y binds X, then X binds Y. A similar point was made by Kenyon and Steinman (1969), who noted that the presence of N,N-dimethylformamide in a solution containing aspartic acid, serine, and dicyanamide increased the yield of α-aspartyl-serine, the dipeptide constituting the center of the active site in the protease chymotrypsin. They point out that a peptide bond is essentially an amide linkage. Thus, in the presence of an amide, an association between amino acids and the potential substrate increased the yield of that portion of the active site of a protease cleaving such bonds. Thus potential substrate could have served as a template about which protoenzymes evolved. In Kenyon and Steinman's felicitous phrase, the lock forms itself around the key. The implication of such shape complementarity, in either peptide or RNA sequence space, is a tendency toward symmetry in catalytic connectivity. If the fact that X catalyzes a reaction involving Y implies that Y has an increased chance of catalyzing either a reaction involving X or a reaction involving peptides near X in peptide space, achieving catalytic closed loops should be more likely in the refined model than in the isotropic model with a constant, independent probability P. In short, since shape complementarity is a reality and not a mere abstract construct, reflexively autocatalytic sets should tend to be populated by polymers which bind one another and hence exhibit a complex web of complementary shapes. Among the reasons this web of complementary shapes is intriguing is this: Any RNA-or-DNA-first view of the origin of life has as one of its dominant advantages the simplicity of point–point complementarity of complementary single-stranded molecules. If we set aside DNA and RNA for a broader look at catalytically active polymers, including peptides, we appear to recover a generalized shape–shape complementarity in a higher-dimensional shape space. In addition, shape complementarity should show up in entropic measures of sequence symbol complexity in mutually catalytic collections of polymers, as noted in Chapter 10.

The basic model ignores inhibition of catalysis. Inclusion of inhibition should permit the formation of autocatalytic sets having complex dynamical attractors. Alternative attractors then become alternative protoorganisms inhabiting the same polymer world. In addition, inhibition of catalysis can cut off unwanted metabolic pathways.

Once a reaction is catalyzed, that catalysis may be inhibited. Indeed, peptide activators and inhibitors of proteases are common in many prokaryotes and eukaryotes (see, for example, Carrell, Pemberton, and Boswell 1987). Because inhibition has been ignored in our model, the dynamics of the autocatalytic sets we have studied is relatively simple. So far, all have gone to a *single steady state* under fixed boundary conditions. If inhibition is included, the dynamics might instead exhibit sustained oscillations or more complex strange attractors. Preliminary analysis including inhibition in idealized Boolean models of such systems suggests that the main conclusions do not alter (Kauffman 1971b). Reasonable-size autocatalytic sets still form. Once a more complex dynamics is allowed, due either to inhibition or to sigmoidal catalytic kinetics, it is intriguing to realize that each alternative dynamical attractor, steady state, limit cycle oscillation, and so on can be thought of as a distinct self-reproducing protoorganism living on the same virtual catalyzed reaction graph.

Inclusion of catalysis inhibition is important for another reason. Such inhibition allows an autocatalytic set to begin to include polymers which inhibit specific reactions; hence the set can begin to cut off useless pathways to unneeded polymers.

The stability properties of autocatalytic sets require examination. Since inhibition of catalysis is likely to open the door to complex attractors, the sizes of those attractors and their stability properties become important with respect to the homeostatic properties of the systems, the ruggedness of their fitness landscapes, and their capacities to evolve. As discussed in Chapter 5, and as we shall see in more detail in Chapter 12, systems of low connectivity tend strongly to have small stable attractors. Later in this chapter, I give reasons to think that autocatalytic systems will evolve from systems containing low-specificity, low-efficiency catalytic polymers to systems exhibiting high molecular specificity and high catalytic efficiency. Since high specificity is the equivalent of low connectivity, we may plausibly project that evolved autocatalytic polymer systems in which the rate of synthesis of each polymer is influenced directly by rather few other polymers will exhibit small stable attractors and homeostasis. Collective self-reproduction and homeostasis may thus be emergent collective properties of polymer chemistry.

EVOLUTIONARY CAPACITIES
OF AUTOCATALYTIC SETS WITHOUT A GENOME

In this section I discuss the capacities of autocatalytic polymer sets to evolve without possessing a genome in the familiar sense of the term. Since possession of a genome capable of heritable modification is considered by many to be the hallmark of Darwinian evolution, this examination is not without general importance. Should the generation of life through collective autocatalysis be demonstrated, we shall need to rethink our theories of heritable information.

The model presented above for the spontaneous emergence of a collectively auto-catalytic set of organic molecules which transform to one another and catalyze the transformations in some statistically prescribed way is the first member of a new universality class. The central ideas are simple, general, and, I believe, important not only in the sense that life on earth might have originated utilizing these principles but also in the sense that the results are robust and withstand a variety of modifications of the assumed distribution of catalytic activities. The combinatorics of the ratio of reactions to polymers is so overwhelming that the argument is insensitive to most of the details. If the model is correct, then the routes to life in the universe are broader than imagined, perhaps boulevards rather than twisted backalleys of thermodynamic improbability. Therefore, it is worth analyzing the issues naturally raised by this new universality class. Foremost among such questions must be the capacity of any protoorganism based on a reproducing "metabolism" to evolve in the absence of template replicating of arbitrary DNA or RNA sequences as the store of heritable properties. Can such evolution occur? What might happen? What might the limits be?

I discuss in outline these issues:

• The surprising expected evolutionary behaviors of supracritical autocatalytic sets under selection

• The capacity of finite autocatalytic sets to exhibit heritable variation and evolve to novel autocatalytic sets without harboring a genome

• A selective transition from supracritical to subcritical autocatalytic sets

• An implied historical supracritical-to-subcritical trend in metabolic evolution

A surprising feature of supracritical autocatalytic sets is that they should be able to use as food essentially any molecular species presented to them. This lack of discrimination implies that such systems should evolve adaptively by tracking a changing food set. This behavior, which appears paradoxical, is rendered less so when we recall from Chapter 4 the concept of catalytic task space and the conclusion that a finite number of protoenzymes is expected to be a universal catalytic toolbox. It follows that such a collection of polymers, able to catalyze almost all reactions on a relevant size scale, would be able to "metabolize" any possible foodstuff.

In a self-reproducing supracritical system, one or more members would be expected to catalyze at least one reaction involving almost any novel molecule presented to the system. In turn, the products of that reaction would likely be subject to further catalysis by other members of the system, and those products in turn would be acted upon. Typically, at least some of the smaller fragments should be built up into preexisting components of the autocatalytic set. Thus those smaller fragments are serving as exogenously supplied food useful to the reproducing system. Further, the new products can exhibit catalytic activity and generate yet new possible reactions; hence they should meld into the set.

This flexibility of supracritical autocatalytic sets suggests that a supracritical system should evolve adaptively from one to another autocatalytic set by tracking its food set as the latter alters as a result of environmental change. Here the system, which might be made purely of catalytic peptides, adapts *without a genome* to an altering environment.

Adaptive tracking of the environment by a single changing autocatalytic set is one issue. But can more than one autocatalytic set live in the same environment, and can one set *evolve to a diversity* of kinds of protoorganisms? We expect so for several reasons. In a finite autocatalytic system having finite concentrations of each polymer and living on an infinite virtual graph, waxing and waning of the number of kinds, concentrations, and rates of feeding should drive an adaptive evolution of the autocatalytic set which shows hysteresis: As the total mass of the system waxes, the number of kinds of polymers present should increase, blossoming outward on the graph. If the system then returns to the same poorer environment after a sojourn on more abundant and richer food sets, it may often yield an autocatalytic system different from that initially found on that same food set. If so, then more than one protoorganism can live in the same environment.

In more detail, hysteresis implies that, in the same environment, the system can exist in more than one dynamical attractor or more than one stable behavior pattern. Each such pattern corresponds to a "different" autocatalytic set, or different protoorganism. There are at least several ways multiple attractors within the same environment might occur. A first is based on the stochastic behavior of self-reproducing metabolisms with small numbers of each kind of polymer in each spatially separated copy of the system. A second is based on sigmoidal enzyme kinetics in deterministic dynamical systems. A third is based on specific phosphorylation. I discuss these next.

First, two major sources of stochastic behavior can be expected. On the one hand, we have included only catalyzed reactions in the model and have ignored the slowly occurring spontaneous reactions which are always possible. Farmer has pointed out, as discussed in Bagley (1991) and Bagley, Farmer, and Fontana (1992), that such spontaneous reactions are a means to create occasional mutant polymers which may then be added to the autocatalytic system. This addition may allow different copies of the system to diverge in different directions. If one considers spontaneous as well as catalyzed reactions and if, as we have now shown, an autocatalytic set with large polymers at high concentrations can sustain itself, then a kind of shadow set, or cloud, of nearby products will exist. This shadow set arises as a result of spontaneous reactions utilizing components of the autocatalytic set. (Here "nearby" means one or two reaction steps away from the autocatalytic set.) Clearly, if any such novel molecule catalyzes its own formation from the initial autocatalytic set, it will be added to the set. More generally, if small collections of such spontaneously formed molecules which fluctuate to appreciable concentrations in the shadow set *jointly permit their own catalysis* from one another and from the original autocatalytic set, they may be added to the initial set. Conversely, addition of new polymers, particularly in the presence of inhibition, may lead to loss of old polymers. Thus, in general, spontaneous reactions and the shadow set of polymers surrounding the autocatalytic set allow evolution to new autocatalytic sets.

These ideas have now been confirmed by simulation. R. Bagley and W. Fontana at Los Alamos National Laboratories and the Santa Fe Institute have implemented a program which tests for the existence of "viable" mutant additions of clusters of new polymers in the spontaneous cloud surrounding the autocatalytic set. They find that such new clusters do exist and can be added to the set. Over a succession of moments, the set evolves by a succession of additions of clusters. These results clearly show that, in principle, autocatalytic sets without a genome can evolve in the space of polymers (Bagley, Farmer, and Fontana 1992).

A further source of stochastic behavior arises from the analogue of cell division. Any beginning account of the origin of life must describe how the protoorganism can segregate reactions into a small region and how it can defend its internal environment. Suggestions here have ranged from Oparin's coascervates (1957, 1971) to Fox's microspheres (Fox and Dose 1977; Fox 1980) to liposomes (Hargreaves, Mulvihill, and Deamer 1977; Deamer and Barchfeld 1982). By whatever mechanism of compartmentation may prove reasonable, the numbers of copies of any single polymer in one copy of a protoorganism will be modest, often small. If the protoorganism compartment is divided into two, the contents of the initial volume will be randomly distributed to the two daughter cells. Any polymer present in low abundance may be asymmetrically distributed to the two daughter cells. Thus, in general, the dynamical behavior of such systems will be governed by stochastic differential equations, and one expects that evolution to new autocatalytic sets will be possible in a population of dividing protocells.

The evolutionary capacity of autocatalytic sets is even greater than discussed. This is because, in one environment, the *same* single pattern of catalytic connectivity may be able to perform in more than one stable and collectively autocatalytic dynamical mode; each alternative mode would constitute an alternative stable dynamical attractor, or protoorganism. Recall from Chapter 5 our discussion of systems in which the activities of enzymes are cooperative sigmoidal functions of the levels of substrates or allosteric modifiers. When biochemical systems whose components realize sigmoidal kinetics exist, the achievement of multiple steady states or multiple attractors is very easy indeed. Thus as soon as sigmoidal kinetics is included in our basic model of self-reproducing metabolisms, the occurrence of multiple protoorganisms in the same external environment becomes virtually inevitable. Such protoorganisms might each exhibit steady-state behavior, limit-cycle dynamics, or chaotic attractors. Thus a single autocatalytic set might have a complex time structure, and more than one attractor might make collective use of the same set of polymers but with different time structures (Kauffman 1971b).

Third, the capacity to evolve in alternative directions is likely to be greater with the advent of specific control over thermodynamic flux by coupling of exergonic reactions to the endergonic reactions synthesizing specific large polymers. As noted above, Fox (1988) has suggested coupling overall pyrophosphate hydrolysis to peptide or RNA condensation reactions. Specific enzymatic control by polymers in an autocatalytic set in which peptides are activated by phosphorylation, and hence form peptide bonds rapidly, may allow much finer control of membership in an autocatalytic set and therefore a larger diveristy of self-consistent autocatalytic sets in a single environment.

Evolution without a genome raises novel questions which do not arise in molecular systems which reproduce arbitrary genomes via a promiscuous polymerase:

1. Since an autocatalytic set reproduces by collective catalysis, only sets which achieve collective catalytic closure can survive. Therefore even in the infinite space of polymers, the number of collectively autocatalytic sets may remain finite.

2. How many neighboring autocatalytic sets exist within some mutation distance from a given autocatalytic set? Here mutation distance must be measured in terms of the number of kinds of new polymers or monomers which must arise as a cluster to become incorporated into the system.

3. What do the pathways of neutral evolutionary change—that is, change in the absence of selection—among such neighboring sets look like? This question seeks the graph of connections among neighboring autocatalytic sets in polymer space.

4. Under given selection conditions—for example, a defined food set—some autocatalytic sets will reproduce more rapidly than others and hence will have higher

Darwinian fitness. In short, a fitness landscape exists over the set of autocatalytic sets. Because of this landscape, an adaptive flow over the space of self-reproducing systems will show historically frozen accidents and hysteresis. Sets will climb to local peaks in a fixed fitness landscape and remain more or less trapped. If the food set or other parameters change, the system may never retrace its steps to an earlier optimal set when it revisits an old environment. What do such patterns look like?

5. Coevolution of autocatalytic sets will occur. If multiple sets live in the same physical environment and exchange molecular variables either by diffusion in a common milieu or by attacking one another, then the polymers in one set may either help or poison a second set. As in Chapter 6, we must consider the coevolutionary dynamics of such systems as the Darwinian fitness landscape of each is deformed by the adaptive moves of its partners.

Notice a point of central interest about coevolution in the current context. In Chapter 6 we used coupled *NK* landscape models to study coevolution, but any given genotype had no natural internal dynamics or function. Here, we have a Darwinian notion of the fitness of any autocatalytic polymer system. In a real sense, any autocatalytic system is a kind of agent; events and molecular inputs can be either useful or dysfunctional to its survival. A natural notion of function, therefore, derives directly from the underlying chemical and collective reproductive properties of the autocatalytic set. Using these properties, we can study the emergence of collaborative functional interaction between autocatalytic sets, such as mutualism and symbiosis, as they coevolve together and come, figuratively, to know one another.

In summary, these evolutionary adaptive behaviors of autocatalytic systems are important. For here we have evolution *without a genome.* Information is stored not in a stable, inert structure such as template-replicating DNA but in the self-consistent web of transformations. But the particular web found is a function of the history of the environments to which the autocatalytic set has adapted. In Chapter 10, I shall generalize these ideas in terms of random grammar models of functional integration.

A Selective Transition from Supracritical to Subcritical Autocatalytic Sets

Physically realized supracritical autocatalytic systems are expected to contain polymers able to catalyze at least one reaction afforded by any novel food molecule and subsequently to catalyze reactions involving the products of that first reaction. Like a universal enzymatic toolbox, such systems would be able to metabolize any substrate in some fashion or another. But current organisms do not appear to be supracritical. If life began with such supracritical sets, then a transition has occurred to subcritical systems. We examine this now.

The model of autocatalytic sets was developed without formal use of the idea of a catalytic task space. Of course, the proper statement of the problem is that an autocatalytic set is a set having the property that each reaction needed to achieve catalytic closure is covered by a ball in catalytic task space and is catalyzed by some member of the set. An autocatalytic set is a self-consistent set of catalyzed balls and has the property that, for the proper substrates offered to each ball and for the proper products of the corresponding reaction, the union of substrates and products catalyzes the same set of balls in task space. Evolution of autocatalytic sets is the concerted evolution of the corresponding set of polymer substrates and products and of the sizes and locations of the catalyzed balls, always maintaining the conditions of self-consistency. A supracritical autocatalytic set has the property that almost any incoming molecule affords at least one reaction which either alters that molecule or else combines it with some other molecule in the set and that this reaction is catalyzed by at least one member of the present set. For this reason, we expect that a supracritical autocatalytic set can cope with essentially any novel food molecule.

Contemporary organisms, even those as versatile as bacteria, can act catalytically on only a subset of the molecular species presented to them. This is so well known as to be thought trivial but is not really so trivial. It implies that cell metabolism is *not* supracritical. If we want to imagine that life began as a supracritical, stochastic reproducing metabolism of polymers having rather weak, nonspecific catalytic activity, then we have to ask how a transition to a *subcritical* system can have occurred.

The immune network leads immediately to a suggestive image: Under selection for protoenzymes with increased catalytic *efficiency,* an autocatalytic collection of catalysts should *simultaneously* evolve enzymes with increased *specificity.* This increased specificity, in turn, will transform a supracritical autocatalytic set into a subcritical one.

According to the immune network theory of Jerne (1974, 1984), antibody molecules have both paratopes (their binding sites) and epitopes (sites recognized *by* the binding sites on other antibody molecules). One picture of this network supposes that the degree to which a B lymphocyte which secretes an antibody which binds an antigen is stimulated to divide is proportional to the affinity of the binding. Conversely, a B lymphocyte whose antibody receptors (identical to the antibody it secretes) are bound by another antibody divides less frequently. The complex shape–space matching of epitope and paratope constitutes the idiotype network (Jerne 1974, 1984; Dwyer, Vakil, and Kearney 1986; Hoffman, Kion, et al. 1988; Perelson 1988). Loops of B lymphocytes form, in terms of a cycle of binding and bound (Chang and Kohler 1986; Perelson 1988). If stimulation of the binding cell is stronger than inhibition of the bound around such loops, each loop can act as a positive feedback loop. In theory (Perelson 1988), those linked loops, and tails, of B lymphocytes having the highest affinity for the antigen are stimulated the most—and hence grow fastest—and dominate the immune response by clonal selection. Subsequent somatic mutation to those cells presumably leads to modified but still coupled loop structures whose member antibodies match and recognize one another with much higher affinity and specificity than initially, and hence mutually stimulate one another more strongly and grow still faster. The looped structures thus might provide an internal memory of the initial antigen and are, in a real sense, *autocatalytic structures* which must maintain cyclic connection while new variant members increase mean affinity for the antigen. In addition, the successive ranks of antibodies form even and odd ranks, each the shape complement of its predecessor and successor. That is, such loops presumably tend strongly to have successive shape–shape complements. In this picture, an initial imprecise loop structure whose members have modest affinity for the incoming antigen *self-tunes* to a focused system (typically one having less heterogeneity of cells) in which the members have higher affinity for the antigen and for one another via generalized shape complementarity.

This idea has a direct implication for autocatalytic polymer systems. We need a single postulate which is plausible, testable, and probably true. Let us suppose that, *on average,* an enzyme which binds its transition-state complex with higher affinity catalyzes the reaction with higher maximal velocity than an enzyme having lower affinity, and that this higher affinity, on average, implies higher *specificity* for substrate(s). That is, let us assume that, on average, higher specificity is correlated with higher maximal catalytic velocity. This average statement need not be true in individual cases. An imprecise site may contain a very reactive group. Presumably, however, if the site became more specific for the transition state of a specific reaction, the affinity for the transition state would increase and the velocity of the catalyzed reaction would increase as well. Indeed, Kacser and Beeby (1984) argue on thermodynamic grounds that in general this relation holds.

The natural consequence of this assumption is that, in a supracritical autocatalytic polymer system—say, in a chemostat with fixed influx of food and overflow—multiple sufficiently connected autocatalytic subsystems are present, some of which catalyze their own formation more rapidly than others. Fixed boundary conditions imply that the total number of monomers free or bound into polymers is constant and thus constitutes selective conditions. Any autocatalytic subsystem which reproduces faster than the other subsystems will outstrip the remainder by familiar Darwinian selection. By the assumption linking specificity and velocity, however, the subsystems which will grow fastest will be those which simultaneously maintain the requirement of catalytic closure *and* have the most highly specific protoenzymes in the system. That is, the natural consequence of natural selection should be to pull such an initial imprecise supracritical autocatalytic system toward subsystems which remain autocatalytically connected but whose mean specificity of coupling and velocity of replication have increased. As this focusing occurs, the increased specificity of catalytic activities means that, simultaneously, extraneous side reactions not needed for autocatalytic closure and reproduction on the given food set are being trimmed away. Thus in this "simple" condition, selection toward minimally complex but still connected autocatalytic systems of ever higher specificity should occur.

Cast onto the image of catalytic task space, under selection, an autocatalytic set remains self-consistent for catalytic closure. The balls grow smaller and the set of reactions and polymers moves across task space in some way, however, and the number of molecular species dwindles toward the level found in minimally autocatalytic systems in which removal of any further polymer is lethal.

But selection toward higher-specificity catalysts must be fundamentally limited. As a very supracritical system with very imprecise protoenzymes begins to undergo selection for slightly more specific and rapidly producing subsystems, these new subsystems should initially remain supracritical. As selection pulls such a system toward more specific catalysts, the task of preserving catalytic closure must become ever more difficult. Less of catalytic task space is being covered, but the tasks accomplished must keep one another mutually in view during the focusing process. At some point, the rate of finding a new polymer which enhances the rate of formation of the system and is itself formed efficiently by the system becomes very low. Another way of saying this is that the selective process toward enzymes which catalyze their target reactions with increased maximal velocities simultaneously increases the specificity of those enzymes. This increase in specificity makes the balls covered in catalytic task space smaller and hence *lowers* the probability that the same enzyme catalyzes other reactions. At some point, if that probability is lowered enough and if the minimal but efficient autocatalytic set is small enough, the reaction graph generated by the set of high-specificity polymers may become finite rather than infinite. Then, under selection, a transition has occurred from an imprecise supracritical finite autocatalytic system living on an infinite virtual reaction graph to a smaller subcritical finite system living on a finite graph.

I summarize the picture we are led to. If there is a correlation between specificity and the velocity with which a protoenzyme catalyzes a reaction, as Kacser and Beeby (1984) have argued on thermodynamic grounds, then, in rapidly growing autocatalytic systems that start from an imprecise supracritical system, natural selection should lead to more focused autocatalytic systems which have increased the specificity of the mutually necessary connected cycles of enzymes, hence trimmed away useless pathways, and perhaps gone subcritical. But selection for ever-increased specificity and velocity is limited. As the system goes subcritical, maintaining catalytic closure becomes harder. Thus there will be a minimal efficient system attainable.

A Supracritical–Subcritical Evolutionary Axis

I close this examination of this new universality class of models by noting the implications of a supracritical-to-subcritical transition in evolution. We are almost forced to believe such a transition occurred by the very picture that life, in the sense of a reproducing coupled catalytic system, arose as a supracritically complex system. We are not supracritical systems now; hence such a transition must have occurred if life started, as here pictured, complex rather than simple. Then there must be a kind of secular evolutionary axis, driven by selection, tuning how supracritical or subcritical free-living entities are. Perhaps invention of the genetic code and DNA was needed not for reproduction and molecular replication but to free autocatalytic systems of a laborious search for catalytic closure each time a new RNA or protein was to be made. Perhaps the evolution of a code was easier than we have pictured it: Perhaps the code crystallized out during the transition from supracritical to subcritical, for supracritical catalytic systems may be astonishingly versatile compared with the systems found in present organisms. Such supracritical systems are both chicken *and* egg.

We may well ask how subcritical current metabolisms are. Three lines of evidence suggest that we may be surprisingly near the edge of supracritical behavior. First, we are now familiar with the potentially universal capability of the mammalian immune repertoire to recognize almost all possible shapes on some size scale, and perhaps to catalyze all possible reactions on the same size scale of active sites. Such a system is essentially supracritical. Second, bacteria and higher organisms evolve very rapidly to cope with an enormous variety of novel metabolic stresses. Bacteria often do so by importing a plasmid, or other vector, producing a novel protein able to deal with whatever agent is causing the stress. But the set of plasmids had to acquire genes coding for proteins with the range of functions required. These plasmid-borne genes may be thought of as a bacterial immune system shared between distinct cells rather than between cells of one organism. The bacterial immune system, too, appears to be nearly complete, nearly supracritical. Virtually any novel stress can be coped with either by an existing plasmid or by the rapid evolution of modified variants from the gene pool already available. And the same considerations apply to higher eukaryotes. Almost any stress will call forth some resistant strain after a modest interval. Thus the gene pool of a breeding population of, for example, fruit flies lies within a short mutational range of proteins or other polymers able to cope with an enormous variety of novel molecular species. We come to the tentative conclusion that our own metabolisms are nearly universal. This thought, initially astonishing, becomes workaday once we realize that a finite number of enzymes, on the order of 10^8 to 10^{12}, ought to be a universal enzymatic toolbox. Our collective genomes as breeding populations within species *are* within range of that diversity.

An Evaluative Summary

I have now discussed a novel model for the origin of life among catalytic polymers based on collective self-organization. The model rests on a phase transition in the hypergraph of polymers, on the reactions which couple them, and on the distribution of their capacities to catalyze the same reactions. Pause to note how simple and general it is. It contains the ideas of (1) a set of entities, (2) the transformations among those entities by which they convert to one another, and (3) the fact that the entities abet those transformations according to some statistically characterizable distribution. A connected whole emerges as a collective property simply because the number of possible transformations grows faster than the number of entities, while the frequency of abetting transformations either stays constant or decreases more slowly than the ratio of transformations to entities increases. At some level of complexity, a percolation threshold is reached and collective autocatalysis emerges.

In an earlier section we discussed the difficulty of achieving template replication of arbitrary RNA sequences. I have suggested that, since template molecules could act as specific ligases, as can ribozymes, reproduction by RNA molecules might be more easily attained by collective autocatalysis. In short, RNA molecules could achieve replication not by templating but by catalytic closure. Thereafter, polymerases able to catalyze replication of arbitrary RNA sequences might arise relatively simply.

If the general ideas are right and truly robust with respect to the idealizations of the model, then the formation of autocatalytic sets of polymer catalysts—peptides, RNA, or otherwise—is an expected emergent collective property of sufficiently complex sets of such polymers. Perhaps the main feature of this model is something like a law of minimum complexity. To obtain connected catalyzed transformations as an emergent collective property, a sufficient complexity is needed. Smaller systems simply fail to achieve catalytic closure.

EXPERIMENTAL CONSEQUENCES

The model of autocatalytic sets of peptides or RNA polymers opens up very large areas for experimental investigation. Foremost among the issues raised is distribution of catalytic capacities in peptide or RNA space (approaches to this issue by cloning of entirely novel genes were discussed in Chapter 4 and will not be further discussed here). Related to this issue is the mapping onto catalytic task space of peptides or RNA sequences considered as reaction substrates and products. Beyond these fundamental issues, we must consider the experimental construction of autocatalytic sets of peptides or RNA ribozymes. I suspect this construction is feasible if we are bold enough to reach the needed complexity and meet the thermodynamic requirements.

Steps Toward Experimental Autocatalytic Sets

If it proves to be the case that at least weak catalysis is widely distributed in peptide and/or RNA space, as now appears likely from the results noted in Chapter 4, then it may soon be possible to create experimental autocatalytic polymer sets. The fact that the basic theory extends comfortably to autocatalytic sets of specific model ribozymes suggests that it may be easier to create autocatalytic ribozyme systems than peptide systems. Either form would be marvelous. The main experimental implication of the body of ideas in the present chapter is this: Start complex, not simple. The goal is a connected catalyzed set of transformations, preferably linking a food set to an autocatalytic set of peptides, RNA, or other catalytically active polymers. A minimum critical complexity is surely necessary to nucleate a connected web. Practical implementation of this possibility confronts three requirements: thermodynamic requirements to drive the synthesis of large polypeptides and polynucleotides from smaller components, the requirement to confine reactants to small volumes, and the requirement for catalytic closure.

Osmotic Drive

The model of autocatalytic sets discussed so far is based purely on polymer autocatalysis and remains innocent of the coupling of exergonic and endergonic reactions in a coupled metabolism. Such coupling is discussed in the next chapter. Before doing so, it is useful to consider possible simple ways to drive synthesis of high-molecular-weight peptides or possibly RNA molecules. Earlier in the chapter, I broached enclosure of peptides in liposomes. Such bilipid membranes are plausibly constructed in prebiotic conditions and, under cycles of hydration and dehydration, can enclose macromolecules, including DNA and RNA (Hargreaves, Mulvihill, and Deamer 1977; Deamer and Barchfeld 1982). Consider peptide or phosphodiester bond formation among peptides or RNA sequences enclosed within a liposome. Each condensation reaction releases one water molecule, which diffuses quite rapidly across the semipermeable lipid membrane. The larger polymers do not diffuse across the membrane, however. If the liposome is placed in a hypertonic medium, the efflux of water, a product of the condensation reaction, should drive the reaction to the right, leading to the synthesis of larger polymers. Conversely, if the liposome is placed in a hypotonic medium, then the influx of water should favor cleavage. Cyclic wetting and drying of the liposome might drive the synthesis, cleavage, and cyclic shuffling of peptide or RNA fragments in the liposome.

A direct experimental approach to this problem would analyze the plastein reaction (Wasteneys and Borsook 1930; Fruton 1982) carried out inside liposomes. You recall that this reaction consists in cleaving large proteins with trypsin, then dehydrating and observing resynthesis of large polymers from small peptides without the need for ATP or other high-energy compounds (Levin, Berger, and Katchalski 1956; Neumann, Levin, et al. 1959). It should be possible to test whether enclosure in a liposome exposed to hypertonic solutions causes a shift of the equilibrium toward synthesis, and whether cyclic wetting and drying drive cyclic condensation and cleavage reactions.

The physical consequences of such ligation and cleavage reactions are also of interest. Large polymers have colligative properties: If cross-linked, they form gels, extrude excess water, and thereafter act as distributed osmometers. Hydrodynamic and other mechanical forces would be expected to deform liposomes in possibly interesting ways. That is, such forces might lead to a liposome with a gel mass at one end and a sol state at the other end, yielding a pseudopod-like protrusion. Sol–gel transitions might even yield the progenitor of amoeboid movement in such a liposome. Finally, confining reactions to a surface in a variety of ways tends to ease the formation of large polymers from smaller components for entropic reasons, and may prove a useful experimental tool.

Autocatalytic Sets: Critical Complexity

If synthetic peptides or RNA sequences show a variety of weak catalytic activities, then it may not be unsurpassingly hard to make autocatalytic sets of polymers. The first steps in any such program require investigating the probability of catalysis by peptides or RNA sequences. Ellington and Szostak's results discussed in Chapter 4 suggest that the probability for RNA sequences may be on the order of 10^{-10} for arbitrary organic reactions. Because of the ease of template complementarity, it might be easier to catalyze cleavage or ligation by RNA sequences of other RNA sequences than to catalyze arbitrary reactions with RNA sequences. Thus the probability of catalysis may be much higher than 10^{-10} among RNA sequences acting upon themselves.

One approach, if focusing on RNA sequences, would generate a complex mixture of perhaps 10^{14} single-stranded RNA sequences of a constant length, say 40 nucleotides, and incubate this mixture by itself in a small volume to ensure high molecular collision rates. If any cleavage or ligation reactions occur, single-stranded RNA sequences shorter or longer than 40 nucleotides will be generated. These can be size-separated on gels, amplified by a variety of techniques, and sequenced. Such sequencing should confirm that sequences longer or shorter than 40 nucleotides have been generated. In Chapter 10, I shall discuss the concept of random grammars by which symbol strings act on and transform one another. RNA sequences catalyzing mutual reactions can be regarded as mediating such a molecular grammar. One measure of the way RNA sequences act catalytically on one another based on template recognition, ligation, cleavage, and splicing is the buildup of nonrandom sequence biases among nearby nucleotides. The resulting biases can be measured as symbol entropy and mutual information as a function of distance between nucleotides. These biases build toward an asymptotic distribution after a sufficient number of molecular recognition events and catalytic transformations have occurred. What is important is that, while the *size* distribution of polymers may be sensitive to thermodynamic factors, such as hydration conditions, the symbol-entropy and mutual-information measures appear to be grossly insensitive to thermodynamic factors but sensitive to site-recognition and sequence-substitution rules. Thus the time course of buildup and of asymptotic deviation from statistical independence as a function of internucleotide distance may yield information on the mutual sequence recognition and molecular catalytic grammar by which RNA molecules act on one another.

The theory described above gives a first hint of the complexity needed in peptide or RNA sequences if these sequences are to contain autocatalytic subsets. Attempts to create autocatalytic sets would require exceeding the minimum complexity. Such experiments might be carried out in a chemostat fed by a rich mixture of organic molecules; pyrophosphate, ATP, and other energy sources; and many small and modest-size peptides or RNA sequences. Analysis of the spectrum of peptides and/or RNA sequences formed could be by column and two-dimensional gel analysis of peptides and polypeptides and by PCR amplification and sequencing of polynucleotides, or by other procedures. If the polymer spectrum settled to a stable, self-regenerating asymptotic state or to a stable core of sequences and a fluctuating periphery, one would surely be entitled to at least a small bottle of champagne, for one would presumably have generated a novel form of life *ab initio.*

If the general theory proves robust, then once such a system formed, the food set should be able to be gradually simplified and the system maintain itself. Rather than carrying out such experiments as spatially homogeneous reactions in a chemostat, work with liposomes may reveal that enclosure and confinement to a small volume drive cyclic ligation and cleavage of large polymers by cyclic hydration and dehydration. In addition, work in a population of liposomes, each containing a unique combination of polymers and perhaps each able to divide, may allow experimental evolution based on heritable variation in model protocells.

It may be the case that random peptide or RNA sequences exhibit catalysis so rarely that assembly of an autocatalytic set from unevolved polymers is possible in theory but impractical. A separate approach is based on use of antibodies and derives from the recent observation, described in Chapter 4, that antibodies against the transition state of a reaction can catalyze that reaction. This is opening the door to the possibility, based on the capacity of antibodies to cover shape space, of finding antibodies to catalyze almost any desired reaction. In particular, such antibodies are likely to be abzymes cleaving or ligating specific peptide bonds. Since the specificity of an antibody molecule is due to its variable regions, consider utilizing entirely random DNA sequences cloned into the variable regions of antibody molecules to generate antibody diversity *in vitro*. Known antibody diversity is about 10^8, and this level of diversity can be matched by use of random DNA sequences inserted into the variable regions of the heavy or light chains. If catalytic function of such antibody molecules requires only reasonably small fragments of the entire molecule around the heavy or light V regions, then a sufficiently large number of such random abzyme fragments may prove to be autocatalytic, in a fashion reminiscent of the self-focusing presumed in the immune network.

Finally, we might note that, if autocatalytic polymer systems are constructable, we might obtain selectable chemical factories, or parallel-processing chemical computers. If supracritical-to-subcritical transitions occur, then in principle, by requiring such a constructed system to evolve using Z as a food source, the system will couple Z, via a sequence of catalysis pathways, to a connected set of other molecules. Then the system will contain the weak enzymes needed to synthesize Z from those other molecules by reversing the flux down those pathways.

SUMMARY

This chapter has discussed a very general theory for the emergence of self-reproducing systems of catalytic polymers, either peptides, RNA, or others. The theory rests on the combinatorial consequences of polymer chemistry. As the maximum polymer length in a system increases, the number of reactions by which polymers can interconvert necessarily rises faster than the number of polymers present. The result is that a sufficiently complex set of polymers has very many potential reactions leading to the synthesis of any one of those polymers. As a consequence, for many possible distributions of catalytic capacity for those reactions among the same set of polymers, autocatalytic sets will emerge.

This new universality class warrants serious exploration both theoretically and experimentally. It seems probable that such systems are capable of evolution in the absence of a genome. It also seems probable that such systems answer, in principle, the question of how peptides might be self-reproducing despite lacking the point–point complementarity found in DNA and RNA. Given the difficulty in achieving template replication of arbitrary RNA or RNA-like polymers but the existence of ribozymes mediating cleavage and ligation reactions, the formation of autocatalytic sets of ribozymes may be both feasible and a more probable route to the origin of life than the familiar hypothesis of nude replicating genes. Indeed, a collective autocatalytic system is a favorable environment from which a general RNA polymerase might have evolved and taken over an integrated metabolism. Further, the hypothesis that life is a collective emergent property of complex polymer systems seems likely to give a first answer to the critical question of why free-living systems exhibit a minimal complexity. Further, whether composed of catalytic peptides, ribozymes, or both, it seems likely that collectively autocatalytic polymer systems recover a generalized shape complementarity in a high-dimensional shape space.

On the experimental front, autocatalytic polymer systems invite exploration, by cloning or other means, of the distribution of catalytic and ligand-binding capacities in peptide space and in RNA space. Further, the theory suggests that achievement of catalytic closure requires a minimal complexity and hence that experiments must be bold enough to reach that complexity. Fortunately, the technological means to do so are just on the point of emerging. We may, in the not distant future, create life anew.

CHAPTER 8

The Origin of a Connected Metabolism

Metabolism is the connected flow of organic molecules through a web of transformations linking exergonic and endergonic reactions by which an organism harnesses the energy needed to drive the synthesis of molecular species. Figure 8.1 shows part of the familiarly complex web of human intermediate metabolism. I find it impossible to view this vast intricate network without wondering what principles might govern its structure and its capacity to have evolved. If life began with a nude replicating gene, either RNA or DNA, how did such a connected web manage to evolve to clothe the nudity? If life were to begin again, would intermediate metabolism look much the same? If so, in what senses and why?

In the present chapter, I shall briefly investigate whether the ideas from the previous chapter on the collective emergence of autocatalytic sets of polymers may shed light on the emergence of a connected metabolism. As in the previous chapter, I shall suggest a heterodox possibility: In a sufficiently complex system of organic molecules mixed with a sufficiently complex set of potential catalytic polymers, a connected web of transformations among those organic molecules will inevitably "crystallize" due to a percolation of a catalyzed reaction network among the organic molecules. Such a metabolism will then couple exergonic and endergonic reactions into the autocatalytic behavior of the catalytic polymers and help drive that system's thermodynamic capacity to reproduce. Thus again I suggest that life may have started with a critical complexity, then *simplified* under selection. After its simplification, it appears to us as a marvelously improbable system, a whole whose chicken-and-egg self-referencing necessity mistakenly mystifies us.

Consider some unfamiliar questions. Organic molecules are complexes of CHNOPS—carbon, hydrogen, nitrogen, oxygen, phosphorus, and sulfur—with small amounts of other elements. If we look at the known map of metabolism, one obvious if unfamiliar question is how many organic molecules with one carbon atom, two carbon atoms, . . . N carbon atoms occur in metabolism? That is, what is the distribution of number of kinds of organic molecules versus their complexity. Whatever the answer may be, we may well ask, Why? Is this particular distribution an accident? Or, if metabolism were to evolve again, would the distribution be sim-

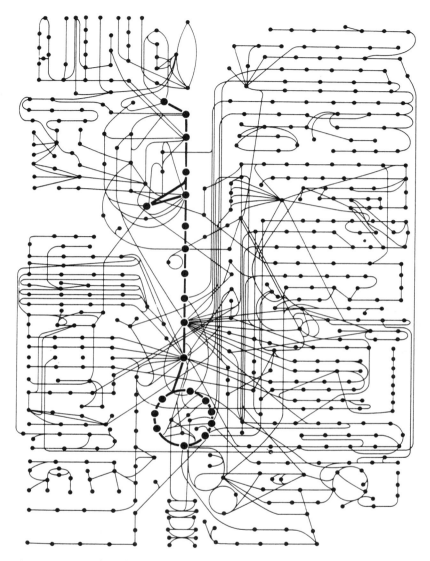

Figure 8.1 Representation of intermediate metabolism. Points correspond to metabolites, edges to chemical transformations. (From Alberts, Bray, et al. 1983)

ilar? Metabolism is replete with branch points and cycles, with some distribution of distances between branch points and number of metabolites in each cycle, hence of cycle lengths. What governs those distributions? Would they recur if metabolism were to originate again? The most overwhelming question, however, is, how could such an integrated system of transformations have originated? As in other aspects of the origin of life problem, we seem again to confront a self-referencing problem. In order to function at all, a metabolism must minimally be a connected series of catalyzed transformations leading from food to needed products. Conversely, however, without the connected web to maintain the flow of energy and products, how could there ever have been a living entity to evolve connected metabolic pathways?

It must not be thought that the requirement that a protometabolism be a functionally connected whole has gone without attempted answers. Indeed, concern over the emergence of such an integrated whole is what lies behind the efforts of Hoyle and Wickramasinghe (1981) and others to "nucleate" a working metabolism in one gigantic leap. You will remember the calculation arguing that the probability of finding a protein able to catalyze a specific metabolic step might be on the order of 10^{-20} but that *2000* such enzymes were needed simultaneously to catalyze a specific metabolic web; hence the probability of attaining a given specific metabolism is $10^{-40\,000}$. By now, one error in this argument is patently obvious. It is not necessary to obtain a given specific metabolism; all we need obtain is some adequate connected metabolism.

As noted above, the main purpose of this chapter is to suggest that the ideas derived in the previous chapter from an analysis of the conditions needed to achieve a complex web of catalyzed transformations might shed light on this problem as well. Connected metabolisms can literally crystallize. First, it is well worth noting the most plausible current orthodox answer.

Horowitz (1945) suggested that metabolic pathways build up backward, proposing that protoorganisms lived in an environment extremely rich in metabolites and therefore initially required little by way of an intricate metabolism. Needed products lay to hand, were used, and then were broken down to waste products. However, gradually, Horowitz supposes, the abundance of exogenous metabolites was lowered. When the availability of any critical metabolite became too low, the evolving protoorganisms had to invent a means of synthesizing that metabolite. The protoorganisms would already possess enzymes capable of acting on that metabolite to produce waste products, however. Further, such waste products would not have been depleted from the environment. Therefore, both the products and the catalysts were available to synthesize the critical metabolite by driving the reaction in the reverse direction. This interesting hypothesis leads to the conclusion that metabolic pathways might have been built up *backward,* gradually over time.

Horowitz's idea is perfectly sensible. Let us pursue the line of thought and, for the sake of argument, forgo the problem of the onset of translation. If one is picturing a single replicating protein or gene as a model of the origin of life, then the problem arises about how such a single entity might come to catalyze a connected web of metabolic transformations. Borrowing from Horowitz, one possibility is at least plausible. Suppose a given enzyme catalyzes a specific reaction converting substrate Sn-1 to product Sn. Then the binding site on the enzyme recognizes Sn-1. A nearby shape mutant of the binding site might therefore be able to recognize a nearby shape on a substrate Sn-2 and hence catalyze the transformation of Sn-2 to Sn-1. Thus, through a succession of gene-duplication and protein-differentiation steps, linked pathways among substrates resulting from a web of similar shapes might be built up from simple beginnings.

I believe Horowitz's hypothesis is plausible because it explains the formation of linked transformations among organic molecules by the existence of successively neighboring substrate shapes along metabolic pathways. Yet the hypothesis has problems. Note first that, where adjacent metabolites are to be linked by a catalyzed transformation but structural shape similarity is *not readily found,* a mechanism of gene duplication and protein differentiation is not readily applicable. Thus for the evolution of enzymes involved with substrates which are simple isomers of each other,

Horowitz's idea may be very reasonable, but many reactions add or remove modest-size functional groups to or from substrates, or cause substantial changes in substrate conformation. It is hard to see how enzymes for such links in the metabolic web might evolve by protein differentiation. The hypothesis is also needlessly limited. Note that it utilizes the plausible idea that a cluster of similar genes/proteins catalyzes a cluster of reactions among similar substrates which therefore can form a linked web. The hypothesis I suggest extends this familiar idea to the crytallization of a linked web of transformations among organic molecules in a sufficiently complex system as a result of percolation of a catalyzed reaction graph. While I state the model without reference to shape similarity among organic molecules, such similarity presumably should abet my general thesis.

CRYSTALLIZATION OF A CONNECTED METABOLISM AS A PERCOLATION PROBLEM

Let us make bold to start again and seek conditions under which a connected metabolism would be expected to emerge. We saw in the previous chapter that a sufficiently complex set of polymers whose members are candidates to catalyze transformations among the polymers should become reflexively autocatalytic as a result of a percolation threshold in the underlying virtual catalyzed reaction graph. Once the set is complex enough, connectivity emerges as a collective property. Then the hypothetical autocatalytic set begins as a supracritical system of imprecise enzymes, each catalyzing a ball of related reactions weakly. We imagine this imprecise system, under selective conditions, focusing down to a tighter catalytic system whose enzymes are of higher specificity but persistently maintain mutual catalytic closure during selection. Selection becomes harder as the system becomes more subcritical.

The basic argument here rests on the combinatorial fact that, as polymer length increases, the number of kinds of polymers increases more slowly than the number of reactions by which they interconvert. It is a separate fact that the polymers are able to catalyze those conversion reactions which allows *auto*catalysis to set in above a threshold. Consider next the space of all possible organic molecules up to a fixed number of CHNOPS atoms per molecule, say M. As M increases, the number of kinds of organic molecules will increase, but the number of reactions by which they interconvert will also increase. If the ratio of reactions to kinds of molecules increases as M increases, then the more complex the set of organic molecules we consider, the more pregnant it shall be with respect to the formation of connected webs of catalyzed transformations. Then we have only to ask the question, How many peptides or RNA sequences, each a candidate to catalyze any one of the many reactions, must be "tried" and be jointly present in order that the number of reactions catalyzed is sufficient for the emergence of connected webs of metabolic transformations? The detailed answer may be long in coming, but the general answer is obvious: A *sufficiently complex system will spontaneously crystallize a connected metabolism.*

A simple example demonstrates this idea. We need only reinterpret polymers made up of two kinds of amino acid as organic molecules made up of two kinds of atoms. Thus a polymer of length M is now to be thought of as a linear organic molecule made up of M atoms. It does not matter for the moment that these model metabolites be reasonable. It matters only that we can calculate the ratio of the number of reactions by which these molecules intertransform to the number of kinds of molecules of length M or less. As before, the ratio is $M - 2$. Then if we consider a separate set of peptide or RNA polymers, each having an independent probability P of catalyzing any of the reactions among the metabolites, we can readily calculate how many peptides or RNA sequences must be tried and remain jointly present in the system in order to achieve a connected metabolism.

More precisely, we can examine this general idea by defining a connected metabolism as one which leads from a defined exogenous food set to a defined internal set of organic molecules. The most stringent requirement is that connected pathways lead from the simplest atoms, A and B, to almost all organic molecules in the set up to length M. But this is just the requirement invoked in the previous chapter as the first stringent criterion for the formation of an autocatalytic set, with the exception that, in that case, the peptides or RNA sequences catalyzed their own formation. The results here are almost the same. If we specify M as the complexity of hypothetical organic molecules, then we can calculate the number of peptides or RNA sequences which must be tried at each probability of catalysis P to ensure that connected transformations lead from A and B to almost all metabolites of length M or less. As before, this analysis bears on the structure of the virtual catalyzed reaction graph among organic molecules and does not yet address the flux of material along such pathways. Above a specific complexity, connected pathways of metabolic transformation crystallize (Table 8.1).

It is critical to note that the crystallization of connected metabolic transformations which emerges here does not yet rely on shape similarity of neighboring substrates. Instead, I have assumed the most difficult case, in which a peptide or RNA sequence which catalyzes one reaction has no increased probability of catalyzing a neighboring reaction which might involve a neighboring substrate. Thus also the nearby mutants of that peptide or RNA have no increased chance of catalyzing such a nearby reaction. Hence the shape similarity argument used to build up a linked set

TABLE 8.1   Critical Number of Hypothetical Organic Molecules Made Up of Linear Chains of Atoms A and B Required for Crystallization of a Connected Metabolism

$$\bar{P} \;=\; e^{-P(5000)(M-1)(1+2^{M+2})} \;=\; \frac{1}{e^8} \;<\; 0.001$$

P	M	2^{M+1}
10^{-4}	1.965	8
10^{-5}	3.81	28
10^{-6}	6.25	152
10^{-7}	8.98	1010
10^{-8}	11.85	7383
10^{-9}	14.83	58 251

Note: Number of potential catalysts present is 5000. Note that, if the number of potential catalysts increases above 5000, a smaller number of hypothetical organic molecules would suffice for the emergence of a connected metabolism. Thus in a two-dimensional space whose coordinates are the number of catalysts and number of organic molecules in the system, each value of P determines a critical curve separating regions with and without connected metabolisms.

of metabolic transformations derived from Horowitz is not necessary for the emergence of a connected metabolism. Rather, such linkage emerges as a threshold property of even random reaction graphs. Presumably, shape similarity will only help in the formation of such connected webs. That intuition, however, remains to be investigated carefully.

A Critical Curve in a Two-Dimensional Parameter Space of Metabolite Set Size and Catalyst Set Size Separates Connected and Nonconnected Metabolisms

A new property which arises in the problem of the emergence of a connected metabolism is that there are twin thresholds, or, more properly, a critical curve in a two-dimensional parameter space consisting of the size of the set of organic molecules and the size of the set of potential catalysts. For a *fixed* number of potential catalysts and a fixed probability of catalysis P, the complexity of organic molecules needed to achieve connectivity can be calculated. As P changes, so does the requisite number of organic molecules. Conversely, for a fixed number of organic molecules containing up to M atoms per molecule and a fixed P, as the number of candidate peptides or RNA sequences increases, a threshold is crossed above which a connected metabolism emerges (Table 8.1). Therefore, there is a two-dimensional space with one axis representing the organic molecules in the metabolic set and the other representing the peptides or RNA sequences in the set of potential catalysts. In that space, for each P, there is a curve dividing the space into two regions. Below the threshold, connected metabolisms are not expected; above it, they are expected. For fixed P, transition to a connected metabolism can take place parallel to either or both axes or at a number of oblique angles.

Thus we arrive at a new point of view. The emergence of a connected metabolism as a supracritical web requires a sufficient complexity of organic molecules and a sufficient complexity of potential catalysts. At that point, such a connected web is an inevitable emergent collective property of the chemical system. Fainthearted experiments with "clean" simple systems of a few components would never uncover these events.

The Reaction Graph Associated with Real Organic Molecules: Unknown but Knowable

Neither the number of organic molecules possible with a given set of atoms nor the number of reactions by which they might interconvert is known. In the present section, I give rough reasons to think that, as the number of kinds of atoms per molecule increases, the number of reactions linking organic molecules rises very much faster than the number of molecules present.

Hypothetical organic molecules which are simple linear strings of two or more kinds of atoms have the virtue that the number of molecules possible and number of reactions by which they interconvert are easy to calculate. The task is very much harder for real molecules. Counting all the possible organic molecules up to some number of atoms per molecule is an unsolved task. Even the more limited task of counting the number of possible molecules having a specific empirical formula—$C_{30}H_{51}N_7O_{15}$, say—is beyond current combinatorial techniques (Read 1976). Counting cyclic and heterocyclic compounds appears to be the hardest task (Dugundji, Gillespie, et al. 1976). Conversely, many acyclic compounds have been counted as a function of the number of carbon atoms per molecule (Read 1976). For example, vastly many acyclic molecules with 25 or fewer carbon atoms exist. The total for only structural isomers, not optical isomers, is beyond 10^{11}. Inclusion of monocyclic and heterocyclic compounds will increase this total enormously.

Consider the space of all organic molecules up to some number of atoms per molecule M. Consider the associated reaction graph, showing for each molecule which reactions it may enter into, subject to the constraint that all product molecules have $2M$ or fewer atoms. That is, as in our considerations in the previous chapter with respect to supracritical graph growth, we again allow reactions to form molecules larger than those in an initial set. We then ask whether, if a fixed fraction of the reactions were catalyzed, a catalyzed metabolic web would grow indefinitely larger. As before, these considerations ignore thermodynamics.

The reaction graph includes the analogue of ligation and cleavage reactions in polymers. In addition, all exchange reactions must be included. Familiar examples are transamination reactions and transfers of a methyl group from one hydrocarbon to another. Further, the graph will include reactions with two substrates and two products. If we limit the complexity of reactions to two simultaneous substrates and two products, we can think of an extremely large two-dimensional space each of whose axes lists all *pairs* of organic molecules, including degenerate pairs in which only one of the two is present. Then any reaction converting one or two organic molecules to one or two other organic molecules can be represented as an occupied cell in this two-dimensional space. The collection of occupied cells in this enormous space is the mapping of substrate(s) onto product(s) by all legitimate reactions, subject to the automatic constraint that the total number of atoms per product molecule is not more than $2M$.

Let us make the simple assumption that virtually any pair of organic molecules X and Y has the property that it can undergo at least one legitimate chemical reaction, producing one or two products via ligation, cleavage, or exchange. This assumption may sometimes fail, but, conversely, we would expect that most complex organic molecules, taken as pairs, would be subject to many more than one possible legitimate chemical conversion, since any molecule containing L atoms has at least $L - 1$ bonds which might be broken. Thereafter, its fragments might be joined to any of the fragments resulting from the cleavage of any of the at least $K - 1$ bonds in the partner substrate, which contains K atoms. The number of potential reactions is then roughly on the order of $L \times K \gg 1$.

Given our assumption, the number of possible legitimate reactions involving molecules containing up to M atoms per molecule increases at least as the square of the number of molecules in the space of molecules containing up to M atoms per molecule. This follows because we are considering pairs of substrates which might transform to pairs of products, and the number of pairs of substrates scales as the square of the number of molecules X up to length M: $X \times X = X^2$. Therefore, the ratio of the number of reactions to the number of molecules present is at least equal to the number of molecules present: $X^2/X = X$. Each molecule might undergo on the order of X reactions. But we also know that, as M increases, the number of possible molecules X explodes enormously rapidly. Thus as M increases, the ratio of number of reactions to number of molecules present also explodes very rapidly.

The reaction graph almost certainly will be nonisotropic because, in the set up to M atoms per molecule, there will be more ways to make small organic molecules than large ones. This is expected from the fact that the number of larger molecules from which any small molecule can be cleaved is larger than the number of smaller fragments from which it can be assembled. The case of linear molecules can be seen from Equation 7.3 and generalizes readily to acyclic branched molecules.

While the reaction graph among the extremely large number of organic molecules containing up to M atoms per molecule is unknown, these crude arguments strongly suggest that the ratio of reactions to molecules up to size M grows very rapidly and that the graph is nonisotopic, having more ways of forming small molecules from within the set than of forming large ones. It seems fully safe to conclude that supracritical growth of a metabolic catalyzed reaction graph can occur if a sufficient number of potential catalysts confront a sufficiently complex mixture of organic molecules.

We turn next to think through a bit more the character of a crystallized connected metabolism.

Crystallization of a Supracritical, Low-Specificity Connected Metabolism

It seems plausible that a protometabolism would crystallize in, integrate into, and support the reproduction of an autocatalytic polymer set. For a metabolism to form and nurture an autocatalytic system, it is necessary that there be formed a connected web of metabolism leading from some subset of readily available organic molecules to many others, some of use in the reproduction of small organic molecules and catalysts larger than the original catalysts. Thus we are led to the picture of a protometabolism with low-specificity protoenzymes, perhaps peptides or ribozymes, emerging spontaneously as a supracritical entity almost inevitably when a sufficiently complex set of organic molecules is brought together with a sufficiently complex set of potential catalysts. In fact, it seems likely that the ratio of number of reactions to number of organic molecules present explodes faster as the number of atoms per molecule increases than does the ratio of number of reactions to number of polymers present as the number of amino acids per peptide increases. Therefore, it should be *easier* to crystallize a connected metabolism than to crystallize an autocatalytic set of peptides or ribonucleotides. This implies, however, that a set of peptides or ribozymes complex enough to become reflexively autocatalytic would probably also have enough members to catalyze a connected metabolism among a reasonably complex set of organic molecules. Initially, both the metabolism and the autocatalytic polymer set would be supracritical. Each lives on an infinite virtual catalyzed reaction graph, and hence the distribution of organic molecules and polymers present in the system depends on boundary conditions. Since amino acids are readily made from simpler organic compounds even in the absence of catalysts, amino acid products of the protometabolism could couple the metabolism to the autocatalytic peptide set and the two could coevolve. A similar story with ribonucleotides arises with the caveat that synthesis of ribonucleotide monomers is harder than synthesis of amino acids.

The same considerations should apply to the coupling of exogenous energy sources into a protometabolism and to the coupling of exergonic and endergonic reactions such that the former help drive the latter. In turn, such couplings would help drive synthesis of large polymers in an autocatalytic polymer system. A coupled protometabolism would be a sprawling thing: imprecise, stochastic, and inefficient, with very many organic molecules coupled into the web of transformations. But many of the molecules—for example, those with ring structures and resonance— would be able to absorb photons and pass to an excited state. In turn, this excitation yields, as targets to couple with endergonic reactions in the supracritical web, the large number of exergonic reactions any such excited molecule might undergo. Alternatively, Fox (1988) and Lipmann (1941) have suggested coupling cyclic redox reactions via iron–sulfur compounds to the synthesis of pyrophosphate in an early metabolism. More generally, many means of coupling redox cycles into such a sprawling protometabolism might exist. In short, a supracritical protometabolism would be a large spider's web set to trap useful reactions. The connected *whole,* the entire coupled system, then must collectively coevolve. There is no mystery in the resulting holism. Torn into pieces, the system is dead.

As in the previous chapter, we must envision, and need to test mathematically and eventually experimentally, that, under selection, the autocatalytic system will select out peptides, polypeptides, or ribozymes of increasing specificity and higher maximal reaction velocity and will also collect inhibitors of catalysis, thereby trimming away useless metabolic branches to leave a core coupled metabolism which is self-consistent: All needed transformations have catalysts of ever-increasing specificity. A metabolism using a specific pathway to form a specific predetermined metabolite Z required for present life, as requested by Hoyle, has not evolved; rather, one of an enormous number of alternative coupled, consistent metabolisms has been focused on collectively by parallel simultaneous selection on the entire integrated, if initially imprecise, autocatalytic system. After the fact, once a subcritical specific metabolism and the requisite enzymes to catalyze it have been achieved by selective focusing, we stand amazed at its integration and complexity. We are thence led to calculate the probability that a set of polypeptide or RNA catalysts would jointly catalyze just that metabolism and led to Hoyle's conclusions. It is the wrong calculation, the solution to the wrong problem. The details of my argument almost certainly are incorrect, but surely the invitation to consider the hypothesis that life began, not simple but imprecise and complex and then *became* simpler by selective focusing to more specific catalysts, deserves serious attention.

Statistical Universals?

The distribution of number of kinds of molecules as a function of atoms per molecule, the connectivity features of metabolic webs, and other features of metabolism may prove to be biological universals. This possibility reflects the fact that these features must fundamentally be governed by the legitimate allowed reactions transforming organic molecules among themselves.

Among the most obvious potential universals which might recur, as remarked on at the start of this chapter, is the expected distribution of numbers of kinds of organic molecules as a function of the number of atoms per molecule. If, as expected, the reaction graph is nonisotropic, so that more reactions are available to form smaller molecles than larger ones, then we expect a unimodal distribution. Figure 8.2 shows the distribution expected for linear polymers as discussed in the previous chapter. Because the number of linear polymers increases exponentially with M but the underlying graph favors synthesis of smaller polymers, the unimodal curve for a finite autocatalytic system living on an infinite virtual reaction graph starts low, reaches a peak at polymers of intermediate complexity, and then drops off exponentially as polymer length increases. Interpreting the entities A and B to be hypothetical organic molecules rather than the polymers of Chapter 7 allows us to view Figure 8.2 as a crude guess at what the *envelope* of organic molecules should look like.

Figure 8.3 shows an interesting tabulation by Morowitz (1968) of the number of known organic molecules as a function of carbon atoms per molecule. The curve is strongly similar to Figure 8.2. Morowitz constructed this graph from chemical index

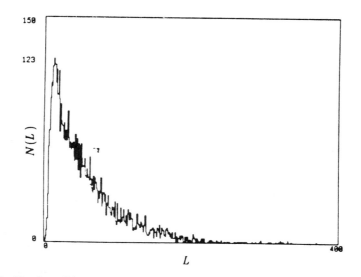

Figure 8.2 Number of kinds of organic molecules $N(M)$ as a function of number of atoms per molecule M. This graph is identical to Figure 7.8*b*; there, however, the entities under consideration were polymers of various lengths rather than organic molecules in general.

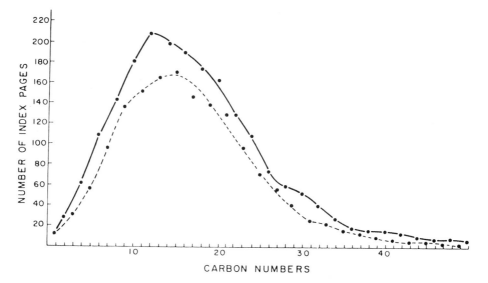

Figure 8.3 Distribution of number of known organic compounds (plotted as number of pages in chemical index volumes) as a function of number of carbon atoms per compound. (From Morowitz 1968)

volumes. It is open to obvious biases. In particular, the numbers of known larger organic molecules may fall off in part because of lack of interest on the part of chemists or because of difficulty of analysis. Further, we are not guaranteed that all these molecules are of direct organic origin. On the other hand, it is pleasing that the two curves are generally quite similar, and the mere fact that Morowitz has called such a distribution to our attention reminds us that there is some true distribution which his curve samples more or less well. We have never had a theory which seeks to explain such curves and now have at least the start of one. Seeing such a curve, it is hard not to wonder how universal the true curve is, and whether near variants would reemerge each time life evolves.

Might the statistical connectivity features of metabolic webs under the action of selection afford universals? Figure 8.1 shows part of intermediate metabolism. Any connected metabolism will exhibit cycles and branch points, with distributions of distances between branch points and distributions of numbers of molecules per cycle. We know from the theory of random isotropic graphs that such properties have robust averages if the ratio of edges to points is given. Thus it becomes a serious theoretical issue to study, as a function of the distribution of catalytic capacities in peptide or RNA space and of the underlying reaction graph of organic chemistry, whether transition from supracritical to subcritical metabolisms yields similarly typical average connectivity properties. Thereafter, selection may attempt to modify the connectivity structure of a metabolic web but be limited by the connections afforded by organic chemistry as well as by the capacity to evolve enzymes capable of mediating those reactions. That the task now looms as extremely hard does not prevent us from supposing that robust statistical properties will be found. If so, they too might be universals which would be expected to recur in almost any evolving metabolism. It would be astonishing if, on closer understanding of the evolution of coupled metabolisms, many other statistical features were not uncovered. All of this then cries out in support of the idea that we must come to understand our particular metabolism as a more, or a less, typical member of the ensemble of coupled metabolisms which might have evolved. The typical properties of that ensemble loom as potential ahistorical universals in biology.

A Final Thought

Autocatalysis by which more complex organic molecules emerge from simpler ones may have played a role in prebiotic chemical evolution. Small organic molecules can catalyze reactions among other organic molecules. Zinc, iron, other metallic ions, histidine, and coenzymes are familiar examples. Then we seem forced to consider the possibility that a sufficiently complex set of organic molecules may become reflexively autocatalytic. In the previous chapter, we explored the growth of the virtual catalyzed reaction graph from a maintained food set with up to L_c amino acids per polymer. We saw that, for ligation and cleavage reactions alone, the number of molecular species in the food set necessary to induce supracritical growth was only $\sqrt{\frac{1}{2}P}$. Were the ratio of reactions to molecules higher, the number of initially maintained molecular species needed for supracritical behavior would be even less. That ratio is almost certainly much higher for organic molecules than for polypeptides. Supracritical behavior of a system of small organic molecules would mean, among other things, that an initial small set of organic molecules might catalyze itself collectively into a complex system of many kinds of organic molecules of use in the later collective crystallization of protoorganisms and their metabolisms. Perhaps such processes occur even now in the organic oceans of Titan.

NEW EXPERIMENTS

What kinds of new experimental work might these ideas portend? Two come immediately to mind:

1. Measure how subcritical the metabolism of a free-living bacterium or other organism is.
2. Make practical use of these ideas in seeking sets of novel peptide or RNA enzymes which might catalyze coordinated sequences of chemical transformation.

In effect, the first type of experiment asks us to measure the probability that an arbitrary organic molecule can be used as "food" by an organism or, more generally, the probability that the molecule will be catalytically transformed into some other products if present in the cell. Restated, the 3000 to 4000 enzymes in a contemporary bacterium cover some fraction of catalytic task space. Suppose we could map all reactions, where a reaction site is bounded by a relevant size scale, onto catalytic task space. Then if we knew the fraction of task space covered by the 3000 to 4000 enzymes and the mapping from molecule substrates onto task space, we would know our answer. While the form of this question seems unfamiliar, it is really the same as asking of the immune system what fraction of arbitrary epitopes on a relevant size scale can be recognized. A fair fraction of task space must be covered for soil organisms to evolve so readily.

Selecting Sets of Novel Peptides, Ribozymes, or Abzymes Which Catalyze a Connected Sequence of Reactions

Among the implications to be derived from the present chapter is the idea that, if enough reactions are catalyzed, a connected sequence of catalyzed reactions should crystallize. This both is testable and has practical implications: Novel enzymes which catalyze a sequence of reactions can be selected.

I have suggested (1) that connected metabolisms emerge because the ratio of number of reactions among metabolites to the number of metabolites present increases as the number of atoms per metabolite increases and (2) that vast numbers of reactions constitute a target web to be catalyzed by potentially catalytic polymers. At some level of complexity, connected webs of catalyzed transformations crystallize. In such a system, very many partially alternative pathways lead from a sufficiently large set of building-block compounds to any target compound Z. In order that Z be synthesized by a connected catalyzed pathway, it is necessary not that any one prespecified pathway to Z be catalyzed but that at least one reasonably high-yield pathway among the many possible pathways be catalyzed.

By this argument, it might prove easier to obtain, from a sufficiently large set of precursors, a set of novel peptides, ribozymes, or *in vitro* cloned abzymes catalyzing an entire sequence of reactions leading to synthesis of an arbitrary target compound Z than to obtain a polymer catalyzing even one prespecified reaction involving Z. For example, suppose the probability of catalysis of a specific reaction is $P = 10^{-6}$. Then on the order of 1 000 000 potential catalysts must be tried to find success. If a prespecified pathway leading to a desired target compound Z has L steps, then on the order of $L \times 1\ 000\ 000$ potential catalysts must be tried if the search is carried out serially to build up the desired prespecified pathway to Z. Thus Ellington and Szostak's results (1990) suggest that the probability of finding a single-stranded RNA sequence which binds to an organic dye is on the order of 10^{-10}. Binding a dye is close to binding the transition state of a reaction. Thus the chance of finding an RNA sequence acting as a ribozyme to catalyze a specific reaction may be about 10^{-10}. A set of L such screenings would be required to find L catalysts for a specified sequence of L reactions.

Instead of building up a set of L enzymes to synthesize a target compound Z, search can be carried out in parallel. Many alternative pathways to Z exist from a properly chosen set of building blocks. What is logically necessary is (1) that a sufficiently large number of candidate catalytic peptides, ribozymes, or abzymes be tried *simultaneously,* by acting on a sufficiently large set of building-block precursors several steps from Z, and (2) that any small yield of Z be detectable. If the target compound Z can be detected, winnowing the initial mixture of candidate enzymes to the subset that yields the target compounds can identify the subset that catalyzes a pathway to the desired target.

SUMMARY

The present chapter has extended the ideas on the emergence of autocatalytic sets of catalytic polymers to the emergence of supracritical connected metabolic webs among organic molecules. The core ideas assert that life began complex, with an autocatalytic polymer system and a metabolic supracritical imprecise web grafted to it from the outset. Such a web would from the start be able to capture energy sources (photons hitting ring structures in organic molecules and inducing resonance, pyrophosphate, cyclic redox reactions, and other possibilities) coupling exergonic and endergonic reactions. Subsequent selective simplification by self-consistent focusing on more specific polymer catalysts covering smaller balls in catalytic task space, plus the advent of inhibition of catalysis, would lead to a focused subcritical metabolism.

This hypothesis is, in a sense, a generalization of the classical idea that metabolisms evolved by gene duplication and differentiation of enzymes to recognize substrates similar to those initially acted upon. Neighborhood relationships among the substrates then engendered linked catalyzed transformations. Our extension of these classical ideas recognizes that, even in randomly catalyzed subgraphs of the reaction graph of organic chemistry, connected metabolic webs will arise as a kind of phase transition when a sufficient fraction of the reactions are catalyzed. Thus the classical answer to the emergence of connected transformations is not logically necessary, although still likely to help in the emergence of connected links in a protometabolism.

It is very much worth noting that selection may often be able to change the kinds of entities on which it acts, yielding more or less correlated fitness landscapes. Its capacity to do so is very limited with respect to metabolism, however, for the connectivity features of metabolism are unmodifiable consequences of organic chemistry. Therefore, more than in other areas, we may well expect that repeated evolution of metabolism would have a good chance of revealing statistical similarities, or ahistorical universals.

The view offered in the last two chapters stands in the sharpest contrast to the nude gene hypothesis. The former asserts that, from the outset, life was holistic and critically complex. Life self-organized with a coherent phenotype, on this view, and evolved to some extent without needing a genome. However, difficulties in maintaining autocatalytic closure would have seriously limited the capacity for evolution.

Coevolution of such a self-reproducing metabolism with RNA and DNA polymers and the evolution of a genetic code were presumably needed to open up wide evolutionary possibilities. We turn in the next chapter to consider models of the origin of life based strictly on RNA or DNA, of the evolution of a genetic code, and of the coevolution of self-reproducing autocatalytic metabolism, replicating RNA, and a code.

CHAPTER 9

Hypercycles and Coding

This chapter considers in more detail the opportunities and difficulties in envisioning life emerging from nude replicating RNA molecules. The possibilities discussed in the previous chapters for the spontaneous formation of autocatalytic sets of peptide or RNA catalysts and of a connected metabolism, even if demonstrable, are no reason to ignore the possibility that template-replicating DNA or RNA was the pre-eminent precursor to life. We recall briefly from Chapter 7 the chemical obstacles to this point of view:

1. Prebiotic synthesis of the components of a nucleotide is rather difficult, and their assembly is yet more difficult.

2. Attempts to use single-stranded DNA or RNA as a surface catalyst to synthesize the complementary strand very often run into the problem that the proper 3′–5′ bond between adjacent nucleotides is not formed; rather, the 2′–5′ bond is formed. The latter is inconsistent with formation of a stable double-stranded helix.

3. There is a marked tendency for some of the nucleotides to form hairpins, looping back on themselves as the second strand.

4. Single-stranded RNA composed of G and C is preferentially C rich can use activated monomers to form a second strand which then melts off, but this second strand is preferentially G rich and not suitable as a template (Joyce 1987).

While these obstacles may appear formidable, they cannot constitute proof that nude single-stranded RNA or DNA cannot have acted with its complement as a template in a replication cycle or, if not RNA or DNA, then some similar but simpler polymer, as recently suggested by Orgel (1987) and Joyce (1987).

THE LOGIC OF HYPERCYCLES

The most systematic efforts to consider the problems which arise when starting with nude replicating RNA or DNA sequences have been carried out by Eigen and Schuster and their colleagues over a number of years (Eigen 1971; Eigen and Schuster

1977, 1978a, 1978b, 1979). Their model is driven by an internal logic worth following. In outline, Eigen and Schuster argue that complex genetic information cannot be built up in single RNA replicating strand pairs; instead, they argue, a cyclic coupling among such template pairs, creating a "hypercycle," is required to achieve the evolution of complex genetic information.

Selection of a Fittest-Mutant Spectrum, or Quasi-species

Eigen and Schuster begin by supposing that arbitrary complementary RNA strand pairs can serve as templates to catalyze each other's formation from free nucleotides by familiar base pairing. Then, as described in Chapter 3, these authors analyze the behavior of a population of template-replicating RNA molecules, taking account of the ideas that each sequence, via its complementary sequence, replicates at its own specific rate and that imprecision in replication substitutes nucleotides at each position with some frequency. Thus each sequence can give rise to a spectrum of mutants altered in one, two, . . . positions. Eigen and Schuster suppose chemostat conditions of a constant influx of nucleotides and energy sources and an overflow rate that dilutes each sequence proportional to its concentration. Thus the system falls to a steady state with a fixed number of nucleotides, either as monomers or bound into polymers, in the chemostat. They then calculate the relative rate of growth of each sequence in the chemostat. For each sequence, this relative rate reflects absolute rate of replication, rate of dilution, and mathematical terms for formation of the sequence by mutation *from* similar sequences minus loss *to* such a spectrum of similar neighbors as a result of mutation of the sequence being monitored. The relative rate of growth of each sequence is then normalized over all sequences in the chemostat. Consequently, only those sequences whose net growth rate is greater than the average growth rate increase in relative abundance. Over a period of time, the single fastest growing sequence—plus an inevitable mutant cloud it drags along with it, the "quasi-species"—comes to predominate in the chemostat. This is Darwinian selection among replicating nude genes.

The Error Catastrophe Again

The next critical logical step is to show that, as the length of a nucleotide sequence increases, selection becomes too weak to pull the population of replicating sequences to a narrow mutant cloud surrounding the single best replicating sequence. As described in Chapter 3, Eigen and Schuster here extend a classical result from population genetics in a new context. Consider an additive model in which the fitness—or the replication rate—of a sequence displaced from the best sequence by K mutants in N positions is proportional to the number of nonmutant nucleotides $(N - K)/N$. Since the mutation rate per nucleotide is fixed but the proportional contribution to fitness declines as polymer length N increases, a complexity "error" catastrophe is reached at a critical value of N, N_c. Beyond this length polymer, N_c, the mutant spectrum associated with the best sequence spreads out widely and most members of the population have many mutants compared with the "canonical" best sequence.

Eigen and Schuster generalize this idea. As described in Chapter 3, the error catastrophe implies that, with a given accuracy of replication per nucleotide and a given fitness superiority of the currently fittest "master sequence" relative to the mutant spectrum around it, there is a maximum sequence length which selection can maintain. Longer sequences degrade stored information. This maximum length is roughly the reciprocal of the error rate per nucleotide. For plausible guesses at possible pre-biotic accuracy, this length maximum would limit sequences to 50 to 100 nucleotides, far from adequate to code for much of a concerted metabolism.

Need for Hypercyclic Organization

This central result leads Eigen and Schuster to their main suggestion: In order to build up systems of replicating RNA sequences which might surpass this error catastrophe, when any one template pair is limited by the catastrophe, conceive of *linking* a set of different template pairs around a closed cycle such that each pair aids in the formation of the next pair. Then the mutually cyclic connections among template pairs, which are themselves already autocatalytic, constitute a higher order *hypercyclic* coupling of such autocatalytic units (Figure 9.1). The central idea is that such mutual coupling allows a complex system of autocatalytic RNA pairs to *coevolve.* Each pair is limited by the error catastrophe, but the *hypercycle as a whole is not.*

Given the logical structure of their statement of the problem, with replicating template pairs and the error catastrophe, surpassing its limits to accumulate complex genetic information appears to require hypercyclic organization. It then becomes reasonable to analyze (1) whether such hypercycles can behave with sufficient dynamic stability to accumulate such information and (2) how such stability depends on the details of the couplings among template pairs in the hypercycle. The basic result rests on the fact that, in such an hypercycle, the rate of concentration increase of any sequence depends on the highest power to which its concentration is raised in the differential equation that describes its concentration increase as a function of its concentration. In turn, this growth rate depends on how richly interconnected the hypercycle is. This dependence follows because the number of ways each template pair contributes to its own concentration increase is given by the number of closed loops that go from that pair through the other pairs in the hypercycle and back to that pair. For the concentrations of all the sequences in the hypercycle to increase stably without the concentration of one pair (or larger subsystem) increasing faster than the whole, thus ripping it apart, the hypercycle must be "homogeneous" in the sense that each component must be equally richly connected. Note that the capacity for a sub-

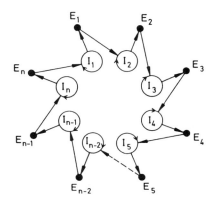

Figure 9.1   Instead of competing, N pairs of replicating RNA sequences $I_1, I_2 \ldots I_N$ cooperate in a hypercycle coupling. Each information carrier I_i codes for a primitive catalyst $E_i,$ which aids replication of the next sequence, I_{i+1}. (From Eigen and Schuster 1979)

system of one or a few template pairs to outgrow the rest of the hypercycle follows from the fact that each pair is, in its own right, an autocatalytic cycle of two complementary strands able to replicate by itself.

Minimum Realistic Hypercycle and Compartmentation

Any such theory must make a choice about the mechanism which couples one strand pair to the next one around the cycle, by which the former helps catalyze the replication of the latter. Eigen and Schuster suppose that each pair makes a product which serves as a specific replicase for the next pair. Further, they assume that the replicase is a protein or peptide. Thus they require that the hypercycle be able to specify at least crudely the synthesis of peptides by each pair to help replicate the next pair. In turn, then, they must suppose that these peptides somehow carry out a coding function, specifying a minimally accurate translation from nucleotide sequences into peptide replicases.

Finally, Eigen and Schuster suppose that compartmentation into protocells might have occurred later in the process.

Some Lessons

The Eigen and Schuster effort is a serious and sustained one. It honestly states and confronts many of the difficulties with a DNA-or-RNA-first view, albeit not all of them. Thus the requirement that all members of the autocatalytic hypercycle have *exactly the same connectivity* so that each proliferates at the same rate is a severe constraint. Granting the assumptions, however, a first question is this: Does the theory work? Niesert, Harnasch, and Bresch (1981) found three stability problems with the model. First, if a single RNA sequence, by mutating, replicates itself well but fails to catalyze the next member of the cycle, the replicating sequence can outgrow the system and will comandeer all mononucleotide resources. Second, a short-circuit catastrophe can occur when one sequence catalyzes a more distant sequence around the loop. The shortened loop may then grow faster than the overall hypercycle, and as a result the hypercycle will contract to a less complex form. Niesert, Harnasch, and Bresch found that these problems increase as the concentrations of molecular constituents increase. The third stability problem these authors found is that the converse catastrophe occurs at low concentrations: Some critical sequence, due to a fluctuation, drops to zero concentration and the hypercycle collapses. These problems suggest that hypercycles, conceived to surpass the limits due to the error catastrophe, may not behave with sufficient stability to surpass it by far.

These three problems may not vitiate the general model. For example, Eigen and Schuster have considered only positive catalytic couplings around the hypercycle. Possibly a mixture of positive and inhibitory couplings would render the coupled system more homeostatically stable to short-circuiting and to selfish sequences and would offer redundancy with stability such that the concentration of any one sequence could drop to zero and the system would still survive long enough to find another sequence to take its place. Perhaps the most important difficulty of the hypercycle model as initially stated is the implausibility of the requisite minimal realistic hypercycle. From *within* the framework of the theory, it is very hard to see how such mutual coupling via specific replicases might have emerged coordinately. For all its rigor, the model seems caught in the familiar chicken-and-egg problem. To be extricated, from within the framework of the theory, one possibility is that the coupling factors between successive sequences need to be something besides peptides—for instance, the RNA sequences might serve as specific replicases. A second possibility is that essentially random peptides having scant coded specificity, each of which is nevertheless "created" by one template pair, can serve as nonspecific replicases acting on a spectrum of "next" sequences to form a coupled hypercyclic system which, despite the sloppiness of coupling, behaves stably as a coevolving whole. This latter condition requires equal coupling for all pairs in order to achieve equal powers in the concentration equations. Even then, evolution of a coordinate metabolism does not follow easily, nor does the theory explain the minimum complexity observed in free-living organisms, beyond historical accident and selective just-so stories.

A new approach to the emergence of hypercyclic organization has been taken by Rasmussen (1989), an approach that in spirit is similar to the autocatalytic polymer set described in Chapter 7. Rasmussen considers template-replicating pairs of RNA molecules. Each such polymer pair may have the capacity to "catalyze" the replication of some other pair. Rasmussen pictures such catalysis as a randomly chosen arrow from the first pair of strands to the second. Then, in a complex system of such replicating pairs, the emergence of a kind of hypercyclic coupling is a question of the expected structure of the random directed graph among the replicating pairs. If the probability of catalyzing replication is high enough, the percolation threshold is passed and connected webs of RNA pairs are formed in which each pair is helped to replicate by others in the web.

Two problems arise in Rasmussen's development. First, the directed graph structure obtained seems typically to have some components receiving more catalytic inputs than others. Thus, as Eigen and Schuster have pointed out, such sequences will replicate faster than other components in the hypercycle and tear it apart. Second, in Rasmussen's theory, each replicating strand pair is a single target whose replication is to be catalyzed by another polymer pair. Thus as the number of kinds of RNA strand pairs increases, the number of target reactions to be catalyzed increases exactly in proportion to the number of polymer species which might catalyze those reactions. In contrast, in the autocatalytic polymer theory developed in Chapter 7, the ratio of reactions to polymers increases as polymer length and density increase; thus the number of target reactions per potential catalyst increases far more rapidly in the autocatalytic polymer model. Therefore, emergence of catalytic closure is harder in Rasmussen's model than in the autocatalytic model. Note that collective reproduction in an autocatalytic system should not suffer the stability failures found by Niesert, Harnasch, and Bresch (1981) because subcomponents of the system are not themselves self-replicating.

Symbiosis of Autocatalytic and Hypercyclic Systems?

As stressed by Dyson (1985), any protein-first theory offers prospect of a dual origin of life in which invasion by parasitic replicating RNA and later symbiosis might occur. Note again that, if single-stranded RNA can act as a catalyst for *ligation* reactions, and if we wish to conceive of an autocatalyic set of RNA molecules based on such catalysis rather than on template replication, then a set of such RNA molecules is an autocatalytic set in the same universality class as an autocatalytic peptide set. Indeed, mixed polymer systems which are autocatalytic are not unthinkable.

In Chapter 7, I stressed that evolution of a protein or ribozyme polymerase might well occur in an autocatalytic peptide–RNA system. For the moment, however, let us assume, as the majority of workers do, that arbitrary RNA sequences can template-replicate. In that circumstance, how might the existence of autocatalytic polymer systems help create stable hypercyclic coupling?

Consider an obvious idea. If a peptide binds to a hairpin loop formed by an RNA molecule, that peptide may help the double-stranded RNA melt to single strands. Peptides which bind to RNA hairpins are known in contemporary organisms. Conversely, if an RNA binds two peptides at two loops, it confines their motions and thus helps catalyze their ligation. Then any sequence specificity in both directions allows an autocatalytic peptide system to coevolve with a replicating RNA system, such that peptides serve as specific replicases for RNA seqeunces, while the latter serve as more or less specific catalysts for the peptides. More precisely, if two peptides are confined by binding to one RNA molecule and then a third peptide catalyzes their ligation, this process converts a bimolecular collision of the substrates in a volume whose frequency varies as the products of the concentrations to an intramolecular rearrangement event whose frequency is independent of the substrate concentrations. Such a conversion always speeds up the reaction.

This is not exactly the Eigen–Schuster hypercycle but is perhaps a quite felicitous marriage nevertheless. It grafts a going metabolic concern to a coevolving set of template-replicating RNA polymers. The details do not have to be right for the general idea to be right. Further, it is at least plausible that such RNA loop binding by peptides is a step in the direction of coding. The closer the loops are to one another, the more closely confined the peptide ends and hence the faster the ligation. In addition, the closer the amino and carboxy ends of the two peptides are to the loops, the more confined their relative motion and hence the faster the ligation. Thus one can envision selective forces for RNA molecules which (1) have hairpin loops that are close to one another and (2) bind to peptides near their carboxy and amino ends, so as to juxtapose those ends closely.

A critical step in the evolution of the genetic code must have been the onset of *colinearity*. Can we envision a plausible sequence of steps leading to at least rough colinearity of protein and RNA molecules from a start with pairs of RNA loops acting as surface catalysts? The first step is to free replication of RNA (or DNA) from specific binding to loops by peptides which are to be ligated. That is, we shall need to suppose the discovery of a general replicase—for example, a ribozyme polymerase—by the coevolving protein RNA system. Next suppose that RNA molecules with loops remain present in the system and remain able to bind peptides. It is easy to suppose that such loops, bearing attached amino acids or peptides, are free-floating and have a crude anticodon allowing attachment to complementary single-stranded RNA. Even nonspecific single-stranded RNA could, by binding two RNA loops to two peptides, serve as a surface catalyst to abet the ligation of the two peptides. Rearranged sequence specificity in the complementary RNA would bring the loops closer together, helping the ligation of each specific peptide pair so approximated.

This is not yet coding. In fact, part of the important point about such a coevolving system is that coding is not needed for this stage to be reached. Transition toward colinearity requires that the peptides which become attached to any loop of RNA (which now become primitive transfer RNAs) become shorter; for such binding between loops and short chains to happen, enzymes which can catalyze the binding of such very short peptides must emerge because short peptides would presumably lose specificity for recognizing and binding to loops. That is, primitive synthetases for charging tRNA loops with very short peptides—and then eventually single amino acids—must emerge. Finally, the proper complementary RNA sequences must come to be adjacent on the complementary crude messenger RNA. These steps toward colinearity would be the precursors of the onset of true coding. (Bedian's elegant model for the onset of coding, which begins with colinearity, is discussed below.)

A slightly modified version of part of this picture has been suggested by Maizels and Weiner (1987). They suppose that the first tRNAs as well as tRNA synthetases (which add amino acids to the tRNA) arose prior to the evolution of mRNA. Rather than using mRNA to locate charged tRNA-like molecules adjacent to one another and so aid peptide synthesis, these authors suppose that charged tRNAs carried amino acids to *peptide-specific ribosomes,* each with its own specific internal template. Base pairing between a specific internal template in such a peptide-specific ribosome and charged tRNA molecules would allow specific peptides to be synthesized. Further, these authors suggest that the persistent failure to isolate a ribosomal protein with peptidyltransferase activity catalyzing peptide bond formation is consistent with the hypothesis that contemporary ribosomes accelerate peptide bond formation primarily by aligning the amino-acid-charged tRNAs. They picture a primitive ribosome with two binding sites: The ribosome synthesizes a peptide longer than a dipeptide by forming a peptide bond, and then releases one tRNA and binds a third charged tRNA before the dipeptide disassociates; this reaction is followed by a second round of peptide bond formation. Over successive rounds, amino acids would form peptide bonds with the growing oligopeptide attached to the remaining tRNA, itself bound to the primitive ribosome. Binding of the amino acid to the tRNA would activate the amino acid for peptide bond formation. Eventual spontaneous hydrolysis of the peptidyl–tRNA bond would release the free peptide.

Maizels and Weiner suggest that each such peptide-specific ribosome would synthesize an amino acid homopolymer which would be functionally useful. I would raise the possibility instead that primitive tRNA-like molecules carried amino acids or small homo- or heteropolymer peptides to specific RNA sequences that, like peptide-specific ribosomes, helped form peptide bonds among the peptides of an autocatalytic set of polymers. In this view, functional oligopeptides already existed in the autocatalytic set of polymers and coevolved with crude tRNA and crude ribosomal-like RNA, each initially helping the synthesis of the other toward a later day when mRNA and coding take over the task of specifying useful proteins.

The origin of life field is littered with idle speculation. It is therefore important to realize that the capacity to utilize cloning methods to generate very large numbers of novel peptides, as well as novel DNA and RNA sequences, means that the kinds of speculation I have allowed myself can be subject to real investigation. Thus if a peptide binds an RNA loop, can that peptide help melt the double-stranded form by affinity to the transiently forming hairpin formed by one strand? How hard is it to find a peptide which will bind a given RNA loop or to find an RNA loop which will be bound by a specific peptide? With what specificities? One can test the hypothesis that binding of two peptides by two hairpin loops on one RNA molecule favors ligation of the peptides. For example, such experiments might be carried out in variants of the plastein reaction in or out of liposomes to help drive synthesis reactions. Can one show that proper distance between hairpin loops and binding of the peptides near the proper carboxy and amino termini abet this catalysis? Can one find novel peptides which carry out a charging function, binding specific small peptides to RNA loops? We may never recover the actual pathway of the genetic code's evolution, but we can hope to test experimentally the feasibility of plausible routes to that coding.

The Anderson–Stein Spin-Glass Model

Another serious conceptual effort to think through systematically how a replicating system based on RNA might have developed is due to Anderson (1983), Stein and Anderson (1984), and Rokhsar, Anderson, and Stein (1986) and grows out of an attempt to avoid the limitation associated with evolution of a single pseudospecies mutant spectrum surrounding a single best replicating RNA sequence. As described in Chapter 3, spin-glasses are models of disordered magnetic materials (Edwards and Anderson 1975; Sherrington and Kirkpatrick 1975; Anderson 1985).

Anderson (1983) and his colleagues (Stein and Anderson 1984; Rokhsar, Anderson, and Stein 1986) draw on this universality class to create a picture of an RNA polymer system with very many alternative locally optimal replicating sequences. The basic ideas are simple. They imagine RNA replicating by templating but include overlapping fragments such that single-stranded tails overhanging at one or both ends of each double-stranded form allow many different double-stranded pieces to aid one another's replication. The serious work of the theory invokes a stability of each sequence which is a spin-glass-like random function of the detailed order of nucleotides within the sequence. One sequence will be highly stable; another—having only a single nucleotide changed—might be very much less stable and hence destroyed rapidly by the environment. In brief, the spin-glass-like theory induces a complex stability landscape over the space of sequences. Selection then has an exponential number of local fitness optima and thus can build up a very large diversity of locally maximally adapted sequences.

In comparison with the Eigen–Schuster model, the Anderson–Stein model has the virtue of proposing a particularly complex fitness landscape and beginning to explore its specific consequences for the spectrum of sequences which might evolve. The Anderson–Stein model has two important limitations however. The first is that the model is still nude genes and requires a means of coupling these genes to a connected metabolism. This coupling mechanism could conceivably grow out of catalytic activity of single-stranded RNA but should more plausibly link to generation of peptides with catalytic activity. Here the difficulty, as for Eigen and Schuster, is *how to achieve sufficiently reliable correlated peptide synthesis by the replicating RNA so that a sufficiently reliable set of active peptides capable of achieving a metabolism can coevolve with the RNA system.* The second limitation is that coding is neither present nor obvious.

The Anderson–Stein model could be married to autocatalytic peptide and/or ribozyme systems as readily as could the Eigen–Schuster model or the Rasmussen model. Since none of the models described in these chapters is correct in detail, there is no reason to be defensive about any or to be shy about borrowing the best suggestions and implications of each.

BEDIAN'S PARADIGM FOR THE ONSET OF CODING

The reader can hardly have missed the point that the onset of coding is a major problem. The literature on the subject is substantial (for example, Crick 1968; Woese 1972; Kuhn and Kuhn 1978) and, as usual in the origin of life field, exceeds the facts. This does not mean that exciting and ultimately testable ideas are not available. I find the conceptual approach taken by Bedian (1982) particularly appealing.

First the framework. Translation, as noted, is mediated by the aminoacyl synthetases. These enzymes bind the proper activated amino acids to the proper tRNA molecules, which in turn bind via their anticodons to the triplet codons in mRNA. The first formulation of the idea of an error catastrophe was based on the notion that, if an error were made in the translation of an aminoacyl synthetase, that error would charge the corresponding tRNA molecules improperly. In turn, these errors would cause further errors in translation, including errors in the synthesis of other aminoacyl synthetases. Errors in the latter would compound the translation errors, leading to a runaway translation error catastrophe (Orgel 1963). Subsequent analysis showed that this qualitative idea is wrong. Instead, there is a zone of tolerance for moderate errors of translation, which tend to die out rather than amplify. Outside that zone, or threshold fraction of errors, the translation system would indeed collapse to disorder. Conversely, the onset of coding would appear to require crossing the threshold from below. This is just what Bedian supplies.

The second part of the framework needed to understand Bedian's ideas is to realize that two strong competing ideas about the origin of the code have argued either in favor of biochemical necessity—that RNA or DNA triplets physically favor specific affinities for specific amino acids—or in favor of a frozen accident (Crick 1968). The former has declined in acceptance over the years. In part this reflects the fact that the amino acid binds tRNA at a site far from the anticodon site. Further strong affinity preferences have not been found experimentally.

The point of departure taken by Bedian (1982) is to ask whether a system of RNA codons and peptides with the capacity to catalyze more or less random binding of amino acids to tRNA-like molecules can coevolve to a coded state. He begins by assuming template-replicating DNA or RNA. He makes the further very strong assumption that, somehow, synthesis of peptides from RNA is already colinear. In fact, he assumes triplet codons, but that is not critical to his central question. Rather, his problem is this: Assume that a set of RNA molecules exists and generates, by colinear polymerization, a more or less random set of peptides. Assume that some of these peptides are capable of acting as crude synthetases, binding amino acids with a *spectrum of preferences* to some or all of the kinds of tRNA molecules. If this binding is completely random, then the system is not coding. For the system to be coding, it must be the case that the peptides specified by the RNA coding sequences can (1) "read" the coding sequences and (2) act as specific synthetases to produce the same peptides. The relationships must be mutually consistent.

Consider for simplicity two amino acid types and two triplet codons. Two coded arrangements are possible. Code 1 assigns amino acid 1 to codon 1 and amino acid 2 to codon 2; code 2 assigns amino acid 1 to codon 2 and amino acid 2 to codon 1. If all four assignments are made equiprobably, no code exists, and random peptide copolymers are produced from any mRNA.

Bedian proceeds by writing down two matrices. An S matrix gives the preferences of each peptide to carry out each of the four assignments; an M matrix shows, for each mRNA, the number of times each code assignment must be used to specify the peptide carrying out one of the present assignments. From these matrices and the assumption that a critical number of amino acids sites are needed for each high-specificity assignment, he derives coupled equations for the joint efficiency with which the set of peptides and their crudely coding mRNA synthesize peptides conforming to code 1 or to code 2 or to random mixtures of the two codes in a noncoded way. By equating such efficiency with the idea of a selective advantage, Bedian is able to show that such a system will often self-organize to one of the coded states. That is, synthetase variants which arise due to mutations and happen to carry out translation more efficiently will be sequentially selected. Thus the system hill climbs. Whether the system evolves to one of the coded states or remains stably in a noncoding state depends upon how fully random the initial preferences of the synthetases are and upon the initial distribution of codons in the quasi-random mRNA species. The coding states and the noncoding state all drain sizable basins of attraction. This suggests that the chances of hitting a coded state are reasonably high from a randomly chosen initial state. While a single evolutionary attempt might fail, similar experiments from different initial states might be likely to have at least one ultimate success.

The spirit of Bedian's central result is that, given a population of random peptides synthesized by fairly random and noncoded colinear polymerization via something like tRNA molecules, and given that these peptides can charge amino acids to the tRNA molecules, selective coevolution of the peptides and underlying coding mRNA to a consistent coded state can occur. How restrictive the conditions are, in terms of the numbers of amino acids for which the code can evolve and so forth, remains to be analyzed. It should be stressed that, in starting with colinear synthesis and tRNA-like molecules, Bedian has already assumed a lot. Nevertheless, his conceptual framework is sensible and encouraging. It does not obviate the possibilities of physical affinity biases or of frozen accidents but attempts to describe conditions sufficient for coevolution of peptides and RNA to a code.

I find Bedian's analysis attractive for a direct reason: He has gone to the center of the conceptual problem in asking how a coded state could evolve from a crude set of peptides carrying out random charging functions. Note that, with the capacity to generate very large numbers of novel RNA and peptide molecules now in hand, it is not beyond imagination to conceive of testing Bedian's ideas directly.

SUMMARY

In the present chapter, I have sketched briefly several alternative views of the origin of life based on template-replicating RNA or DNA molecules. We have followed Eigen and Schuster through the necessity for some kind of hypercyclic coupling forming a mutually catalytic community of replicating RNA polymers, given that the RNA polymers can replicate and individually suffer the error catastrophe. Anderson and Stein follow a slightly different avenue, seeking a natural account of fitness landscapes with very many alternative optima. Their spin-glass-like model of polymer stability achieves this and is closely analogous to the NK family of rugged landscapes discussed in Chapters 2 and 3.

Hypercyclically coupled systems of replicating RNA polymers may have emerged more easily than Eigen and Schuster initially supposed, given the graph theoretical arguments of Rasmussen. As in the autocatalytic polymer theory, Rasmussen suggests that, if enough RNA polymers catalyze one another's replication, connected hypercyclic webs will emerge. However, the stability of such coevolving systems appears to be delicate. Those with higher numbers of catalytic inputs replicate faster and therefore tear the coupled system apart. Such instability does not arise in an autocatalytic system of peptide or ribozyme polymers, since no single polymer individually replicates; all are jointly and collectively necessary to self-reproduction. Thus we explored the plausible hypothesis that autocatalytic polymer systems, peptide and/or ribozymes, might coevolve with template-replicating RNA systems. Such systems appear capable of some potential steps toward a coded state. Finally, we discussed Bedian's elegant attack on the core problem of how a system of peptides with random synthetase activities and partially random mRNA might evolve to a coded state.

A general feature of the topics and approaches in the three chapters on the origin of life is the emergence of ordered properties in sufficiently complex systems of interacting polymers and organic molecules. This is a very new subject. While the details I have discussed may prove wrong, it is apparent that unexplored potential exists for emergent order in complex biochemical systems as a result of (1) crystallization of connected pathways of catalyzed reactions, (2) dynamic stability in complex reaction systems, and (3) selective modification of such systems by increase in the specificity and efficacy of catalysis. The theories presented are merely the beginnings of a new area of thought and investigation in biology, chemistry, and physics—perhaps even in economics and other areas of social sciences. I turn to such expanded topics in the next chapter.

The spirit of all the ideas discussed in this and the previous two chapters is a kind of unrepentant holism and a sense of synthetic biology rather than the familiar reductionistic analytic mold. The task here is to understand in detail not just *what now is* but *the ways what now is might plausibly be expected to have arisen.* We want such theories to be testable, we want them to be explanatory. If we were ever to synthesize an evolving autocatalytic peptide-cum-ribozyme-cum-metabolic-hypercycle system, synthetic biology would surely have much to say to complement analytic biology. We shall have to come to understand what "explanatory" might mean in this new context.

CHAPTER 10

Random Grammars: Models of Functional Integration and Transformation

In this chapter I extend the basic model of autocatalytic polymer systems in an effort to develop a new class of models for functional integration, transformation, and coevolution in biological systems. The ideas appear far more general, however, and may well extend to social systems normally considered in fields such as economics, anthropology, and perhaps even history. The topics covered in this chapter are first efforts. Only further work will reveal whether the approaches I discuss will fulfill their apparent promise.

The generalization of autocatalytic polymer systems is based on the realization that polymers can be regarded as strings of symbols. For example, a protein is a string of 20 kinds of amino acids. The catalytic and other chemical rules governing the ways enzymes catalyze ligation and cleavage among proteins can be thought of as a kind of *grammar.* In this grammar, strings of symbols act on strings of symbols to yield strings of symbols. An autocatalytic set is a type of *collective identity operator* in this space of symbol strings, an operator which produces at least itself. Given this basic idea, we may generalize to *random grammars.* Each such grammar is a kind of hypothetical set of chemical laws. Each will yield a world of symbol strings and their joint transformations. Such symbol strings can be thought of as polymers in a prebiotic soup, molecules in an organism, goods and services in an economy, and perhaps even conceptual elements in a cognitive web or mythic elements in a cultural system. As we see below, the kinds of compositional sets of symbol strings that emerge in systems in which strings act on strings to produce strings become models of functional integration, transformation, and coevolution.

Analysis of random grammars is one approach to the general problem which arises in investigating the capacities of complex systems to adapt. That problem lies in understanding both the functional and the dynamical order which integrates these systems. *Escherichia coli* "knows" its world. A wealth of molecular signals pass between a bacterium and its environment. The signals entering the bacterium are harnessed to its metabolism and internal transformations such that, typically, the cell maintains itself, replicates, and passes its organized processes forward into history. Similarly, a colony of *E. coli* integrates its behavior. The organisms of a stable ecosystem form a functional whole. The niches occupied by each organism jointly add up to a meshwork in which all fundamental requirements for joint persistence are met. Similar features are found in an economic system. The set of goods and services making up an economy form a linked meshwork of transformations. The economic niches occupied by each set allow the producers of that set to earn a living and jointly add up to a web in which all mutually defined requirements are jointly met. Both biological and technological evolution consist in the invention of slightly or profoundly novel organisms, goods, or services which integrate into the ecological or economic mesh and thereby transform it. Yet at almost all stages, the web retains a functional coherence. Furthermore, the very structure and connections among the entities set the stage for web transformation. In an ecosystem or economic system, the very interactions and couplings among the organisms or among the goods and services create the conditions and niches into which new organisms, goods, or services can integrate. The web governs its own possibilities of transformation (Kauffman 1988).

Similar functional integration of roles, obligations, and institutions applies at societal levels. The revolution that occurred in Eastern Europe and the former Soviet Union in these *anni mirabili* was accompanied by a sense that the Soviet system was an integrated whole with the property that if one or a few features were removed or altered, the entire system must transform to something quite different—and whole. In June 1989 the Communist leaders in China tragically saw fit to kill their students in Tienamin Square. Why those leaders did so is clear: The students were demonstrating for increased democracy, and the government feared the consequences would transform Chinese Communism. In short, the puzzle is to understand not what China's leaders did, but what they *knew*. In a real and deep sense, the Chinese government knew that, were a few features of their system altered, the entire edifice stood in danger of dramatic transformation. What, indeed, did they know?

In the biological and social sciences, we badly lack a body of theory, indeed even a means of addressing these issues: What is a functional whole and how does it transform when its components are altered? As remarked, a new approach to answering this question is based on the use of random grammars. The objects of the theory are strings of symbols which may stand for chemicals, goods and services, or roles in a cultural setting. Symbol strings act on one another, according to the grammar, to yield the same or other symbol strings. Thus the grammar specifies indirectly the functional connections among the symbol strings. It defines which sets of strings, acting on other sets of strings, produce which sets of output strings. These mappings are the functional couplings among molecular species in a proto-organism, among a population of organisms in an ecosystem, and among the linked production technologies in an economy. Diverse grammars model diverse possible chemistries or possible production technologies. By studying the robust features of functionally integrated systems which arise for many grammars, we should find that grammars fall into a few broad "grammar regimes." Using these robust features, it should be possible to build toward a new theory of integration and transformation in biological and social sciences. Among the features we shall find are phase transitions between finite and potentially infinite growth in the diversity of symbol strings in such systems. As we have seen, this phase transition may well underlie the origin of life as a phase transition in sufficiently complex sets of catalytic polymers. Similar phase transitions may underlie "takeoff" in economic systems, such as the Industrial Revolution, once the systems attain a critical complexity of goods and services that allows the set of new economic niches to explode supracritically, and may provide models for the conceptual explosion wrought by the redevelopment of science three centuries ago.

The relationship between such random-grammar models and more familiar dynamical systems, such as those discussed in Chapter 5, must be made explicit. In familiar dynamical models, such as Boolean networks, the "meaning," "connections," and "dynamical laws" governing the interpretation of each variable, its couplings to other variables, and how it changes as they change are all defined externally to the theory. That is, we have no underlying theory showing why a given pair of variables are coupled. Granted the structure and equations for the dynamical system, its dynamical behaviors are studied as flows in its state space. Precisely because they offer no account of the ways couplings among the variables arise, such models cannot readily be models of functional integration. In sharp contrast, however, models based on grammars afford an underlying account of the particular couplings among symbol strings. Thus an entity such as an autocatalytic set of symbol strings *announces* its functional integration! The members of the system collectively *make one another.* It is in just this sense that random-grammar models allow us to begin to study the relation between the structure of the grammar rules and the kinds of functional integration and transformation that emerge in systems of symbol strings which act on one another according to those rules.

The chapter is organized into three sections. In the first section, some of the kinds of finite or infinite sets of symbol strings which emerge as algorithmic transformations of one another are described. In the second section, I suggest two approaches to studying a denumerably infinite space of grammars. One approach is based on infinite-dimensional Boolean networks; the second, on random grammars. The third section discusses applications to biological, neural, and social sciences. First, however, the remainder of this introductory material describes briefly Walter Fontana's generalization of the models of autocatalysis in polymer systems to an "algorithmic chemistry" which, strikingly, yields autocatalytic sets of symbol strings.

Fontana's AlChemy

In Chapter 7, when discussing the emergence of autocatalytic sets of polymers, we considered two basic rules for the assignment of catalysis. The simplest was based on a constant probability P that any polymer catalyzed any reaction. The second, modeling RNA molecules, required that a putative RNA enzyme template match the right and left terminal nucleotides in its potential substrates; then that RNA sequence had a chance governed by its "matchstrength" to be an enzyme able to catalyze the reaction. In these models of the origin of life, we are describing mappings of symbol strings into strings, mediated by strings. Thus, as remarked above, this mapping is some kind of algebraic or algorithmic transformation in which autocatalytic sets of symbol strings are certain kinds of collective identity operations of a cluster of the objects acting on themselves. The use of a constant probability of catalysis P and the matchstrength rules yields two specific grammars assigning transformations among symbol strings to the action of symbol strings. Such transformations are algorithmic in the precise sense that the action of one string on another is defined by the grammar and that recursive actions occur. Fontana devised a third grammar mapping strings into strings.

Walter Fontana (1990) borrowed a powerful algorithmic language derived from the lambda-calculus, invented to be as powerful as universal Turing machines, and the progenitor of Lisp. His idea is to exploit the general idea of strings acting on strings as algorithms. In this hypothetical chemistry, Fontana does not require that mass be conserved. Two strings collide; the first is the program, which acts on the second as an input. By construction, most strings are legal both as program and as input. Thus most collisions between strings transform the second string into some single new string. Fontana defines a "Turing gas" in which a random collection of strings is placed in a "chemostat." After each productive collision between strings, the number of strings has increased by one. To supply a selective condition, Fontana removes one string randomly chosen from all present, hence holding the number of strings in the chemostat constant.

Fontana has carried out three kinds of numerical experiments. In the first, a set of 700 strings is allowed to interact by random collisions. At first only new strings are generated, but over time more and more of the strings generated already have identical copies in the chemostat. Eventually a closed set of strings, an *autocatalytic set,* emerges. In this first set of experiments, the autocatalytic set is dominated by a general replicase, which is a sequence that can copy itself and any other sequence. Such a replicase is equivalent to a ribozyme which might copy itself and all others. Harkening back to our concern in Chapter 7 about nude replicating genes, here the replicating system based on a general replicase *does not* build up a complex metabolism about itself. It remains a simple system in string space.

In the second set of experiments, Fontana disallows copying strings. Nevertheless, *closed collectively autocatalytic* sets of strings emerge. Thus one set contained 45 kinds of strings which mutually transformed into one another. These sets are the direct analogue of the collective autocatalytic polymer sets discussed in Chapter 7.

In the third numerical experiment, Fontana injected sets of 20 random strings into an evolving chemostat. He found that the terminal autocatalytic set differed from that which would have occurred without exogenous perturbation. Often a core metabolism is stable, while a penumbra of string types come and go under such perturbations as the injection of new strings.

JETS AND AUTOCATALYTIC SETS: TOWARD A NEW STRING THEORY

Whether we are considering the transformation of molecules in a prebiotic soup or goods and services in an economy or a variety of other cases, it seems useful to consider the infinite set of binary strings as the objects under analysis. Then, in general, strings or sets of strings act on strings or sets of strings to yield strings or sets of strings. In general, such transformations are just mappings specified by a kind of grammar. The set of strings operated upon can be one or many. The set of strings carrying out the operations can be one or many.

In general, the set of transformations will increase more rapidly than the set of strings. Thus the general question is this: *For various kinds of random or nonrandom mappings of strings into strings, what kinds of sets of strings emerge?* What we need, in general, is a way of generating families of grammars, or finite state automata which realize those grammars, and of discovering the kinds of functionally generative sets we obtain. I turn next to some intuitions about those sets, then return in the next section to consider ways of studying the space of possible grammars.

String Set Geometries: Jets, Lightning Balls, Mushrooms, Eggs, Filigreed Fogs, and Pea Soups

Consider first a Jet. Imagine a rule for polymer sets which states that any string can catalyze the ligation of two other strings only if both merging strings are larger than the catalyst string. Then, by construction, no feedback loops can form. All catalyzed transformations lead to ever larger strings. Let me define a Jet as a set of transformations among strings from some *maintained "founder" set of strings* (analogous to the food set S_0) having the property that, under the algorithmic transformations among the strings, each string is produced only by a unique set of parent strings and is produced in a unique way. This definition is probably sufficient but more than necessary. In any case, it leads to a Jet of string productions which never cycles back on itself.

Note that a Jet might be either finite or infinite.

A Lightning Ball is a Jet cut free from its founder set, free to propagate through string space until it either dies out (if the Jet were finite) or propagate forever (if the Jet were infinite). Presumably a periodic or quasi-periodic Lightning Ball which orbits either back to the starting set of strings or back to *near* the starting set of strings is possible. The orbits in string space; defined as the succession of sets of strings in the Lightning Ball, might be periodic, raylike, chaotic, or ergodic.

Let me define a Mushroom. The first example is an autocatalytic set of polymers growing forth from a maintained food set. First a set of transformations jets up via a kind of stem which is free of feedback loops, and then feedback loops begin to form, creating the head of the Mushroom. In effect, a Mushroom is a Jet from a maintained founder set, with feedback loops.

Mushrooms are models of functional "bootstrapping." An immediate example is an autocatalytic peptide set with a sustained metabolism of coupled transformations from the food set. Perhaps another example is the technological evolution of machine tools. For example, the first tools were crude stones; then came shaped stones, which enabled formation of better tools, which could be used to dig ore for making metal tools, the ultimate development being machine tools which themselves generate tools such as axes, chisels, and machine parts for other machine tools. Presumably the onset of agriculture among hunter-gatherers is a similar example. Many more must exist in economic and cultural evolution, as well as in organic evolution.

Like Jets, Mushrooms can be finite or infinite.

Next consider the Egg. The hexamer and trimer RNA sequences which reproduce only themselves in RNA sequence space (Chapter 7) are Eggs, whole in and of themselves. Eggs are self-sufficient sets of algorithmic transformations with no need of a stem from a maintained founder set. In an egg, strings can produce arbitrary strings and hence can enter a closed set which finds only itself, free of all other strings. Let me reserve the term "Egg" for *finite closed autocatalytic sets.* An unchanging finite Egg is precisely an identity operator in the process algebra or grammar by which strings act on strings such that the collection of processes produces precisely and only itself. Eggs may prove to be useful models of self-confirming mythic or even scientific conceptual systems by which the outside world is parsed. They may also prove useful as models of cultural identity, integration, and wholeness.

Presumably Eggs come in several types. We have already considered the Fixed Egg, which maps into itself. Eggs might move through string space, however, creating Traveling Eggs. Fixed Eggs correspond to autocatalytic sets which are closed and hold to a fixed set of consequences and transformations. Traveling Eggs are like Lightning

Balls but contain feedback loops and change composition in sequence space in various ways. Presumably, Wobbly Eggs, which orbit among either a period or a quasi-periodic set of sequences, might be possible; so might Chaotic Eggs and Ergodic Eggs exist. The set of strings in an Ergodic Egg would wander randomly over string space as the Egg traveled. In addition, Hairy Eggs would be finite objects from which Jets or Mushrooms may extrude, perhaps stochastically if the production rules are activated probabilistically. Fontana may have found such structures with a stable core metabolism sending out a fluctuating flare of other strings.

The Filigreed Fog is an infinite supracritical autocatalytic set which either may have a stem to a sustained founder set or, like an Egg, may float free. Unlike an Egg, however, a Filigreed Fog is not bounded. Nevertheless, it is limited in that there are at least *some strings which can never be generated by the set.*

Finally, there is the Pea Soup, defined as an infinite set which, in principle, will eventually include all possible strings. It is intuitively plausible that the autocatalytic set generated by the model in which each polymer has a fixed chance of catalyzing each reaction will form a Pea Soup if the set is supracritical. Ultimately all strings should have their formation catalyzed by some string.

Evolution and Stability of Functional Sets

Among the obvious questions about such sets are their stability and capacity to evolve. Consider an Egg. How many Eggs does the algorithmic set contain? A few? Many? Given a definition of one-mutant variants, is an Egg stable to all one-mutant variations in its composite strings? All two-mutant variants? Thinking of Eggs as attractors, how many are accessible from any other Egg, by how much of a mutation in the set of strings present? For example, Fontana began to study this question by injection of exogenous strings. Can one jolt an Egg to another Egg? Similar questions apply to all the kinds of structures depicted. Such questions bear on the stability of Jets, Mushrooms, and so on and on their capacity to evolve as a result of "noise." To return to a theme from Chapter 7, the capacity of such systems to evolve constitutes evolution without a genome. Hence such examples strike at an over-narrow view of the basis of heritable variation and the capacity for selective evolution based on DNA template complementarity.

Note that transformation from one Jet to another, one Mushroom to another, one Egg to another, or from Jet to Egg or Egg to Mushroom and so forth begins to get at our intuitions that the Soviet or Chinese political system is fragile, that a few minor changes in the coherent structure must lead to the replacement of many or most functional linkages.

Decidability Problems

A number of issues may be undecidable. For example, whether a given set of founder strings in a given algorithm is subcritical or supracritical might be such an issue. It appears intuitively related to the halting problem: Will an algorithm halt with the answer or continue forever? Similarly, in a Filigreed Fog, it may not be formally decidable that the grammar cannot produce a given string from the initial set of strings. I suggest below in an economic context that, where strings are goods, such formal undecidability may map into the logical requirement for bounded rationality in economic agents and into an equal logical requirement for incomplete markets. Thus such models may invite modification of neoclassical economic theory.

Size Distribution of Avalanches of Change

In autocatalytic polymer sets, addition of a new polymer may trigger the formation of many new strings and the elimination of old ones. In a technological web, addition of the automobile drove out the horse and many horse trappings. When Fontana injected random strings, a peripheral component of the autocatalytic metabolism tended to change. What is the size distribution of such avalanches? For example, in Chapter 6 I discussed the sand pile model and self-organized criticality of Bak, Tang, and Wiesenfeld (1988); at the critical state, there is a power-law distribution with many small and few large avalanches. In Chapter 5 we saw a similar distribution in Boolean networks at the edge of chaos, and in Chapter 6 we found the same distribution of avalanches in model ecosystems which have optimized joint fitness. Thus we are led to ask what such avalanches of "damage" or changes look like in our various objects. For example, it might be that in finite Jets early avalanches are large and late ones are small. Or avalanches might show a common distribution regardless of when they are unleashed in the lifetime of a Jet. Similar questions arise with respect to Fixed Eggs and Traveling Eggs. Perhaps a power-law distribution obtains just at the phase transition when sets go supracritical.

These questions may allow us to begin addressing such issues as the *sensitivity of history* to small perturbations. For history, too, is an unfolding of transformations among some indefinite set of possibilities. Similarly, the evolution of autocatalytic sets in a world of polymers, with coevolution among the sets, captures both historical accidents and a kind of entropic exploration of the world of the possible.

Sets of Strings Acting on Sets of Strings: Aggregated Transformations as Machines Tune the Ratio of Transformations to Strings

Consider Fontana's Turing gas. A thousand strings interact with one another by random collisions. This system parallels the studies on autocatalytic sets. Suppose that 100 different types of strings are present. Then the chance that any specific string will undergo an ordered set of five of the transformations mediated by these strings is low. Consider instead, a "machine," by which I mean a complex made of a sequential aggregate of five kinds of strings, such that any string which encounters the machine undergoes sequentially all five transformations. Thus the machine, which is made of *simple* transformations, ensures a *complex set* of transformations. Since we are supposing 100 kinds of strings in the gas, there are $100^5 = 10^{10}$ of these combined fivefold transformations.

One implication of the use of an aggregate as a machine which acts on a single string or on a set of strings is that the number of machines is very much larger than the set of single strings. Since each machine carries out a compound transformation on an input string, construction of complex machines increases the ratio of potential transformations mediated by one machine to strings. Hence achieving phase transitions to more complex supracritical sets becomes easier.

Another implication is that coordination of five strings in an ordered way into a machine alters effective time scales. Thus we can think of the joining of primitive strings into aggregates, the *invention of machines,* as a means of mediating specific compound transformations at high frequencies. Clearly, the emergence of machines will alter the functional sets formed. This example makes it clear that *time scales matter.* The members of a set of strings interact by some dynamics, as in Fontana's random-collision dynamics. Altering the probabilities of string interactions profoundly alters which sets of composite transformations occur and which sets of strings arise.

The image is not a poor one. The machines in our economy form specific complex objects among a set of many other possible ones.

Compound machines made up of sets of strings also afford models of downward causation. More lion proteins abound these days than do trilobite proteins largely because lions as whole organisms are now more abundant that trilobites. This abundance of leontine proteins reflects the recent success of lions as integrated wholes. Similarly, the abundance of pistons in landfills reflects the usefulness of motors as integrated functional wholes. Neither lion proteins individually nor pistons individually vie for abundance in their respective historical records; it is the wholes embedding them that vie for abundance.

From One Chemostat to Many: Coevolution and Phase Transitions

By introducing a multiplicity of chemostats which operate on strings internally, and which may also exchange strings between chemostats, we can explore models of coevolution, the emergence of competition, mutualism, or symbiosis in biology, or trade between economic agents or units. In addition, phase transitions among the kinds of sets generated—Jets, Mushrooms, and so forth—may take place as a function of the number of chemostats which come to interact. As that number increases, the joint complexity of strings being operated on can pass critical thresholds. Such transitions may model "takeoff" in an economy or even intellectual community. I now discuss this in more detail.

The autocatalytic models and Fontana's model so far take place in a stirred reactor. All strings can interact with all strings. Consider instead a set of chemostats, or boxes. Each box, to be concrete, begins with a sustained founder set, which constitutes the box's sustainable natural resources. Each set of strings proliferates purely internally. So far this is nothing but the stirred reactor within one chemostat. Now let some of the strings be made for "export only." These exported strings may pass to other boxes. Those other boxes may be identified by spatial location, or some strings may bind to the box surface and serve as address strings. Either way, the invention of multiple chemostats identifies individual regions of local processes which may then coevolve with other such regions.

Among the first questions to consider are these:

1. Imagine that each box, granted its sustained founder set, yields only a finite Jet. It may be the case that, when strings can be exported between boxes, some or all of them are lifted to a more complex level of activity. For example, the collective system might form an infinite Jet, a finite or infinite Mushroom, a Filigreed Fog, or even a Pea Soup! The point to stress is that collaborative interaction may transform a system from one type of functional set to another. In particular, there may be a critical level of complexity for any given set of algorithmic transformations, a critical level that leads with high expectation to each of these transitions. If so, what are these thresholds like? Might they, for example, bear on economies which are unable to expand in diversity of goods versus those which can explode? Do they bear on the consequences to isolated cultural systems when brought in contact with other isolated systems or with a larger world culture? Do they bear on the scientific explosion following the Renaissance?

2. Consider the question of functional integration between the boxes. Each box can be thought of as a kind of country with natural resources, or a firm interacting with other firms, or perhaps an integrated functional organism. String inputs from other countries, firms, or organisms may perturb the internal dynamics of each box. In response, the box may die; that is, the Jet or other process may collapse to a sustained founder set, or it may transform to some other more or less constant functional set. In the latter case, we have an image of entities which alter their internal structure in response to external couplings such that each entity is internally a stable sustained flux of collaborative processes in conjunction with couplings to other boxes. It is an image of stable signal relations among bacterial cells or perhaps, as we see in more detail below, trade relations among nations endowed with different natural resources and different histories of technological development. Are there many alternative attractors to such a system, given the same founder sets to each box? How history-dependent is the system? How stable to perturbations?

Such coevolving boxes literally come to know one another and to know their worlds. We must consider when and whether such systems are competitive and when and whether they coevolve mutualisms which optimize mutual growth rate or, equivalently, utility. Indeed, I suspect that these processes must occur in biologic and economic evolution.

Dynamical Stability as Well as Compositional Stability

The sets we considered above—Jets, Mushrooms, Eggs, Fogs, and Soups—deal with the string composition generated by different rules by which strings interact. In addition to the composition of such generated sets, however, it is also important to consider their dynamical aspects in terms of the "concentrations" of strings of each type over time. For example, an Egg might reproduce itself compositionally at a dynamical steady state or along a limit cycle or chaotic orbit in string space along which its constituent strings were successively produced. Presumably other orbits might suffice for an Egg to persist. Similar questions arise for Jets, Fogs, and other potential objects.

A critical difference between string systems and familiar dynamical systems is that the former operate in an indefinitely large state space, and the latter in a fixed state space. The functional sets we are considering are, in a sense, evolving in an open state space of strings.

INFINITE BOOLEAN NETWORKS AND RANDOM GRAMMARS: APPROACHES TO STUDYING FAMILIES OF MAPPINGS OF STRINGS INTO STRINGS

In order to study Jets, Eggs, Fogs and functional interactions, we require mathematical models of the algorithmic interactions by which strings act on strings to produce strings. The autocatalytic polymer set with fixed probability of catalysis P, the RNA string match rule Bagley and I have investigated, and Fontana's alchemy are three choices of rules by which strings act on one another. The aim of this section is to consider alternative approaches to generate in some ordered way the set of "all possible" mappings of strings into strings. In fact, this cannot be done in an ordered way because the set of all such mappings involves the infinite power set of binary strings of infinite length acting on itself to produce the infinite power set of binary strings. This class of objects is not denumerably infinite. It maps to the reals. Consequently, any ordered approach to this problem requires simplifying at least to a denumerably infinite set of objects, categorized in terms of some parameters such that mappings of increasing complexity can be studied and such that these mappings fall into useful classes.

The aims of this endeavor should be stated clearly. I believe such mappings sending strings or sets of strings, operated on by strings or sets of strings, into strings or sets of strings may provide useful models of several things: molecular interactions or molecular machines in organisms; production technologies in economic systems; and conceptual linkages in psychological, scientific ideational, or cultural systems. We surely do not, at this stage, have detailed understanding of such functional couplings in any of these areas, and the hope is this: by exploring large tracts of "grammar space," we may find rather few "regimes" in each of which the same general behavior occurs in the sets of strings generated by the specific grammar. That is, just as exploration of random Boolean networks has revealed three broad regimes—ordered, complex, and chaotic—so too may exploration of grammar space reveal rather few broad regimes. We can then hope to map these broad regimes onto biological, economic, conceptual, or cultural systems. Thereby we may obtain models of functional couplings among biochemical, technological, or ideational elements without first requiring detailed understanding of the physics or true laws governing the couplings. We may find, in short, the proper universality classes.

I next discuss two approaches to this task. The first explores the representation of mapping of strings into strings via potentially infinite Boolean networks. The second considers the use of random grammars with definable parameters which allow grammar space to be explored.

A Natural Infinite-Dimensional State Space Representation of the Mapping of Strings into Strings via Infinite Boolean Networks

One representation for strings mapping into strings is a potentially infinite-dimensional state space of symbol strings which are finite but of arbitrary length. Consider binary strings of length L, where L can increase up to infinity. Order these in counting to infinity, beginning with the two "monomers" 0 and 1, then the four "dimers" 00, 01, 10, 11 . . . , followed by the eight trimers, and so on. At each string length L, there are 2^L types of strings. This infinite list of string types, whose lengths also increase to infinity, can be ordered from a starting point: the monomer 0. Create two infinite matrices. The first, the input matrix, is ordered such that each column denotes one specific binary symbol sequence, and the columns begin with the monomers at the rightmost column, the dimers to the left, the trimers to the left of the dimers, and so forth, stretching to infinity in the left direction. Thus each possible symbol string is assigned one column in the input matrix. The second matrix, the response, is simply the mirror image of the input matrix. The response matrix lists the monomers in the leftmost two columns, the four dimers to their right, and so on, stretching to infinity in the rightward direction. The input matrix has as its rows all possible combinations of the presence or absence of the possible types of symbol strings, starting with the row (. . . 00000) on top. There are an infinite number of rows in the input matrix. Lower rows, for example (. . . 01100), represent states with longer symbol strings.

The positions of 1 values in each row of the input matrix represent which strings are present in that state of the world. The response matrix will show the next state of the world as strings act on strings to produce strings. By construction, the input and response matrices are mirror symmetric; hence in order to read the next state formed from each input state, the reading must be flipped from right to left for the input matrix to left to right for the response matrix.

Alternative mappings from input to response represent alternative mappings of the set of strings into itself. In order to proceed, some further definitions are required. Let a machine M^* be an ordered collection of M strings. Let an input bundle I^* be an ordered set of I strings. The action of M^* operating on I^* will yield an ordered output set of strings O^*.

Any row of the input matrix has a finite number of cells having the value 1, representing the fact that each row represents a unique combination present or absent in the world of strings up to some length. The possible machines built of these strings might be limited to a specific maximum number of strings—five, for example—or might range up to the finite total number of strings present in that state of the world. Call the latter maximum size machine unbounded in the sense that, as rows farther down the infinite input matrix (which represent the presence of longer and often more symbol sequences) are considered, still more complex machines can be built. Because of the ordered way the input matrix is constructed, it is possible, given constraints on which ordered sets of strings count as legitimate M^* or I^*, to uniquely number each machine and input bundle. In a moment I shall use such unique numberings to produce a deterministic mapping from current state to next state.

The choice to include all possible unbounded machines as legitimate machines or all possible unbounded input bundles as legitimate bundles specifies the power set of strings operating on itself as the mathematical entity of interest. This is clearly the widest interpretation. It allows us to generate, in the presence of a fixed set of strings, the maximal number of strings possible under *any* interpretation of the kinetics in which strings collide with and act on one another. (Other choices are more limited. For example, we might wish to assume that the largest machine any string might enter was the only machine which formed and acted. In that case, transformations mediated by smaller machines would be inhibited by the capacity to form the larger machine.) Here, instead, we assume that all legitimate machines able to be formed are present and carry out their transformations on all possible legitimate input sets. Thus, for the moment I therefore choose the widest, full power set interpretation of machines and input sets.

If a string is acted on and transformed, we need to choose whether the initial string remains in the system or not. The natural interpretation, adopted here, is that the string is used up, or disappears, in the transformation. (Note that, in chemistry, back transformation always occurs. This is not in general the case, however.)

With these assumptions, the next state of the world is a mapping from the present state, a mapping given by some Boolean functions in the response column. Such a system can be thought of as a discrete-time, autonomous, synchronous, potentially infinite automaton. The dynamics of this automaton tells us the way strings engender strings in the potentially infinite space of strings.

Three further assumptions lead us to a canonical and ordered way to generate a denumerable family of transformations:

1. We can parameterize such transformations by the largest machines allowed, M^* and the largest input sets to a machine allowed, I^*. Thus the system might at the

moment have 1000 strings, but only machines or input bundles with five or fewer strings might be allowed. In any row in the input matrix, there are a finite number of sites with 1 values, corresponding to strings present in the system, say N^*. The maximum number of machines is N^{*5}. The maximum number of input bundles is similarly N^{*5}. The product of these is the maximum number of pairs of machines and input bundles. Determinism demands that for each pair, there is a unique outcome.

2. In order to retain denumerability, we need a rule which limits the number of output strings given the number of input strings. Call this a limit on the "output spray" S^*. Given this limit and then given N^* in the input row, we have a maximum limit on the number of strings present in the next state of the world.

3. Next, again to retain denumerability, we can choose to delimit the maximum length of a new string produced by machines whose maximum member number is M and whose maximum input bundle string length is L to some finite bound which increases with M and L by some bounded function. Thus in the origin of life model from a food set, at each iteration, the maximum string length doubles because one imagines ligating two strings present in the system. Any such bounding choice is a third parameter, which in effect creates an expanding cone down the rows of the response matrix. The cone asserts that maximum string length can grow only as fast as a function of lengths of strings already present in the system.

Given these bounds, we have, for each row of the input matrix, a bound on the maximum number of strings which can be present in the next state of the world and a bound on the maximum length of those strings. (Note parenthetically that a constraint exists between the maximum rate of cone expansion of string lengths and the total number of output strings from any input state. There must be enough possible strings in the space allotted to accept the new strings.)

The Quenched Deterministic Version: Mapping the Infinite Power Set of Binary Strings into Itself

The mathematical object we are considering in the Boolean idealization where all allowed transformations occur is really a mapping of the now at most denumerably infinite power set of N strings into itself. That is, consider a row of the input matrix. It contains a set of N strings. The power set of ordered pairs of strings, ordered triads of strings, ... , ordered Nads of strings, however, is just the set of all possible machines M^* constructible from those strings. Similarly, the sets of single strings, ordered pairs of strings, and so on are the set I^* of possible input bundles. Thus I^* and M^* are the same power set in the limit when machines and bundles having N strings are allowed. As remarked above, identify each unique ordered set of strings which is a machine M_i with a unique number. Similarly, identify each unique ordered set of strings which is an input bundle I_i with a unique number. Then the pair of numbers ij specifies a unique machine input pair and hence must always have a fixed output bundle O_{ij}. The output, of course, is bounded by the output spray S.

Given the unique identification of I_N, it is possible to define for each machine input pair—regardless of *which state of the world it occurs in,* hence which row of the input matrix—a unique output bundle. This ensures both determinism and constraints within the family M, I, and S. Since both the length and the number of strings making up machines, input bundles, and output sets are finite and bounded for any unique machine input pair, we can generate *all possible finite number and length legitimate output sets* which might be generated by the pair. Thus it follows that, in terms of the parameters giving (1) maximum machine, bundle, and output spray sizes, and (2) cone angle or string length amplification factor, and (3) the deterministic machine input pairs mapping into unique output bundles, we can consider all possible mappings of any finite set of strings into itself. Hence within these parameterizations, we can explore all possible dynamical behaviors of this family of systems.

The Annealed Model

Given these bounds on the response matrix, we may consider a simpler, "annealed" model which may prove useful. Consider, for each row of the input matrix and the bounds on amplification, output spray, and so forth, all possible ways of filling rows in the response matrix with 1 and 0 values, consistent with those constraints. Each way corresponds to a well-defined transformation of the set of strings into the set of strings and allows an expanding cone of complexity. However, this model does *not* preserve deterministic dynamics. It is an *annealed* approximation to a deterministic grammar of the infinite set of strings into itself. The lack of determinism is easy to see. Consider two input rows in which strings S1 . . . S5 are present, but in the second row string S6 is also present. By determinism, all the machines, input bundles, and transformations which might occur in the first row are also present in the second row and hence must occur in the corresponding next state of the world. Under the annealed model, however, filling the response rows in all possible ways, such determinism is not guaranteed. Instead, this model is an annealed approximation to a deterministic dynamics whose statistical features it may prove useful to analyze, as has proved to be the case with the Boolean networks discussed in Chapter 5 (Derrida and Pomeau 1986).

The concepts of Jets, Lightning Balls, Eggs, Filigreed Fogs, and Pea Soups are all clear in either the deterministic or the annealed picture as either trajectories from a maintained source set (Jets, Mushrooms) or "free" dynamics (Eggs, Lightning Balls, Filigreed Fogs, Pea Soups).

Since these systems are just infinite Boolean networks explored from an initial state invoking relatively short and few symbol strings, the concepts of *dynamical attractors and ordered, chaotic, and complex behaviors* carry over directly. In addition, however, in comparison with a fixed state space, we have here the idea of attractors in *composition space*—the sets of strings which make up the system as well as the dynamical behaviors evidenced among the sets of strings, be they Eggs, Mushrooms, or Filigreed Fogs.

We can begin to guess at the relation between dynamics and composition space. The Boolean idealization shows the set of all possible transformations from the current set of strings into the next set. By contrast, in other dynamics not all machines and input bundle pairs will interact at each moment, and hence only a subset of all transitions will occur. In particular, as remarked above, we might want to model the presence of an inhibitor string which, when present, unites with a machine and reliably blocks its action, just as repressor molecules bind to *cis*-acting DNA sites and block transcription. Note that this kind of dynamics departs from the choice to allow all possible transformations induced by machines on inputs. Were that most general choice made, then in the presence of the inhibitor string, *both* the machine without the string and the machine with the string would be present. The former would carry out its prefigured transformation; the latter would not. In the modified dynamics, the inhibitor string uses up all copies of the machine and prevents the transformation which would otherwise occur. The important point is this: Once we allow inhibition of transitions in this way, the dynamics can be chaotic, ordered, or complex.

Consider the case in which, on the infinite graph, graph growth creates a leaky Egg which emits a narrow infinite Jet. Will the infinite Jet actually occur? The *dynamics itself can control the subset of the composition set explored.* For example, the dynamics of the system might cut off all transformations at the base of the infinite Jet, so that there will be no strings to flow up and create the Jet. Clearly, this process is easier to control if the dynamics are in the ordered regime rather than the chaotic. In the ordered regime, all string processes from the Egg to the Jet entrance might be inhibited. Under chaotic dynamics in the Egg, firing of strings at the base of the Jet would be hard to prevent. Thus which subset of the composition set actually occurs is clearly more readily controlled if there are ordered dynamics. Conversely, achievement of ordered dynamics in Boolean networks requires control over the number of inputs per variable and over the biases in the Boolean functions. Both controls will be easier to maintain in a *finite Egg than in an infinite Mushroom, Filigreed Fog, or Pea Soup.* In these latter cases, the elaboration of feedback connections to each string is roughly unbounded. Thus these systems are more likely to exhibit chaotic dynamics, and thus to explore fuller reaches of their possible composition set, than are finite Eggs with orderly dynamics.

Obviously, finiteness in physical systems is also controlled—by thermodynamics in chemical systems, for instance, and by costs of production, aggregate demand, and budget constraints in economies. However, in the worlds of ideas, myths, scientific creations, cultural transformations, and so on, no such bound may occur. Thus it is of interest to see how such algorithmic string systems can control their own exploration of their possible composition set by dynamic control over the processes they undergo.

The generalization to the case with multiple chemostats is obvious. It is equivalent to a set of linked Boolean nets—that is, those which share some external variables.

Random Grammars

While infinite Boolean systems may prove useful, random grammars may be more readily studied. Grammars range from simple regular languages to context insensitive and context sensitive to recursively enumerable. The most powerful grammars are known to be as powerful as universal Turing machines. A grammar can be specified by a list of pairs of symbol strings, with the interpretation that each instance of the right member of the pair in some input string is to be substituted by the corresponding left member of the pair. Thus, were the sequences (110011) and (0011) such a pair, then starting with a given input string, any instance of (0011) would be replaced by (110011). Effectively carrying out such a transformation on an initial string requires a precedence order among the pairs of symbol sequences in the grammar plus a means to limit the depth to which such substitutions are allowed. For example, replacement of (0011) with (110011) creates a new (0011) sequence. Shall it be operated on again by the rule? If so, recursion will generate an infinite string by repeated substitutions at that site. If not, the depth has been limited. Limiting depth limits the length of the transformed string with respect to the input string.

Recursively enumerable grammars, which can be defined by a finite list of pairs where the partner on the left can be shorter or longer than the partner on the right, are as powerful as universal Turing machines. Tuning the number of pairs of symbol strings, the lengths of those strings, and their symbol sequence complexity tunes the power and character of the grammar. A further "amplification" parameter specifies by how much and whether always or on average substituted symbol sequences are longer or shorter than the original sequence. Additional rules allow strings to be cleaved or ligated. In short, a few simple parameters can be used to specify a grammar space. Using them, random grammars within each set of values of the parameters can be chosen and the resulting string dynamics studied.

A simplest approach is this: Use a random set of pairs of strings as the random grammar. Begin with a set of strings and operate on each string according to the grammar. Here, however, strings do not act on one another.

A more useful approach, suggested by Albert Wong (personal communication 1991) and closely related to Fontana's work as well as to our own origin of life model, is to define grammars of substitutions plus ligation and cleavage operations, but require that strings contain "enzymatic sites" such that the strings themselves are carriers of the grammatical operators. Thus if the grammar specifies that string ab is replaced by string cddcde, then an enzyme string with an ab enzymatic site would search target strings for a matching ab site and, if found, substitute cddcde in the target string at that site. Or the enzymatic string might cut or glue strings at sites. Clearly, such grammars can be implemented in binary strings, with matching as complements or as identities. Any such grammar-chemistry must also make definite choices about the precedence order in which rules are applied and in which depth of recursive substitution or other actions at one site are pursued.

More complex machine and input bundle sets can also be built up by generalizing on the idea of enzymatic sites. Real proteins often cooperate with one another by forming multimeric enzymes carrying out the same or even a succession of biochemical transformations. Here the constituent monomer proteins recognize one another and self-assemble within the cell to form the ordered protein aggregate which is the cooperative complex enzymatic machine. Similarly, we might extend our grammar rules to specify how ordered collections of strings self-assemble and act as machines or input bundles to yield unique output sets of strings.

The use of grammars is likely to be very important in analyzing the emergence of functional adaptive systems. The Boolean idealization allows the set of all possible next strings to be followed, but it does not readily allow for growth in the numbers of copies of each string, for inhibitory interactions and hence competition between strings, and so forth. In contrast, just such features emerge readily in models where strings interact with one another via grammatical rules. I return to this model below in considering the implications of these ideas for mutualism, community structure, and economics.

The relation between grammar complexity and the kinds of Jets, Eggs, Fogs, and so forth which arise is a central object for analysis. Some points already seem plausible. A simple grammar may be likely to give rise to finite Sets, Jets, Mushrooms, or Eggs. A complex grammar may be likely to give rise only to infinite Mushrooms or Filigreed Fogs. The reason is intuitively clear. The first finite autocatalytic sets found were the hexamer single-stranded RNA and its two trimer substrates, as noted above. The point–point complementarity due to base pairing allows this system to make first an exact complement and then itself in a closed cycle *which need not expand out into sequence space*. It is possible for this autocatalytic set to remain a two cycle and finite. Once overlapping sticky ends and ligation are allowed, this more complex grammar can give rise at least to infinite Filigreed Fogs. Now consider a very complex grammar: the fixed probability rule for autocatalytic sets. Here, each string has a fixed probability of catalyzing any reaction. The grammar is complex in the sense that, after catalytic interactions are assigned, each enzyme can typically act on many unique substrates. In due course, in supracritical systems under the probability of catalysis P rule, the formation of all strings will be catalyzed; hence this system creates a Pea Soup. It seems highly likely that, for a fixed founder set, the more complex the grammar, the less easy to limit string generation to finite sets.

The Growth and Asymptotic Form of Mutual Information as Strings Act on Strings

The action of strings on strings to produce strings according to a grammar should, over time, build up constraints in symbol sequences in the strings produced. As described briefly in Chapter 7, such constraints should show up in a measure of relations between symbols called mutual information. The mutual information between pairs of symbols a distance S apart is defined as

$$M(S) = \sum_{a,b} P_{ab}(S) \log_2 \frac{P_{ab}(S)}{P_a P_b}$$

where P_a or P_b is the frequency of value $a = 1$ or $b = 0$ in the set of symbol sequences and $P_{ab}(S)$ is the frequency of symbol value a at position 1 and symbol value b at position 2 at distance S from position 1. The factor $P_{ab}(S)$ is averaged over all pairs of positions S apart in the set of symbol sequences under consideration.

In natural language texts, $M(S)$ typically decreases as a power law as S increases (Li 1989). Thus nearby symbols tend to be more strongly correlated than distant symbols.

Consider now a system of 1000 binary strings, each chosen at random among strings having length 100. Because the set is chosen at random, the mutual information between sites at any distance S will be 0. Let the strings act on one another in a chemostat such that 1000 strings are always maintained in the system. As these mutual interactions occur, the action of strings on one another creates correlations and hence mutual information. Preliminary studies with David Penkower in my laboratory at the University of Pennsylvania indicate that in these systems mutual information begins very close to 0 and builds, as interactions take place, to an asymptotic form which depends on the grammar. Figures 10.1 and 10.2 show examples of these results for grammars of different complexity. The following features have been observed. The time course of the buildup is complex and can vary for different distances $_D S_1$ in the same grammar and for grammars of different complexity (Figure 10.1). Surprisingly, the establishment of correlations need not be monotonic in time. For some values of S, mutual information can first increase and then decrease (Figure 10.1b). Mutual information patterns do appear to build to a final asymptotic form as a function of number of grammar transformations. Simple grammars appear to build to their asymptotic patterns faster than complex grammars. The asymptotic pattern of mutual information is high for adjacent symbols and appears crudely to fall off exponentially, rather than in a power law, as S increases (Figure 10.2a). Finally, the mutual information patterns which build up can be strongly statistically significant (Figure 10.2b).

These preliminary results suggest that two factors—(1) the time course in which mutual information builds to the asymptotic form as a function of numbers of string interactions and (2) that asymptotic form—give information about the complexity of the grammar.

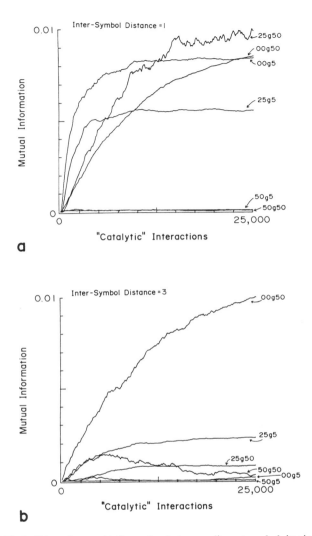

Figure 10.1 (*a*) The buildup of mutual information between adjacent symbol sites in random grammars with 5 or 50 pairs of symbol strings defining the grammar. Grammar rules also varied the fraction that mediates ligation and cleavage from 50 to 25 to 0 percent. The *x* axis plots the total number of interactions between symbol strings that have occurred. A system contains up to 1000 interacting symbol strings. (*b*) Similar to (*a*), except that the symbols compared are three apart along symbol strings rather than adjacent.

As remarked in Chapter 7, one can envision experiments in which random single-stranded RNA molecules of perhaps length 100 are allowed to interact with one another. If these molecules interact via specific sites that mediate ligation, cleavage, and splicing reactions (as do hexamers and ribozymes), then over time the sequences in the system should build up mutual information as a function of internucleotide distance S. This hypothesis should be testable by using PCR amplification, cloning,

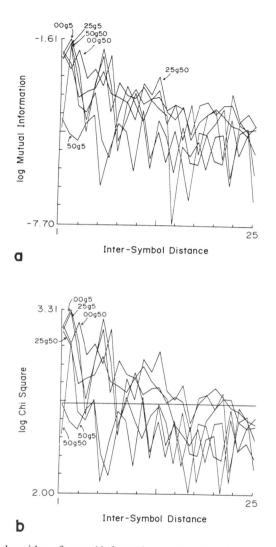

Figure 10.2 (*a*) The logarithm of mutual information as a function of intersymbol distance for the different grammars shown in Figure 10.1 and after 250 000 interactions between symbol strings have occurred. (*b*) Replotting of the data shown in (*a*), showing the logarithm of the chi-square value for each intersymbol distance after 250 000 interactions between symbol strings have occurred. The horizontal line represents the 5 percent statistical significance level.

and sequencing of the interacting RNA sequences over time. In turn, estimates of grammatical complexity are bulk estimates of enzymatic-site complexity as RNA sequences act catalytically on one another. Further, the buildup of short-range sequence biases implies that some sequences on a short range are far more common than are fully random sequences. This process might be detectable by the buildup of the capacity of one such sequence to hybridize with many complementary sequences as sequence biases build up. Analysis of the buildup of mutual information may prove useful for mixtures of initially random polypeptides or other potentially catalytic polymers which act on one another. While the length distribution of RNA or peptide polymers in such a system will be sensitive to thermodynamic factors, it seems likely that measures of mutual information among nucleotides or amino acids as a function of distance apart is sensitive only to the grammatical complexity with which RNA or peptide sequences act on one another.

Stochastic Generalization

The model above is deterministic. It can be expanded to a stochastic form in two ways. First, it is possible to include random bit mutations in strings to yield a stochastic version of the same basic model. Second, note that the grammar rules, when applied to strings without reference to use of other strings as enzymatic tools, are the analogue of spontaneous reactions occurring without an enzyme in the autocatalytic polymer set model. Hence spontaneous applications of grammar rules are the natural form of spontaneous mutations in these systems. This provides a second direction in which to make grammar systems stochastic. Analysis of the capacity of such systems to evolve will presumably be aided by study of stochastic versions of the basic grammar models.

APPLICATIONS TO BIOLOGICAL, NEURAL, AND ECONOMIC SYSTEMS

Random grammars and the resulting systems of interacting strings will hopefully become useful models of functionally integrated, functionally interacting molecular, biological, neural, psychological, technological, and cultural systems. The central image is that a string represents a polymer, a good or service, an element in a conceptual system, or a role in a cultural system. Polymers acting on polymers produce polymers; goods acting on goods produce goods; ideas acting on ideas produce ideas. The aim is to develop a new class of models in which the underlying grammar implicitly yields the ways in which strings act on strings to produce strings, to interpret such production as functional couplings, and to study the emergent behaviors of string systems in these contexts. I consider first some implications for biological models.

Waiting for Carnot: Biological Integration and the Evolution of Open Self-Constructing Systems to the Subcritical–Supracritical Phase Transition

Carnot gave us equilibrium thermodynamics and the general laws which have been extended to statistical mechanics. Yet we have no general laws concerning the behavior of open, far-from-equilibrium systems. Indeed, for apparently good reasons, we could not hope to have such laws: Turing machines, we know, can carry out universal computation. Thus *any* well-specified algorithm can be carried out by some machine. Such machines can be made of real physical materials and can carry out the algorithm if the system is displaced from equilibrium, open to matter and energy. Thus physical systems displaced from equilibrium and properly constructed can carry out any well-specified sequence of behavior. We know that some computations cannot be described in a more compact form than carrying out the computation and observing its unfolding. Thus we could not, in principle, have general laws, shorter more compact descriptions, of such behavior. Thus we could not have general laws about the behavior of arbitrary, far-from-equilibrium systems.

This argument, however, contains a vital premise: It is *we* who construct the non-equilibrium system in some arbitrary way. Having specified its structure and logic, we find the system capable of arbitrary behavior. Perhaps if we focus instead on *self-constructing* far-from-equilibrium systems, we may hope to find general laws. Among these, I now suggest, are laws of functional integration and the real possibility that coevolving far-from-equilibrium systems may evolve to the phase transition between subcritical and supracritical behavior.

Part of the interest in models of autocatalytic polymer systems, beyond the serious hope that they bear on the origin of life on earth and presumably elsewhere in the cosmos, lies in the fact that such systems afford a crystalline founding example of functional wholeness, hence functional integration. Given the underlying model of chemical interactions, once an autocatalytic set of polymers emerges, it is a coherent whole by virtue of achieving catalytic closure. Given the underlying model chemistry and catalytic closure, the functional role of each polymer or monomer in the continued existence and proliferation of the autocatalytic set is clear. Note that we here feel impelled, almost required, to begin to use functional language. This requirement reflects the fact that such a self-reproducing system allows a natural definition of the "purpose" of any polymer part, a purpose which is subservient to the overarching purpose of the autocatalytic set, which is of course, abetted by natural selection, to persist and prevail. In this nonconscious sense, an autocatalytic set becomes a locus of agency.

Model autocatalytic sets are natural testbeds for studying the emergence of collaborative or competitive interactions. We need merely specify how such systems may export or import strings to one another, and we shall find out how they cope with such exchanges. As remarked above, such interactions and their internal consequences within each set literally are what it means for such systems to come to know one another. By studying these properties across grammars, it should be possible to understand how grammar structure as well as the structure of interacting autocatalytic sets governs the coupled coevolutionary structures which emerge. The ways model autocatalytic sets build internal models of one another may well mimic the ways *E. coli* and IBM know their worlds. In addition, these models may well yield insight into the onset of mutualism, symbiosis, and competition in the biological realm.

In Chapter 5, I discussed the bold hypothesis that knower and known evolve to the edge of chaos. This hypothesis seems to me to be very attractive indeed. Like Boolean networks, model autocatalytic systems are parallel-processing systems that might lie in the ordered regime, chaotic regime, or complex regime in terms of their dynamical behavior. If deeply ordered or deeply chaotic, such systems would be expected to coordinate complex sequences of molecular events poorly and to adapt poorly. One might expect such systems to evolve their internal organization to the edge of chaos.

Autocatalytic polymer systems overcome a powerful limitation of the use of Boolean networks. In such networks, the binary elements themselves have no internal "meaning." It is just such internal meaning that is afforded by random grammar models. The strings which interact to produce one another exhibit their mutual meanings to the evolving system precisely by their functional couplings. It is in just this sense that grammar models are the natural extension of our interest in the properties of complex systems which allow them to adapt. Here we can deduce such properties based on hypotheses about the structure of the underlying "laws of chemistry" which govern functional couplings among symbol strings. Meaning, immanent, becomes explicit to the coevolving systems.

I now sketch some simple, preliminary grounds to support a candidate general law concerning the evolution of *self-constructing* open, far-from-equilibrium systems: Such systems may evolve and coevolve to the phase transition between subcritical and supracritical behavior. I find even the possibility of such a general law fascinating.

Prior to the origin of life, the earth was an open thermodynamic system. As noted in Chapter 8, some form of complex chemical evolution from simple organic molecules to a far higher diversity of organic molecules must be presumed to have occurred during the 100-million-year interval between the cooling of the crust sufficient to support liquid water and the first known signs of life 3.8 billion years ago. Thus we must seek the laws which govern such far-from-equilibrium chemical evolution. We already have some hints. In Chapters 7 and 8, we investigated the subcritical–supracritical phase transition. Once a supracritical diversity of organic molecules was present, such that the molecules themselves catalyzed the formation of new organic molecules, that very system must have been able to catalyze an open-ended explosion of diversity. In this sense, open chemical systems can be *self-extending*. We also have a clue that this view might be correct. Recall from Chapter 8 that the distribution of the numbers of kinds of organic molecules as a function of carbon atoms per molecule yields a single peaked distribution with a roughly exponential fall-off for larger organic molecules. This, I argued, reflects twin combinatorial facts about chemistry: As larger organic molecules are considered, with more atoms per molecule, the potential diversity of kinds of molecules increases exponentially or faster. Conversely, in the reaction graph among such organic molecules, there are always *more* pathways to create small molecules by cleaving them off larger molecules than there are pathways to build large molecules by combining smaller ones. The implications of these two combinatorial factors seems to be that if a supracritical fraction of reactions is catalyzed, then the diversity of molecules whose formation is catalyzed will reach a peak at an intermediate number of atoms per molecule and will fall off in an exponential tail for larger molecules. In short, the observed diversity distribution appears to be consistent with the self-extending features of organic chemistry.

Our next step is to try to extend Darwin's principle of natural selection to the prebiotic earth. Suppose that red and blue bacteria compete for resources, and red bacteria grow faster than blue bacteria. Eventually, the red bacteria overgrow the blue. A petri dish inoculated with red and blue bacteria becomes red. The fastest replicators inherit the earth. But is cell division essential for this process? Suppose that a *nondividing* red bacterium grows faster than a nondividing blue bacterium and that neither lyses from obesity. Eventually, the petri dish will be red. The fastest eaters inherit the earth. In short, cell division is not essential to Darwin's argument about *selection* leading to the overgrowth of one form compared with another. Continued evolution, however, does require heritable variation. Cell division is essential because it allows persistent independent heritable variation and thus persistent evolution. Thus let us, as a thought experiment, extend Darwin's concept to what might be called *producing organizations,* rather than reproducing organisms. We consider next hypothetical vesicles on the prebiotic earth which transport materials across their boundaries and carry out metabolic transformations on those materials. Such vesicle organizations may be imagined to grow in mass and size, even to divide by budding. Those organizations which increase in mass the fastest will inherit the earth. However, even if vesicles bud, no organization need be literally self-reproducing. Our hypothetical vesicles are, instead, loci of coordinated chemical productions.

Such vesicles can, in fact, even form reproducing organizations. Bachmann, Luisi, and Lang (1992) have created a micelle which transports an organic molecule across its boundary from a bulk medium. The organic molecule is transformed within the vesicle into a product molecule which itself forms the vesicle or micelle boundary. Thus the vesicle grows and ultimately divides as a result of surface tension.

Following Bachmann et al. in this thought experiment, I wish to make use of the fact that quite simple organic molecules can form such hollow micelles, or hollow lipid bilayer vesicles such as liposomes. Such vesicles divide an interior environment from an exterior environment. Thus they allow a system of organic molecules to *regulate* which molecules have access to one another. Therefore, while the entire collection of molecules located in many vesicles might be strongly supracritical if each molecule could interact with all the others, vesicle boundaries can *prevent* specific molecular interactions from occurring. In consequence, such a system of vesicles can, in principle, remain subcritical or can be maintained on the boundary between subcritical and supracritical behavior.

The central possibility I want to sketch is that open self-extending molecular systems *able to regulate which molecules interact* will do so such that each vesicle, or other type of compartment, is just subcritical. Two reasons support this possibility. The first asserts that vesicle organizations at the phase transition can evolve more successfully than those which are very supracritical or very subcritical. The second suggests that vesicle organizations at the phase transition can grow fastest.

Imagine two extreme alternative kinds of nested vesicle systems: "onions" and "pomegranates." Onions are nested concentric shells of vesicles. Each vesicle contains a single vesicle nested within it, which, in turn, contains still another. Multilammelar liposomes are an example. Pomegranates are nested sets of vesicles, each of which contains a large number of vesicles at the next nested level. In the onion, each vesicle is alone in its environment. In the pomegranate, each vessicle shares its environment with many others whose metabolic activities may impinge on it. Imagine next that each membrane layer contains molecules which control vectorial transport across the membrane, either inward or outward. Since vesicles can grow only by inward vectorial transport, such membrane sites allow the vesicle to "eat." Further, imagine that the molecules within a vesicle can interact only with those within that vesicle or in the membranes forming the boundaries of that compartment. Molecules in a membrane can interact only with those within that membrane or in the outside and inside compartments divided by the membrane. Finally, imagine two extremes of supracritical and subcritical organization. In the supracritical extreme, each vesicle alone contains a sufficient diversity of organic molecules—perhaps including polymers—that it is already supracritical. In the subcritical extreme, each vesicle contains a very small, hence very subcritical, diversity of molecular species.

Following Bachmann et al. in our thought experiment, consider first a subcritical vesicle organization where the outer boundary membrane transports an organic molecule, A, to the interior. The interior contains three types of organic molecules, one of which is a catalyst. The incoming organic molecule is transformed to yield three products: one forming the outer membrane, one forming the molecule which acts catalytically on the incoming organic molecule, and one that will form and be transported into the next inner vesicle. A series of such reactions might transport material to the innermost vesicles in the organization. Thus we might imagine that each compartment of the organization and its boundary membranes are able to grow and bud. Each daughter vesicle would contain, at random, the molecules and the vesicles nested within the mother vesicle. Since different vesicles in the organization would bud at different rates, the resulting organizations would explore a variety of more or less elaborate arrangements of nested vesicles. Some arrangements would accumulate mass and would bud faster than others. If stably produced, such organizations would come to predominate.

But consider the further evolutionary potential of these very *subcritical* organizations. Suppose that the initial food molecule A disappears from the environment, while a new food molecule, B, arrives. Alas, our subcritical organizations are ill-equipped to make use of B. If B encounters the outer membrane, it is likely not to be transported across the membrane or, once inside, to be acted on by the molecules within the compartment.

Precisely because such organizations are profoundly subcritical, they are likely to be *blind, or nonreactive,* to incoming new molecules. Hence they cannot make use of such novel molecules in either of two ways:

1. The new molecules cannot be used as food for further metabolism and growth.

2. The new molecules do not interact with existing molecules to unleash a cascade of still more new molecules whose formation is catalyzed within the compartment and which might be of further use to the organization.

But it is just such cascades which afford an analogue of metabolic mutations in which new kinds of molecules are generated and tried out by the organization. If adaptive evolution requires heritable variation, subcritical organizations cannot vary enough to evolve well. This would be expected to create a selection pressure for organizations which are less profoundly subcritical. Such organizations should be able to evolve more efficiently.

Consider next profoundly supracritical vesicle organizations. Each compartment is already supracritical. Then suppose that a new molecule arrives in the environment and is transported into the first, outermost vesicle. Once inside, that novel molecule will unleash a massive cascade of novel types of molecules, never before experienced by the organization. Many of these molecules would be expected to disrupt any useful coordination of processes within the organization. For example, some might destroy membrane sites, thus preventing the outer vesicle from taking in material across the boundary and killing the organization. In short, supracritical systems are so radically altered by any novel molecular input that they cannot stop varying. If evolution requires heritable variation, profoundly supracritical systems vary too much and too unendingly. An adapting population should suffer the error catastrophe noted by Eigen and Schuster (1979): Any useful information or coordination will be lost by an evolving population. This would be expected to create a selection pressure for less supracritical organizations.

If, in fact, selection pressures tend to lower the degree of supracritical diversity and increase the degree of subcritical diversity, the balance struck might lie very near the phase transition, such that any vesicle is just subcritical in the face of the molecular diversity impinging on it from its environment. Such poised systems would seem to be those both best able to adapt and best able to make use of a diversity of resources to grow.

A very crude calculation supports this idea: The biosphere as a whole appears supracritical. A typical human cell appears to be just subcritical. The human genome encodes about 10^5 genes. There may be on the order of 10^8 species. Estimate that the total diversity of proteins in the biosphere is somewhere between 10^5 and perhaps 10^{13}. A conservative estimate of the diversity of organic molecules in the biosphere might be about 10^6. As in Chapter 8, estimate that, on average, any pair of organic molecules with several atoms per molecule can serve as a pair of substrates which undergo at least one chemical reaction. Then the total number of reactions possible is the square of the organic molecule diversity, or 10^{12}. Estimate, based on the probability that a random peptide binds a monoclonal antibody and on the ease of finding catalytic antibodies, that the probability that a randomly chosen peptide or polypeptide binds a randomly chosen transition state and catalyzes the corresponding reaction is on the order of 10^{-9}. Then if the entire molecular diversity of all the organisms in the biosphere were placed in a beaker, in a kind of "Noah's vessel" experiment, such that all molecular species could interact with one another, the minimum expected number of reactions catalyzed would be on the order of $10^{12} \times 10^5$ divided by 10^9. Hence 10^8 reactions would be catalyzed, exploding the diversity of organic molecules from about 1 million to about 100 million. Since those novel organic molecules would be available for further reactions, further supracritical explosion would occur. The biosphere as a whole appears to be vastly supracritical!

Is a human cell, by these arguments, supracritical? The 100 000 genes encoded by our genome control both ontogeny and metabolism. Let us estimate the intracellular organic molecular diversity of metabolism at 1000 species. The genes and proteins within us have already evolved to handle the flow of reactions possible among these 1000 to create the ordered flux of metabolism. Thus any "unwanted" side reactions among the perhaps 1000×1000 possible reactions have been eliminated by selection acting on the enzymes that catalyze metabolic flux. However, in addition to its evolved active site, any protein unavoidably has a variety of epitopes, nooks, and crannies that might catalyze reactions with novel substrates. Imagine injecting a novel organic molecule, Q, into a human cell. Q might interact with each of the 1000 organic molecules to form a two-substrate pair for some novel reaction. By our earlier estimates, the probability that such a reaction is catalyzed is $1000 \times 100 000$ divided by 10^9. Thus the probability is 0.1! Only one in ten such types of novel molecules would be expected to undergo a reaction. Were a novel product molecule, R, formed, it too would have a 10 percent chance of being a substrate for a further reaction, yielding a still further molecule, S. Thus by this very crude reckoning, the human cell is just subcritical!

This argument is far too crude to be taken seriously in detail. But it provides a powerful hint. The biosphere appears to be highly supracritical. The human cell clearly appears to be somewhere near the boundary between subcritical and supracritical behavior. If a human cell is near the boundary, then so too are all other known cells—from bacteria to higher metazoans. Bacteria have smaller genomes, but may experience a somewhat higher diversity of intracellular organic compounds. The human genome probably encodes as many distinct proteins as any other cell. Indeed, the supracritical boundary would appear to be of fundamental importance and might limit the total molecular diversity within present-day cells. Suppose that a eukaryotic cell harbored 10^6 organic molecules and 10^6 types of proteins. If the probability that an arbitrary protein catalyzes an arbitrary reaction is actually about 10^{-9}, then the cell would be quite supracritical. A novel molecule that entered the cell would unleash the formation of about 1000 new types of molecules at the first step, and thereafter the diversity would explode. Thus such a cell would require very precise control over its membrane in order to prevent intrusion. Far easier, it would seem, to live just subcritically and be prey to cascades of novelty from only the rare molecular invader.

The boundary between subcritical and supracritical diversity within vesicles or cellular compartments emerges as a plausible candidate attractor for an evolutionary dynamics. If contemporary cells are bound by this boundary, then the phase boundary probably has constrained molecular evolution for 3.8 billion years. Total diversity within a single vesicle, compartment, or cell cannot transgress the frontier, but total prebiotic and biosphere molecular diversity can and apparently has continued to expand along this phase-transition boundary by creating novel vesicle organizations, other forms of compartmentation, and novel organisms whose own physical boundaries sharply regulate molecular interactions. The fact that the biosphere as a whole is supracritical serves, I believe, as a fundamental wellspring for a persistent increase in molecular diversity. There are perhaps two reasons for the increase in the molecular diversity of the biosphere since life came to rely on genes and proteins. First, genetic mutations may lead to an enzyme which diverts the metabolic flux of existing small molecules down some new pathway to a novel organic molecule. Here the proliferation of organic molecular novelty is driven by the formation of novel proteins. Second, novel molecules arising from *other organisms* may impinge on an organism as toxins or food sources, requiring the adaptive evolution of an enzyme to transform the novel molecule into another novel molecule. Both mechanisms are but other expressions of the fact that the entire biosphere appears to be supracritical. Novel molecules produced in one venue ultimately impinge on another and afford the possibility of novel reactions leading to a further increased diversity of new molecular species.

If these ideas have merit, we are led to think that self-constructing, far-from-equilibrium molecular systems, whose molecular components can regulate the interactions among themselves, may actually flow to the phase transition between subcritical and supracritical behavior, and then diversify further along that frontier. If true, then Carnot may have whispered to us about the shape of general laws governing the behavior of self constructing, far-from-equilibrium molecular systems.

Potential Neural and Psychological Implications

Artificial intelligence has long harbored a debate between those who favor models of the mind based on sequential inference, as exhibited by sequential computer programs, and those who favor models based on parallel-processing neural networks. The former models are widely used in expert systems, in analyses of linguistic and inferential webs, and so forth. As described in Chapter 5, parallel-processing neural networks have reemerged more recently as models of content-addressable memories. Here a dynamical attractor is thought of as a memory or as the paradigm of a class. All initial states flowing to that attractor achieve the desired memory or class. Hence such systems generalize from attractor to basin. Learning consists in sculpting attractor basins and attractors to store desired patterns of neural activity (Hopfield 1982a, 1982b; Rummelhart, McClelland, and PDP Research Group 1986).

Random grammars and the consequent models of strings acting algorithmically on strings to form Jets, Eggs, Mushrooms, or Fogs may be a new and useful marriage of the two classes of AI models. Like sequential-rule-based models, where one action or classification triggers downstream cascades of actions, one string or a set of strings creates downstream cascades of strings. Like parallel-processing networks, many strings can act on one another in parallel to create Jets, Mushrooms, Eggs or Fogs.

Unlike AI models, though, where the couplings among the elementary processes are defined by external criteria, the coupling in grammar string models is defined internally by the grammatical rules which determine how strings generate one another. There is an important sense in which the "meaning" of one elementary process with respect to others is given by local production transformations and the global structure—Jet, Egg, Mushroom—and its natural dynamics. Random grammars are somewhat like Holland's (1986) problem-solving computer algorithms called classifier systems, in which rules cast as binary strings trigger the firing of other rules, attain a fitness or "strength" dependent on payoff in a mock economy, and coevolve with other rules by mutation, recombination, and selection. In Holland's case, the couplings are governed by match criteria by which the action part of one rule acts on the message condition part of another rule.

Another feature of grammar models is that the set of neural network processes is open and potentially infinite, unlike the case with parallel-processing models. Such open systems may remain perpetually changing, always out of equilibrium, always adapting, rather than falling to simple dynamical attractors.

It is not entirely implausible that such grammar-string models may prove useful in thinking about the "schemas" by which personality elements are constructed. Consider, for example, the stunning phenomenon of multiple personalities. Typically, each "self" has only faint or no awareness of the alternative personalities. The situation is like a gestalt shift when regarding a Necker cube. When seeing the cube in one way, one cannot simultaneously perceive it in the second way. The two views are mutually exclusive perceptual organizations of the visual world. It seems of interest to consider an Egg able to interact with an external world as a kind of self which knows and organizes its world in some self-consistent way. The same system may harbor more than one Egg, however, each mutually exclusive of other Eggs, each living in its own self-consistent world.

Parallel-processing grammar models may help relate holism in science to stability of ego structures and centrality in the web of string processes. First, consider the thesis of holism in science. Suppose I hold the earth to be flat, and you hold it to be round. We perform a critical experiment at the seashore, watching a ship sail out to sea. I predict it will dwindle to a point. You predict the hull will lapse from sight before the superstructure. Your prediction is confirmed. "The world is round, admit it!" you claim in jubilation. "No," I respond, "light rays fall in a gravitational field, so of course the hull disappears first." The point, first stressed by Quine (1961) is that any hypothesis confronts the world intertwined in a whole mesh of other hypotheses, laws, and statements of initial conditions. Given disconfirming evidence, consistency requires that *some* statement(s) of the premises be abandoned, but we are free to choose which premise we shall abandon and which we shall save. I can save my hypothesis that the earth is flat at the price of a very bizzare and convoluted physics. We cannot avoid Quine's point. Typically, we choose to save those hypotheses that are the most central to our conceptual web and give up peripheral hypotheses or claims about initial conditions. But that very choice renders those central claims very hard to refute, indeed, almost true by definition. Now the interesting point to add is that the hypotheses we choose to save are those which, in a graph theoretic sense, are central to the conceptual web. Let us use, as our model of a conceptual framework, a string process that creates an Egg, a Mushroom, or another object connected via string exchange to an outside world. That Egg entity will have more central and less central elements. If an Egg is a self knowing its world, preservation of self becomes preservation of the central elements in the Egg while a peripheral metabolism fluctuates into and out of existence. Indeed, one wonders if the concept of resistance in psychotherapy, a phenomenon familiar in practice if hard to quantitate, can in part be made sense of in terms of preservation of core elements of the Egg. One can consistently continue to maintain that the world is flat despite apparently enormous evidence to the contrary.

Models of Cultural Coherence and Transformation

What did China's leaders know in the summer of 1989? What occurs when an isolated culture comes into contact with a world culture? What constitutes the integration and coherence of a culture and how do new ideas, myths, or production techniques transform the culture? Just as it is a vast jump from grammar-string models to models of personality structure, so too is it hubris to leap to cultural models. Yet the phenomena feel the same. New strings are injected into an Egg. It transforms to something different and coherent, perhaps another Egg, another closed, coherent culture. Conversely, modern society is open, explosive, changing, indefinitely expanding in ideas, goods, services, myths. Have we now become culturally supracritical? Can we construct models in which cultures can be stable Eggs but then transform into a different kind of object, perhaps a Fog? It seems worth considering.

Application to Models of Technological Evolution of Economic Webs

Grammar models may prove useful in developing a new class of theories about technological coevolution. It is quite surprising that, although technological evolution is thought by many economists to be a major, perhaps the preeminent, factor driving modern global economic growth, economists lack a coherent theory of the phenomenon. The problem is that the issue is not merely economic; it is technological. In a way which requires understanding, the goods and services in an economy themselves offer new opportunities to invent yet further goods and services. In turn, new goods and services drive older goods and services out of the economy. Thus the system transforms. For example, the invention of the automobile led to the requirement for a host of other goods and services, ranging from paved roads, traffic lights, traffic police, and courts to oil refineries, gasoline stations, motels, automobile-repair facilities, parts manufacturers, and emission-control devices. And the advent of the automobile led to elimination of the horse for most transport. With the horse went stables, public watering troughs, blacksmiths, the Pony Express, and a host of other goods and services.

This example states the problem faced by the economist. In order to understand the current web structure of the goods and services of an economy, and how that structure governs its own possibilities of transformation by the invitation to invent new goods which intercalate into the web, transform it, and eliminate other goods, one needs a theory for which goods and services fit together technologically.

Economists call such fitting "complementarity." Thus nut and bolt are complements, hammer and nail are complements, and so forth. Complements are sets of goods or services which are used jointly to produce a given other good, service, or consumer product. Substitutes are sets of goods which might substitute for one another in a given production technology or consumption good. Screws can substitute for nails; potassium chloride can substitute for ordinary salt at dinner.

We have no theory of which goods are substitutes or complements of one another, but grammar models provide a novel approach. Just as any grammar mapping the power set of strings acting on itself into itself is a mock-up of the laws of chemistry, so too any such grammar is an "as if" model of the unknown laws of technological complementarity and substitutability. Goods and services are modeled as symbol strings, and machines, input bundles, and output bundles become production technologies. We have no idea what those technological laws are, but if we can find that large regions of grammar space yield model economies which behave much as do real economies, we shall have grounds to map real economic technological growth to the same universality class.

A concrete way to build grammar-string models of economic growth is the following. First, specify a grammar by which strings act on one another to produce strings. The set of strings which are a machine M and jointly act on a string or set of strings to produce an output set are complements. All parts of M are needed to make the product. Alternative strings or sets of strings which, as input to M, yield the same output set are substitutes. Weaker senses of complements and substitutes arise if output sets which are overlapping but not identical are considered. The transformations specified by the production technologies specify the numbers of each type of string required as input or machine part to make a specified number of each kind of output string.

Economist Paul Romer pointed out that the grammar implies an input–output matrix. Using this framework, a simple, formal economic model can add constraints on exogenous inputs to the economy, such as raw material mined from the ground. These constraints might be supplied by a founder set of strings maintained at a constant "concentration." To carry economic analysis further, the utility of each string must be specified. Given these constraints, the equilibrium for the current economy specified in terms of the linked set of goods and services is that ratio of production of all goods and services which maximizes the total utility of all the goods and services in the economy subject to the constraints. That ratio can also be thought of as the price of the goods relative to one another, taking any single good as the unit.

The growth of the economy over time in terms of the introduction of new goods and services can be studied as follows. Start at the current equilibrium with the current set of goods and services. Use the grammar rules to construct all possible new goods and services derivable by allowing the current goods and services to act on one another in all possible ways. Doing so generates all possible new goods which are technologically next to those in the current economy. (Alternatively, you could choose some random or nonrandom subset of these new goods as potential new goods.) Now construct the next economy, which contains the potential new goods and services plus all the current goods and services. These new and old goods specify, via the grammar, a new input–output matrix for the economy. Now assess the equilibrium of the new economy, as derived from its modified input–output matrix. At that equilibrium, some of the new potential goods may make a profit and hence are produced at a positive rate. Others may make a loss and hence are not produced at a finite rate. Similarly, some old goods will still make a profit, and others will now make a loss. The new economy comprises only those old and new goods which jointly make a profit. Hence over time, cascades of new goods enter the economy, and cascades of old goods are driven from it.

I have begun to investigate very simple economic models of this type in order to explore the implications for technological evolution. In these models, a grammar specifying all possible input bundles, machines, and output bundles is chosen. The grammar implicitly defines the possible technological evolution of the economy. All goods are assumed to be consumable. In addition, some goods can be inputs or machines yielding further goods. For simplicity, there is a single consumer. The utility of each good is specified by a spin-glass-like function. In addition, there is an infinitely rational social planner whose task it is to decide which possible production functions are to be carried out and at what rate, at each period of time, in order to maximize the overall utility of the consumer.

At each period, a defined set of diverse symbol strings can be harvested from the ground. These strings serve as renewable resources. The social planner begins at period 0 and either thinks ahead or looks ahead for some finite number of periods T. Her first task is to conceive of all possible goods which might be created in the first period, given those growing from the ground. Then she considers all the goods which might be created in the second period from those created in the first period and so on for T periods ahead. For sufficiently complex grammars and sufficiently diverse types of strings emerging from the ground, the set of possible goods and services expands over look-ahead periods. The planner's next task is to create an optimal plan for the T periods. To do so, she optimizes a specific utility function which discounts the utility of future goods by a constant fraction $B < 1.0$. If an apple today is worth 100, the same apple tomorrow is $B \times 100$. The same apple in T days is $B^T \times 100$. The total utility to the consumer is just the sum of the discounted utilities over the T periods of all the goods and services provided at each period. This simple utility function yields a linear programming optimization problem which the planner solves. The solution specifies an optimal plan showing which goods are consumed, which are produced, and which are carried over to the next period, for all periods from 0 to T. Thereafter, the planner implements the optimal plan for the first period. She again plans T periods into the future, hence from period 1 to $T + 1$, creates a new optimal plan, and carries out the first period of that optimal plan, thereby carrying the evolving economy to period 2. Over iterations, the economy evolves.

Economic models of this type are of interest in a number of regards.

First, they model economic growth due to the growth in niches afforded by goods to create new goods.

Second, such systems afford novel models of economic takeoff. The behavior of the economy depends on grammar complexity, the diversity of renewable resources, the discount factor B, and how far into the future the social planner looks at each period. Economies which have too few renewable goods or too simple a grammar may not be technologically able to be supracritical and hence may never take off. If the grammar is complex, however, or if more types of renewable resources are available, or if several economies come into contact and exchange goods and services, the coupled system may jump from one in which each separate economy makes a small finite Jet to a supracritical Mushroom which explodes into the space of potential goods and services. Hence this is a model for economic takeoff. This behavior suggests that technological diversity is a major factor in abetting economic growth. Recent evidence appears to support this. Growth in cities appears most strongly correlated with industrial diversity and not with concentration within single industries (Schenkman et al. 1991). In turn, this correlation carries policy implications: to engender growth, it may be better to support a web of cottage industries than the Aswan dam.

The extent to which the planner looks into the future governs whether the economy grows at all, slowly, or rapidly. Figure 10.3 shows a model economy plus the growth of the total number of goods and services over time as a function of the plan-

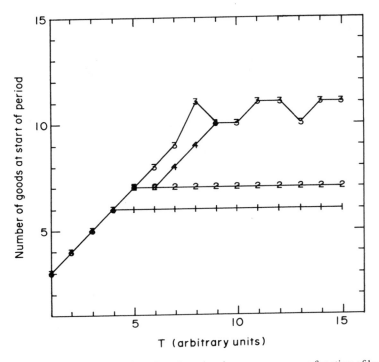

Figure 10.3   Increase in diversity of goods and services in an economy as a function of lookahead. The *x* axis represents number of periods through which the economy has evolved. Each curve is labeled with the planning horizon of the social planner.

ning horizon. The economy rapidly stagnates in terms of diversity when the horizon is only one or two periods ahead, but diversity increases when the horizon is longer. This result is general in this class of models. Naturally real economies have no social planner, but the expectation that technological growth is strongly correlated with the capacity to see its implications is reasonable. Finally, growth in diversity of goods does not occur as rapidly if *B* is close to 0. If the consumer places little value on the future, diversity of goods and services remains small.

Third, these model economies can be expected to exhibit enormous historical contingency coupled with lawlike behavior. If, at each stage in the growth of the kinds of goods and services in the economy, a random subset of potential new goods is tried and some are accepted, that sampling process will strongly bias the future directions of growth. Hence the goods which emerge and integrate into the system will become frozen accidents guiding the future evolution of the system in ways similar to biological evolution. Yet the statistics of the process—the size distribution of new sectors, the numbers of new goods entering and old goods leaving, the changes in richness of interconnection within the web—may all be stable given membership in a regime or class of grammars.

Fourth, several features of these models invite modifications of the core of neoclassical economics, which rests on general competitive equilibrium theory. This beautiful theory demands infinitely rational economic agents and what are called "complete markets." Such markets allow exchange of all possible dated contingent goods. An example would be a contract to deliver apples tomorrow if it rains in Nebraska today. In the absence of infinite rationality and complete markets, we see failure of the beautiful theorems guaranteeing an equilibrium such that all markets clear. If we take grammar models seriously as models of technological evolution, then the decidability problems in Filigreed Fogs and other objects imply that it may be logically impossible to deduce that a given good is ultimately producible from the current technologies. This implies that markets must be *incomplete*. The same failure of decidability may imply that economic agents must logically be boundedly rational. Both these latter two points cut at the core of neoclassical economics and hence may invite its extension.

Fifth, grammar models hold the hint of that extension, a hint which suggests the possibility of a theory of bounded rationality and a new nonequilibrium solution concept: Markets may not clear. Rather, boundedly rational economic agents may achieve the edge of chaos, where markets come close to clearing. At the edge, bankruptcies, like extinctions, would propagate through the economy from time to time.

Figure 10.4 shows the grounds for these hints. Maintaining the fiction of a social planner, the figure examines the behavior of the optimal plan for the first period as a function of lookahead (10.4a) and as a function of a slight change in the utility function (10.4b). In Figure 10.4a, as the planner extends the planning horizon further into the future, the plan for the first period at first changes a lot and then gradually settles down and stops changing. Suppose, however, that the planner is a bit unsure of the consumer's utility function. Then let her assume a slightly modified utility function and generate a new optimal T period plan. Figure 10.4b shows the difficult results. The further into the future the planner thinks, the more dramatically the optimal plan for the first period under one utility function diverges from that under the slightly modified utility function! In short, given uncertainty over the consumer's utility function, the further the planner thinks into the future, the less certain she becomes about the actions to be taken in the very first period. Thinking into the distant future has become counterproductive! Yet the planner is tempted to think into the future: The further she thinks ahead, the more an optimal plan can take account of the highly valuable novel goods and services which can be constructed from the renewable resources. The total expected utility to the consumer goes up as the planning horizon extends. The consequence of these countervailing effects is this: As the planner extends the planning horizon, expected utility to the consumer rises but the risks of being drastically wrong increases if she is slightly wrong about the consumer's utility function.

It seems plausible, but I have not yet shown, that an optimal planning horizon is the consequence of the countervailing advantages of thinking further into the future but risks increasingly catastrophic misallocation of resources if a mistake is made about the consumer's utility function. Rather than thinking into the infinite future and basing action on an optimal plan, the planner should take account of uncertainty over utility functions and think an optimal distance into the future, balancing risk and reward. Such an optimal planning horizon offers a clear sense of a kind of bounded rationality: It pays to plan only so far ahead.

A further potential implication of this model is that a multiplicity of economic agents may attain the edge of chaos. In this model, lacking markets, the analogue of market-clearing equilibrium is the optimal plan where the flows into and out of all production functions mesh perfectly. Consider a modified economic model in which there is no social planner and each production technology belongs to one of four

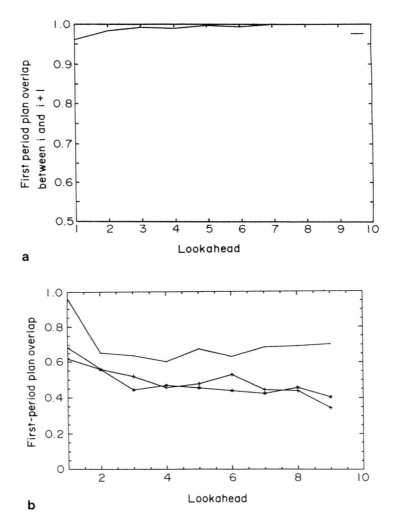

Figure 10.4 (*a*) Overlap of optimal plan for the first period as a function of planning horizon. As the social planner thinks further into the future, the first-period plan changes less and less. Hence the overlap between successive lookahead periods increases to 1.0. (*b*) Overlap between the optimal plan for the first period as a function of planning horizon when two slightly different utility functions are used to generate the optimal plans. A plan generated by an unperturbed utility function is compared with plans having small, modest, and large random perturbations of the utility function. As the planning horizon increases, optimal plans for the first period diverge.

firms—Red, Blue, Green, and Yellow. Suppose each firm is trying to decide how much of each of its production technologies to run, under uncertainty about the consumer's utility function. Let each firm reason as did the social planner. Suppose each assumes a slightly different utility function for the consumer, constructs an optimal plan for some period T, and attempts to execute the first-period plan. Alas, Red assumed four kilograms of butter supplied by Green to allow Red to make cakes, but

Green planned to produce only three kilograms. The mismatches of flows between firms are losses in total utility produced and losses to each firm. Yet even given uncertainty over the consumer's utility function, each firm would be tempted to think far into the future to take account of coming technological evolution. It is plausible that, in this world, economic agents will balance how far into the future they plan against the increasing risks of discoordination if they are slightly wrong. Note that the *further into the future each agent plans, the more rugged is his fitness landscape!* I would like to believe, but have not shown, that generically such agents tune how far into the future they plan such that the set of agents achieves the edge of chaos. Here, the analogue of markets would not clear but would fluctuate near clearing in a characteristic way. Small and large bankruptcies would propagate through the system.

An even broader basis suggests that in order to optimally predict one another's behavior, complex adaptive agents will build optimally complex, and hence boundedly rational, models of one another. Such adaptive agents might well coevolve to the edge of chaos. The framework of ideas has several steps.

First, given *finite data,* models which can optimize the capacity to generalize accurately must be of optimal, intermediate complexity. For example, given a set of 20 prices of oil versus time over the past 20 years, an overcomplex model connecting all the data points precisely is merely a lookup table, and hence is incapable of generalizing. Overfitting the data is captured by the idea of using such a high-order polynomial equation that there are as many parameters as data points. In the neural net field, overfitting of data arises when too many hidden units are used. Conversely, an oversimple model—for example, fitting a straight line through the data—also generalizes poorly, since it ignores major trends in the data. In general, given finite data, optimal models best able to generalize are of an intermediate complexity tuned to the data available. The complexity of a model can be roughly defined as the number of free parameters needed to specify it. Thus one might constrain models to use seven Fourier modes in the attempt to fit the data.

Second, when adaptive agents make models of one another as part of their mutual ongoing behavior, the *eventual failure* of any finite, approximate model of another's behavior drives substitution of a "nearby," optimally complex model of the other's behavior which now appears to be the best fit to the other's behavior. Here the point is that, given any finite set of data, multiple models of about the same complexity will fit the data roughly as well. As the data stream evolves, overlapping patches of the data are optimally fit by nearby models drawn from a set of models of the same complexity.

Third, adaptive agents may persistently alter their models of one another's behavior. Once an agent adopts a changed model of another agent, then his own decision rules, and hence behavior, will change. It follows that such agents must *coevolve* with one another using changing models of one another's behavior.

Fourth, presumably, such coevolving behavior can be chaotic, ordered, or at the edge of chaos. Chaotic behavior would correspond to rapidly changing models of the other agents. Ordered behavior would correspond to converging on a mutually consistent set of models of one another. This yields full behavior coordination and is close to the theory of rational expectations in economics. At the edge of chaos, models of one another would be poised, tending to change, unleashing avalanches of changes throughout the system of interacting agents.

Fifth, a qualitative argument suggests that, in a persistent attempt to optimize prediction about the behavior of other agents, adaptive agents will alter their finite, optimally complex models of one another so that the entire system approaches the edge of chaos. If the dynamics are very stable and mutually consistent, then each agent has an abundance of reliable data about the behavior of the other agents. Given more data, each agent naturally attempts to improve his capacity to generalize about the other agents' behavior by constructing a *more complex model* of the others' actions. This more complex model is necessarily more sensitive to small alterations in the other agents' behavior. It lives on a more rugged landscape. Thus as agents adopt more complex models to predict better, the coevolving system of agents tends to be driven from the ordered regime toward the chaotic regime. Conversely, in the chaotic regime, each agent has very limited reliable data about the other agents' behavior. In part, the absence of reliable data reflects the fact that in the chaotic regime each agent adopts successive models of the others, thus actually changing the "law" or decision rules governing his own behavior. Given the small amount of reliable data, each agent, in order optimize the capacity to generalize, is driven to build a *less complex model* of the other agents' behavior. These less complex models are less sensitive to the behavior of the others, and thus live on smoother landscapes. The substitution of simpler models therefore drives the system from the chaotic regime toward the ordered regime. The process, I hope, will generically attain the edge of chaos.

This framework is entirely untested. It seems to hold promise, however. At present, there appears to be no established framework which successfully characterizes the coordination of interacting agents. The classical solution in economics is the invisible hand, which leads to price equilibrium at which markets clear. However, the invisible hand can fail if more than one good is in the economy. The equilibrium price at which supply and demand balance can be unstable to price fluctuations. Nash equilibria in game theory provide a second framework for coordination. However, in general there is no dynamics among the agents which ensures convergence on a Nash equilibrium. Further, Nash equilibria, as in the prisoner's dilemma, are often not particular good solutions. Rational expectations in economics argues that it can be rational to speculate on bubbles because of a convergence of beliefs among agents. But given any finite data, multiple models or expectations fitting that data are always possible. The size of the set of alternative models which might fit the data increases with the complexity of models that the agents might construct. Thus convergence of all agents on any single expectation or vector of expectations presumably cannot be guaranteed (Grammont, personal communication). In place of these established attempts to find a general framework by which agents coordinate behavior, the approach outlined here hopes to find coordination to the coevolving edge of chaos among agents bent on optimal prediction of one another's behavior. If correct, it may help us understand that *E. coli* and corporate executives build optimally complex, boundedly rational, models of the other agents constituting their worlds.

Why do such economic models matter in this book? Economics is the study of allocation of resources to optimize something like utility. If the bold hypotheses of Chapters 5 and 6, that complex systems achieve the edge of chaos internally and collectively, were to generalize to economic systems, our study of the proper marriage of self-organization and selection would enlist Charles Darwin and Adam Smith to tell us who and how we are in the nonequilibrium world we mutually create and transform.

SUMMARY

We are interested in how complex systems come to exist and adapt. Current theories, as discussed in Chapter 5, are based on dynamical systems in which the elementary variables may be molecules, genes, organisms, neurons, cognitive elements, or cultural roles. Yet in these theories there is no underlying microscopic theory or account of the functional couplings among the variables. Rather, these couplings must be postulated or defined based on some external criteria which might range from observation to ad hoc.

In this chapter, I introduce a new class of models based on random samples from grammar space. In contrast to standard dynamical theories, grammar models yield a theory of the functional couplings among the variables. The approach is a generalization of studies of autocatalytic peptide and RNA systems and of Fontana's recent extension of that work to AlChemy, an algorithmic chemistry. In this further generalization, I consider finite power sets of symbol strings whose lengths may be arbitrary. Ordered sets of strings act as machines on ordered sets of strings as input bundles to yield unique output bundles of strings. Limitation on the number and length of output strings as a function of the number and lengths of input and machine strings ensures that the system, if started with a finite number of finite-length strings, remains finite after finitely many iterations. Each grammar specifies a unique mapping of this set into itself. The transformations mediated by symbol strings on one another *are* the functional couplings among the symbol strings.

In this string world, new kinds of dynamical objects—here dubbed Jets, Lightning Balls, Mushrooms, Fixed or Traveling Eggs, Filigreed Fogs, and Pea Soups—arise and inhabit string space. In addition to the compositional character of such sets, we may consider their dynamical behavior. An Egg, which is a collective identity operator in string space, may re-create its strings in a steady state, in a limit cycle, or on a chaotic attractor.

Eggs and autocatalytic Mushrooms and other sets seem natural models of integrated function and even of agency. In these models, strings of symbols can stand for molecules, neural activities, cognitive elements, goods or services, or cultural roles. Thus such objects may be useful models of prebiotic molecular evolution, of ecosystems and organisms among which mutualistic or competitive interactions may coevolve, and of parallel-processing neural systems, ego systems, economic systems, and cultural systems. In all cases, we attain models of functional integration and transformation.

Biology is a deeply historical science which may yet be the locus of law. An old debate wonders what it might be, in a historical science, to exhibit laws. Grammar models promise answers for such issues. The specific regions of string space inhabited by any such system may often be critically dependent on frozen accidents, and yet phenomena such as phase transitions, the sizes of cascades triggered by minor perturbations, and the statistical connectivity features of Jets, Mushrooms, Eggs, and Fogs may all show powerfully lawlike behaviors. This may be just the conceptual scheme we need: a locus of law, accident, design, selection, ever unfolding and transforming in novel functionally integrated forms.

It is as if we were waiting for a new Carnot. Grammar models allow us to study model worlds of pure linked processes. These processes, in general, are not reversible and hence can be seen as arbitrarily far from equilibrium systems. We have no general theory about the behavior of such systems. Indeed, if we construct such systems, they can function as universal computers about which no predictions can be made. Perhaps if we model worlds of coupled processes which govern their own transformations, however, worlds in which autocatalytic sets emerge and coevolve in some wavefront of increasing complexity, building models of one another, ever transforming, ever subject to frozen accidents and avalanches of change, perhaps then we shall find general principles.

In this chapter, I have suggested one candidate general principle: Open far-from-equilibrium, self-constructing systems whose molecular variables can regulate which variables interact with one another may form organizations which evolve to the phase transition between subcritical and supracritical behavior. Each vesicle, compartment, or cell would be slightly subcritical. The system of interacting vessicles, compartments, or cells would, as a whole, be supracritical. Crude estimates of the total organic molecular diversity and protein diversity of the biosphere strongly suggests that the biosphere as a whole is very supracritical. By the same estimates, human cells, and thus all other cells, appear slightly subcritical. If so, then 3.8 billion years of life have respected the subcritical–supracritical boundary. Cells may have evolved to that boundary. The total diversity of molecules may have been driven to increase, in part, by the supracritical character of the entire biosphere. If so, supracritically is an ultimate wellspring for the growth of molecular diversity. The phase transition as an evolutionary attractor for open, self-constructing molecular systems may hint at the answer Carnot would wish to suggest.

The largest intellectual agenda of this chapter is based on the presumption that—by analyzing a variety of grammars from regions of a parameterized grammar space, each grammar a kind of hypothetical set of laws of chemistry or functional complementarity—a few broad regimes will emerge. Where we can map such generic behaviors onto molecular, organismic, neural, psychological, economic, or cultural data, we may have found the functional universality class needed to explain phenomena in these areas of chemistry, biology, and the social sciences. In Chapters 5 and 6, I stressed that complex systems might adapt and coevolve to the edge of chaos both in terms of internal structure and in terms of mutual interactions. The approach of this chapter points toward the proper extension of these efforts in understanding the emergence of functionally coupled, coevolving systems. Indeed, we found grounds for another bold hypothesis. Coevolving adaptive agents attempting to predict one another's behavior as well as possible may coordinate their mutual behavior through optimally complex, but persistently shifting models of one another. Again, we suspect, the edge of chaos will be attained. We may find that *E. coli* and IBM do indeed know their worlds in much the same way.

PART III

Order and Ontogeny

The remainder of this book looks at ontogeny in current organisms. In these introductory pages preceding Chapter 11, I outline the two main problems in ontogeny: cellular differentiation and morphogenesis (for a good general discussion, see Gurdon 1974). The former includes all the processes by which cells become different from one another; all the latter, the processes by which either cells of the same type or cells of different types become coordinated into organized tissues, organs, and morphologies. Chapters 11, 12, and 13 discuss the control of cellular differentiation, and Chapter 14 discusses morphogenesis. These are vast topics. My aim throughout is to attempt to characterize two features about ontogeny—those aspects which may reflect the self-organized properties of the underlying genomic and biochemical system and those which reflect selection—and to determine a way of recognizing the marriage between the two. If our look at ontogeny is to be useful, we must find not only ways of thinking about these issues but also ways of turning them into research programs.

Multicellular organisms have existed at least since the late Precambrian. There is no reason to think that the major features of ontogeny have changed in 600 million years or more. Almost all multicellular plants and animals begin as a single cell. Typically this single cell is the zygote, formed by union of the male and female gametes. During development, this single cell undergoes 10 to 50 mitotic divisions, creating $2^{10} = 10^3$ to $2^{50} = 10^{15}$ cells, which form the adult. Perhaps the most critical single fact about ontogeny in almost all organisms is that the set of genetic instructions is identical in all cells of the organism. The familiar example—which, despite its familiarity, should enthrall us all—is the carrot (Steward 1958; Steward, Mapes, et al. 1964). The mature carrot can be dispersed into single cells; under appropriate conditions, any single cell type—whether from root, cambium, or leaf—can be induced to divide, form a rudimentary tissue mass, undergo organogenesis, and re-form an entire organized carrot. From this it follows that each cell of the carrot carries the genetic instructions needed to form an entire carrot. In other words, ontogeny is an overwhelming example of self-organization.

It is not entirely true that all cells of each multicellular organism retain identical information, however. In some cases, entire chromosomes sets are lost, or maternal or paternal chromosomes are lost. In other, rare cases, specific genes in the genomic set are amplified to form extra copies in only some cells. For example, the chorion genes in the fruit fly, Drosophila melanogaster, *responsible for proteins in the chorion covering of the egg, are amplified in the cells which make the chorion (Spradling and Mahowald 1980). Such alterations are rare, however. Almost universally, the DNA appears to be identical in all the different cells of an organism. This uniformity leads to the* central dogma *of developmental biology: Cells are different in an organism because different genes are expressed in different cell types. In turn, the genetic identity of different cell types focuses attention on understanding the mechanisms which control and coordinate differential gene expression in the organism. It is this* genomic regulatory system *which orchestrates cell differentiation and morphogenesis. This view is, in fact, the contemporary expression of Weismann's doctrine (1885, 1904), discussed in Chapter 1. The genomic system constitutes the developmental program.*

An overview of the probable complexity of this genomic system is a worthwhile exercise. A higher mammal has enough DNA to encode on the order of 2 million average-size proteins. Based on the diversity of messenger RNA species found in one or another cell type in the organism, the total number of different kinds of proteins may be on the order of 20 000 to 100 000 (Alberts, Bray, et al. 1983). In turn, the synthesis or lack of synthesis of each protein is almost certainly under the control of a variety of regulatory genes and processes, described in detail in the next few chapters. Focusing *attention on structural genes—that is, those coding for proteins—and simplifying to imagine each protein to be either present or absent in a cell, we find that there are at least $2^{20\ 000}$ possible combinatorial patterns of gene expression. This is 10^{6000}, a number vastly larger than the number of hydrogen atoms in the known universe. It is the genomic regulatory system which constrains into useful behavior the patterns of gene activity during ontogeny. The problem, simply stated, is to understand how such coordination is achieved and how it could have evolved. The central tenet of the ensuing chapters is that many of the highly ordered properties of genomic regulatory systems are* spontaneous, self-organized *features of complex control systems which required almost no selection at all. Clearly, if much of the order we see in ontogeny reflects the natural features of complex control systems, we must rethink evolutionary biology. Some of the sources of order lie outside selection.*

Cellular Differentiation

During mitotic divisions, the progeny cells of the zygote become different from one another via the process called cellular differentiation. *In virtually all known organisms, differentiation proceeds down* branching pathways *such that one cell type gives rise to two or a few new cell types and they in turn branch to give rise to a few further cell types, until the spectrum of cell types characterizing the adult organism is eventually formed. While ontogenies in Cambrian metazoans and metaphytons cannot be studied directly, presumably they too followed branching pathways of differentiation. Thus presumably such pathways are a deep property of ontogeny.*

There are probably three fundamental ways differentiation occurs (Gurdon 1974):

1. *A single cell becomes different over time. Equivalently, a single cell and its daughters become different over time, but all daughters remain identical to one another. Only mitotic division has intervened in this pattern of change.*

2. *Two daughter cells become different from each other. If one remains the same as the mother cell, it is called a* stem cell. *Alternatively, two daughter cells differ from each other and from the mother cell. Almost certainly, such branching differentiation is controlled internally by the cells. In general, one supposes that the two daughter cells become different because they are placed in different states at division, perhaps by differential distribution of key molecules. Examples include asymmetric divisions in the development of the molluscs (Dohmen and Verdonk 1979), in the nematode* Caenorhabditis elegans, *in yeast, and in neural development in the grasshopper.*

3. *Cells are induced to differentiate by exposure to external factors derived from the environment or from other cells. The classical example here is induction of the ectoderm to form neurectoderm by contact with the underlying mesodermal mantle in the early vertebrate embryo. Similar examples are rife in vertebrate development and include sequential inductive events in the differentiation of the eye and the ear. In most cases, it seems clear that specific substances, probably small molecules, transfer from one cell to a neighbor and are the normal inductive agents.*

A particularly striking fact that emerged from the initial attempts to isolate the normal inducer molecules for neurectoderm development is that cell types are poised *among only a few alternative pathways of differentiation. A variety of compounds, including pure chemical substances, even changes in pH and pure water, were found to induce embryonic ectoderm to differentiate into neurectoderm. Thus the critical conclusion is that the ectoderm is, as noted, poised between only two alternative pathways and can be triggered to differentiate into neurectoderm by a variety of normal and abnormal stimuli.*

The property of being poised between only two or a few alternatives, called com-petence, is almost certainly directly related to branching developmental pathways. In turn, branching pathways almost certainly reflect the fact that each cell type is a highly constrained pattern of gene expression. At each stage in ontogeny, each cell type has only a few accessible neighboring cell types. The specificity of the inducer-created differentiation step lies in the responding cell and in the constraints in the pathways of change open to it. Thus branching pathways and the poised properties of competent cells are both a reflection of an idea we can state qualitatively: Cell types must be very constrained patterns of gene expression among the $2^{20\,000}$ to $2^{100\,000}$ possibilities. Then exogenous inductive signals can trigger each such constrained pattern, or cell type, to change to only a few other constrained patterns.

In the following chapters we shall see that the achievement and maintenance of the poised character of cell types and of the universality of branching pathways of differentiation have not required extensive selection since the Cambrian. Rather, highly constrained, poised cell types and ordered patterns of gene activity, each able to change to only a few others, are gratuitously present in a vast class of genomic regulatory systems. Recall from Chapter 5 our discussion of parallel-processing Boolean networks. These may lie in the ordered regime, the complex regime, or the chaotic regime. The phase transition from one regime to another is governed by simple parameters of the system, such as richness of coupling among the variables. The order seen in ontogeny, I shall suggest, is just that which arises spontaneously in the powerfully ordered regime found in parallel-processing networks. Selection, I shall further suggest, by achieving genomic systems in the ordered regime near the boundary of chaos, is likely to have optimized the capacity of such systems to perform complex gene-coordination tasks and evolve effectively.

One striking fact about the ordered regime is that the generic properties of complex parallel-processing systems in it also appear to account for many other characteristic features of ontogeny and cell differentiation. These properties range from the number of cell types in an organism as a function of genomic complexity, to the avalanches of gene activity alteration in a cell when the activity of one gene is transiently altered, to the similarities and differences in gene activity patterns in different cell types of one organism, and even to the existence of archaic cell types which can be reexpressed in an organism in unusual circumstances. The very nonintuitive point of the next three chapters, in short, is that much of the order seen in ontogeny is due to a previously unsuspected and powerful tendency for order to emerge in massively complex systems. Selection had such order to work with from the outset. And, wonderful if true, selective adaptation toward the boundary between order and chaos, the true marriage of self-organization and selection, may prove to be an ahistorical universal in biology.

Morphogenesis and Pattern Formation

Ontogeny unfolds into an ordered organism: cells in tissues, tissues in organs, organs in proper array. Spatial organization, or pattern formation, must reflect the coordinated, coupled behaviors of cells. This topic, morphogenesis, is the focus of Chapter 14. Our general framework for thinking about morphogenesis includes the ideas that cell types may adhere specifically to certain other cell types, may divide in oriented ways and controlled rates to generate ordered spatial patterns, may move relative to one another, may exert physical forces on one another which literally generate shapes, and may emit and receive, from neighboring or distant cells, chemical or physical signals which control the differentiation, growth, or shape of individual cells at specified positions within a maturing tissue. A great deal has been learned about these processes in a number of organisms. Just as we may wonder to what extent the deep features of cell differentiation are achievements of selection, so too we may ask whether there are underlying general properties of morphogenesis and, if such properties do exist, to what extent do they reflect more or less inevitable properties of coupled cells, guided by a complex genomic regulatory system?

The problems we confront here include old ones. D'Arcy Thompson (1942), for one, made famous many examples of organismic forms which closely approximate simple physical forms. Radiolaria, for example, distribute spicules in the interstices separating protoplasmic bubbles which are organized as if to minimize surface tension, just as are soap bubbles in a child's loop. How are we to recognize such apparently physical properties in morphogenesis? And, in our search for the relation between spontaneous order and selection, how are we to understand the ways in which the genome and its evolution can use, modify, and be limited by those abiological morphogenetic processes? Morphogenesis is not just the genome's "doing"; rather, it is the consequence in time and space of the structural and catalytic properties of proteins encoded in time and space by the genome, acting in concert with nonprotein materials and with physical and chemical forces to yield reliable forms. Our problem is to understand not merely how genes influence morphology, but how they marry to the entire morphogenetic process, and how that process may evolve.

In Chapter 14 we discuss aspects of morphogenesis ranging from induction, to pattern duplication and regeneration, to pattern formation and morphogenesis in such varied organisms as slime molds, fruit flies, molluscs, and vertebrates. A central concept we shall discuss is that any developmental mechanism gives rise to a well-defined and natural family of forms. *Shifting among members of such a family occurs as parameters of the developmental mechanism are altered. The morphologies which arise as parameters change smoothly define what we must mean by morphological neighbors. This is the structuralist theme struck by Webster and Goodwin (1982), mentioned in Chapter 1. The proper marriage of self-organization and selection with respect to morphology interweaves two major strands. First, some developmental mechanisms lie to hand in the evolution of morphogenesis. The evolutionary discovery of such mechanisms appears almost inevitable. Consequently, some families of form lie to hand. Second, among any family of forms, selection is unlikely to be able to avoid those forms or morphologies which correspond to large volumes of the parameter space of the developmental mechanisms. It is just these forms, the ones naturally generated by the governing mechanisms, which we must expect to find.*

In summary, Part III searches for and finds deep signs of self-organization in ontogeny underlying the wondrous coordination of gene activities governing cell differentiation in systems with tens or even a hundred thousand genes. That order reflects the unexpected, emergent order in a vast class of parallel-processing regulatory systems. Similarly, the fundamental chemical and physical mechanisms which underwrite morphogenesis were, in many cases, natural emergent features of interacting cells. The natural families of forms which follow from those mechanisms lay ever available to selection. Selection, I suggest, has molded but was not compelled to invent the native coherence of ontogeny.

CHAPTER 11

The Architecture of Genetic Regulatory Circuits and Its Evolution

Cellular differentiation in plants and animals is a consequence of the capacity of genes to modify the activity of other genes. Thus different cell types, as noted in the introduction to Part III, differ largely because different sets of genes are active in them. As known since Jacob and Monod published their seminal work on the lactose operon (1961, 1963), however, and as stressed by Monod in his elegant *Chance and Necessity* (1971), the existence of allosteric interactions allows a molecule to control the activity of enzymes or other effector molecules bearing on the synthesis of molecules entirely unrelated to the controlling molecule. That is, this freedom allows the "cybernetic" aspect of genomic control systems to achieve arbitrary complexity.

We shall, in the present chapter, define and develop a body of theory to think about the wiring diagram of these cybernetic control systems and about their evolution. The background expectations we bring to this enterprise, which reflect the present conceptual tradition, should be noted again. I shall suggest that this tradition is inadequate. We have come to think of the genomic system as a kind of biochemical computer which executes a developmental program leading to the unfolding of ontogeny. We know from experience with genuine computer programs how complex, intricate, and often fragile to minor errors they can be. We know that programs can be written to compute literally *any* algorithm. We suppose that selection has filtered out useless programs and has achieved rare, intricate, balanced yet buffered programs guiding development. The rough idea that programs can compute virtually any algorithm and that selection alone determines which among a vast number of developmental programs are fit fits our conception of the genome as the master causative agent in ontogeny and the common idea that selection is the sole source of order. The idea that even in the absence of selection something like orderly developmental programs might exist seems absurd. After all, what would a "random" program do?

I strongly believe this set of views to be ill-founded and, worse, misleading, Thus the major purpose of this chapter and of Chapter 12 is to develop a theory concerning random developmental programs based on the *expected wiring diagram, or architecture,* of genetic regulatory circuits and on the *expected dynamical behavior* of such networks. I shall develop an initial theory for the robust statistical structure and behavior of these networks in the face of mutational forces. As we shall see, abundant order prior to the action of selection exists. In prospective summary: The genome is a system in which a large number of genes and their products directly and indirectly regulate one another's activities. The proper aim of molecular and evolutionary biology is not merely to analyze the structure and dynamical behavior of genomes but also to comprehend *why* they have the architecture and behavior observed and *how* they may evolve in the face of continuing mutations. I shall suggest that we must build statistical theories of the expected structure and behavior of such networks. Those expected properties then become testable predictions of the theory. If discovered in organisms, those properties then find their explanation as the typical, or generic, properties of the ensemble of genomic regulatory systems which evolution is exploring.

INDEPENDENCE OF THE MOLECULAR EVOLUTIONARY CLOCK AND MORPHOLOGICAL EVOLUTION

In Chapter 4 we briefly discussed the data roughly supporting the existence of a molecular "clock." For a number of different proteins or DNA sequences, the rate of amino acid or nucleotide substitution appears approximately constant (Zuckerkandl and Pauling 1965; Ohta and Kimura 1971; Kimura 1983). The idea of a molecular clock ticking at a fairly uniform rate in protein evolution causes a certain shock when the protein sequences of chimps and humans are compared: we are nearly identical. Nevertheless, the morphological variation is considered, at least by human classifiers, to be great. In contrast, divergent species of frogs are morphologically similar but exhibit strong differences at the DNA and protein levels. This important set of facts led Wilson and co-workers (1974, 1977), and subsequently many others, to suppose that the major changes in morphological evolution were due to mutations not in *structural* genes but rather in *regulatory* genes, whose actions determine the action of other regulatory and structural genes. The general idea is that a single regulatory mutation can cause very large alterations in patterns of gene expression by disrupting the coordinating behavior of the genomic regulatory system (Wilson, Sarich, and Maxson 1974; Valentine and Campbell 1975; Bush, Case, et al. 1977; Stanley 1979; Bush 1981; Campbell 1982).

COMPONENTS IN THE GENETIC REGULATORY SYSTEMS OF PROKARYOTES AND EUKARYOTES

In this section, I review the major molecular mechanisms controlling gene expression at the DNA, RNA, and protein levels. These mechanisms collectively constitute the genomic regulatory networks governing ontogeny. I also review the types of chromosomal mutations which persistently alter genomic regulatory networks in the course of evolution. This material, familiar to biologists but less familiar to other readers, forms the basis of the ensemble theories of genomic regulatory networks developed below.

Cis-*Acting Loci at the DNA Level*

The term "*cis*-acting" refers to regulatory genes which act on more or less nearby genes on the same DNA molecule. *Cis*-acting regulatory loci known at present are promoters; operators; enhancers; boxes, or hormone responsive elements; chromatin folding domains and loops; and facultative heterochromatin. I discuss each briefly.

In both prokaryotes and eukaryotes, promoters are small DNA regions, typically on the order of 10 to 30 base pairs (Alberts, Bray, et al. 1983; Ptashne 1986), which serve as the sites of attachment for the appropriate RNA polymerase enzymatic complex required for transcription of adjacent structural genes. For example, in *Escherichia coli,* the lactose promoter is the site where first sigma factor and core RNA polymerase bind each other to make holoenzyme, then cyclic AMP and its binding protein, CAP, bind each other, and finally all bind to the promoter site to form a competent transcription complex (Zubay and Chambers 1971). Promoters on eukaryotic genes that transcribe relatively large amounts of mRNA have similar structures. They have an ATA sequence (sometimes called the TATA box) located about 30 base pairs upstream from the site where transcription begins plus one or more upstream promoter elements (Maniatis, Goodbourn, and Fischer 1987). The upstream promoter element is often a variation of the sequence CAAT, but other upstream promoter elements have been found (Grosschedl and Birnstiel 1980; McKnight and Tjian 1986). The CAAT and TATA boxes have been found to be critical elements in numerous eukaryotic promoters (Efstratiadis, Pasakony, et al. 1980). Table 11.1 shows a number of common eukaryotic promoter elements (from Gilbert 1988).

Operators, well established in bacteria and viruses (Jacob and Monod 1961; Vogel 1971; Ptashne 1986), are typically DNA sites between a promoter and adjacent structural genes which can regulate transcription of the adjacent structural genes. Often operators have the property that, when bound by a regulatory protein, they block transcription initiating at the promoter from reading through to the structural genes. Such operators are examples of negative regulators, shutting off transcription.

TABLE 11.1　Basic Promoter Elements Common to Several Genes

Gene	CAAT region[a]	TATA region[a]
Histone		
Sea urchin H2A	GGACAATTG (-85)	TATAAAA (-34)
Sea urchin H2B	GACCAATGA (-92)	TATAAAA (-26)
Sea urchin H3	GACCAATCA (-75)	TATAAAT (-30)
Drosophila H2A	AGTCAATTC	TATAAAT
Drosophila H3	CGTCAAATG	TATAAGT
Globin		
Mouse α	AGCCAATGA (-88)	CATATAA (-29)
Human $\alpha2$	AGCCAATGA (-70)	CATAAAC (-28)
Collagen		
Chick $\alpha2$ type 1	GCCCATTGC (-78)	TATAAAT
Insulin		
Human	GGCCAGGCG (-73)	TATAAAG (-29)
Rat I	GGCCAAACG (-78)	TATAAAG (-30)
Ovalbumin	GGTCAAACT (-74)	TATATAT (-31)
Conalbumin	GGACAAACA (-81)	TATAAAA (-30)
Silk fibroin	GTACAAATA (-93)	TATAAAA (-29)

[a]Numbers in parentheses correspond to the position upstream from the point where transcription is initiated.

Source: Efstratiadis, Pasakony, et al. 1980; Vogel 1971.

Similar sequences can also serve as positive regulators, activating transcription when bound by a regulatory protein.

Another kind of *cis*-acting regulatory elements are the enhancers (Maniatis, Goodbourn, and Fischer 1987), initially found in viruses but also occurring in the eukaryotic genome. They are DNA regions which may be a modest distance from the genes whose transcription they influence, transcriptionally either upstream or downstream from those genes. Enhancers act via *cis*-regulation of promoters which themselves are *cis*-regulators of adjacent structural genes. In addition, enhancers can function equally well if removed and reinserted in the opposite 3'-5' orientation. One of the first enhancers found appears to control the cell specificity of immuno-globulin gene transcription (Gillies, Morrison, et al. 1983; Potter, Weir, and Leder 1984). Tissue-specific enhancers have also been found in the pancreas (Boulet, Erwin, and Rutter 1986), where the enhancers for the exocrine protein genes (chymotrypsin, amylase, and trypsin) are different from the enhancers for the endocrine protein insulin (Walker, Edlund, et al. 1983). Other enhancers appear to regulate the timing of gene expression. For example, Krieg and Melton (1987) found that certain frog DNA sequences contain an enhancer which triggers transcription at the mid-blastula transition. When B-globin genes were placed adjacent to the enhancer, the globin protein, usually first seen in the tadpole, was turned on during the mid-blastula stage. Other enhancers are hormone-responsive. For example, the enhancers for ten exocrine proteins of the pancreas share a 20-base-pair consensus sequence, suggesting that coordinate expression of all ten is mediated by this similarity (Boulet, Erwin, and Rutter 1986; Gilbert 1988). The general idea is that first hormones bind to a specific protein receptor and then this complex is able to bind to the enhancer (Miesfeld, Rusconi, et al. 1986).

In eukaryotes, DNA is folded into chromatin. At the lowest structural level, the DNA winds twice about a complex of four types of histone proteins to form a *nucleosome* (reviewed in Alberts, Bray, et al. 1983). The nucleosomes, which are separated from one another by about 60 bases, are packed into condensed higher order structures and form a 30-nm chromatin fiber. The fibers are organized into a series of looped domains each containing 20 000 to 80 000 base pairs. (Human chromatin might have about 2600 looped domains.) These looped domains may be functional domains as well. There is evidence that eukaryotic transcription is initiated in subregions of decondensed chromatin, at *cis*-acting DNA sites specified within the decondensed domain. The decondensed state appears to be stable once it forms.

A general view of eukaryotic DNA is that the default condition of chromatin is a condensed state due to packing into nucleosomes. In this repressed state, regulatory factors cannot gain access to the DNA unless the tight packing is relieved by specific activation factors (Schlissel and Brown 1984; Zaret and Yamamoto 1984; Weintraub 1985). Thus, in this general view transcription is activated in two stages. First, the chromatin region must be unfolded so that the gene and its promoter are accessible to transcription factors and RNA polymerase. Second, these transcription factors must be present in the nucleus and must bind the exposed DNA sites.

Trans-*Acting Regulatory Elements at the DNA Level*

By definition, *trans*-acting elements on one chromosome are able to influence genes located on other chromosomes (Alberts, Bray, et al. 1983; Ptashne 1986). Typically, these elements can act on other chromosomes because they create diffusible products, such as RNA, protein, or metabolites. Figure 11.1 shows the lactose operon in *E. coli*. Transcription is blocked if the protein product of a distant gene, the repressor protein R, is present and binds to the operator DNA site. In turn, a metabolic transformant of lactose, allolactose, binds to the repressor protein at a second "allosteric" site, alters the conformation of the repressor, and hence releases it from the operator. Therefore, the operator is under the control of two *trans*-acting molecular inputs. Four other *trans*-acting molecules interact with the promoter to stimulate transcription.

Similar *trans*-acting regulatory proteins have now been well established in a number of eukaryotes, ranging from yeast through nematodes, *Drosophila,* maize, and mice (McClintock 1956; Paigen 1979; Dickinson 1980a, 1980b, 1980c). For example, the 5S rRNA gene in *Xenopus* is transcriptionally regulated throughout development (Korn 1982). Transcription requires the presence of a 32 500-dalton protein, called TFIIIA, which binds to the control region of the gene and directs RNA polymerase III to bind and begin transcription (Ng, Parker, and Roeder 1979; Engelke, Ng, et al. 1980). Similarly, transcription of eukaryotic mRNA is mediated by *trans*-acting factors that enable RNA polymerase II to initiate transcription. For example, Parker and Topol (1984) isolated from *Drosophila* nuclei two factors required for the transcription of two histone genes and an actin gene. One factor appears to bind an upstream promoter element; the other binds to the TATA box. Davison, Edgly, et al. (1983) have isolated mammalian TATA-box binding proteins.

Some *trans*-acting nuclear proteins are operative in a limited set of cells and mediate tissue-specific gene expression. An example is GHF-1 in the anterior pituitary cells. This protein binds to an upstream promoter element of the human growth hormone and can activate transcription of growth hormone when added to nonpituitary nuclear extract (Bodner and Karin 1987).

Among the most interesting putative *trans*-acting genes are homeodomains, found first in a number of homeotic mutants in *Drosophila* and subsequently in other organisms. These mutants will be described in detail in Chapters 12 and 14. Briefly, homeotic mutants convert one tissue or organ to another: eye to wing, for example, or antenna to leg. A number of protein-encoding homeotic genes have been found to encode a 180-base-pair region that becomes part of the functional protein. The DNA region is called the homeobox, the protein region is the homeodomain (McGinnis, Garber, et al. 1984; Scott and Weiner 1984). Evidence from several sources suggests that the homeobox and homeodomain are regulatory site and *trans*-acting factor, respectively. First, the homeobox looks like a nuclear-localization sequence. Second, the homeobox-encoded proteins return to the nucleus after synthesis and are able to bind DNA (O'Farrell, Desplan, et al. 1985). Indeed, the homeodomain of one gene in *Drosophila,* called *engrailed,* binds to the 5' flanking region of its own gene and to that of another homeotic gene, the *fushi tarazu (ftz)* gene (Desplan, Theis, and O'Farrell 1985). Similarly, in mouse the homeodomain of a protein is known to bind to a DNA site upstream of its own homeobox (Fainsod, Bogarad,

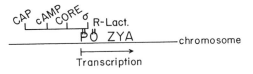

Figure 11.1   The lactose operon in *E. coli.* Z, Y, and A are structural genes; O is the operator, P the promoter site, and R a *trans*-acting repressor protein which binds O and blocks transcription unless itself bound by lactose or allolactose. P is regulated by four *trans*-acting factors: cyclic AMP; a core enzyme; σ, a factor which combines with the core enzyme to form holoenzyme RNA polymerase for transcription; and CAP, a factor which binds cAMP. All must be present for transcription to initiate.

et al. 1986). This evidence strongly suggests that the homeodomain is a *trans*-acting factor able to regulate gene expression on genes at distant loci and to act in a feedback loop for regulating its own transcription via its own *cis*-regulatory sites. The conservation of homeoboxes in evolution has received a great deal of attention and caused much excitement. Homologous sequences have been found in frogs, mice, humans, echinoderms, annelids, and molluscs but not in tapeworms, roundworms, or slime molds (Holland and Hogan 1986). The dramatic effects of homeotic genes in insects has suggested to many workers that these highly conserved sequences play critical roles in development of all the organisms in which they occur.

In addition to all these modifications mediated by *trans*-acting factors, it is important to mention a quite different mechanism of gene regulation: methylation of one of the DNA bases, cytosine, when this base is adjacent to guanosine. About 5 percent of the cytosine residues in mammalian tissues are methylated. Methylation is generated after DNA replication. In some organisms, methylation appears to be associated with diminished transcriptional activity (Groudine and Weintraub 1981). Organ-specific methylation patterns are seen in some cases. For example, the chick ovalbumin gene is unmethylated and expressed in the oviduct but is methylated and not expressed in other chick tissues (Mandel and Chambon 1979). When the DNA replicates in the cell division cycle, two hemimethylated sites are created, one on each sister chromatid. A methylase enzyme then regenerates the fully methylated state on each hemimethylated site. The novel feature of gene control by methylation is that, once created, such methylated sites are passively heritable over cell division cycles (Stein, Gruenbaum, et al. 1982). Unlike specific controls of transcriptional activity, such as the lac operator, whose repression depends both on the continued presence of respressor molecule and on the specificity of the repressor for the operator DNA sequence, maintenance of patterns of methylation is nonspecific. Essentially, any site which can be methylated will be maintained passively and stably thereafter by the general enzyme which methylates any hemimethylated site. This passive mechanism cannot be necessary for ontogeny. A number of organisms, including *Drosophila,* exhibit all the typical features of ontogeny in a complex metazoan without detectable methylation of cytosine.

RNA Loci of Control: Processing, Transport, Translation

In bacteria, the product of transcription is messenger RNA (mRNA). In eukaryotes, however, the nuclear DNA is first transcribed into short-half-lived heterogeneous nuclear RNA (nRNA) (for a review, see Lewin 1980). Then several additional steps are required to produce mature eukaryotic mRNA. Nuclear RNA contains many more sequences than those which ultimately appear as cytoplasmic message. The unexpressed sequences are called *introns,* and they are spliced out of the mature mRNA before it leaves the nucleus. The regions between introns, called *exons,* are the parts expressed in the protein being made. Thus the same nRNA transcript can produce different mRNA molecules by splicing out and adjoining different subsets of exons (Maki, Roeder, et al. 1981; Amara, Jonas, et al. 1982; Breitbant, Andreadis, and Nadal-Ginard 1987; Crenshaw, Russo, et al. 1987). That is, one "gene" can code for more than one protein, each gene made up of a different subset of the exons contained in its transcript. Differential splicing may play a role in creating different proteins in different cells at different times and even in sex determination (Baker, Nagoshi, and Burtin 1987; Boggs, Gregor, et al. 1987).

In addition to splicing, maturation of mRNA often, but not always, requires addition of a large tail of A bases, creating a polyA 3′ tail. Next, modification of the 5′ end of the RNA, called capping, must occur. Finally, the mRNA must be transported from the nucleus to the cytoplasm, where it becomes attached to ribosomes and is translated into protein.

The formation and breakdown rates of any specific mRNA are further obvious control points in gene expression. For example, the cellular oncogene (see Chapter 12) *c-fos* encodes a nuclear protein needed for normal fibroblast cell division. The mRNA for this gene contains a large 3 ′ untranslated region rich in AU sequences. Deletion of this region creates a longer messenger half-life and results in more *c-fos* protein, which in turn signals the cell to divide. The result is a tumor (Meijlink, Curran, et al. 1985). Among the most important ways control of mRNA half-life acts in development are the well-established stable "storage" forms for mRNA, found in particular as maternal mRNA stored in the oocyte.

Finally, the protein product of translation is itself subject to modification. It may be cleaved to leave a mature subfragment, as occurs when insulin is cleaved from proinsulin. The activity of many proteins are modified by phosphorylation or dephosphorylation of lysine and serine residues, mediated by specific kinases and phosphatases. In addition, once synthesized, proteins are often modified by addition of sugar sequences to form glycoproteins.

In the remainder of this book, I shall use either "genetic regulatory network" or "genomic regulatory system" inclusively to point at the integrated system underlying homeostasis of cell types, differentiation of cell types, regulation of developmental pathways, stability of those pathways, and unfolding of morphology. This system is made up of all the complex feedback webs and cycles of control acting at all these levels.

At this point, it is worth pausing again to ask, How many distinct molecular species, kinds of sites, and elements are components in the genomic system in a higher eukaryote? The answer is unknown, but plausible guesses are available. As noted earlier, a human genome has sufficient DNA to code for about 2 000 000 average-size proteins. Almost certainly this is a large overestimate because much of the DNA may be "noncodonic." The estimated number of mRNA species in the cytoplasm or of the number of heterogeneous nRNA species found in all the nuclei of one organism is on the order of 20 000 to 100 000 (Bishop 1974; Hough-Evans, Smith, et al. 1975; Axel, Feigelson, and Shultz 1976; Chikarraishi, Deeb, and Sueoka 1978; Brown 1981; Alberts, Bray, et al. 1983). (This estimate does not include nontranscribed regulatory loci such as promoters and operators.) Suppose roughly one regulatory gene per structural gene exists, an estimate which might be off either way, perhaps by a factor of 10 to 100; then the number of genes might range from 101 000 to 10 000 000. Guess 200 000 to be conservative. In addition, all the different RNA molecules and different splicing opportunities offer different points of control, as do the different phosphorylation and cleavage sites on the proteins. Perhaps the regulatory system has a total of 250 000 components? If this were a tenfold overestimate, which seems unlikely, then the genomic system would have "only" 25 000 distinct components. Not only do we wish to understand how any one such system works, but we also wish to understand how it might have occurred at all in evolution.

Chromosomal Mutations and the Fluid Genome

The genomic regulatory system of 25 000 to 250 000 components can be thought of as a kind of chemical computer. Each component is a node in the computer and receives inputs from those other components which directly regulate its activity. Whenever I refer to the "architecture" of this genomic regulatory system, Monod's "cybernetic system," I simply mean the "wiring diagram" showing which component affects which component. Of course, the wires are not physical connections; they are merely representations of the fact that, for example, the gene synthesizing the repressor protein for the lactose operator acts on that operator. Thus the wires are better thought of as directed arrows pointing from the regulating to the regulated component. This is no mere abstract image. Although we do not yet know the architecture of the genomic system, it surely has one. On the other hand, that architecture is surely not stable. A variety of chromosomal mutational events literally scramble the wiring diagram.

In addition to point mutations, in which one base is substituted for another, there are regional chromosomal mutations. A whole region may be deleted, moved, or duplicated (Markert, Shaklee, and Whitt 1975). The duplicated regions may then amplify into a long, multiply repeated set of identical sequences by unequal exchange between homologous chromosomes or other mechanisms. In turn, these redundant sets may diversify by accumulating point mutations to produce a family of similar sequences called a multigene family (G. P. Smith 1974; Kafatos 1983; Ohta 1983; Ohta and Dover 1983). Some of these sequences may accumulate stop codons and hence form nonfunctioning pseudogenes (Li 1983). Other types of chromosomal mutations, called inversions, translocations, and transpositions, invert a segment of DNA or move it to a new chromosomal location (McClintock 1956; Cameron, Loh, and Davis 1979; Young 1979; Berg 1980; Chaleff and Fink 1980; Green 1980; Corces, Pellicer, et al. 1981; Peterson 1981; Spradling and Rubin 1981; Finnegan, Will, et al. 1982). Conversions involve a process whereby two different but very similar genes become identical to one of the two (Dover 1970; Dover, Brown, et al. 1982; Dover and Flavell 1982).

The general consequence of such chromosomal mutations is summarized by a picture in which genes are duplicated, modified, and dispersed to novel positions around the chromosome set. Every biologist who has remarked on this process has immediately remarked as well that such mutations obviously can alter the regulatory connections in the genomic system (Wilson, Sarich, and Maxson 1974; Gould 1977; Wilson, Carlson, and White 1977; Sherman and Helms 1978; Stanley 1979; Bush 1981; Dover, Brown, et al. 1982; Flavell 1982; Campbell 1985). That is, a *cis*-acting gene can be moved to near a novel set of structural genes and hence bring the latter under its control and thus under the control of those upstream genes acting on that *cis*-acting gene. Similarly, a *trans*-acting gene may move into the domain of a new *cis*-acting gene, bringing to the *cis*-acting gene an influence over the downstream cascade of that *trans*-acting gene. In other words, chromosomal mutations rearrange genomic circuitry.

It must be stressed that, in evolution, rapid chromosomal alterations often occur, leading a number of workers to suppose that such alterations may be the predominant factors in speciation (Bush, Case, et al. 1977; Dover, Brown, et al. 1982; Flavell 1982). Despite the plausibility of this claim, it remains controversial. In particular, a clear link among chromosomal mutations in sibling species, alterations in the genomic wiring diagram, and speciation is lacking. Nevertheless, the rate of chromosomal mutations is sufficiently great to have displaced the picture of the well-ordered, static genome in favor of a rapidly changing, fluid genome in evolution.

AN ENSEMBLE THEORY BASED ON
RANDOM DIRECTED GRAPHS

If the genome is fluid, how shall we conceive of its structure and evolution? Let us recast the problem as follows: Duplication and dispersion of *cis*- and *trans*-acting loci around the scrambling genome explores an ensemble of regulatory architectures. We do not know yet what that ensemble is, but one way to define it is to consider any constraints in the mutational process by which the regulatory connections are rearranged. Whatever the proper description of the scrambling process, we can use it to define the one-mutant variants of a given genomic system and then iterate from each variant to *its* one-mutant variants, and so on, to generate an enormous ensemble of genomic systems. In the case of peptide space, each point is a protein. Similarly, in genomic network space, each point is a genomic system, and that point's neighbors are its possible one-mutant neighbors. Obviously, we do not yet know enough about the structure of the genomic system in any complex eukaryote to carry out this gedanken experiment. We can reasonably begin, however, by building simplified models of genomic cybernetic systems, capturing some of the correct features, and then asking what occurs when chromosomal mutations arise (Kauffman 1985a, 1985b, 1986c).

In Figure 11.2 I show two haploid genomes, each with four hypothetical chro-

CHROMOSOME 1   C1 T1 S1—C2 T2 S2—C3 T3 S3—C4 T4 S4—

CHROMOSOME 2   C5 T5 S5—C6 T6 S6—C7 T7 S7—C8 T8 S8—

CHROMOSOME 3   C9 T9 S9—C10 T10 S10—C11 T11 S11—C12 T12 S12—

CHROMOSOME 4   C13 T13 S13—C14 T14 S14—C15 T15 S15—C16 T16 S16—

a

CHROMOSOME 1   C1 T2 S1—C2 T3 S2—C3 T4 S3—C4 T5 S4—

CHROMOSOME 2   C5 T6 S5-C6 T7 S6—C7 T8 S7—C8 T9 S8—

CHROMOSOME 3   C9 T10 S9—C10 T11 S10—C11 T12 S11—C12 T13 S12—

CHROMOSOME 4   C13 T14 S13—C14 T15 T14—C15 T16 S15—C16 T1 S16—

b

Figure 11.2 (*a*) Hypothetical set of four haploid chomosomes with 16 kinds of *cis*-acting (C1,C2, . . .), *trans*-acting (T1,T2, . . .), and structural (S1,S2, . . .) genes and regulatory domain boundary markers (dashes). Each *cis* gene acts to its right until the first boundary marker. Each *trans* gene acts on the *cis* gene having the identical index. Structural genes code for proteins which are assumed to play no regulatory roles. (*b*) Similar to (*a*), except that here the regulatory domains contain C(*x*), T(*x* + 1), S(*x*).

mosomes containing four types of genetic elements: structural genes, *cis*-acting genes, *trans*-acting genes, and domain markers showing the boundaries of action of adjacent *cis*-acting genes. For concreteness, I have assumed that each *cis*-acting gene acts to its right and controls all *trans*-acting and structural genes until the first boundary marker. I leave out those parts of the genomic system depending on RNA control points, and so forth. The meaning of the numbers is simply to capture the idea that a *trans*-acting gene is *targeted* to act on specific *cis*-acting genes. Numbers on structural genes just distinguish them. For simplicity, I have assumed that structural genes play no regulatory roles.

Each arrangement of the four types of genes or loci implies a specific regulatory architecture. In Figure 11.3, I have drawn the arrows representing these control interactions for the chromosomes of Figure 11.2. The result in Figure 11.3*a* is a set of separate small circuits, each independent of the others. In Figure 11.2*b* I permuted which *cis* gene is next to which *trans* gene. The result, Figure 11.3*b*, is a large loop passing through all the regulatory genes. I include structural genes because they are assumed only to be regulated, not to regulate. These simple pictures should convey the obvious result: Altering which genes are neighbors changes the regulatory architecture.

What happens if chromosomal mutations duplicating and transposing chromosomal regions around the genome occur at random? To investigate this, it is simple to write a computer program which decides at random, for each small region, if it is to be duplicated or transposed and, if the latter, to what site. Figure 11.4 shows the results of 2000 generations of such random scrambling of the two initial genomic systems of Figure 11.2. Although the results look like scrambled spaghetti, an important point is immediately obvious to the eye: Figures 11.4*a* and 11.4*b* look much more similar to each other than do Figures 11.3*a* and 11.3*b*. What your eye has picked up effortlessly is that the *statistical connectivity features* of the two parts of Figure 11.4 are very similar to each other, while those of the two parts of Figure 11.3 are very different from each other. Your eye is correct: Two "well-scrambled genomic architectures" look very much alike, although, of course, the detailed positions of the genes differ. Restated, in the course of scrambling, two initially highly improbable architectures have "fallen" toward what must be the statistically expected, typical connectivity features of such architectures.

It is useful to define several connectivity features more precisely. Because such wiring diagrams use arrows, not lines, to connect points, they are *directed graphs* (Berge 1962; Harary 1969; Hararay, Norman, and Cartwright 1975). We can define the ratio of arrows to points as well as the mean number of direct inputs and outputs per point. These latter two are the *local connectivity* properties of the network. In addition, each gene regulates, either directly or indirectly, all those genes which can be reached by following arrows, tail to head, from that initial gene. Therefore, we can define the number of *descendants* from and *antecedents* to each gene as well as the mean number of descendents and antecedents of each gene. Any gene can reach all of its descendents in a finite number of steps by following a sequence of arrows tail to head. The *radius* from each gene is the minimum number of steps which allows it to reach all its descendents. Thus we can define the mean radius of the network, which measures how spread out the network is. Some genes lie on *feedback loops*, and thus we can define the fraction of genes lying on feedback loops. Any such gene may lie on a number of intermeshed feedback loops. We can study the length of the smallest loop any such gene lies on and the mean minimum loop length. We can define as a *strong component* any set of genes which can mutually reach one another,

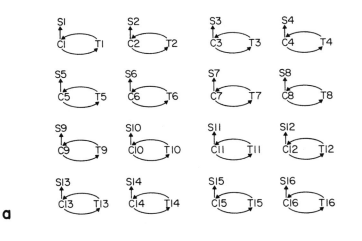

Figure 11.3 (*a*) Regulatory interactions among genes in the chromosomes of Figure 11.2*a*. (*b*) Regulatory network of interactions implied by the chromosomes of Figure 11.2*b*.

directly or indirectly, by following arrows tail to head. A variety of other connectivity features can be defined, but these just mentioned will suffice to indicate the spirit of the enterprise.

The natural question becomes this: What are the expected connectivity properties in the well-scrambled genome? This problem, simply stated, is too hard to answer, but the flavor of an answer can be obtained by considering a simple problem.

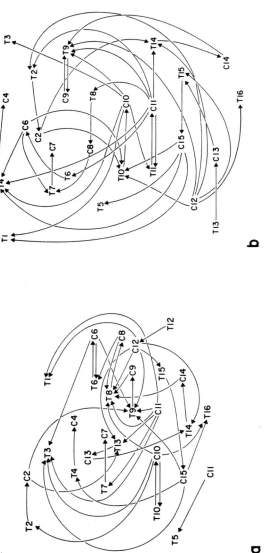
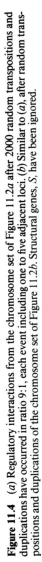

Figure 11.4 (*a*) Regulatory interactions from the chromosome set of Figure 11.2*a* after 2000 random transpositions and duplications have occurred in ratio 9:1, each event including one to five adjacent loci. (*b*) Similar to (*a*), after random transpositions and duplications of the chromosome set of Figure 11.2*b*. Structural genes, *S*, have been ignored.

Random Isotropic Directed Graphs

Consider a system of N genes, shown as points and connected entirely at random by M arrows. Our intuitions are already tuned by the results of Erdos and Renyi (1959, 1960) on random undirected graphs, where lines, not arrows, connect points at random. These results were discussed in Chapter 7 and exhibit threshold phenomena. As the ratio of lines to nodes increases, critical thresholds are crossed at which the connectivity properties change abruptly. Similar transitions occur in directed graphs.

Figure 11.5 shows 20 genes connected at random by more and more arrows. As

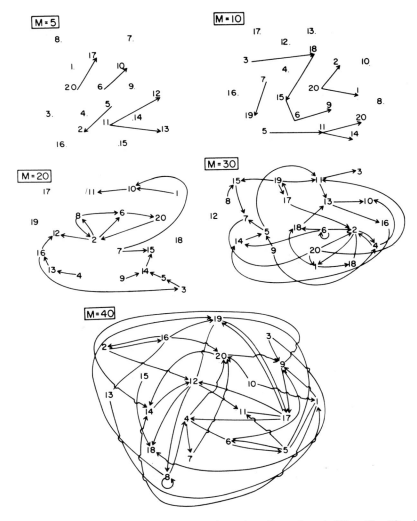

Figure 11.5 Evolution of connectivity in an isotropic random directed graph. When $N = 20$ points are connected at random by an increasing number of arrows M, a large connected structure begins to crystallize when $N = M$.

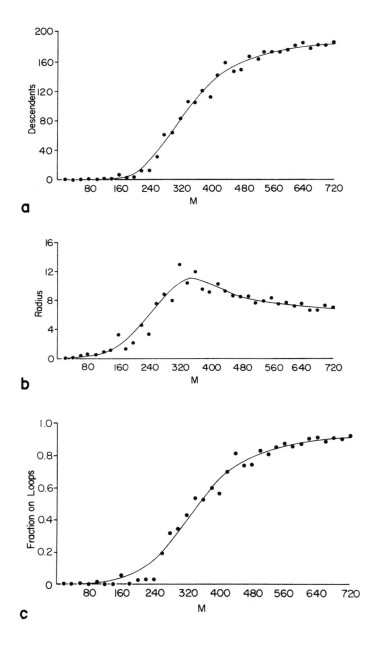

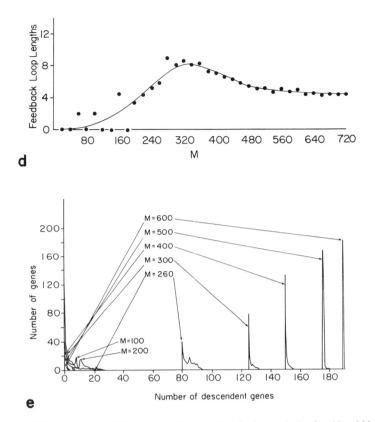

Figure 11.6 (*a*) Mean number of descendants from each point in graphs having $N = 200$ nodes, as the number of directed arrows M increases from 0 to 720. Mean descendants are mean number of genes any gene directly or indirectly influences. (*b*) Radius distribution showing mean number of steps via intermediate genes for influence to propagate to all descendants of a gene, as a function of M. (*c*) Average number of genes lying on feedback loops, as a function of M. (*d*) Average length of shortest feedback loops genes lie on, as a function of M. (*e*) Descent distribution within a network, showing how many descendants each gene has. Note bimodal character of distribution for $M > N$.

the ratio of arrows to genes increases, small, isolated, branched tree circuits form, coalesce into fewer large, sparsely connected circuits, then finally become multiply connected in a richly cross-coupled web having many interlocked cycles. Figure 11.6 shows mean connectivity features in numerical trials for $N = 200$ genes, connected at random by $M = 0$ to 720 arrows. In Figure 11.6*a*, a threshold is reached for the crystallization of connected circuits when the ratio of arrows to genes equals 1. Below that value, on average each gene influences no other gene, or at most very few; beyond that value, genes begin to have many descendants. The mean radius (Figure 11.6*b*) is at a maximum when the ratio is about 2, corresponding to the steepest part of the curve in Figure 11.6*a*, when the descent distribution is increasing most rapidly and the underlying graph is most diffuse and large. This is just what we might expect. As the number of connections increases beyond this level, shorter pathways to all descendents emerge, and hence the radius falls. Since feedback loops are likely to form when any gene has a large number of descendants, each of which might feed back to act on that gene itself, the curve in Figure 11.6c parallels that in Figure 11.6a; eventually almost all genes lie on feedback loops. The mean minimum size of feedback loops (d) parallels the mean radius and increases, then grows shorter as the ratio of arrows to genes increases. This makes sense, for the minimal loop lengths should be less than the radius from any gene to all its descendents, and the radius increases, then decreases as the ratio of arrows to nodes increases. Finally, Figure 11.6e shows the descendant distribution for all the genes in the network as the ratio of arrows to genes change. The interesting result of Figure 11.6e is that the descendant distribution is bimodal. Many genes have few descendants, a vast number have an identical and large number of descendants, and a few have yet more descendants. The reason is simple. Once interconnected loops form, all the genes on the loop(s) have one another as descendants, as well as all genes lying on tails descendant from the loops. The set of genes on these interconnected loops can each reach all the other genes on those loops, and so these genes constitute the strong component of the genomic regulatory network. The genes on tails have only a few descendants. Genes antecedent to genes on loops have a few more than the very large number of descendants of the genes on loops. The antecedent genes and tail genes are not members of the strong component. The former cannot be affected by the genes of the strong component and the tails; the latter can affect only their few descendants.

Generic Properties of the Ensemble, Statistically Robust Null Hypotheses, and, If Unavoidable, Potential Universals

What we have just described is a toy model. The real scrambling genome is not an isotropic random graph. In order to use the insights gained from this toy model, we must certainly develop far more accurate models of how real genomic systems are scrambling in evolution. We return to this below. Nevertheless, the analysis helps us see quite a bit. What is most obvious is that, given any well-defined prescription on how the real genomic system is scrambling, there is a well-defined ensemble of regulatory architectures being explored. That ensemble is enormous. As in any other large ensemble, the generic features of this ensemble will be very robust. Then the following statements appear absolutely reasonable:

1. *If* we can correctly identify how the genome is scrambling, the average connectivity properties of the resulting ensemble of systems constitute the proper *null hypothesis* about the properties we would expect in the absence of outside forces.

2. Thus, such mean properties are our reasonable best first guess as to regulatory architecture.

3. Selection is the most important outside force. It may pull populations of adapting organisms to odd regions in the underlying ensemble, where the genetic regulatory architectures are grossly untypical of the ensemble as a whole. If so, we cannot predict which features will occur on the basis of ensemble average properties.

4. On the other hand, if selection is too weak or otherwise cannot escape the typical properties of the ensemble, those properties will be widespread in organisms not because of selection but despite it. Then the null hypothesis will be explanatory of features actually found.

5. Further, we shall be able to predict statistical features without carrying out the full reductionistic analysis to work out all 250 000 components and their interactions.

6. Insofar as the typical properties are ordered in some way, we do not have to account for that order on the basis of selection. It lies to hand in the ensemble.

7. We stressed in Chapter 5 that selection may alter the kinds of entities upon which it is acting. The corollary here is that selection may have built up specific *ways* it is scrambling the genome. If so, it will follow that the ensemble being explored is constrained in some ways. It will not follow that such an ensemble does not exist, for the ensemble claim is merely a restatement of those aspects of genomic scrambling which are random, subject to the constraints. Insofar as we can define the current constraints, and hence the current ensemble being explored, we can legitimately consider the average properties of members of that (selected) ensemble as proper null hypotheses and assess the extent to which selection *within that ensemble* can attain entities which deviate significantly from its robust mean properties.

In short, we may need selection to account for the particular ensemble selection is exploring but not to account for the fact that organisms remain typical members of that ensemble. We here confront a new pattern of evolutionary inference: generic properties shared among organisms due to common membership in an ensemble, not by virtue of common descent.

While appeal to the simple model of a scrambling genome as a serious description of genomic systems is premature, some of the features of the model seem likely to apply in real genetic cybernetic systems. In particular, the rich results on connectivity and sizes of descent distributions are known to exhibit sharp thresholds as the ratio of arrows to points increases (Erdos and Renyi 1959, 1960; Kauffman 1985a, 1985b, 1986c; Cohen 1988). It would be quite astonishing if those typical graph properties did not carry over to the wiring diagrams of current genomic systems. In the current context, we suppose that a eukaryote has on the order of 20 000 to 100 000 structural genes and a variety of other control points in the regulatory network. We suppose each gene is regulated somehow, probably through multiple regulatory inputs to account for tissue-specific expression. Suppose for simplicity that we consider a plausible picture with 70 000 structural genes and on the order of 200 000 regulatory interactions via *cis*- and *trans*-acting elements. Given a typical directed graph with 200 000 regulatory connections and 70 000 genes, we can make strong predictions about the connectivity features of such a system. For example, it would be astonishing if a very large number of genes were not coupled to one another via loops on a gigantic strong component such that each could, in principle, influence all the rest. (I stress "in principle" because, as we know from Ashby [1960] and examine further in Chapter 12, potential influence may not propagate in such a network.) Equally, it would be astonishing if many genes did not have a very large number of descendants. Thus it would be astonishing if such a system were not richly webbed, with many feedback loops interlocked in complex ways and with minimal-length feedback loops from each gene to itself which were reasonably short. Conversely, the probability that such a system has a wiring diagram which falls apart into truly independent separate circuits is infinitesimal.

Genomic Regulatory Architectures Are Likely to Be Rich in Feedback Loops Rather than Purely Hierarchical

A familiar conceptual model of genetic regulatory architectures has been a military hierarchy, with a few generals commanding overlapping battalions of officers and troops below them (Britten and Davidson 1969; Davidson and Britten 1976). In these models, a few master genes control overlapping downstream cascades, or batteries, of genes. Control of alternative developmental programs lies in unleashing the different possible cascades of activities from the master control gene at the top of each hierarchy. The overlapping activations spreading through the overlapping downstream cascades would create different cell types, each with a unique battery of gene activities. In one extreme form, such regulatory networks are entirely free of feedback loops. The terminal structural gene "twigs" of the hierarchial cascades, once activated, might maintain a stable pattern of activity by virtue of a stable unmethylated state.

The hypothesis that genetic regulatory systems are hierarchical and free of feedback loops is an important one. If true, propagation of control is particularly simple. Yet even with only modest insight into the statistical properties of scrambled wiring diagrams, it is clear that such a hierarchical system is highly improbable. Genomic regulatory networks are very likely to be rich in feedback loops. In turn, the existence of such loops raises complex issues about the control of cascades of gene activities in such webbed networks, issues which will be discussed in the following chapter. For now, focus attention again on the genomic regulatory network, ignoring structural genes playing no further control roles. The hierarchy of control from master genes to lowest level structural genes is assumed to pass via intermediate regulatory genes, presumably *cis*- and *trans*-acting factors. Suppose there are N such regulatory genes. Then in order that they be connected into the simplest possible *directed-tree* hierarchy, we must assume that each regulatory gene (except perhaps the master genes) has one regulatory input from above in the hierarchy. Thus M, the number of regulatory connections, is at least equal to N, the number of regulated regulatory genes. If cascades of genes downstream from master genes *overlap*, as commonly assumed, then some or many of the regulated genes must have more than a single regulatory input from above. Then M is clearly larger than N. Recall from the analysis of Figures 11.5 and 11.6 that, when M is significantly less than N, the generic structure of connections is small, directed trees. As M increases toward N, the trees get larger and begin to merge. As large trees begin to merge, however, creating just the kinds of hierarchy envisioned, they typically also begin to form feedback loops. Precisely when M equals N, *both* large descendant trees *and* richly interconnected loop structures almost surely begin to form. As M becomes greater than N, feedback loops become ever richer. The genome is, in fact, scrambling its architecture and logic. Thus in order that hierarchical command structures among regulatory and structural genes be generic, and hence readily attained and maintained by selection, M must be less than N. If M is to be less than N, many regulatory and structural genes must have no regulatory inputs. Conversely, if all or almost all regulatory and structural genes have one or more regulatory inputs, then, generically, the genomic architecture is rich in feedback loops. Thus to attain and maintain an improbable hierarchical wiring diagram in the face of mutations tending to randomize connections and creating a richly webbed architecture, selection would have to struggle against powerful forces. We see in the next section of this chapter that selection is very likely to fail in such an endeavor. However, in Chapter 12 we shall find conditions based on Ashby's percolating walls of constancy and frozen components, discussed in Chapter 5, which permit evolution of hierarchial command structures even in genomic systems rich in feedback loops.

This brings us, inevitably, to a fundamental question: Can selection build arbitrary genomic wiring diagrams?

Simple Population Models and the Error Catastrophe Again

Selection acting to attain optimal wiring diagrams confronts the same limitations we encountered in Chapter 3: On smooth fitness landscapes, as mutation rate increases, populations fall from fitness peaks and exhibit the error catastrophe; on rugged landscapes, populations become trapped in small regions of the space of possibilities. Recall the error catastrophe: As the number of elements in a system increases in the face of a constant mutation rate, a maximum complexity is reached. Beyond this complexity level, selection is no longer able to hold the population at the globally optimum phenotype, errors begin to accumulate, and the population flows away from the optimum toward less fit combinations. We now reexamine this phenomenon in the context of selection acting to mold the wiring diagram of a genomic regulatory system. The results repeat and extend the familiar ones. In the simple model I consider next, I shall assume that a specific wiring diagram is the globally optimal diagram and that no other local optima exist. More realistic models of genetic regulatory networks and their evolution, discussed in Chapter 13, admit a multiplicity of local optima.

We now define a fitness landscape with a single optimal wiring diagram. Let us, for simplicity, continue to model a genetic wiring diagram by a set of N genes, shown as points, wired together by a total of T directed regulatory connections, shown as arrows connecting the points. For complete simplicity, let us suppose that a mutational event can alter the wiring diagram arbitrarily, by changing either the tail or the head of any arrow such that it starts from or terminates at a different gene. That is, in a stepwise fashion, we can change any wiring diagram into any other wiring diagram. Further, for extreme simplicity, let us suppose that some specific wiring diagram is optimally fit. Since all wiring diagrams are equally likely, for our modeling purposes we can pick any one as optimally fit. For example, we might as well pick the diagram that has $N = T$ and is connected in a single loop of genes, $1 \rightarrow 2 \rightarrow 3 \ldots N \rightarrow 1$.

Next we need to define a simple fitness function for wiring diagrams which deviate from the optimal diagram. One choice is to measure fitness by G_x/T, the fraction of "correct" connections in a given network X. A slight generalization admits that the fitness might be linearly proportional to this fraction but might also change in some nonlinear fashion. A further generalization allows the possibility that a network with no correct connections might even be slightly fit. This leads to a simple form:

$$W_x = b + (1 - b)\left(\frac{G_x}{T}\right)^{\alpha} \tag{11.1}$$

where W_x is the fitness of a given wiring diagram; b, for $0 \leq b \leq 1.0$, is a basal fitness even if no connections are correct; and α is a parameter measuring how fitness changes as the ratio of good connections ranges from 0 to T. If $\alpha = 1.0$, fitness is just proportional to the fraction of good connections and the model corresponds to additive fitness. If $\alpha > 1.0$, fitness drops off rapidly at first and then more slowly as the number of incorrect connections increases. The fitness peak is a spike, which corresponds to saying that the components act cooperatively in fitness. If $0 < \alpha < 1.0$, fitness falls off slowly at first but then more rapidly as the number of incorrect connections increases from 0 to T. Now the fitness peak is round-topped and has steepening sides, a typical form in genetic models which assumes that, near the optimum, most mutants have only very minor deleterious effects.

The first thing to notice about this fitness landscape is that it has a *single* peak—the global optimum—and every wiring diagram which is less than optimal has one or more adaptive walks via one-mutant fitter neighbors to that peak. That is, no trapping on local optima below the global optimum can occur, and so the problem is excessively simple.

We can next define a haploid-population genetic model by assuming a population of "organisms," each consisting in a specific genetic wiring diagram. At each generation, each connection in each wiring diagram is subjected to mutations with a given probability, altering heads or tails of regulatory connections at random. At each generation, each wiring diagram produces progeny in the next generation proportional to its current fitness. The program proceeds by fixing a population size starting with an initial population of identical networks, mutating each at random as determined by a mutation rate, determining the fitness of each network, and then allowing each network a chance to be a parent with a fixed low probability. For each network afforded a chance to be a parent, the program decides with a probability *equal to network fitness* whether that parent has an identical offspring. Each potential parent can be sampled more than once, and sampling continues until the fixed number of networks needed for a constant population size is chosen for the next generation. This procedure normalizes fitness each generation, relative to the fittest in the population. The networks are subjected to new mutations at each generation, and the process is iterated.

For sufficiently high mutation rates, the error catastrophe shows up as populations falling from the fitness peak to an intermediate fitness level. Figure 11.7 shows the results of two numerical trials at parameter values where the error catastrophe discussed in Chapter 3 has set in. In one case, the entire population began as the optimally fit network, the loop. In the other case, the entire population began as a network with very few correct connections—the number which are correct by chance in a random network. In the former case, the population accumulates errors over time, and the mean fraction of correct connections per network decreases. No net-

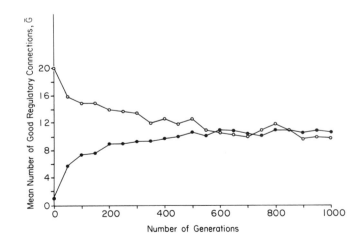

Figure 11.7 Population selection for 1000 generations after initiation at $G = T = 20$ (○) and $G = 1$ (●), where G is the mean number of correct connections per organism. For both graphs, μ 0.005, $\alpha = 1$, $b = 0$. (From Kauffman 1985a, 1985b, 1986c)

work which is the global optimum remains. The entire population slides down the side of the fitness peak to an intermediate height and remains there. Conversely, the initially poorly fit population climbs in fitness to the same intermediate height on the fitness hill. That is, regardless of the initial position of the population, it ends up, in the presence of continuing selection and mutation, at a stable *intermediate* level of fitness. If perturbed above or below that fitness, the population returns to the same stable distribution. Clearly, selection is here unable to hold the population on the global optimum, even if the population had been released there, and surely cannot pull a population to that peak from suboptimal positions, despite the fact that connected adaptive pathways to the peak abound. Also, obviously, the position of the population midway between the peak and very low fitness networks is consistent with an enormous number of alternative wiring diagrams, all having the same ratio of correct to total connections. Thus Figure 11.7 is misleading. In the proper high-dimensional space, the population may be located at any point on a shell surface that is at a constant distance from the global optimum and has a finite thickness. Since evolution occurs by small mutations which accumulate and since all points in the shell have essentially the same fitness, derivative lineages of networks will show branching random (literally neutral) walks in the shell. As we described in Chapter 3, as a result of finite population size and of founder effects such that after sufficient generations all descendants in the population derive from one (random) initial member, the walks tend to be coherent rather than resulting in diffusion of the population over the entire shell. Rather obviously, the spherical symmetry of this shell derives from the fact that all loci are assumed to contribute equally to total fitness. If that symmetry were broken, the shell would be more or less distorted.

Bifurcation Behaviors

Altering the parameters leads to sharp changes, or bifurcations, in the behavior of the adapting population. In some regimes, the population climbs to and remains clustered about the fitness peak. In others, it falls from the peak to a sustained intermediate level. Numerical experiments have been carried out varying the mutation rate μ, the peak shape parameter α, the basal fitness b, and the total number of arrows T. Qualitative insight into the bifurcation behavior of this system follows from approximate equations for the rate of change of the mean number of correct connections $\overline{G}$ per network in the population (Kauffman 1985a, 1985b, 1986c).

$$\frac{d\overline{G}}{dT} = \frac{\alpha(1-b)\overline{G}^{\alpha-1}\sigma^2 G}{(1-b)\overline{G}^\alpha + bT^\alpha} - \mu(TP + \overline{G} - T) \tag{11.2}$$

where σ is the variance in G and P is the probability that a given connection is correct. The second term reflects mutations, which tend to destroy correct connections already established. Mutations provide a restoring force which pushes the adapting population toward the mean number of correct connections expected in the absence of selection in an entirely random wiring diagram. This restoring force is proportional to the deviation of the population above that unselected mean. That is, the mutational restoring force is like the restoring force in a simple spring, proportional to deviation above rest length. By contrast, the first term in Equation 11.2 is the selective force tending to increase the mean number of correct connections per network in population members. Recall Fisher's fundamental theorem (Ewens 1979):

$$\frac{d\overline{W}}{dT} = \frac{\sigma^2 W}{\overline{W}} \tag{11.3}$$

where $\overline{W}$ is the mean fitness of the population. This general result states that the rate of change of mean fitness in the population is proportional to the variance of fitness and inversely proportional to the mean fitness. That the rate of change depends on the numerator is immediately reasonable since selection can increase fitness only if there is a fitness range, allowing the fitter to outgrow the less fit. That the rate is inversely proportional to the present mean fitness, a result following simply from the difference equations for the change in fitness over one generation, means that selection is a less effective force as fitness increases.

The relation between effective selection and mutational restoring force shows up in the two terms of Equation 11.2 (so long as $b = 0$). In Figure 11.8, I graph these two terms separately. The restoring force lowering fitness increases proportional to mean fitness. The selection force trying to increase fitness falls off first rapidly and then more slowly as fitness increases. If these two curves cross in the interval where $0 < \overline{G} < 20$, that level of mean fitness is a steady state, and the tendency to increase is the same as the tendency to decrease it. Thus the steady state is stable, and the population adopts a stable distribution around it. Conversely, if the selection force is greater than the restoring force throughout the interval of mean fitness from 0 to 1.0, the population is pulled from any initial distribution to the global optimum and remains there. It should be stressed that Equation 11.2 is inadequate. We really need a companion equation showing how the variance of fitness in the population changes in time. This is a hard problem. The idealization in Equation 11.2 is based on the

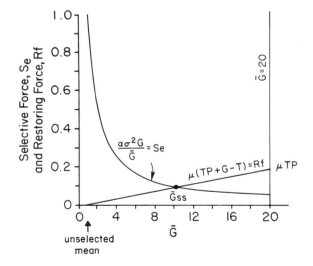

Figure 11.8 Graph of the two terms in Equation 11.2. The effective selective force S_e increases the number of correct connections, and the mutational restoring force Rf tends to reduce this number toward the unselected expected mean (1). Both forces are plotted as a function of the mean number of correct connections $\overline{G}$ in networks in the adapting population. Curves cross at steady state level $\overline{G}_{ss}$, attained when selection balances mutation. ($T = 20, P = .95, \alpha = 1, b = 0, \mu = .01$ were used in Equation 11.2, and $\mu TP = Rf$ at $\overline{G} = T$.) (From Kauffman 1985a, 1985b, 1986c)

assumption that the variance settles down to a constant in a few generations and remains stable thereafter. This is borne out by simulations.

Despite its inadequacy, Equation 11.2 can help us gain some insight into how behavior depends on the parameters of the adapting system. If α increases, the selective curve shifts upward but the restoring curve is constant (Figure 11.8). Thus if the two curves cross inside the 0 to 1.0 fitness range for $\alpha = 1$, the point of crossing will shift toward higher fitness as α increases. For sufficiently large α, implying that fitness falls off very steeply from the optimal network, the selection curve is guaranteed to be above the restoring curve; hence overwhelming selection for an optimum wild type can always dominate over mutational forces. The limit of infinite α is identical to hand-picking the fittest variant at each generation and seeding the entire next generation from that hand-picked network. Very high values of α, or their biological counterparts, seem implausible in general. In reality, the steepness of fitness gradients appears bounded.

As the mutation rate increases or decreases, the mutational restoring force becomes steeper or less steep. Thus, for a fixed selection curve, decreasing the mutation rate will move the steady-state crossing point inside the range 0 to 1.0 toward increasing fitness. If the mutation rate is low enough, selection is able to pull the population to the global optimum. Conversely, if the mutation rate is higher and the population is released on the optimum, mutations accumulate gradually and the population will first remain at and then flow away from the optimum.

Perhaps most important, all these properties depend on the complexity of the system under selection. Recall the classical population result for additive models which leads to the error catastrophe. Let each correct gene connection make a contribution to fitness inversely proportional to the total number of connections. That is, let α equal 1.0. Then as the number of connections T increases, the contribution of each to fitness decreases inversely proportional to T. However, the mutation rate from a correct to an incorrect connection is constant. Therefore selection becomes a weaker force per connection as T increases, while mutation remains as powerful as ever. Eventually, selection becomes weaker than mutation—in other words, too weak to hold a population at the global optimum.

The classical result shows that the expected number of incorrect connections will increase proportional to the square of the total number of connections T^2 (Chapter 3). This rate is very rapid. It can be compensated by decreasing the mutation rate inversely to T^2. For any real system, however, there is a lower bound on the mutation rate. Eventually, as T increases, selection is unable to hold the population at the global optimum. The error catastrophe sets in.

Simulation results on model genetic wiring diagrams strongly suggest that these results extend from the *additive fitness* case, $\alpha = 1$, in which each correct connection makes a proportional and independent contribution to fitness, to values of α substantially greater than 1.0. This result reflects the fact that, for a range of values of α, the population variance in fitness increases only as $\sqrt{T}$ while the mutational restoring force increases as T. Therefore, the classical population genetic result appears to extend fairly widely to nonadditive models, where $\alpha > 1.0$.

A Novel Bifurcation When b Is Greater than Zero

The behavior of this system changes dramatically when basal fitness $b > 0$, for now the population can jump discontinuously from a stable high-fitness state to a stable low-fitness state. This bistable behavior reflects the fact that the effective selection curve can increase rather than decrease as fitness increases. Consequently, as shown in Figure 11.9, the selection and mutation force curves can cross in two places in the interval of mean fitness between 0 and 1.0. In this case, the upper crossing (point G_2) is an unstable steady state. If the population is released above that steady state, selection is stronger than mutation and the mean fitness increases toward the global optimum. However, the global optimum, $G/T = 1.0$, is a *reflecting barrier*. Thus the population hovers in a highly fit steady-state distribution between the G_2 steady state and the global optimum. Conversely, if the population is released below G_2, mutational forces lowering fitness are stronger than selectional forces, and the population falls to G_1, which is stable. Thus the system will remain in either a high-fitness state or a low-fitness state unless a fluctuation drives the population across the threshold unstable steady state.

In fact, such fluctuation-driven transitions do occur (Figure 11.10). Further, the system exhibits striking behavior if the mutation rate begins low and increases gradually. At first, the mutational restoring force curve is low and crosses the selection curve only once (Figure 11.9). Selection is always stronger than mutation, and hence the population climbs to and remains at the optimal wiring diagram. As mutation rate increases, however, and the mutational curve becomes steeper, the second upper crossing is first created, then moves to the right, and then hits the boundary where mean fitness is 1.0 and disappears. At that mutation rate, the population suddenly falls abruptly from the global optimum to a wiring diagram only slightly better than random and corresponding to the lower steady state created by the crossing of the two curves.

Increase of total complexity for the interesting case where $b > 0$ leads to a similar dramatic result. The population distribution of high fitness near the global optimum is at first stable, but as T increases, the mutation curve becomes so steep relative to

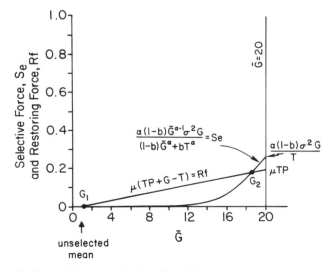

Figure 11.9 As in Figure 11.8, except that here $b = 0.5$, $\alpha = 10$, $T = 20$, $P = .95$, $\mu = .01$. These values allow the curves to cross at two points in the fitness interval between 0 and 1. This double cross creates two domains of attraction: one near the optimal state $\bar{G} = T$, and one near the fully unselected state $\bar{G} = 1$. The point G_1 represents a stable steady state, G_2 an unstable steady state; $\mu TP = Rf$ at $\bar{G} = T$; $[\alpha(1 - b)\sigma^2 G]/T$ is S_e at $\bar{G} = T$. (From Kauffman 1985a, 1985b, 1986c)

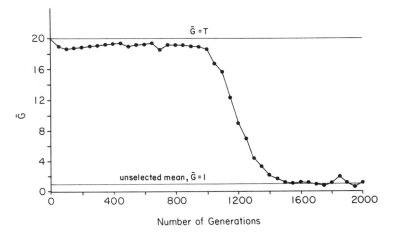

Figure 11.10 Fluctuation-driven transition of adapting population from stable upper, near-perfect $\overline{G} = T$ state to stable low-fitness $\overline{G} \approx 1$ state. Here $T = 20$, $\alpha = 10$, $b = .5$, $\mu = .005$. (From Kauffman 1985)

the selection curve that the upper stable state disappears and the population crashes to a low-fitness state. Thus if a simple loop of genes is the maximally adapted architecture, and if "evolution" begins small with a perfect loop and attempts to integrate new genes into the loop, at first selection can maintain perfect loops. Eventually, however, the addition of one more gene crosses the bifurcation threshold and the loop architecture abruptly becomes highly scrambled (Figure 11.11).

The General Qualitative Result

The general qualitative result of these simulations is that, as the complexity of the system under selection increases, for a fixed mutation rate and a range of additive

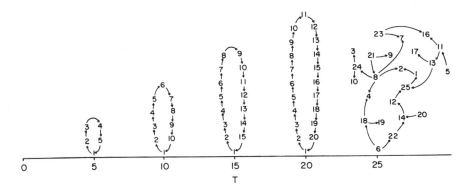

Figure 11.11 Selection for a perfect-loop structure in a genetic wiring diagram. The five diagrams represent the maximal adaptation achieved and maintained in a population as the number of genes in the loop T increases. Eventually, addition of one more gene causes the loop to cross the bifurcation threshold and its architecture abruptly becomes highly scrambled. Here $\alpha = 10$, $b = .5$, $\mu = .005$, from Equation 11.2. (From Kauffman 1985)

and nonadditive fitness rules, selection can at first either maintain a population at a global or a local optimum to which pathways exist or pull a population to such an optimum. At some critical complexity value, however, selective forces become weaker than the mutational forces driving the population toward the mean properties of the underlying ensemble of systems. As complexity increases further, the population falls away—either smoothly or in a large jump—from its optimum toward the mean properties of the ensemble. The population then arrests at some stable distance from a single global optimum and occupies a shell neutrally, wandering in it in coherent, branching, ergodic ways.

This model has only a single optimum. It represent the simplest case to consider. Already it shows that complex systems are likely to be suboptimal, depending on size and mutation rate. The existence of a global optimum in the present case is associated with the existence of connection changes via fitter variants to reach that global optimum. In real genomic systems, single-optimum landscapes may not be the case for two reasons:

1. The space of genomic regulatory architectures virtually certainly will have many local optima. Recall the *NK* model of correlated rugged landscapes discussed in Chapters 2 through 5. We noted in general that conflicting design constraints in complex systems lead to multipeaked rugged landscapes. Recall, too, that on such landscapes an adapting population tends to become trapped on local optima in small subregions of the very rugged landscape.

2. We have assumed that *any* rearrangement of connections is a possible mutation, but this may be drastically incorrect. That is, it may typically be the case that non-random "hot spots" for genomic rearrangement exist while other connections rarely, if ever, change. For example, insertion of transposable elements into a site typically generates small repeated sequences which remain after the element transposes from that site. The repeated elements make that locus a hot spot for further transposition (Spradling and Rubin 1981). Such nonrandomness suggests that some connections are almost inviolate, while others are modified often. In turn, this means that the way the genome scrambles is nonrandom and constrained. From a given wiring diagram, therefore, not all other diagrams are accessible to the mutational move generator.

In sum, the real fitness landscape in genomic architecture space may be highly rugged and multipeaked. There would probably be extensive trapping on local optima even if all mutational moves to adjacent genomic wiring diagrams were possible, but the problem is exacerbated by nonrandomness in the mutational move generator across the space of wiring diagrams. Thus more realistic assessment of real genomic systems will force us to ask which transitions among diagrams are even legitimate sequences of mutational moves, let alone which pass via neutral or fitter variants.

In examining the *NK* model in Chapter 2, we found that a second complexity catastrophe sets in when K increases proportional to N. The fitness peaks decay toward the mean features of the space of systems. In the present case, we have no coherent way to guess the distribution of fitness values across the space of genomic architectures. Thus we cannot be sure whether this further complexity catastrophe sets in. In Chapter 13, however, we shall see that this complexity catastrophe does set in in models of cell differentiation. Thus, decay toward the mean features of an ensemble as complexity increases, even with strong selection, must be a very general phenomenon. On the other hand, it does seem likely that the fitness landscape for genomic architecture is quite rugged. Therefore, trapping of the evolutionary adaptive process in small subregions of network space seems likely. In turn, this possibility suggests that, even after strong selection acts, networks trapped in local subregions due to landscape ruggedness are likely to exhibit fairly typical features of network architecture space as a whole. If so, the wiring diagram architecture of complex networks, even in the face of strong selection, will be generic members of the class of systems explored by evolution.

In short, in the real biological world with all its complex organisms, it is not foolish to ask how "precise" such organisms are likely to be as a result of selection. Can we guess anything about the structure of regulatory networks? Is an organism with 70 000 genes and 200 000 regulatory connections likely to be graced by an arbitrary improbable wiring diagram?

To be concrete, let us consider how precise the genomic wiring diagram in a mammal might be. Suppose that the genomic system has on the order of only 10 000 genes and two regulatory connections per gene, and that the mutation rate of a regulatory connection is 10^{-6} per genome per generation. The classical population result for additive fitness with a single global optimum suggests that on the order of 25 percent of the connections will be "wrong." With a genome of 100 000 genes and two connections each, the expected fraction of wrong connections is about 80 percent. In either case, it is obvious that highly nontypical architectures, such as isolated subcircuits or strictly hierarchical wiring diagrams, simply could not be maintained. Any architecture whose statistical connectivity features were remarkably rare in the underlying ensemble would not be maintainable. Conversely, as noted already, if the landscape is very rugged, selection cannot pull an adapting population to an arbitrary correct wiring diagram from an arbitrary initial point in the space of systems because selection becomes frozen into a small subregion of the space. In either case, genetic networks having thousands of components seem likely to be rather typical members of the available ensemble.

A Research Program

The models presented in this chapter are oversimple examples of the connectivity properties of a scrambling genomic regulatory system under the drives of mutation and selection. The purpose of the models is to make us realize that such systems have well-defined, robust statistical features which we can characterize. Further, we can develop theories about the extent to which selection can pull adapting populations away from the average properties of ensemble members and toward rare, presumably useful wiring diagrams.

The general theory is an injunction to find out precisely how real genomic systems scramble and thereby to attempt to build a well-informed picture of the true ensemble of genomic regulatory systems selection is exploring. As noted, we already know about a number of features which bias the ways the genomic system scrambles. For example, in the evolution of real chromosomes, once a given gene has undergone tandem duplication, the existence of the duplicates renders further expansion of the duplicated genes via unequal crossing-over during meiosis more likely (G. P. Smith 1974; Dover, Brown, et al. 1982). Thus some sequences amplify tandem duplicates by what may be a selectively neutral mechanism. As this duplication occurs, extra copies of *cis-* and *trans-*acting genes may be being generated into multigene families. Later translocations, inversions, transpositions, and so forth will then disperse this randomly amplified *particular set* of *cis* or *trans* loci around the genome. Consider a *trans*-acting gene which acts on the duplicated and dispersed *cis*-acting gene. That *trans* gene will now have a very large numbers of descendants via those particular *cis* loci. Thus large biases in the number of regulatory descendents are almost surely introduced by such adventitious and biased amplification of copies of an initial *cis* or *trans* gene. In short, the wiring diagram is probably not isotropic. Further, as noted above, transposition occurs by specific transposable elements whose terminal sequences typically leave a small duplicated trace when they hop out for a new location. Those traces become hot spots for the reentry of another transposing element. Hot spots imply that the moves among genomic regulatory systems by which the systems rewire are not random but strongly biased. Translocation of genes to noncoding, junk DNA which is not transcribed may often be silent. The bits of genes so translocated may themselves fall silent, as do pseudogenes, and play no further role in the evolution of the connected regulatory network.

Selection has had unknown wisdom in protecting genomic systems from the ravages of chromosomal mutation. Close linkage of *cis*-acting loci to their regulated loci undoubtedly reflects, in part, protection against disruption by chromosomal mutations. This protection plan is a form of the familiar idea that recombination builds up closely linked coadapted gene complexes. Experiments in *Drosophila* using a specific transposable P element have shown that structural genes plus their adjacent *cis* regulatory elements can be cloned into the P element and the whole unit then transposed and integrated into many positions of the *Drosophila* chromosome; yet the structural gene is often activated in proper tissue-specific and time-specific manner in development (Garabedian, Shepherd, and Wensink 1986). This implies that influences from outside the boundaries of the P element do not often "reach inward" to influence the action of the structural gene. Limiting the range of action of *cis*-acting elements along the chromosome limits the sensitivity of the genomic system to chromosomal rearrangements.

Whatever these boundaries may be, whatever the nonrandomness may be, ultimately we should be able to characterize the way the genomic regulatory system is scrambling and build a good statistical theory of its expected structure. That that prospect lies in the future means only that it awaits our efforts.

Can we, in the meantime, build up information about the overall wiring diagram in a genomic system? This surely is the aim of contemporary developmental genetics. Here a major point to emphasize is that our experimental procedures are well tuned to uncover *local features* of the overall genetic regulatory system plus *global consequences* of that organization. That is, we can find out in detail which genes directly regulate one another, via which molecules. This knowledge yields insight into the local structure of the wiring diagram. As I emphasize in Chapter 12, studies of the overall similarities in patterns of transcription in different cell types and other studies provide information about the global, large-scale behavior of the genomic system. What we need to invent is a means of linking clues about the local architecture and clues about global behavior into a coherent theory for the probable structure and behavior of genomic systems. The ensemble theories of this and the next chapter are meant to provide a toehold on this vast problem.

In attempting to understand the overall wiring diagram of genomic systems, we confront the experimental difficulty that alteration in the activity of one gene in the network may typically not propagate to all its descendant genes. Recall that if a gene is activated by the prior activity of either of two input genes and if one input gene is now active, the activity of the second has no influence on the activity of the regulated locus. Ashby warns us that influence may not propagate. In order to study the downstream cascade of events, I shall return in Chapter 12 to testing the structure of genomic regulatory systems by means of inclusion in a cell of an arbitrary cellular gene, and its willful activation by a controllable *cis*-acting element. Similarly, any single gene can be deleted and the downstream consequences of the deletion analyzed. In this way, we can reasonably hope to build a picture of such small-scale features of the genomic system as small feedback loops and small patches of the wiring diagram. Limitations to the cascading effects may make it hard to pick up a strong component linking thousands of genes, however. Contrast this problem with one in which signals propagate to all downstream components. In that case, activation of one element reveals, by the subsequent cascade, all descendants. The fact that signals may not propagate in genetic regulatory networks means that the detailed wiring diagram structure may have to be teased apart bit by bit.

In order to study the evolution of genomic regulatory systems, we need to understand how rapidly and easily connections vary. This promises to be a fruitful direction, for we can already study the variability in genomic regulatory systems in closely related species (Dickinson 1980a, 1980b). Such variability may reflect either selection for different microhabitats or random drift in sloppy, suboptimal genomic systems. Discriminating which will be difficult, but at least establishing substantial regulatory variability is very important. I discuss this further in Chapter 13.

SUMMARY

The genome is a system in which a very large number of genes and their products directly and indirectly regulate one another's activities. The proper aim of molecular and evolutionary biology in considering such systems is not merely to analyze their architecture and dynamical behavior but also to understand *why* they might have more or less the architecture and behavior observed and *how* they may evolve in the face of mutational effects. To this end, we must build statistical theories of the expected structure and behavior of these complex systems. Structure is discussed in this chapter; behavior, in the next.

In order to understand the probable structure and behavior of evolving genomic regulatory systems, we must confront a new pattern of reasoning in biology: We must invent and investigate ensemble theories because an ensemble houses our current best hypotheses about the local or global structural or behavioral characteristics of genomic systems. The ensemble is just the set of all regulatory networks which are consistent with the constraints we currently know. A principle of adequate reason then enjoins us to suppose that, in the absence of further forces, genomic systems will be typical of the ensemble in question. Thus, for any genomic system, the typical properties of such ensembles emerge as the proper null hypotheses describing the structural or behavioral features we would expect in the absence of outside forces.

The most obvious outside force which may cause genomic systems to deviate from the typical features of the ensemble is natural selection. Thus we are enjoined to develop ways of testing what genomic systems might look like in the absence of selection, comparing predictions with features observed, and then attempting to deduce the effects of selection. Such an analysis makes explicit the interacting roles of self-organization and selection in accounting for the features we see.

Unlike more familiar theories in biology, ensemble theories, by their very nature, predict distributions of features, not particulars. This in no way implies that such theories are not testable, for they *are* testable against the corresponding observed distributions. In developing such theories, however, care must be taken to garner sufficient facts to generate the proper ensemble whose distributions constitute the null hypothesis. And selection is always operating, we may presume, both in tuning the ensemble and in locating organisms within it. That this be so appears unavoidable; that we must develop theories of parallel subtlety is the consequence.

CHAPTER 12

Differentiation: The Dynamical Behaviors of Genetic Regulatory Networks

The aim of this chapter is to develop a framework for thinking about the integrated behavior of the genomic regulatory systems underlying ontogeny. In the preceding chapter, we examined the evidence showing that genes directly and indirectly regulate one another's activity. In this chapter, we shall look in detail at such regulatory interactions.

In overview, the problem is to find a way to think about genetic systems containing thousands of genes whose products turn one another on and off. It is the integrated dynamical behavior of this regulatory system which coordinates the expression of different genes in each cell type of the organism and underlies the orderly unfolding of ontogeny. From what we now know, it appears this regulatory system is extremely complex and, as suggested in Chapter 11, likely to be a rich web of feedback loops rather than a simple hierarchial command structure. Further, chromosomal and point mutations are continuously "scrambling" the "wiring diagram" and the "logic" of the regulatory system. The obvious question is this: How can such a genomic system manage to behave with sufficient order to control ontogeny? The answer I propose in this chapter is controversial: I suggest that much of the order required for ontogeny is spontaneously present in a vast class of complex genetic regulatory systems. Selection may have less to accomplish than we have thought.

Among the reasons my suggestion is controversial is that we have come to think of ontogeny as controlled by a "genetic program." The very thought that a "program" might have self-organized properties appears utterly foolish, but this sense of bewilderment is due to our underlying picture of a genetic program as being similar to conventional programs which operate on conventional computers. Conventional computers are von Neumann architectures which carry out sequential computation. In such computation, a central processing unit performs a single operation at a time—a multiplication, for example, or a comparison of two values to find which is larger. A program is an effective algorithm which arrives at a final solution in a step-by-step fashion and uses Turing's concept of recursively computable functions. Anyone familiar with programming is acutely aware that minor changes in the sequence of instructions in a program typically alter output catastrophically. Hence the question "What would a random computer program do?" seems nearly meaningless.

It is therefore a major initial point to realize that, in whatever sense the genomic regulatory system constitutes something like a developmental program, it is almost certainly not like a serial-processing algorithm. In a genomic system, each gene responds to the various products of those genes whose products regulate its activity. All the different genes in the network may respond at the same time to the outputs of those genes which regulate them. In other words, the genes act in parallel. The network, insofar as it is like a computer program at all, is like a *parallel-processing network* (Rummelhart and McClelland 1986). In such networks, it is necessary to consider the simultaneous activity of all the genes at each moment as well as the temporal progression of their activity patterns. Such progressions constitute the integrated behaviors of the parallel-processing genomic regulatory system. And, as we shall soon find, it makes very precise sense to conceive of and analyze the expected self-organized behaviors of such parallel-processing networks. Doing so, however, requires that we develop insight into some of the main construction features of the networks.

The spirit of the adventure is first to characterize two local features of the system: the number of genes or gene products which directly regulate any one gene and the ways that gene responds to those regulatory inputs. We shall find that any gene appears to be directly regulated by only a few other genes. Put another way, the regulatory network is sparsely connected. In addition, almost all known genes turn out to be regulated according to a special class of rules that govern their activity as a function of the activity of the genes acting on them. This class, which I call "canalyzing," was introduced in Chapter 5.

The second step is to ask whether these two local features by themselves carry implications about the integrated behavior and structure of genomic regulatory systems constrained to have these local features. There are an enormous number, an entire ensemble, of possible genomic regulatory systems constrained to share these two local properties. Therefore, this adventure naturally leads us to study the average properties of members of such an ensemble and to assess whether those typical properties are due to the local features.

As we shall see, it is even now possible to demonstrate that almost all members of the enormous ensemble of genomic regulatory systems having these two properties exhibit many of the orderly properties seen in ontogeny. This unexpected but powerful spontaneous order must naturally lead us to suspect that the local features we have identified may harbor principles of order on which evolution has relied.

In Chapter 5, I introduced the idea of a dynamical attractor, such as a steady state or limit cycle, in the integrated behavior of a system whose variables influence one another. Any such attractor is a recurrent pattern of states of the variables of the system and typically occupies a small subvolume of the system's state space. In this chapter, I shall interpret a *cell type* as a recurrent pattern of gene activity, and *hence an attractor,* in the integrated dynamical behavior of a coupled system made up of thousands of genes and their products, all of which turn one another on and off. This identification is controversial but testable. It is also, I believe, the most natural and sensible image to hold of a cell type, given our picture of genetic regulatory systems and genetic circuits turning one another on and off in complex ways. For example, consider a stem cell population able to undergo repeated mitotic divisions and sustain itself as the same stem cell population while at the same time splitting off daughter cells which follow some further pathway of developmental change. One is almost forced to think of such stem cells as exhibiting some form of stable recurrent patterns of gene expression—in other words, as an attractor. A similar clear example is cells in culture which maintain a stable phenotype over hundreds of generations. It may be somewhat more controversial, but still reasonable and testable, to characterize cell types along differentiation pathways as attractors. Given this single interpretation of a cell type as an attractor, I shall show that mathematical models of very large genetic networks in which each gene is regulated by only a few other genes and is constrained to use the canalyzing rules which occur in known real genetic circuits yield model cell types which fit many known features of real cell types.

The network ensemble theory allows us to predict a number of fundamental features of cell differentiation in ontogeny:

- The expected number of gene-expression patterns which recur in one cell type
- The distribution of cell cycle times
- The number of cell types in an organism
- The similarity and difference in gene-expression patterns in different cell types in one organism
- The existence of a large core of genes active in all cell type of the organism
- The distribution of downstream cascading events resulting from a hormonal signal or mutation of a single gene
- The inherent homeostatic stability of cell types
- The fact that each cell type can differentiate directly into only a few other cell types
- The fact that ontogeny must be organized around branching pathways of differentiation from the zygote to all the ultimate cell types in the adult

Branching pathways of differentiation, present in all multicellular organisms presumably since the Precambrian, are but one candidate example of a universal feature which may hardly reflect selection at all. Instead, such pathways may reflect self-ordered features of complex genomic regulatory systems so powerful selection cannot avoid them. At a minimum, the powerful self-order exhibited by the models we shall consider invites our attention to the central theme of this book: Order in organisms may largely reflect spontaneous order in complex systems.

This chapter is organized into six sections. The first examines genetic circuits and attempts to define the two local properties noted above: low connectivity and use of canalyzing functions. The second considers "large-scale" aspects of cell differentiation, such as the existence of regulatory cascades, the similarity in protein and gene-expression patterns in different cell types, and the number of cell types in an organism. The third section develops the need for ensemble theories giving the expected structure and dynamical behavior of genetic regulatory systems. There I reintroduce the Boolean switching network models used to study complex genetic regulatory systems, first considered in Chapter 5. The purpose of these efforts is to find the conditions required for orderly behavior in such large, complex networks. To do so, the fourth section considers both implausible and plausible models of genetic regulatory systems. By doing so, we consider the entire range of possible models of genetic regulatory systems within the idealization and framework of Boolean networks. This analysis reveals that biologically plausible networks built with our two observed local properties behave with marked spontaneous order. In the fifth section, I ask whether the statistically expected behaviors of the biologically plausible class of genetic networks match known features of differentiation and ontogeny. I shall claim that the answer is "yes." The theory appears to account at least for the features of ontogeny alluded to above. In the sixth and final section of the chapter, I discuss the famous metaplasias—homeosis and transdetermination—seen in *Drosophila,* other insects, arthropods, and even humans. Evidence abounds for combinatorial features and an epigenetic code in ontogeny, precisely as expected from our network theory.

SIMPLE GENETIC CIRCUITS
AND THE BOOLEAN IDEALIZATION

In this section, we examine simple genetic circuits known in viruses, bacteria, and multicellular eukaryotes. We shall uncover two major generalizations. First, most genes are regulated directly by rather few other molecular variables. Second, regulated genes appear overwhelmingly to be regulated by a special and critical subset of the possible Boolean functions: the canalyzing functions. Because of the central importance of Boolean networks when we are thinking about self-organization and evolution in genetic regulatory networks, I recall here some material originally presented in Chapter 5.

Throughout much of this chapter, I shall use the idealization, introduced by Sugita (1963), that a gene is either active or inactive, and its product either present or absent. Given this on–off idealization, the effects of molecular signals controlling a gene are described by a logical switching function, or Boolean function, giving the activity state of the regulated gene as a function of the presence or absence of the regulating variables. This "binary" idealization is extremely valuable, for we shall shortly be considering model genetic regulatory systems comprising 10 000 or 100 000 such genes, and the idealization allows us to obtain mathematical results. Like the physicist's idealization of gas molecules as hard, elastic spheres, the binary idealization is meant to capture the essential behavior of the entity in question, a gene and its activities. We must remember, however, that the idealization is false. "Inactive" genes in bacteria exhibit a low level of transcriptional activity such that an average of fewer than 1 but more than 0 product molecules per cell are present. Furthermore, an active gene can exhibit graded levels of activity. Thus while the binary picture captures important features, all it can do is give us the logical dynamical features of the regulatory circuitry and tell us how the circuitry behaves over time. (I shall discuss the uses and limitations of this idealization shortly.)

Figure 12.1*a* shows again the now familiar lactose operon. As described earlier, the operator is bound by the repressor molecule unless the metabolic product of lactose, allolactose, binds to a second site on the tetrameric repressor, altering the repressor's conformation and thus pulling the repressor off the operator site (Muller-Hill, Rickenberg, and Wallenfels 1964; Burstein, Cohn, et al. 1965; Bretscher 1968; Zubay and Chambers 1971). When we use 1 to represent the bound state of the operator and 0 to represent the free state, the corresponding Boolean function is the "Not If" function. The adjacent promoter P is regulated by CAP, cAMP, core enzyme, and sigma factor. All four must be present for an active transcription complex to form (Zubay and Chambers 1971). This is the Boolean "And" function of four input variables. The structural genes themselves—Z, Y, and A—are regulated by the operator; the promoter, by the "Not If" function (Figure 12.1*b*).

The lactose operon displays an interesting kind of "differentiation" into two alter-

CAP cAMP CORE σ R-Lact.

P O ZYA ——chromosome

Transcription ——→

allo-lactose	Repressor	Operator
0	0	0
0	1	1
1	0	0
1	1	0

"NOT IF"

CAP	cAMP	CORE	SIGMA	PROMOTER
0	0	0	0	0
0	0	0	1	0
0	0	1	0	0
0	0	1	1	0
0	1	0	0	0
0	1	0	1	0
0	1	1	0	0
0	1	1	1	0
1	0	0	0	0
1	0	0	1	0
1	0	1	0	0
1	0	1	1	0
1	1	0	0	0
1	1	0	1	0
1	1	1	0	0
1	1	1	1	1

"AND"

a

P	Op	ZYA
0	0	0
0	1	0
1	0	1
1	1	0

b

Figure 12.1 (*a*) *First panel:* The lactose operon in *E. coli.* The Z, Y, and A are structural genes; O is the operator, P the promoter site, and R *trans*-acting repressor protein which binds the operator and blocks transcription unless itself bound by lactose or allolactose. The promoter is regulated by four *trans*-acting factors: cyclic AMP; core enzyme; sigma factor, which combines with core enzyme to form holoenzyme RNA polymerase for transcription; and CAP, a factor which binds cAMP. All four factors must be present for transcription to begin. *Second panel:* Boolean function describing regulation of the operator by repressor and allolactose. For the operator site, 0 = free, 1 = bound. For repressor and lactose, 0 = absent, 1 = present. Boolean function "Not If" specifies activity of operator at next moment, given each of the four possible current states of regulatory inputs. *Third panel:* Boolean function for promoter activity as a function of the four *trans*-acting variables regulating it. Boolean function is "And." All four variables must be present for the promoter to be active. (*b*) Boolean function for transcription of the structural genes in the lactose operon. Transcription requires that the promoter be bound and active (1) and the operator be free (0). This is the "Not If" Boolean function.

445

native stable states, and this differentiation is readily captured, even in this simple binary picture. If *E. coli* is placed in a medium containing a high level of lactose, lactose enters the cell, converts to allolactose, and binds to the repressor, freeing it and allowing transcription of the adjacent structural genes Z, Y, and A. Gene Z corresponds to beta-galactosidase, the enzyme that metabolizes lactose. Gene A's function is not known. Gene Y, however, codes for a protein which is a specific permease for lactose. In the presence of this enzyme, lactose enters the cell far more easily. Therefore, once the operon is derepressed and Y is present, the external supply of lactose can be lowered to an intermediate level and, thanks to the permease, lactose will still enter the cell rapidly enough to keep the operon derepressed.

This simple example exhibits three very important properties:

1. Think of external lactose as a control parameter. If we begin an experiment with external lactose at a very low concentration, then smoothly raise it to a high concentration, then smoothly lower it back to a low concentration, the response of the *E. coli* shows hysteresis. Specifically, at the same intermediate level of external lactose, the cell can be in either one of two alternative states, with the lactose operon either activated or inactive. Which state the cell is in depends on its history of external perturbation.

2. This simple system can exhibit two alternative steady states, operon active or inactive, for an appropriate constant intermediate value of the external lactose concentration. Thus this tiny system is already a genetic feedback circuit with something like two cell types or, more modestly, two stable and different states of gene expression.

3. Notice that if exogenous lactose were confined to vary from an intermediate to a high concentration (which it is not), then by shifting the concentration up, we could switch the operon to an active state. If we thereafter lowered the concentration to the intermediate level, the operon would remain active. It is obviously easy to make a genetic circuit which has two alternative steady states and in which the system can be switched to one of those states virtually irreversibly.

Here is the start of a picture of irreversible differentiation: the attainment of a stable pattern of gene activity which, because of hysteresis, cannot be reversed by the chemical signals that triggered the step, despite the identity of the genome.

The Number of Boolean Functions of K Variables Is $(2^2)^K$

We shall be seriously concerned with Boolean functions, and thus characterization of them is essential. The operator is controlled by two molecular variables: R and allolactose. Since each regulating variable can be either present or absent, there are $2^2 = 4$ possible combinations of their presence or absence. For each such state of the two inputs, the Boolean function must specify the state of activity which the regulated gene will adopt. Any Boolean function of two input variables therefore must specify four response values, chosen to be either 0 or 1. Consequently, the total number of possible Boolean functions of two input variables is $(2^2)^2 = 16$ (Table 12.1). More generally, if a gene is regulated by K variables, the number of possible combinations of their presence or absence is 2^K, and for each of these states, the response might be 0 or 1. Hence the number of possible Boolean functions of K variables is $(2^2)^K$.

The rapid explosion of possible Boolean functions as K increases is important. As Monod (1971) stressed, the existence of allosteric enzymes having regulatory sites

TABLE 12.1   The 16 Boolean Functions of $K = 2$ Inputs

1	2	3
0	0	0
0	1	0
1	0	0
1	1	0

1	2	3		1	2	3		1	2	3		1	2	3
0	0	0		0	0	0		0	0	0		0	0	1
0	1	0		0	1	0		0	1	1		0	1	0
1	0	0		1	0	1		1	0	0		1	0	0
1	1	1		1	1	0		1	1	0		1	1	0

1	2	3		1	2	3		1	2	3		1	2	3		1	2	3		1	2	3
0	0	0		0	0	0		0	0	1		0	0	1		0	0	1		0	0	0
0	1	0		0	1	1		0	1	0		0	1	0		0	1	1		0	1	1
1	0	1		1	0	0		1	0	0		1	0	1		1	0	0		1	0	1
1	1	1		1	1	1		1	1	1		1	1	0		1	1	0		1	1	0

1	2	3		1	2	3		1	2	3		1	2	3
0	0	1		0	0	1		0	0	1		0	0	0
0	1	1		0	1	1		0	1	0		0	1	1
1	0	1		1	0	0		1	0	1		1	0	1
1	1	0		1	1	1		1	1	1		1	1	1

1	2	3
0	0	1
0	1	1
1	0	1
1	1	1

Note: In each function, the four possible combinations of activity of inputs 1 and 2 each determine the activity of variable 3 at the next moment.

distinct from the normal catalytic site of the enzyme implies that utterly arbitrary cybernetic or logical control systems can, in chemical principle, be built. Therefore, we are not, or at least not yet, prohibited from conceiving of genes, or enzymes, having five regulatory variables which might realize any of the $2^{32} = 5.9 \times 10^8$ possible Boolean functions. It is therefore of interest to ask the following central question: Can we find any constraints on the Boolean functions which appear to be used in genetic regulatory systems? The answer appears to be "yes" and of considerable importance.

Canalyzing Boolean Functions

Consider again the lactose operon in Figure 12.1 before passing to others. The operator, regulated by the Boolean "Not If" function, has the following property. If allolactose is present, the operator is free regardless of whether the repressor is present or absent. If the repressor is absent, then the operator is free regardless of whether allolactose is present or absent. That is, either regulatory input has one value which suffices by itself to guarantee one operator state. Notice that the opposite value of either input is unable by itself to guarantee the operator state. If allolactose is absent, the

Prom 1	Prom 2	Structural gene Transcription
0	0	0
0	1	1
1	0	1
1	1	0

"EXCLUSIVE OR"

Figure 12.2 Hypothetical gene regulated by two promoters and transcribed if promoter 1 is bound or if promoter 2 is bound but not transcribed if neither or both are bound. This corresponds to the Boolean "Exclusive Or" function and is not canalyzing. No single state of either input 1 or 2 can determine the activity of the regulated gene.

operator may be free or bound, depending on whether the repressor is absent or present. If the repressor is present, the operator may be free or bound, depending on whether allolactose is present or absent. As in Chapter 5, I here define a *canalyzing function* as any Boolean function having the property that at least one of the regulating variables has one value which alone suffices to guarantee one state of the regulated locus. (The term "canalyzing" [Kauffman 1971a, 1974] was chosen to honor the fine biologist C. H. Waddington [1957], who liked to think about such things with respect to entire epigenetic "landscapes.")

Notice next that the "Not If" Boolean function has two canalyzing inputs, since either allolactose alone, if present, or repressor alone, if absent, can guarantee the value of the operator (free). Also, the canalyzed, or guaranteed, state of the operator (free) is the same for both canalyzing inputs. That is, the canalyzed state is *necessarily the same* for both canalyzing inputs.

Not all Boolean functions are canalyzing. Consider a hypothetical gene which is regulated by two promoters. The gene is transcribed if promoter 1 is bound or if promoter 2 is bound but not transcribed if neither or if both are bound (Figure 12.2). This is the Boolean "Exclusive Or" function. Note that no single state of the first promoter, free or bound, guarantees whether the gene shall be transcribed or not. Similarly, no single state of the second promoter guarantees whether the gene shall be transcribed or not. In all cases, the free or bound states of *both* promoters must be known in order to know whether the gene is transcribed or not. Thus the noncanalyzing "Exclusive Or" function has no gene state which can be guaranteed by any single state of any single input gene.

The Fraction of Boolean Functions that Are Canalyzing Decreases as K Increases

We shall see shortly that almost all known regulated genes are governed, in the Boolean idealization, by canalyzing functions. Thus it shall become critical to ask whether this is due to chance. Almost certainly the answer is "no," for the fraction of the $(2^2)^K$ Boolean functions of K input variables which are canalyzing is a maximum for $K = 2$ and decreases rapidly as K increases.

As K increases, an upper bound on the number of the $(2^2)^K$ functions which are canalyzing is given by $(4K)(2^2)^{K-1}$ (Gelfand and Walker 1984). Dividing this number by $(2^2)^K$ gives the *fraction* of Boolean functions of K variables which are canalyzing. This is a maximum for $K = 2$ and drops to less than 5 percent of the $(2^2)^4$ Boolean functions of four inputs. We shall see in the next sections that, even where regulated genes have more than two inputs, the genes typically are regulated by canalyzing functions. This presumably reflects either selection or other causes.

Boolean Functions as the Logical Skeleton
of Continuous Control Rules

Boolean functions idealize a gene to be either active or inactive as a function of the concentrations of the molecular variables which regulate the gene. A more accurate description notes that gene output can be a graded response to input. Nevertheless, as we saw in Chapter 5, the binary idealization yields results which can validly be extended to homologous systems of genes whose activities vary sigmoidally over a continuous range as the activities of their inputs vary smoothly (Walter, Parker, and Ycas 1967; Glass and Kauffman 1972, 1973; Glass 1975, 1977b, 1985; Glass and Pasternack 1978a, 1978b; Thomas 1979, 1984; Hopfield and Tank 1986a, 1986b; Kaufman 1988). The binary idealization is the limiting case of such sigmoidal response functions. Recall that such functions are characteristic of positive cooperative responses in allosteric enzymes and other cellular constituents, such as tetrameric hemoglobin or multimeric cell receptors.

Figure 12.3 shows the continuous function for the Boolean "Not If" function of Figure 12.1 and plots activity of the regulated gene vertically (ranging from 0 to 1) as a function of the activity level (minimum 0 and maximum 1) of its inputs, which are an inducer I and a repressor R. The function is a surface in this unit cube. The surface is at height 0 where the two input activities are 0 and rises to unit height 1 where I activity is maximal and R activity is minimal. I have drawn a sigmoidal surface, flat near the corners and rising steeply in the middle of the unit cube. Clearly, the values at the four corners of input activities correspond directly to the Boolean idealization. The interior of the unit cube is the entire function of which the corners are the skeleton. The sigmoidal surface reflects typical cooperative kinetics. In the limit of high cooperativity, the steep part of the sigmoidal curve rises to a vertical cliff, or explicit threshold. As input activities cross above or below this threshold, the regulated gene switches on or off. In Chapter 5 we examined a specific small network coupling two variables which were governed by sigmoidal response functions of one another's concentrations. Recall that the system exhibited two stable steady states corresponding to the "extermal" steady states of the homologous Boolean system but also contained an unstable steady state in the interior of the Boolean cube. That unstable steady state lay on the separatrix dividing the basins of attraction flowing to the two stable steady states.

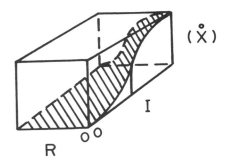

Figure 12.3 The continuous sigmoidal response of a gene to inducer I and repressor R is the analogue of the "Not If" Boolean function. Corner values of activity in the unit cube, where I and R take on minimum or maximum values, correspond to Boolean values for off and on states of I and R. Note that either an increase in R or a decrease in I alone can ensure decreased activity of the regulated gene. This is the continuous analogue of canalyzing in Boolean functions.

The general issue of the relation between the dynamical behavior of discrete Boolean networks and that of homologous systems of continuous differential equations governed by sigmoidal functions deserves extensive investigation. Traditional mathematical procedures allow study of the local stability of the steady states of nonlinear dynamical systems but are not easily used to obtain insight into the global behavior of complex nonlinear dynamical systems. In contrast, the Boolean idealization now appears, in a number of instances, to provide excellent qualitative insight into the numbers, locations, and stability of attractors in systems whose components are governed by a sigmoidal response function (Newman and Rice 1971; Glass and Kauffman 1972, 1973; Glass 1975, 1977b, 1985; Glass and Pasternack 1978a, 1978b; Thomas 1979; Kaufman, Urbain, and Thomas 1985; Hopfield and Tank 1986a, 1986b; Kaufman 1988). This usefulness of the idealization carries three implications. First, if analysis of Boolean networks finds ensembles which exhibit orderly dynamics with small attractors, we have reasonable grounds to expect the same qualitative behavior in a wide class of continuous nonlinear dynamical systems. Second, the Boolean idealization is robust with respect to our basic questions. Third, the idealization may prove a powerful mathematical technique for global analysis of the complex nonlinear dynamical systems.

The concept of canalyzing extends to continuous functions (Newman and Rice 1971). Continuous versions of "Exclusive Or" of one of the two noncanalyzing Boolean functions of two inputs correspond to a saddle-shaped surface in the unit cube. In contrast, Figure 12.3 corresponds to one of the canalyzing Boolean functions, which are characterized by the fact that, in the unit cube, the continuous function is monotonic with respect to at least one regulating variable. Thus in Figure 12.3, corresponding to the "Not If" function, if the activity of input I decreases while that of input R is held fixed at any level, the activity of the regulated gene is guaranteed to decrease. In contrast, for the noncanalyzing "Exclusive Or" function, the effect of the saddle-shaped surface is that, if the activity of one input increases, the activity of the regulated gene may increase or decrease, depending on the fixed activity of the second input. Such functions are not monotonic. Therefore, canalyzing Boolean functions correspond to continuous functions which are monotonic on at least one regulating variable. The fact that the fraction of Boolean functions which are canalyzing decreases as the number of inputs K increases carries over directly to the claim that the probability density of such continuous functions which are monotonic on one or more variables decreases as the number of regulating variables K increases.

Note also the interesting fact that the continuous canalyzing class of functions of several variables offers the maximum possible control to at least one control variable. For any fixed activity level of the other regulating variables, unidirectional change in the activity of this one control variable is sure to yield unidirectional change in the activity of the regulated gene. It is worth a moment to stress this point. Either the canalyzing functions in their Boolean idealization or the fully continuous functions afford the *maximum control possible by at least one input variable*. In general functions of two or more variables, no single "input variable" has even this much control. It may be for this reason, as we are about to see, so many regulated genes are governed by canalyzing functions.

Many Simple Circuits Use Canalyzing Functions

Consider again the lac promoter, which is regulated by the "And" function on four inputs. This is a canalyzing function, since absence of any single variable guarantees that the promoter is inactive. Each input is a canalyzing input, and again the canalyzed value of the promoter (off) is necessarily identical for all canalyzing inputs.

The structural genes in the lactose operon, Z, Y, and A, are directly regulated by the two *cis*-acting sites: promoter and operator (Figure 12.1*b*). Transcription occurs only if the promoter is activated and the operator is free. This is the "Not If" Boolean function, which is canalyzing. If either the operator alone is bound or the promoter alone is inactive, transcription does not occur.

Figure 12.4 shows components of the arabinose operon for *E. coli*. The probable sequence of genes is COIBAD. Genes B, A, and D are structural (Zubay, Gielow, and Englesberg 1971), O is an operator, and I is a promoter. The product C, a regulatory protein, appears to exist in two forms, P1 and P2, which attach respectively to O and I. The complex P1 at O inhibits transcription of BAD; P2 at I is required for transcription. L-arabinose is a specific effector, probably binding to either form of C and stabilizing P2 over P1. This binding frees the operator from inhibition by C. The operator is therefore regulated by C and L-arabinose (Figure 12.4*a*). The operator is bound only if C is present and L-arabinose is absent. Again, this is the canalyzing "Not If" function. Activation requires CAP and cAMP.

Figure 12.4*b* shows BAD transcription as a function of O and I. Transcription occurs only if O is free (0) and I is bound (1), the canalyzing "Not If" Boolean function. The presumptive molecules controlling the I locus are RNA polymerase core, sigma factor, CAP, cAMP, C protein, and L-arabinose. This is the "And" function of six input variables. All six are canalyzing, since absence of each determines that polymerase is not bound at I.

In the regulatory circuitry of bacteriophage lambda, the left operator O_L is complex but canalyzing. Lambda repressor C_I binds to O_L and is removed during lytic induction, presumably by a substance (X). However, the product of the lambda gene cro (also called tof) represses leftward transcription by binding at O_L even in the presence of X (Kumar, Calef, and Szybalski 1970; Szybalski, Bovre, et al. 1970; Eisen and Ptashne 1971; Ptashne 1986). Thus O_L is governed by the Boolean function in Figure 12.5. This function of three regulatory variables is canalyzing: In the presence of cro protein, O_L is bound regardless of the values of the remaining variables.

Two features of all these regulated loci are (1) the number of regulatory variables per locus is low, from 0 for constitutive mutants to 6, and (2) when regulated, each locus is regulated by a canalyzing Boolean function. Both properties appear to generalize to eukaryotes.

C	L-Arabinose	Op		O	I	BAD
O	O	O		O	O	O
O	I	O		O	I	I
I	O	I		I	O	O
I	I	O		I	I	O
a				**b**		

Figure 12.4   The Boolean functions governing the activity of each component of the arabinose operon in *E. coli*. (*a*) Regulation of the operator. (*b*) Regulation of the structural genes. (From Kauffman 1974)

Cro	C_I	X	O_I
O	O	O	O
O	O	I	O
O	I	O	I
O	I	I	O
I	O	O	I
I	O	I	I
I	I	O	I
I	I	I	I

Figure 12.5   Regulation of the left operator OL in bacteriophage lambda. (From Kauffman 1974)

Genetic Circuits Controlling Cell Type and Cell Lineage in Saccharomyces cerevisiae

Genetic control of cell differentiation is currently better understood in the simple eukaryotic yeast system than elsewhere. Yeast demonstrate canalyzing functions. Cells of yeast, *Saccharomyces cerevisiae,* are of three types: A, a, and A/a. The three types differ in their ability to undergo transitions in the life cycle. The view presented here is summarized from Sternberg (1986). Haploid A and a cells are specialized for mating, each with the other. Mating produces diploid A/a cell, capable not only of mitotic division but also of meiosis and spore formation, yielding two A and two a cells. Both A and a cells can undergo mating-type conversion to the other type. One consequence is that the two types can generate a colony of A/a cells.

   Cell specialization results from differential transcription of cell-type-specific genes—A-specific genes, a-specific genes, and haploid-specific genes, the last being transcribed either in A or in a but not in A/a cells. Genes specific for A/a cells have not yet been found but may exist. Expression of cell-type-specific genes is controlled at the level of transcription, largely by the mating-type locus, MAT. The two alleles MATA and MATa encode *trans*-acting proteins that regulate unlinked cell-type-specific genes. MATa encodes two regulatory proteins, a1 and a2, while MATA encodes at least one, A1. The a1 protein is a positive regulator of a-specific genes, and a2 is a negative regulator of A-specific genes. Thus in a cells, the set of a-specific genes is turned on and the set of A-specific genes is turned off. In A cells, the set of a-specific genes is silent because a1 is absent and the set of A-specific genes is expressed because a2 is absent. A third type of control is clearly combinatorial, with more than one *trans*-acting regulatory input. A combination of A1 and a2, present only in A/a diploids and probably a DNA binding dimer, represses a wide variety of haploid-specific genes that are otherwise expressed in A or a cells. Thus the haploid-specific genes are regulated by A1–a2. In addition, MATa1 expression in A/a cells is turned off by A1–a2.

   In the Boolean idealization, the logical functions seen so far are "Yes" for activators, "No" for inhibitors, and "Not And" for inhibition by A1–a2. Note that "Not And" is canalyzing, with absence of either A1 or a2 alone sufficing to veto inhibition. In fact, the a-specific and A-specific genes are regulated by more than a single input and hence are more complex than "Yes" or "No."

   In mating-type interconversion, genetic information is unidirectionally transferred to the MAT locus from either of two silent mating-type loci, HML and HMR, leaving a silent copy in place. This is the cassette model. A change in MAT yields a switch in cell type because the proteins encoded by MAT regulate cell-type-specific genes. Mating-type interconversion requires the action of the HO gene and a group of DNA repair genes, all of which must function normally. Thus inaction of any of these genes blocks conversion. Therefore, again, the corresponding Boolean function is canalyzing. Mating-type switching is probabilistic but highly regulated. It occurs only in a or A cells and not in A/a cells, preferentially in the mother cell, and only over a specific short interval early in the cell cycle between a Start event, defined by a mutant's blocking of cells at an early stage in mitosis, and the onset of DNA replication marking the start of the S phase.

Regulation of HO expression has revealed some of the *cis*-acting and *trans*-acting control mechanisms. Canalyzing functions abound. The HO gene encodes a site-specific endonuclease which cuts MAT. Transcription of HO is precisely regulated, occurring only in A or in a cells, in the G1 phase of the mitotic cycle between Start and S. *Cis* regulation of HO involves two positive sites and two families of negative sites upstream (5′) of the gene. These families lie in two major regions of upstream regulatory sequences, URS1 and URS2. Removal of these regulatory sequences removes the constraint of HO to be synthesized only in the G1 phase of mitosis. *Trans*-acting positive and negative regulators of HO expression have been identified. These regulators together strongly support the existence of *three* pathways of negative control, all of which must be alleviated if HO is to be expressed; thus the overall Boolean function is canalyzing. The three pathways are all controlled by canalyzing functions:

1. Cell-type control. The HO gene is expressed only in a or in A cells, not in A/a cells, thanks to repression mediated by a1–A2. The Boolean rule is the canalyzing "Not And" function.

2. Asymmetric control favoring mating-type switching, resulting from an interaction of a positive regulator, SW15, which probably relieves repression by a negative factor, SIN. The Boolean function is the "If" function. Again, it is canalyzing, for HO is not repressed by this negative pathway if either SIN is absent or SW15 is present.

3. Cell-cycle control, probably mediated by another positive regulator, SW14, acting on the CACGAAAA consensus sequence to repress HO transcription. This is the Boolean "Not" function.

Finally, examples of "And" and "Or": Both SW15 *and* SW14 are necessary to activate HO expression (the "And" function). In contrast, the CYC1 gene requires either *or* both of two conditions to be met, each realized by the presence of either of two activators that act as distinct promoter elements (the "Or" function).

Other Eukaryotes

The ubiquity of canalyzing functions seen in yeast appear to be typical of those higher eukaryotes whose regulatory connections have been worked out. These include steroid receptors which bind steroids and regulate transcription of specific genes through DNA binding (Miesfeld, Rusconi et al. 1986). These receptors realize the Boolean "And" function: Transcription requires both the steroid and its receptor. The function is canalyzing, since absence of either input alone blocks transcription. Enhancer-like sequences have been found in *Drosophila.* For example, two distinct enhancers flank and regulate joint-tissue-specific expression of two yolk proteins, YP1 and YP2. One enhancer appears sufficient to cause synthesis in the ovaries; the second, to cause synthesis in the fat body (Garabedian, Shepherd, and Wensink 1986). Each is thought to be regulated by a *trans*-acting factor. These yolk genes appears to realize the "Or" function with respect to these enhancers, since either enhancer alone suffices, when activated, to activate both adjacent genes. The "Or" function is also canalyzing: either input alone suffices to ensure transcription. Analysis of other regulated genes in mammalian, maize, insect, nematode, and other systems, so far reveals similar features.

Local Features: Genes Appear to Be Regulated by Few Direct Inputs and by Canalyzing Boolean Functions

The general summary of work over 20 years analyzing regulated genes in viruses, bacteria, plasmids, and now eukaryotes is simple. So far, almost all known regulated genes—structural and *cis*-acting—are directly controlled by rather few variables. The number ranges from zero for constitutively active genes (discounting TATA boxes and so forth) to perhaps six or ten for genes such as HO, which appear to be the focal point for control. Also, virtually all regulated genes appear to be regulated by canalyzing Boolean functions. Only further work will establish whether this pair of local features holds widely in eukaryotic systems.

As noted earlier, the fraction of Boolean functions which are canalyzing decreases rapidly as the number of input variables K increases. Thus, since canalyzing functions are widespread in organisms, we must explain their prevalence. Two plausible explanations are selection and chemical simplicity. That is, it may be that selection has ensured that, even when genes are regulated by several inputs, the functions are constrained to the canalyzing subset. Alternatively, it is almost certainly *easier* to construct molecules to implement canalyzing Boolean functions. For example, consider a repressor protein binding to an operator site. It is easy to imagine that the repressor has an allosteric site which is bound by either of two different molecules, X or Y, each of which suffices to alter the conformation of the repressor and pull it off the operator. The allosteric site need merely recognize X or Y. Conversely, it is hard to make a repressor which is pulled off the operator if X or if Y is present but not if *both* are present. That would presumably require two different allosteric sites—one for X and one for Y—and a complex interaction between the two. Thus this noncanalyzing "Exclusive Or" function is more complex molecularly than the simpler canalyzing "Or" function. In short, I strongly suspect that the prevalence of canalyzing functions largely reflects simple chemistry.

The possibility that widespread use of canalyzing functions may reflect chemical simplicity is part of the evidence for a tentative and rather astonishing possibility: As we shall see, the two local features of few inputs per gene, which reflects *molecular specificity,* and canalyzing functions, which reflect *chemical simplicity,* already appear to account for much of the large-scale order we find in cell differentiation and ontogeny. If so, then *we must entertain the hypothesis that much of that large-scale order is a direct reflection of fundamental features of polymer chemistry.*

LARGE-SCALE FEATURES OF CELL DIFFERENTIATION

Before exploring the consequences of these two local features, it will be useful to consider three well-known large-scale features of cell differentiation: the existence of regulatory cascades, the overlap in gene expression pattern at the RNA level and protein level in different cell types of one organism, and the number of cell types in an organism. All three obviously are large-scale consequences of the integrated behavior of genetic regulatory systems.

Regulatory Cascades

In Chapter 11, we used the general idea of random directed graphs to consider the architecture of connections in genetic regulatory systems. We saw that, in a system containing 10 000 regulated genes connected at random by 20 000 regulatory arrows, there is a very high probability that almost every gene will have almost all other genes in the system as direct or indirect descendants via connected pathways emanating from the gene in question. The situation does not alter substantially if we become slightly more realistic and assert the existence of a large number of structural genes which are themselves regulated but have products that play no regulatory role. We may simply remove these structural genes from consideration when we are thinking about the *regulatory* structure by which regulatory genes control the action of regulatory genes. We retain, of course, the *trans*-acting structural genes whose products act on other *cis*-acting loci and hence are part of the regulatory system.

In thinking about the regulatory architecture in Chapter 11, we temporarily suppressed concern about the response of each regulated gene as a function of the presence or absence (or activity or inactivity) of its regulating variables. However, mere connections between two genes, direct or indirect, do not ensure that a change in behavior of one will alter the behavior of the second. We have already seen this. If gene C is regulated by genes A and B by the Boolean "Or" function and if A and B are both active, then C is active. If A is inactivated, the behavior of C will not change. Thus alteration of A's behavior does not propagate through C to C's descendant genes. It follows obviously that the architecture of the regulatory system showing the connectivity of the directed graph structure is a necessary but not sufficient condition for the existence of regulatory cascades. The proper connections and the current state of activities of all the participants in a multi-input/multi-output network determine the ways in which alteration of the activity of one gene propagates through the network. Conversely, and obviously, if we alter the activity of a single gene and witness a cascade of alterations in activities of some number of other genes, we have at most a *minimum* estimate of the number of genes influenced by the initial gene. Other genes may in principle be influenced, but the current state of genes in the network may keep changes from propagating to them.

Regulatory cascades are well known in viruses as well as in higher eukaryotes. In bacteriophage lambda, infection of *E. coli* results in either lysis or lysogeny, as noted earlier (Ptashne 1986). The lytic cascade involves a complex unrolling of 10 to 15 *cis*-acting and *trans*-acting loci which coordinate the expression of structural and regulatory genes leading to replication of the phage DNA, synthesis of coat proteins, packaging, and cell lysis. A map of lambda and the main aspects of the regulatory cascade are summarized in Figure 12.6. Inspection of the *cis*-acting promoters and operators, as well as of antitermination sites, reveals that virtually all of them are guided according to canalyzing functions (Kauffman 1974; Ptashne 1986).

The alternative developmental pathway open to lambda, lysogeny, consists in the integration of the viral DNA into the host chromosome by precise integration at specific sites. The lysogenic state is maintained by synthesis of the lambda repressor C1, which feeds in a positive-feedback circuit to maintain its own synthesis from the maintenance promoter. Thus C1 forms a small feedback loop able to hold itself in a

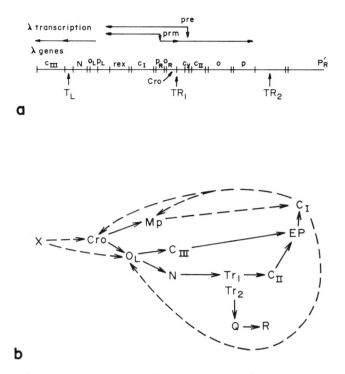

Figure 12.6 (*a*) Schematic representation of bacteriophage lambda genome and major transcription control points. (From Kauffman 1974; see Ptashe 1986) (*b*) Regulatory cascade triggered during lysis by transient activation of cro. This cascade is an example of a forcing structure, discussed in Chapter 5 and in this chapter.

steady state of activity. In fact, C1 is involved in a circuit of considerably greater interest: It represses the gene called cro, while cro is able to repress C1, each acting on the other indirectly by competition at a promoter. During the onset of lysis from the lysogenic state, cro is activated. If the product is abundant enough, cro suppresses C1 and thereafter allows the lytic cascade to unroll (Figure 12.6*b*).

C1 and cro: A Mutually Inhibitory Circuit with Two Alternative Steady States.
The C1–cro feedback system supports two states of differentiation. Each gene indirectly represses the other. Thus it seems reasonable that this little circuit be open to two alternative steady states: C1 active, cro inactive or C1 inactive, cro active. Normally the former state is the lysogenic state, or immunity+ state, where C1 activity also renders the host *E. coli* immune from superinfection. The latter case, the immunity− state, cannot be seen because lysis is unrolling. In lambda missing N and P genes, however, lysis cannot occur. In that case, both steady states can be seen. In fact, a kind of stochastic switching between the two states occurs. If lysogens are grown from an immunity+ cell, a small number spontaneously convert to the immunity− state. The reverse transition also occurs. Thus this small circuit supports two "cell types," each able to differentiate into the other (Neubauer and Calef 1970).

The Polytene Puffing Cascade in **Drosophila melanogaster.** *Drosophila melanogaster* is a holometabolous insect. That is, it has four life stages. The fertilized egg hatches as a wormlike larva which undergoes three molting cycles separating three larval instars. The third instar forms a pupa, undergoes metamorphosis, and emerges as an adult. Male and female adults mate and begin the cycle again. Among the many features which have properly drawn geneticists and developmental biologists to *Drosophila* are its immense polytene chromosomes. These are 100- to 2000-fold lateral duplications of the normal diploid set of four pairs of homologous chromosomes. Each polytene chromosome has all lateral copies synapsed in proper register, such that even light microscopy reveals prominent banding patterns. Over the years, debates have raged concerning the number of genes per band. *Drosophila* has about 5000 bands. It now appears that the insect has more genes than bands, but grounds persist to think that one band might be a unit of functional organization.

Among the tissues in *Drosophila* having the most prominent polytene chromosomes are the salivary glands, which secrete a specialized set of glue proteins during the onset of pupariation. Long before their products were known, however, these salivary-gland chromosomes invited attention. Metamorphosis in *Drosophila* is triggered by an increased level of the molting hormone ecdysone. Examination of the salivary-gland chromosomes during the late-third-instar and early-to-mid pupal stages revealed that the diameter of specific chromosomal bands waxed and waned dramatically (Becker 1959; Ashburner 1970, 1972, 1974). Subsequently it was shown that addition of exogenous ecdysone to salivary glands cultured *in vitro* induced the same "puffing" cascade (Ashburner 1972; Ashburner and Berendes 1978). Thereafter it was established that puffing is associated with transcription (Bonner and Pardue 1977). Good evidence has long suggested that this cascade is a causal one during which a succession of genes are sequentially activated and repressed via protein products of the cascade (Ashburner 1974). The cascade, which consists of about 150 distinct puffs, each presumably containing at least one gene, is organized into early, middle, and late puffs. The early puffs are directly responsive to ecdysone. The middle and late puffs presumably depend on the activity of the early ones, since inhibitors of protein synthesis during the activity of the early puffs prevent the puffing and accumulation of RNA in the middle and late puffs.

A few of the early genes have now been cloned, and mutants at one specific locus, B2 on the X chromosome, suffice to block the entire cascade induced by ecdysone (Chao and Guild 1986). The intron/exon structure of the B2 gene has been worked out, and work is now under way to uncover *cis*-acting sites in its vicinity which might play a role in its regulation (Chao and Guild 1986). This gene presumably acts on later genes in the cascade. According to one model (Ashburner 1974; Zhimulev, Vlassova, and Belyaeva 1982), transient activity by an early gene X, perhaps activated by B2, feeds back on B2 activity, even in the continued presence of ecdysone, and feeds forward to middle and late genes to potentiate their activities. In this hypothetical feedback loop on B2, the B2 gene is regulated by the canalyzing "Not If" function, for the gene is active only if ecdysone is present and X is absent. Thus absence of ecdysone alone or presence of X alone inhibits B2 activity.

If it is legitimate to conclude that the salivary-gland puffing cascade is a control cascade, then a single trigger molecule—ecdysone—unleashes alteration in the behavior of on the order of 150 "genes." While we do not know exactly how many true genes this number represents, we do know that at least 150 bands puff. Since *Drosophila* has about 5000 bands, this implies that on the order of 150/5000 bands, or 3 percent, alter their behavior in the presence of ecdysone. Thus 3 percent is a rough minimum estimate of the number of genes directly or indirectly influenced by ecdysone in the salivary gland and probably by B2, since B2 deletion blocks the entire cascade.

Similar cascades are well known in other insects, such as *Chironomus* (Beermann 1952) and other species of *Drosophilae* (Berendes 1966). In *D. melanogster* similar cascades, not yet worked out in detail, occur during the onset of metamorphosis of several tissues. Many of these tissues have been analyzed at the protein level and reveal at least a dozen or more novel proteins whose synthesis appears to be unleashed by ecdysone. Since different cascades unroll in different tissues, it has long been supposed that ecdysone interacts, via specific receptors, with different DNA control regions in different cell types.

Conclusions About Regulatory Cascades.   The conclusions to draw from these studies are that regulatory cascades certainly exist, that the consequences of altering the behavior of a single gene in a cascade can easily propagate to a few percent of the genes in the system, and that we can hope to eventually work out the statistics of the sizes of such cascades. More precisely, we can ultimately learn, for any arbitrary gene, how many other genes alter their activities when the activity of that first gene is altered. Experiments along these lines are now completely feasible. First, transformation of specific cell lines containing a desired gene which is under the control of an arbitrary inducible promoter allows the desired gene to be activated at will in the host cell. By utilizing deletions of the gene of interest in the host cell and by controlling the timing of activation of the introduced copy, we can study downstream effects on other genes at the RNA and protein levels. Conversely, and in principle, we can study the cascading effects of deleting the activity of any single gene at a defined time by injecting antisense RNA to the mRNA of that gene. In principle, the antisense RNA binds to the sense mRNA and blocks its translation, thereby functionally deleting the gene product from the cell.

It is important to analyze such statistical features of regulatory cascades. Current efforts to understand the detailed molecular mechanisms by which specific genes are regulated are highly useful, as are efforts to work out specific cascades. However, we must develop approaches to learn about the structure of the overall architecture of genomic regulatory systems. Sampling to establish the sizes of descent cascades from many points within the network is clearly a useful way to proceed. Furthermore, once the distribution is established, we shall want to be able to *deduce* those statistical features from some deeper theory about the expected architecture and dynamic of genetic regulatory systems, not merely list them. Development of such a deeper theory is the aim of this chapter.

Cell Types as Constrained Patterns of Gene Expression

RNA Data.   By comparing the gene sets transcribed into heterogeneous nuclear RNA in different cell types within one organism, workers have been able to examine the overall behavior of the genomic regulatory system underlying cell differentiation in a number of systems (Alberts, Bray, et al. 1983). The same studies carry over to analysis of the mature mRNA populations which appear in the cytoplasm. These analyses have yielded a number of major conclusions:

1. Only a small fraction of the total DNA is transcribed.

2. The total number of distinct sequences transcribed ranges from about 17 000 for *Drosophila* (Levy and McCarthy 1975; Arthur, Weide, et al. 1979) up to perhaps 70 000 to 100 000 for mammals (Bishop 1974; Bantle and Hahn 1976).

3. The abundances of different RNA sequences vary enormously, from highly repetitive sequences present in millions of copies per cell to genes whose transcripts are present in one or fewer copies, on average, per cell (Britten and Kohne 1970; Alberts, Bray, et al. 1983; Davidson, Hough, et al. 1983; Darnell, Lodish, and Baltimore 1986).

4. A very large common set of genes is transcribed into heterogeneous nuclear RNA in virtually all cell types in the organism; this common core may range up to 70 percent of the transcribed complexity (Hough-Evans, Smith et al. 1975; Hastie and Bishop 1976; Chikarraishi, Deeb, and Sueoka 1978; Hough-Evans Ernst, et al. 1979; Alberts, Bray, et al. 1983).

5. The differences between different cell types is rather small, typically on the order of a few percent of the total number of different sequences present.

6. A subset of the sequences transcribed into heterogeneous nuclear RNA is processed to mature mRNA and transported to the cytoplasm.

7. Qualitatively, the conclusions about heterogeneous RNA hold up for mature messenger diversity: A common set of 60 to 70 percent of the genes is expressed in all cell types as mRNA, the abundances vary enormously, and different cells typically differ in a few percent of the message sequences. Thus a typical plant cell might have on the order of 20 000 distinct mRNA sequences and differ from another cell type in the same plant in perhaps 1000 sequences (Kamalay and Goldberg 1980; Alberts, Bray, et al. 1983). Some evidence suggests that differences in mRNA sequences from one cell type to another may often be greater than differences in nuclear RNA sequences.

Protein Data. Analysis of the protein sets synthesized in different cell types in one organism has revealed essentially the features shown at the RNA level. Technical differences in the ways RNA sequence diversity and protein diversity can be analyzed must be borne in mind. The bulk of the RNA work has relied on hybridization of single-stranded RNA to single-stranded DNA. Competition experiments with radioactively labeled RNA from one tissue and unlabeled RNA from another tissue allow assessment of the overlap between the two tissues in kinds of RNA sequences transcribed and of their relative abundances. However, detailed analysis of the time course of any one mRNA sequence is almost impossible by these techniques. Newer procedures for examining diversity at the RNA level rest on first cloning specific genes or fragments of genes (Grunstein and Hogness 1975) and then determining the concentration of the corresponding mRNA sequences over time. The concentration analysis entails hybridization of each kind of RNA sequence to its corresponding DNA sequence's complementary strand by dot-blot analysis (Benton and Davis 1977). These techniques, while powerful, allow investigation of only several to several dozen genes and gene products at a time. Thus they naturally focus effort on finding and studying a few genes of interest. In contrast, two-dimensional gel electrophoresis (O'Farrell 1975; Garrels 1979, 1983) has opened the door to parallel analysis of up to 2000 proteins at once.

Typically, analysis of different cell or tissue types taken from the same organism and taken at about the same developmental stage reveals surprisingly few differences.

For example, Rodgers and Shearn (1977) examined the different imaginal discs of *D. melanogaster* during the third-larval-instar stage and found only a few differences among many hundreds of identifiable protein spots. Similar results are reported by Greenberg and Adler (1982).

Since RNA data suggest that the total number of different RNA sequences is 20 000 or more (Levy and McCarthy 1975), the 1000 to 2000 proteins which can be visualized on two-dimensional gels are a 10 percent sample of the total diversity. One guesses, but does not know, that the 10 percent is representative.

The major point to emphasize is that, in most higher eukaryotic organisms, a common set of proteins is synthesized in all cell types. Different cell types differ in only a few percent of the proteins synthesized (Alberts, Bray et al. 1983; Darnell, Lodish, and Baltimore 1986). This generalization may not hold up for very simple multicellular organisms, such as the cellular slime mold *Dicteostylium discoidum,* where the protein or RNA differences between the few cell types may be a larger fraction of the total diversity (Firtel 1972).

In addition to studies of different cell or tissue types at one stage in development, two-dimensional gel electrophoresis has been used to analyze the time course of protein expression patterns during development (that is, the ontogeny in organisms). Such work in *Drosophila* (Gutzeit and Gehring 1979; Sakoyama and Okubo 1981; Summers, Bedian, and Kauffman 1986) reveals that, over the whole developing embryo, hundreds of proteins translated very early in development later wane, while hundreds of others either wax or first wax and then wane. Thus a complex temporal pattern in which many proteins alter synthesis levels unfolds from the initial state of the zygote. Similar studies have been carried out in other organisms (for instance, Brandhorst 1976; Bedard and Brandhorst 1983).

The general summary of these results is that, typically, a large core set of genes is transcribed in common in all cell types of an organism and that cell types differ in a "penumbra" of genes or proteins that represent only a few percent of the total number of genes expressed in the organism. While it is clear that these cell-type-specific genes must be regulated, it is less clear whether and how the common-core genes are regulated. That core might be constitutively active, or it might form linked circuitry that mutually ensures its conjoint activity. It is also not clear what the functions of the core might be. One obvious proposal is that these genes represent "housekeeping" genes whose products are needed in general cell metabolism. However, this seemingly plausible hypothesis is not that plausible after a moment's thought. *Escherichia coli* has 2000 to 3000 structural genes which suffice for a rigorous life demanding metabolic flexibility. Why should cells in a multicellular organism, each cell living in a buffered internal environment, require an order of magnitude more diversity of structural genes to handle housekeeping tasks? And we must bear in mind that we have not measured the activities of nontranscribed *cis*-acting regulatory loci in this story.

A rather unexpected possible role for the core genes will emerge shortly. I shall suggest that they correspond to a frozen component in a fixed state of mutually ensured activities within the genomic system. Recall from Chapter 5 that such a frozen component is a precondition for orderly dynamics in complex Boolean networks.

The Number of Cell Types in an Organism

Different organisms have different numbers of cell types. While any biologist, not to mention butcher, would agree with this qualitative statement, it is difficult to quantitate the differences. The problem is the criteria we should use in distinguishing cell types. The qualitative point is so simple and critical, however, that it often passes without comment. Since the earliest days of the cell theory and histology, it has been quite apparent that cell types are distinctly different from one another and do not intergrade. Over the years, biologists have used morphological and histological staining procedures, as well as behavioral criteria, to analyze the distinct cell types of a wide variety of organisms. If we choose to utilize those procedures and criteria for counting cell types, we must bear in mind that analysis at the protein or RNA level may lead us to distinguish cell types which biologists now count as the same. On the other hand, we might reasonably hope that any discoveries we made at the protein/RNA level would lead us to multiply the number of cell types in each organism by some fairly constant factor. Thus refinement of criteria will alter the number of cell types per organism but may not alter the proportional way that number changes as we move from simple organisms to more complex ones.

Simple organisms have fewer cell types than complex organisms. For example, yeast has three, as distinguished above. Bacteria may have two: vegetative and spore forms. The mold *Neurospera* has about five (Baldwin and Rusch 1965). By similar criteria, the hydra, a simple coelenterate, has 13 to 15 (Macklin 1968), as do sponges (Borradaile, Potts, et al. 1958). Jellyfish are more complex, with perhaps 20 to 30 cell types (Borradaile, Potts, et al. 1958). Annelids have about 60 cell types (Borradaile, Potts, et al. 1958), and adult humans about 254 (Alberts, Bray, et al. 1983). Therefore, there must be some relation between the number of genes in an organism and the number of its cell types. If it is hard to count numbers of cell types, it is also hard to count numbers of genes. The simplest way to count genes is to count total DNA per cell, a not too unreasonable choice. If this approach is used, we recall that many plants are polyploid, having extra copies of all chromosomes. Since strains virtually identical to polyploid strains have lower ploidy, we should probably use the minimum diploid DNA content per cell of members of that variety of plant. The virtue of using DNA content per cell to measure the number of genes is that very reliable data are available. The vice is that much of the DNA appears to be noncodonic and may play no genetic role at all. If we adopt the alternative approach and use measures of the number of distinct RNA sequences transcribed into heterogeneous nuclear RNA, we miss the nontranscribed *cis*-acting regulatory loci, plus possible control loci in RNA itself, and so on.

In Figure 12.7, I show a graph of the number of cell types, based on histological criteria, versus the total DNA content per cell, across a wide range of phyla. The data, on a log–log plot, are quite close to linear, indicating that the correlation between amount of DNA and number of cell types is a power law. Specifically, the data suggest that the number of cell types in an organism is crudely proportional to the square root of the DNA content per cell. This means that, in evolution, adding the next cell type has required ever greater amounts of DNA. Suppose instead that we consider, for any organism, the number of cell types versus the estimated number of different RNA sequences transcribed in that organism. Thus for yeast, the estimated number of cell types is 3 and the number of transcribed genes is 4000. In *Drosophila,* as in annelids, the number of cell types might be about 60 and the transcribed complexity about 17 000. In humans the number of cell types is 200 to 300 and the transcribed complexity is about 100 000. Thus on the basis of measured complexity of RNA sequences, the number of cell types increases roughly linearly with, or perhaps a bit faster than, the number of genes. It certainly does not increase much more rapidly than linearly. That is, the number of cell types does not increase as the number of genes squared or as an exponential function of the number of genes. Since total DNA

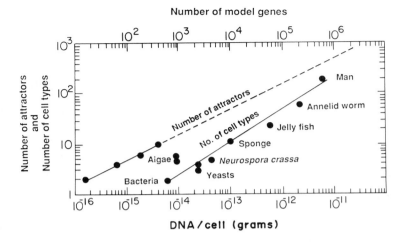

Figure 12.7 Logarithm of the number of cell types in organisms across many phyla plotted against the logarithm of the DNA content per cell. Plot is linear with a slope of 0.5, indicating a power-law relation in which the number of cell types increases as the square root of the amount of DNA per cell. If total number of structural and regulatory genes is assumed proportional to DNA content, then the number of cell types increases as a square-root function of the number of genes. Number of attractors refers to predictions of numbers of model cell types in model genomic regulatory systems having $K = 2$ inputs per gene.

is an overestimate of the number of genes and transcribed complexity is an underestimate, we might safely guess that the true relation is somewhat less than linear. This statement says that, in evolution, progressively more genes are required for each additional cell type. Although it fits the general notion that more regulatory genes are needed to coordinate gene expression, this image is quite naive. It is trivial to construct model genetic networks in which the number of model cell types increases *exponentially* as the number of genes increases. Exponential growth is faster than any power law.

In short, it is very much a nontrivial observation, across many phyla, that the number of cell types increases at something like a linear or square-root function of the total number of genes in an organism. Obviously, we would like to know why that might be true. And we might begin to wonder whether a property such as this, splayed across many phyla, represents selection or some deeper feature of genomic regulatory systems.

THE CONCEPTUAL FRAMEWORK: CELL DIFFERENTIATION IN BOOLEAN NETWORKS

This third section of the chapter develops a new conceptual framework to describe, analyze, and think about the coordinated behavior of integrated genomic regulatory systems containing tens or hundreds of thousands of genes. In this section and the next, I reintroduce Boolean networks as models of genetic regulatory systems and summarize a wide range of classes, or ensembles, of such networks. Because the material is unfamiliar to many readers whose background is in the biological sciences, I should make the motivation clear: As we discuss many of the possible classes of Boolean networks, some of which are clearly *not* biologically plausible, our intuitions will become tuned to the network *construction requirements* that allow the spontaneous occurrence of the kind of orderly dynamical behavior which must underlie ontogeny. Indeed, such emergent order arises in model regulatory systems which have local structure similar to that found in real genetic regulatory systems: Each gene is regulated by few other genes and is governed by a canalyzing function.

Need for an Ensemble Theory

We have now concluded with fair conviction that, in bacteria and viruses as well as in eukaryotes, most genes whose regulation is established are directly controlled by only a few regulatory inputs and appear to be regulated by one of the canalyzing Boolean functions in the binary Boolean idealization. In addition, it is clear that the Boolean function by which one gene is regulated may differ from the function describing another gene's behavior. Our ultimate aim is to understand how 10 000 to perhaps 250 000 genes and their products are organized into a coupled network by which the activity of each gene is orchestrated into some form of harmony with the activities of all the others. We ultimately want to understand, as well, how such systems evolve. The latter necessarily involves understanding how the behavior of genomic systems changes as new structural and regulatory loci evolve and as new regulatory connections are made, old ones broken, and regulated genes come to respond differently to their regulating inputs, thereby altering the logic of the developmental program. Thus it shall not suffice merely to understand one such network; we shall need a theory about how changes in the network change its behavior.

How can we sensibly proceed? First, of course, we can continue to work out the molecular mechanisms of gene regulation. Second, we can discover local, small-scale properties of the regulatory network, such as those described above. With more tenacity, we can discover large scale features of patches of the regulatory architecture. Indeed, proposals to examine regulatory cascades, both upstream and downstream, by cloning *cis*-acting loci, by identifying *trans*-acting components as suppressor mutants able to suppress in *trans,* and so on are now rife. It is genuinely marvelous, in just the proper sense of that word, that such experimental work is now feasible. Based on the results we have obtained at each stage in this process, however, we shall want to begin to frame hypotheses about the overall organization and behavior of the integrated system of 100 000 or so genes. Then we need to be clear about the kind of problem we must pose to ourselves.

If all we know about the genomic system is some set of small-scale features—such as numbers of inputs per gene, use of canalyzing functions, and perhaps local patches of cascades and small feedback loops—then the rational way to proceed, based on those local features, is to study the ensemble of all large regulatory systems, each constrained to have those local properties. That ensemble comprises, in a real sense, all our guesses about possible large-scale genomic systems which can be derived from the known local features. From this, it inevitably follows that the *typical* large-scale properties of members of the ensemble must stand as our current best guesses about the large-scale features we expect to find in genomic systems, based on the local features we have discovered to date and used to construct the ensemble. As stressed in Chapter 11, the generic features of typical members of the ensemble are the proper *null hypothesis.* In the absence of outside forces and if we have generated the correct ensemble given known local properties, the null hypothesis gives the expected features in genomic systems. Of course, the most obvious outside force is selection; we shall return to selection acting on genomic regulatory systems, mentioned already in Chapters 3 and 5, later in this chapter and in Chapter 13.

Methodological, Epistemological, and Ontological Consequences of an Ensemble Approach

Scientists tend to eschew philosophy. Nevertheless, the task of understanding the integrated behavior of a genomic system in the continuing process of selection and drift may necessitate changing how we manage this part of biological science. First a methodological issue. We are not likely to analyze an entire genomic system of 100 000 genes and their interacting products in detail soon. Bacteriophage lambda, with a few dozen genes, took many people many years of work. Manfred Eigen playfully defines a unit of work as one Ph.D. dissertation. Many such units can be expected to elapse before all the details are worked out. Brute force reductionism is likely to either fail outright or else be very slow methodologically. Therefore, I contend, we genuinely have no alternative to a mixed methodology, based on discovering local features, analyzing the corresponding ensemble of genomic systems embodying those known features, and attempting to use the predicted features of ensemble members to predict corresponding large- and small-scale features of cell differentiation and ontogeny. In turn, however, any use of an ensemble theory leads us inevitably into a new pattern of thinking, for the answers which an ensemble theory can provide are necessarily statistical distributions. No ensemble theory can be expected to assert that, in fact, gene A regulates gene B. Therefore, we can hope to predict means and variances of properties—in other words, qualitative aspects of development rather than specific cases. Therefore, testing and improving an ensemble theory reside in testing expected distributions and, where they fail, in attempting either to improve the ensemble by better understanding of local and larger features or to understand how outside forces such as selection modify the features we see from those predicted by the null hypothesis.

These methodological implications of an ensemble approach immediately carry epistemological implications. From a practical point of view, we may never be able to work out the details of the genomic regulatory system in a single inbred higher eukaryote. Even were we to succeed, the fluidity of the genome means that the regulatory system in neighboring organisms, sibling species, and so forth, are dancing away from us faster than we may ever be able to grasp them. The practical epistemological problem, in short, is that we may never to able to carry out the reductionistic dream of complete analysis but will want nevertheless to understand how these systems work. An ensemble theory offers hope of such understanding. To the extent that known local properties engender an ensemble whose typical members exhibit many of the properties found in differentiation and ontogeny, then to that extent those properties do not depend upon the details of the genomic system. Therefore, we can legitimately *explain* those properties by understanding why they are typical of the ensemble in question. Indeed, the typical properties of such ensembles emerge as the new "macroscopic" observables of this new kind of statistical mechanics.

These same issues arise in biology at levels other than genomic regulatory systems. For example, we want to understand the immune system. The immune response depends on the integrated action of the B-cell and T-cell network, including the anti-idiotype network (Jerne 1974, 1984; Kaufman, Urbain, and Thomas 1985; Kaufman and Thomas 1987; deBoer 1988; Hoffman, Kion, et al. 1988; Kaufman 1988;

Perelson 1988; Sieburg 1988). Yet the structure and logic of the immune network in each organism are unique not only because the organism's immune repertoire is genetically unique but also because the structure and logic change as a function of both antigenic history and the maturation of the immune response. Despite the fact that the immune network in each organism is shifting connections and local logic rapidly, we need theories about the integrated behavior of such systems. Brute reductionism seems likely to fail by itself: Given any immune network, we shall need to know how minor variations in anti-idiotype coupling and repertoire expression alter dynamics. Some kind of ensemble theory over classes of immune networks seems an inevitable necessity. Again, the generic features of such ensembles emerge as the macroscopic observables which can be explained by the theory.

The epistemological issue leads to an ontological one. If membership in some ensemble of genomic systems virtually ensures some spectrum of ordered features, then to account for the evolutionary origin of these features, we need only account for the evolutionary occurrence of the ensemble itself. Thus I shall try to show that membership in the ensemble of genomic systems characterized by low connectivity and use of canalyzing functions accounts for much of what we see in ontogeny. If so, we need only to account for the evolution of such classes of regulatory systems in order to explain the spectrum of properties. And, as hinted at repeatedly, if we can show that selection or drift is unable to avoid those generic properties, then they should be widespread in organisms by virtue of membership in a common ensemble, not by virtue of common descent.

The Marriage of Self-Organization and Selection: Selection Sustains the Useful Ensemble

The ontological issue points to the marriage of self-organization and selection. Selection constrains the ensemble being explored in evolution. In Chapter 5, we found conditions for orderly dynamics; order requires the percolation of unchanging frozen components across the network. We found three kinds of behavior: "solid," "liquid," and "gas," where solid corresponds to frozen unchanging states and gas to chaotic dynamics. Between them is the liquid interphase, corresponding to the most intricate behavior in such networks, where the most complex "computations" can occur. Previewing the discussion below, genomic systems appear poised in the solid regime near the edge of chaos, but this positioning is unlikely to be happenstance. The most plausible hypothesis is that selection achieves this poised state. Thus the natural marriage of self-organization and selection first discovers the powerful order inherent in complex systems in the near-liquid regimes and then appeals to selection to achieve and sustain membership in this ensemble. The generic properties of this useful ensemble emerge as the quasi-universals characterizing ontogeny.

Boolean Regulatory Networks as Logical Skeletons of Genomic Regulatory Systems

Consider again a small Boolean network comprising three on–off genes (Figure 12.8a). I assume each gene receives regulatory inputs from the remaining two. Gene 1 is governed by the Boolean "And" function and is activated the next moment only if genes 2 and 3 are active at the present moment. Genes 2 and 3 are each governed by the Boolean "Or" function and hence are active at the next moment if either or both regulatory inputs are active at the present moment (Figure 12.8a). In Figure

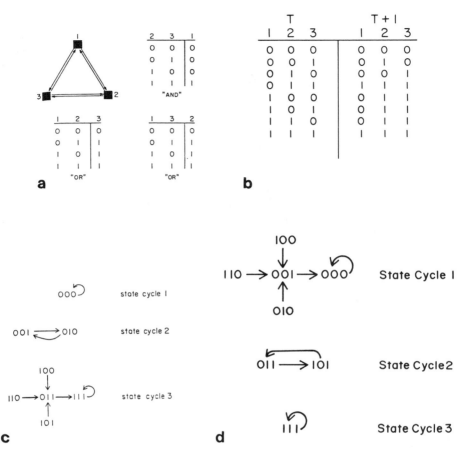

Figure 12.8 (*a*) The wiring diagram in a Boolean network containing three binary elements, each an input to the other two. One element is governed by the Boolean "And" function, the other two by the "Or" function. (*b*) The Boolean rules of (*a*) rewritten to show, for all $2^3 = 8$ states of the network at time *t*, the activity assumed by each element at the next time moment *t* + 1. Read from left to right, this figure shows the successor state for each state. (*c*) The state transition graph, or behavior field, of the autonomous Boolean network of (*a*) and (*b*), obtained by showing state transitions along trajectories through successor states connected by arrows. This system has three state cycles. Two are steady states, (000) and (111); the third is a cycle with two states. Note that (111) is stable to all single Hamming unit perturbations—[for example, to (110), (101), or (011)—while (000) is unstable to all such perturbations. (*d*) Consequences of changing the rule governing element 2 from "Or" to "And." Such mutation alters some state cycle attractors and basins of attraction leading to old attractors. Here (000) and (111) remain steady state attractors, (000) becomes stable to all one-unit perturbations, and (111) becomes unstable to all one-unit perturbations.

12.8*b*, these rules are rewritten. Each of the 2^3 possible combinations of activities of the three genes corresponds to one *state* of the network. Each state at one time moment causes all the genes to assess the values of their regulatory inputs and, at a clocked moment, to assume the proper next activity. Thus at each moment, the system passes from one state to a unique successor state. Over a succession of moments, the system passes through a succession of states, called a *trajectory*. Figure 12.8*c* shows these trajectories.

The first critical point is this: Since there are a finite number of states, the system must reenter a state previously encountered; thereafter, since the system is deterministic and must always pass from a state to the same successor state, the system will cycle repeatedly around this *state cycle*. Therefore, each state cycle is an attractor in this dynamical system, and the set of states flowing into one state cycle or lying on it constitutes the basin of attraction of that state cycle attractor. The state cycle is called an attractor because it "attracts" trajectories flowing from all states which lie in its basin of attraction. The state cycles are also the "asymptotic," or "long-term," behaviors of the genomic system because the system ultimately reaches one state cycle and cycles about it forever afterward if undisturbed by outside perturbations.

The length of a state cycle can range from 1, a steady state which transforms to itself, to 2^N. In the latter case all the states lie on a long cycle through which the system passes repeatedly.

Any such network must have at least one state cycle attractor but may have more than one, each draining its own basin of attraction. And, as should now be familiar, each state either lies on or flows into one state cycle. Hence the state cycle(s) of the system are its dynamical attractors, and their basins partition the state space. Figure 12.8c shows that our three-gene system has three state cycles. Each is a discrete alternative recurrent asymptotic pattern of gene expression in the entire model genomic system. Left to its own, the system eventually settles down into one of its state cycle attractors and remains there.

The Central Interpretation: A Cell Type Is an Attractor

I shall make a single central interpretation of such genomic models: A *cell type corresponds to a state cycle attractor* in the dynamical behavior of the genomic system. The immediate question to assess is whether this interpretation is sensible. It distinguishes between two classes of patterns of gene expression: *transient* and *recurrent asymptotic*. Thus among the $2^{100\,000}$ possible patterns of gene expression among 100 000 genes in the on–off idealization of a eukaryotic genomic system, not *all* count as cell types. Rather, only those patterns which are recurrent asymptotic behaviors of the genome count as cell types. Note that an immediate consequence of this identification is that cell types are *discretely* different. Attractors lie in different basins of attraction and do *not intergrade.*

My interpretation seems reasonable on a second ground. Cell types are constrained and apparently stable recurrent patterns of gene expression. Attractors are precisely the kinds of dynamical objects which have the possibility to be constrained because only a subset of states occurs on an attractor. Attractors are likely to be reasonably stable to perturbation, since each typically drains some basin of attraction. Thus if perturbed from some attractor into the attractor's basin, the system will return to the same attractor. Small attractors located inside a volume of states constituting their basins of attraction are the natural image of stable systems exhibiting homeostasis. Homeostatic return of a perturbed cell to its previous pattern of gene activities after many possible chemical perturbations seems likely to be a fundamental property of most cell types.

The assertion that cell types are attractors is a hypothesis, not yet an established fact. In a moment, I shall discuss the beginning implications of this new view. First, though, it is well to consider some possible concerns and objections. The first concern is mere unfamiliarity. Developmental biologists are perfectly familiar with genetic circuits and cascades of genes switching one another on and off. The puffing cascade in *Drosophila* polytene chromosomes is but one example. Much less familiar is the idea of thinking of the behavior of *all* the genes at the same time and hence following a trajectory of their switching on and off in a state space. This state-space description is merely the most natural complete description of the trajectories of the genomic system, however. The transient trajectories capture transient cascades. The attractors capture the final regions of state space into which the genomic dynamics traps itself.

While we are all familiar with the idea of a chemical system which might have one or more steady-state behaviors, we are less familiar with the idea that each such steady state is a point attractor draining some surrounding basin of attraction. Once one is comfortable with the image of all genes simultaneously changing activities, however, and with the image of a trajectory of states through state space, the most natural distinction to make is between attractors and the transient states which lie on trajectories flowing to the attractors. The attractors—whether steady states, oscillations, or even chaotic regimes—are the only *persistent* behaviors of the genomic system. Thus it is natural to identify persistent cell types with attractors rather than with transients.

As noted in the introduction to this chapter, the hypothesis that cell types are attractors is almost forced on us in certain circumstances. For example, in an adult organism, many tissues are maintained by the existence of a proliferating stem cell population. At some divisions a stem cell may give rise to both a stem cell daughter and a second daughter which differentiates to a further state. Since stem-cell populations persist over hundreds of cell divisions, those cell types must be persistent, recurring states. Similarly, many kinds of metazoan and metaphyten cells can be grown *in vitro* and maintain a characteristic persistent differentiated state which may or may not differ from the state *in vivo*. The stable persistence of such cell types over hundreds of divisions again almost certainly demands a recurrent stable pattern of gene expression—an attractor. Similar examples are found in the imaginal disc tissues of *Drosophila melanogaster,* described in detail below, where larval disc tissues can be cultured by serial transfer in adult abdomen over numbers of years and maintain a cell heritable committed state which is finally expressed in terminal differentiation to adult cuticular structures after exposure to ecdysone (Hadorn 1966; Nothiger 1972; Nothiger and Gehring 1973; Gehring 1976).

Some cell types which make their appearance during embryogenesis along branching pathways of differentiation may prove to be transitory. Such types might simply be transient trajectories flowing to an attractor. It seems more probable, however, that many such transitory cell types are *transitory attractors* whose existence reflects the fact that, in real genomic systems, different molecular variables may change concentrations or activities on different natural time scales. Thus half-lives may be long for some mRNA sequences and proteins and short for others. The existence of a heterogeneity of time scales complicates the simpler picture based on synchronous updating in model Boolean networks. In general in such systems, the slow variables can be thought of as parameters to the subsystem of fast variables. The latter fall rapidly to their attractor(s), but the locations of those attractors in state space may either change gradually or bifurcate abruptly into two attractors as the slower variables change. The importance of such a heterogeneity of times scales lies, in part, in understanding cellular differentiation. In the picture I shall focus on, based on synchronous Boolean networks, each attractor is a cell type. Differentiation then consists in a transition from one attractor to another caused by some external perturbation. In a fuller theory based on a heterogeniety of time scales, cell types along branching developmental pathways would sometimes be attractors of the faster variables which change and bifurcate into different fast-variable attractors as slower variables change. Here differentiation along such branching pathways can reflect both response to external inductive perturbations and gradual changes in the slow variables which are internal to each cell. The probable bearing of the simpler to the fuller theory is likely to be similarity in the number of cell types into which each cell type can differentiate.

A further complication may arise. Suppose we find evidence that in some organism's development a specific gene is in some way deleted, amplified, or rendered heritably inactive. Then such alteration merely alters the genomic dynamical system and alters the behavior of the remaining genes, which are free to be active or inactive. The modified system still has attractors, perhaps different from those of the initial system. It remains natural to think of those attractors as the cell types, but now we must take account of the fact that an added regulatory process has switched the genomic regulatory system irreversibly to a modified version with new cell type attractors. The entire process of development, then, would need to include a theory about control of such irreversible alterations and the behavioral options open to each alternative version of the genomic system in different cell lineages. Therefore, understanding such complications as irreversible activation or inactivation of specific genes will require an analysis of the dynamical behavior and attractors of the remaining genomic system. In short, development and cell types may be more complex that the simplest attractors we discuss here, but the former almost certainly include the latter as fundamental features.

Natural Properties of Cellular Differentiation Conceived Within the Framework

Given the identification of an attractor as a cell type, we have achieved a preliminary conceptual framework in which a number of properties are naturally expressed:

1. If an attractor is a cell type, then we are interested in how many states occur on the attractor; that is, how constrained is the pattern of gene expression per cell type? Are the attractors small compared with the possible 2^N patterns of gene expression?

2. Turning genes on and off takes time. How long would it take the genomic system to cycle through a cell type attractor? Again, are the attractors small?

3. If a state cycle is a cell type, then the genome contains only a certain number of different state cycle attractors, and these are the different cell types within the genomic repertoire. How many cell types are there?

4. If different attractors are the different cell types, how similar are the patterns of gene expression on them? Is a core of common genes active in all cell types? Do cell types differ in a penumbra of gene activities outside that core?

5. If an attractor is a cell type, then how stable is that cell type if the activities of one or a few genes are transiently reversed as a result of chance fluctuations in the chemical milieu of the cell? For example, the small genomic network in Figure 12.8 has three cell types. Note that model cell type 1—that is, state cycle 1—responds to transient reversal of the activity of any single gene by flowing to either state cycle 2 or state cycle 3. In other words, model cell type 1 is unstable to any small perturbation. Conversely, state cycle 3 is stable when the activity of any one gene is reversed. After each such perturbation, the system flows back to state cycle 3.

6. If an attractor is a cell type, then differentiation is passage from one attractor to another. This passage can occur either as a result of asymmetric distribution of gene products or other regulatory variables to daughter cells or as a result of exog-

enous inductive influences. The minimal number of genes such an inductive stimulus (a hormone, say) can directly influence is one. If each gene can have its activity transiently reversed by such a stimulus, how many different cell types can one cell type differentiate into directly? Pictorially, how many basins of attraction are close to a given attractor cell type located inside its own basin? For example, in the genetic network of Figure 12.8, cell type 1 can differentiate directly into cell type 2 or cell type 3 by altering the activity of a single gene. Cell type 2 can differentiate into cell type 1 or cell type 3 by properly timed reversal of the activity of a single gene. Cell type 3 is irreversibly locked into remaining the same cell type with respect to transient reversals of any one gene's activity.

7. If one cell type can differentiate directly into a subset of cell types, how many of the cell types in the genomic repertoire can the first type ultimately reach by repeated differentiation along branching pathways? Can at least one cell type reach all the other cell types in the genomic repertoire? Such a cell type might be the zygote, able to reach all other cell types by branching differentiation. If no cell type can reach all other possible cell types, what might the existence of unreachable cell types imply about ontogeny?

8. If the activity of a single gene is transiently altered—by a hormone signal, for instance—how many downstream genes alter their activities in an avalanche of changes? How many steps does it take for that influence to propagate? What are the mean and variance of these properties for genes in the network?

9. If a single gene is deleted, a single regulatory connection altered, or a single Boolean rule changed, how many downstream genes, on average, alter their behaviors? How many cell type attractors are altered? Do some cell types literally disappear and become transient patterns of gene activity which flow to other preexisting attractors, such that the mutant causes cells to differentiate into normal cell types, but by an aberrant pathway? For example, in Figure 12.8d, I show the consequence of altering the rule regulating gene 2 from "Or" to "And." The resulting basins of attraction have the same cell types (1 and 3), but note that now cell type 1 lies in the center of a large basin of attraction and is stable to transient reversal of the activity of any single gene, while cell type 3 is now unstable to any such perturbation. If the developing system regulated by the small genome in Figure 12.8a were subjected to persistent occasional reversals of the activities of one gene at a time, cell type 1 would not be seen for long and cell type 3 would persist. The mutation of "Or" to "And," however, means that cell type 3 would not persist and cell type 1 would. Thus the mutation has left cell types intact but altered pathways of differentiation between them.

As you may now imagine, this list only begins to frame the questions which naturally emerge. Naturally, as well, the answers depend critically on the particular ensemble of genomic regulatory systems being considered. Therefore, the next, and fundamental, task is to explore the expected behaviors of diverse ensembles of genomic systems. The aim is to understand the requirements for the *natural emergence of ordered properties in complex regulatory systems.*

ENSEMBLES OF GENETIC REGULATORY SYSTEMS: GENERIC PROPERTIES

I now turn to the task of discussing properties of genetic regulatory systems drawn from different ensembles. I shall do so by recalling the main points of the detailed discussion of Boolean networks from Chapter 5. Recall first the major result we found: Random Boolean networks, parallel-processing systems which are examples of massively disordered systems, exhibit three main regimes of behavior—ordered, chaotic, and complex.

In the ordered regime, a frozen subnetwork, or component, of genes, each gene in a fixed state of activity 1 or 0, spans, or percolates, across the network, leaving behind functionally isolated islands of genes which may twinkle on and off in complex ways but which cannot communicate with one another through the frozen component. Attractors are small, few, and stable. Alteration in the activity of single genes typically does not unleash a large cascade of changes in the activities of other genes. That is, typically damage does not propagate beyond one functionally isolated unfrozen island of genes. Further, mutational alteration in network structure or logic does not cause massive changes in dynamical behavior.

In the chaotic regime, it is the unfrozen component of genes—those not in fixed states of activity—which percolates across the network, leaving behind isolated frozen islands of genes in fixed states of activity. In this regime, attractors lengthen exponentially as N increases, and alterations in the activities of one or a few genes unleash cascades of change, or damage, which propagate through the unfrozen component to very many of the genes. Hence chaotic networks exhibit sensitivity to initial conditions. The hallmarks of chaos in large parallel-processing networks are attractors whose lengths scale exponentially with the number of genes and whose cell types, rather than exhibiting homeostasis, show massive sensitivity to initial conditions.

The complex regime occurs at the boundary between the ordered and the chaotic regime. The transition from order to chaos is, in fact, a phase transition, driven by alterations in control parameters such as K, the number of inputs per gene; P, the internal homogeneity of Boolean functions as redefined below; or alterations in the fraction of canalyzing functions utilized in construction of the network. The fact that such simple parameters can tune where networks lie on the order–complexity–chaos axis suggests that selection might easily determine whether genomic networks are ordered, complex, or chaotic.

The phase transition occurs when the unfrozen percolating component is just breaking up into isolated islands separated by genes in the frozen component. Because unfrozen islands are large but do not percolate across the entire system, damage propagates on all scales in a power-law distribution, with many small and few large avalanches of alterations in gene activities. As we have seen, the most complex internal communication and computation can be attained in the complex regime. For example, mutual information (MI) is defined as the sum of entropy of the activity states of two genes minus their joint entropy. MI is 0 if both genes are in fixed states or if the twinkling pattern of activity of the first gene is random with respect to the second gene. Hence MI is 0 deep in the ordered regime and deep in the chaotic regime. In the complex regime, MI is maximum, even among distantly connected genes. For this and other reasons, it is plausible to think that parallel-processing systems in the solid regime but near the boundary of chaos could perform the most complex controllable behaviors and also adapt optimally in a fixed or changing world. One approach to testing this hypothesis is via adaptive games played between Boolean networks. If networks in the solid regime near the edge of chaos adapt and perform best, we shall have to suppose that selection attains and maintains such poised systems, whose generic properties would emerge as quasi-universals in biology.

Chapter 5 introduced the results on Boolean networks but did not do so in the context of genomic regulatory systems. Since our interest here is in discussing how to think about genomic systems comprising thousands of coupled genes and their products, it is important to reconsider diverse ensembles of Boolean networks as specific models of genomic systems. I emphasize that the ensembles in the chaotic regime are *not,* in my view, biologically plausible. They are studied merely to illustrate their typical dynamical behaviors and, by the stark contrast, to underscore the construction requirements for, and implications of, spontaneously ordered dynamics in massive genomic networks.

K = N: *The Chaotic Grand Ensemble*

Networks in which $K = N$, where each gene is directly regulated by all other genes, are the most important chaotic ensemble. This is the benchmark Grand Ensemble. Its importance is twofold. First, all other ensembles are specialized subensembles of it, obtained by constraints either on the numbers of inputs per gene or on the set of Boolean functions among the $(2^2)^N$ possible Boolean functions of N variables. Second, the Grand Ensemble is *maximally disordered.* Thus *any* order which may emerge as typical in such systems is surprising and, more important, may prove very robust and general. Indeed, signs of order with marked biological implications emerge even in this most chaotic of ensembles: they have remarkably few alternative attractors.

Conceive, then, of a hypothetical genome with N binary genes, each directly regulated by $K = N$ genes. In other words, each gene is directly regulated by all genes. There is only a single wiring diagram. To study the typical properties of members of this or any other ensemble, it is necessary to sample a number of members at random. Mean properties of these members identify the typical properties of the ensemble. Thus the way to proceed mathematically is to construct, entirely at random, a large number of genetic networks having $K = N$ and then analyze their properties. Figure 12.9a shows an example with five genes, each with an input from all $K = N = 5$ genes. Figure 12.9b shows the random assignment of a Boolean function to each of the five genes. This assignment is the equivalent of randomly assigning to each state its successor state. Figure 12.9c shows the single state cycle and basin of attraction of this network.

This class of maximally disordered systems, sometimes called the random-map model because each state of a network passes at the next moment to a randomly chosen, but then fixed successor state, has received increasing attention (Kauffman 1969, 1971a, 1971b, 1974, 1984a, 1986a, 1986c; Wolfram 1983, 1984; Gelfand and Walker 1984; Coste and Henon 1986; Derrida and Flyvbjerg 1987b; Derrida and Bessis 1988). In part, the interest reflects the relative analytic simplicity of this extreme case.

In $K = N$ networks, the expected median length of state cycles is $0.5(2^{N/2})$. Thus median cycle lengths increase exponentially as N increases (Kauffman 1969, 1971, 1984a, 1986a, 1986c; Wolfram 1983, 1984; Gelfand and Walker 1984; Coste and Henon 1986). The implication is striking. In a model genetic system containing only 200 binary genes, the expected state cycle length is $2^{100} = 10^{30}$. Suppose genes turn on and off in a microsecond. then traversing the state cycle would require 10^{24} seconds, or 3.17×10^{16} years. The estimated age of the universe is only on the order of 14×10^9 years. Thus, since time began, the tiny system of 200 binary genes would have covered only one-millionth of the traverse around its attractor. This is enough to make a galaxy seem friendly, but consider these hyperastromical numbers in the context of realistic cell biology. Attractors are the natural candidates to think about

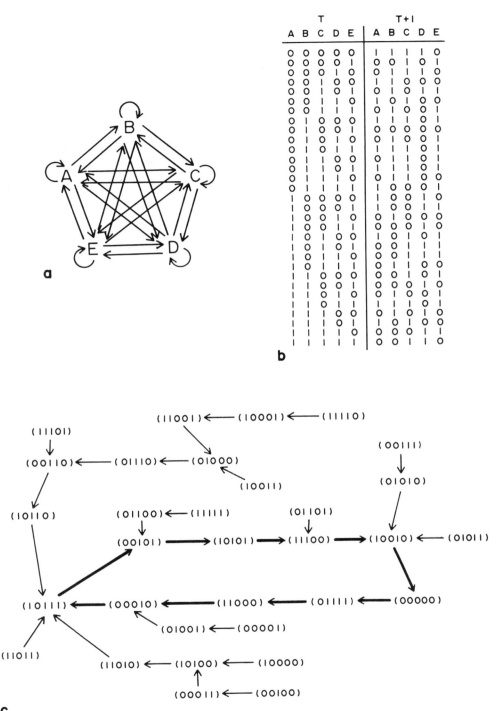

Figure 12.9 (*a*) Wiring diagram for a network containing $K = N = 5$ genes, each receiving regulatory inputs from all five. (*b*) A different random Boolean function of $K = 5$ variables is assigned to each gene. (*c*) State transition diagram of the network leading to a single state cycle.

as model cell types. Thus the time scale for traversing an attractor becomes important. Natural cell time scales, such as the cell cycle, are on the order of hours. A microsecond is a vast underestimate of the time required to activate a gene or replicate a replicon. Nevertheless, the resulting time scale to explore an attractor is very much longer than the possible history of life. Clearly, $K = N$ networks, deep in the chaotic regime, are very poor candidate models of real genomic systems.

Despite their massive chaotic behavior, $K = N$ networks exhibit striking order in one respect: The number of attractors is only N/e (where e is the natural logarithm base) and thus a linear function of the number of genes. A tiny genome containing 200 genes might take eons to traverse an attractor but would have only about 74 attractors. A system containing 1000 binary genes would have a state space of 2^{1000}, or 10^{300}, but only 370 attractors (Rubin and Sitgreave 1954; Kauffman 1969, 1984a, 1986a, 1986c, 1986d; Derrida 1987a, 1987b; Derrida and Flyvbjerg 1987b; Derrida and Bessis 1988). Within one system, however, the basin sizes of these attractors differ sharply. A few drain huge basins of attraction, the rest typically lie in tiny basins.

This result should immediately strike a chord. We have seen that, as a maximum estimate, the number of cell types in organisms increases linearly as the number of genes increases. If we wish to think of an attractor as a cell type, then even in these maximally disordered systems we have already found the emergence of unexpected order. Even these maximally disordered systems begin to yield a scaling law that predict the relation between genomic complexity and numbers of cell types across phyla in evolution. We shall see that this slow increase in number of attractors as N increases is a very robust property, holding for $K < N$ and reaching a minimum for $K = 2$, where the number of attractors increases as $\sqrt{N}$.

In $K = N$ networks, alterations in the activity of any single gene propagate changes to most other genes. A mathematically more precise statement of this property is that, as the number of genes in the system increases toward infinity, changes propagate to a finite and roughly fixed fraction of the genes. This sensitivity to changes in initial conditions reflects the fact that no frozen component percolates across the system. Avalanches of changes propagate widely through the enormous unfrozen component. This property, too, is very unlike the biologically observed cascades of alterations in gene activities unleashed when the activity of some single gene is altered by a hormonal signal. In reality, typically a small fraction of genes alter their behaviors.

In $K = N$ networks, attractors are unstable to minimal perturbation. Thus unlike real cells and genomic systems, *homeostasis is absent* in $K = N$ networks. Further, each attractor can differentiate into all other attractors by single minimal perturbations. This point is easy to understand qualitatively. Since the successor to each state is chosen at random from among the set of 2^N states, neighboring states along trajectories show no similarities in patterns of gene expression. It follows immediately that, if the system is perturbed off its present state cycle attractor by the transient reversal of the activity of any gene chosen at random, the probability that the perturbed state returns to the perturbed cycle is simply proportional to the relative size of the basin of attraction of that attractor. Further, the probability that the perturbed state flows, or differentiates, to any other attractor is proportional to the size of the other basin. Consequently, considering all possible ways any single gene can have its activity transiently reversed on each state of each state cycle, we expect that each attractor has low stability and can differentiate directly into every other attractor. In short, each attractor is next to every basin of attraction in the sense that reversing the activity of some gene at some point on its cycle causes the system to jump into each of the possible basins of attraction. This property, too, clearly does not occur in real cells, however. In the ontogeny of all metazoans and metaphytens, any cell type differentiates directly to only a few other cell types. This deep property of ontogeny is not found in chaotic networks.

Networks in which $K = N$ "live in" uncorrelated adaptive landscapes. Deletion of any gene in such a network or, equally, considering it to be constitutively inactive changes the successor state assigned to fully half the 2^N states and causes havoc with all attractors. Randomly changing the Boolean function assigned to any gene to one of the $(2^2)^N$ other Boolean functions will equivalently alter half the successor state assignments and almost completely rearrange the dynamical attractors. We shall return in Chapter 13 to consider the adaptive capacities of genomic regulatory systems. We can presage that discussion here by noting that, if we consider attractors as cell types and then model cell-type evolution by either deletion of mutants or alterations of the Boolean rules assigned to genes, then we shall not be surprised to find that profoundly chaotic $K = N$ networks have nearly uncorrelated fitness landscapes with respect to such mutations.

Chaos in $K = N$ *Boolean Networks of High Internal Homogeneity*

The first specialization from fully random $K = N$ networks is meant to explore the obvious intuition that increasing convergence in state spaces might suffice to achieve short, orderly attractors. However, this intuition is wrong. We have just seen that fully random $K = N$ networks have only N/e attractors (Kauffman 1969, 1984a, 1986a, 1986c, 1986d) but that cycle lengths increase exponentially in N and equal the square root of the total number of states, $2^{N/2}$. Obviously, to be remotely reasonable models of genomic regulatory systems, attractor lengths, which reflect how constrained a pattern of gene activity is on a cell type, must be very much smaller than this number.

Recall from Chapter 5, where its implications for network behavior were discussed, that the *internal homogeniety P* of a Boolean function is defined as the deviation of the fraction of 1 values or of 0 values from 50 percent of the 2^K states (Gelfand and Walker 1984). For example, a Boolean function with an internal homogeneity of 0.8 has either 80 percent 1 values and 20 percent 0 values or 80 percent 0 values and 20 percent 1 values in the 2^K positions specifying the Boolean function (Figure 12.10).

In $K = N$ networks, increasing P markedly increases convergence in state space and decrease state cycle length. However, the expected median state cycle length still increases exponentially as N increases (Kauffman 1984a):

$$\text{Expected median cycle length} = 0.5 \left(\frac{1}{\sqrt{P}} \right)^N \tag{12.1}$$

Let $B = 1/\sqrt{P}$; then, since $B > 1$, cycle lengths increase as $0.5B^N$. Thus, according to these results for expected median cycle length, even $K = N$ networks with fixed high P remain chaotic and do not exhibit biologically short and plausible state cycle attractors. If $P = 0.8$, a small genomic system comprising only 200 genes would have state cycles requiring 4.9 billion years to traverse at a microsecond per state transition.

The critical implication here is that *no fixed P* alone suffices to ensure that state cycles remain small as $K = N$ grows large. Here I shall mean by "small" that state

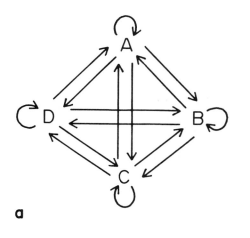

		T				T + 1		
	A	B	C	D	A	B	C	D
0	0	0	0	0	0	1	1	1
1	0	0	0	1	1	1	1	0
2	0	0	1	0	1	0	1	1
3	0	0	1	1	1	1	0	1
4	0	1	0	0	0	1	1	1
5	0	1	0	1	1	1	1	0
6	0	1	1	0	1	1	1	1
7	0	1	1	1	1	1	1	1
8	1	0	0	0	1	0	1	1
9	1	0	0	1	1	1	1	0
10	1	0	1	0	1	0	1	1
11	1	0	1	1	1	1	0	1
12	1	1	0	0	1	1	0	1
13	1	1	0	1	0	1	1	1
14	1	1	1	0	1	1	1	1
15	1	1	1	1	1	1	1	1

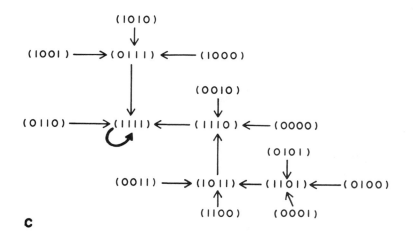

Figure 12.10 (*a*) Wiring diagram for a network containing $K = N = 4$ genes, each receiving regulatory inputs from all four. (*b*) Biased Boolean functions in which the probability of a 1 response is nearly 0.8 and the probability of a 0 response is nearly 0.2. (*c*) State transition diagram of the $K = N = 4$ network leading to a state cycle length 1 in which the central state (1111) transforms to itself.

cycles do not grow faster than, say, linearly as N increases. Exponential growth is explosively faster than linear growth as N becomes large. Convergence due to fixed P alone is not sufficient principle of order as N and K increase.

Boolean networks with $K \geq 5$ and randomly chosen Boolean functions remain in the chaotic regime. They have attractors whose lengths increase exponentially as N increases and exhibit sensitivity to initial conditions. The mean-field argument on which these analytic results depend (Chapter 5) are based on P_K, the mean value of P, for Boolean functions of K inputs:

$$P_K = \frac{1}{2} + 2^{-(2^K+1)} \times \binom{2^K}{2^{K-1}}$$

The value of P_K is a maximum for $K = 2$ and falls thereafter. For $K \geq 5$, the mean-field assumption is reasonable and implies that, for fixed K, attractors increase exponentially as N increases. Expected median cycle lengths as a function of P_K is found by substitution into Equation 12.1.

The important result is this: Genetic networks having a modestly high number of inputs per gene, say $K \geq 5$, and randomly chosen Boolean functions remain in the chaotic regime. They have state cycle attractors whose lengths increase exponentially as N increases and show sensitivity to initial conditions in the form of very large avalanches of change which propagates throughout the network. For these chaotic networks, the cycle lengths increase far too fast as N increases for such attractors to be plausible models of cell types.

While state cycles are long for $K > 4$, disordered networks have few attractors, and $K = N$ fully random genomic systems have only N/e distinct attractors. Recent analytic results strongly indicate that, for $N \geq K > 4$, the number of attractors is at most a linear function of N and increases to N/e as K approaches N. Letting $a = P_K - \frac{1}{2}$ (that is, a represents the deviation of the mean internal homogeneity above 0.5), Coste (reported in Kauffman 1986d) was able to obtain bounds on the number of attractor cycles:

$$N^{\left(\dfrac{\log \dfrac{1}{\frac{1}{2} + a}}{2}\right)} \leq \text{number of cycles} \leq N^{\left(\dfrac{\log \dfrac{1}{\frac{1}{2} - a}}{2}\right)}$$

Since the number of cell types in organisms is between a linear function and a square-root function of the number of genes, chaotic genomic systems are reasonable biological models with respect to predicting number of cell types as a function of genomic complexity.

On the basis of the time needed to traverse chaotic attractors, the major instability of chaotic attractors in lack of homeostasis, and the high ruggedness of the fitness landscapes on which systems must evolve, we can tentatively conclude that real genomic systems are very unlikely to lie in the chaotic regime. Hence if genomic regulatory systems do in fact have high mean connectivity, the kinds of Boolean functions present in the systems must be constrained in order that the systems have small attractors which might be plausible models of cell types. Two reasonable candidate biases are (1) sufficiently high values of P relative to K and N to ensure a phase transition to the ordered regime and (2) use of a high enough fraction of canalyzing functions to ensure a phase transition to the ordered regime in genomic systems.

Sufficiently High Internal Homogeneity in the Boolean Functions in Boolean Networks Yields a Frozen Component and Dynamical Order

Random networks with $K = 5$ or a similar modest number of inputs per gene and no constraint on the Boolean functions used lead to chaotic systems which have exponentially long state cycles as N increases, as just seen. It is therefore very important that, in large networks in which K remains reasonably small—between 4 and 20, for example—*sufficiently high internal internal homogeneity P leads to Boolean networks having small attractors and stable dynamical behavior.*

Recall the discussion in Chapter 5 of P_c, a critical value of P above which the phase transition to order takes place. Derrida and Weisbuch (1986) studied how altering P affected behavior networks which were regular lattices in which each binary element was influenced by its four or six immediate neighbors according to a biased Boolean function. They discovered the critical P_c value such that, for $P > P_c$, there arises in the lattice a large frozen component in which many binary elements become fixed either in the 1 state or in the 0 state.

In the ordered regime, as a result of the presence of such a frozen component, the lengths of state cycles increase *less than linearly* with respect to the number of elements in the lattice. Because such attractors are very small and homeostatically stable to perturbations, such systems exhibit dynamical order.

Frozen components are a precondition for the occurrence of short state cycles in large Boolean networks. When networks lack frozen components, chaotic dynamics, in the sense defined, ensues. Frozen components arise by at least two means: by percolation of fixed states due to sufficiently high P ($> P_c$) in the Boolean functions relative to the number of inputs to each site and by the formation of forcing structures built from canalyzing Boolean functions, as emphasized later in this chapter. I shall suggest that cells exhibit orderly behavior as a result of the formation of forcing structures whose genes fall to fixed activities.

These results raise the possibility that orderly dynamics in cells may arise as a result of Boolean functions in genomic regulatory systems which have $P > P_c$ rather than as a result of any constraint to canalyzing functions. Testing this possibility would require, in part, discriminating whether, in the Boolean idealization, regulated genes are typically governed by high-P Boolean functions or by canalyzing Boolean functions. Among the Boolean functions of few inputs, there is a marked overlap between these two classes of functions. The canalyzing "Or" function is also one with high internal homogeneity: $P = 0.75$. The overlap between these two classes of Boolean functions dwindles as K increases (Gelfand and Walker 1984; Stauffer 1978b). Thus as the control rules governing genes with modest numbers of regulatory inputs become known, it will become possible to discriminate between these hypotheses. Since the evidence does not discriminate and does support the claim that genetic systems are rich in canalyzing Boolean functions, I shall continue to base our discussion on the generalization that real genetic networks are built of genes regulated by canalyzing functions in the Boolean idealization. The biological implications of genomic order due to percolations of frozen components from $P > P_c$ or forcing structures are similar in many but not all respects.

K = 2 and Canalyzing Ensembles of Genomic Systems: The Crystallization of Order

Massively disordered networks having random connections and logic among their N elements crystallize order under the simple constraint that $K = 2$. So too do networks which have a larger number of inputs per gene but are constrained to the canalyzing functions (Kauffman 1969, 1971a, 1971b, 1974). As noted on page 198, a number of workers have studied these systems.

Three major approaches to this problem were discussed in Chapter 5, but here I shall merely redescribe the numerical studies which detail the emergent order. Numerical simulations of randomly chosen members of the ensemble of $K = 2$ input networks in which the number of binary genes N ranges up to 10,000 have been carried out (Kauffman, 1969, 1971a, 1971b, 1974, 1984a, 1986a, 1986c, 1986d). The following are the dominant results:

1. The expected median state cycle length is $\sqrt{N}$. That is, the number of states on an attractor scales as the square root of the number of genes. A 10 000-gene genomic regulatory system which is utterly random within the constraint that each gene is regulated by only two genes would therefore have a state space of $2^{10\ 000} = 10^{3000}$ but settle down and cycle recurrently among a mere $\sqrt{10\ 000} = 100$ states. Thus in contrast to genomic networks having $K > 4$, where cycle lengths increase *exponentially* in N and rapidly become hyperastronomical, randomly assembled genomic systems having $K = 2$ spontaneously confine their dynamical behavior to truly tiny subvolumes of their state space. A system of 10 000 genes which localizes its dynamical behavior to 100 states has restricted itself to 10^{-2998} parts of its state space.

2. The distribution about this median is skewed. Most networks have short state cycles, while a few have very long ones. If replotted as the logarithm of cycle length, the distribution remains skewed rather than becoming a familiar bell-shaped Gaussian distribution.

3. The number of state cycle attractors is also about $\sqrt{N}$. Therefore, a 10 000-gene system would be expected to have on the order of 100 alternative attractors. A 100 000-gene system would have about 317 alternative asymptotic attractors. Then a genomic system containing 100 000 genes would flow to about 317 state cycle attractors, each having about 317 states.

4. If the stability of each state cycle attractor is probed by transiently reversing the activity of each gene on each state of the state cycle, then, for 80 to 90 percent of all such perturbations, the system flows back to the same state cycle. Thus state cycles are inherently stable to most minimal transient perturbations.

5. When it flows back to the same state cycle, the perturbed system typically reaches that state on the state cycle which it would have reached in the same time interval had the system not been perturbed. Thus if one thinks of position around a state cycle as carrying phase information, the system tends to return to the same cycle after perturbation and maintain phase.

6. For perhaps 10 to 20 percent of the minimal perturbations (transiently reversing the activity of a single gene), the system leaves the state cycle from which it was perturbed and flows to another.

7. By such perturbation or signal-induced transitions, each state cycle can directly change to only a small number of other state cycle attractors in the system. Thus if the system has 100 000 genes and 317 alternative attractors, typically each can be triggered to flow to only a few of the 317 whenever the activity of any single gene is altered. It follows that many of the 10 to 20 percent of the perturbations which cause the system to change from attractor A induce change to the same neighboring attractor B. In short, a variety of stimuli acting on different genes in the system induce the same specific response.

8. A large fraction of the N genes, typically 70 percent or more, fall to either a fixed active or a fixed inactive state, and that state is *identical* on all the alternative attractors of the genomic system.

9. The mean difference in patterns of gene activity on different attractors is a few percent.

10. Transient alternation of the activity of a single gene typically propagates and causes alterations in the activity of a small fraction of the total number of genes in the system.

11. Deleting any gene or altering its Boolean function typically causes only modest changes in attractors and transients.

The high order seen in random $K = 2$ networks extends to networks having more than two inputs per gene, provided the rules controlling behavior, in the Boolean idealization, are confined to canalyzing functions. Further, the constraint to synchronous updating of gene activities can be relaxed; order persists in networks in which the timing of the changes of activities of the different genes is asynchronous (Thomas 1979; Fogelman-Soulie 1984, 1985a, 1985b). This is of central importance: Real genomic systems are limited neither to $K = 2$ inputs nor to simultaneous updating of the activities of all genes in the genome. Thus the fact that the order found in $K = 2$ networks extends to a far wider class of systems, those lying in the ordered regime, means that we can attempt to account for the order in organisms on the basis of the two observed local properties of genomic systems: few inputs per gene and a preponderance of canalyzing functions.

Genomic Systems in Which K = 1: Harbingers of Genetic Subcircuits in Functionally Isolated Islands

Genomic systems in the ordered regime having a modest number of inputs per gene limited to canalyzing functions shall emerge as our central interest. Because the genetic subcircuits within the functionally isolated islands of such canalyzing networks are similar to $K = 1$ networks, it is particularly important to analyze this simplest set of networks.

Consider a simple system containing N genes, each regulated by one gene. This constraint means that the wiring-diagram structure of such a network must fall into feedback control loops and descendant tails. Such tails may hang off the loops, since we have allowed one gene to regulate more than one gene. However, since no gene can have more than one regulatory input, loops cannot be interconnected. Since connections are made at random, the wiring diagram of $K = 1$ networks has the same statistics as the state transition diagrams of random $K = N$ networks, with nodes interpreted as genes rather than states and the direction of arrows representing state transitions reversed. This interpretation yields loops with descendent tails of regulated genes. In $K = N$ networks, the length of state cycles is of order square root of the number of states 2^N. Similarly, in $K = 1$ random networks, the lengths of the wiring-diagram loops are on order square root of the number of elements N. Jaffe (1988) has shown that the total number of genes lying on feedback loops scales roughly as $\sqrt{\pi/2} \sqrt{N}$.

There are four Boolean functions of one input: "Yes," "No," "Tautology," and "Contradiction." If each gene is assigned one of these four at random, the consequences are quite trivial. "Yes" means that the input activates the regulated gene. "No" means that the input inhibits the regulated gene. Any gene assigned "Tautology" is constitutively active, regardless of the activity of its regulatory input. Those assigned "Contradiction" are permanently inactive. Each such fixed gene fixes the activities of those genes downstream of it governed by "Yes" or "No" until the next fixed gene is encountered. Thus if at least one gene per feedback loop is constitutively either active or inactive, that loop falls to a fixed state, as does its hanging tails.

To make the system more interesting and applicable to the functionally isolated islands in more realistic genomic systems, let us require that only the Boolean "Yes" and "No" functions are used. Then each gene either copies the activity of its regulatory predecessor or reverses that activity. Here is the simple result. Any feedback loop must have either an even or an odd number of "No" Boolean functions. If the loop has an odd number, it must oscillate; it has no steady states of gene activities. If the loop has an even number, it has exactly two steady-state patterns of gene activity. Loops with even numbers of "No" functions can, however, oscillate through a state cycle. It is easy to show in general that the number of ways a loop of length L can oscillate is given by the factors of $2L$ (Holland 1960; Kauffman 1971a, 1971b). More recently, $K = 1$ networks have been analyzed by Flyvberg and Kjaer (1988) and Jaffe (1988). In particular, Jaffe shows that the number of attractors in $K = 1$ networks which do not use "Tautology" and "Contradiction" is an exponential function in N, while cycle length increases slowly (Table 5.1).

Genomic systems in which $K = 1$ exhibit "combinatorial" attractors. A similar combinatorial character also arises from the joint activity combinations of functionally isolated islands in Boolean networks in the ordered regime. Since $K = 1$ networks fall apart into separate wiring-diagram loops, each of which may have descendent tails of genes, it is obvious that the different *unconnected loops* must behave independently of one another. Therefore, if each loop by itself has some number of different attractors—say, two steady states, or J modes of oscillation—then the number of alternative attractors of the entire network is given by the product of the J_1 attractors of the first loop times the J_2 attractors of the second loop, and so forth. This is our first introduction to a property which shall become increasingly important. It shows that, for $K = 1$ networks, any attractor of the entire network can be thought of as being made up of a specific *combination* reflecting the choice of one among the J_1 alternatives for the first loop, one among the J_2 choices for the second loop, and so on. Then, if we want to think of a cell type as an attractor of the genomic system, it follows that $K = 1$ networks naturally have the property that each cell type can be described in a kind of *combinatorial epigenetic code.* The network has a specific number of genetic decision-taking "circuits," the different independent feedback loops. Each can make $J_1, J_2, \ldots$ alternative choices, which can be thought of as numbered from 1 to J_1 for the first loop, 1 to J_2 for the second loop, and so on. Then each cell type is an attractor in this code, reflecting the combination of choices made by each decision-taking loop. Remember the Cl–cro feedback loop in lambda with the immunity $+$ and immunity $-$ alternative steady states. A genomic system with ten such loops would have $2^{10} = 1024$ cell types, each reflecting a specific combination of choices by the ten decision loops. We shall see below that such combinatorial behavior almost certainly occurs in metazoan ontogeny. Since real genomic systems have more than $K = 1$ input per gene, however, we shall find combinatorial behavior arising from the combinations of activities of functionally isolated islands.

Not all networks which contain multiple attractors have combinatorial attractors. For example, genomic networks with $K = N$ have N/e alternative attractors. It will by now be obvious to the reader, however, that these attractors are *not* due to the combinatorial behavior of unconnected and independent subsystems. Rather, they are basins of attraction which occur in systems of the highest possible wiring-diagram interconnectivity.

With this background about the behavior of diverse ensembles of model systems, we can now turn to genomic systems that are based on known local properties.

IMPLICATIONS FOR ONTOGENY

We have now examined different ensembles of model genetic regulatory systems as a preface to asking whether any such ensemble might exhibit spontaneous order which accounts for aspects of the order seen in cellular differentiation and ontogeny among real organisms. Known genomic regulatory systems are graced by two local features:

1. Most genes are directly regulated by few other molecular variables.
2. Within the Boolean idealization, most genes are regulated by canalyzing Boolean functions.

Such constraints appear sufficient to specify an entire ensemble of model genomic systems whose typical members lie in the ordered regime. It would simply be foolish not to entertain the hypothesis that the ordered properties spontaneously present in this ensemble may bear on the ordered properties found in cell differentiation and ontogeny. In this section, we examine the match between theory and fact.

State Cycle Attractors as Cell Types

I introduced above the central interpretation I shall make: A cell type is an attractor of the genomic regulatory system. Whether we are thinking of Boolean models of genomic systems or of later, more sophisticated models, this identification distinguishes between transient patterns of gene expression lying on trajectories which converge to dynamical attractors and patterns of gene expression which are "on" the attractor. In a loose picture, if you will, cell types are the stable, recurrent patterns of gene expression in the dynamical flow driven by the couplings among genes and their products. In the Boolean network models, then, the natural identity is that a state cycle attractor is a cell type. The immediate consequence is that the patterns of gene expression corresponding to one cell type are constrained to those which constitute the recurrent cyclic pattern. On the other hand, as we have seen, the attractors for $K = N$ networks have on the order of $2^{N/2}$ states each. For even 100 genes, a cell type is localized to only $2^{50} = 10^{15}$ patterns of gene expression. Even for $K = 5$, or perhaps $K = 3$, length of state cycle attractors increases exponentially as N increases. In contrast, for $K = 2$ and canalyzing Boolean networks, state cycle attractors are on the order of only $\sqrt{N}$ in length. Thus a genomic system having as many as 100 000 genes would localize its patterns of expression to 317 states through which it cycled repeatedly. A set of 100 000 genes is big enough to be a reasonable and serious model of a genome; a set of 317 patterns of gene expression is small enough to be biologically completely reasonable for a cell type. Thus $K = 2$ and canalyzing networks exhibit adequate order in this respect without selection.

How restricted is the pattern of gene expression in one cell type? The best available data which bear on this question examine, via two-dimensional gel electrophoresis, the patterns of protein synthesis for many hundreds of proteins simultaneously, in cell populations which are synchronized in the mitotic cycle. The general observation is that the synthesis rate is constant throughout the cell cycle for almost all the hundreds of proteins and waxes and wanes for a few percent of them. Thus, based on these data, a cell type is a very constrained pattern of gene expression. It should be borne in a skeptic's mind, however, that the variability in patterns of gene or protein expression in one cell type over its cycle might be substantially higher. Data from two-dimensional gels—indeed, most similar data—average over the behavior of millions of individual cells. If transcription or translation were twinkling on and off, or up and down, in complex temporal oscillations over short intervals, so that the different cells were not doing exactly the same dance, then data averaged over many such cells would smooth out that variability. In short, on a microlevel, within one nucleus or cell, the patterns of gene expression or protein translation may well cycle through a set larger than what we can currently detect.

The identification of a cell type as an attractor raises a fundamental issue: Are cell types really attractors? That is, if we could reach into a cell and at any moment arbitrarily switch on or switch off the activity of any single gene, or modulate translation of its mRNA, and so forth, would we find that the system *typically returned* to the *recurrent pattern* of expression experienced prior to the perturbation? The answer is that we do not know. The supposition appears absolutely reasonable, since cell types must be stable to a variety of biochemical fluctuations in their microenvironments, but the direct experimental proof is, to my knowledge, currently lacking. How might we obtain such evidence? As mentioned in the preceding chapter, it is now feasible to clone any cellular gene adjacent to an arbitrary controllable promoter and to introduce that construction into living cells. Thus it is possible to switch on or off the cloned gene at arbitrary moments in synchronized cell populations. If the cell's own copies of the gene remain present but inactive, then the exogenously introduced copy can be activated at will, and the cascading consequences upon other genes studied at the levels of transcription, translation, and beyond. Two-dimensional gel analysis is currently good enough to indicate how often such a perturbation relaxes back to the previously observed pattern. And if a transition occurs to another recurrent pattern, does that pattern correspond to a known cell type pattern of gene expression?

Similarly, injection of complementary RNA which hybridizes to the mRNA of a specific gene can, in principle, block expression of the gene, allowing study of the cascading consequences. Does the pattern return to that perturbed? Does the system flow to another normal pattern of expression? to some new pattern? I should stress that the theory we are discussing suggests that cell types are attractors that are stable to *most but not all* perturbations in which the activities of any single gene are altered. Were a cell type stable to all such perturbations, then inducing differentiation from one cell type attractor to another by single hormonal signals to some specific sensitive gene would be impossible.

Interestingly, although direct data do not yet firmly show that a cell type is an attractor, aspects of phase resetting in the cell cycle do suggest that the mitotic cycle is an attractor. We discuss this briefly next.

Expected Cell-Cycle Times

The expected state-cycle length for a $K = 2$ or canalyzing synchronous switching Boolean network is on the order of $\sqrt{N}$. Thus, as noted, the recurrence time for a given state of gene activity is 100 for a genome of 10 000 and only 317 for a 100 000-gene genome. This length does not change dramatically for asynchronous models, as just noted. Turning transcription or translation on or off requires time. For example, the time may be on the order of a few minutes in bacteria and on the order of a few tens of minutes in eukaryotes. Therefore, in a eukaryote cycling through 100 patterns of gene expression, the expected recurrence time is on the order of 1000 minutes, or 16.6 hours. The first implication of the hypothesis that genomic networks lie in the ordered regime, then, is that cells can traverse their attractors in *biologically reasonable lengths of time.* Doing so, given that genomic systems have the established local features of low connectivity and canalyzing functions, requires no selection per se and is a self-organized property of this class of genomic regulatory systems. If cell types are attractors, those attractors must be small enough to allow the attractors to be traversed on a short time scale.

The most obvious cycling process in cells is the mitotic cycle. Therefore, let us ask if we can use the distribution of state-cycle lengths for each size network, plus the change in that distribution as network size increases, to predict anything about the known distribution of mitotic-cycle times in organisms. Figure 12.11 shows, as a function of DNA content per cell, the distribution of mitotic-cycle times for organisms ranging from bacteria to mammals, for a diversity of cell types within each organism and under a broad range of conditions. The distribution of state-cycle lengths for networks containing 200 genes is shown, along with the square-root relation of state-cycle length to numbers of binary genes. A surprising number of features are parallel. First, the time scale, on the order of hours, is correct. It surely need not have been. Second, median mitotic time does increase as a square-root function of the DNA content per cell across this wide range of phyla. Third, even in a logarithmic plot for cycle time, both Boolean networks of the same size and cells of the same genomic complexity show a similarly skewed distribution, with the bulk of the cases having short cycle times and a few scattered examples having very long cycle times.

The observed time scale of cell division (hours to tens of hours), the observed relation across phyla between genomic complexity and cell-division time, and the skewed distribution about the mean cycle time at each level of genomic complexity are basic features of cellular life. The ensemble theory of the distribution of cycle times in complex Boolean networks in the ordered regime fits the observed distributions. It would be premature to conclude that the theory accounted for the phenomena but even more premature to dismiss the theory. Indeed, at present, this approach is the only one able to predict the observed distributions. Other models, discussed next, consider in more detail the control of the mitotic cycle but do not bear on the distribution of cycle times across phyla.

The oscillatory-network view is broadly consistent with, and can unite, much of the extant work done on the mitotic cycle. The cycle is typically analyzed in four periods: mitosis, the G1 period between mitosis and the onset of DNA replication, the S period of DNA replication, and a subsequent G2 period leading to mitosis. There are three dominant lines of thought on control of mitotic time:

1. A G1 interval of indeterminant length which leads stochastically into fixed S, G2, and mitosis intervals.

2. A model of the mitotic cycle based on a collection of mutations which block cells in particular phases of the mitotic cycle. The best work has been carried out in yeast (Hartwell, Culotti, et al. 1974). The initial naive picture of the mitotic cycle was that of a simple cyclic sequence of events, each necessary and sufficient for the occurrence of the next. Then a mutant blocking one event would collect all cells at that stage of the cycle. The situation is more complex, however. There exist in yeast mutants which block DNA replication but allow rhythmic rounds of budding to continue. Conversely, mutants which block budding but allow rhythmic rounds of DNA synthesis exist. Evidently, the cyclic causal hoop of events is a *multiplexed web* which has more than one cyclic pathway through it.

3. A mitotic cycle governed by a central clock which is a limit-cycle oscillator, with a threshold concentration of one or more dynamical variables of the limit cycle needed to trigger downstream mitotic events (Kauffman and Wille 1975; Shymko, Klevecz, and Kauffman 1984). Limit cycles are attractors describing stable oscillations (Winfree 1980, 1987). If perturbed off the limit cycle, the oscilla-

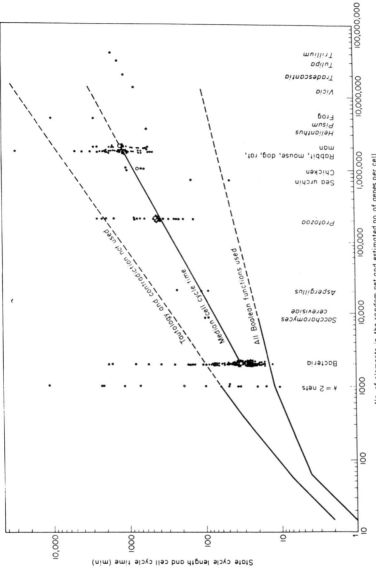
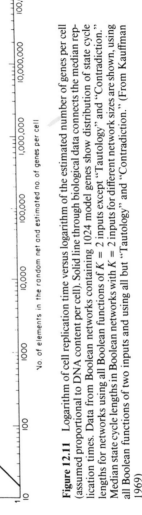

Figure 12.11 Logarithm of cell replication time versus logarithm of the estimated number of genes per cell (assumed proportional to DNA content per cell). Solid line through biological data connects the median replication times. Data from Boolean networks containing 1024 model genes show distribution of state cycle lengths for networks using all Boolean functions of $K = 2$ inputs except "Tautology" and "Contradiction." Median state cycle lengths in Boolean networks with $K = 2$ inputs for different network sizes are shown, using all Boolean functions of two inputs and using all but "Tautology" and "Contradiction." (From Kauffman 1969)

tory system *returns* to the limit cycle but may do so at a perturbed phase. Thus for any limit cycle, one can derive a mapping which shows the new phase attained as a function of the phase the system was at prior to perturbation, and the intensity of the perturbation. Such a mapping is a *phase-resetting curve* (Winfree 1980, 1987). One virtue of a limit-cycle model is its capacity to account for observed phase-resetting behaviors when cells are subjected to temperature and other perturbations, which are thought of as transiently driving the dynamical system off the limit cycle (Kauffman and Wille 1975; Willie, Scheffey, and Kauffman 1977; Shymko, Klevecz, and Kauffman 1984). The position regained on the limit cycle relative to where the system would have been measures the phase resetting due to the perturbation. Limit cycles in continuous state spaces, with continuous differential equations, naturally show smooth phase resetting. For most small changes in the perturbation applied, the new phase is typically only slightly modified.

The fact that dividing cells show this smooth phase-resetting behavior strongly suggests that the cells are cycling along some kind of underlying attractor and return stably to it when perturbed. That is, smooth phase resetting suggests that cell cycles, at least, are attractors. In fact, the most recent work on components of the eukaryotic-cell cycle—including cyclin, kinases, and dephosphorylases—is entirely consistent with a limit-cycle theory having at least several biochemical variables.

Qualitative predictions derived from older limit-cycle models remain to be tested with our new understanding. For example, if the mitotic cycle is controlled by a limit cycle of biochemical variables, one or more of which must reach a threshold to trigger mitosis or other cellular events, then appropriate perturbations (such as heat shock, inhibition of protein synthesis, or transient inactivation of specific genes) should cause one or more cycles of *subthreshold oscillations.* These subcycles would cause skipping of the normally triggered event, which would occur one or more cycles later plus or minus some phase resetting also induced by the perturbation. Evidence for such skipped mitoses and subthreshold oscillations does exist in the older literature (Wille, Scheffey, and Kauffman 1977; Shymko, Klevecz, and Kauffman 1984). In a similar vein, a resting G0 state might be thought of as a bifurcation in a central clock from an oscillatory regime to a steady-state regime. Further, if mitosis is governed by a limit cycle, then one might try to understand the transition in many early embryos from synchronous cleavage divisions to asynchronous cleavage divisions as a consequence of the increased diffusion barriers between cells which drive systems from spatially homogeneous behavior to behaviors which propagate phase waves.

Since real cellular systems have hundreds or thousands of linked biochemicals, however, would one expect smooth phase-resetting behavior in such systems? Strikingly, even in the extreme form of discrete state on–off switching networks, the $K = 2$ or canalyzing ensemble shows this smooth property as well. If perturbed off the state cycle, the system flows back onto it nearly in phase with an unperturbed copy (Gelfand and Walker 1984). Presumably, this behavior generalizes to continuous systems. Smooth phase resetting should occur even in large coupled nonlinear oscillatory systems in the ordered regime.

The image of a cell type as a cyclic attractor in a complex network is a marriage of the limit-cycle picture and the multiplexed cyclic web of events picture. Imagine replacing the phrase "gene A turns on" with "event A occurs in the cell" and the idea of *cis* and *trans* control of gene expression with the more general notion of the entire genomic system. Then as long as such a system has few inputs per event and as long as events are governed by canalyzing functions in the Boolean idealization, the number of sequential states on attractors in the Boolean idealization will be about the square root of the number of processes each of whose presence or absence constitutes an event. The causal webs in such networks are multiplexed. Blocking one event or process often allows cycling to continue via other causal routes. Thus the similarity in the statistical distributions of cycle times in Boolen models and real cells can reasonably be taken seriously as a first step toward a theory of the distribution of these phenomena is cells across phyla.

Ahistorical Universals?

Consider our hypothesis. We now have two consequences:

1. Cell types should be constrained patterns of gene expression.
2. Mitotic cycle times should show the distributions which are in fact observed across many phyla.

We may then ask, Does the observed distribution reflect selection acting directly to attain cell cycle times which, for higher eukaryotes, typically range from several hours to around a day? Is the similarity across diverse phyla to be accounted for by descent from common ancestors? Or might it possibly reflect the spontaneous distribution of properties in genomic systems with few inputs per gene (or event) and use of canalyzing functions? I have repeatedly raised the point that, in our current world view, where the only source of order is Darwinian selection, common ordered features across organisms are always accounted for by selection and descent from a common ancestor. Here, however, we have candidate properties which may be biological universals of a different type. Granted that organisms are in the ordered regime due to membership in the canalyzing ensemble, which may itself reflect selection, then these other properties may simply follow as otherwise ahistorical universals. Such a picture does not violate the letter of Darwin, who noted that selection for some property might "carry along" other properties which were correlated with the first, but it does violate the Darwinian spirit, for the correlated properties are enormously ordered and so deeply generic to the ensemble of genomic systems in question that selection may not be able to avoid them. Thus the order becomes universal and entrenched, part of the way organisms must henceforth be.

The Number of Cell Types Expected

If a cell type is an attractor, then the number of attractors in the dynamical repertoire is the number of cell types of which that genome is capable. Therefore, an ensemble theory affords us a means to *predict the expected number of cell types in an organism.* For the canalyzing ensemble in the ordered regime, the number of cell types should grow as about a square-root function of the number of genes and processes in the genomic regulatory system, reaching 317 for humans with about 100 000 genes. For $K > 5$ to $K = N$, the number of attractors grows roughly linearly with the number of genes and processes which are coupled. Thus a robust conclusion is that this vast ensemble, from $K = 2$ to $K = N$, predicts that the number of cell types should be between a square-root and a linear function of the genomic complexity.

Figure 12.7 plots the numbers of cell types against DNA content per cell across many phyla, up to about 254 for humans. As noted earlier, the number of cell types increases as a square-root function of DNA content. If rough estimates of the number of structural genes are used instead of DNA content, the number of cell types increases roughly linearly with the number of structural genes. This number leaves out nontranscribed regulatory sequences and other components of the control cascades; thus, as concluded above, it seems safe to say, using the histologist's criteria, that the number of cell types does increase between linearly and as a square-root function of the genomic complexity. Were finer-grained criteria used to discriminate cell types, the number of cell types per organism should increase fairly proportionally and hence not alter the logarithmic slope reflecting the power-law relation between genomic complexity and number of cell types across phyla.

In summary, our best guess must be that the number of cell types increases as a *fractional* power of the number of genes. Since the number of attractors increases as the 0.5 power of the number of genes in the ordered regime and increases as the 1.0 power of the number of genes in the chaotic regime, we may conclude that our genomic-network models in the ordered regime and perhaps near the boundary between order and chaos do in fact match and predict the relationship between number of cell types and genomic complexity across phyla.

The C Value Paradox

One of the puzzles in developmental biology is called the C value paradox (Alberts, Bray, et al. 1983). There are really two separable paradoxes:

1. The complexity of any given genome seems to be poorly correlated with the complexity of the organisms generated by that genome. Thus mammals have nearly 1000-fold the DNA per cell found in very simple metazoans, yet do not seem 1000-fold as complex.
2. Different metazoans of apparently the same complexity can have very different amounts of DNA.

The ensemble theory we have now explored has bearing on the first paradox. In $K = 2$ and canalyzing ensembles, the number of cell types increases only as a square-root function of the number of genes. Restated, as genomic complexity increases, it takes ever more new genes to add the next new cell type.

Another Universal?

The distribution of number of cell types per organism versus genomic complexity across many phyla may reflect selection sifting. It is hard, in light of the predictive capacities of the class of theories we are investigating, not to consider an alternative hypothesis. Build genomic systems constrained to the canalyzing ensemble and, without further selection, organisms will exhibit the observed distribution of cell types. The scaling law may be a universal.

Homeostatic Stability of Cell Types to Perturbation: Toward a General Theory of Homeostasis

We need a deep theory of homeostasis. What are the requirements in complex dynamical systems such that the systems settle down to constrained behaviors and, in addition, those behaviors are stable to perturbations? How did such ordered behavior evolve? Homeostasis, at root, is the buffered capacity of a system to return after a perturbation. We now have the start of such a general theory, for we know that dynamical systems in the ordered regime exhibit attractors which are stable to most minimal perturbations. In $K = 2$ and canalyzing networks, the systems return to the same state cycle attractor for 80 percent or more of the possible minimal perturbations achieved by transient reversal of the activity of any single gene. Figure 12.12a shows a system of 30 state cycles. After perturbations, the system typically returns to the state cycle from which it was perturbed. If it passes to another state cycle, typically one or a few of the other 29 cycles are accessible from each state cycle. Thus each model cell type is stable to most perturbations and can "differentiate" into only a few others by transient reversal of the activity of any single gene.

But what is this canalyzing ensemble? Low connectivity ($K = 2$) is just high molecular specificity. As emphasized above, canalyzing Boolean functions are just the simplest molecular kinetic rules to build. It is easy to make an enzyme which is activated by one or another allosteric effector acting on the same allosteric site. The site needs merely to bind one or the other indiscriminately. To create an enzyme which realizes the *noncanalyzing* "Exclusive Or" function, however, the enzyme needs two allosteric sites and must be active if either one but not both are bound. This latter condition is a difficult one to achieve. Then we are led to the hypothesis that large-scale dynamical homeostasis is an almost inevitable consequence of molecular specificity and simplicity. This general conclusion should apply to the evolution of autocatalytic peptide or RNA polymer systems. It should also apply to cell types, with a number of important implications. Foremost among these is that, even without further selection, the molecular specificity and simplicity which engender membership in a canalyzing ensemble already purchase homeostasis for cell types. A liver parenchyma cell will remain a liver parenchyma cell after most insults.

Restricted Pathways of Differentiation

If a cell type is an attractor, then differentiation is passage from one attractor to another, driven either by exogenous signals or by asymmetric distribution of cellular constituents at division. In the canalyzing ensemble, any cell type can be triggered to differentiate to only a few neighboring cell types by hormonal or other signals altering the activity of single target genes. This is a powerful property with many consequence. First, it necessarily follows that any single cell type can differentiate into many cell types only by following branching pathways of differentiation from the first cell type to its few immediate neighbors, from them to their few additional neighbors, and hence ultimately to a large set of cell types. In short, *ontogeny must be organized around branching pathways of differentiation!* And in fact, as far as we know, all metazoans and metaphytens since the Paleozoic have ontogenies organized around such branching pathways (see, for instance, Conklin 1905). This is so commonly known that it has not even constituted a question. Yet there is nothing evidently necessary about its entrenched feature of ontogeny. Consider a sponge. If disaggregated into individual cells which are already differentiated, the sponge will gladly reassemble itself into a working whole. We can easily imagine a sponge embryo proliferating into a large number of identical cells which might then differentiate into many different cell types. The many cell types might then arrange themselves into a sponge. The sponge does not work this way, however. Instead, it follows its rehearsed sequence of branching pathways of differentiation (Wilmer 1970). Why? Why, indeed? Is the

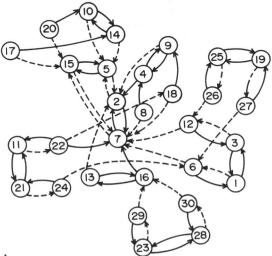

	1	2	3	4	5	6	7	8	9	10	11	12	13	14	15	16	17	18	19	20	21	22	23	24	25	26	27	28	29	30
1	103	2	16			8	1																							
2		289			2		11																							
3	92	20	976			9	2	1				46	5																	
4		49		987				8	32	69				11																
5	6	24	6		1076										39															
6	46	1	2			324	6								3	1														
7		40					340								4															
8		8		32			11	329									3													
9		49		64			8		987	12								31												
10		3			46					21	966	6			64	8			5	31										
11	6		27	2						18	9	949	3									85	42	5	1					
12	2	1	46			6						324			3	2														
13	9	18	9										1080			56														
14		3		21	46					64				966	8		32				6				5					
15					40	2	8						2		332															
16						3	6						3	56		316														
17					8	1	7							32	10		322						2		1					
18		8					11			32								329		4										
19	37		12	2		6				18	9		2					1	939				6		80		39			
20				8	1					32					10			7		322		2					2			
21	27		6	18		3			1	2		86									949			43					5	
22		3								1		43						6		3	2	315								1
23										18		9		35			1		9				917					108	53	3
24	3		1		9		6					2					3			43			315							2
25	12		37	18		2			1	2			6		9					80					939	40		6		
26	2		6	1				6				11				3									40	313				2
27	6		2			11				1								6	40								313		2	
28			18					0						35			1			9		106		9				917	4	53
29														1			11	6			3	53					3	4	303	
30						6								1			11				4	3		3				53		303

a

Cycle Perturbed (vertical axis label)

Cycles Found

b

490

modus operandi of the sponge a consequence of selection? Has Darwin's force labored for 600 million years or more to achieve and maintain this feature of ontogeny because it is essential? Or is it an unavoidable universal, so deeply generic to genomic systems that for 600 million years selection has been constrained to make do with this self-organized property?

Further, if each cell type has only a few neighboring cell types and responds to any possible reversal of the activity of any single gene either by homeostatic return to the same cell type or, occasionally, by flow to one of these few neighbors, then each cell type is poised between few alternatives. Then a variety of exogenous stimuli, acting on different genes in the system, will necessarily trigger the same transition. This implies that the specificity of the response lies largely in the cell tickled and only slightly in the feather doing the tickling. As I discuss in more detail in Chapter 14, inductive phenomena are vital in ontogeny. Thus, in early vertebrate development, the infolded mesoderm in the gastrula comes to lie beneath and juxtaposed to the overlying ectoderm which has not invaginated. Inductive influences from the mesoderm cause the overlying ectoderm to differentiate into neurectoderm (Saxen and Toivonen 1962). This phenomenon led early investigators to search for the normal inducing agent. It was rapidly discovered that a variety of abnormal agents (pure chemicals, pH alterations, even pure water) could induce the ectoderm to form neurectoderm. That is, the embryonic ectoderm is poised, ready to be tipped into neurectoderm, by a variety of specific and nonspecific stimuli. Being poised, or competent to respond in a restricted way (Waddington 1957, 1962), is not an accident; it is another expression of the fact that each cell type has only a few neighbors. Furthermore, the capacity of a cell to be triggered by an inductive influence depends on the cell type. Hence the competence to respond changes during development as cells pass branch points along developmental pathways. All these are deep structures in the logic of ontogeny. All, I believe, are consequences of membership in the canalyzing ensemble. Selection has these properties to work with, like it or not. They, too, may be universals.

Does an Organism Use All Its Cell Types: Implications of Restricted Global Reachability

If a cell type is an attractor and if differentiation is passage between attractors induced by perturbations or signals, then it is natural to ask if any cell type can ultimately

---

Figure 12.12 (*a*) A matrix listing the 30 state cycles of one network and the total number of times one unit of perturbation, transient reversal of the current activity of a single gene at a single state of a state cycle, shifted the network from each cycle to each cycle. The system generally returns to the cycle perturbed and hence exhibits homeostasis. Division of the value in each cell of the matrix by the total of its row yields the matrix of transition probabilities between state cycle modes of behavior under the drive of occasional random perturbations and constitutes a Markov chain. The transition probabilities between two cycles are often asymmetric. (*b*). Transitions between cycles shown in (*a*). The solid arrows are the most probable transition to a cycle other than that perturbed; the dashed arrows are the second most probable. The remaining transitions are not shown. Cycles 2, 7, 5, and 15 form an ergodic set into which the remaining cycles flow. If all transitions between cycles are included, the ergodic set becomes 1, 2, 3, 5, 6, 12, 13, 15. The remainder are transient cycles leading into this single ergodic set. Under the drive of occasional reversal of the activity of any single gene, cell types within the ergodic set can reach one another but cannot reach cell types not in the set. (From Kauffman 1969)

differentiate into *all* the cell types in the genomic repertoire. The generally expected answer is "no." Typically, in the canalyzing ensemble, each cell type can directly differentiate into only a few other types. The directed graph showing this (Figure 12.12) typically has the following features:

1. No cell type can reach all cell types.
2. A large number of cell types can mutually reach one another via one or more steps. Call this the "strongly connected set of cell types."
3. Some cell types can reach this strongly connected set but cannot be reached from it.

If we imagine that all cell types are attractors, therefore, one consequence is that typically no cell type can differentiate into all legitimate cell types. Thus the genome must typically harbor *unused* cell types.

We shall examine the implications of the strongly connected set of cell types below. Briefly, the existence of this set implies that other sets of cell types should exist which can transform into one another in a more or less complex web. Just such metaplasias are observed. The implication of the fact that the genome uses only a subset of cell types in ontogeny is more immediate, however: Unused cell types may include cancers.

What might it mean if an organism uses only a subset of its cell types? At least one obvious hypothesis is this: Among the unused cell types, some may behave poorly in the integrated organism. For example, they might continue to undergo mitotic divisions without cessation. That is, some unused cell types might be cancers. Almost inevitably, then, we are led to suppose that improper combinations of activation of normal cellular genes can cause cancer. That is, we would not be surprised if there existed genes whose improper activation yields transformation (Braun 1968, 1969; Kauffman 1971c). Such improper activation would not logically require that the genome be mutant, in which case the malignant behavior would be a "disease of differentiation." In fairness to this body of theory, I note that this prediction was made in 1971, long before the discovery of first viral and then cellular oncogenes, which now is revolutionizing cancer research (Shih and Weinberg 1982). Placed in this framework, the oncogene story has some implications which have only been partially recognized. Obviously, chromosomal mutations such as translocations (Leder, Battey, et al. 1983), rearrangements (Croce 1987), or promoter insertion (Hayward, Neel, and Astrin 1981; Adams, Harris, et al. 1985), which bring oncogenes under aberrant control, often play a role in carcinogenesis. Alternatively, abnormal differentiation in genetically normal cells may lead to aberrant activation of cellular oncogenes. Without or with such somatic mutational events, however, the resulting genomic system is capable of very many alternative cell types. The question, then, is whether it might be possible to trick cancer cells into differentiating into either normal or benign cells. The general answer to this may be affirmative.

Embryonal carcinomas are particularly interesting. In appropriate circumstances, they can be tricked into behaving benignly. Embryonal carcinomas derive from testicular or ovarian cells which form progenitors to the germ cells. Such tumors form teratoma, in which tissue derivatives of the three major classes—ectoderm, mesoderm, endoderm—are found. Thus teratomas may literally contain muscle cells, nerve cells, and hair follicles in haphazard array. In a beautiful set of experiments, Mintz and Illmensee (1975) constructed chimeric (mixed cell) mouse embryos in which some cells were from a normal mouse embryo and other, genetically marked cells were from a testicular teratocarcinoma. The chimeric embryos were transplanted into pseudopregnant females, and fully normal young mice were born. These young mice were a composite of cells derived from the normal embryo and the tumor tissue. In fact, over a set of such mice, the genetically marked tumor-cell derivatives were found normally integrated into almost all tissues of the progeny chimeric mice. In brief, when exposed to the normal sequence of inductive and regulatory interactions, the tumor cells reverted from malignant to utterly benign behavior.

These tumor reversion results can be found in other systems. For example, tumors can be induced in the tail of the newt by tar carcinogens (Braun 1968, 1969). If such tumors are induced on the basal region of the tail and the distal tail tip is cut off, the newt obliges by regenerating its distal tail. If this cycle of cutting and regeneration is repeated several times, the *tumor regresses and vanishes* (Braun 1968, 1969). Presumably, the normal regulatory factors brought into play during regeneration mediate this effect. Differentiation of a number of juvenile cancers, such as neuroblastoma and retinoblastoma, to nonmalignant mature cell types is well known *in vivo* (Braun 1968, 1969; Kauffman 1971c; Klein 1987).

The differentiation from malignant to benign behavior is underscored by the large number of cases where somatic hybridization of spontaneous, virally induced, and chemically induced tumors results in low-tumorigenic or nontumorigenic cells (summarized in Klein 1987). Reversion from malignant to benign behavior has also been demonstrated by negative selection that killed transformed cells (Rabinowitz and Sachs 1970) and, more recently, by positive selection (for example, Noda, Selinger, et al. 1983). It seems clear that some genes act as tumor suppressor genes. For example, Knudson (1987) has suggested that retinoblastoma typically arises by the loss of both alleles of a specific gene (RB-1). Normally the retinoblast differentiates into a retinocyte that has irreversibly lost the ability to divide. A child who inherits a deletion for one copy of RB-1 risks developing a retinoblastoma during its first years of life, but by age five, all retinocytes have differentiated. The child is no longer at risk even if the remaining copy of RB-1 is lost. Apparently, a single copy of the gene suffices to suppress transformation to malignant behavior. The genetics of retinoblastomas suggests that abnormal inactivation of normal RB-1 genes might suppress their suppression and hence allow transformation.

As Klein (1987) points out, a number of malignancies can be induced to differentiate terminally *in vitro*. In some clear instances, suppression is mediated by small molecules produced by normal cells that diffuse in solid tissues through gap junctions and damp tumor cell precursors (Paul 1988).

All these results suggest that, more often than we now suppose, it may be possible to induce malignant cells to behave benignly (Kauffman 1971c; Klein 1987). Any interest in this approach, however, must be tempered with the realization that most tumors undergo progression during which a variety of chromosomal mutations build up. The hope to trick such aneuploid cells into benign behavior may prove faint indeed. On the other hand, integration into such cells of a properly chosen set of normal cellular genes, properly activated either together or in sequence might succeed in inducing a transformation to benign behavior even in badly aneuploid malignant cells.

It should be noted that the hypothesis that the normal genome can generate cell types which are not normally utilized in ontogeny is clearly open to direct experimental investigation. As remarked above, high-resolution two-dimensional gels even now allow a fine-grained portrait of the pattern of protein synthesis in a large number of genes and are reliable enough to distinguish different cell types. One can hope eventually to build up data on the normal cell types of any organism and then ask whether transient activation or inhibition of one or several genes causes a transition to a new stable recurrent pattern of protein expression which differs from known normal patterns in other cell types. Obviously one would have to rule out somatic mutations, but in general such studies are even now on the verge of feasibility.

Can Mutations Alter Developmental Pathways and Not Cell Types?

The existence of such atavisms as hens' teeth (Kollar and Fisher 1980) and whales' legs (Andrews 1921) and the evolution of differentiation raise an obvious question: Can there occur mutants which alter developmental pathways but not cell type? Such possibilities would be useful, and, indeed, Figure 12.8 shows an example. It is particularly easy to accomplish such limited alteration in the canalyzing ensemble of genomic systems. A large fraction of the genes, up to 70 percent, fall to fixed active or fixed inactive states and then remain in the same fixed state on all cell types of the organism. Therefore, deletion of a gene which is normally fixed inactive on all cell types will *not* alter the cell types. However, that gene need not be inactive during the differentiation from one cell type to another when the differentiation is induced by exogenous stimuli or by asymmetric distribution of material at cell division. Consequently, deletion of such a gene can alter which stimuli cause which differentiation steps between which cell types, without altering the cell types themselves. In fact, the ways in which this occurs has surprising features in these model genomic systems: in general, a subset of differentiation pathways is selectively altered by a class of random mutations. Pathways of differentiation can be sculpted independently of cell types.

Consider a Boolean genomic system containing L cell types. Suppose that 70 percent of the N genes are fixed either active or inactive; hence about 35 percent are fixed inactive. Numerical simulations were carried out deleting each fixed inactive gene one at a time and testing how the deletion affected the differentiation pathways between the L cell types. The possible transitions between the L cell types are conveniently shown in a square $L \times L$ matrix whose left-hand column labels the cell type which is perturbed by all minimal perturbations to the N genes of each state of each state cycle and whose top row labels the cell type to which the perturbed cell type changes; Figure 12.12 shows an example. Since most often a cell type returns to the same cell type, the main diagonal in Figure 12.12a has the most entries. Since cell type A might be able to differentiate to B but B not to A, the matrix is not symmetric. The frequencies in the boxes can of course be normalized to show the probability, for all possible minimal perturbations, that cell type i changes to cell type j. The row totals are then 1.00, and the matrix is just a Markov chain. That is, the matrix shows the transition probabilities between any cell type and each of the other possible cell types, under repeated occasional random transient reversal of the activity of single genes. In this context, a mutation which deletes an inactive gene does not alter cell types but may alter the entries in the matrix and hence change the transition probabilities between two cell types in the face of fluctuations.

The striking results of the numerical simulations are:

1. Any such deletion alters only a few of the $L \times L$ entries in the matrix and hence *alters only a few pathways of differentiation.* Thus mutations can tune one or a few transitions between cell types without altering all pathways at once.

2. Most of the fixed inactive genes caused an overlapping small subset of the $L \times L$ entries in the matrix to be altered. Thus the same few differentiation steps are affected by most of the mutants in this class.

3. Any particular entry in the matrix was much more likely either to increase or to decrease, for all or most of the mutants in the 35 percent tried! This is particularly interesting. It says that the next random mutation in this class of mutants is much more likely to increase or more likely to decrease specific transition probabilities between particular cell types. Thus not only are there mutants which can tune up or tune down differentiation transitions between specific cell types and hence open or close pathways, but the generic property of complex genomic systems is that those alterations do *not* change up or down with equal probability. Whether one is comfortable with the idea or not, the existence of preferred "directions" of alteration of developmental pathways implies something like "orthogenesis"—a tendency of evolution to occur in preferred directions not because of selection constraints but because the underlying system has preferred directions of change in the face of *random* mutations.

These results make an experimental prediction. We know that a large number of genes are actively transcribed in all cell types, but we are led to the prediction that the genome should also contain a large number of genes which are *not* transcribed in any stable cell type. Just these nontranscribed genes are the prime candidates to play roles in transitions between cell types—that is, in differentiation between stable alternative cell types. The results fit a second set of facts. The emergence and disappearance of clusters of "whole-cloth" features in ways unrelated to phylogenetic lineages—homoplasy, in short—are just what one would expect of genomic systems which allow differentiation pathways to be tuned in such a way that flow to particular cell types and tissue types is either cut off or opened up.

Collective Order: Forcing Structures and the Percolation of Frozen Components

Canalyzing Boolean networks crystallize orderly dynamics—in other words, they spontaneously lie in the ordered regime—because a specific kind of subnetwork, called an *extended forcing structure,* literally crystallizes out of the dynamics of the network (Kauffman 1971a, 1974, 1984; Fogelman-Soulie 1984, 1985a, 1985b). The genes making up the forcing structure fall to fixed active or inactive states; the forcing structure percolates through the genomic network and typically leaves behind one or more unfrozen islands of genes free to turn on and off in complex patterns. These islands are functionally isolated from one another, and the combinations of alternative behaviors of the set of isolated islands correspond to the alternative cell types exhibited by such a genomic system.

The percolation of forcing structures and Derrida's annealed model are two analytic approaches to understanding the emergence of order in random $K = 2$ Boolean networks and in the canalyzing ensemble. Both approaches were described in Chapter 5. Here I recall the features of forcing structures, describe some of their properties, and consider their biological implications.

Forcing Structures and Their Fixed States. Percolating forcing structures fall to a fixed, "forced" state of gene activity. Consider a Boolean network in which each gene receives inputs from $K = 2$ other genes and each gene is governed by the "Or" function, which is canalyzing. If a given gene is placed in the active state, then we are assured that each of its descendant genes in the network will be in the active state at the next moment. In turn, all immediate descendants of these daughter genes will certainly be active the third moment. That is, if any gene is placed in the active state, that active value propagates to all descendants regardless of the values of other inputs to those descendants (Figure 12.13). This is the simplest example of a forcing structure.

Genes A and B will be coupled by a forcing connection if three conditions are met:

1. Gene A must be a canalyzing input to B.
2. Gene A must itself be regulated by a canalyzing function on its own inputs.
3. The canalyzed value of gene A (the value of A which can be guaranteed by its own inputs) must be the value which guarantees B.

Thus, in this case, A is governed by the "Or" function, as is B. The value of A which can be canalyzed by its own inputs is 1, and this is simultaneously the value of A which ensures that B shall be active the next moment. These conditions create a transitive relation such that, if A forces B and B forces C, the forcing value at A propagates in two steps and forces C to its own forced value. Thus the first important point about forcing structures is this: When a forced value is propagating down a forcing structure, the propagation is impervious to perturbations by other regulatory inputs to the cascade.

Not all connections between two canalyzing genes are forcing. If A were controlled by the "And" function and B by the "Or" function, then A would not force B. The value of A which can be canalyzed is 0, not 1, since A in inactive if either of its inputs is inactive. Since 0 at A does not ensure any value at B, the connection is not forcing.

By the appropriate choice of Boolean functions, the forced value at different points in a forcing structure can be 0 or 1. Figure 12.14 shows a forcing structure with different functions for the different genes; here a 1 value for A forces a 0 value for B, which in turn forces a 1 value for C.

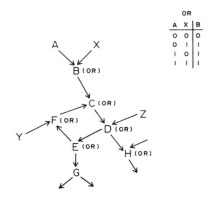

Figure 12.13 Forcing structure among binary elements governed by the Boolean "Or" function. The forcing 1 value propagates down the structure and around the forcing loop, which eventually is frozen into the forced state with 1 values at all elements around the loop. The loop then radiates fixed forced values downstream.

Forcing Structure

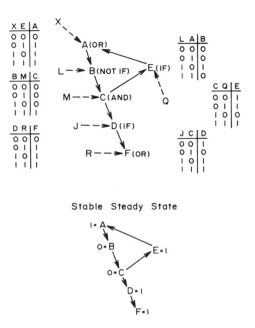

Figure 12.14 Forcing structure among binary elements governed by a variety of Boolean functions. Forced values propagate downstream through the structure and around the loop, which eventually falls to a frozen forced state. The loop then radiates fixed forced values downstream into the forcing structure.

Forcing structures can include genes which are regulated by only a single gene. If A is governed by a single input gene B and if B is governed by genes C and D via a canalyzing function, then C can indirectly canalyze the behavior of A via B. Gene B can pass along the forced value from C to A.

A second important feature of a forcing structure is that it can form forcing feedback loops. The salient feature of such a loop is that it must have a steady state, in which all members are in their forced values. In that steady state, the loop is completely impervious to perturbations impinging on it from outside. For example, the forcing loop in Figure 12.14 has four genes (A, B, C, E) and the forced values (1001). If placed in that state, each gene at each moment is forced, by the value of its forcing predecessor around the loop, to maintain that state regardless of the values assumed by its other input from outside the loop. Notice too that any forcing loop has three other properties:

1. It has a second steady state, the complement of the first, with the value of each gene reversed, which is unstable to any perturbation.

2. This instability reflects the fact that, if any input to genes on a forcing loop causes a loop member to assume its forced value, that value propagates ceaselessly around the loop thereafter and cannot be dislodged. Therefore the loop will tend to accumulate forced values and fall to the fixed forced steady state.

3. Once in that fixed forced state, the loop will radiate those fixed forced values to all forced descendants in the forcing structure. Therefore, the forcing loop and all its descendants will fall to a fixed steady state—the forced state—and remain in that state thereafter, regardless of the behavior of other genes in the network.

This *propagation of fixed behavior* is the key consequence of forcing structures. The fixed behavior, percolating across the network, is precisely the crystallization of the ordered regime.

An added property arises if the forcing loop is multiply connected, with subloops whose lengths or factors are relative primes. In that case, if any single gene is ever in the forced value, then at some fixed time later, given by the product of those primes, all genes in the multiply connected loops will be forced. Such a multiply connected forcing loop is said to be strongly connected (Kauffman 1971a, 1971b, 1974).

Forcing Structures Are Expected to Percolate in* K = *2 and Canalyzing Networks.
Whether forcing structures do or do not percolate is a phase transition governed by the ratio of forcing connections to genes in the genomic system. Above a critical ratio, forcing structures form. They form spontaneously in $K = 2$ networks and in $K > 2$ networks biased to use canalyzing functions. Thus the ratio of canalyzing to total functions in a genomic system is a control parameter which can tune whether genomic systems lie in the ordered, complex, or chaotic regime.

The forcing structures in a genomic network are a random subgraph of the directed graph showing the full architecture of the wiring diagram of the genomic system. Our experience with the connectivity properties of random directed graphs in Chapter 11 leads us to expect phase transitions as the ratio of forcing connections to genes increases gradually. If forcing connections are imagined as being colored red, then as the ratio increases, only a few isolated connections are red at first; then small red trees emerge. We know that a threshold will be reached when the number of red arrows equals the number of genes, however, for suddenly large connected red forcing structures will emerge. That is, for a high enough ratio of forcing connections to total number of genes, such a large connected structure will suddenly crystallize, or percolate. This can be seen in Figures 5.5 and 5.6.

Percolating forcing structures crystallize spontaneously in $K = 2$ networks and do not form when $K > 2$ unless choice of functions is biased in favor of canalyzing functions. Eight of the Boolean functions of $K = 2$ inputs are canalyzing in the central sense, and four "Yes" or "No" on one or the other input can propagate a forcing value. Two functions, "Tautogy" and "Contradiction," are trivially canalyzing. Only "Exclusive Or" and its complement, "If and Only If," are genuinely noncanalyzing. The probability that a regulatory connection between two genes each governed by a canalyzing functions is forcing is ½, and since each gene has $K = 2$ inputs, the expected number of forcing connections is close to N. Forcing structures are expected to, and do, percolate. In $K = 3$ networks, the fraction of Boolean functions which are canalyzing is far smaller than in $K = 2$ networks and the expected number of forcing connections is substantially less than N. Here, forcing structures are not expected to, and do not, percolate. As K increases, the fraction of Boolean functions which are canalyzing decreases dramatically and very few forcing connections are formed; hence, in general, forcing structures do not crystallize. If instead of the full set of Boolean functions of $K = 3$ or $K = 4$ variables, only canalyzing functions are used, then the number of forcing connections is again N or greater and forcing structures again percolate.

Very pretty numerical experiments by Fogelman-Soulie (1984, 1985a, 1985b) demonstrate this. She built $K = 2$ networks in a particularly simple way: on a lattice where each gene had four neighbors, two of which serve as input lines, and two as output lines. In this pre-fixed wiring diagram, Fogelman-Soulie assigned random Boolean functions of two inputs to genes in the lattice and followed the subsequent dynamics. As expected, large connected sets of genes fall to fixed 0 or 1 values. This is the forcing structure. Added to the forcing structure are all those genes which receive *all* their inputs from within the structure. When the latter is fixed, so are these genes. Fogelman-Soulie calls the extended forcing structure augmented by the elements which receive all inputs from it the *stable core*. The stable core is functionally similar to the frozen component arising as a result of high internal homogeniety in the network ($P > P_c$, Figure 5.10). Like the high-internal-homogeniety clusters, the stable core percolates across the lattice. Figure 12.15 shows the forcing structure and stable core in a random $K = 2$ Boolean network.

Because functionally isolated islands of unfrozen genes exist, cell types governed by genomic networks in the ordered regime should typically exhibit a combinatorial character. The random network in Figure 12.15 has one such nonforced cluster, and in Figure 5.10 there appear to be four nonforced clusters. Since these pockets of interconnected genes are not forced, they are free to vary even when the forcing structure has fallen to its fixed values. Typically, a large system will have several isolated pockets, each free to behave in several ways. That is, each isolated unfrozen pocket will have its own alternative attractors as a small cluster of interconnected genes.

Once the forcing structure falls to its fixed state and pockets are functionally iso-

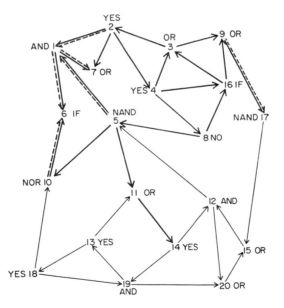

Figure 12.15 Forcing structure found in random Boolean network containing $K = 2$ inputs. Thick arrows are forcing connections; thin connections are not forcing. Dashed arrows lead from binary elements which fall to forced values to further elements which fall to fixed values because all their inputs become fixed. Such secondarily fixed elements become frozen in active or inactive values when the forcing structure falls to its fixed frozen state.

lated from one another, the dynamical behavior of the full system is given by the fixed values of the percolating forcing structure and by the *combinations of alternative behaviors* of the functionally isolated pockets. In short, such systems inherently behave combinatorially. If the first pocket has x alternative attractors, the second has y alternative attractors, the third has z, and so forth. Then the possible alternative attractors of the entire system are the product of the modes, xyz. Each attractor of the whole system can be thought of as a particular combination of choices among the x possibilities of the first pocket, among the y possibilities of the second, and so on for all the functionally isolated pockets. These properties lead to a number of predictions. In particular, cell types might be expected to exhibit a similar combinatorial character. And, indeed, they appear to do so.

Functionally isolated pockets must be headed by recurrent genetic circuits with the property that each gene in the recurrent circuit is "alive." Such recurrent circuits, the analogue of master genes (Garcia-Bellido 1975), are the loci of developmental choice in the genomic system. Notice that, in a network in the ordered regime, any gene which is not part of the forcing structure must have at least one input which is also not part of the forcing structure. Define such an input as *alive* (Aleksander 1973), a term meant to convey the fact that the "live" input is potentially free to be either active or inactive. Thus each live gene must have at least one live input, and this gene–input combination must eventually form a live feedback loop. Therefore, the functionally isolated pockets of genes each must contain at least one $K = 1$ input live feedback loop, which may have tails descendant from it. The live loop may or may not be as simple as a single gene which feeds back upon its own activity. There may be cross-couplings between more than one such loop, or it may have partial $K = 2$ connections within it. The conclusion to be borne away is that such pockets are headed by more or less simple feedback loops. Since these live loops can have more than one attractor each, they are the *loci of developmental choice in the genomic system.*

Reemergence of Hierarchial Command Structures?   In Chapter 11, we considered the military command structure made familiar by the Britten and Davidson (1969) model of genetic regulatory systems. In the simplest version of that model, master genes stand at the head of a descendant army of regulatory and structural genes, and control structural batteries of genes via a hierarchial command tree. I argued that such hierarchical command structures would be extremely hard to achieve and maintain in genomic regulatory systems having more regulatory connections M than regulatory genes N, since richly webbed structures are generic when $M > N$. In richly webbed networks which *lack* a frozen component and behave chaotically, it is quite clear that nothing like a hierarchial command structure exists. Altering the behavior of any gene will propagate changes to most genes in the system. This connectedness is not found in real genetic systems.

Conversely, the existence of a percolating frozen component raises again the possibility of more or less hierarchial command modules. This possibility arises in two ways. Consider a small isolated island of genes headed by a single $K = 1$ live feedback loop which has descendant tails of regulatory and structural genes. Such a loop is logically identical to the master gene in charge of the descendant battery of genes. Alternative attractors in one loop, like alternative states of a master gene, drive alternative cascades in the descendant tails. However, the isolated islands need not be fully isolated from one another. In a random genetic network, a frozen component may percolate throughout the network and at the same time the remaining islands may be more or less sparsely connected with one another. Suppose that island A is fully cut off by the frozen component from inputs emanating from other islands but that a gene in A happens to regulate a gene is an otherwise isolated island B. Then the alternative attractors of A can modify and constrain the dynamical behavior of B. Similarly, B may be connected to an otherwise isolated island C. In this way, hierarchial commands without (or with) feedback can be built up among the nearly isolated islands. The critical point is this: When $M > N$, attaining and maintaining structurally hierarchical military command trees seem extremely difficult because the overwhelming majority of networks with more regulatory connections than genes are rich in feedback loops. If some form of hierarchical regulation is often useful, which seems plausible, but must occur in genomic systems rich in feedback loops, then the existence of a large frozen component is almost certainly an essential requirement. Given such frozen components and given the generic existence of isolated or near-isolated islands, selective sculpting of sparse connections among these islands should permit the easy adaptive evolution and maintenance of hierarchical command structures embedded in the frozen component.

Harking back to the theme of *adaptation to the edge of chaos* and recalling that *chaos emerges when frozen components just melt,* it is also clear that, if genomic regulatory systems are just on the "solid" side of the edge of chaos, when frozen components are just large enough to begin to percolate, then the *ease* of molding such hierarchial connections among the nonfrozen, almost isolated islands should be optimized.

The Expected Number and Sizes of Functionally Isolated Islands in Genomic Systems.
If functionally isolated or nearly isolated islands are the carriers of alternative developmental decisions via their alternative attractors, then it is important to have some crude idea of the expected number of such islands. Figure 12.16 shows numerical results giving the distribution of numbers of fully isolated unfrozen islands as a function of the number of the total of N genes which are not frozen. Thus 0 unfrozen corresponds to a fully frozen network at a steady state. This analysis ignored the direction of connections among the unfrozen genes; hence more or less complex hierarchial dependent connections between subregions may exist within one fully isolated island.

Figure 12.16*a* examines 1000 networks, each containing $N = 100$ genes; Figure 12.16*b* examines 1000 networks, each having $N = 1000$ genes. The following results are clear. As the unfrozen fraction of the system increases, the number of isolated unfrozen islands first increases to a maximum and then decreases. This behavior reflects the fact that, when almost the entire system is frozen, few and small isolated islands exist. When the frozen component is nearly melted, the isolated islands merge into one large unfrozen island, which thereafter becomes the unfrozen sea. Thus the maximum number of isolated unfrozen islands occurs at an intermediate fraction frozen, which appears to be about 75 to 80 percent for these values of N. The total number of islands is quite small. The maximum was 13 in the $N = 100$ networks and only 27 in the $N = 1000$ networks. The mean numbers were, respectively, 2.7 and 5.2. A small number of $N = 5000$ networks were examined, and the mean number of isolated unfrozen islands increased to only 6.0. While the scaling laws for such clusters are not known, I would guess that a genomic system containing 100 000 genes might have on the order of 10 to 12 functionally isolated unfrozen islands of

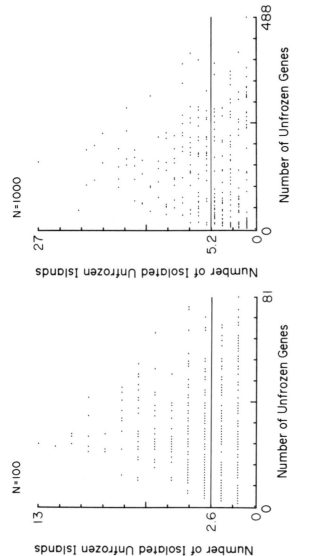

Figure 12.16 The number of functionally isolated unfrozen islands as a function of the number of unfrozen genes in 1000 $K = 2$ Boolean networks. In this analysis the directions of connections between genes in an unfrozen island are ignored. (a) $N = 100$; (b) $N = 1000$. Solid horizontal line is mean.

genes. If true, this result is remarkably encouraging. If isolated island are the loci of developmental choice, then genomic systems comparable to those in humans may have only on the order of 10 to 12 functionally isolated clusters of genes.

The number of cell types in a genomic system may bear a relatively simple relation to the number of functionally isolated islands. Make the undoubtedly oversimple assumption that each island is capable of only two alternative attractor modes of behavior. Then the expected total number of attractors of the system is roughly 2 raised to the power of the expected number of functionally isolated islands. For $N = 100$, this yields $2^{2.7} = 6.5$; for $N = 1000$, $2^{5.2} = 36.7$; and for $N = 5000$, $2^{6.0} = 64$. Recall that, for $K = 2$, the expected number of cycle attractors is roughly $\sqrt{N}$. The values 6.5, 36.7, and 64 are quite close to the square roots of 100, 1000, and 5000. Thus understanding the scaling laws for the number of isolated islands may yield scaling laws for numbers of attractors in real genomic systems.

The expected sizes of functionally isolated unfrozen islands can also be estimated, given the fraction of genes which are frozen. For example, assuming that about 80 percent of the genes are in the frozen component, a genome containing 100 000 genes, such as the human genome, would have about 20 000 unfrozen genes. Assuming that such a system would have 10 to 12 islands, each might contain about 2000 genes. A plant having about 20 000 genes would have about 4000 of them unfrozen. If a system with 20 000 genes enjoys about eight islands, then each island might contain about 500 genes. We see in a moment that these cluster sizes come close to predicting the observed differences between cell types.

Stable Core and Combinatorial Penumbra. In $K = 2$ and canalyzing networks, 70 percent or more of the genes fall to fixed active or inactive states; these genes are members of the forcing structure. The different state-cycle attractors, because of different activities of the functionally isolated islands and perhaps because of the overlapping downstream unfrozen genes which the islands may jointly control, typically differ from one another in the activities of a few percent of the genes. Therefore, these ensembles predict the existence of a large core of genes in fixed states of activity or inactivity in all cell types of the organism and differences between cell types in the activities of only a few percent of the genes.

Both the unchanging core and the rather small differences between cell types are found. As described earlier, a large core of genes is transcribed into heterogeneous nuclear RNA (hnRNA) in all cell types (Hough-Evans, Smith, et al. 1975; Kleene and Humphreys 1977, 1985; Chikaraishi, Deeb, and Sueoka 1978; Darnell, Lodish, and Baltimore 1986). A modest fraction of the hnRNA is processed and transported to the cytoplasm. While differences in cytoplasmic mRNA from one cell type to another may be greater than differences in hnRNA, the mRNA differences typically are in the sequences present in high numbers of copies (100 to 50 000) in one but not the other cell type. There is substantial cross-hybridization for cytoplasmic RNA sequences which are present in lower copy numbers. Thus it appears that different cells typically contain markedly overlapping sets of message sequences in the cytoplasm as well as in the nucleus (Goldberg, Galau, et al. 1973; Axel, Feigelson, and Shultz 1976; Galau, Klein, et al. 1976; Hough-Evans, Ernst, et al 1979; Alberts, Bray, et al. 1983; Darnell, Lodish, and Baltimore 1986).

Let us now turn our attention from the RNA level to the protein level. Most proteins are synthesized in all cell types. As noted earlier in this chapter, two-dimensional gels typically exhibit a qualitative difference in some small percentage (perhaps 15 percent) of the proteins in different cell or tissues of an organism. Some caution is needed here, however. Examples of myeloid differentiation are known in which about 50 percent of the abundant proteins alter during differentiation (Lieberman, Hoffman-Lieberman, and Sachs 1980). Whether these differences extend to the rarer proteins is not clear.

With caution, we may conclude with Alberts, Bray, et al. (1983) that, in plant cells having about 20 000 different RNAs transcribed, the typical difference between two cells is on the order of 1000 kinds of RNA. These differences are a kind of penumbra surrounding a stable common core of RNA and proteins which appears in most or all cell types of the organism. Note that the observation that plant cell types typically differ in the expression of about 1000 kinds of RNA is remarkably close to the estimates one would get for the mean sizes of functionally isolated islands in genomic systems containing 20 000 genes. If the stable core comprises 80 percent of these and about eight unfrozen clusters partition the remaining 4000 genes, each cluster should contain about 500 genes.

This predictive success of the canalyzing ensemble in the ordered regime is not trivial. There is good evidence for a frozen core and, as we see below, good evidence for combinatorial behavior due to unfrozen gene clusters. The ordered-network ensemble predicts the distribution of these properties in organisms. Order emerges in disordered Boolean networks, as far as is now known, either because of extended forcing structures or because of frozen percolation clusters. Otherwise, chaotic dynamical behavior is found in Boolean networks. Presumably, such chaos will be found in continuous homologues to Boolean networks. In networks with $K > 3$ and no bias in the choice of Boolean functions, neither orderly behavior nor a core and penumbra are found. In such networks, no genes are in fixed states of activity or inactivity in any cell type, and the typical differences in patterns of gene activities between attractor cell types is larger than for $K = 2$ or canalyzing networks.

The hypothesis of a frozen stable core makes a number of predictions:

1. The forcing structure must *fall* to its fixed state. Thus if the zygote is not already in a state such that the frozen component is yet frozen, early embryogenesis should witness a rapid cascade of alterations of activities of very many genes. Note that until the stable core falls to its frozen state, genomic islands which shall later become functionally isolated are still functionally coupled. Influences can propagate widely across the genomic network. At some point, the frozen core becomes established. Thereafter, unfrozen islands are functionally isolated. If this view is correct, temporal differences in gene activity cascades early in development should be larger than differences between cell types that occur later in development, after the stable core is frozen. Analysis of developmental profiles of protein synthesis in *Drosophila melanogaster* (Summers, Bedian, and Kauffman 1986) revealed a very large number of proteins which are actively translated in early development and fall silent later in many or all cell types. This pattern is not rare. Undoubtedly many of these proteins are uniquely required in early development, but a more subtle role of genes and proteins which fall quiescent in early development might be predicted.

2. Such fixed inactive genes include those which play a role in controlling later differentiation between cell types via *transient activation.*

3. Genes which are fixed active in all cell types are not merely constitutively active; although some might be constitutively active, many others are regulated by other genes in fixed active or fixed inactive states.

4. A function of such fixed active and inactive genes is to *isolate* the pockets of unfrozen genes which are loci of developmental choices.

5. Insofar as those pockets are isolated, different cell types should exhibit somewhat combinatorially different patterns of gene expression. However, since the unfrozen clusters can be tenuously interconnected in a hierarchial command structure such that alternative choices of live recurrent loops high in the hierarchial structure constrain the alternative choices of live recurrent loops lower in the hierarchial structure, the overall behavior may not be fully combinatorial. The failure of simple complete combinatorial behavior will also occur because more than one live-loop master-gene circuit may converge in the control of overlapping subsets of unfrozen downstream genes. Further data bearing on some of these features will be discussed shortly.

Cascading Effects of Transiently Activated Single Genes Are Limited. A further consequence of forcing structures is that the effects of altering the activity of any single gene are usually limited and do not propagate widely throughout the network (Kauffman 1974, 1984a; Fogelman-Soulie 1984, 1985a, 1985b; Stauffer 1987a). If the gene whose activity is altered is part of the forcing structure, the alterations typically propagate either not at all or to only a few descendant genes. This limited propagation is due to the fact that the forcing structure is typically multiply connected—that is to say, many genes are forced by several of their inputs. Transiently altering the activity of the gene does not alter the activity of the descendant locus, whose other forcing input remains in its forced and forcing value. Genes which are part of the isolated pockets not within the forcing structure may propagate influence to many or all genes in the isolated cluster. Members of the head loops can potentially influence all genes in the cluster. Members of descendant tails propagate influences to their fewer descendants. Numerical data show that, typically, alteration in the activity of a single model gene cascades consequences to 0 to 15 percent of the genes in $K = 2$ and canalyzing networks. This value is fairly close to that seen in higher eukaryotes. For example, in the puffing pattern in polytene chromosomes, ecdysone unleashes a cascade of 155 puffs among about 5000 bands, or in about 3 percent of the potential puff regions (Berendes 1966; Ashburner 1970). If we use an estimate of six for the expected number of functionally isolated islands in a genome containing 5000 genes and 80 percent for the frozen component, then the expected maximal size of puffing cascades in unfrozen islands should be about $(0.2 \times 5000)/6 = 166$. This number is remarkably close to the observed 155. Conversely, the data surely do not fit the expectations of genomic systems in the chaotic regime. There, cascades would propagate to thousands of other genes across the unfrozen sea (Stauffer 1987a, 1989).

The Effects of Deletion Mutations Are Limited. The effects of deletion mutants are similarly limited. If a gene which is active in some attractor cell types but inactive in one or more other cell types is deleted, the consequences of such a mutation include transformation from the former cell type(s) to the latter. We shall see that this transformation parallels homeotic transformations. Other effects include the formation of novel cell types—that is, attractors which did not exist before. The typical number of genes affected by the deletion of one gene ranges from 0 to about 15 percent (Kauffman, unpublished results). Alterations do not spread widely.

Comparable experimental data are available on deletion mutants in cell lines and on cell lines transformed with a tumor virus. Typically a small fraction of the proteins analyzed on two-dimensional gels alter their synthesis patterns (Strand and August 1977, 1978). Rarely, a larger number is observed (Lieberman, Hoffman-Lieberman, and Sachs 1980).

I stress again that these properties are not trivial. In chaotic $K = N$ networks, deleting any single gene will alter about half the state transitions in the dynamical system, dramatically altering all attractors. Any mutant will therefore propagate alterations widely throughout the network. Presumably, the factor which limits propagation in $K = 2$ and canalyzing networks is the forcing structure. This is clearly a *sufficient* condition to limit propagation, but recall the results of Derrida and Weisbush mentioned above and in Chapter 5, which show that high internal homogeniety alone can lead to frozen gene activity when the genes are present on a regular lattice. Thus there may prove to be principles other than forcing structures which limit propagation of alterations in genomic systems.

Forcing Loops and Structures Exist in Lambda, E. coli, *and Higher Organisms.*

Forcing loops and structures are not hypothetical, mathematical objects. They occur in known genetic circuitry. Consider again the now overfamiliar lactose operon. Recall that the structural genes Z, Y, and A include the permease A. Once lactose enters the cell and switches on the operon, the permease permits the operon to remain active in the presence of a *lower concentration* of external lactose than would have initially sufficed to induce the operon. Thus let us, as before, constrain lactose concentration to vary from this intermediate level to high concentration levels. In this case, the lac operon is a self-reinforcing feedback loop. Once external lactose concentration is high enough to activate the operon, the concentration may either remain high or fall to its intermediate value; in either case, the operon remains active, thanks to the induced permease. The active state is the forced value of this small forcing loop.

Another familiar example of a forcing loop arises in phosphorylation cascades, where a growth factor receptor which acts as a kinase and phosphorylates a downstream cascade of proteins must itself be specifically phosphorylated in order to be active as a kinase. Once activated by binding the growth factor, the receptor is able to phosphorylate itself. Thus once into the forced phosphorylated state, the system maintains that active state on its own despite absence of the original growth factor whose binding to an allosteric site induced the initial phosphorylation. The forced active state of such receptors can, however, be overridden by additional molecular inputs—enzymes called phosphatases, which dephosphorylate the receptor. This fact urges me to note that *conditional forcing structures* can readily be constructed. A Boolean function may be conditionally forcing on some inputs if some other input is held in a specific value but not be forcing otherwise. Little theoretical work has been done on the structure and behavior of nested conditional forcing structures.

Not only do forcing loops occur, but so do entire forcing structures. Figure 12.6*b* shows the regulatory cascade in lambda underlying the lytic response (Hershey 1971; Kauffman 1974; Ptashne 1986). Lambda is held as a lysogen, integrated into *E. coli* DNA, until the cell is damaged and then leads to the synthesis of cro. In turn, cro is the head of a forcing cascade which ensures sequential activation and repression of the lytic cascade, regardless of the presence of lambda repressor, C1, or other regulatory inputs acting on lambda in the lysogenic phase. Mating-type specific-gene expression in yeast, described above, is a forcing structure. It will be interesting to see whether the complex genetic system underlying sex determination in *Drosophila* (Baker, Nagoshi, and Burtin 1987; Gilbert 1988) turns out to be a forcing structure.

CELL TYPES AS A COMBINATORIAL EPIGENETIC CODE

The final theoretical implication of percolating frozen components is that the resulting functionally isolated islands which are the loci of alternative choices necessarily afford a more or less combinatorial character to cell types. Thus insofar as there are isolated islands in the genomic system, each with two or more alternative attractors, we can think of a cell type's attractor as being specified in a combinatorial code consisting in the alternative attractor choice made by each such unfrozen gene cluster. Thus a kind of epigenetic code is to be expected, in which each cell type is specified by a unique combination of choices made by the isolated clusters of genes. It is striking indeed, therefore, that very substantial evidence for such combinatorial epigenetic codes now appears to exist in *Drosophila* and elsewhere.

Lessons in Metaplasias: Uncommon Developmental Pathways Suggest an Underlying Binary Combinatorial Epigenetic Code in Drosophila *and* Caenorhabditis

Normal ontogeny unfolds along the normal branching pathways of differentiation. Metaplasias, by contrast, are rare deviations from those pathways, deviations in which one normal cell or tissue type transforms to a cell or tissue type proper to another region of the organism. To give a first example, crayfish will regenerate an antenna which is cut off. On occasion, however, a leg is formed in place of the antenna, arousing both social comment among the congeries of crayfish and proper interest among biologists. Such *homeotic regeneration,* as this example is called, and other metaplasias show the uncommonly expressed developmental opportunities which the genome affords. Darwin taught that anomalies are central in biology. The anomalies of metaplastic transformations hint that cells and tissues which normally develop along given pathways have unused neighboring pathways. Thus the *structure,* or patterns of transitions, along those unused pathways may reveal something about the underlying logic of development.

Nowhere are such metaplasias more abundantly known than among the arthropods and, among these, the Insecta, and among these, the best studied is the fruit fly, *Drosophila melanogaster.* The lessons drawn from *Drosophila* appear to extend to other organisms and will lead us to the nematode *Caenorhabditis elegans,* to humans, and to plants. Metaplasias in many of these organisms hint at an underlying combinatorial logic in ontogeny.

Drosophila *and Its Metaplasias.* *Drosophila* development passes through egg, larval, and adult stages. Adults (Figure 12.17) have a head; three thoracic segments called prothorax, mesothorax, and metathorax, each having a pair of legs; eight abdominal segments; and terminal genitalia. Three of the four major types of metaplasias occur in *Drosophila:* (1) homeotic mutants; (2) phenocopies of homeotic mutants in normal flies; and (3) transdetermination of a cell-heritable determined or committed state in certain organ cultures. The fourth, homeotic regeneration, is not well documented in this insect.

"Homeosis" originally was a term reserved for metaplasias in which parts normal to one body segment were replaced with parts normal to another segment, such as the replacement of antenna by legs. More generally, the term now means replacement of one normal body part by another essentially normal body part. Among the thousands of mutants which have been recovered in *Drosophila* (Lindsley and Grell

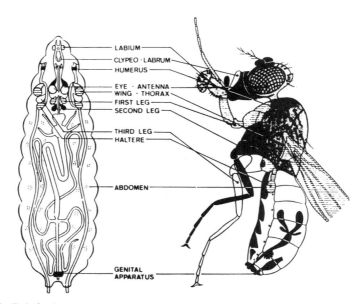

Figure 12.17 Relation between larval organs, called imaginal discs, and the adult epidermal regions to which they give rise during metamorphosis.

1968), a class of 40 to 50 are homeotic mutants (Gehring 1973, 1975, 1976; Postlethwait and Schneiderman 1974; Ouweneel 1976). An example is the dominant mutant *Nasobemia,* which transforms the antenna to a mesothoracic leg. Thus one finds flies with legs where antenna ought to be. *Nasobemia* sometimes also transforms the eye to a wing (Gehring 1966b, 1976). Similarly, *opthalmoptera* transforms eye to wing (Postlethwait 1974). Another surprising mutant is *tumorous head,* which produces growths in the head which are a transformation of head tissues into adult genitalia (Postlethwait, Bryant, and Schubiger 1972).

Homeotic mutants can be either dominant or recessive, requiring either one or both parental homologues to be mutant in order for the transformation to be expressed. Some homeotic mutants are point mutations; others are deletions of chromosomal regions; others are due to transposition of extraneous DNA into the normal, or wild-type, gene.

Phenocopies of homeotic mutants can be produced by exposing normal fly embryos to chemical physical perturbations during development. For example, normal flies have halteres on their metathorax and wings on their mesothorax. A famous cluster of homeotic mutants, called the *bithorax* series, transforms thoracic and abdominal segments into each other. In particular, the *bithorax* mutant transforms the anterior part of the halteres into wings (Lewis 1978, 1981). However, it is possible to expose 1- to 3-hour-old normal *Drosophila* embryos to heat shock or to ether and recover adult flies in which the halteres have been transformed into wings (Villee 1945; Gloor 1947; Maas 1948; Capdevilla and Garcia-Bellido 1974; Ho, Bolton, and Saunders 1983; Ho, Tucker, et al. 1983; Ho, Matheson, et al. 1987). The heat or ether treatment has *mimicked* the effect of the homeotic mutant. Nevertheless, the adult fly subjected to treatment is genetically perfectly normal and gives rise to progeny which are normal, not homeotically transformed. Thus the chemical or physical perturbation has caused apparently the same transformation as that caused by the *bithorax* series, but in a wild-type fly.

It is true—but perhaps a bit shocking—that it is also possible to phenocopy the *normal morphology* in a homeotic mutant. Thus *Nasobemia* can be made to yield progeny having antenna, not leg, where antenna ought well be, yet the subsequent progeny of that pleased, phenotypically normal fly will express the *Nasobemia* phenotype that has leg in place of antenna. A variety of other treatments, such as salts (Sang and McDonald 1954), can phenocopy other mutants, and many homeotic mutants have been phenocopied (Villee 1943, 1944; Woolf 1949).

Transdetermination refers to a change in the heritable committed state of saclike organs in the larva, called imaginal discs, from a commitment to mature into one kind of adult ectodermal tissue—for example, leg—to a commitment to produce instead a different adult cuticular structure—for example, wing (Hadorn 1967, 1978; Gehring 1973, 1975, 1976; Gehring and Nothiger 1973). Cells which shall form the imaginal discs are set aside early in normal development, either in the embryo or in the first larval instar molting cycle. During the three larval instar phases, each disc grows, reaching about 60 000 cells in the late-third-instar wing–mesothorax disc and 20 000 or so cells in the other discs. During metamorphosis, the entire ectoderm of the adult is formed by the terminal differentiation of these discs. During this differentiation, the cells change shape and neighbors; deform into legs, wings, antennae, head parts, and so forth; and secrete proper cuticle and pattern elements in proper spatial arrangement. The larva contains left and right discs committed to form eye–antenna; clypeolabrium; humerus; mesothorax and wing; metathorax and haltere; pro-, meso-, and metathoracic legs; and the bilaterally symmetrical genitalia. In addition, the ectoderm of the eight abdominal segments is formed by a burst of growth and maturation of small nests of cells, called abdominal histoblasts, in the walls of the larval abdomen.

Imaginal discs afford the cleanest demonstration both of a heritable committed state and of the operational separability of that committed state from terminal differentiation. A third larval instar disc—say, the wing–mesothorax (wing for simplicity) disc—can be dissected free and cultured as an organ in the abdomen of an adult female. In that environment, the disc cells divide but do not differentiate. Continued proliferation over years can be maintained by removing the disc from the adult abdomen, cutting the disc in half, and reinjecting each half into a different female host. At any stage, the capacity of disc tissue to differentiate into adult tissue can be tested by removing the disc from the adult abdomen and injecting it into a third-instar larval host. When the host undergoes metamorphosis, the molting hormone ecdysone acts on the injected tissue as well. In favorable cases, the resulting adult has in its abdomen a ball of cuticle formed by the injected disc tissue. This ball can be removed and studied.

The first, and still impressive, result of such studies is that wing disc tissue exhibits a cell-heritable determined state. When cultured for up to *ten years,* wing disc tissue can still give rise to adult wing structures (Hadorn 1967, 1978; Gehring and Nothiger 1973). Over ten years, the initial cells have undergone hundreds of divisions. Therefore, the commitment to form wing cannot be due to the synthesis, once and for all, of some special RNA or protein molecules. Such molecules would be diluted out over hundreds of divisions. Thus the determined state must be a heritable state which constantly regenerates itself over cell-division cycles. A second inference from this work is the operational distinction between this determined state, which can be indefinitely maintained, and terminal differentiation, which can be indefinitely delayed and then induced by ecdysone. While the astonishing heritability of the determined state is important, no less so is the pattern of its jump changes.

Transdetermination has occurred when cultured wing discs metamorphose into perfectly fine adult *antennae!* That is, the cultured disc tissue can jump to a new heritable determined state specifying another disc type adult tissue (Hadorn 1966; Nothiger 1972; Gehring 1976). Good evidence shows that this transdetermination jump is not due to a somatic mutation and that it is heritable once it has occurred (Hadorn 1967; Nothiger 1972; Kauffman 1973, 1975; Nothiger and Gehring 1973; Gehring, 1976). Figure 12.18 shows the transdetermination steps which have been observed, with arrow lengths representing relative frequencies. I confess that I have long found this figure fascinating. Notice four features. First, there are allowed and forbidden single-step transdeterminations. Thus genital can change to antenna or to leg but not to wing, eye, haltere, proboscis, or mesothorax; antenna can change to genital, eye, wing, or leg but not to mesothorax, haltere, or proboscis. Second, there are pathways of sequential transdetermination. Genital can change to wing but must do so by way of antenna or leg. Proboscis can change to wing, again by way of antenna or leg. Third, number each tissue with the minimal number of transdetermination steps needed to reach mesothorax. Thus mesothorax is zero steps; wing is one step; antenna, leg, eye, and haltere are two steps; and genital and proboscis are three steps. Each step *closer* to mesothorax is more probable than its inverse. Thus genital changes to antenna more readily than antenna to genital. Antenna changes to wing more readily than wing to antenna. It is as if mesothorax were a developmental sink and the remaining tissues tended to fall back into the sink.

The fourth feature of Figure 12.18 cannot be seen from what I have explained to date. However, we shall see in Chapter 14 that, on the early embryo, the positions of cells which shall give rise to different parts of the adult are well specified in a *fate map*. In that map, cells forming adult head are at the anterior end of the egg and cells forming adult thorax are somewhat posterior to those forming head. Next come cells which will form the abdomen, and finally, in the posterior, come the cells which will form the adult genitalia. In other words, the order of tissues in the adult is laid out in a homologous two-dimensional fate map in the early embryo. The fourth fascinating feature of transdetermination is that it leaps long distances over the fate map. Even a rank nonbiologist will appreciate that genitalia and head are at opposite ends of the fly—and the fate map. Yet genital discs transdetermine readily to form head. Why is this important? Because one would like to believe that the metaplasias revealed in transdetermination show which developmental programs are neighbors of one

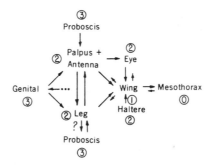

Figure 12.18 Transdetermination between different imaginal-disc-determined states in *D. melanogaster*. Arrows show observed transdetermination steps. Arrow lengths reflect relative probabilities of transitions. Dotted arrow into genital disc indicates that the transdetermination source to genitalia is not certain but is thought to be antenna. Circled numbers indicate minimum number of transdetermination steps separating a disc from the mesothorax determined state. (From Kauffman 1973)

another. In some sense which we want to understand, genital and antenna must have neighboring developmental programs, while genital and mesothorax do not. Then the implication of this jumping over the fate map is simple but very important: Whatever sense may exist to the phrase "neighboring developmental program," distant regions on the fate map can have neighboring programs. Here is a puzzle worth some struggle, for these metaplasias must be telling us something about the underlying logic of development in this slight fly.

The final point to make in this overview of metaplasias in *Drosophila* is that the different processes exhibit very much the same transformations (Ouweneel 1976). For example, *Nasobemia* converts antenna to mesothoracic leg, and eye to wing–mesothorax (Gehring 1966a, 1966b; Stepshin and Ginter, 1972). These are both transdetermination steps in Figure 12.18. *Tumorous head* transforms genitalia to antenna (Postlethwait, Bryant, and Schubiger 1972), and *eyeless opthalmoptera* converts eye to wing (Postlethwait 1974). *Bithorax* alleles convert haltere to wing and wing to haltere (Lewis 1978, 1981). The haltere-to-wing conversion has been seen in transdetermination, but not the reverse. Thus almost all the transdetermination steps are mirrored by some homeotic mutant, and vice versa. Similarly, the capacity to phenocopy homeotic mutants causes transformations which are also seen in transdetermination. The major difference between the homeotic mutants and transdetermination is that each mutant causes a restricted subset of the transformations seen in transdetermination and in fact a restricted set of the transitions from any one disc seen in transdetermination. Thus, *Nasobemia* converts antenna to leg and eye to wing but does not convert antenna to wing or to genitalia. It is important to comment here that *one* homeotic mutant can cause *parallel* transformations in more than one tissue. *Nasobemia* transforms antenna to mesothoracic leg *and* eye to wing. Another mutant transforms genitalia to antenna *and* haltere to wing. The implication of this will be brought out below.

It is truly impossible to review these data and not come to at least the tentative conclusion that all these metaplastic transformations are showing that specific developmental programs are neighbors of one another. The same neighbor relations are exhibited by all the forms of metaplasia in *Drosophila*.

A Binary Combinatorial Epigenetic Code. The most natural model to account for the features of transdetermination noted above posits that each imaginal disc is specified by a "binary combinatorial epigenetic code" (Kauffman 1973, 1975). First, I consider a nonpredictive (because noncombinatorial) model in the general context of cell types as dynamical attractors. An entire disc is not a single cell type, but let us relax a bit and think of a tissue as an attractor. Transdetermination steps occur between at least eight separate tissues. Suppose as the most general hypothesis that we conceive of a genomic system with eight different attractors. Let us assign one attractor to each disc type; attractor 1 to genitalia, attractor 2 to antenna, 3 to leg, and so on. Can we use this model to make any predictions about which disc transdetermines to which disc with what frequency? Clearly not. Having specified an arbitrary dynamical system containing eight attractors, one per tissue type, we are *entirely free* to imagine transitions between any of these attractors in any frequency distribution we wish. Thus consider an 8×8 matrix whose elements reflect the transition probabilities between attractors. The probabilities per row must add up to 1.0, but that is the only constraint. Our hypothesis is a mere universal, and hence empty, redescription.

Suppose instead we imagine that the determined state of each tissue is due to the combination of choices made by several independent bistable (binary) genetic feed-back loops. There are eight disc tissues; thus if we wish to imagine that the deter-mined state in one tissue is specified by commitments carried by feedback loops with only two steady states each, we need at least three independent loops. Then we shall specify the determined state in one disc, say genitalia, by the combination of states on the independent feedback loops. Consider, to be concrete, the four discs genital, antenna, leg, and wing. Let 1 stand for one of the steady states of a bistable genetic feedback loop, such as the im+ state of the C1–cro circuit in lambda (Neubauer and Calef 1970; Ptashne 1986; Reinerts 1987), and let 0 stand for the other steady state, such as the im − state of the C1–cro circuit. Then consider two independent feedback loops and specify genital as (00), antenna (10), leg (01), and wing (11) (Figure 12.19).

There are three immediate consequences of this *combinatorial hypothesis:*

1. Since each feedback loop is assumed to be independent and since transitions between the 0 and 1 states of each loop occur rarely but independently, it follows that the one-step transdetermination moves from each tissue reflect changes in only one of the two loops. Therefore, each disc can undergo only some of the possible transformations. Put another way, there are *allowed and forbidden one-step transitions.* Thus genitalia can change in one step to antenna or to leg but not to wing.

2. *Sequences* of transformations occur, by way of intermediates. Genitalia can change to wing via antenna or leg.

3. The consequence is subtle: In a bistable system, if one choice is more stable than the other, the less stable transforms to the more stable more readily than the more stable transforms to the less stable. This inequality implies a directionality toward a combination of choices on independent circuits toward a developmental sink with all circuits in their more stable state. This directionality is lost if the circuits have more than two alternative developmental choices. Consider a 2×2 matrix showing the transition probabilities between the two states, calling those states 1 and 0 (Figure 12.20). Again, each row sum of probabilities must add to 1.0. Therefore, if the 1 state is less likely to change to the 0 state than the 0 state is to change to the 1 state, it must follow that the 1 state is more likely to remain the 1 (more stable) state than is the 0 state. In short, because the feedback loop has only two steady states, if 1 is more stable than 0, then the 1-to-0 transition is less probable than the 0-to-1 transition. This constraint is missing in an 8×8 matrix because in that larger system there are too many degrees of freedom. The 2×2 matrix has only one degree of freedom per row. Let us, by convention, let 1 represent the

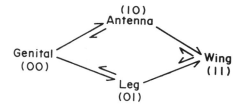

Figure 12.19 Four disc tissues, each assigned a determined state in terms of the combinations of steady states of two bistable circuits, 1 and 0. Transdetermination is assumed to require flipping one circuit to the other steady state.

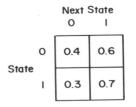

Figure 12.20   The 2 × 2 transition matrix between states 0 and 1. If 1 is more stable than 0, then 0 transforms to 1 more easily than 1 transforms to 0.

more stable state of each bistable feedback loop. Then the 0-to-1 transition is more probable than the reverse. Thus the third consequence of our binary combinatorial hypothesis is that, with the assignment of combinatorial code words to define the committed states in gentalia (00), antenna (10), leg (01), wing (11), it follows that each step toward wing is more probable than its inverse. In brief, an oriented flow toward a developmental sink emerges inevitably with a binary combinatorial code.

A binary combinatorial epigenetic code model makes a number of independent predictions. Notice that the circuit transition transforming genital to antenna, in the first of the two bistable circuits, is exactly the same transition as that occurring in the transition from leg to wing. Therefore, we must predict that if the gential-to-antenna transition is more probable than its reverse, it follows that the leg-to-wing transition is more probable than its reverse. The prediction is independent of the assumptions and is in fact true. Similarly, the same 0-to-1 transition in the second circuit which occurs in the genital-to-leg transition occurs in the antenna-to-wing transition. Thus if gential to leg is more probable than its reverse, we must predict that antenna to wing is more probable than its reverse. Again independent prediction, again true.

In Chapter 14, I present a detailed model for the sequential establishment of the combinatorial code for the different imaginal discs shown in Figure 12.21. The model is based on sequential compartmentalization, discussed in Chapter 14. As you can see, four bistable circuits are used (Figure 12.21*d*), more than enough for eight imaginal discs but not yet enough for the known alternatively committed subregions of the *Drosophila* embryo. The data which led to those particular binary epigentic code word assignments are independent of the predictions listed in Table 12.2, most of which derive from the postulate that, if transdetermination from *x* to *y* involves switching one bistable circuit while transdetermination from *x* to *z* involves the same switch for that circuit *plus* switching another circuit, then transdetermination from *x* to *y* should be more frequent than transdetermination from *x* to *z*. There are 37 independent predictions, and all but two are true. The *a priori* probability of such success is very small. I do not wish to draw the conclusion that the combinatorial code in this model is correct in detail; it cannot be with two false predictions. However, the very strong capacity to account for the ordering relations among the observed transdetermination transformations must be taken as encouraging support for some similar form of a combinatorial model. In fact, such a model receives substantial further support, as we see next.

A combinatorial epigenetic code model should also account for homeotic mutants. Destabilization of a master gene "choice" circuit might transform many disc tissues in parallel, and tissues which are transformed into one another by home-

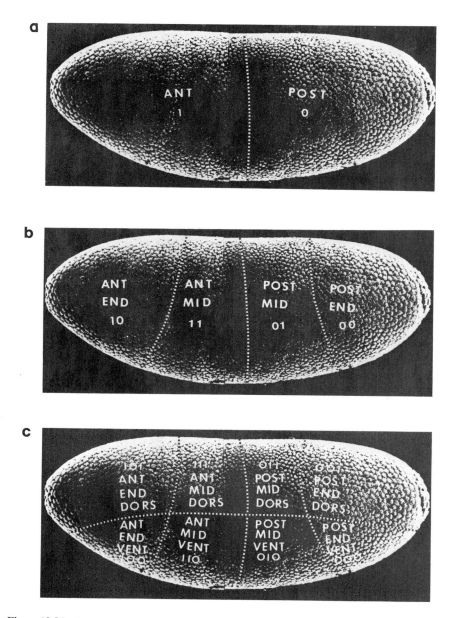

Figure 12.21 (*a–c*) Hypothetical sequence of compartmental boundaries subdividing the progressing blastoderm. Each boundary triggers a binary development commitment to one of two developmental fates.

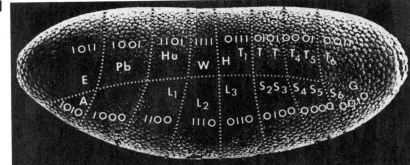

Figure 12.21 (*d*) The four-digit combinations show, for each compartment, a proposed binary combinatorial epigenetic code word specifying the determined state of each major imaginal disc. (From Kauffman, Shymko, and Trabert 1978). E = eye, A = antenna, Pb = proboscis, Hu = humerus, L1–L3 first to third leg, W = wing, H = haltere, T1–6 first to third abdominal tergite, S1–6 first to sixth abdominal sternite, G = genitalia.

otic mutants should be specified by neighboring epigenetic code words. In the simplest case, let us assume that a homeotic mutant acts on a single bistable circuit and destabilizes one of the two states, increasing the frequency of transition to the other. A concrete example would be a temperature-sensitive C1 mutant in lambda, which could be induced to switch to the im− state by high temperature. It follows that such a mutant will cause the affected disc to undergo a transition which is only one possibility from the larger set of transdetermination transitions possible from that disc. Thus a mutant changing the second circuit from 0 to 1 in Figure 12.21*d* will change antenna (1010) to mesothoracic leg (1110). This is only one of the transdetermination steps open to antenna, which can transdetermine to genitalia and wing as well

TABLE 12.2 Predicted Relative Transdetermination Frequencies Derived from the Model Shown in Figure 12.21*d*

Prediction	Status	Prediction	Status	Prediction	Status
$H \to W > H \to A$	T	$A \to W > A \to H$	T	$L \to W > L \to E$	T
$H \to W > H \to L_{1,2}$	T	$A \to L > A \to W$	F	$L_{1,2} \to W > L_{1,2} \to H$	T
$H \to W > H \to E$	T	$A \to P > G \to P$	?	$L \to A > L \to E$	T
$H \to W > H \to P$	T	$A \to E > A \to W$	F	$L_{1,2} \to A > L_{1,2} \to G$	T
$W \to A > H \to A$	T	$A \to G > L_{1,2} \to G$	T	$L_2 \to G > L_3 \to A$	?
$W \to E > H \to E$	T	$A \to E > E \to A$	T	$L_1 \to P > L_1 \to G$	?
$W \to L_{1,2} > H \to L_{1,2}$	T	$A \to L_2 > L_2 \to A$	?T	$G \to A > G \to P$	T
$W \to L > W \to A$	T	$E \to W > E \to H$	T	$G \to A > G \to W$	T
$W \to L > W \to G$	T	$E \to A > E \to G$	T	$G \to L_{2,3} > G \to W$	?T
$W \to A > W \to G$	T	$E \to A > E \to L$	T	$G \to A > G \to L_{2,3}$	?T
$W \to E > W \to P$	T	$E \to W > E \to L$	T	$G \to A > A \to G$	T
$W \to E > W \to G$	T			$G \to L > L \to G$	T
$W \to E > W \to A$	?			$G \to H > G \to W$	?F

Note: $L_{1,2} \to A > L_{1,2} \to G$ means that the model predicts that transdetermination from the first or second leg is greater to antenna than to genital.

A = antenna	L = leg	T = true
E = eye	P = proboscis	F = false
G = genital	W = wing	
H = haltere		

as to leg. Furthermore, the supposition that there is a combinatorial code of independent bistable circuits implies that, if the mutant changes the second circuit from 0 to 1 in antenna, it might also change the *same circuit* from 0 to 1 in any or all other disc tissues where the same circuit is in the same 0 state. Thus, in particular, it might change from 0 to 1 in the eye disc, transforming the eye (1011) to wing (1111). This is precisely the change due to *Nasobemia*.

The essential point to emphasize is that the hypothesis that there are *independent feedback loops* having more than one alternative state, say bistable, implies that each disc tissue is specified by a unique *combination* of the alternative states of each loop. Therefore, the same state of any one loop occurs in more than one disc. Therefore, if a mutant destabilizes that state and tends to convert it to another state of the loop, the destabilization might occur in all tissues where the weakened state of that loop occurs. Then the hallmark of combinatorial behavior is the occurrence of *parallel* transformations induced by the same mutant. On the other hand, full independence of isolated loops, whether functionally isolated or otherwise, is probably an idealization. Hence lapses from complete fulfilling of these stringent conditions mean not that the combinatorial picture has no correct pieces to it but that full independence may not exist.

As expected by a combinatorial epigenetic code, *Drosophila* is replete with parallel transformations caused by one mutant. *Nasobemia* has been mentioned several times. Another mutant (Shearn, Rice, et al. 1971) transforms genitalia to antenna and haltere to wing, and is captured in the 0-to-1 transition of the first circuit in Figure 12.21*d*. The *ultrabithorax* locus transforms mesothorax to metathorax and mesothoracic leg to metathoracic leg (Lewis 1978, 1981). The *engrailed* mutant converts the posterior half of each segment of the entire body into a mirror-image anterior half (Morata and Lawrence 1975, 1977, 1978). Thus *engrailed* acts in half-segment units in parallel throughout the body. A large number of mutants, called the *even-skipped* or *odd-skipped classes* and discussed in detail in Chapter 14, act in half the segmental tissues of the body.

If the combinatorial code in Figure 12.21*d* is at all close to correct, it should also be the case than the known homeotic transformation in *Drosophila* are between tissues which require the changing of only a single feedback loop from 1 to 0 or 0 to 1. That is, if each tissue has a binary combinatorial code word, then different tissues differ by one or more bits, and the known homeotic transformations should strongly tend to occur between tissues with neighboring code words that differ in a single bit. Table 12.3 shows the known transformations and the required bit changes. Most changes are either one or, rarely, two bits. Since a random code predicts a mean of two bits, this model is very much better than chance.

Evidence for Binary Decisions and Parallel Homeotic Transformations in Caenorhabditis elegans

Caenorhabditis elegans, a small nematode which has made major contributions to developmental genetics in the last few decades (see, for instance, Kimble and White 1981; Greenwald 1987), also shows evidence of binary developmental decisions and parallel homeotic transformations. The adult worm has only about 1000 cells, which arise in extremely precise branching lineages. Ablation experiments demonstrate that almost all cell fates are determined cell-autonomously, without reference to neighboring cells. Each lineage comprises sequential divisions in which a mother cell may give rise to two daughter cells which are either identical to or different from each

TABLE 12.3 Observed Homeotic Transformations and the Code Changes Required for the Code Scheme in Figure 12.21*d*

Mutant	Transformation		Coordination	Code change	Switches required
Antennapedia[1]	Antenna →	leg 2	—	1010 → 1110	1
Pointed wing	Antenna →	wing	—	1010 → 1111	2
Nasobemia	Antenna →	leg 2]	Parallel	[1010 → 1110	1
	Eye →	wing		1011 → 1111	1
Dachsous	Tarsus →	arista	—	1110 → 1010	1
Opthalmoptera[2]	Eye →	wing	—	1011 → 1111	1
Hexaptera	Prothorax →	mesothorax	—	1101 → 1111	1
Pedoptera	Wing →	leg		1111 → 1110	1
Tetraltera[3]	Wing →	haltere	—	1111 → 0111	1
Contrabithorax	Wing →	haltere]	Parallel	[1111 → 0111	1
	Leg 2 →	leg 3		1110 → 0110	1
Ultrabithorax	Haltere →	wing]	Parallel	[0111 → 1111	1
	Leg 3 →	leg 2		0110 → 1110	1
Tumorous head	Eye →	genital]	Parallel	[1011 → 0011	1
	Antenna →	genital]	Divergent	1010 → 0010	1
	Antenna →	leg		1010 → 1110	1*
Lethal(3)III-10	Haltere →	wing]	Parallel	[0111 → 1111	1
Lethal(3)XVI-18	Genital →	antenna]		0010 → 1010	1
	Genital →	leg	Divergent	0010 → 0110	1
Lethal(3)703	Antenna →	leg]	Parallel	[1010 → 1110	1
Lethal(3)1803R	Genital →	leg]	Divergent	0010 → 0110	1*
	Genital →	antenna		0010 → 1010	1*
	Haltere →	wing]	Parallel	0111 → 1111	1
Proboscipedia	Proboscis →	antenna]	Divergent	[1000 → 1010	1
	Proboscis →	leg		1000 → 1100	1*
Extrasexcombs[4]	Leg 2 →	leg 1]	Convergent	[1110 → 1100	1
	Leg 3 →	leg 1		0110 → 1100	2
Polycomb	Antenna →	leg 2		1010 → 1110	1
Lethal(4)29	Leg 2 →	leg 1]	Convergent	[1110 → 1100	1*
	Leg 3 →	leg 1		0110 → 1100	2*

Note: A set of homeotic mutants causing the same transformation is represented by one member. (1) Antennapedia, antennapedix, aristapedia, aristatarsia; (2) opthalmoptera, opthalmoptera, eyes-reduced; (3) tetraltera, metaplasia, haltere mimic; (4) extrasexcombs, reduplicated sex comb, sparse arista. (From Kauffman, Shymko, and Trabert 1978)

*More than one circuit must change to account for *all* the transformations due to this mutant.

other but both different from the mother; alternatively, the mother cell may give rise to two daughters in which the mother cell is reiterated in one of the daughters but not in the other. Occasionally, cell fate is determined by cell–cell interaction, and an *equivalence group* of cells can replace any ablated cell. A number of mutants have been uncovered which transform fates among pairs of cells. Many of these mutants convert a sister cell such that it becomes identical to its normally different sister. Mutants causing the reverse transformation are known as well. Each of these pairs suggests a binary-decision circuit underlying the different states in the pair. Note that the same binary behavior is expected in simple "live" decision loops in canalyzing networks. If such a loop is a one-input live loop with an even number of "no" elements, then it has two alternative steady states and is driven to either one or the other by proper deletion mutants.

The telltale signal for a combinatorial code, however, lies in evidence that the *same state* of a decision circuit occurs in more than one cell type, such that a single mutant converts a *set* of cell types to another set *in parallel.* In favorable cases, a second mutant should cause the second set to transform in parallel to the first set. The *lin*-12 mutant in *C. elegans* appears to be just such a gene. Let us characterize two cell fates, A and B, for each of a number of quite different pairs of cells. In the wild-type worm, both A and B occur. In a semidominant *lin*-12(d), both members of each pair adopt the A fate and so AA is formed. In the converse *lin*-12(o) allele, both members of each pair adopt the B fate, forming BB. It is now clear that the semidominant allele is an overproducer and the (o) allele is an underproducer of the gene product. In all, *lin*-12 controls homeotic transformations between 11 pairs of cell types in distinct lineages, drawn from four development stages and in quite different cell types—for example, neuroblasts and myoblasts (Greenwald 1987; summarized also in Gilbert 1988).

Lin-12 shows that at least one mutant in *C. elegans* acts in parallel in many cell types, causing reverse transformations between binary alternatives. Apparently, in each cell type, the same binary decision occurs but carries different implications in each as a result of other differences. In summary, *C. elegans* clearly has several binary-decision systems, each acting in at least one pair of cells and *lin*-12 acting in 11 pairs of cells. These data begin strongly to suggest that any cell type will be specified at least partially by a binary combinatorial epigenetic code word.

Metaplasias in Mammals

While best studied in *Drosophila,* startling metaplasias exist in many other organisms, including humans. Those in humans demonstrate largely reversible transformations among 14 tissues in two major sets, one belonging to endodermal germ layer derivatives, the other to mesodermal germ layer derivatives. The patterns of transformation parallel those expected of genomic systems in the ordered regime.

Homologues for homeotic mutants in humans and other mammals have long been doubted, but Slack (1985) has published an elegant account of metaplasias and heterotopias in humans. "Heterotopia" refers to the formation during embryogenesis of a patch of tissue proper to one region, *in situ,* in an improper position. "Metaplasia" here refers to transformations taking place postnatally and hence probably due to injury and repair. Slack notes that the ectoderm of humans is not very richly embued with distinct regional characteristics, as it is in the arthropods, and that much internal tissue is muscle, cartilage, or bone, whose exact regional specificity in a very small patch cannot be determined. On the other hand, the epithelial lining of the gut, urinary system, and female genital system is highly specific and the kinds of cell types which occur are highly distinct in distinct regions. Slack carefully disallows cases where endodermal epithelium characteristic of one region might be present in another region solely as a result of cell migration. He considers only cases where the ectopic tissue patch is well integrated into the surrounding epithelium and can be assumed to have arisen *in situ* by an altered behavior of the committed stem cells at that point. Figure 12.22 presents a global overview of the transformations seen. Four features of this directed graph of transformations are immediately apparent. First, each tissue transforms to only a subset of the 14 tissues. Second, sequences of transformations occur. Third, there are two *isolated basins,* joined by urinary bladder and ovaries. Within each basin, different tissues can transform to one another metaplastically. One of these basins corresponds to the endodermal germ layer and includes epithelia of the intestine, stomach, esophagus, pancreas, oral cavity, gall bladder, and nasal cavity. The second basin corresponds to mesodermal germ layer derivatives

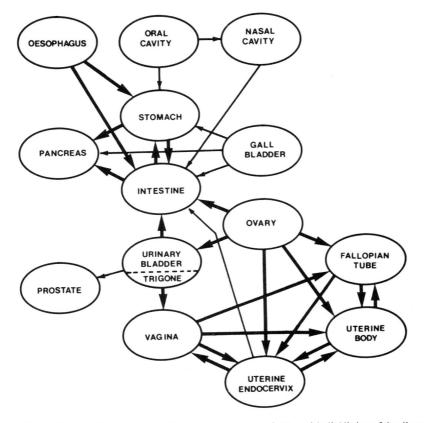

Figure 12.22 Graph of homeotic transformations in humans in the epithelial lining of the digestive, urinary, and female reproductive systems. An arrow from tissue A to tissue B means that patches of B epithelium can be found in the epithelium of A. Thick arrows denote relatively common events, and thin arrows denote very rare ones. Only the epithelial component of each organ is transformed. (From Slack 1985)

and includes the urinary and female genital systems. Within the endodermal basin, the pancreas is a developmental sink. No similar single sink exists in the mesodermal basin. Fourth, there is a tendency for directionality. Gall bladder and ovary, for example, can be left but cannot be returned to, and only four of the transformations are reversible, only one of these in the endodermal basin.

It is uncanny that the endodermal basin is almost identical to the *Drosophila* transdetermination graph (Figure 12.18) when the identities pancreas = wing, stomach = antenna, intestine = leg, gall bladder = genital, and esophagus = proboscis are used. These identities merely point to the same directed graph structure, of course. Thus a simple combinatorial code fits these phenomena. On the other hand, the mesodermal basin is far more complex. I can find no simple combinatorial scheme for this basin, although one may well exist.

What implications can be drawn from the metaplastic transformations of Figure 12.22? Recall the behavior of model genetic networks having a very large number of inputs per gene and random choice of Boolean function for each gene. In particular, reconsider the limiting case of $K = N$ networks. We know that such a system has N/e attractors. Since the successor to each state of gene activities is drawn at random from among the 2^N possible states, however, successive states along state-cycle attractors are not similar to one another at all. Thus if the network is perturbed off the state cycle by transient reversal of the activity literally of any single gene at *all states* along the state cycle, the chances approach 1 that the system will undergo a transition to each of the other state-cycle attractors for some of those perturbations. If all possible perturbations of any single gene on all states of the state cycle are considered, it is virtually certain that each state cycle can be transformed to all other state-cycle attractors. That is, it is characteristic of $K = N$ networks that each cell type can transform by some minimal perturbation directly to *all* cell types. This clearly is not what is occurring in the transformations see in Figure 12.22. Each cell or, more properly, tissue type can transform to only a few of the possible alternative types. This characteristic strongly implies that, if cell types are attractors, each has only a few other basins of attraction as its neighbors. These properties, as we have abundantly seen, are generic self-organized properties of genomic systems built with few inputs per gene and/or limited to canalyzing Boolean functions (Figure 12.12). It is well, therefore, to rest with the obvious features of the transformations in this mammalian endodermal epithelial system. All are explicable as natural consequences of the framework we have developed throughout this chapter: Cell types can be thought of as basins of attraction in the dynamical behavior of integrated genomic systems underlying ontogeny. Canalyzing genomic systems have the property that any cell type can differentiate into only a few cell types. These properties obtain not only in normal development, viewed along normal developmental pathways, but also in all the known metaplasias. That the properties obtain in the metaplastic transformations is extremely unlikely to reflect direct selection. It seems very much more probable that the property of having only a few neighboring cell types into which each cell type can differentiate is a fundamental consequence of membership in the canalyzing ensemble. That membership, not direct selection, ensures that ontogeny has been organized into branching developmental pathways for the past 700 million years or so.

SUMMARY

The genomic regulatory system can in part be characterized by its local features. These include the number of genes or processes directly regulating any given gene or process and the constraints which may exist on the ways the regulated locus responds. Almost without exception, known genomic systems have few inputs per gene. Almost all the genes are governed, in the Boolean idealization, by canalyzing functions. Constraint to a few inputs is the molecular equivalent of high molecular specificity, however, and constraint to canalyzing functions reflects chemical simplicity. Thus these basic constraints in large part reflect aspects of organic chemistry.

Asking what those local properties imply about the overall organization and behavior of large genetic regulatory systems led us to study ensembles of genetic systems whose members had the known local features. In fact, this entire ensemble spontaneously exhibits many highly ordered dynamical features which closely parallel known features of cellular differentiation.

The parallels uncovered are far from trivial. They allow us to make predictions about

- The complexity of recurrent patterns of gene expression corresponding to single cell types
- The distribution of cell-cycle times
- The number of cell types in an organism
- The homeostatic stability of cell types
- The ability of any cell type to differentiate directly into only a few other cell types
- The organization of ontogeny around branching pathways of differentiation

We were also able to predict that cell types share a core of expressed genes and differ in the expression of a small fraction of the genes, that hormonal or other signals propagate to a small and characteristic fraction of the genome, and that mutations propagate alterations to a similar fraction. We found evidence supporting the prediction that an organism uses only a subset of its cell types and evidence for a combinatorial character to alternative developmental commitments in the metaplasias of *Drosophila* and other organisms.

The ensemble approach has methodological, epistemological, and ontological implications. The genomic system underlying ontogeny is vast in its complexity. Complete analysis of any such system lies ever before us. Meanwhile, practical progress must use those features of this system which can be teased apart to build up an overall theory for the system's structure and behavior. Even were genetic regulatory systems fixed in structure rather than fluid over short evolutionary times, we would need to use some form of ensemble theory to state our current best conjectures of large-scale organization and behavior.

If there is a single image to bear away from these efforts, it is that random genetic programs, defying our earlier prejudices, can exhibit very great order. To ignore the possibility that such order underlies the order of ontogeny is simply foolish.

In Chapter 5 we found a "liquid" interface separating parallel-processing networks into three classes: solid, liquid, and gas. "Solid" corresponds to the ordered regime due to percolation of large frozen components via forcing structures or internal homogeneity clusters. "Gas" corresponds to chaotic dynamics having (1) exponentially long attractors as a function of the number of genes in the system and (2) spreading avalanches of alterations in gene activity patterns following minor perturbations. Both these hallmarks of chaos arise when no frozen component percolates. The liquid region is the boundary between order and chaos. All the data and models we have reviewed in this chapter are consistent with the hypothesis that genomic systems across phyla are in the ordered, solid regime, perhaps quite near the edge. Grant that genomic systems rich in canalyzing functions will tend to exhibit spontaneous order due to the percolation of frozen components. Self-organization interacts with selection at all times. Thus the hypothesis that selection tunes the detailed structure of genomic networks toward the edge of chaos is highly plausible. Such a poised state would yield stable attractors as cell types, combinatorial features to development, evolvable hierarchial command structures among the unfrozen regions, cascades of alterations in gene activity which propagated to some (but not a vastly large) number of other genes, and the capacity to evolve cell types by altering the behavior of single isolated unfrozen islands without propagating alterations to all regions of the genome. The testable intuition is that the genome, if poised near the edge of chaos, maximizes dynamically ordered complexity and evolvability.

This chapter has explored the strengths of an ensemble theory. If evolution is itself exploring an ensemble of systems, then the generic properties of that ensemble stand as null hypotheses to account for the order we see. We need merely explain membership in the ensemble in question. If it proves true that selection tunes genomic systems to the edge of chaos, then evolution is persistently exploring networks constrained to this fascinating ensemble of dynamical systems. The generic properties of this perhaps most important ensemble emerge as the best hypotheses to account for the remarkable order in organisms. If this overall view proves useful, then the manifold marriage of self-organization and selection consists in constraint of genomic systems to this well-wrought ensemble as a result of selection's achieving systems best able to adapt and the emergence of the generic properties of that ensemble as quasi-universals in biology.

CHAPTER 13

Selection for Cell Types

My purpose in this chapter is to explore some of the issues which arise in considering either the evolution of novel cell types or the potential "perfection" of existing cell types. In Chapter 2, 3, 5, and 12 we examined the character of adaptive walks on rugged fitness landscapes, Ashby's *Ansatz* for adaptive evolution in integrated dynamical systems, and the emergence of order in disordered Boolean models of genomic regulatory systems. In particular, in Chapter 12 we found that the spontaneous order in the canalyzing ensemble of Boolean networks closely mirrors a number of ontogenic features observed across many phyla. However, the "cell types" in such random networks are themselves random attractors in the dynamics of the regulatory systems. These preliminaries lead us to ask (1) whether natural selection can mold the genomic regulatory system in order to achieve particular patterns of gene expression in specific cell types and (2) whether and when the rugged character of adaptive landscapes may limit the capacities of selection.

Caveats are immediately in order. The evolution of cell types since the Cambrian has included the generation of new structural genes coding for new proteins with new or modified catalytic or structural functions. As described in Chapter 4, formation of such genes has involved duplication of structural genes and their subsequent divergence in gene families; it has involved recombinations that united functional domains from each of the proteins being recombined; it has involved local mutational search either perfecting a given catalytic or structural function or marching toward good performance of some neighboring function. I shall not discuss all these fundamental aspects of the evolution of cell types. Instead, I shall focus on a simpler but still fundamental subproblem: part of the evolution of cell types has undoubtedly involved alterations in the purely cybernetic aspects of the genomic system, such that different combinations of activities of the *same set* of structural genes come to exist as attractors. That is, the evolution of cell types has in part involved alterations in the regulatory system such that new cascades and new patterns of expression of the same structural genes arise.

Thus the fundamental question examined in this chapter is the extent to which selection can mold cell types by altering the cybernetic regulatory system controlling their conjoint activities. A major conclusion reached, echoing the results from Chapters 2, 3, and 5, is that selection is sharply limited by the ruggedness of the adaptive landscapes on which the system evolves. Selection, we shall see, almost certainly cannot mold the cybernetic system to achieve arbitrary patterns of gene expression in a given cell type; instead, adaptation becomes trapped on local optima. A second major conclusion, however, is that such trapping is less of a hindrance in genomic regulatory systems with few inputs per gene than in those with many inputs per gene. That is, the same class of model genetic regulatory systems which best captures the behavior of known genomic systems simultaneously appears best tuned to adapt well. Such networks lie in the ordered regime and hence adapt on highly correlated landscapes. Conversely, networks with many inputs per element typically lie in the chaotic regime and adapt on very rugged landscapes. We shall therefore have to ask whether the occurrence of such ordered cybernetic systems in organisms may not reflect "second-order" selection for the class of regulatory systems which adapt well.

The results critical to this chapter were discussed in Chapter 5 in the general context of the adaptive capacities of Boolean networks and draw on work done with my colleague R. Smith (Kauffman and Smith 1986). Here I present the broader context into which those results fit and briefly recapitulate the findings of Chapter 5.

THE FRAMEWORK

As we noted in Chapters 1 and 2, chromosomal mutations which move *cis*-acting regulatory sequences to novel positions adjacent to structural genes literally alter the regulatory wiring diagram of the genomic system. Similarly, either point or grosser mutations readily alter the behavior of a regulated locus. The simplest examples include mutation of an operator locus to a constitutively free state (Jacob and Monod 1963; Monod 1971; Vogel 1971; Zubay and Chambers 1971). Examination of the capacity of selection to mold cell types requires, at the level of theory, an appropriate abstraction of our knowledge of genuine genomic regulatory systems. Given such an abstraction, our reasonable aim must be to understand some of the major issues which arise in the evolution of coordinated patterns of gene expression. Later, guided by those insights, we may turn back to experiments. Therefore, I shall persist in utilizing Boolean-network models of genomic regulatory systems and in using the central interpretation of a "cell type" as a dynamical (state-cycle) attractor in the repertoire of a given genetic network. Given this identification of a cell type, the framework for this chapter is to examine whether selection can act on a population of mutating genomic regulatory systems (in which connections and Boolean rules assigned to individual genes are altered) and can thereby achieve genomic regulatory systems which have *desired* patterns of gene expression as the dynamical attractor cell types of the system.

Recall from Chapter 5 the similarity of this question to Ashby's (1960) *Ansatz*. For Ashby, adaptation is based on identifying a subset of essential variables in a system. Those variables must be kept in bounds by the coordinated dynamical behavior of the system coupled to its environment (called the System). In any initial state, the System flows to some atrractor. On that attractor, the essential variables either are or are not kept in bounds. In the former case, Ashby alters nothing. In the latter case, he in effect introduces a jump mutation in some parameter setting, thereby altering basins of attraction. With the new basins, the System may flow from its current state, with some essential variables out of bounds, to a new attractor which keeps all essential variables within bounds. If so, Ashby stops the adaptive process. If not, he allows another step change in some parameter to the system. Clearly, if we are examining the extent to which selection can alter regulatory connections and Boolean rules in a mutating population of genomic regulatory systems in order to achieve some desired pattern of gene expression as a dynamical attractor, we are embarked on a version of Ashby's *Ansatz*.

Another caveat is in order. What kind of question is being posed in asking whether selection can mold a model genomic regulatory system to achieve some desired pattern of gene expression as a dynamical attractor? Obviously we are in no position yet to guess which particular patterns of gene expression may be useful, nor do the Boolean models we shall consider assign real properties to individual genes. The underlying task of interest is to *characterize the adaptive landscape for genomic regulatory systems* such that, *whatever* the patterns of gene expression which might be advantageous, we shall understand the extent to which selection for such advantageous patterns is either aided or limited by the statistical features of the adaptive landscape underlying the evolution of cell types. For this purpose, it is fully legitimate to choose any arbitrary pattern of gene expression as a "good" cell type, ask whether selection can begin by operating on an arbitrary genomic network, and ultimately achieve a network exhibiting the desired target pattern of gene expression as an attractor.

I note that efforts to evolve parallel-processing networks to have attractors with desired properties follow in a tradition explored by Fogel, Owens, and Walsh (1966), who were among the first to attempt to apply evolutionary methods to evolving computer programs with desired behavior. More recently, Lenat (1977, 1980) has studied the capacity of evolutionary procedures to prove theorems and has suggested heuristics for biological adaptation. In these cases and in our current context, a major issue is to understand the structure of the fitness landscape and the limits this structure imposes on adaptive evolution.

GENOMIC NETWORK SPACE

In Chapters 2, 3, and 5 we discussed the concept of an ensemble of entities and their arrangement in a high-dimensional space such that each entity was next to its one-mutant neighbors. In the simplest case, an entity is a peptide of defined length and its one-mutant neighbors are other peptides of the same length but differing by a single amino acid. In the case of NK Boolean models of genomic regulatory systems, the ensemble consists of all networks having N binary genes and K inputs per gene. Each such network is a one-mutant neighbor of all networks differing in a single regulatory connection or single Boolean function. More precisely, any regulatory connection can be thought of as an arrow from the regulating gene or process to the regulated gene or process. Thus either the regulating or the regulated gene might be altered. A Boolean function of K variables assigns a 1 or 0 value to each of the 2^K possible combinations of activities of the K input variables. Thus we can define the one-mutant neighbors of any Boolean function to be those which change a single bit in the 2^K positions either from a 1 to a 0 or from a 0 to a 1.

Consider a genomic system with $N = 1000$ and $K = 2$. The number of regulatory connections is $KN = 2000$, and each is subject to mutation in either the regulating or the regulated gene. If the origin or termination of any regulatory connection can be changed to any other connection, there are 2×1999, or 3998, one-mutant neighbors with respect to regulatory connections. Similarly, each Boolean function of $K = 2$ inputs has four bit positions, and hence each admits of four neighboring Boolean functions. Thus any such NK network has 4000 one-mutant neighbors with respect to Boolean functions. Therefore, we may consider two one-mutant spaces: that induced by alterations of regulatory connections and that induced by mutation in Boolean functions. It may or may not be the case that the adaptive landscape is statistically similar for these two distinct "move" generators. The full space, of course, allows both kinds of mutations, so that, in the full space, each network has almost 8000 neighbors.

In Chapters 2, 3, 4, and 5, I discussed the idea of an adaptive walk which is constrained to start at some initial entity and pass via *fitter* one-mutant or *j*-mutant variants to a local optimum. Such an optimum is any entity which is fitter than all its one-mutant or *j*-mutant neighbors. We saw that the one-mutant case allows us to characterize the statistical properties of the adaptive landscape in a constant environment (such that the fitness of each entity is constant in time). Those properties include number of local optima, expected lengths of adaptive walks to local optima, number of alternative local optima accessible from any initial entity, similarity of local optima, and fitness of the accessible local optima with respect to the mean fitness of entities in the space. It is natural, in assessing these statistical properties, to idealize the adaptive behavior of a real population driven by mutation and selection to a limiting case in which mutation frequency is low relative to population size and fitness differentials. Then, on a slow time scale, fitter variants are encountered. Any such variant either dies out or, once above a rough threshold frequency in the population, invades the population essentially deterministically and on a fast time scale. Recall that Gillespie (1983, 1984) has shown that such a process can be considered a continuous-time, discrete-state (here "entity") Markov process. The adaptive process first is "resident" at a single entity over long time periods and then jumps probabilistically to any of its fitter one-mutant neighbors on a rapid time scale. Thus this limiting case of low mutation rate samples the one-mutant fitness landscape in question and proceeds via one-mutant fitter variants to or toward local optima.

I simplify this behavior of an adapting population in the models below and consider an adaptive process in which, at each generation, all copies but one of the current fittest network are subject to a single random mutation in connections or in Boolean function. Each copy is examined to assess whether or not it has a state-cycle attractor best matching a predetermined target pattern of gene expression. The entire population of networks then moves in a single step to the fittest variant, and the adaptive walk iterates from that new network. For concreteness, we may normalize the fitness of any network to lie between 0.0 and 1.0 by taking as its fitness the *fractional match* between its closest attractor and the predefined target pattern of gene expression.

It is important to stress that it is trivially possible to build model genomic networks in which any single desired pattern of gene expression by N genes is achieved as a stable steady-state attractor. It suffices to assign the Boolean function "Tautology," corresponding to constitutively active, to any gene which is to be active in the target pattern; similarly, it suffices to assign the Boolean function "Contradiction" to any gene which is to be inactive in the target pattern. There may of course be very many other genomic networks having any specific single target pattern as a steady-state attractor. Thus failure of a selection regimen to achieve networks capable of matching a single desired target pattern cannot be due to a mathematical impossiblity of constructing such networks. By contrast, it might be the case, were we to seek networks which had some number of predefined target patterns as steady-state attractors while the class of networks under selection were constrained to, say, $K = 2$ inputs per gene, that such requirements were mathematically inconsistent. In the analysis below, therefore, I limit attention to cases where selection attempts to find networks matching a *single target pattern of gene expression.*

Adaptation in the Space of Genetic Networks

The results reported in Chapter 5 show the following:

1. Long-jump adaptation confirms the Universal Law. In Boolean networks, a long jump consists in mutating half the connections in the networks, or one-quarter of the bits in the Boolean function of the networks, and then assessing the population for fitter variants. In the long-jump case, the expected waiting time to find the next improved variant doubles after each improvement step. Further, the cumulative number of improvement steps S increases as a logarithmic function of the number of trials, or generations, G:$S = \log_2 G$ (Figure 5.11).

2. In long-jump adaptation the *complexity catastrophe* described in Chapter 2 arises. As the complexity of the network N grows larger, the mean fitness attained at any fixed generation *declines* toward the mean fitness in the space of genetic networks, 0.5. This relationship implies that, as genomic systems grow larger, adaptation by long-jump searches in genomic space becomes an increasingly poor strategy because such searches jump beyond the correlation spaces of the landscapes.

3. Adaptation via one-, two-, or five-mutant variants in $K = 2$ and $K = 10$ networks (Figures 5.12 and 5.13) exhibits six general features:

   - Adaptation never reaches the global optimum of a perfect match to the desired target pattern of gene activities.
   - The waiting time to find fitter variants increases as fitness increases.
   - As the mutant search range increases from 1 to 5, the fitness attained at local optima increases. Searching is therefore optimized by tuning the search range to the structure of the landscape.
   - In $K = 2$ networks, the fitness attained at local optima *falls* as N increases (Table 5.5). Thus even adaptation via fitter one-mutant neighbors confronts the complexity catastrophe. One expects that, as N increases, the fitness attained at local optima will dwindle to nearly that found in unselected networks in the space of genetic networks.
   - The correlation structure of $K = 2$ landscapes is much smoother than that of $K = 10$ networks (Figure 5.14).
   - If the activities of only a subset of the N genes are considered important and measured against a target pattern of desired gene activites while the rest of the N genes are hidden variables not directly affected by selection, selection still does not achieve the global optimum of a perfect match for the subset. The fitness level attained seems roughly the same regardless of the fraction of important and hidden variables.

These results carry a number of important implications. Foremost, in this limit of strong selection relative to mutation rate, where the population can always move "uphill' to the fitter variant, the general failure of adaptive walks via fitter one-mutant variants to find networks which match a target pattern of gene expression strongly suggests that real natural selection cannot (often or at all) mold genomic systems to achieve arbitrary desired patterns of gene activities as cell types. Instead, trapping of the adaptive walk on local optima prevails. The results go further. The multimutant cases also fail to reach networks with a fitness of 1.0 in 100 generations. The number of neighbors having two or five mutations is extremely large. Because all neighbors cannot be searched, we cannot be sure that any given network is a local optimum with respect to all two- or all five-mutant neighbors. Thus although multimutant search does not attain perfect networks, it remains conceivable that, in a sufficient number of generations, a two-mutant or five-mutant search might achieve a perfect network. It seems unlikely, however.

Thus for complex genomic regulatory systems, in the limit of long-jump adaptation and in the limit of one-mutant adaptive walks, it seems virtually certain that even strong selection cannot begin with an arbitrary genomic regulatory system and follow adaptive walks to achieve precisely defined cell-type patterns of gene expression. It remains conceivable, but does not appear too hopeful, that either an adaptive process with an intermediate search range or one with a spectrum of mutants per individual might typically be able to achieve networks having any arbitrary desired pattern of gene expression as attractors. Only further mathematical work will answer this question.

In short, we must begin to reckon with the fact that the adaptive landscape in which the cybernetic aspects of the genomic system evolve is very rugged, reflecting the fact that such networks pose complex combinatorial optimization problems. Adaptive walks in the simplest case of constant fitness landscapes and very strong selection march uphill to a local peak which, generically, is substantially below the global optimum fitness.

The second general result is that the rate of finding fitter variants slows, implying that the number of fitter j-mutant variants decreases as fitness increases. Thus adaptive walks can in principle branch in different directions, but the rate of branching decreases as fitness increases. This relationship very strongly implies that, from an initial state of moderate fitness and in a fixed fitness landscape, strong selection can follow adaptive walks to a fairly large number of *alternative local optima.* Multiple, if suboptimal, solutions to the cybernetic task of coordinating gene expression are to be expected. As the fitness of the initial entity increases, the number of alternative optima accessible decreases rapidly. Highly fit entities can walk only uphill to a single local optimum. Thus bushy radiation from initial genomic systems of moderate fitness which move toward alternative solutions to the adaptive problem and quiets to stasis as fitness increases in a fixed fitness landscape is essentially built into the structure of complex combinatorial optimization processes.

A third feature of adaptation is that, as the mutant search range increases from 1 to 5, the fitness achieved increases. While increasing the range from 1 to 5 helps the adaptive process, increasing it very far toward the long-jump limit hurts. Intuitively we know that the system is jumping beyond local good hills and wasting time searching far away. This constraint is related to the concept of an optimal search range, tuned by the rate of uncovering fitter variants, which can be readily obtained by assuming that the *mutation rate is heritable.* Then if it currently pays to search farther away, such mutants will find fitter variants and carry the high mutation rate with them. Alternatively, if it currently pays to search nearby, then entities with low mutation rates will find fitter variants faster and will carry the low mutation rate with them. A heritable mutation rate in principle allows selection to tune the search range to the statistical features of the adaptive landscape as they are encountered. As noted in Chapter 3, this is a mechanism which is familiar in evolutionary biology and even in engineering applications of mutation selection search procedures but whose use in evolution has been doubted by serious evolutionary biologists (Crow 1987).

The fourth important feature is that even $K = 2$ networks suffer the complexity catastrophe. Because they adapt on more correlated landscapes than do $K = 10$ networks, however, $K = 2$ networks appear to suffer that catastrophe more gradually. The complexity of $K = 2$ networks was varied from $N = 20$ genes per network to $N = 100$ genes, and the adaptive process was restricted to one-mutant search. The salient result is that, as N increases, the fitness achieved decreases from 0.90 for $N = 20$ to 0.78 for $N = 100$, a decrease which is not as great as in the long-jump adaptation, where the fall is from 0.88 to 0.67 as N increases from 20 to 100. The fact that limitation of selection to a subset of important genes while the rest remain hidden variables does not help attain higher optima strongly suggests that, as N grows, the match of *any fixed "important" fraction* of the N to a desired pattern of activity will also fall toward 0.5.

These results imply that, however adaptation is carried out—via fitter neighboring mutants or via long-jump adaptation—as the complexity of the genomic cybernetic system under selection increases, the expected capacity to match even a single desired target pattern of gene expression falls to the match expected by any randomly chosen unselected network (0.5). The complexity catastrophe limits selection.

In Chapters 2 and 3, we examined the NK family of landscapes and found that, for $K = 2$, the fitness of optima attained appeared to be independent of N. On those landscapes, the complexity catastrophe could be completely avoided as long as K remained fixed small as N increased, perhaps even if K increased slowly as N increased. In the present case, we are considering the dynamical behavior of disordered Boolean switching networks. The results suggest that, even for the $K = 2$ case where order emerges spontaneously, the complexity catastrophe still arises.

If these results are generally applicable, they suggest that attaining specific coordinated patterns of gene expression becomes progressively harder as N increases. Truly large genomic systems may therefore be constrained to cell types in which, at best, the activities of rather *few* genes can be prescribed with precision. Guessing ahead to the capacity of selection to build specific patterns of gene expression in a number of *different* cell types in one organism, we would expect that, in any one cell type, a modest number of genes might have their activity specified with precision, but those same genes might well express themselves "adventitiously" in a number of the other cell types of the same organism. We return below to evidence which suggests that this may well be true.

The fact that both genetic networks of low connectivity and those constrained to utilize canalyzing Boolean function spontaneously exhibit a highly correlated adaptive landscape, while high-connectivity networks do not, raises the very important possibility that the current apparent prevalence of regulatory systems of low connectivity and/or rich in canalyzing functions may reflect selection itself. If it is the case that low-connectivity networks are better able to adapt and can climb through fitter one-mutant variants to rarer optima in spaces of complex genomic systems, then it requires no group selection argument to note that natural selection will tend to enrich populations founded by entities which, thanks to the correlation structure of their adaptive landscapes, can better adapt. Therefore, we are forced to confront the idea that selection can *alter the very ensemble* of genomic systems in which adaptive evolution is occurring. Therefore, as discussed in Chapter 5, we must conceive of a kind of *second-order theory:*

1. What kinds of systems—genomic, protein, hierarchial—have correlated landscapes which circumvent in part the general tendency for local optima to fall toward the means of their spaces as the complexity of entities under selection increases?

2. Since in coevolution or in the face of high mutation rates, maintained fitness can be higher on more rugged than on less rugged landscapes, optimal adaptation may typically require a compromise between smooth and rugged landscapes. What kinds of genomic systems match this optimum ruggedness of landscape structure?

3. When can Darwin's selection, acting predominantly on individuals, yield organisms having the property that their mutant landscapes are well correlated? Protein domains, low-connectivity genomic networks, and hierarchial construction in organisms, modular in each case but in different ways, all suggest that selection reliably molds entities to have "good" landscapes. We obviously have much to learn about this problem.

Can Recombination Aid Adaptation in Regulatory Networks?

In Chapters 3 and 5, we considered the power of selective hill climbing based on recombination rather than mere accumulation of single advantageous point mutations. In protein space, for example, recombination allows new combinations of functionally independent domains on two initial proteins to be joined in a single mutational step. By contrast, achieving this on any amino acid substitutions would require that many point mutants accumulate in either parental protein alone. Recall that the usefulness of recombination almost certainly depends on the ruggedness of the fitness landscape. For a fully uncorrelated landscape, it seems unlikely the recombination is a useful strategy. For recombination to be useful, it is plausible that it must be possible to marry two partial solutions to a task with reasonable chances that the recombined trial solution is as good as or better than the initial partners to the recombination event.

Recombination certainly occurs in real genetic systems. Insofar as the genomic systems in two parents are not identical, such recombination may generate an offspring system whose detailed wiring diagram and logic differ from those of either parent. Indeed, presumably this differing from one generation to the next is an utterly common phenomenon. Why does it not lead to chaos? Presumably, such recombinational alterations in the genomic system do not often drastically alter the system's behavior.

No studies have yet been carried out with $K = 2$ networks in the ordered regime or $K = 10$ networks in the chaotic regime to test whether recombination can be a useful strategy for adaptive hill climbing toward desired patterns of gene activity. However, I believe it is reasonable to anticipate that such recombinational moves are more likely to be useful for $K = 2$ networks than $K = 10$ networks. I anticipate this on two general grounds. First, $K = 2$ networks adapt on more correlated landscapes. All alterations yield, on average, smaller effects on dynamical attractors than in $K = 10$ networks. Further, in $K = 2$ networks in the ordered regime, a large fraction of the genes fall to a fixed active or inactive state. This frozen core isolates islands of genes which influence one another within an island but do not influence genes in other islands. Consider recombining two similar genetic networks having a large overlap in the frozen component. Break each system into two fragments, and glue the complementary fragments together to form the recombined system. If the broken regulatory connections fall largely in the frozen component, the rejoins will hardly alter the dynamical behavior of the recombined system; only recombined connections to the isolated islands will have marked influence. Thus such recombination events will not create mass havoc. In contrast, the same operation on two fairly similar chaotic systems having $K = 10$ inputs per gene and no frozen component can be expected to alter dynamical behavior drastically. Therefore I surmise, but do not know, that recombination may often be useful in $K = 2$ networks or, more broadly, in networks in the ordered regime which result either from percolation of forcing structures or from networks which achieve frozen components by virtue of high internal homogeneity in their Boolean functions.

Population Selection as Complexity Increases: The Error Catastrophe Again

The simulations and analysis I have reported concern the character of adaptation only where selection is always strong enough to pull an adapting population to the next fitter variant. The structure of the fitness landscapes for $K = 2$ and $K = 10$ Boolean networks clearly is multipeaked and quite rugged. The flow of an adapting population across such landscapes as a function of population size and mutation rate remains to be examined. Nevertheless, experience with the NK family of landscapes makes it clear that, as the mutation rate increases—or, for a fixed mutation rate, as the complexity of the system increases—populations will eventually suffer from the error catastrophe. Selection will become too weak to hold the population at or near local optima. Thus the population will flow down from the peaks and across more or less vast tracts of network space among large numbers of nearly neutral mutants.

Conclusions and Caveats from the Theory

The conclusions to be drawn from even these preliminary models are these:

1. Complex genomic systems adapt on rugged fitness landscapes which have many local optima. It is very unlikely that selection for any arbitrary overall pattern of gene expression as a cell type can achieve such a goal. Adaptation arrests on optima far short of any such goal. While the image of selection trying to achieve a specific arbitrary pattern of gene expression is obviously improper, the conclusion is not. It asserts that, whatever may be advantageous, trapping on local optima will block attainment of arbitrary useful patterns of gene expression.

2. As the complexity of genomic systems increases, long-jump adaptation alone would lead to systems whose attainable fitness falls toward mean properties of the ensemble of systems under selection. The same holds true as complexity increases, if the fitness landscape is sufficiently uncorrelated.

3. These limitations can be delayed only if the landscape is sufficiently well correlated. Genomic systems which have few inputs per gene and those which have canalyzing functions appear to meet these requirements. Both have small attrac-

tors, and hence both exhibit homeostasis and thereby also adapt on correlated landscapes. While systems having few inputs per "gene" slow the rate of onset of the complexity catastrophe, they cannot avert it. As the genomic system becomes more complex, it appears that the match of any fixed fraction of genes to a desired target pattern falls toward chance levels.

4. It remains possible that, as N increases, an ever-larger number of genes can be selected to match a target pattern but the number which matches increases less than linearly as N increases. If so, then larger genomes could specify with precision the activities of more important genes, constituting a smaller fraction of the genome.

5. Our analysis has been limited to the oversimple case of selecting on Boolean networks to achieve only a *single* target pattern of gene expression as an attractor. If attractors are the proper model for cell types, and if an organism has several hundred cell types, then selection for specific patterns of gene expression on each cell type must juggle the joint requirements of this regulation. Even selection of a single attractor to match a single target pattern meets with very limited success, and even such modest success may degrade drastically if selection must attain a single genomic system which has many attractors each matching a different desired and arbitrary pattern of gene expression.

6. Boolean networks in the ordered regime but near the edge of chaos may be those having both the capacity to perform the most complex computations and the optimal landscapes for adaptive evolution. Because most mutants cause minor changes in behavior but some cause drastic changes, such networks live on landscapes which are typically smooth but harbor directions of sharp change. These features allow such networks to rapidly alter behavior in drastic ways if the environment or task changes rapidly.

7. Selection may be able to act in a second-order way to achieve systems which adapt on properly correlated fitness landscapes. Such action does not require group selection; it requires only that entities with properly correlated landscapes more often find fitter variants and hence are selected and carry with them those constraints which ensure a correlated landscape.

8. Molecular specificity and simplicity appear to be sufficient conditions to achieve molecular dynamical systems—genetic, autocatalytic, antiidiotype networks—of low connectivity. These molecular features would be expected to be selected for by virtue of the increased catalytic and recognition efficiency due to that specificity. Thus selection for molecular efficiency will purchase, at the same time, both globally ordered dynamics and integrated systems able to adapt. This duality should be a strong part of the answer to Dyson's (1985) request for a theory of emergent homeostasis.

9. Selection for cell types is limited not only by the abundance of local optima but, as complexity increases, by the encroaching failure of selection to overcome the effects of mutation. We recall from Chapters 3 and 6 that, in the face of high mutation or in the face of nonfixed landscapes resulting from environmental changes or coevolution, high maintained fitness may require the entities to adapt on more rugged landscapes. Such landscapes may have lower peaks than smoother landscapes do, but the gradients toward the peaks are steeper, thus helping to offset mutational forces or landscape deformation. Thus there may be an optimal landscape ruggedness which preserves attainable high optima as complexity increases

but has steep enough loss of fitness in enough directions to allow selection to hold entities at such high optima in the face of mutation. In short, additive fitness, or gently rounded optima, yield the error catastrophe as complexity increases: Mutational forces overwhelm selection forces. Conversely, very uncorrelated landscapes confront the first complexity catastrophe: Attainable optima fall toward the mean of the ensemble as complexity increases. Selection may thread a path between these new Scylla and Charybdis in constructing systems whose landscapes are rugged in the "right way."

EXPERIMENTAL AVENUES

Two lines of thought spring to mind when we consider ways of testing our hypotheses. The first revolves around two questions: Are genomic regulatory systems likely to be highly variable? Is such variability open to experimental verification? If selection for regulatory network structure and logic is based on additive models, high variability seems inevitable. In this case, selection in the face of known mutation rates appears unlikely to hold adapting populations very close to possible local optima. In Chapter 11 we used an additive model and calculated that, based on estimated rates of chromosomal mutations moving *cis*-acting regulatory loci, a genomic system with on the order of 25 000 regulatory connections would have about 10 percent "incorrect" with respect to single global optima. This result should carry over to accumulation of mutants affecting the regulated behavior of genes and thus their Boolean functions in the binary on–off idealization of gene expression. If each of 100 000 genes is regulated by a Boolean function on some regulatory inputs and if mutations modify such rules at about 10^{-6}, then the same argument leads to the expectation that 25 percent or more of such rules should be incorrect with respect to spaces with single global optima. If so, then the regulatory system should drift and vary considerably.

The additive-landscape idealization that underlies such a calculation is, of course, unrealistic. More sensible work is needed—as a function of mutation rate, landscape ruggedness, population size, and extension to diploid models—to achieve reasonable insight into the expected precision in the genomic architecture and logic. Nevertheless, a plausible first guess is that considerable variability should be found.

This guess should be directly testable by exploration of the variability in the regulatory architecture and behavior of actual populations. Proper experimental approaches to this line of testing can be based on analysis of tissue-specific patterns of protein expression or gene expression in different isolates of one inbred strain of a species or in sibling species.

Analysis of sibling species is showing very high regulatory varibility. Work on the rich sibling species of Hawaiian picture wing *Drosophila* has analyzed tissue-specific patterns of enzyme expression for eight enzymes in 13 tissues across 23 species and is summarized in Dickinson (1980a, 1980b, 1980c, 1988). The general observation is that each enzyme chosen is expressed in a core set of tissues in all 23 species but twinkles on and off—present and absent—in the remaining tissues. Among the most interesting features of this twinkling is that it does not fit neatly into the best guesses for the phylogenetic branching lineage of the sibling species, nor do the twinkling patterns for different enzymes across the 23 species fit parsimoniously into the same branching lineage. In analyzing these patterns, Dickinson (1988) has been led to suppose that selection is capable of ensuring the expression of each gene in the core of tissues, where it may be critical, but that the complex structure of genetic regulatory systems with combinatorial regulation of many genes implies that the same gene will be expressed more or less randomly in a variety of other cell types of the same organism. Cavener (1987) has reached somewhat similar conclusions.

The failure of the expression patterns of enzymes to fit naturally into a phylogenetic lineage is reminiscent of the quixotic disappearance and recurrence of small bristle clusters and other pattern elements in the same set of *Drosophila* species (Garcia-Bellido 1983). In all these cases, it can be argued that the different sibling species have enzyme patterns or bristle and pattern elements which reflect maximally adapted phenotypes in the local microenvironment. Conversely, the constancy of enzyme expression in a core of tissues and the quixotic expression in the remaining tissues suggest that selection, at best, can control gene expression in the core but not in the remaining tissues, which exhibit imprecise behavior. Ruling out selective interpretations will be forever difficult in sibling species studies. Ultimately, testing isolates of inbred strains directly and assessing the rate of mutation of regulatory elements will help establish where and when selection can maintain cybernetic systems and with what precision.

The second line of thought asks how we might test the premise that, if many local optima exist, all suboptimal for any target pattern of gene expression, selection for any specific pattern of gene expression should become trapped on a local optimum well short of the target pattern. The experimental difficulty lies in conceiving of procedures to select for combinations of expression of a set of genes embedded in the genomic system of an organism. Such selection schemes are on the edge of feasibility. For example, a number of structural genes coding for enzymes for a connected metabolic pathway not present in a given cell, and in which the terminal metabolite was utilizable by the cell as food, might be inserted at random locations in the genome of the host cell, and selection for their joint synthesis sought by supplying substrates for various enzymes in the pathway. Such studies might reveal that sequential selection starting with expression of the enzyme required at the bottom of the pathway and followed by expression of its predecessors could succeed, or such selection might show that the coupled genomic system typically makes joint expression extremely difficult to achieve. Either answer is interesting.

SUMMARY

This chapter has begun to explore the capacity of selection to modify cell types or to coordinate patterns of gene expression. We have uncovered two familiar limitations. First, selection is limited by the structure of the fitness landscape. Second, we expect that selection in the face of mutations cannot hold a population at optima. With respect to the first limitation, adaptation becomes trapped on local optima. Even strong selection for arbitrary desired patterns of gene expression is unable to achieve those patterns via fitter mutants which alter the wiring diagram and the logic of the genomic regulatory system. Adaptation climbs to local optima quite far below the global optimum of a perfect match to the desired pattern. However, landscape ruggedness for genetic networks in the ordered regime, where there are few inputs per gene, differs from ruggedness in those in the chaotic regime, where each gene is regulated by many other genes. In the former case, the landscape is highly correlated and the fitness of attainable optima recedes only slowly as N increases. This echos the results in Chapters 2 and 3 for the NK family of rugged landscapes, where, for small K, the fitness of attainable optima does not decrease at all as the number of traits in the organism N increases. In both cases, low connectivity (low K) corresponds to systems whose fitness landscapes are highly correlated. Both are the kinds of highly complex systems which can adapt well because each adapts on a good fitness landscape with high and attainable optima that do not recede as N increases.

The second limitation is due to the mutation/selection error catastrophe. We have not carried out numerical studies of this problem for $K = 2$ regulatory networks, but from experience reported in Chapter 3, we anticipate that, in the face of continuous mutation, selection will be limited in its capacity to hold adapting regulatory networks at the attainable good local optima. As the number of genes and hence the complexity of the network increase, the fitness of such optima declines gradually. Further, as N increases, the average fitness loss due to a single mutation becomes smaller. Correspondingly, the selective force tending to restore the mutation to the optimal state becomes less. In parallel with results on the mutation/selection error catastrophe discussed in Chapter 3, mutation ultimately becomes a stronger force than selection as N increases, and the population falls from the attainable local optima toward the mean of the underlying space of genetic systems. The population wanders the foothills of the adaptive cybernetic landscape. Given plausible numbers for the rate of mutation and the number of genetic regulatory connections in a higher eukaryote, it is reasonable to guess that selection may well not hold populations closely clustered about optima in the space of genetic networks.

Results for fixed landscapes must be extended to ever-deforming landscapes due to environmental changes or coevolution, as discussed in Chapter 6. There we saw that, up to a point, the more rugged the landscape, the better a system may fare. Optimization of network architecture and logic for the capacity to adapt and coadapt appears to be a subtle problem.

If indeed it is the case, and can be shown experimentally, that selection does not hold populations clustered very narrowly about a wild-type genomic regulatory system, then actual genomic systems are diffusing in a space of possible genomic systems, perhaps well *below* local optima. Then actual genomic regulatory systems are imprecise, drifting, and suboptimal. If this is true, then such systems should remaining fairly *typical* members of the space of genomic systems in which evolution is occurring. Further, if optimal networks, both in terms of complexity of task performed and in terms of capacity to adapt, lie in the ordered regime at the edge of chaos, then selection must hold networks in this poised ensemble. The generic features of this ensemble emerge as possible quasi-universals. Thus, again, we return to the recurrent general theme of this book: Complex systems, contrary to our naive beliefs, exhibit self-organized behavior. Insofar as selection tunes the ensemble explored but is unable to avoid its generic properties, those quasi-universal features may be expected to shine through across the eons and across phyla.

CHAPTER 14

Morphology, Maps, and the Spatial Ordering of Integrated Tissues

The two fundamental problems of developmental biology are cell differentiation and morphogenesis. In the preceding several chapters, we examined some of the ways in which cell differentiation may exhibit ordered properties based on the spontaneous order inherent in a wide class of genomic regulatory systems. The intellectual theme was an exploration of the possibility that such spontaneous order may account for the origin and persistence of basic aspects of ontogeny, such as the existence of branching developmental pathways and the homeostatic stability of cell types. We repeatedly encountered the fundamental problem of understanding the relation between such "spontaneous order" and the effects of selection. In this chapter, we turn to an equally vast topic: How can we understand the genesis of ordered tissues and organs in multicellular organisms? This problem is indeed vast, for it encompasses more than the familiar issues of understanding the developmental mechanisms which faithfully unroll in ontogeny and the ease with which those mechanisms were "found" in evolution. In addition, the problem also involves asking whether any such mechanism can properly be thought of as having generated a distinct *family of natural forms* and, if so, whether and to what extent selection can modify those natural forms. That is, we must again ask (1) whether the morphologies of organisms to some extent represent the "self-organized," or natural, forms readily constructed via known developmental mechanisms and (2) how we must conceive of the interaction between such forms and selection.

This problem is an old one, yet one never well formulated. Let me be explicit. Cells are bounded by a bilipid membrane. Such membranes form readily in an aqueous environment and, for simple surface-energy reasons, readily form a closed spherical surface enclosing an aqueous interior. Many properties of cells depend critically on this simple self-organized property of lipids in water. Are the formation of such bounding membranes and their spherical form "achievements" of the genome and natural selection? To ask this question is to realize that many aspects of organismic form must reflect the natural properties of the building blocks from which organisms construct themselves. How much of what we see reflects such properties, and how does selection enter into the picture? D'Arcy Thompson (1942) wrote a justly famous book, *On Growth and Form,* in which he showed myriad examples in which organismic forms appeared closely analogous to forms generated on purely physical bases. For example, soap bubbles, as a result of surface tension and the contraction of each bubble to a minimal energy surface, cling to one another and form corners having very specific angles. Figure 14.1 shows two views of a species of *Radiolaria,* whose surface is covered with protoplasmic bubbles meeting at angles almost exactly like the angles seen in soap bubbles. *Radiolaria* deposit calcium crystals in the interstices between protoplasmic bubbles, forming spicules in a regular "cage."

Thompson argues, of course, that this morphology can be explained "purely" on physical grounds. In so arguing, he is making two distinct points, one correct, the other inadequate. First, he properly draws our attention to the fact that the genome can generate an organism *only* by making use of a variety of developmental mechanisms which depend on very many physical and chemical principles beyond the genome's raw capacity to coordinate, in time and space, the synthesis of specific RNA and protein molecules. Thus, surely, the natural ways such mechanisms construct forms must impinge on the forms we see. However, Thompson eschews much mention of selection. Yet selection occurs. Thus our fundamental problem is how to think about the relation between selection and the natural form generated by different classes of developmental mechanisms. We understand design principles, more or less, but we have virtually no coherent research program in biology attempting to understand this other utterly basic problem. As if all we see reflects selection alone, accidents, and remnants of past designs.

My aim in this chapter, therfore, shall be to discuss a number of aspects of pattern formation. Each aspect appears to be underwritten by a mechanism or class of mechanisms which can be expected to arise readily in evolution. Each generates either a family of related forms or a set of well-ordered properties in ontogeny. In examining these natural consequences of a variety of developmental mechanisms, we may begin to discern how to formulate the question of the relation between something like natural forms and selection.

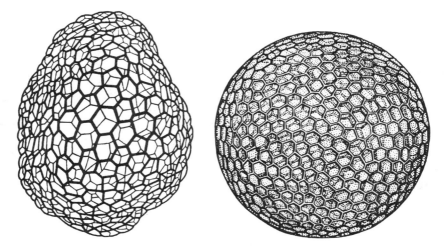

Figure 14.1   *Radiolaria,* whose surface is covered with protoplasmic bubbles meeting at angles similar to the angles seen in such minimal-energy surfaces as soap bubbles. Calcium crystals in the interstices form a regular cage of spicules. (From Thompson 1966)

The problem of pattern formation can be teased into at least two separate issues. First, how does the zygote give rise to different cell types in a *spatially ordered way?* Second, how does a collection of one or several cell types give rise to *morphology?* In the first part of this chapter, I explore whether we can begin to identify sources of order lying behind the ways a zygote may give rise to a spatially ordered arrangement of different cell types in a tissue, an organ, or an organism. In particular, in the first section I discuss the evidence for intercellular inductive interactions which lead the responding tissue to differentiate in new directions. Many of the ideas which came up in earlier chapters viewing cells as "poised" between only a few alternative pathways of differentiation apply here as well. Induction appears to be one of the most fundamental means of engendering spatial heterogeneity in cell types in one or several interacting tissues. Beyond such local interactions, phenomena of pattern duplication and regeneration following wounding or grafting experiments give evidence of long-range order in tissues. All this evidence leads us, in the second section, to introduce the concept of "positional information" and the related idea that regeneration is organized by the "smoothing out" of positional-information discontinuities caused by grafting or wounding. Limitations of the positional-information paradigm lead us, in the third section, to the famous Turing model of pattern formation. This class of models is able to generate beautiful standing wave patterns of the biochemicals in a tissue in specific spatial patterns. Astonishingly similar patterns appear to arise in the development of several organisms, including *Drosophila melanogaster,* discussed in detail in the fourth through eighth sections, where we also consider similar mechanochemical models which lead to formation of morphologies. Throughout, we shall find clues that many developmental mechanisms may lie to hand for evolution. Simultaneously, each mechanism generates a family of forms, some readily and others with difficulty. Selection may be constrained to the former. These themes are brought together in the final, ninth section.

Let us be Kantian for a moment. What must cells be such that a collection of them can organize into a spatially ordered array of distinct cell types? A first prerequisite is that the cells must be able to become different from one another. Second, the formation of a spatially ordered array of cell types requires that, whatever the cell types, their arrangement in space be provided for. Third, if the formation of the spatial pattern of cell types initiates with a single cell, rather than as a process of aggregation of preexisting cell types, then that cell must undergo mitosis. Two major alternatives for mitosis immediately arise: Either each cell, at division, autonomously "computes" the kind of cell type each daughter cell is to become and the position each should take by virtue of the past history of the clone of cells from which it, the mother cell, derives or, alternatively, each cell consult the surrounding cells within some range and bases its future differentiation and mitotic behavior on the information it receives from its neighbors. The first alternative leads us to consider mechanisms whereby cell might count mitotic divisions and partition different instructions to two daughter cells when appropriate conditions are met. The second alternative leads us to consider how cells might talk to their neighbors, what they might learn, and what they might be able to do with what they learn. Obviously, organisms can and do utilize both strategies.

I intend to focus on the second strategy for the bulk of this chapter. However, organisms such as *Caenorhabditis elegans* (Chalfie, Horovitz, and Sulston 1981; Kimble and White 1981) forcefully demonstrate that "clonal computation" is a powerful and pervasive general means of producing tissues, organs, and organisms. Yet formation of an ordered array of cell types by clonal computation, in which each cell ignores its neighbors and bases its behavior on its present state and presumably on its recording of the past history of the clone, is a logical subproblem of the broader problems which arise when cells can also talk to their neighbors. In the latter case, cells may carry clonal histories as well as consult their neighbors.

INDUCTION AS A BASIC INTERCELLULAR CONVERSATION

In this section, I show the parallels between the phenomena of induction and the natural properties of cells when modeled by canalyzing Boolean networks. The deepest features of induction appear to be generic to this class of regulatory networks.

If cells are to talk with one another and, by virtue of the exchange, coordinate in space and time the proper formation of different cell types relative to one another, then the most basic requirement of such interaction must be that one of the interacting cells should change its behavior. In the simplest case, two cells can influence each other by being in very close proximity; in more complex cases, one cell may act on the other at a distance by virtue of long-range chemical signals, such as hormones, electrical signals, or, presumably, mechanical force signals. We begin with the simplest case and consider direct cell–cell interaction by virtue of either membrane contact or short-range chemical signals. Such interactions, mediating alteration in cellular behavior and differentiation, are the well-known *inductive* interactions.

Since Spemann and Mangold's work (Spemann and Mangold 1924; Spemann 1938) on the dorsal lip of the newt blastula-stage embryo demonstrated the capacity of the transplanted dorsal lip to organize the formation of a new primary embryonic axis, developmental biologists have devoted enormous attention to such inductive processes. Figure 14.2 shows the newt blastula and the subsequent gastrulation and neurulation in the early newt embryo. The newt egg is mesolecithal, having a mod-

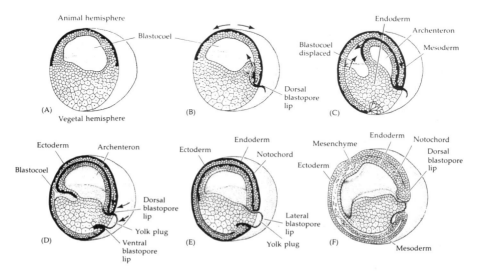

Figure 14.2   Gastrulation sequence in the newt. (From Gilbert 1988)

erate amount of yolk. Cleavage divisions are complete but unequal. This imbalance results in a blastula having a thinner animal wall and a thicker vegetative one. The animal pole region is darkly pigmented, the vegetal pole is clearer, and the marginal region between them, broader on one side of the equator than on the other, is the gray crescent. These three regions correspond roughly to the future three main germ layers: the animal region will constitute the outer germ layer, or ectoderm; the vegetative zone will form the inner germ layer, or endoderm; the middle zone will form the middle germ layer, or mesoderm. The future ectoderm comprises two main areas: the presumptive central nervous system and the presumptive epidermis. In the anlage, or formative zone, of the mesoderm is the material for the notochord, occupying a large middle part of the gray crescent; on each side of the notochord are located the materials for the myotomes. The lateral and ventral parts of this marginal zone correspond to the mesodermal linings of the body cavity, the kidneys, and so on. The endoderm forms the primitive gut, the lining of the gut, and the glands of the digestive system.

Segregation of these regions of the blastula into three germ layers occurs in gastrulation. As shown in Figure 14.2, the cell layer forming the surface of the hollow blastula invaginates through the blastopore to form the gastrula; the inward-migrating sheet of cells, the mesodermal mantle destined to form the primary mesoderm of the embryo, extends forward just beneath the overlying presumptive neurectoderm. Eventually, this mantle extends around the ventral, vegetal half of the embryo as well. Gastrulatin is followed by the formation of the neural tube, comprising the presumptive forebrain, midbrain, hindbrain, and spinal cord. The initial ectodermal cells overlying the inward-migrating mesodermal mantle, if cut free of the embryo before contact between these two cell layers and cultured independently, remain simple primitive ectoderm. This and other experiments demonstrate that the mantle acts on the overlying layer of cells which has not invaginated and triggers the conversion of those cells from progenitors of ectodermal cells to neurectodermal cells. This triggering constitutes primary induction (Spemann 1938; Saxen and Toivonen 1962; Jacobson 1966).

The inductive capacities of different regions of the invading mesodermal mantle differ. The most advanced, or leading tip, induces the overlying ectoderm to form neurectoderm which will form forebrain. More posterior regions, closer to the blastopore, induce the overlying ectoderm to become neurectoderm which will form more posterior regions of the primary nervous system. The most posterior region of the invaginating tissue, which folds inward through the blastopore last, induces the overlying ectoderm toward mesodermal fates.

This primary induction in vertebrate development is merely the most well known. Another example is the cascade of three major inductive events between two adjacent sheets of cells which occurs in the formation of the vertebrate eye. Briefly, the optic stalk originates as an evagination from the posterior part of the forebrain. The stalk contacts the overlying ectoderm and induces the latter to form lens, while the stalk is induced to form the optic cup, which invaginates to form a double-layered structure whose inner layer becomes the retina and whose outer layer becomes the pigmented epithelium of the eye and iris. In turn, the lens induces the ectoderm which re-forms over it to differentiate into cornea.

Formation of most glandular tissues in vertebrates requires inductive interactions on the endodermally derived cell which will form the tubes and secretory cells of the given gland from the surrounding mesodermally derived mesenchyme cells (Alberts, Bray, et al. 1983; Gilbert, 1988). In most of cases, the inductive capacity is specific to the particular local congregation of mesenchyme cells and is presumed, based on the capacity to induce when tissues are separated by small-pore filters, to be due to one or more small molecules. Specificity is sometimes shared, however. If salivary gland mesenchyme is grafted in place of mammary gland mesenchyme around the developing tubes which normally form the mammary gland, then the branching morphology of the tubes comes to resemble that of the salivary gland, but the glands still secrete milk proteins upon appropriate hormonal stimulation.

Analysis of primary and secondary induction in normal embryogenesis has led to the development of two major concepts: the *inductive capacity* of an inducing tissue and the *competence* of the responding tissue to be induced (Waddington 1940, 1957, 1966). It is important to stress that each capacity shows a characteristic spatial extent in the embryo, a distinct time course during which it may first wax and then wane, and a spectrum of specificities in action or response.

The spatial extent issue is interesting. The general picture arrived at by classical embryologists is that the areas which are either inductive or competent to respond first to primary induction and later to secondary induction become progressively smaller. That is, the primary competent field of cells for primary induction is the entire presumptive ectoderm and, as we know from Mangold's demonstration of formation of a secondary embryonic axis, extends even to the presumptive endoderm. Once the neurectoderm has formed, however, it creates restricted fields of cells underlying the ectoderm which forms over the neural tube. Thus only subregions of the neural tube are able to form the optic stalk, or otic stalk. In secondary induction of the overlying ectoderm, induction is acting on cells which have already been induced to form ectoderm. In the induction of nephric tubules, the responding tissue is already determined as mesenchyme and so on. Similarly, the time course of inductive ability, or capacity to respond, exhibits increases and decreases. Thus the capacity of the ectoderm to respond to the primary mesoderm by forming forebrain wanes prior to the capacity to form hindbrain or spinal chord. Cells not yet determined to form ectoderm cannot yet respond to the optic stalk to form lens. Finally, specificity of action by a given inducing tissue is not complete, nor is the specificty of response of a given responding tissue. A region of the invading mesoderm will induce *a spectrum of responses from the same overlying tissue.* This latter point is typical and very important. It states that, in response to a given inductive stimulus, a *limited range* of responses is found. For example, classical work by Saxen and Toivonen (1962) examined the response of the newt ectoderm to induction by guinea-pig liver. Tissues ranging from forebrain, nose, eye, and balancer, which are archencephalic derivatives, to spinal cord, fin, myotomes, and notochord are found. While a small area of responding tissue may be induced in several directions, not all directions are accessible. Thus no endodermal structures are formed.

Analysis of normal and heterogeneous inducers has revealed another fundamental feature of primary and secondary induction: Many different inductive agents trigger the same response in the induced tissue. Saxen and Toivonen (1962) analyzed such diverse inductive agents as retina, bacteria, liver, bone marrow, kidney, and skin. Even a simple chemical agent, such as methylene blue, or an alteration in pH was found sufficient to induce ectoderm to form neurectoderm. The capacity of a wide range of biological and nonbiological stimuli to evoke the identical transformation is evidence that the channeling of the response into a few alternative inductive responses lies in the responding tissue. Further, the wide stimulus range is strong evidence that there is unlikely to be some single "trigger gene" or process which constitutes the final common pathway of an inductive response and on which all these disparate stimuli act.

Very much worth noting is the profound similarity between the variety of heterogeneous inducers which can cause the same transformation and the phenomena of phenocopying and genetic assimilation (Waddington 1942, 1956; Ho, Bolton, and Saunders 1983; Ho, Tucker, et al. 1983). Recall that a number of homeotic mutants convert one tissue or structure to another in *Drosophila melanogaster*. In particular, members of the *bithorax* complex of genes transform third thoracic segment into second, or second into third, and also convert different abdominal segments to one another. Treatment of normal early embryos with ethyl ether results in some adults having third thoracic segment transformed to second, or second to third (Gloor 1947; Capdevila and Garcia-Bellido 1974; Ho, Bolton, and Saunders 1983; Ho, Tucker, et al. 1983; Ho, Saunders, et al. 1987). This procedure is called *phenocopying* a homeotic mutant. That the fly is genetically normal is revealed by its normal progeny. Waddington (1942) carried out selection experiments for ease of phenocopying *bithorax* transformations. After a number of generations, a population of flies was achieved which exhibited the transformations in the *absence* of ether. Thus genetic modifiers had been collected in such a population, modifiers which mediated the same transformation as that induced by ether. This accretion of modifiers has been called *genetic assimilation*. Genetic analysis showed that these modifiers were not members of the *bithorax* complex of genes. Different selection experiments yielded different populations exhibiting the same transformation, typically due to the assembly of a different collection of modifier genes. Meanwhile, Maas (1948) showed that heat shock in early embryogenesis in *Drosophila* also phenocopies the *bithorax* transformations.

What does phenocopying imply? First, it is clear that a variety of external stimuli—here ether and heat—can cause the identical transformation in the developing organism. Second, those transformations which are readily evoked by external stimuli appear readily open to genetic assimilation, assembling a constellation of genes which thereafter causes the same transformation in the absence of the external stimulus. Third, often the set of modifier genes so assembled is not identical to a single major gene which causes the same transformation as a dominant or recessive. For example, a recovered mutant mimicking some of the *bithorax* transformations is in a totally different gene, now known to code for an RNA polymerase (Greenleaf, Weebs, et al. 1980)!

These facts have curious consequences. Clearly the fact that a restricted number of transformations are caused by a variety of external stimuli and by a variety of mutants, acting either alone or in combination, demonstrates that the cell types or tissues in question are poised between few alternatives. Since cell and tissue types are complex systems, this poised character very strongly suggests, but does not prove, that variations at *many different points* in the system cause the same transformation. This raises two important questions: What kinds of genomic cybernetic regulatory systems have this poised property? Does the existence in organisms of these poised properties reflect a *selective achievement* of adaptive evolution?

In Chapters 12 and 13, we examined the dynamical behavior of Boolean model genomic regulatory systems. In that analysis we identified a cell type as an attractor, a state cycle, in the dynamical repertoire of one genomic regulatory system. Given that identification, we have a way of thinking about induction and part of the process of differentiation. The natural sense of "neighboring cell type" derives from asking which other cell types a given cell type can differentiate into by *transient* reversal of the activity of any single gene. One imagines here that an exogenous inductive stimulus—a hormone or an ether, say—transiently reverses the activity of some single gene. Any such single perturbation may leave the cybernetic system in a state which lies in the original basin of attraction and hence returns to the same state-cycle-attrac-tor cell type, or it may be left in a state which lies in a different basin of attraction and hence flows to a different state-cycle-attractor cell type. Figure 12.12a shows the transformation induced among 30 model cell types in a typical genomic regulatory system having $K = 2$ inputs per gene. The salient facts are that, after most pertur-bations, the cell returns to the same attractor and hence remains the same cell type. After some perturbations, however, the cell flows to a small set of different cell types. Thus, as stressed in Chapter 12, any cell type is stable to most perturbations and can directly differentiate into only a few other cell types. With respect to our current inter-est, note also that very many different stimuli cause the identical transformation from cell A to cell type B. Thus cell type A is poised between few alternatives, and many stimuli cause transformation of A to the same one of its few accessible neigh-bors. Indeed, given that the genomic system has many genes and has attractors, each having only a few neighbors, it is almost necessarily the case that a number of stimuli acting on different genes in the cybernetic system will cause the same transformation.

Genetic assimilation shows that exogenous perturbations can cause the same transformation of developmental pathways as those caused by selection on sets of modifier genes. This, too, suggests that the cell types of the genomic system are poised such that either minor external perturbations or alterations of internal components cause the same transformation. In our model studies of the effects of deletion of model genes frozen in the inactive state, discussed in Chapter 12, the same phenom-enon was observed. Many such mutations increased or decreased the probability of the same few transitions among those model cell types which already existed in the wild-type genetic network. In short, the model cell types are poised to differentiate in restricted directions, and a class of mutants can increase or decrease the probability of just those transitions.

We reach an important conclusion. Poised cell types, having only restricted ave-nues of differentiation open to them and the capacity to integrate and channel diverse stimuli acting at divergent points in the regulatory system, are *inherent, generic* prop-erties of genomic systems which have few regulatory inputs per gene. Such systems are poised in the sense that genetic assimilation on a variety of modifier genes will increase the same transitions. In short, the existence of genomic regulatory systems possessing the proper dynamical features for exhibiting the known patterns of induc-tive transformations is "built into" the class of genomic systems which best models real genetic cybernetic systems. As with other self-organized properties in this class of genomic regulatory systems, we may ask whether these constrained patterns of inductive transformations per se reflect selection or not. I stress that our question is not whether the particular inductive transitions observed are the consequence of selection, but whether the fact that the genomic system has the property that such transitions occur between poised cell or tissue types is itself selected. Once asked, as in parallel questions in earlier chapters, the answer is not obvious. What is obvious, however, is that selection to hold genomic regulatory systems in the "good" class of those governed by few inputs and rich canalyzing functions is sufficient to account for this fundamental property of developing systems.

Induction of New Cell Types and "Dedifferentiation"

The cases of induction discussed above required transient interaction between induc-ing and induced tissues at specific stages of development when each tissue was com-petent to either induce or be induced. In contrast, there is substantial evidence that persistent interaction between cells is often required for maintenance of each cell's morphology and function. Evidence supporting this first came to light in the analysis of cells in tissue culture. As Weiss (1939) noted, many explanted cells gradually lose their former differentiated aspect and distinguishing cytological features:

> Eventually, all cultivated cells appear in only three forms: closely packed epithelial cells, loosely connected mesenchyme, and free amoeboid cells. Originally, it had been thought that this abandonment by the cells of their specialized aspects meant a real reversion to a more primitive level of differentiation, and that identical appearances were indicative of identical character. Continued studies, however, have made it clear that the character of a cell cannot be judged by external appearances. Although superficially they resemble one another, the *explanted cells obstinately preserve many of their functional distinctions* even over prolonged periods of cultivation *in vitro*. . . . Cultivated glandular cells of the intestine and pigment cells of the eye may become very similar in shape and general conduct, but under the proper conditions the former still resume the production of digestive enzymes and the latter of black pigment. . . . Thus, . . . *differentiation of cellular character in a vertebrate is irreversible . . . ; only modulations are reversible.*

The "proper conditions" to which Weiss refers are those in which the different cell types in a tissue are cultivated in close proximity, forming an aggregate inside of which the diverse cell types reverse their modulation to take up again their normal cytological and synthetic patterns.

If inductive interactions can occur between tissues in embryogenesis, and if heterogeneous inducers such as guinea-pig bone marrow can act to induce neurulation in the newt, it would be astonishing if cell–cell interactions did not also engender modulations in cytology and gene expression as a result of persistent cross-talk. What would it mean to develop a *theory* about such modulations? I consider next some initial steps.

Tissues as Sheets of Interacting Cells Exchanging "Microhormones"

In Chapter 12 we considered Boolean models of genomic regulatory systems. Each model network stands for the genome in a single cell. We presume that, in almost all metazoans and metaphytens, the genomic system is identical in all cell types of the organism. Conceive, then, of a tissue as a two-dimensional sheet of cells, each cell endowed with the same genomic regulatory system, modeled as a Boolean network. Add to our previous discussion the general idea that *products of a subset of genes can reach neighboring cells in the tissue.* There are indeed a number of cellular mechanisms which do mediate such transfer. Most simply, a product can diffuse out of the initial cell and is small enough to pass via gap junctions to neighboring cells. Alternatively, a product cannot pass out of the initial cell, but a metabolite whose synthesis the product controls can diffuse to neighboring cells. In a third alternative, a product is vectorially excreted from the initial cell and binds to surface receptors on a neighboring cell; from there it may act via second messengers or via pinocytosis into the nieghboring cell.

Let us call these migrating gene products "microhormones," where the name is meant to imply that such products act only locally, on their immediately neighboring cells. Thus "microhormone" is a renaming of untold many local inductive agents.

The set of genes whose products can reach neighboring cells can be thought of as a kind of external, or *positional,* regulatory network, for it is these genes whose products mediate cell–cell interaction and begin to control the arrangement of cell types in the tissue.

Emergence of Novel Cell Types in a Spatial Pattern in Tissues

What happens in a tissue, or sheet of cells all of which contain the same genomic regulatory system, when a subset of the genes in the network make products which can reach neighboring cells? The general answer turns out to be threefold. First, the interactions generate a variety of novel cell types whose continued existence as those cell types requires persistent contact and exchange with the neighboring cells. Second, the number of novel cell types which arise is strongly dependent on the fraction of genes whose products can reach neighboring cells—in other words, on the fraction of genes producing microhormones—and appears to reach a peak at about 15 to 20 percent microhormones. Third, primitive spatial order of cell types in the tissue arises spontaneously.

The first two results were evident in my own early unpublished numerical studies, and all three results are clear in work carried out by Jackson, Johnson, and Nash (1986). An organism is modeled as a one-dimensional growing line of cells. Inside each cell is a copy of the same small genomic regulatory system. In addition, each cell is provided with a schematic cell cycle which governs the conditions under which the cell will divide or cease division. This oversimple model of the cell cycle is not important in its own right; it merely provides internally generated rules such that a group of adjacent cells can divide and ultimately form a tissue which may or may not cause cell division to cease. The critical questions concern the consequences of microhormones. Jackson, Johnson, and Nash further constrained microhormones to be vectorial, that is, to act only on neighboring cells and not the synthesizing cell.

Mathematically, the problem is simple to state. Each copy of the genomic regulatory system, one copy resident in each cell, has a set of attractors, or cell types, to which it will settle if the cell is left in isolation. A tissue of interacting cells—either a one-dimensional line of cells or a two-dimensional sheet of them in which all cells "compute" the next activity value of all genes at the same synchronous instant—is just a large iterated array of the identical Boolean network with synchronous state transitions. Thus a state of the tissue is just the current activity of all genes in all cells at one instant. Over a succession of moments, the tissue will pass from state to state. Ultimately, the tissue falls onto an attractor. That is to say, the tissue settles down to some recurrent pattern of gene expression in which each cell necessarily exhibits its own recurrent pattern of gene expression.

We can now ask whether any individual cells in the tissue are exhibiting novel cell type state cycle attractors which were not possible in an isolated cell containing the same genomic regulatory network. That is, we can ask whether cell–cell interactions generate new cell types. The generic answer is "yes." In the study carried out by Jackson, Johnson, and Nash (1986), each attractor was scored by the fraction of time each gene was active, in 20 percent intervals. Thus these workers may have categorized as identical many cell types which differ upon more detailed analysis. Nevertheless, they found that the formation of new cell types occurs and is most common when about 20 percent of the genes are microhormones (Figure 14.3). Further, they found that, as the number of genes in the network increases, the probability that new cell types are formed also increases. In my own earlier work, using detailed state-cycle identity

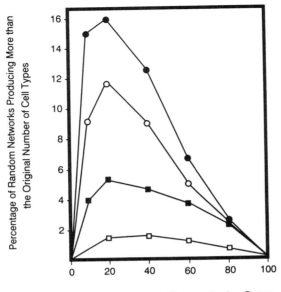

Figure 14.3 Probability of induction of novel model cell types in a one-dimensional organism as a function of the fraction of Boolean genes whose products (microhormones) communicate with neighboring cells. Increasingly high curves reflect increasing numbers of genes in the model genome. (From Jackson, Johnson, and Nash 1986)

to characterize each attractor cell type and fairly large genomic systems (50 to 100 genes), I found that virtually all networks with 15 to 20 percent of the genes taken as microhormones formed new cell types by virtue of tissue interactions. The total number of novel cell types of which a given network was capable of producing by such interactions was not noted to be very large. Good numerical data are no longer available, but, crudely, an isolated copy could double the number of cell types.

These model Boolean networks yield three fundamental conclusions:

1. Tissue interactions typically yield new cell types in model tissues which couple adjacent cell via microhormones.

2. The number of new cell types generated is maximally on the approximate order of the number of cell types of which an isolated cell is capable.

3. Most surprising, the capacity to engender new cell types is coupled to the fraction of the genes whose products can reach neighboring cells. The maximum occurs when around 20 percent of the genes produce microhormones.

These results powerfully suggest that *attaining novel cell types by cell–cell interactions lies to hand in evolution.* The results also hint that, as a result of the generic properties of the class of genomic regulatory systems under selection, selection may be constrained in the numbers of induction-dependent cell types which can be created through the use of microhormones.

Jackson, Johnson, and Nash found that primitive spatial order of two types arises.

Given the assumptions that the organism is a growing line of cells and that cells can divide and both daughters will be in the identical state, it is not terribly surprisingly that adjacent blocks of cells in the same state, and cell type, emerge in the growing tissue models. It is far more interesting that simple cell patterning arises. Thus a specific gene is active in every third cell along the posterior half of the model organism. Presumably, this patterning reflects a cascade of mutual inductive effects along a line of cells, effects which set up a repeating pattern of alternative cell types in the tissue. It is not grand, but it is spatial order. Indeed, as Jackson and his colleagues hasten to point out, the spatial order is reminiscent of regularly spaced bristle patterns in *Drosophila*, where it is known that formation of the adjacent bract from one epidermal cell is dependent on an inductive signal passing to that cell from the nearby cells forming the bristle and its socket (Bryant 1984).

The Boolean network models are idealizations for a more general class of model cybernetic systems utilizing continuous nonlinear equations. The analogues of state-cycle attractors in Boolean networks are dynamical attractors of the system of ordinary differential equations describing the behavior of the intracellular variables. The continuous-variable analogue of the on–off "positional variables," or microhormones, passing between cells in the iterated network model of a tissue consists in describing those spatially distributed subsets of variables via partial differential equations. Thus the entire system is a linked dynamical system in which a subset of variables pass between cells and hence are described by partial differential equations. Note that the general concept that the *entire tissue* falls to an overall dynamical attractor which may be spatially heterogeneous will carry over to continuous nonlinear models. I shall return to this idea in the final section, for the spatially inhomogeneous pattern of the external, or positional, variables naturally constitutes what we might want to call a *map*. Further, the attractor to which each cell flows, governed by its initial conditions and by the values of the positional variables constituting the map, corresponds to the *interpretation* that cell makes based on location in the tissue. The interpretation made by the cell constitutes its "decision" based on location.

What lessons can be drawn? Most fundamentally, induction of novel cell types, establishment of spatial heterogeneity, and setting up of simple spatial order are virtually inherent in almost any genomic system which has highly localized attractors as cell types when those cells are coupled to one another via a subset of their products. These are deep properties of genomic systems which probably require little or no selection other than membership in the proper class of genomic systems and the wit to allow a subset of products to move between cells. Adaptive selection to achieve "good" patterns of "good" cell types is a different matter entirely.

Summary: Inductive Interactions Are Generic in the Canalyzing Ensemble

Local inductive interactions are critical features of ontogeny in higher plants and animals. Identification of the kind of underlying cybernetic regulatory structure which permits the types of inductive interactions that we know to occur can be attempted by asking what kinds of genomic regulatory systems exhibit induction, competence, and simple spatial order as generic properties. We have succeeded in identifying sufficient conditions. Genomic systems in the *ordered regime* have the requisite properties. Genomic cybernetic systems of high specificity, such that each gene is regulated by few other genes and products, generically are in that ordered regime. Consequently, cell types are poised. Many alternative external stimuli cause the same transformation. Many alternative mutations increase the probability of the same transformation. Selection of such mutations will cause genetic assimilation for those transformations induced by the external stimuli. Microhormonal interaction between cells carrying the same genomic system generates new cell types whose continued maintenance requires continued inductive interactions. Simple spatial patterning arises spontaneously.

In contrast, genomic systems in the *chaotic regime* do not exhibit these properties. Here, almost any perturbed attractor has access to all other attractors. In other words, each cell-type attractor is not poised between few alternatives. Thus the fact that cells are poised strongly suggests that cell genomic systems are in the ordered regime.

EVIDENCE FOR LONG-RANGE ORDER IN TISSUES: DUPLICATION, REGENERATION, AND POSITIONAL CONTINUITY

The philosopher Ludwig Wittgenstein once remarked that avoidance of errors was significantly enhanced by consideration of more than one theory for the same phenomenon. His advice is worthwhile. In this section, I review the concept of *positional information* and describe some of the basic phenomenology of epimorphic pattern regulation. I then consider the relative merits of three alternative possible "coordinate systems" for that positional information—polar, Cartesian, and spherical—and show that none of these models can account for all the current data. Despite the inadequacies, however, it shall emerge that many features of pattern regulation can be accounted for by a very simple general developmental mechanism for achieving "positional smoothing" which almost certainly "lies to hand" in evolution and has probably recurred many times. Thus the kinds of spatial order entailed by such positional smoothing appear hard to miss in evolution. The incapacity of this mechanism to account for all the data leads to a critique of an entire class of theories, however, and emphasizes the need to focus on theories which link the geometry of a tissue to the presumptive profiles of positional fields within the tissue. This approach leads, in the subsequent section, to the development of a large class of theories, all of which derive from the famous English mathematician Alan Turing. These theories naturally have the property that spatially organized patterning arises spontaneously, and the expected pattern accord remarkably well with those seen in a variety of developing organisms. Nevertheless, even Turing's broad ideas appear to be too simple, and in a subsequent section of the chapter I consider ways to extend his fundamental ideas in testable ways.

Positional Information

In the past two decades, a resurgence of interest in spatial patterning in developing organisms has followed Wolpert's (1969, 1971) reformulation of this fundamental problem in terms of the concept of positional information. Prior to Wolpert's introduction of this concept, the dominant theory guiding research postulated the existence of developmental fields, or regions of cells, possessing "prepatterns"—in other words, nonuniform spatial distributions of hypothetical biochemical substances in a tissue. Local concentration peaks of these substances would induce the formation of specific pattern elements, such as digits, sensillae, or bristles (Stern 1968).

In contrast to Stern, Wolpert proposed the more abstract idea that, through access to a underlying positional coordinate system, cells within a developmental field possess positional information about their location with respect to the boundaries of the field. The behavior of each cell in the field was assumed to be due to two independent processes. The cell first *assesses* its positional information, then *interprets* this information according to the type of cell it is, and finally forms a specific structural element in the overall pattern (I note parenthetically that Wolpert's distinction between positional information and interpretation lay behind my use of these terms in the previous section in reference to linked partial differential equations and ordinary differential equations, where the former constitute a map and the latter drive the cell to an attractor.)

Among the kinds of data which initially lent support to Wolpert's idea are the much-noted homeotic mutants in *Drosophila melanogaster*. An example is the mutant *Antennapedia*, which converts antenna to second leg. A quite surprising additional feature of this mutant is that it converts distal antenna to distal leg (Postlethwait and Schneiderman 1971). In Wolpert's terms, the cells in the prospective distal area of the antennal imaginal disc, which metamorphoses to an adult antenna, "know" both that they are "distal" and that they are "antenna." The homeotic mutant converts the cells' determined state from antenna to leg, but the cells continue to assess their local positional information, continue to know that they are "distal," but now interprete that information in light of their new "leg" state—and so form distal leg. Similar results are known for many other homeotic mutants.

The chief difference between Wolpert's positional information and Stern's prepattern is that positional information is free of assumptions about the existence of specific biochemical "morphogen" peaks underlying the subsequent differentiation of specific pattern elements. This freedom in one sense makes the theory of positional information less predictive; yet it allows for two important possibilities:

1. The positional information in all the developmental fields of one organism might be identical.

2. The positional information system *in all organisms* might be identical!

The general success of the positional-information concept led to a search for the coordinate system which supplies the information. At present, polar (French, Bryant, and Bryant 1976), Cartesian (Cummins and Prothero 1978; Kauffman 1978; Winfree 1980, 1984; Kauffman and Ling 1981; Lewis 1981, 1982; Kauffman 1984b; Totafurno and Trainor 1987), and spherical (Russell 1978) coordinate system models have been proposed. The differences among these models are not trivial. Although it is always possible mathematically to transform from one coordinate system to another, the "forces," or tissue properties, which must be postulated to explain the observed features of pattern regulation differ sharply in the different models. In general, the "morphogens" have not been found. Nevertheless, one task in this area of biology consists in efforts to discover the simplest theory and postulates about cells to account for the known data. A realistic hope has been that the proper formulation will both provide macroscopic laws describing the behavior of integrated tissues and aid in the discovery of the underlying molecular variables. It is, furthermore, of particular importance to our efforts in this book to assess the extent to which ordered properties of organisms are readily accessible to adaptive evolution, to discover—as we shall in this section—that beautifully ordered features of pattern formation may well rest on very simple general laws.

The Phenomena

To assess the relative success of the alternatives that have been proposed, it is necessary to review briefly at least some of the major phenomena of pattern generation and regeneration.

Intercalary Regeneration of Intervening Structures.   If an amphibian limb capable of regeneration is transected proximal and distal to the elbow and the distal wrist fragment then is grafted to the proximal shoulder stump, cells proliferate in the wound area to form a blastema, followed by eventual regeneration of the missing elbow region, with the normal bone and pattern elements in the proper proximal–distal order. This process is called *epimorphic pattern regeneration.* The defining feature is that the old tissues, here proximal shoulder and distal wrist fragments, remain intact, while new cells *grow and intercalate* the missing pattern elements. The fact that the intercalated pattern elements are regnerated in the proper proximal–distal order is ubiquitous and fundamental (Slack 1980). Juxtaposition of normally nonadjacent tissues from a single developmental field is generally followed by regeneration, in the proper spatial order, of the structures normally lying *between* the juxtaposed tissue edges (Mittenthal 1981). This notion of betweenness is necessarily central to any theory of pattern formation. Another fundamental feature of this example is its suggestion of long-range order. That is, the reestablished pattern elements require specification of the fates of a reasonably large number of cells in a spatial order lying between the bounding proximal and distal stumps. The fact that the order carries linearly over a reasonably large number of cells (perhaps 100 or so) in the blastema between apposed stumps does *not* entail that the positional signals pass otherwise than between adjacent cells. Surely, however, this fact does hint that the process differs from local induction of a transition of cell types A to cell type B in the presence of cell type C. It is, in fact, the serially ordered betweenness of such intercalary regeneration which points to the existence of a system mediating long-range order.

The simplest physical model to account for betweenness in intercalary regeneration postulates the existence of one or more chemical concentration gradients spanning the tissue, with the various concentration levels specifying the positional information of cells at each point in the domain. As shown in Figure 14.4, in which a proximal–distal gradient along an amphibian limb is envisioned, surgical removal of the elbow and grafting of wrist to shoulder create a discontinuity in the gradient at the graft junction. If one imagines that gradient concentrations are held fixed in the

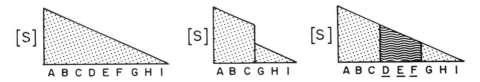

Figure 14.4   [S], the concentration gradient of substance S, provides proximal-distal positional information in amphibian limbs. Serial threshold levels specify pattern elements A, B, . . . I. Removal of the limb midregion, denoted by D,E,F, and grafting create a discontinuity which stimulates cell proliferation. Diffusive smoothing of gradient discontinuity regenerates the missing gradient levels (wavy lines) and structures D,E,F. (From Kauffman 1984)

old tissue fragments while diffusion occurs in the new cells of the wound blastema, then simple diffusive averaging of the concentration discontinuity at the graft junction smooths over the discontinuity, re-creating all the intervening gradient values in proper spatial order. In fact, the utter simplicity of this argument does much to explain the persistence of faith in morphogen gradients, despite the difficulty developmental biologists have had in finding the presumptive biochemical morphogens.

For the remainder of this section, I adopt the postuate that position is specified by graded scalar properties in tissues, such as chemical concentrations, although it is important to stress that discrete models, such as the genetic networks forming a tissue discussed in the previous section, require attention. In fact, we shall return to such networks later. Given the postulate of positional gradients, a fundamental question is the extent to which the simple property of diffusive-like averaging of gradient discontinuities can account for pattern formation and regeneration. This simple property turns out to be very powerful indeed. So much so that it may stand as another deep property of organisms which is very nearly inevitable.

Sequential Formation of Positional Axes in Development.   In several systems, positional axes appear to be established sequentially during development. In classical experiments, Harrison (1918, 1921) removed the right forelimb bud of the amphibian *Ambystoma* and grafted in its stead the left forelimb bud. Such grafts must either invert the anterior–posterior limb axis while keeping the donor and host dorsal–ventral axes aligned or invert the dorsal–ventral donor and host axes while keeping the anterior–posterior axis aligned. Harrison found that, if very early left limb buds were grafted onto the right, they developed into normal right limbs. If late left limb buds were grafted, they formed normal left limbs with the axis which was inverted at surgery still inverted with respect to the host. But if left-to-right grafts were made at an intermediate stage, the outcome depended on which axis was inverted at the graft junction. If the anterior–posterior axis remained normally aligned and the dorsal–ventral axis was inverted, the donor left limb bud formed a *right* limb; if the dorsal–ventral axis remained aligned and the anterior–posterior axis was inverted at the graft junction, the donor left limb bud formed a left limb which remained inverted at the donor–host junction. Harrison inferred from his results that the donor anterior–posterior axis becomes autonomously self-sustaining before the dorsal–ventral axis does. Similar data suggest that the amphibian eye axes are established sequentially, although the status of the data on the eye is in dispute (Hunt 1975).

Distal Transformation.   If an amphibian limb is transected—for example, at the elbow—the proximal stump can form a regeneration blastema and regenerate the distal limb (Harrison 1918). If the digits of the transected distal fragment are implanted into a host flank to establish an adequate blood supply to the distal fragment and the limb is then cut, the cut surface at the elbow, which initially faced proximally, forms a regeneration blastema and regenerates a second distal wrist and hand structure which is *mirror-symmetric* to the implanted distal limb (Harrison 1918, 1921). That is, if the implanted hand is a left hand, the regenerated second hand is a right hand. Except for handedness both the proximal stump and the implanted distal limb fragment regnerate the same set of distal limb structures from the cut surfaces at the elbow, identified as a regenerate hand on the proximal stump and a duplicate hand on the implanted distal limb fragment. The fact that both fragments form distal limb has been called the "rule of distal transformation" (Rose 1962). Similar results have been found in many insect legs and in the imaginal discs of *Drosophila,* as discussed further below.

Supernumerary Limbs. Among the most striking observations in pattern regulation is the induction of supernumerary limbs following grafting. After both the anterior–posterior and dorsal–ventral axes of amphibian limbs are fixed, transplantation of a left distal limb to a right proximal stump which reverses the anterior–posterior axis of the donor relative to the host but leaves the dorsal–ventral axis aligned typically results in the formation of two super numerary limbs at the anterior and posterior margins of the donor–host junction. If, instead, the anterior–posterior axes of host and graft are aligned but the dorsal–ventral axes are inverted, the two supernumeraries emerge from the dorsal and ventral margins of the host–donor junction. These supernumerary limbs generally have the handedness of the proximal stump (Harrison 1918, 1921; Bryant and Iten 1976). Similar results have been found in transplantation of cockroach limbs (Bulliere 1970; Bohn 1972).

Rotation of a left distal limb by 180 degrees and regrafting to its own stump give a more variable range of results. After such a rotation, the limb may partially rotate back toward its normal alignment; sometimes zero, one, two, or more supernumerary limbs are formed at the graft site and have either the same or opposite handedness (Bulliere 1970; Bohn 1972; Bryant and Iten 1976; French, Bryant, and Bryant 1976).

Duplication and Regeneration by Complementary Tissue Fragments. Distal transformation by both proximal and distal amphibian limb fragments is one example of duplication and regeneration by complementary fragments of a developmental field. The phenomenon is common, however, and has been studied in greatest detail in the imaginal discs of *Drosophila.* As we noted earlier, *D. melanogaster* is a holometabolous insect with egg, larva, pupa, and adult stages. During metamorphosis, the larval ectoderm lyses and the ectoderm of the adult is formed by the terminal differentiation of special larval organs called imaginal discs (Gehring and Nothiger 1973). In the late-third-instar larva, each imaginal disc is a two-dimensional sheet of cells forming the surface of a hollow sphere. The columnar cells on one hemisphere form the imaginal disc proper, while thin squamous cells on the other hemisphere form the peripodial membrane, which is lost during metamorphosis. Imaginal discs are found as bilaterally symmetric pairs, each destined to form specific left and right regions of the adult ectoderm: the left and right first leg discs from the two prothoracic legs; the two wing–thorax discs form the left and right mesothoraces and wings, and so forth.

By injecting specific fragments of each disc into host larvae which undergo metamorphosis and then recovering the metamorphosed implanted tissue from the abdomen of the now-adult host and characterizing the adult cuticular structures formed by that disc fragment, it has been possible to construct a fate map of each part of each imaginal disc. The fate map of the wing–thorax disc (hereafter wing disc) is shown in Figure 14.5*a*. Note that the upper and lower margins of the disc along its longitudinal axis forms ventral and dorsal thoracic structures, while the midregion of the disc forms wing structures. During metamorphosis, the wing disc folds along an arc running from the anterior to the posterior disc margin. This folding apposes ventral and dorsal thorax areas and also apposes ventral and dorsal wing hinge and wing blade areas, creating a bag that everts through the peripodial membrane. The center of the disc forms the distal wing tip, while an arc running from anterior to posterior disc edge corresponds to the wing margin.

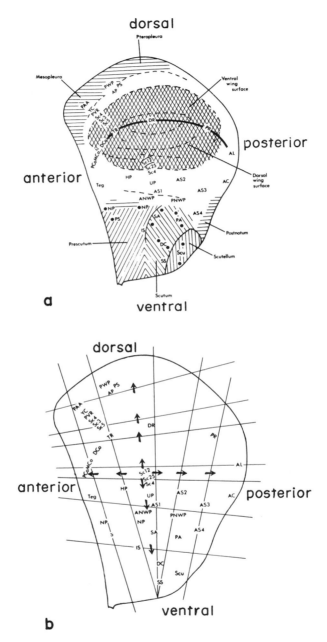

Figure 14.5 (a) Fate map of *Drosophila* wing disc. AC, axillary wing process; AL, alar lobe; ANWP, PNWP, anterior and posterior notal wing processes; AP, axillary pouch; AS1–4, first to fourth axillary sclerites; DC, dorsocentral bristles; DR, double bristle row (distal wing margin); HP, humeral plate; NP, notopleural bristles; PA, postalar bristles; PCo, MCo, DCo, proximal, medial, and distal costa; PR, posterior row of hairs; PS, pleural sclerite; SA, supraalar bristles; Scu, scutellar bristles; Sc4, Sc3, Sc5, sensilla campanifomia on ventral radius; Sc4, Sc25, Sc12, sensilla campanifomia on proximal dorsal radius; Teg, tegula; TR, triple bristle row; UP, unnamed plate; YC, yellow club. (Fate map from Bryant 1975) (b) Fate map as in (a). Lines show positions of single cuts. Arrow across each line points from the fragment which regenerates to the fragment which duplicates. (From Kauffman 1984b)

Grafting experiments are not yet feasible in *Drosophila,* but analogous experiments can be performed by cutting the wing disc into known fragments and injecting each fragment into the abdomen of an adult female. In that environment, the disc fragment heals its cut edge. In the healing process, tissue regions which are normally nonadjacent are apposed, and new cells grow in the wound area. After a week in culture, a disc fragment's mass typically doubles. After such culture, the fragments may be recovered and injected into host larvae for metamorphosis, then recovered from the emerged adult. By comparison of the patterns of hairs, sensillae, and bristles which form when a known disc subfragment is injected directly into larvae for immediate metamorphosis, it is possible to characterize the pattern regulation which occurs in the cultured fragment. The following are the dominant results (Bryant 1975, 1978):

1. If the wing disc (Figure 14.5*b*) is cut into two fragments of unequal size by a straight cut, the smaller fragment duplicates some or all of its pattern elements. In favorable cases, a mirror-symmetric duplicate is generated whose symmetry axis lies along the position of the cut. The larger complementary fragment regenerates the pattern elements normally formed by its smaller complement. Thus complementary fragments exhibit complementary behavior: One regenerates, one duplicates (Figure 14.6). Therefore, it is possible to draw an arrow across each such single straight cut on the disc, pointing from the fragment which regenerates to that which duplicates. As shown in Figure 14.5*b*, such arrows point *radially outward* from a small region in the interior of the disc. The polarity of regeneration and duplication reverses around this interior point.

2. If the disc is cut into arbitrary three-quarter and one-quarter pie sectors, the former regenerates, and the latter duplicates.

Figure 14.6   Duplicated wing disc fragment following one week in culture. (From Kauffman 1984b)

3. If an interior "distal" circular region containing the region from which arrows radiate is cut out and cultured, it duplicates. If the corresponding outer "proximal" annulus is cultured, it regenerates the central distal region. Similar results have been found in the leg disc (Schubiger and Schubiger 1978).

4. If two narrow normally duplicating crescent fragments cut from opposite edges of the disc are mixed, they regenerate the intervening pattern elements spanning the disc (Haynie and Bryant 1976).

Polar, Cartesian, and Spherical Coordinate Systems

The first major advance in predictive use of the positional-information hypothesis lay in the formation of the polar coordinate or clockface, model for pattern regulation in epimorphic fields by French, Bryant, and Bryant (1976). The initial model was based on results described in amphibian limbs, cockroach limbs, and imaginal discs and is well illustrated by application to the wing disc of *Drosophila.*

The existence of an apparently special region in the wing disc about which the direction of regeneration reverses, called the *high point,* suggested that cells might measure their distance in the tissue from this special point. This conjecture raised the possibility that the position of cells in the wing disc is specified by a polar coordinate system having the high point as its origin. Since the wing disc is a two-dimensional surface, an azimuthal angle must be measured. Were an angle specified by a single scalar variable, that variable would necessarily be discontinuous along some radial line from the high point, but Bryant found than any one-quarter pie wedge fragment duplicated, while its three-quarter complement regenerated. If an azimuthal discontinuity were present, the one-quarter fragment containing it should behave differently, and regenerate. In other words, if cells measure angle, they do so seamlessly. Thus the model postulates that cells measure radial distance from a distal high point and angle seamlessly, modulo 2π.

In order to account for the bulk of the data on epimorphic regeneration, the polar coordinate model initially proposed two rules of intercalary regeneration and a third special rule for distal transformation:

Rule 1. If cells having different radial values are apposed, cell proliferation will be stimualted and the missing intervening radial values will be restored to a resting radial gradient; then proliferation will cease.

Rule 2a. If cells having different angular values are apposed, cell proliferation will be stimulated and the missing angular values will be intercalated back to a resting angular gradient.

Rule 2b. Since two angular arcs around a 2π circle of values join any two juxtaposed angular values, a choice rule is needed. The simplest postulates that angular intercalation occurs along the *shorter* arc.

Special Rule 3. The complete circle rule: If a complete circle of angular values at a proximal radial level is exposed, distal regeneration occurs. The special nature of rule 3 will be discussed later.

Duplication and regeneration by complementary fragments of the wing disc are explained by rules 2a and 2b. Figure 14.7 shows a single straight cut on the wing disc, yielding a narrow anterior fragment and a broad posterior fragment. During culture in an adult abdomen, the narrow anterior fragment folds over, apposing the cut edge

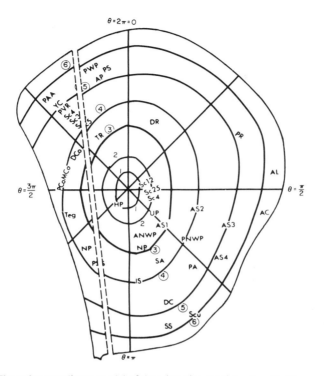

Figure 14.7   The polar coordinate model of the wing disc. Radial values (1–6) are measured from the high point in the middle of the disc. Angle θ is measured without discontinuity, modulo 2π. A straight cut anterior to the high point creates a narrow anterior fragment and a wide posterior fragment. Wound closure juxtaposes discordant angular values, leading to duplication of the narrow fragment and regeneration of the wide fragment. (From Kauffman 1984b)

such that the ventral and dorsal thoracic regions heal together and ventral and dorsal wing blade regions also heal together. This healing juxtaposes cells having similar radial values but discordant angular values. This discontinuity stimulates cell proliferation and smoothing of the angular discontinuity along the shorter angular arc. Since this shorter angular arc is the arc already present in the original narrow anterior fragment, the positional values in the new cells form a mirror-symmetric duplicate of those in the original anterior fragment, and the fragment duplicates.

The positional values present along the cut margin of the large posterior fragment are *identical* to those along the cut margin of the narrow anterior fragment. If the wound on the broad fragment heals in a similar way, the pairs of positional values apposed in the posterior fragment must be similar to those apposed in the anterior fragment. Therefore, the posterior fragment must intercalate, along the shorter angular arc, the same intervening positional values as did the anterior fragment. Therefore, the posterior fragment regenerates.

The polar coordinate model demonstrates a more general result. Whatever the coordinate system specifying position in a developmental field may be, the positional values along the two margins of a cut are identical. If the two fragments heal in similar ways, both will appose essentially identical pairs of positional values. Therefore, if subsequent pattern regulation is governed by diffusive-like averaging of positional disparities, both complementary fragments must re-form the same set of structures. If one fragment duplicates, the second, complementary fragment must regenerate. *The prediction of complementary behavior in complementary fragments* of a field is a *coordinate-free property* which follows directly from the postulate that regeneration is determined solely by the apposition of discordant pairs of positional values.

Since the polar model is symmetric about the high point, a narrow posterior fragment will duplicate and its complement will regenerate. Also, any one-quarter wedge will heal its cut margins and duplicate, and its three-quarter complement will regenerate.

Intercalary Regeneration, Betweenness, and Convex Sets

A central feature of the postulates of intercalary smoothing is that only those positional values lying *between* the apposed values can be re-formed. This constraint leads to a critical restriction in the predictive consequences of any given coordinate system, since the constraint implies that diffusive-like smoothing can re-create *only* positional values lying in the convex set bounded by the positional values in the apposed tissue edges (Kauffman 1984b). This restriction in turn implies that different coordinate systems may demand different special cellular behaviors beyond simple diffusive-like smoothing to account for the data.

The concept of a convex set and the limitations it imposes can be brought out in the polar coordinate model. Radial positions can be visualized without loss of generality as a radially symmetric gradient whose conelike peak is at the distal high point. Figure 14.8 shows a wing disc from which the distal high-point region has been removed, thus removing the radial gradient's peak. In the remaining outer proximal annulus, only lower values of the radial gradient are present. Therefore, no juxtaposition of tissue edges in the proximal annulus can lead to diffusive-like filling in of the missing high-points radial peak. In a polar model, the region containing the origin does not lie between the positional values in the proximal annulus. That is, the region around the origin is not in the convex set of all those positional values derivable by averaging any pairs of positional values present along the cut margin of the proximal annulus. Thus if the "high point" is cut out, it cannot be regenerated by the proximal annulus.

The implication of this feature of any polar coordinate model is that averaging of positional discontinuities cannot lead to re-formation of the distal peak, and some special rule is needed to re-create the missing peak by processes other than positional smoothing. In the initial formulation of the polar model, special rule 3, the complete circle rule, was proposed. According to this rule, exposure of a complete circle of angular values at a proximal level leads to regeneration of missing distal radial values. With the assumption of this rule, the model accounts for the capacity of a truncated amphibian limb to undergo distal transformation and regenerate distal wrist and hand structures from a proximal shoulder stump. The same rule accounts for distal regeneration by a proximal wing disc annulus and duplication of the high-point region when that region is cut out and cultured. Finally, rule 3 accounts for the striking observation that grafting a left hand to a right stump yields two supernumerary limbs at the positions of maximal discord in the angular values. Such a graft creates two complete circles of angular values at the radial level of the graft. These circles undergo distal transformation and yield two supernumeraries having the handedness of the host.

The polar coordinate model has been successful in the best sense. It has stimulated

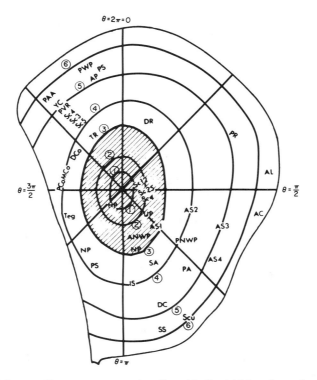

Figure 14.8 Polar coordinate system on wing disc with distal high point region (shaded area) removed. Remaining proximal annulus lacks missing distal radial values. Thus tissue contact and positional averaging cannot intercalate the missing values. (From Kauffman 1984b)

a great number of experiments which have helped shed light on pattern regeneration and its macroscopic lawlike properties and, inevitably, have uncovered weaknesses in the model.

Note that special rule 3, or any modified form of it, is formally equivalent to postulating a special mechanism *beyond* diffusive-like averaging of positional values to regenerate a missing radial gradient peak. While such a postulate is not a flaw, its status should be made explicit. We shall see shortly that distal regeneration does not require such a special process in Cartesian and spherical coordinate systems. Further, if special mechanisms exist to re-create missing gradient values, then those mechanisms are of central importance: They are likely to play a role in the initial establishment of positional gradients as well as in subsequent pattern regulation. The forms of special mechanisms suggested depend on the choice of coordinate system. Thus different coordinate systems suggest that different cellular properties beyond diffusive averaging are required to account for the data on pattern formation.

Transverse Gradients, or a Modified Cartesian Coordinate System

Several workers have independently suggested a modified form of a Cartesian coordinate model to account for the data on epimorphic pattern regulation (Cummins and Prothero 1978; Kauffman 1978, 1980; Kauffman and Ling 1980; Winfree 1980, 1984; Lewis 1981, 1982). Figure 14.9*a* shows the *Drosophila* wing disc with roughly orthogonal monotonic anterior–posterior and ventral–dorsal gradients of two chemicals, X and Y. Lines of constant concentration are bowed outward on the disc, symmetrically about the high point. Figure 14.9*c* shows the "image" of the cross gradient in XY morphogen space, or tissue specificity space (TSS) (Winfree 1980, 1984). In this image space, lines of constant concentration are straight, and so what is a convex line in (*a*) becomes a straight line in (*c*). Similarly, a straight cut on the actual wing disc corresponds to a *concave* line in (*c*). The model assumes that a cut wing disc fragment heals its cut margins, apposing nonadjacent positional values, and that simple diffusion smooths discontinuities in X and Y and fills in the convex set bounded by the apposed XY pairs along the cut margin. As shown in Figures 14.9*a* and 14.9*c*, the bowing of lines of constant X and Y concentration on the disc implies that diffusive smoothing of X and Y discontinuities in new cells of a large posterior fragment of a single straight cut will fill the shaded convex set and *regenerate* anteriorly to the anterior-most value present along its cut margin. The complementary anterior fragment apposes the same pairs of discordant values and duplicates to the same anterior-most value.

Symmetric convex bowing of X and Y concentrations about the high point ensures that the direction of regeneration reverses about the high point. A large anterior fragment from a straight cut will regenerate posteriorly, its anterior complementary fragment will duplicate. Similarly, any one-quarter fragment which apposes its two cut margins will duplicate, its three-quarter complement will regenerate. Therefore, a transverse gradient Cartesian model can yield reversal of the direction of regeneration about a high point without the assumption that the high point is a special locus from which cells measure position.

Notice next that distal regeneration is a direct consequence of simple diffusive-like smoothing of discontinuities. As shown in Figure 14.9*b*, deletion of the high-point leaves a proximal annulus. Wound healing apposes tissues around the circular cut margin, creating discontinuities in X and Y concentration gradients. Since it lies in the convex set reached by diffusive smoothing from the proximal annulus, the distal high-point region is regenerated. Similarly, the high-point region itself closes like a purse string, heals, and duplicates. By the same argument, a truncated amphibian limb will regenerate the distal wrist and hand from the proximal shoulder stump, but if the fingers of the distal fragment are implanted into the flank of the host and regeneration occurs from the cut elbow surface, a second hand, mirror-symmetric to the implanted hand, will be formed via distal regeneration.

Figure 14.9 (*a*) A Cartesian coordinate system with two monotonic gradients showing concentration profiles on wing disc. Chemical Y has ascending concentration levels 1–8; X has ascending concentration levels A–H. A straight cut creates a large posterior fragment which heals by folding in half and juxtaposing opposite ends of the cut margin. For this cut, such folding leads to a discontinuity in Y values but not in X values. The convexity in lines of constant concentration (isocones) ensures that simple diffusive smoothing of X and Y concentrations leads to regeneration out to the anterior-most X isocone (B) contained along the cut margin of the posterior fragment (stippled area). Similarly, the smaller anterior fragment heals the same way and duplicates to the B isocone. Symmetry of convexity in the X and Y isocones about a central region implies that the polarity of regeneration will alter about an apparent high point in this region. (*b*) Removal of central high-point region leads to regeneration by outer proximal annulus and duplication by central region as purse-string-closure juxtaposes the positional values around the cut margin and diffusive smoothing re-creates missing distal positional values. (*c*) Tissue specificity space (TSS) assigning each point on wing disc a unique

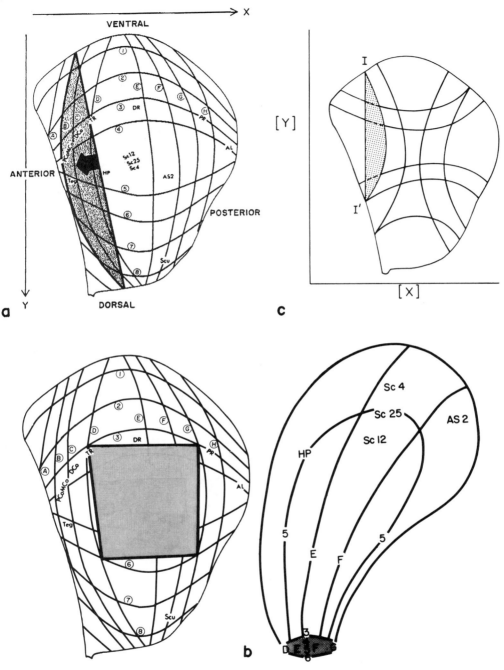

concentration value of chemicals X and Y. Concave lines are images in XY morphogen space, or TSS, of straight cuts on wing disc from (a). A single straight cut from I to I' on disc lies along the concave arc from I to I' in TSS. Wound healing apposes the wound margin in the posterior fragment, causing positional discontinuities which are smoothed by diffusion to fill and hence regenerate the convex set (shade area) bounded by the cut margin. Anterior fragment heals in a similar fashion, juxtaposing the same sets of positional values, and duplicates to the straight edge of the shaded area. (From Winfree 1984)

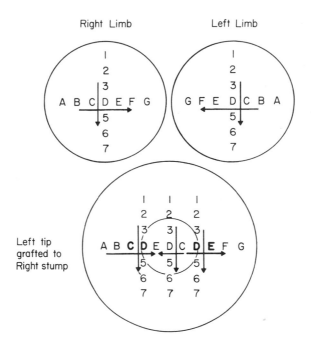

Figure 14.10 For the concentrations of chemicals X and Y, a schematic picture of anterior–posterior and dorsal–ventral Cartesian coordinates for left and right limbs, projected onto a plane. Position at the middle values of the two variables (4,D) corresponds to the distal limb tip. Grafting a left distal limb to a right proximal stump leads to smoothing of discontinuities in X and Y (**CD** and **DE**) and the formation of two supernumerary distal limb tips having the handedness of the host proximal stump, at the positions of maximal disparity of host and graft axes. (From Kauffman 1984b)

Finally, a simple transverse gradient model explains the striking observation that grafts of distal left limbs to proximal right stumps generate two supernumerary limbs which have the handedness of the host. (Figure 14.10).

Comment

Analysis of the Cartesian coordinate model, which need not be truly Cartesian in the sense that the x and y axes need not be strictly orthogonal, shows that very many of the phenomena mentioned above with respect to epimorphic pattern regulation can be accounted for quite simply by assuming *preexisting transverse gradients* of morphogens and assuming that juxtaposition of discordant positional values leads to a diffusion which simulates averaging of gradient discontinuities to fill in the convex set of accessible gradient values. Thus an extremely simple set of postulates leads to regeneration in proper spatial order, to duplication and regeneration by complementary fragments of a developing field, even to the marvelous formation of supernumerary limbs. All that are needed are unique positional information in two dimensions and averaging.

But the phenomena covered occur in organisms as disparate as cockroach, newt, and *Drosophila*. Does this broadness imply that the *identical* molecular system mediates positional information in all such organisms? That would be remarkable if true. It seems far more likely that a variety of molecular mechanisms may behave *as if* they were scalar morphogens, capable of smoothing discontinuities. Indeed, I find myself strongly tempted to say that these widespread properties of regeneration are *macroscopic, lawlike, quasi-universal* aspects of tissue organization (quasi-universal simply because they follow so simply from so many mechanisms that may act between cells to coordinate cell differentiation in space). Our analysis of genetic regulatory systems in Chapter 12 and earlier in this chapter already shows that such systems typically have many cell types in their repertoire and that, given almost any kinds of interactions between cells, local inductive interactions will generate different cells types at different spatial positions. Microhormones which pass between cells are almost bound to arise. Pick your favorite microhormone morphogen, let it communicate between cells, and graded levels are almost bound to occur. If different genes respond at different threshold levels of a microhormone, different graded levels readily elicit different responses from different cells. Juxtapose distant cells, smoothing of those gradients occurs, and the story follows. This part of the ordering in space quite lies to hand.

Spherical Coordinate Model

The final model we consider was proposed by Russell (1978) and consists in a spherical coordinate system based on three orthogonal gradients to form a tissue specificity space. Position in a tissue is specified by a solid angle, ϕ and θ, corresponding to latitude and longitudinal angles on the earth. Each positional value is a unique *ray* at a constant $x:y:z$ ratio emanating from the origin. That is, each ray is a *line* of equivalent positional values in xyz space. The longitudinal angle ϕ is defined by the ratio of x and y in the equatorial plane, and the latitude angle θ is defined by either the $x:z$ or the $y:z$ ratio. The image in xyz space of a two-dimensional tissue is a two-dimensional spherical surface pierced by a set of solid angle rays emanating from the xyz origin, which does not normally lie in the physical tissue (Figure 14.11).

The spherical coordinate model can account for almost all the data based only on

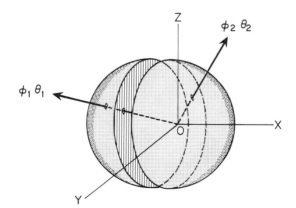

Figure 14.11 Russell's (1978) spherical coordinate model. The variables x, y, and z are orthogonal gradients. Position is specified by ratios of $x:y:z$ with respect to the origin. Each ratio is a ray at a unique solid angle $\phi\theta$. The diagram shows a disc as a spherical surface (dotted region), with an anterior fragment cut off. Wound healing in the anterior fragment creates an additional, new xyz surface (parallel lines). Rays $\phi\theta$ pierce both surfaces of the anterior fragment, and hence the anterior fragment duplicates. The posterior fragment forms the same new xyz surface and regenerates. (From Kauffman 1984b)

the concept of diffusive smoothing. This is most easily visualized in the wing disc by remembering that the disc is topologically almost a spherical surface, the disc proper, backed by the peripodial membrane. (The sphere has a hole where the traceal tube enters, but this detail does not matter for the theory.) Let this tissue spherical surface be embedded in *xyz* space such that the surface surrounds the origin. Then each solid angle ray pierces the closed two-dimensional surface of the image once (Figure 14.11), specifying the positional angles ϕ and θ. Suppose now that a narrow anterior fragment is cut from the wing disc and cultured. The cut margin purse-string-closes the wound and heals. Then smoothing of X, Y, and Z discontinuities in the new cells in the wound area forms a *second surface* in *xyz* space, a surface whose edges join those of the original anterior fragment image (Figure 14.11). Rays from the origin pierce both the new surface and the original surface; hence the narrow fragment duplicates completely. The broad posterior fragment wound heals similar and hence forms the same new image surface in *xyz* space in the new cells of the wound area. These cells are pierced by the same rays that pierce the anterior fragment; hence the posterior fragment regenerates completely.

The model is spherically symmetric. Therefore, not only do narrow anterior, posterior, ventral, or dorsal fragments duplicate while their broad complements regenerate, but a distal fragment containing the high point will purse-string-close, creating a second image surface in the new cells in the wound area. This second surface will be pierced by the same rays which pierce the high point region and therefore will duplicate the high point. Equally, the proximal annulus will regenerate distally. Finally, this model directly explains distal regeneration proportional to the proximal arc cultured and the incidence of supernumerary limbs.

The spherical coordinate model is very elegant. Indeed, I shall return to a version of it called the four color wheels model later in this chapter, where I try to account for a number of mutants which cause deletion and mirror-symmetric duplications of major body plan elements along the anterior–posterior axis of the embryo. Despite the elegance of Russell's attempt, however, it suffers the same failing that the polar and Cartesian models do: It cannot account for all the data based solely on diffusional smoothing.

The Inadequacy of Simple Diffusive-like Smoothing of Positional Discontinuities

The lawlike behaviors just described presume the preexistence of morphogen gradients. Those hypothetical spatial inhomogeneities must come from somewhere in development. Thus any body of theory resting solely on smoothing must be inadequate. It is worth asking, however, whether pattern regeneration by itself provides direct evidence of the inadequacy of this simple view. The answer is a strong "yes." A direct implication of these models is that any two complementary fragments of the same developmental field must regenerate the *same* new pattern elements. If one fragment regenerates, the other must duplicate. This consequence is coordinate-free and follows from the postulate that regeneration is driven entirely by the set of positional values apposed along the wound margin. Since a single cut unveils the same positional values on both fragments, both enjoy the same convex set of possibilities. Thus, with only positional smoothing, it is impossible for both fragments to regenerate *all missing pattern elements*. Specifically, the duplicating fragment must not regenerate as well.

In a number of experiments using long-term culture, however, it has proved possible to obtain essentially complete regeneration by small disc fragments which normally only duplicate. For example, results in our own laboratory showed that narrow anterior wing disc fragments, which normally duplicate, can regenerate pattern elements reaching as far as the posterior wing disc margin (Kauffman and Ling 1980). Similarly, Karpen and Schubiger (1981) and Schubiger and Karpen (1981) showed that a small fragment of leg disc, which normally duplicates, can regenerate an entire leg! Similar results were reported in a different system by Slack and Savage (1978).

These critical results, because they are entirely sufficient to prove that processes beyond mere smoothing of positional discontinuities are involved in pattern regeneration, inevitably drive us to ask what kinds of processes might be capable of *setting up* positional gradients in the first place. For any such process would presumably be capable of re-creating during pattern regeneration, missing gradient peaks or valleys which were not in the convex set accessible by mere diffusive smoothing. This questions leads us to the beautiful ideas of Alan Turing (1952) and the topic of *dissipative structures* (Nicolis and Prigogine 1977).

A Brief Critique of the Concept of Positional Information

Before taking up ways of thinking about the spontaneous generation of spatial patterns, a pause to reconsider the basic ideas of positional information is warranted. At root, the idea rests on a severe distinction between positional information itself, which a cell assesses, and the subsequent *interpretation* the cell places on that information, which reflects the type of cell it may be. Further, the abstract idea of positional information requires no necessary relationship between the spatial profiles of the molecular variables which carry the information and the subsequent patterns which emerge. Thus a positional-information system might be used by a tissue to set up three stripes of cells—red, white, and blue—to form a French flag, but the same information might be reinterpreted to form a Wolpertian flag of Jackson Pollack complexity. No constraints are imposed. This severely idealized model can be relaxed in two quite distinct directions. The first would suppose that the hills and valleys—the actual profiles in space—of the putative morphogens bear some natural relation to the geometry of the tissue in which pattern is being formed and perhaps even guide the growth of that geometry. That is, we might well seek theories in which the expected distribution of morphogens in space and time foretold the subsequent morphological patterns which arise. Indeed, this approach is, in a sense, a retreat to the earlier prepattern concept of Stern (1968). We take up this strand of possibilities in the next sections.

The second direction raises the fundamental concern that both the prepattern concept and the latter-day theory of positional information rely on a conceptual separation between positional-information assessment and the subsequent interpretation of that information by the cell, with no provision whatsoever for the obvious possibility that the very interpretation made by the cell might feed back and modify the information. Yet our familiarity with cascading induction events and with reciprocal inductions, which play a role in generating spatial order, suggests that such phenomena are common, not rare. Already we have seen that, in Boolean models of genomic regulatory systems splayed out in a two-dimensional tissue and with exchange of a subset of gene products between cells, induction-dependent cell types in a spatial ordering emerge almost inevitably. It is not clear at all that, in describing such situations, it would be natural to draw a sharp distinction between positional maps and the independent interpretation of those maps. Rather, the entire "tissue" settles to a spatially heterogeneous and more or less ordered attractor pattern in which the cell types which arise modify the local microhormones which are exchanged and hence the local positional map. We return to this theme in a later section. Meanwhile, suppose we consider a tissue all of whose cells are entirely identical. How might they ever become different from one another and set up a spatial pattern?

THE SPONTANEOUS GENERATION OF SPATIAL PATTERNS: TURING MODELS

Alan Turing, the same man who helped discover the German high command's ultrasecret for encrypting codes during World War II and the same man who laid the basis for modern computers with his analysis of the logical requirements for algorithms in his Turing machine, also formulated a fundamental idea concerning the onset of patterns in organisms (Turing 1952). Here I develop not his exact model in detail but only its central ideas, which have recurred, as we shall see, in many minds and been directed to many problems.

Turing's question is simple. If an organism starts from a single cell and grows to some mass of cells, all of which are identical, how do those cells ever become different and set up spatially ordered patterns? Two answers which Turing ignored are biologically reasonable and, indeed, occur. First, specialized molecules may be prepackaged in specific areas of the egg and hence partitioned to specific daughter cells during cleavage, such that different daughter cells contain different subsets of critical molecules. Molluscs are but one example of this phenomenon, where the polar lobe in the egg contains material which will later direct cells which come to contain it in particular developmental directions (Dohmen and Verdonk 1979). Related to this first answer to Turing's question, it might be supposed that, at cell division, either a nonrandom or a random but unequal partitioning of cellular materials to daughter cells caused them to enter different developmental pathways. Presumably, such unequal partitioning occurs in organisms such as *Caenorhabditis elegans.*

Turing wished to imagine the worse case and show that spatial patterning can arise. Thus consider a ring of cells, all utterly identical, each connected to its left and right neighbors around the circle. Could one imagine a mechanism whereby the complete identity—that is, the complete spatial homogeneity of chemical state around the ring of cells—would *spontaneously* break and patterns of maximal and minimal chemical concentrations would arise around the ring?

Turing could. His model builds up spatially heterogeneous patterns of chemicals from an initial uniform distribution. He supposes a chemical system containing two important chemicals. In outline, one chemical, say X, autocatalyzes the formation of itself from some precursor molecule. In addition, X catalyzes the formation of a second chemical, Y. In turn, Y *inhibits* the formation of X and also *inhibits* the formation of Y. Gierer and Meinhardt (1972) and Meinhardt and Gierer (1974) call X and Y "activator" and "inhibitor" for obvious reasons.

The next idea is that both X and Y can diffuse in the tissue but that the inhibitory Y can diffuse more rapidly than the activating X. Suppose, speaking qualitatively, that the chemical system of reactions and diffusion linking X, Y, and their precursor and product molecules has the property that the entire system has a steady state in time which is spatially homogeneous. If such a chemical system were placed in a petri dish (Figure 14.12), the concentration of X and that of Y would each be constant

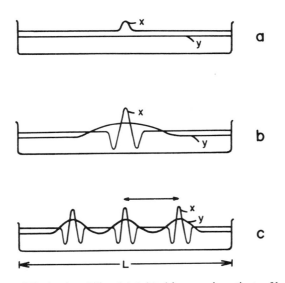

Figure 14.12   Onset of Turing instability. (*a*) A local increase in activator X above the spatially homogeneous steady-state level leads to autocatalytic increase in X concentration and to increase in inhibitor Y concentration. (*b*) Because the inhibitor diffuses more easily than the activator does, the ratio of Y to X is low in the center of the peak and high in the lateral flanks, where Y inhibits an autocatalytic accumulation of X. (*c*) Far enough away from the first peak and out of range of the diffusing inhibitor, other peaks may form. Thus there is a natural "wavelength" between peaks in the system.

throughout the dish, unchanging in time. One such state, of course, is thermodynamic equilibrium. In the absence of precipitation patterns, such as the famous Liesgang rings, no pattern can emerge at such an equilibrium (Nicolis and Prigogine 1977). Therefore, we must imagine that the chemical system is displaced away from thermodynamic equilibrium—for example, because the precursors to X and Y are present in high concentrations relative to the products and relative to the expected equilibrium ratio of precursors to products. Persistent displacement of such a system from thermodynamic equilibrium implies both that the system is open to matter, energy, or both and, as we shall see, that ordered chemical patterns can arise. Because such systems are open to the flow of matter and energy, and because they use energy continuously, they are called *dissipative systems* (Nicolis and Prigogine 1977). Such systems can exhibit the "spontaneous" onset of spatially ordered patterns.

The next issue to understand is that such a system will break symmetry and build up a macroscopic pattern if the homogeneous spatial distribution is perturbed a little bit. Suppose that a few extra molecules of X were added to the petri dish at a specific point (Figure 14.12*a*). Since X autocatalyzes the formation of itself, this addition will lead to the buildup of a local peak of X, above the steady-state level. In turn, X catalyzes the synthesis of Y and hence a local peak of Y centered on the same spot in the dish will build up. Now the first critical idea: X diffuses less well than Y. Therefore, after a little time, the X concentration will remain a narrow peak about the initial location of extra X, while the Y concentration will be a flatter, broader peak centered at the same point (Figure 14.12*b*). Consequently, in the center the level of X is high relative to Y, while in the lateral flanking regions the concentration of X is low relative to Y. Since X catalyzes its own and Y's formation, while Y inhibits its own and X's formation, it follows that, in the central region of the peak, X may continue to dominate and increase the concentration of X and Y, while on the two lateral flanks, Y may dominate and inhibit the synthesis of X and of Y, causing the concentrations to fall. Thus a local peak of X and Y concentration can build up and *suppress* the formation of similar peaks in the vicinity of the lateral flanks. Obviously, at a location far enough away from this local peak, the inhibitory influence due to excess Y relative to X wanes and a second peak can arise. This formation of a second peak leads to the next critical idea: There is some *natural minimum spacing, or "wavelength," between peaks* which such a system can form. Therefore, intuitively, we come to the conclusion that such a system might form spaced peaks and troughs of X and Y concentration around a ring of cells (Figure 14.12c). This is the heart of Turing's idea.

Mathematical Analysis

Before we go on, we need to understand the central ideas behind this class of models in more detail. As a particular model, we shall postulate a biochemical system comprising two chemical components, X and Y, with concentrations $X(r,t)$ and $Y(r,t)$ at position r at time t; the two components are being synthesized and destroyed at rates $f(X,Y)$ and $g(X,Y)$ at each point in the spatial domain and are diffusing throughout a tissue. The partial differential equations for this system are

$$\frac{\delta X}{\delta t} = f(X,Y) + D_X \nabla^2 X \qquad (14.1a)$$

$$\frac{\delta Y}{\delta t} = G(X,Y) + D_Y \nabla^2 Y \qquad (14.1b)$$

where D_X and D_Y are diffusion constants. These equations are chosen to have a spatially homogeneous temporal steady state X_0, Y_0. Since the state is steady, the rate of change is zero for both X and Y; hence $f(X,Y)$ and $g(X,Y) = 0$ for X_0, Y_0. Furthermore, in that spatially homogeneous state, no inhomogeneity exists, and so the ∇^2 term, or Laplacian operator, is zero.

The Laplacian operator is critical to our discussion. Picture a line of cells, with different concentrations of a chemical in the different cells. The rate at which the chemical enters a cell in the middle of the line (Figure 14.13) is proportional to the difference between the concentration in that cell and the concentration in each of its neighbors. Figure 14.13 shows a concave monotonic gradient which becomes flatter toward cells at the right end of the line. Therefore, for a middle cell, say cell 4, the rate at which a diffusing substance S enters from cell 3 is greater than the rate of efflux of S to cell 5. Therefore, S accumulates in cell 4. However, the difference in concentration between cells 3 and 4 and between cells 4 and 5 is just the derivative, or slope, of the concentration of S (denoted by [S]) along the line of cells at those three points. The fact that S accumulates in cell 4 is shown by the slope into 4 being greater than the slope out of 4. The difference between these two slopes is the second derivative of the concentration profile [S], which in a single spatial dimension is just the Laplacian operator. The Laplacian therefore gives the rate of change of concentration, [S], at each point in the line of cells. Since, by hypothesis X_0, Y_0 is a *spatially homo-*

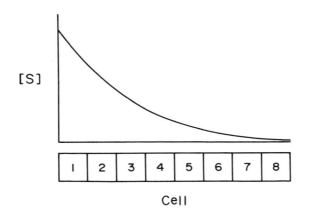

Figure 14.13 Monotonic concave morphogen gradient along a line of cells. The difference in concentration between cells 3 and 4 causes morphogen to diffuse into 4 from 3. The difference in concentration between cells 4 and 5 causes morphogen to diffuse from 4 to 5. The fact that the concentration gradient into 4 from 3 is steeper than the gradient out of 4 into 5 implies that morphogen concentration builds up in 4. This simple example shows intuitively why, over time, the change in morphogen concentration at a point in space due to diffusion depends upon the second derivative of the spatial concentration, as given by the Laplacian operator.

geneous steady state, there is no gradient profile in either X or Y and hence the Laplacian term in each equation is zero.

Analysis of the system described by Equation 14.1 begins by linearizing the equations about this spatially homogeneous steady state. This is carried out by using new imaginary chemical variables x and y, which measure the *deviation* of X and Y from the steady-state levels X_0 and Y_0. Thus while X and Y cannot have less than zero concentration, x and y can be less than zero, meaning less than the steady-state concentrations. Linearization means merely that the *local* way the synthesis or destruction of X or Y changes as concentrations deviate a bit above or below the steady-state levels is approximated by simple constants, numbers which show the tangent approximation to the curves $f(X, Y)$ and $g(X, Y)$ at the point X_0, Y_0. That is, the linear term in the Taylor expansion of $f(X, Y)$ and $g(X, Y)$ is kept.

We need next the idea of an eigen function of the Laplacian operator. This operator simply expresses the idea of local averaging of concentration differences due to diffusion. In fact, the evidence for averaging of positional discontinuities in epimorphic pattern regulation strongly argues that a process analogous to diffusion is occuring—the diffusive-like positional averaging positional values. An *eigen function* is a spatial distribution of a diffusing chemical having the property that, over time, the only effect of diffusion is to alter the amplitude of the pattern but not its basic shape. One eigen function for diffusion is a flat, homogeneous distribution, since a flat pattern remains constant in time. More interesting, in a one-dimensional domain, such as ink in a glass capillary tube, a perfect sinusoidal distribution is an eigen function. Under the action of diffusion, the sine pattern will flatten but not change shape. The proof is simple. The first derivative of a sine pattern is a cosine pattern. The first derivative of a cosine pattern is the negative of a sine pattern. Thus the second derivative of a sine pattern is the negative of a sine pattern, while the second derivative of a cosine pattern is the negative of a cosine pattern. This means that the shape stays the same, but the amplitude falls.

The next straightforward idea we need is that the spatial wavelength of the sine pattern of an ink or of any other chemical can be short or long. The reciprocal of the wavelength is the wavenumber k, which denotes the number of wavelengths which can be fit into some fixed interval.

Linear Stability Analysis

Armed with these ideas, we can approach the classical task of linear stability analysis. The question of interest is this: If the homogeneous steady-state pattern X_0, Y_0 is perturbed by sinusoidal deviations of X and Y above and below the steady-state levels, with the deviations having any specified wavenumber k, will that sinusoidal pertubation *die away* and dampen back to the flat, spatially homogeneous steady state or will it *amplify* and create a high-amplitude pattern of wavenumber k and the corresponding wavelength? If, for all wavenumbers k, the system damps back to the spatially homogeneous pattern, the system is unable to create a pattern. If for some values of k or for some ranges of values of k, the system can amplify a small perturbation, then it can form a pattern of high amplitude. The way in which the system will form the new pattern requires a next critical idea: At the spatially homogeneous steady state, the real system is subjected to *fluctuations* in the concentrations of X and Y at all points in the spatial system. Any fluctuation which has small peaks and valleys can be decomposed by Fourier techniques into the weighted *sum* of very many short- and long-wavelength sine patterns of X and Y or, more precisely, of x and y deviations above and below X_0, Y_0. Thus, in the noisy spectrum, almost all wavelengths occur at low amplitude *simultaneously,* each distributed across the *entire tissue domain.* Thus if the chemical system can amplify a specific wavelength, a bit of that wavelength is bound to be present in the noisy fluctuations. The system will pluck that bit out and amplify it, thereby creating a macroscopic pattern of X and Y peaks and valleys throughout the tissue.

The linearized equations for our two-chemical system are

$$\frac{dx}{dt} = K_{11}x + K_{12}y + D_x \nabla^2 x \qquad (14.2a)$$

$$\frac{dy}{dt} = K_{21}x + K_{22}y + D_y \nabla^2 y \qquad (14.2b)$$

where the constants K_{ii} are the slopes of $f(X, Y)$ with respect to X and Y and $g(X, Y)$ with respect to X and Y at X_0 and Y_0. The linear stability of the system described by Equation 14.2, after sinusoidal pertubations of wavenumber k, is analyzed by evaluating the determinant of the matrix

$$\begin{vmatrix} K_{11} - k^2 D_x - \lambda & K_{12} \\ K_{21} & K_{22} - k^2 D_y - \lambda \end{vmatrix} = 0 \qquad (14.3)$$

where λ is one of two *eigen values* in the system comprising the two chemical species X and Y. If an eigen value is positive, the associated pattern will grow in amplitude; if negative, the associated pattern will decay. The wavenumber k enters squared because the second derivative of sin kx is $-k^2$ sinkx. This determinant leads to a relation between the two eigen values, λ_1 and λ_2, and the remaining parameters of the linearized system:

$$\lambda = f(k)$$
$$\lambda_{1,2} = \tfrac{1}{2}[K_{11} + K_{22} - k^2(D_x + D_y)]$$
$$\pm \tfrac{1}{2}\sqrt{[(K_{11} + K_{22} - k^2(D_x + D_y)]^2 - 4[k^4 D_x D_y - k^2(D_y K_{11} + D_x K_{22} + K_{11}K_{22} - K_{21}K_{12})]}$$
$$(14.4)$$

With appropriate constraints on the linearized reactions and diffusion constants, this *dispersion relation* represented by Equation 14.4 between one of the two eigen values and wavenumber is positive for a restricted range of wavenumbers and hence for a restricted range of wavelengths, L_1 to L_2 (Figure 14.14). In this case, as noted, the chemical system acts as an amplifier and plucks from thermal noise the spatial wavelength which arises.

It is also true that a specific *ratio* of the two underlying chemical concentrations x and y, measured as deviations from the steady state X_0 and Y_0 ([X] = X_0 + x, [Y] = Y_0 + y), occurs in the pattern. This fixed ratio is called an *eigen vector* and can be determined mathematically by substituting the eigen value for λ in the determinantal equation and then solving the pair of algebraic equations for x and y, which are thus determined to be a fixed ratio. Thus the pattern might amplify $1.5x:1.0y$ everywhere in the spatially sinusoidal distribution.

The next concept needed is a *boundary condition*. Imagine a one-dimensional (in other words, narrow) capillary tube in which this reaction occurs, closed in a ring. Now, only sinusoidal patterns can grow in the ring, and therefore an *integral number* of patterns must fit in around the ring. That is, the pattern might be flat or it might have a single sinusoid (one peak and one trough) or a double sinusoid or a triple (Figure 14.15a). A ring has periodic boundary conditions, meaning simply that the peak–valley patterns must mesh smoothly to maintain a sinusoidal pattern around the ring. The importance of such boundary conditions is that intermediate wavelengths do not fit into the ring. Therefore, if the length around the ring is fixed at L, only a sequence of distinct wavelengths can fit, having wavelengths L, $L/2$, $L/3$, $L/4$ … L/n. Therefore, in a ring structure, the amplification of a sinusoidal pattern by the chemical system requires that two conditions be met simultaneously: the system must amplify some range of wavelengths, and at least one of those wavelengths must fit into the ring domain. If both conditions are met, a pattern will form.

The same ideas carry over to a straight capillary tube on which the ends are sealed. No diffusion can occur across the ends. This no-flux boundary condition implies that, at the boundary, the gradient is entirely flat. In turn, this flatness implies that the pattern in the tube must be a cosine (Figure 14.15b). Again, for any tube length L, the patterns which can fit in are cosines of lengths L, $L/2$, $L/3$. …

Linear stability analysis is useful but limited. I shall describe more fully the nonlinear analyses made of this important class of models. Before doing so, however, I turn to initial applications of Turing's ideas to biological pattern formation in a cellular slime mold and in the development of the fruit fly. I note that the mathematical material described above has received detailed attention by a number of workers, who may be consulted for further information. Gmitro and Scriven (1966), Babloyantz and Hiernaux (1975), Nicolis and Prigogine (1977), and Kernevex (1980) are useful sources.

SUCCESSIVE EIGENFUNCTION MODEL OF COMPARTMENTALIZATION

NONLINEAR REACTION DIFFUSION SYSTEM:

$$\frac{\partial X}{\partial T} = F(X,Y) + D_x \nabla^2 X$$

$$\frac{\partial Y}{\partial T} = G(X,Y) + D_y \nabla^2 Y$$

LINEARIZE ABOUT SPATIALLY HOMOGENEOUS STEADY STATE AND PERFORM STABILITY ANALYSIS

$$\begin{vmatrix} A_{11} - K^2 D_x - \lambda & A_{12} \\ A_{21} & A_{22} - K^2 D_y - \lambda \end{vmatrix} = 0$$

UNDER CONDITIONS:

(i) $A_{11} + A_{22} < 0$

(ii) $A_{11} A_{22} - A_{12} A_{21} > 0$

(iii) $(A_{11} - A_{22})^2 > -4 A_{12} A_{21}$

(iv) $D_x A_{22} + D_y A_{11} > 0$

(v) $\left(\sqrt{\frac{D_x}{D_y}} A_{22} - \sqrt{\frac{D_y}{D_x}} A_{11} \right)^2 > -4 A_{12} A_{21}$

THE REACTION-DIFFUSION SYSTEM HAS SPATIAL PATTERNS WITH A "NATURAL" WAVELENGTH L_*

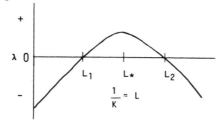

Figure 14.14 Linear analysis of nonlinear coupled reaction-diffusion system in the two-chemical Turing class leads to a dispersion relation between the two eigen values of the system, λ_1 and λ_2, and the wavelengths L or, alternatively, wavenumbers $k = 1/L$, which are unstable and grow in amplitude. For proper choices of parameters, one eigen value is greater than zero between a restricted range of wavelengths L_1 and L_2; hence the system amplifies any eigen function pattern within that wavelength range.

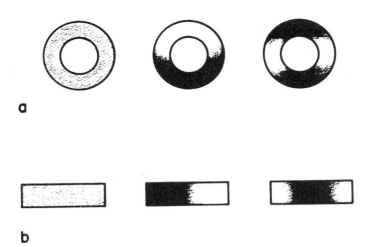

Figure 14.15   (*a*) Eigen function sine patterns in a ring domain with periodic boundary conditions which demand an integral number of peaks and troughs. (*b*) Eigen function cosine patterns, flat at boundaries where no flux of reactants across the boundary occurs.

Application to Radial Branching Patterns in Polyspondilium discoidum

Reaction-diffusion models, the familiar name for this class of systems, have been applied to a large variety of systems. The first application, by Turing himself, was to the formation of the tentacles which emerge from the circular mouth region of hydra. The first example I want to discuss, however, is based on very current work. *Polyspondilium discoidum* is a cellular slime mold which is a first cousin of the more famous *Dicteostyleum discoidum. Polyspondilium* lives as free individual amoeba which feed on bacteria. Under appropriate conditions of starvation, large numbers of cells orient toward an initial signaling cell or small cluster of cells, which emit cAMP. This chemical signal causes the amoebae to migrate toward the signaling source and form a mass of cells which creates a sorogium. This mass erects itself vertically and forms two major types of cells: stalk cells, which support the mounting column and its later branching structures, and spore cells, which are the germ cells for the next generation. The morphology of interest to this discussion concerns the branching pattern which arises (Figure 14.16*a*). As the column of the sorogium mounts, it leaves behind globular masses of cells, called whorls, at spaced intervals along the column. Each whorl in due course gives rise to radial branches oriented perpendicular to the main stalk column. The number of branches from any one whorl ranges from one to ten and is uncorrelated from one whorl to the next. The angular spacing between radial branches is quite regular in each whorl. Inspection of Turing-like patterns on a ring (Figure 14.15*a*) and of the patterns of radial branches around a whorl suggested to Cox and his colleagues (Byrne and Cox 1986, 1987; Cox and McNally 1989) that this class of models might well account for the observed pattern.

In order to test this general hypothesis, Cox and co-workers have raised monoclonal antibodies, seeking those which might react, in the developing organism, with spatially specific antigens which reflect this branching pattern. Their hope was that

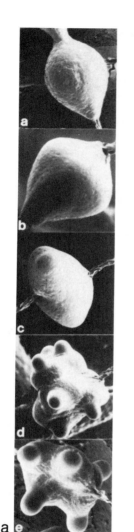

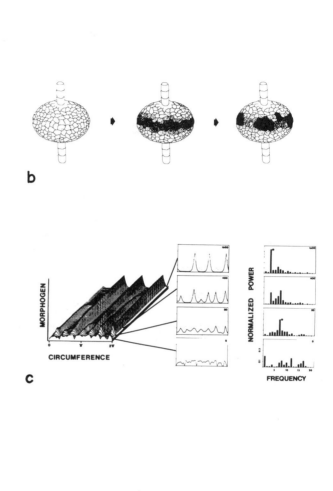

Figure 14.16 (*a*) Distribution of young and old whorls in *Polyspondilium discoidum*. (*b*) Distribution of antigen around sorogium over developmental time. Pattern begins diffuse, then gathers into peaks located over loci where whorl tips form. (*c*) Predictions of time course of antigen pattern by Turing class reaction-diffusion model. (From Byrne and Cox 1987)

by being able to visualize the nascent pattern at the molecular level via antigen markers which can be taken to reflect the prepattern before subsequent morphology is obvious, they might have a better clue to the way that pattern originates.

Figure 14.16*b* shows characteristic patterns of antigen distribution on young and old whorls. The authors analyzed the distribution of antigens based on Fourier decomposition of the observed patterns into a spectrum which had weightings on the 0,1,2,3 . . . peak sinusoidal distributions around the equator of each whorl. The observed distributions show, roughly, the following phenomena. First, each antigen is first visible diffusely throughout the early whorl. Slightly later, the antigens appear to be concentrated in small clumps scattered at random throughout the whorl. Slightly later, the antigens are preferentially located at the whorl equator, forming a rough ring around the main stalk. Thereafter, the antigen ring breaks up into a number of almost evenly spaced clumps, or peaks, around this ring. The location of these peaks essentially foretells the location of the subsequent tip of each radial branch. The Fourier analysis confirms this impression. The initial pattern has contributions from many "modes," or sinusoidal wavelengths, around the whorl. As the whorls mature, the power spectrum concentrates almost all the weight on a single mode (sometimes two modes) with a general shift in weights toward lower wave numbers— that is, toward longer wavelengths. Thus, as the pattern matures, high spatial frequencies damp away and the lower wavelengths dominate, to set up the branching pattern.

Cox and co-workers point to the similarities between these data and reaction-diffusion models (Figure 14.16c). In a Turing-like system, our expectation is that initially the pattern will be seen as random fluctuations away from a diffuse middle state (X_0, Y_0), with short and long wavelengths present. If the ring circumference is large enough so that several different amplifiable wavelengths can each fit an integral number of times around the ring, then each wavelength can be amplified. Higher- and lower-wavenumber patterns (shorter and longer wavelengths) will damp out. Thus in the midterm of the pattern's formation, many different sinusoidal modes should be present. In the longer term, one or a few wavelength patterns should dominate. Thus the power spectrum should simplify. Indeed, in the purely linear analysis, the wavelength having the largest eigen value must eventually dominate entirely, since each wavelength mode grows exponentially as $e^{\lambda t}$. Thus, eventually, the largest value of λ wins. For a fully nonlinear analysis, discussed below, the amplitudes of sine patterns cannot grow indefinitely; hence this simplification of pattern to a single dominant wavelength may be limited. Finally, the radial spacing ought to be, and is, quite regular, as expected from the natural spacing of a sinusoid wave around a ring.

The similarities between theory and observation are obviously gratifying. Before criticism, what further points can be made? Cox and his colleagues argue, quite convincingly I believe, that this pattern would be very hard to account for on the basis of a clonal "computation" or a cell-lineage mechanism. Further, this is a clear case in which the mere hypothesis of positional information is far less predictive, hence far less interesting, than a more detailed theory in which the putative spatial distribution of morphogens accounts for the pattern seen. Despite these successes, the results do not establish that a reaction-diffusion mechanism in fact accounts for the observations, for several reasons. First, the theory is incomplete. The authors argue about the formation of peaks and valleys of morphogens around a one-dimensional ring, whereas the actual whorl is a three-dimensional globular mass of cells. This discrepancy does not rule out a Turing-like model; it is, however, a request that the theory be carried through. A second reason for caution, is even more important and even more interesting, as we see next.

Mechanochemical Models Give Much the Same Patterns

Spatial patterning can be based not only on instabilities in reaction-diffuse systems but also on mechanochemical systems. Equation 14.4, the dispersion relation between eigen values and wavenumber, $\lambda = f(k)$, is the critical underlying mathematical form which governs the kinds of macroscopic patterns that emerge. Many dynamical systems having a dispersion relation with a single hump where an eigen value is positive for a restricted range of wavelengths $0 < L_1 < L_2$ give rise to very much the same patterns. Therefore, it is exciting that recent work investigating mechanochemical models reveals very similar potential patterning. In the Turing class of models, the key idea is short-range activation and long-range inhibition (due to low diffusion of X and high diffusion of Y). Similar mechanochemical models are based on calcium-release-mediating alteration of stress and strain in fibrillar proteins forming the cytoskeletal matrix. Mechanical strain acts as the long-range inhibitory force, while local calcium release, which stimulates further local calcium release at a point in the cytoskeletal matrix, acts as the short-range activator. The resulting equations are different from Turing models in detail; the dispersion relation is the same (Oster and Odell 1984; Goodwin and Trainor 1985). The consequence is that we can now begin to envision an unknown range of mechanochemical models, in addition to the Turing class, all of which generate similar eigen function patterns. A clearly interesting aspect of the mechanochemical models, however, is that, by generating mechanical forces, these models suggest cellular mechanisms by which actual morphologies are built. Below we return to these developments and consider a model for the sequential formation of bone elements in the vertebrate limb. Cox and co-workers comment favorably on this class of models as well, as potentials to account for the phenomena they observe. Figure 14.17 shows an example these authors analyzed.

Similar eigen function patterns are not limited to even these mechanochemical models. For example, consider either the modes of bending of a uniform beam or the vibrations of a stretched string or a plate. Vibrating strings have natural harmonics with $1, 2, 3 \ldots n$ wavelengths fitting onto the string. Vibrating plates exhibit a variety of resonant modes, analagous to those we shall discuss shortly in trying to account

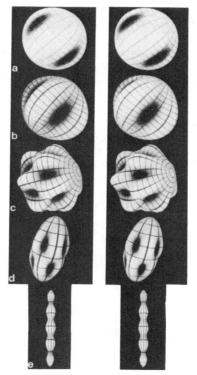

Figure 14.17 Mechanochemical model of whorl formation in *Polyspondilium discoidum.* (From Cox and McNally 1989)

for phenomena in *Drosophila*. Indeed, the patterns found on vibrating strings and plates would be expected to arise in growing two-dimensional epithelia. Such an epithelium—forming a spherical closed surface, say—will buckle inward or outward if excess cell proliferation occurs in subregions of the spherical surface and creates stress. The spatial patterns of that buckling will strongly tend to follow the natural buckling eigen function modes for the overall geometry of the tissue. All these examples have similar eigen functions and hence generate a similar family of possible "natural" patterns.

From our present point of view, a very important message is that quite different underlying chemical, mechanochemical, and mechanical systems are governed by similar mathematical principles. Just as positional smoothing accounts for many phenomena in pattern regeneration in a coordinate-free way, many similar themes of tissue organization may all derive from modified forms of Turing's essential, and central, idea.

Analysis of Turing and related models reveals a further issue of central importance to our thinking about the relation between self-organization and selection: Evolution may often be constrained to form those patterns which are "easily" generated by any developmental mechanism. Our analysis already shows that, within any such developmental mechanism, some patterns are easy while others are hard. The latter require that parameters be held in tiny volumes in the corresponding parameter space. For example, in a reaction-diffusion model, obtaining a system with short-wavelength sinusoidal patterns is very difficult because diffusive smoothing of concentration differences over short distances is very rapid. Short-wavelength *macroscopic* sinusoidal patterns generate just such steep concentration gradients, however. Therefore, diffusive smoothing which drives incipient short-wavelength sinusoidal patterns back to the homogeneous state is very powerful. Thus selection would have a hard time building a reaction-diffusion system which generated short-wave length patterns, in the precise sense that the parameters of the system which amplify a short wavelength must be confined to very small volumes of paremeter space. In short, the sizes of the volumes in parameter space corresponding to specific members of a family of patterns provide a way to think about the relation between natural selection and the pattern found. Evolution ought typically to exhibit the "typical" patterns easily generated by any developmental mechanism.

COMPARTMENTAL AND SEGMENTAL PATTERNS IN *DROSOPHILA MELANOGASTER*

I return now to *D. melanogaster*. This organism offers stunning evidence of wavelike phenomena in its pattern formation. We shall discuss this topic first with respect to the beautiful phenomenon of sequential compartmentalization in the wing disc and other imaginal discs of the larval stage. We shall find that the symmetries, time course, and spatial patterning of the sequential formation of compartmental boundaries all appear open to explanation in Turing's terms. Thereafter, I shall discuss the onset of pattern formation in the early embryo. We shall find complex spatial-temporal patterns of gene transcription which are hauntingly similar to those which Turing might have foretold. Those patterns hint at a mechanism of positional specification which I shall call the *four color wheels model*. It, too, follows with pleasing directness from Turing's fundamental insights.

To set the stage for our consideration of *Drosophila*, I now sketch its development. *Drosophila's* egg (Figure 14.18*a*) is about 600 micrometers long and 100 micrometers from dorsal to ventral, forming a slightly distorted ellipsoid whose dorsal surface is straighter than its ventral surface. Early cleaveage division in *Drosophila*, as in

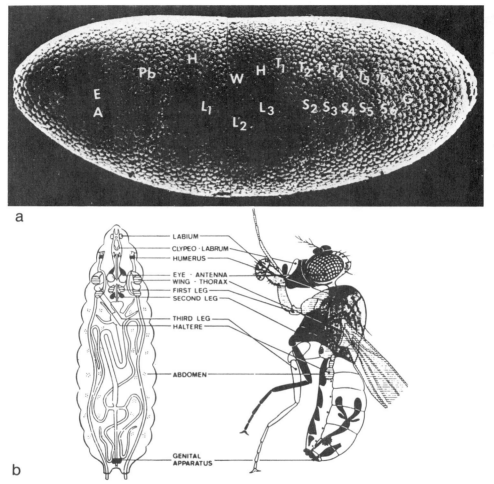

a

b

Figure 14.18 (*a*) Fate map of *Drosophila* egg. See key to Figure 12.21 (p. 515) for abbreviations. (*b*) Correspondence of imaginal discs to parts of the adult ectoderm.

other insects, is unlike that in amphibians and other vertebrates. In the former, the egg does not divide, while the zygotic nucleus undergoes 14 rounds of division to create a *syncytium* in which many nuclei coexist in a common cytoplasm. The first few nuclei to be formed are roughly on a spherelike surface inside the egg but migrate to the egg cortex at the eighth and ninth divisions. After four more divisions, most nuclei lie adjacent to the cortex while some remain in the interior. Following the last cleavage division, which creates a stage called the *synctial blastoderm,* cell membranes extend first downward and then beneath each subcortical nucleus, creating the first cells after the egg itself. This stage is *cellular blastoderm,* which comprises about 6000 cells. The nuclei which remained in the yolky interior of the egg cellularize as well (Foe and Alberts 1983).

Subsequent to cellular blastoderm formation, at about three hours at 25°C, the movements of gastrulation ensue. Invagination along the ventral midline of mesodermal and neurectodermal progenitor cells starts. Simultaneously, two furrows, the cephalic and hindgut furrows, running obliquely dorsal to ventral in the anterior and posterior halves of the blastoderm, arise. As gastrulation proceeds, the germ band extends: The region along the ventral half of the embryo lengthens, pushing the anterior and posterior terminal masses first dorsally and then backward along the dorsal part of the embryo toward its midregion. At about this stage, clear evidence of segmentation is visible. Thereafter, the germ band retracts, and by about 12 hours, the fully segmented embryo is evident (Counce and Waddington 1972; Turner and Mahowald 1976, 1977, 1979; Underwood, Turner, and Mahowald 1980). *Drosophila* has—by one counting criterion, at any rate—17 segments. The anterior-most five of these form the head segments of the larve and adult, the next three form the three thoracic segements—prothorax, mesothroax, and metathorax—the next eight form abdominal segments, and the final one forms part of the internal and external genitalia. At about 24 hours, the first-instar larve hatches, undergoes two further molting cycles to create second- and third-instar larvae, pupates at about 96 hours at 25°C, metamorphoses over a few days, and emerges as an adult.

As noted in Chapter 12 and above, the entire ectoderm of the adult derives, during metamorphosis, from special nests of cells in the larva, called either *abdominal histoblasts* or *imaginal discs* (Gehring 1973, 1976; Nothiger and Gehring 1973). The imaginal discs initiate as nests of cells present in the first-instar larva, each disc committed to form a unique part of the adult ectoderm. Each disc grows during the larval-instar stages to about 10 000 to 60 000 cells.

A variety of techniques, including ablation by hot needles or lasers, ligation experiments, and genetic techniques marking cells and their progeny, have made it possible to constrct a *fate map* of the adult or of the larva, cast back onto the blastoderm stage of the embryo (see, for instance, Poulson 1950a, 1950b; Garcia-Bellido and Merriam 1969; Bownes and Sang 1974a, 1974b; Janning 1974, 1978; Bownes 1975; Lohs-Schardin, Cremer, and Nüsslein-Volhard 1979). Such a map is shown in Figure 14.18a. A number of points must be stressed. First, a fate map shows that cells in a specific area will ultimately give rise to specific parts of the adult; it does *not* demonstrate that the cells of the blastoderm have become *committed* in any sense to that prospective fate. The cells might in fact be fully labile, such that if transplanted elsewhere in the embryo they would adopt another fate, depending on location, and yet the fate map would still hold. We return to this point below. Second, if we compare the fate map on the blastoderm with the geography of the adult (Figure 14.18b), it is startling how similar the two are. Cells which will form head regions lie in the anterior of the blastoderm; those which will form the three thoracic segments lie in linear order along the fate map, anterior to posterior; those destined to form the three thoracic legs lie ventral to those destined to form the dorsal parts of each thoracic segment; cells which will form successive abdominal segments lie in serial order along the blastoderm. This topological near-identify restates Wolpert's issue: Somehow cells at the proper places in the blastoderm come to "know" what segmental fate to assume.

A next fascinating feature of the fate map on the blastoderm is brought out by recalling the marvelous phenomena of transdetermination and homeotic mutants discussed in Chapter 12 (Hadorn 1966, 1967, 1978; Gehring 1973; Gehring and Nothiger 1973; Kauffman 1973, 1975). *Transdetermination events and homeotic mutants jump long distances across the fate map!* Eye tissue can change and form genital tissue. Even a casual biologist will recognize that the eye and gentialia are located at opposite ends of the fly and of the fate map. Recall that the evidence showing a committed, or determined, state in each imaginal disc rests on culturing a fragment of that disc in an adult abdomen for a week, removing the tissue and transplanting it to a successive adult abdomen, and so on, over a series of transfers. The determined state of the cultured disc tissue can be assayed by implantation into a host third-instar larva. When the host goes through metamorphosis, so does the implant.

The resulting cuticular patterns identify the adult tissue, if any, formed by the implanted disc. Recall that the first important result is that wing disc may be cultured for up to ten years by serial transfer and still form adult wing. Hence the determined state must be cell-heritable. Occasionally, however, the cultured tissue jump-change to a new heritable state, a process called transdetermination. Thus wing disc can jump to an eye-heritable state, therefore propagatable, and form parts of the adult eye on differentiation. Figure 12.18 shows the pattern of transdetermination seen among the major imaginal discs. Recall also that most of these transdetermination steps also are known homeotic mutations. For example, eye transdetermines to form wing. Similarly, the *eyeless opthalmopteria* homeotic mutant transforms eye to wing (Ouweneel 1976). Antenna can transdetermine to genitalia. The *tumorous head* mutant replaces parts of the head and antenna with genitalia (Postlethwait, Bryant, and Schubiger 1972; Ouweneel 1976). Antenna can transdetermine to form leg. Members of the *Antennapedia* complex transform antenna to leg (Gehring 1966a, 1966b; Ouweneel 1976; Kauffman and Ling 1980). All these tissues are distant from one another in the adult and on the fate map.

What is the implication of the fact that transdetermination steps and homeotic mutants can jump long distances across the fate map? The importance of these metaplastic transformations is that, in some sense, they show us that those tissues which can transform one to another have neighboring developmental programs. The question is: In what sense of "neighboring"? Since transdetermination steps and homeotic mutants leap over large regions of the underlying fate map, *whatever* the sense of "neighboring developmental programs" may be, that sense is not trivially related to the fate map. Clearly, we would like to find a deeper sense of neighboring programs and, if we are lucky, a way to relate that sense to the geometry of the fate map. Possible important clues to this relation arise in the seemingly unrelated phenomenon of *sequential compartmental boundary formation* in the developing wing disc, discussed next. We shall find a clear sense, based on the combinatorial epigenetic code broached in Chapter 12, in which nonneighboring tissues can have neighboring developmental programs.

Sequential Compartmentalization in the Wing Disc

Compartmental boundaries in *Drosophila* are lines separating regions of tissue whose cells have progeny which do not cross the boundary. Thus compartmental boundaries are lines of *clonal restriction.* Discovery of such boundaries rested on a genetic technique called *mitotic recombination,* which is useful for our discussion.

In *Drosophila,* maternal and paternal homologous chromosomes are synapsed, a state of affairs which allows breakage and recombination between homologues. In particular, this mitotic recombination occurs, after replication of each chromosome, between sister chromatids (Garcia-Bellido 1975; Becker 1978). Suppose the fly is heterozygous for a recessive mutant y which, when both copies of the mutant gene are present, causes bristles to be yellow rather than the normal brown. That is, y/y gives yellow bristles, while either y/+ or +/+ gives brown bristles. Recombination can yield two daughter cells having the property that one is +/+ and the other is y/y. The progeny cell which is homozygous for y, as well as *its* progeny, remains y/y. Hence any cells which form bristles from this y/y clone of progeny form yellow bristles. Using such mitotic recombination techniques, it is possible to mark a cell and all its progeny. Therefore, it is possible to determine the spatial location of the yellow clone on the adult cuticle and hence determine what part of the adult is formed by progeny of the initial y/y cell. (Since not all y/y cells create bristles, of course, not all progeny are visualized.) It is not possible to decide which cell will be genetically "marked," but by utilizing X rays to induce mitotic recombination, the developmental stage at which a cell is marked can be defined.

Using this mitotic recombination technique, Garcia-Bellido, Ripoll, and Morata (1973), 1976) and Garcia-Bellido (1975) found lines of clonal restriction. These workers marked cells early in development, during midembryonic life. They examined the wings of the resulting adults and were surprised to find that large clones of marked cells appeared in either the posterior half or the anterior half of the wing. Such clones occupied overlapping areas in wings of different flies. What was striking, however, was that no clone crossed a particular boundary separating the anterior and posterior wing surfaces into two distinct compartments. Instead, a large posterior clone might abut this boundary and run along it for hundreds of cells; alternatively, an anterior clone might do the same. Otherwise stated, the marked clones appear to be respecting this boundary and not crossing it. This behavior is the operational definition of a compartmental boundary (Crick and Lawrence 1975; Garcia-Bellido 1975).

By irradiating at different times, Garcia-Bellido and colleagues found that, at successively later times, a succession of compartmental boundaries arise and sequentially subdivide the maturing wing disc into finer and finer spatial subdomains. Figure 14.19a shows the compartmental boundaries on the adult wing. Figure 14.19b projects the same boundaries onto the third-instar wing disc. The first anterior–posterior boundary is formed in early larval life. A second boundary, the dorsal–ventral one, arises in the first instar stage and divides the dorsal and ventral wing surfaces from each other. This boundary comprises what will become the wing margin, since the adult wing has two epithelial surfaces. Prior to the formation of the dorsal–ventral boundary, clones can extend from the dorsal to the ventral surface but do not stray across the anterior–posterior boundary. After the dorsal–ventral boundary arises, clones are constrained to respect both the anterior–posterior *and* the dorsal–ventral boundary. Thus cells are now confined to finer subregions. Shortly after the formation of the dorsal–ventral boundary, a third boundary arises, dividing the thoracic regions of the disc from the wing blade area. This wing–thoracic boundary is particularly interesting. The fate map of the wing disc (Figures 14.5a and 14.19b) shows that the wing blade derives from the midregion of the disc, while the ventral and dorsal thoracic areas derive from the two separate ends of the wing disc. Note that *distant regions,* here the two ends of the disc, adopt fates which appear to be more similar to each other than to the midregion between them. By the second-instar stage, a further compartmental boundary divides the dorsal–thoracic region into subregions. In the third-instar stage, a final boundary divided the distal wing blade from the proximal wing hinge area.

Projection of the boundaries onto the third-instar wing disc (Figure 14.19b) makes it clear that the wing is successively divided into subregions in a particular order. We discuss next the possible significance of compartmental boundaries, then take up possible mechanisms which control the sequence and geometry of these boundaries.

The Compartmental Hypothesis and a Binary Combinatorial Epigenetic Code—Again

A central hypothesis advanced by Garcia-Bellido (1975) and by his co-workers (Morata and Lawrence 1975, 1977; Lawrence and Morata 1977) is that a compartmental

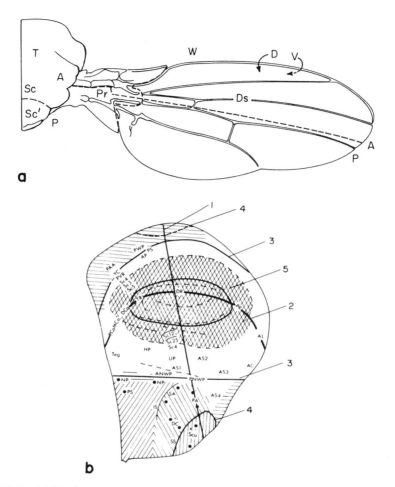

Figure 14.19 (*a*) The five compartmental boundary lines on the thorax and wing. A, anterior; D, dorsal; Ds, distal wing; P, posterior; Pr, proximal wing; Sc, scutum; Sc′, scutellum; T, thorax; V, ventral; W, wing. (From Garcia-Bellido 1975) (*b*) The five compartmental boundaries which arise successively on the growing wing–thorax disc: 1, anterior–posterior; 2, dorsal–ventral; 3, wing (midregion of disc)–thorax (two end regions); 4, scutum–scutellar; 5, proximal–distal wing. Abbreviations as in Figure 14.5. (Fate map from Bryant 1975)

boundary separates domains of cells which have take alternative developmental commitments. Thus the anterior–posterior boundary is taken to reflect the fact that anterior and posterior cells have adopted alternative decisions. The first point to be made about this hypothesis is that the existence of a compartmental boundary cannot, in itself, prove that the isolated domains of cells have adopted different developmental commitments. The line might represent a zone of cell death, for example, separating cells which were identically committed when the boundary formed. Despite the logical insufficiency, however, the hypothesis is obviously plausible. We know that commitment of cells to form a particular adult appendage is a cell-heritable property in the third-instar imaginal disc. Thus if cells in the anterior and in the posterior compartment adopted alternative commitments but each remained able to form arbitrary parts of the anterior or posterior compartments respectively, then we would expect that the progeny of such cells would be constrained to form structures in the respective compartment and never those in the other compartment. Thus cell heritability plus a restricted equipotentiality implies compartmental boundaries.

Grant the hypothesis that compartmental boundaries separate domains having different heriatable commitments. Then the implication is that cells in any one final compartment have adopted a *sequence of alternative commitments*. Thus an individual cell has become anterior, not posterior; dorsal, not ventral; wing, not thorax; proximal, not distal. Then we can think of each of these decisions as one between *binary* alternatives and hence think of cells in each final compartment as defined by a combination of binary decisions. That is, we come again, as in Chapter 12, to a picture in which cells can be described by a *binary combinatorial epigenetic code!*

Other workers have examined the other major imaginal discs—leg, eye, proboscis, and genital (Steiner 1976; Dubendorfer 1977; Baker 1978; Morata and Lawrence 1978; Struhl 1981a). Not only do compartmental boundaries form sequentially on the wing disc, but analogous boundaries sequentially subdivide the other major discs. In fact, sequential compartmentalization appears to occur in the early embryo as well, as we shall discuss further below.

Evidence that Compartments Are Alternatively Committed Domains of Cells

Several lines of evidence have been adduced to attempt to show that compartments separate groups of cells which have taken alternative developmental commitments. The hypothesis suggests the possibility that all cells of a given compartment might share the expression of some gene or set of genes or, similarly, some set of antigens. Brower, Piovant, and Reger (1985) screened monoclonal antibodies raised against imaginal tissue and succeeded in finding one which specifically reacts with all dorsal tissue in the wing disc and another which reacts with all ventral tissue in the wing disc. The two compartments containing these two types of antibodies abut along a crescent line which is presumed to match well with the future wing margin and with the dorsal–ventral compartmental boundary. If accepted as evidence, it is interesting that these two antibodies each react with dorsal or ventral regions comprising more than one final compartment—all the dorsal or ventral ones. This is just the kind of common expression across a set of compartments which a combinatorial code hypothesis leads to.

A second line of evidence that compartments are alternatively committed domains of cells is the claim that certain mutants act on compartment domains—that is, that the transformed region may be smaller than, but is delimited by, a compartmental boundary. For example, Morata and Lawrence (1975) and Lawrence and Morata (1976a) claim that the mutant *engrailed-1* converts part of the posterior compartment in the wing disc to a mirror-image anterior compartment. Indeed, as described below, *engrailed* appears to convert posterior compartments in each segment of the entire body plan to anterior compartments. The mutant *bithorax* (Lewis 1978, 1981) appears to convert variable parts of the anterior compartment of the metathoracic haltere to anterior wing, while the mutant *postbithorax* converts the posterior compartment of the haltere to posterior wing (Lawrence and Morata 1976a).

These data are important but cannot be fully persuasive. As Karlsson (1984) points out, the transformation by *engrailed-1* is less than complete. Nevertheless, the evidence is very suggestive. Insofar as we may accept the hypothesis, it also strongly suggests two important consequences:

1. The wing disc, and perhaps other aspects of *Drosophila,* are sequentially divided into finer and finer overlapping subdomains of alternative developmental commitments.
2. Those commitments can be represented in a combinatorial epigenetic code.

We saw in Chapter 12 that transdetermination phenomena also led us to both these conclusions.

Application of Reaction-Diffusion Models to the Sequential Formation of Compartmental Boundaries

The geometry of compartmental boundaries is fascinating. Notice first a twofold symmetry. The four quadrants defined by the anterior–posterior and dorsal–ventral boundaries are regions which are essentially mirror-symmetric (Figure 14.19*b*). Ignoring for the moment the imprecision of the lines on the wing disc, let us idealize them on an ellipse, as shown in Figure 14.20. The twofold symmetry is here evident. Why should the compartmental boundaries exhibit such symmetry? Why should they arise in the observed sequence? Why are they located, even roughly, where they are?

It is a striking fact that Turing's model or, in general, the natural "eigen function" patterns of the Laplacian operator on a two-dimensional surface come very close to explaining the sequence and symmetries of the compartmental boundaries on the wing disc and other discs of *Drosophila.* The chemical patterns which a reaction-diffusion system generates in two spatial dimensions are more complex than those generated in one spatial dimension. With respect to the expectations derived from linear stability analysis, however, the two-dimensional patterns which arise are again eigen functions of the Laplacian operator in the two-dimensional spatial domain,

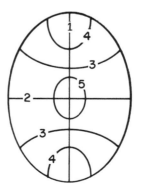

Figure 14.20   The five compartmental boundaries on a wing disc projected onto an ellipse. Lines are also predicted nodal lines of sequential eigen functions of reaction-diffusion model on a growing ellipse whose axis ratios change to elongate the major axis.

with appropriate boundary conditions. On a circular domain, the eigen functions are given by Bessel functions; on an ellipse, they are given by Mathieu functions (McLachlan 1947; Morse and Feshbach 1953; King and Wiltse 1958; Abramowitz and Stegun 1972). On a circular domain, the nodal lines of eigen functions form radial spokes and concentric circles, the natural coordinates of a circle. The homogeneous state is one eigen function. The first spatial pattern (Figure 14.21a) is a gradient which is low on one side of the circle and rises to a peak on the opposite side. This distribution has the property that, traveling around the circumference of the circle, a cosine pattern is experienced, with a single maximum and minimum. The same is true at any radial distance from the center, but the amplitude of the single sinusoid is smaller nearer the center and zero at the center. The steady-state chemical values X_0, Y_0 form a *straight-diameter nodal line* across the circle. The second pattern which arises is similar, but now two sinusoidal undulations occur around the circumference of the circle, given by the cosine of twice the angle around the circle and creating a *saddle-shaped surface* which curves down from the origin along one axis and up from the origin along the orthogonal axis (Figure 14.21b); the steady-state nodal line chemical values X_0, Y_0 both occur along two perpendicular crossed lines. Higher modes, analogous to higher sinusoidal frequencies in one spatial dimension, are other eigen functions on a circle. In addition to these radial-spoke patterns, which vary angularly around the circle and have an integral number of peaks and valleys, there also exist a different set of patterns which do not vary angularly but instead form concentric nodal line circles at different radial distances from the center to the edge of the circular domain. The simplest, shown in Figure 14.21c, is a simple hill-shaped pattern having a peak at the center and falling to identical minimal value around the edge. The steady-state values X_0, Y_0 form a circular ring around the center of the circle. More complex radial patterns have a hill in the center and a circular valley which rises to a circular rim at the margin of the circular domain. In general, the nodal lines of the eigen function patterns, corresponding to the loci where X_0 and Y_0 occur, lie along the natural polar coordinates for a circle—along radii or along concentric circles.

The eigen functions on an ellipse similarly have nodal lines which lie on the natural coordinates for an ellipse (McLachlan 1947) (Figure 14.20). These are confocal ellipses, the analogue of concentric circles, and pairs of parabolas, the analogue of radii. For example, simple patterns include a monotonic gradient along the long or short axis of the ellipse, a gradient which is high in the midregion of the ellipse and low at both longitudinal ends, and hill patterns having no angular components. Notice, in prospect, that the twofold symmetries of the nodal lines on an ellipse are just those seen in the compartmental boundaries on the wing disc. In any of the complex patterns, the four quadrants of the ellipse are mirror-symmetric patterns about the major and/or minor axes. Thus patterns derived from the Laplacian operator will naturally have these symmetries.

Sequential Generation of Compartmental Boundaries: A Bifurcation Sequence

The nodal lines of eigen functions on an ellipse have much the same symmetries as the compartmental boundaries on the wing disc. But those boundaries arise in succession. To see how naturally a reaction-diffusion model accounts for such a succession, we return to the idea of bifurcations, raised in Chapter 5. Recall that, as a parameter of a dynamical system is slowly varied, the dynamical behavior, such as

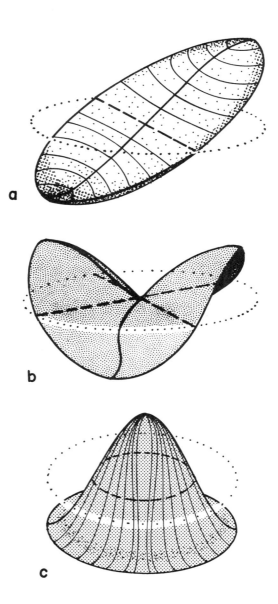

a

b

c

basins of attraction, will be little changed over some range. At critical values of the parameter, however, the behavior may change dramatically. Earlier examples included the onset of eddies as stream velocity past an obstruction increases and the onset of Benard cells as the temperature differential from bottom to top of a heated layer of liquid increases. Thus the set of parameters affecting a dynamical system themselves constitute a *parameter space,* which can be divided into subvolumes. Inside any subvolume the behavior of the dynamical system is much the same, but crossing the bifurcation boundary between subvolumes causes sharp changes in behavior. In Turing-like models, one of the two critical parameters is the ratio of the diffusion constant D_x/D_y and the other is the overall length or size of the physical domain in which the reaction system is placed. We now see that, as these parameters are changed slowly, jump changes in the pattern formed by the system occur.

Consider again the dispersion relation $\lambda = f(k)$ (Equation 14.4). For the curves of interest, in which there is a single interval of wavenumbers, and hence of wavelengths, having positive eigen values $0 < L_1 < L_2 < \infty$, the chemical system will amplify patterns in that range of wavelengths. But the boundary conditions must *also* be satisfied. Therefore, consider a ring whose overall length is less than L_1. The longest wavelength which can fit into that ring is less than L_1, and such wavelengths are damped out by the dynamical system. Therefore, for sufficiently small rings, the homogeneous state is fully stable. No pattern can emerge. Let the ring lengthen slowly. When it reaches length L_1, the periodic boundary condition can be met by a wavelength pattern which can be amplified, and a first mode arises. The critical implication is simple: As a small system grows gradually larger, no pattern is present at first; then a first pattern, corresponding to the longest wavelength which can fit into the domain, spontaneously arises.

As the length of the ring system increases, a succession of patterns of increasingly complex waveform will amplify and decay in succession. By appropriate tuning of the parameters of the Turing system, the dispersion relation $\lambda = f(k)$ can be tuned so that only a very narrow range of wavelengths $0 < L_1 < L_2 < \infty$ are amplified. Consider this limiting case and, to be precise, consider only small-amplitude patterns for which the linear stability analysis applies strictly. Then when the ring length increases beyond L_2, the single sinusoidal around the ring is too long and so is damped out by the dynamical system. However, the two-peaked pattern, $\sin 2x/L,$ is too short a wavelength for two tandem patterns to fit into the domain and so is also not amplified. Thus the initial pattern, $\sin x/L,$ decays back to the spatially homogeneous steady state. When the ring length is $2L_1$, this second pattern, which contains two peaks and troughs, can fit in and hence is amplified. As the ring grows further, this pattern becomes too long to be amplified and hence again decays. Thus as the ring lengthens, a *succession* of eigen function patterns arise and decay. That is, the *same* chemical variables, X and Y, form a succession of different gradient patterns with different geometries, each an eigen function of the Laplacian operator on the

---

Figure 14.21   (*a*) Wave pattern generated on a circle with a scaled radius of 1.82. The pattern is the product of a radial part, J1(kr) (the first-order Bessel function), and an angular part, cos ϕ. The dashed nodal line of zero (steady-state) concentration runs along a diameter of the circle. The dotted circle outlines the circular radius. (*b*) Wave pattern from J2(kr) cos 2ϕ, generated at a scaled radius of 3.1. The dashed lines are crossed nodal lines on two perpendicular diameters. (*c*) Pattern generated at a scaled radius of 3.8, where the zero in the derivative of Jo(kr) matches the radial boundary condition. The pattern is Jo(kr) cos 0ϕ, which has no angular variation. The nodal line is concentric with the outer radius.

ring domain. Therefore, a natural property of this class of systems is that parameter space is divided into distinct regions. As a parameter is tuned slowly, the system passes through distinct regions such that a succession of patterns arise and decay. In this simplest case, because of the stricture that the range of allowed wavelengths, L_1 to L_2, be very narrow, only a single pattern can grow at a time. Even here, though, the ring will eventually be long enough that N of the longer-wavelength patterns, length L_2, can fit into the ring, while $N + 1$ of the shorter-wavelength patterns, length L_1, can fit in. At that point, linear stability analysis predicts that both patterns will grow at once, with the pattern having the larger eigen value dominating the ultimate pattern. More generally, when the window L_1–L_2 is wider, several modes may grow simultaneously as the domain becomes larger.

The sequential formation of a succession of patterns by reaction-diffusion models as a parameter such as size is increased gradually is robust. Such a sequence which arises as a parameter is tuned is called a *bifurcation sequence*. In addition to size, the absolute values of the diffusion constants can be changed while their ratio is held constant. A decrease in diffusion constants is equivalent to an increase in length. Indeed, size scales as $\sqrt{D}$ in a line, as D in two spatial dimensions, and so on (Arcuri and Murray 1986). Similar properties arise in mechanochemical models (Odell, Oster, et al. 1981; Goodwin and Trainor 1985).

Compartmental boundaries on the wing disc arise sequentially as the disk grows. Therefore, it is of interest to try and use a bifurcation sequence to account both for the geometries and for the sequence in which compartmental boundaries form on the wing disc (Kauffman, Shymko, and Trabert 1978; Kauffman 1979; Kernevez 1980). The first, the anterior–posterior boundary, is present in the blastoderm prior to formation of the wing disc. Thus we must account for it on the early embryo. The remaining compartmental boundaries arise on the wing disc proper and are thus to be accounted for by the model. Figure 14.20 shows the succession of nodal lines for eigen function patterns on a simple ellipse as the ellipse grows and the ratio of major to minor axes changes slowly. Figure 14.22 shows the first five eigen function patterns on the wing disc as it grows. Figure 14.22a shows that the first pattern forms along the long axis, with a peak at one end and a valley at the other. The nodal line X_0, Y_0 lies along the minor axis. We take this line to model the dorsal–ventral boundary transversing middle of the disc. The second pattern (Figure 14.22b) arises with a single maximum and minimum oriented the short way across the ellipsoid; the nodal line would repeat the anterior–posterior axis were it to arise. The third pattern (Figure 14.22c) is again longitudinal, low at both ends and high in the midregion. The nodal lines are a *pair* of parabolas roughly parallel to the first nodal line of the previous pattern. We take this pair to correspond to the pair of wing–thorax boundary lines, separating the two thoracic ends of the disc from the wing region in the middle. The fourth pattern (Figure 14.22d) creates two arc nodal lines oblique to the major and minor axes; one line we take to create the scutum-scutellar (fourth) compartmental boundary. The fifth eigen function (Figure 14.22e) creates a concentric hill pattern whose nodal line is the analogue of the proximal–distal wing boundary.

It is natural to seek a connection between the chemical patterns which are imagined to arise and developmental commitments. The compartmental hypothesis suggests that each boundary separates domains where cells have adopted different com-

---

Figure 14.22   First five eigen functions on wing disc shape as the disc grows larger but maintains its relative shape.

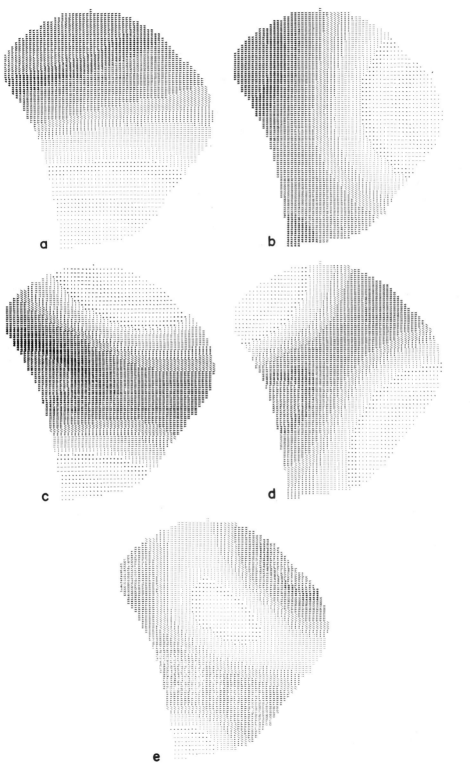

mitments. Thus it is attractive to suppose that the nodal line X_0, Y_0 is a threshold chemical concentration. Cells above this concentration adopt one commitment, those below it adopt the other. On this view, a succession of differently shaped gradients of the *same* morphogens drives a succession of different developmental commitments by which space is sequentially subdivided, and each terminal domain is characterized by a combinatorial epigenetic code word.

Critique

In a moment we shall consider defects with this model, but first, its successes are not trivial. It predicts that compartmental boundaries should arise in a sequence. In addition, those boundaries should and do exhibit the inherent twofold symmetry of the ellipse. The model comes close to explaining the locations of compartmental boundaries on the wing disc. Errors may reflect the fact that predictions relate to the growing disc while observations are made on the adult wing and projected back onto the late-third-instar disc. Note, too, that, in predicting a sequence, symmetries, and location, the model also predicts which specific members of the sequence are earlier than others. Thus it predicts that the dorsal–ventral boundary should precede the wing–thorax and proximal wing–distal wing boundaries and that the wing–thorax boundary should precede the proximal wing–distal wing boundary. Indeed, the model also fits well the boundaries which arise on the haltere, eye-antenna, and genital discs. Data on the leg discs suggest that formation of the initial anterior and posterior compartments is followed by subdivision of only one of those two compartments by a further boundary (Steiner 1976). If true, this second subdivision would be difficult to explain by this mechanism. I stress again that a variety of mechanisms, some mechanical or chemomechanical, have the same eigen functions. Hence perhaps the most important idea to hold onto is that a variety of mechanisms imply that space should be sequentially subdivided as size or other parameters change and that a family of related but discretely different eigen function patterns should arise. Those patterns are at least closely similar to this set of observed phenomena.

Despite its successes, this linearized Turing model is not perfect. First, the transverse mode repeating the anterior–posterior boundary is not needed. In our initial formulation of this model (Kauffman, Shymko, and Trabert 1978), we suggested that, because of nonlinearities in the full model, this mode might be skipped. (I return to this point in a moment.) Second, within the limits of linear analysis, the polarity of each pattern is determined by initial conditions. Thus whether it is the top or the bottom of the ellipse which shall have a peak when the first longitudinal mode arises is due to initial conditions. Similarly, the second logitudinal pattern might have a peak in the midregion of the disc or at both ends. If one wishes to imagine that the nodal line chemical values are a threshold, then the polarity of the chemical patterns matters critically to proper development. Third, the first several eigen function patterns which arise are robust with respect to slight distortions in the shape of the spatial domain, but the higher modes become very sensitive (Nicolis and Prigogine 1977; Bunow, Kernevez, et al. 1980; Kernevez 1980; Arcuri and Murray 1986). This sensitivity is to be expected since higher modes arise on larger domains, where a distortion that is small with respect to the overall domain is long enough to accommodate much of an entire wavelength. A fourth problem is that the final compartmental pattern is not very sensitive to wing disc size. This is a general problem in models of pattern formation. In many cases, the same pattern is formed over a considerable range of organism or organ size. The straightforward Turing class of models, in the linear stability region, have difficulties coping with this insensitivity to size. Some solutions, such as scaling the diffusion constants to cope with changes in domain size, seem too ad hoc to live with.

Nonlinear Analysis of Reaction-Diffusion Models: Skipping of Modes, Sequentially Determined Polarities, and Spatial Frequency-Doubling Bifurcations

A number of the defects which arise when reaction-diffusion models are applied to *Drosophila* and other systems may well reflect limitations of linear analysis. We see next that a more adequate analysis of the full nonlinear equations redresses some, and perhaps all, of the difficulties I have raised. Several lines of evidence suggest that the patterns which arise can be far more robust than expected. These studies indicate that the polarity of the patterns is insensitive to initial conditions, that certain "unneeded" modes are skipped as domain size or other parameters increase, and that there is a strong tendency to spatial frequency-doubling bifurcations.

A particularly careful analysis of the general class of Turing models in a single spatial dimension has been carried out by Arcuri and Murray (1986). They consider both the linear case, common to all members of the class, and a specific fully nonlinear model. They analyze a variety of boundary conditions on an open line: (1) no flux; (2) flux across the ends with an environmental bath set at the steady-state levels X_0, Y_0; (3) flux with the bath set above or below that steady state; and (4) spatially inhomogeneous initial conditions. In particular, they analyze the behavior of a growing system in which the bath is set at the homogeneous steady state but the first mode is already present in the domain as it grows large enough to accommodate higher modes.

The authors chose parameters for their model such that a modest window of wavelengths L_1–L_2, can be amplified and analyzed in terms of the two fundamental underlying parameters: $\beta = D_x/D_y$, the ratio of the diffusion constant for the two chemicals, and γ which scales with the size of the domain as $\sqrt{\gamma}$ in one dimension, γ in two dimensions, and so forth. This latter parameter is also interpretable as being inversely proportional to the absolute size of the diffusivities. As noted, decreasing diffusivity is the same as increasing domain size. Alternatively, γ represents the strength of the overall reaction terms. An increase in enzymatic activity can therefore increase γ.

As parameters are tuned, the sizes of parameter domains associated with some patterns change as a result of nonlinearities in the model. Figure 14.23 shows the straightforward predictions of linear stability theory for regions in parameter space, or "Turing space," where either single modes or combinations of modes arise. Figure 14.23*b* shows the regions in which each mode or combination of modes will arise and Figure 14.23*c* shows regions of parameter space where a single nonlinear pattern dominates. Notice that as γ increases, the domain associated with the longer wavelengths—that is, lower numbered modes—becomes *larger*. This relationship implies that there is a tendency, at large β, for longer wavelengths around a ring to dominate. This tendency correlates well with the data on *Polyspondylium* (Byrne and Cox 1986, 1987), where the power spectrum preferentially simplifies to sinusoids having few peaks.

In the nonlinear model, modes can be skipped. In Figure 14.23*d*, simulations for one fully nonlinear dynamical system are shown for no-flux boundary conditions and for a single initial state which is a random fluctuation away from the homogeneous steady state. The interesting features are (1) that a single mode dominates in

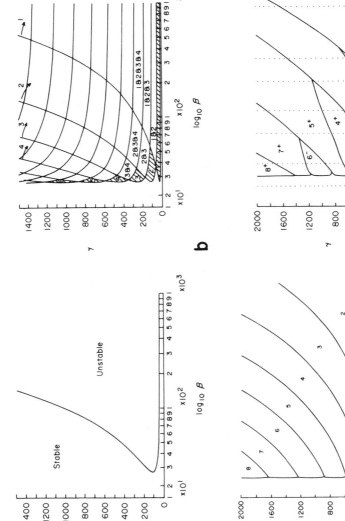

a

b

c

d

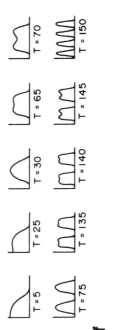

Figure 14.23 Two-dimensional "Turing space" given by the two parameters γ and β, where γ reflects the absolute size of the domain or diffusion constants and β reflects the ratio of diffusion constants. (a) Region of instability of the second mode in a one-dimensional spatial domain, determined by linear stability analysis. When unstable, the mode will amplify in time and establish a macroscopic pattern. (b) Linear predictions of domains in which single modes or linear combinations of modes are unstable and therefore amplify. (c) Regions where linear theory suggests that single modes will have the largest eigen value and hence dominate the long-term pattern. Note that as β increases, the longer-wavelength patterns dominate. (d) Nonlinear simulation results for no-flux boundary conditions show that the domains in parameter space associated with some modes are bounded. This figure shows regions where even or odd nonlinear patterns arise. Note that for large β only the odd modes arise. (e) Nonlinear simulation results on a growing one-dimensional domain having no-flux boundary conditions, early modes remain as later modes become established. The two sets of graphs show that different initial patterns evolve to the same pattern. Thus final pattern can be independent of initial conditions. T = time. (f) Spatial frequency-doubling bifurcations arise in the *Drosophila* embryo. (From Arcuri and Murray 1986)

each parameter domain and (2) that a number of domains are large, while others have become small. Here the odd modes have large parameter domains, the even modes small ones. Note that the even modes occupy parameter domains which are *bounded as β increases*. Thus, for β at a fixed large value—that is, a high ratio of D_x/D_y—only the odd modes arise, as size increases; all the even modes are skipped.

In nonlinear models, pattern polarity can be determined uniquely by the existence of prior patterns. As size changes slowly from an initial small domain, a first monotonic pattern arises. If the window L_1–L_2 is of reasonable size or if the decay rate of the first mode is not too fast relative to domain growth, then the first monotonic mode will still be present as the domain becomes larger and the second mode becomes admissible and begins to grow. Therefore, the initial condition for the second mode will be not the homogeneous steady state but the preexisting pattern. This case leads to the expectation that the first mode can dictate the polarity of the second mode.

The expectation that a first pattern can dictate the polarity of a second pattern is correct. Figure 14.23*e* shows the parameter domains where either the first or the second mode persist once established. Sizes of domains are enlarged relative to the linear theory and relative to one another. Figure 14.23*e* is of basic importance to our further discussion. It shows the time succession of patterns on a line for two sets of initial conditions, with the full cosine mode having either a peak or a trough in the middle. As the one-dimensional spatial domain enlarges, both patterns merge to the *same polarity!* That is, the final pattern is independent of initial conditions!

A third dramatic feature is *spatial frequency-doubling bifurcations.* Figure 14.23*f* shows this behavior for a large value of β, well above a critical ratio $\beta_c = 27$. Note that the odd modes, those having maxima at one end of the line and minima at the other, do not occur. The pattern goes from the monotonic-gradient one mode, to the single-peaked two mode, skips the third mode, and jumps to the two-peaked four mode. These peaks then split to form the four-peaked eight mode. Thus, for large β, this model naturally give spatial frequency-doubling. We shall see strikingly similar phenomena in the early *Drosophila* embryo.

The nonlinear analysis offered by Arcuri and Murray (1986) goes some distance toward meeting objections to the Turing class of models for pattern formation. In particular, the fact that pattern polarity can be independent of initial conditions is a critical feature if concentrations above and below threshold levels are to set developmental decisions. Further, the capacity to skip unneeded patterns is obviously important.

Summarizing, nonlinear models of the Turing type have the properties that pattern polarity can be made relatively insensitive to initial conditions, that modes can be skipped, and that spatial frequency-doubling bifurcations can occur. The power of this class of models is undoubtedly great and carries over to a variety of mechanochemical models. Once out of the linear range, the simplest, hence most general, predictions fall away a bit. Different nonlinear models have as yet poorly understood differences in their behaviors.

PATTERN FORMATION
IN THE EARLY *DROSOPHILA* EMBRYO

We now turn from imaginal discs to the wondrous phenomena known to occur in the early *Drosophila* embryo. These phenomena and a body of theory are discussed in four stages. First I shall consider the simplest application of Turing's general idea of reaction-diffusion instabilities to the problem of pattern formation in the early embryo. This analysis leads to a model for the hypothetical sequential subdivision of the early embryo by the successive formation and decay of longitudinal and dorsal–ventral eigen function patterns on the embryo. Each successive pattern of the *same* hypothetical morphogens is assumed to trigger alternative heritable commitments, depending on whether that morphogen pattern is above or below a threshold concentration. The succession of patterns triggers a succession of different developmental commitments leading to the binary combinatorial code model examined in Chapter 12. This model has major strengths but is inadequate to account for additional fundamental features of pattern deletion and pattern duplication which occur on a variety of length scales, from half egg to an approximate $\frac{1}{32}$ egg length along the anterior–posterior axis of the embryo.

In the second part, I consider a number genes falling into four classes: maternal, gap, pair-rule, and segment-polarity. Weak mutants of these genes delete parts of a longitudinal pattern, while stronger alleles delete more pattern elements and yield mirror-symmetric duplications of the remainder.

The third part considers dramatic data demonstrating that genes belonging to these four classes exhibit increasingly complex, multipeaked patterns of transcription and translation during the earliest stages of development. Many members of these systems of genes pass through what appears to be a bifurcation sequence very similar to that already envisioned for the wing disc and through the hypothetical sequence of eigen functions subdividing the egg into successively finer subdomains.

In the fourth part, I propose that longitudinal position is specified by the *ratios* of these phase-offset gene-transcription patterns in a "four color wheels" model of positional information. This model, a generalization of the binary code combinatorial model we discuss first, accounts naturally for the deletions and mirror-symmetric duplications considered in the second part.

The four color wheels model, developed with B. Goodwin (Goodwin and Kauffman 1989a, 1989b; Kauffman and Goodwin 1989), appears quite powerful. However, I should stress that, at present, *no single adequate theory* of pattern formation in *Drosophila* is in hand (Sander 1975, 1977, 1980, 1984; Meinhardt 1977, 1986; Kauffman 1983; Anderson and Nüsslein-Volhard 1984; Gergen, Coulter, and Wieschaus 1986; Akam 1987; Nüsslein-Volhard, Frohnhofer, and Lehmann 1987).

Before beginning, an overview. Pattern formation in the early embryo is concerned with assigning defined segmental and other committed fates to nuclei or cells at specific positions on the blastoderm (Gehring and Nothiger 1973; Gehring 1976) (Figure 14.18a). That is, there is good evidence to believe that early syncytial blastoderm nuclei are developmentally unrestricted but by the late syncytial blastoderm stage, and certainly by the cellular blastoderm stage, a number of *nuclear- or cell-heritable commitments* have occured. Transplantation of nuclei from one region to another in early-cleavage-stage embryos showed that the genetically marked nuclei formed larval and adult tissue proper to the location at which they were placed (Illmensee 1972, 1976; Okada, Kleinman, and Schneiderman 1974). Thus such nuclei remain uncommitted to specific fates. Conversely, transplantation of anterior cells at the blastoderm stage to the posterior results in formation of anterior-specific structures by the transplanted cells (Chan and Gehring 1971). Further, transplantation at blastoderm of cells from the prothorax to the metathoracic area results in formation of prothoracic-leg-specific bristles in the patch of transplanted cells later found residing in the host's metathoracic leg. Thus such blastoderm cells clearly have a cell-heritable commitment.

It is also now clear that *late syncytial blastoderm nuclei plus cytoplasm can carry heritable commitments.* Thus when late syncytial anterior nuclei and surrounding cytoplasm were transplanted to the posterior flanks of late syncytial hosts, the transplanted material formed anterior-specific adult cuticle. Similarly, posterior nuclei and cytoplasm injected anteriorly formed posterior-specific structures (Kauffman 1980). These results strongly suggest that, by the late syncytial blastoderm stage, *regional biochemical differences controlling heritable commitments exist in the syncytium* and are *regionalized* despite the fact that the syncytium remains an open domain with respect to diffusion.

Early embryogenesis is concerned with setting up regionalized commitments of nuclei and cells. Some of these heritable commitments clearly form sequentially and are related to sequential subdivision of the embryo into successively finer regions.

The first evidence for sequential cell-heritable developmental commitments at about the cellular blastoderm stage came from mitotic recombination experiments. Clones marked at the blastoderm stage are confined to form adult cuticle derived from a single segment. However, blastoderm clones can cross from the mesothorax to the mesothoracic leg in the adult (Lawrence and Morata 1977). Thus, at the blastoderm stage, cells may be committed with respect to the mesothorax segment but are uncommitted with respect for formation of its dorsal or ventral adult cuticular structures. Shortly after blastoderm, clones become restricted to wing or to leg. Similarly, shortly after blastoderm, clones are restricted to anterior or posterior compartments in each segment (Morata and Lawrence 1977, 1978). These data suggest that at least some ectodermal compartmental cell-heritable developmental decisions are taken sequentially in the embryo and subdivide it into successively finer subdomains. Dramatic further evidence for sequential subdivision of the early embryo is discussed further below.

Theoretical Sequential Subdivision of the Embryo Based on Reaction-Diffusion Instabilities and the Triggering of a Binary Combinatorial Epigenetic Code

Reaction-diffusion models of the Turing type account for many features in the sequential compartmentalization of *Drosophila* imaginal discs. Therefore, it becomes of interest to study the predictions this class of models might make concerning early pattern formation in the *Drosophila* embryo. The predictions of sequential eigen function patterns I now describe were made prior to data demonstrating the actuality of spatially sinusoidal gene transcription in the syncytial embryo. The simplest expectations of a Turing type model applied to the syncytial cleavage states, as we shall see, come close to predicting what has come to be observed. In addition, I here use these patterns to build up the combinatorial epigenetic code described in Chapter 12.

The egg is an ellipsoid. It does not increase in size during cleavage, but three events—the increased mass of cytoplasm relative to yolk, the migration of this cytoplasm to the cortex, and the subsequent increased amount of membrane material and formation of cell boundaries as membranes migrate inward beneath cortical nuclei—are all likely to decrease diffusion constants, at least in the cortical area. As described above, a decrease in diffusion constants is equivalent to an increase in length. The eigen function patterns which arise on the surface of an ellipsoid are given by generalized Lamé functions and again follow the natural coordinates for an ellipsoid. These functions consist of concentric ellipsoidal surfaces, on each of which a straight major and minor axis along and across the hemi-ellipse are present. Flanking the minor axis are pairs of roughly paraboloid lines, analogous to those on a planar ellipse. The analogues to confocal ellipses in a planar ellipse occur as well, oriented along the long axis of the hemi-ellipsoid, at its equator, and along its "saggital" midline.

Figure 14.24 shows the first four expected eigen function patterns for an ellipsoid having about the same axes ratio as the *Drosophila* egg. The first mode is longitudinal, high at one end, low at the other, and increasing sigmoidal in the midregion. Its threshold nodal line X_0, Y_0 divides the egg, at half egg length measured from the posterior pole, into anterior and posterior halves. Following the hypothesis with respect to compartments, let us assume that this division triggers an anterior versus posterior commitment in each half egg, coded for by either the 1 state or the 0 state of a genomic circuit like the C1–cro loop in bacteriophage *lambda*, which has two steady states, im+ and im− (Neubauer and Calef 1970; Ptashne 1986). These alternative states represent the heritable, memorized, determined commitments throughout each half embryo.

The second pattern which arises (Figure 14.24*b*) is low at both ends and high in the middle, creating two nodal lines separating the ends from the midregion. Let a second decision circuit record commitments to end versus middle states in these regions. Note that this hypothesis assigns a similar state, "end," to both ends of the embryo even though they are separated by the midregion. This double assignment is reminscent of the fact that the two ends of the wing disc form thorax while the midregion forms wing blade. This assignment begins to create a combinatorial code. We might think of the combinations

Anterior + end = head
Anterior + middle = thorax
Posterior + middle = proximal abdomen
Posterior + end = distal abdomen and genitalia

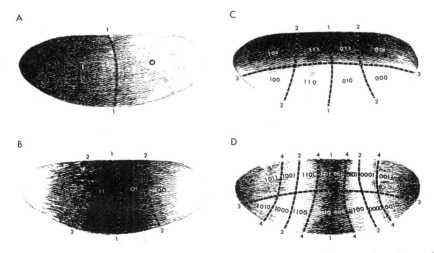

Figure 14.24 Hypothetical sequence of four eigen-function patterns on preblastoderm *Drosophila* egg. Nodal lines were used to trigger alternative commitments to create combinatorial epigenetic code. The 1 and 0 for each decision stand for alternate states of a genetic circuit which records each developmental commitment. The final subdomains shown in (*d*) are specified by combination of the four binary alternative commitments taken. See Figure 12.21d.

Because the second pattern is *non*monotonic, with a peak in the middle, this procedure makes distant regions similar to one another. Foreshadowing: A single switch from posterior to anterior will convert gentalia to head, jumping over the midregion of the embryo.

The third pattern (Figure 14.24c) runs dorsal to ventral, drawing a compartmental boundary separating dorsal and ventral ectodermal structures along the lateral equator, recorded in a third commitment: dorsal versus ventral.

The fourth predicted pattern (Figure 14.24d) is based on the expectation of a spatial frequency-doubling bifurcation. The argument, confirmed by Arcuri and Murray (1986), that the second longitudinal mode, once established, makes transition to the third mode difficult and causes skipping to the fourth mode yields the fourth pattern. This pattern has two peaks, recorded by yet another decision-taking circuit into even versus odd.

Later, as a result of frequency-doubling, longitudinal modes yield an eight-peak pattern containing 16 monotonic domains, and then double to a 16-peak pattern. We ignore these more complex patterns for the moment but return below to examine data showing that just such spatial frequency-doubling in gene-transcription patterns occurs in the preblastoderm embryo while it is still a single syncytium open to diffusion. The now-known transcription patterns are stunningly close to those expected on a Turing model.

The hypothetical sequence of modes envisioned in Figure 14.24 successively subdivides the embryo into a set of domains, each of which carries a specific binary combinatorial epigenetic code word, shown in (d). This is the model alluded to in Figure 12.21d, where we tried to account for the latticelike patterns seen in transdetermination. Recall that this model predicted rather well most major features seen in the transdetermination and homeotic mutant data.

According to this combinatorial code model, transdetermination is due to a switch of one circuit from a 1 to a 0 or vice versa. Then, as noted in Chapter 12, the following predictions can be made. First, consider two transdetermination steps, such as haltere to wing and haltere to antenna. Both require changing the first circuit from 0 to 1, but haltere to antenna requires changing an additional circuit, the second, from 1 to 0. Thus the model predicts that transdetermination from haltere to wing is more probable than transdetermination from haltere to antenna. Table 12.1 shows 36 independent predictions made by this code assignment. All but two are true. The *a priori* probability of this success is less than one in a billion. Thus the combinatorial code is genuinely capturing major features of the transdetermination flow. Restated, the combinatorial code gives an interpretation to the sense in which any two tissues have "neighboring developmental programs." Each imaginal disc is characterized by the *combination of binary choices* made during its ontogeny. Two discs are neighbors if they differ by a single choice.

Next, the code accounts for the fact that any disc transdetermines to only a few other discs and that transdetermination sequences occur. Each sequence follows from the fact that only a single circuit is likely to change state at one time; hence each disc can jump only to neighbors whose code differs in a single decision. Further, if the 1 state is more stable than the 0 state, the oriented flow toward mesothorax ensues. The fact that such a binary combinatorial code comes very close to capturing the transdetermination lattice suggests quite strongly that the combinatorial theory has something right about it.

Because the code arises from monotonic and nonmonotonic chemical patterns, distant tissues can transdetermine into one another, jumping across the fate map—

genitalia to antenna, for instance. Indeed, this model commends itself, despite difficulties which we shall ennumerate below, in part because it is the only one available which accounts for such jumps in a rational way.

Each disc is coded by a number of decision circuits. Since transdetermination might allow any one of them to switch, but any homeotic mutant might be assumed to alter the stability of only one state of one circuit, the model correctly predicts that the set of tissues to which any disc can transdetermine should be *broader than but include* the effects of any single homeotic mutant (Hadorn 1966; Kauffman 1973; Ouweneel 1976).

The simple assumption that some but not all homeotic mutants may act to destabilize a single memory circuit leads to further predictions:

1. Known homeotic mutants should preferentially cause transformation between tissues whose code words differ in a single decision. Table 12.2 shows that this is largely true. This success, too, is unlikely by chance. That is, if random four-bit code words were assigned to *Drosophila* domains, the mean number of bits changed per mutant would be 2. Again the model exhibits a strong success.

2. If a homeotic mutant alters the stability of a memory circuit in one disc tissue, it might act in another in the same way. Thus *parallel transformations* should arise. For example, were the second end versus middle decision weakened in the antenna, the antenna would jump over intervening tissue to the mesothoracic leg. The same effect in the eye would convert eye to wing. In fact, *Nasobemia*, a member of the *Antennapedia* complex, mediates just this pair of transformations (Gehring 1966a, 1966b; Stepshin and Ginter 1972). A mutant which converted the initial posterior decision to an anterior would transform haltere to wing and genital to antenna or head. A mutant found by Shearn, Rice, et al. (1971) mediates this pair of parallel transformations. Best known, but not in this combinatorial scheme, the *engrailed* gene appears to be needed in the posterior compartment of each segment of the organism. Its deletion appears to convert each posterior compartment, wholly or in part, to an anterior compartment proper to that segment (Morata and Lawrence 1975).

Thus a combinatorial model predicts parallel homeotic transformations, and a number of such transformations are known. This model comes close to accounting for most known homeotic transformations as one-step changes in the underlying code. Both the combinatorial concept and the idea that the spatial arrangement of the code may be set up by a sequence of monotonic and nonmonotonic positional signals are encouraged.

Critique

This sequential-morphogen-mode, combinatorial-code model clearly has a number of attractive features. First is the idea that space is successively subdivided by the fitting in of higher "harmonics" on the egg domain; as we shall see, there is direct evidence in mutant data and gene-transcription patterns to support this idea. Second is an attempt to link those harmonics to a binary combinatorial coded state in each domain. Such coding allows distant regions to be developmental neighbors in an attractive way that fits homeotic and transdetermination data. This model was formulated before evidence for multipeaked gene-transcription patterns in the syncytial blastoderm egg and early embryo became available. The model is worth recalling precisely because it still offers one of the best accounts of why distant regions are developmental neighbors and because the concept that monotonic and nonmonotonic gradients of gene products and other morphogens divide space into regions having unique positional identities seems very likely to be correct.

On the other hand, this early theory appears to be wrong in important ways and to harbor difficulties. In particular, the model assumes, for simplicity, that successive gradient patterns of the *same* morphogen system wax and wane in the eigen function bifurcation series. This simplest hypothesis then requires that space be divided by a series of different developmental "switches," each triggered to alternative states by the proper member of the eigen function series. It now seems clear that at least four gene systems—maternal, gap, pair-rule, and segment-polarity—pass through a bifurcation series, setting up complex, multipeaked longitudinal transcription patterns. The different systems have different wavelength patterns, and within each wavelength transcript patterns are phase-shifted. Therefore, it now seems likely that cells might specify position by the *simultaneous* values of the patterns of at least four gene systems on four different wavelengths.

Any reaction-diffusion theory confronts the fact that Turing models are sensitive to the shape of the egg. A fairly wide range of changes in axial ratios causes no apparent alteration in the adult (see, for instance, Bunow, Kernevez, et al. 1980). However, recent data, discussed below, suggest that the spatially localized products of maternal, gap, pair-rule, and segment-polarity genes along the longitudinal axis may act as an internal framework governing which patterns form. Such a framework would reduce sensitivity to changes in egg shape.

Further difficulties arise with respect to the gap phenomenon (Schubiger 1976; Schubiger and Wood 1977; Newman and Schubiger 1980). If the egg is ligated during cleavage divisions, a gap of missing longitudinal segments appears flanking the location of the ligation. The gap grows narrower if the ligation is made closer in time to the blastoderm stage and is not due to a loss of total number of cells. Further, the ligation shifts the prospective fate map of the early embryo; if the trans-egg bicellular layer which forms after ligation is pierced after the cellular blastoderm stage, then the gap of missing elements is eliminated. Thus after the blastoderm stage, the total requisite information can be restored even if it was deformed earlier. This gap phenomenon is hard to account for on the basis of a sequence of transient wave-shaped patterns which arise and decay on the embryo, as posited in the present model. This difficulty, however, appears to be met by the four color wheels model, described below, which has grown out of and incorporates much of this Turing model.

The most important difficulties of this particular form of a combinatorial-code model, however, arise with respect to experimental data on mutants which cause deletions and mirror-symmetric duplications of anterior—posterior segmental pattern elements. These phenomena strongly support a Turing-like model fitting successively shorter-wavelength, multipeaked morphogen patterns on the egg but require extending the binary epigenetic code model and lead us to the four color wheels model. In that model, cells measure longitudinal position along the embryo by assessing "phase" with respect to each of four spatially periodic morphogen patterns of successively shorter wavelength. As in the combinatorial-code model, the monotonic and nonmonotonic character of the wave patterns again allows the four color wheels model to account naturally for transformations between distant regions of the embryo.

SPATIAL HARMONICS SUGGESTED BY MUTANTS AFFECTING SEGMENTATION: LONGITUDINAL DELETIONS AND MIRROR-SYMMETRIC DUPLICATIONS

Perhaps the most interesting mutants discovered in *Drosophila* are a number of maternal-effect mutants, which affect the embryo only if present in the mother, and maternal-plus-zygotic mutants, and zygotic mutants, which affect the primary organization of longitudinal segmental patterning or dorsal–ventral patterning in the embryo (Anderson and Nüsslein–Volhard 1984). In the following body of ideas, developed with B. C. Goodwin (Goodwin and Kauffman 1989, 1990; Kauffman and Goodwin 1990), I focus on longitudinal organization. As noted above, these mutants fall into four major classes: maternal and zygotic genes, which are concerned with the entire longitudinal axis; gap genes, whose mutation deletes a contiguous longitudinal region of the primary axis; pair-rule genes, whose mutation deletes alternative regions on a double-segment periodicity; and segment-polarity genes, whose mutation deletes parts of the longitudinal pattern on a single-segment periodicity. Analysis of these four classes of mutants which affect embryo organization in the anterior–posterior axis has revealed four fundamental properties:

1. Mutants can delete part of the anterior–posterior longitudinal pattern.
2. Mutants can delete part of the longitudinal pattern and form mirror-symmetric duplications of the remaining pattern elements.
3. Among those mutants which can produce duplications, weak effects produce only deletions but stronger effects produce deletions plus duplications.
4. Such phenomena occur on a variety of length scales along the embryo, ranging from half the longitudinal axis to deletions and duplications within each segment.

Bicephalic *and* Bicaudal *Create Mirror Symmetries on a Half Egg Length Scale*

Among the mutants which cause mirror-symmetric duplications extending over half the embryonic body are the maternal-effect mutants *bicephalic* and *bicaudal* (Bull 1966; Nüsslein-Volhard 1977; Nüsslein-Volhard, Frohnhofer, and Lehmann 1987). During oogenesis, the oocyte derives from a nest of nurse cells which remain connected to the nest by cytoplasmic bridges and transfer cytoplasmic material into the oocyte. Outside the egg proper is the complex chorion. In *bicephalic* mothers, the oocyte has nurse cells at both poles of the egg and the chorion has anterior stigmata disposed at both poles. The embryo, as the name implies, forms a mirror-symmetric double anterior set of structures, with the plane of symmetry in the thorax. The *bicaudal* mutants, much better studied (Bull 1966; Nüsslein-Volhard 1977; Mohler and Wieschaus 1986), show a beautiful range of phenotypes (Figure 14.25). The chorion is normal, and the egg appears normal. As the name implies, the dominant phenotype is a double-abdomen embryo. In favorable cases, the plane of mirror symmetry is as far anterior as the second abdominal segment and the embryo is entirely symmetric. That is, the duplicate anterior half has as many segments as does the posterior half. The symmetric class can vary in the position of mirror symmetry as far posterior as the seventh abdominal segment. In that case, the entire embryo has four abdominal segments, ordered 8-7-7-8. It is important to stress that the plane of sym-

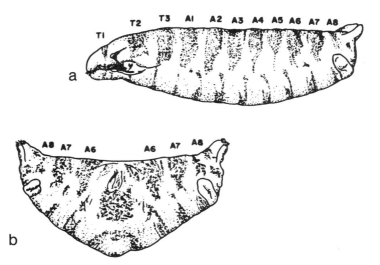

Figure 14.25 (*a*) Wild-type larva. (*b*) Mirror-symmetric double abdomen. Asymmetric double abdomens have more posterior segments than anterior segments. (From Nüsslein-Volhard 1977)

metry can pass through *any* position within a segment. Therefore, there is no position within a segment which is uniquely the location about which anterior–posterior mirror symmetry occurs in all cases. Furthermore, double-abdomen embryos typically have more segments along the ventral side than along the dorsal side; therefore, the line of mirror symmetry from dorsal to ventral runs at an oblique angle with respect to segments. As pointed out by Gergen, Coulter, and Wieschaus (1986), this phenomenon poses serious difficulties for the compartmental-domain, binary-code picture described above. According to the compartmental hypothesis, a compartment is an equivalence class of cells having identical fates, specified by some heritable committed state. Yet the positions of mirror symmetry in *bicaudal* and *bicephalic* embryos do not respect compartmental boundaries at all. Any position within a segment can be the symmetry line, and that line is oblique across two or more segments along a dorsal–ventral transect. The same features occur with mutants causing mirror symmetry on shorter length scales.

In addition to the symmetric class, a large asymmetric class exists, with fewer segments in the mirror-symmetric anterior region than in the posterior region. Finally, the *headless* class of mutants has all segments in normal serial order but the head segments are either missing or poorly involuted. Note that the range of phenotypes— from *headless* to symmetric *bicaudal* embryo—suggests that weak effects delete part of the longitudinal pattern, while stronger effects beyond some critical extent of deletion yield mirror-symmetric duplications.

Bicephalic and *bicaudal* and their phenocopies in other systems (Kalthoff and Sander 1968; Kandler-Singer and Kalthoff 1976; Kalthoff 1983) show that several insects can exhibit mirror-symmetric duplication on a rough *half egg length scale.*

Bicaudal embryos also exhibit an additional striking phenomenon. It is not uncommon to find embryos in which the *dorsal* aspect is a symmetric bicaudal on the left and right halves of the embryo, with posterior spiracles on left and right halves at the anterior pole. Ventrally, however, one side is a *headless* phenotype, while the other side is a mirror-symmetric double abdomen. This astonishing form means that, if one considers a transect from dorsal to ventral along the *headless* side near the anterior pole of such an embryo, the segment phenotype will jump discontinuously from posterior abdomen dorsally to thorax ventrally. This phenotype is found in other insects. For example, van der Meer (1984) induced *bicaudal* phenotypes in *Callosorbruchas* and recovered embryos having a longitudinal stripe of *bicaudal* segmental pattern elements embedded within a normal left or right half embryo. Obviously, we need to ask what kind of system can readily account for such a strikingly discontinuous pattern in the embryo. As we shall see, a significant virtue of the four color wheels model is that it naturally accounts for such discontinuities.

The Gap Genes and Mirror Symmetry on Quarter Egg Length Scales

Among the zygotic mutants affecting pattern, a number are gap mutants which delete a set of adjacent segments (Nüsslein-Volhard and Wieschaus 1980; Akam 1987; Lehman and Nüsslein-Volhard 1987; Nüsslein-Volhard, Frohnhofer, and Lehmann 1987): *hunchback, knirps, Kruppel,* and *giant.* The *hunchback* mutant deletes meso- and metathorax. *Knirps* deletes several adjacent abdominal segments, yielding a continuous ventral setal belt. *Kruppel,* the best studied (Weischaus, Nüsslein-Volhard, and Kluding 1984; Knipple, Seifert, et al. 1985; Jackle, Tautz, et al. 1986), deletes the entire thorax and proximal abdomen and produces an embryo which has an abnormally involuted head at the anterior end, adjacent to a mirror-symmetric duplicate of up to the last three abdominal segments, 8-7-6-6-7-8. Often, only the sixth abdominal segment is present in mirror-duplicated fashion. Again, the symmetry line can pass through arbitrary points in the sixth segment. As in the case of *bicaudal,* weak alleles of *Kruppel* cause loss of meso- and metathoracic pattern elements but no mirror symmetry (Weischaus, Nüsslein-Volhard, and Kluding 1984). Thus, again, weak alleles or phenotypes cause deletion, whereas stronger ones cause successively more deletion and then the onset of mirror-symmetric duplications. Notice that the mirror symmetry occurs on roughly a *quarter egg length scale.*

A similar progression occurs in *hunchback.* Weak alleles or phenotypes cause deletions, and stronger ones cause more extensive deletions and then the onset of mirror-symmetric duplications. The familiar phenotype deletes the second through sixth "parasegments," corresponding to the maxillary through third thoracic segment, as well as the thirteenth parasegment, corresponding to parts of the seventh and eighth abdominal segments. Stronger mutants of *hunchback,* however, deleting both maternal and zygotic contributions, also cause mirror-symmetric duplications of several posterior abdominal segments (Jackle, Tautz, et al. 1986). As with *Kruppel, hunchback* duplication is on roughly a quarter egg length scale.

Pair-Rule and Segment-Polarity Mutants Cause Deletions and Mirror-Symmetric Patterns on Double-Segment and Single-Segment Spacings

Saturation mutagenesis on all three major chromosomes of *Drosophila* (Nüsslein-Volhard and Wieschaus 1980; Akam 1987: Nüsslein-Volhard, Frohnhofer, and Lehmann 1987) has yielded a number of other mutants which fall into two further classes: pair-rule and segment-polarity mutants. The pair-rule mutants, roughly said, delete every other segment along the embryo. Thus *even-skipped* deletes the even-numbered segments, and *odd-skipped* deletes the odd-numbered ones. The domains deleted are often not exact segments but rather overlap segmental boundaries in characteristic ways for each mutant. Among the pair-rule mutants, strong alleles of *runt* form an embryo having half the normal number of segments, but each segment is present with longitudinal mirror symmetry (Nüsslein-Volhard and Wieschaus 1980). Here the mirror symmetry is on roughly a sixteenth egg length scale. Like *runt*, weak alleles of *even-skipped* delete parts of alternative segments in a two-segment spacing, but strong alleles cause deletion of more than an entire segmental domain and mirror-symmetric duplication of the remaining segmental domain in the double-segment interval. It is not the case that all pair-rule genes cause mirror-symmetric duplications in their strongest alleles, but striking new evidence suggests that some double mutants may cause such duplications: Moderate alleles of *even-skipped* delete the denticle bands of even-numbered segments, while *odd-skipped* deletes the denticle bands of odd-numbered segments; thus one might expect the double mutant to delete all denticle bands. Instead, the double mutant forms eight mirror-symmetric denticle bands on a double-segment spacing (Gergen, Coulter, and Wieschaus 1986)! Here deletion of the second gene *replaces part* of the pattern normally deleted by loss of the first gene. As we shall see, the four color wheels model naturally accounts for these striking phenomena.

The pair-rule mutants were the first persuasive evidence that, during embryogenesis, a double-segment interval occurs prior to and is perhaps causally necessary to the ultimate formation of segmental patterning.

A number of *segmental* mutants delete part of each segment along the embryo and form mirror-symmetric duplicates of the remaining region in each segment. The deleted region can include the segmental border; equivalently, the duplicated region can include the segmental border. Thus the segmental border is not a preferred location. Further, while each mutant deletes and duplicates a characteristic zone, it typically overlaps two segments. Thus the length scale is segmental, and therefore the mirror symmetry here occurs on roughly a $\frac{1}{32}$ egg length scale.

These astonishing data show that mirror-symmetric mutants occur in *Drosophila* on a variety of length scales. Furthermore, the general character of all these length scales is that weak alleles delete parts of a longitudinal pattern, while stronger alleles cause more extensive deletion and the onset of mirror-symmetric duplication. Thus it appears that, on all length scales, a critical amount of the longitudinal pattern must be deleted before mirror-symmetric duplications arise in the remaining pattern elements. The mutants which have mirror-symmetric patterns on successively shorter length scales suggest the possibility that the embryo is successively subdivided on those length scales into subdomains. Further, the longitudinal mirror-symmetric patterns hint at wavelike properties on the egg: If positional information is specified along the egg, or within a tissue, by a morphogen gradient which has multiple peaks and valleys—that is, if the gradient is nonmonotonic—then it is necessarily multivalued. Further, domains with mirror-symmetric gradients necessarily abut one another. In order to "cope with" the existence of such repeating spatial positional maps and generate a unique pattern, the cell needs some further mechanism to individuate each monotonic subdomain in the morphogen pattern. If that mechanism fails, the underlying mirror symmetry may be revealed. Again, as we shall see, the four color wheels model offers a simple interpretation of these phenomena. Passage from mirror-symmetric mutant patterns to wavelike properties on the egg with successively shorter wavelengths is a theoretical step. We turn now to direct and stunning evidence for sinusoidal, sequentially shorter spatial patterns of gene and protein expression exhibited by some of these genes.

SINUSOIDAL TRANSCRIPTION AND PROTEIN PATTERNS: A BIFURCATION SEQUENCE OF HIGHER HARMONICS ON THE EGG

The *Drosophila* syncytium is open to diffusion until formation of the cellular blastoderm at about three hours. Therefore, it is terribly striking that a number of the maternal, gap, pair-rule, and segment-polarity genes do actually come to exhibit complex, multipeaked longitudinal patterns of RNA transcripts and protein abundance in the syncytial egg. Whatever the mechanism governing the patterns, the phenomena are truly beautiful.

Three features of these data are critical in supporting a Turing reaction-diffusion model. First, many of the gene systems to be described exhibit spatial expression patterns which are close to eigen functions of the Laplacian. Second, the spatial patterns of each gene system become increasingly multipeaked during early development in a way which either is extremely close to a spatial frequency-doubling bifurcation sequence or otherwise is close to a natural bifurcation sequence in which higher harmonics arise as diffusion or some other parameter is tuned. Third, different gene systems which appear to have different natural wavelengths appear to pass through similar bifurcation sequences, although at different rates.

The Long-Wavelength System: Bicoid, Oskar, and Caudal

The longest-wavelength system includes the maternal genes, such as *bicoid* and *oskar,* and another gene, the maternal-plus-zygotic gene *caudal* (MacDonald and Struhl 1986; Akam 1987; Nüsslein-Volhard, Frohnhofer, and Lehmann 1987). In the oocyte, egg, and early embryo, RNA for *caudal* and other gene products is localized by *in situ* hybridization with radioactively labeled DNA or RNA which is complementary to the RNA (Hafen, Levine, et al. 1983). Protein products of the gene are localized with tagged monoclonal antibodies specifically binding the products.

The maternal genes *bicoid* and *oskar,* as well as others, are established in the syncytium in stable monotonic gradients: *Bicoid* is high in the anterior; *oskar,* high in the posterior. Neither passes to a more complex waveform during the later stages of cleavage. In contrast, *caudal* exhibits an increasingly complex waveform. Its transcript is first found in the nurse cells adjacent to the oocyte and appears to migrate into the oocyte. In early cleavage divisions, this maternal RNA is uniformly spread throughout the egg, with less in the yolky interior and more in the cytoplasmic islands. During cleavage, as these islands and their nuclei migrate toward the cortex, the transcript migrates as well. A dramatic change occurs between the ninth and the thirteenth divisions, a change which leads to the formation of an anterior–posterior gradient which is low in the anterior, high in the posterior. Note that this monotonic gradient is flat at both poles and steep in the middle, forming a half-cosine-like pattern. Such a distribution is as expected for the first longitudinal Turing-like mode and the Laplacian operator (Figure 14.24*a*). The mechanism underlying the shift from a uniform to a monotonic gradient is unclear (MacDonald and Struhl 1986).

Analysis of the *caudal* protein abundance has also been carried out by staining whole-mount embryos with monoclonal antibodies against a β-galactosidase-*caudal* fusion protein. This analysis reveals that the protein is initially uniform and at the tenth division shifts to a monotonic pattern parallel to the RNA abundance, low in the anterior, high in the posterior, and rising sigmoidally mid-egg. Thus both RNA and protein pass from a uniform distribution to a pattern remarkably close to the first longitudinal eigen function mode on the ellipsoidal egg, as predicted by the Turing model and other field equations.

Thereafter, the RNA and protein pattern becomes still more multipeaked. Shortly before cellularization, the level drops at both the anterior and posterior poles but increases in the posterior part of the egg, yielding a rough ring of high RNA abundance in the posterior third of the embryo. During germ-band extension, the transcript is present throughout the longitudinal extent of the germ band but waxes and wanes in abundance in a *double-segment* spacing.

The *caudal* RNA and protein therefore progress through the following stages: (1) a uniform distribution; (2) a monotonic pattern like the first longitudinal eigen function of the Laplacian; (3) a third pattern which is remarkably like the *superposition* of the first longitudinal mode plus the second mode, low at both poles and high in the midregion; and (4) a highly multipeaked pattern, abundant in alternative segmental spacing along the axis.

Further evidence of transient multipeaked patterns comes from a detailed study of *caudal* mutant phenotypes. Eggs from *cad−* females presumed to be fertilized by a *cad+* sperm from *cad+/cad−* males often develop into larvae that lack the eighth abdominal segment and sometimes show partial deletions of the fourth abdominal segment, hinting at a four-segment spacing. In addition, a few larvae show partial deletion of even-numbered abdominal segments, in correlation with the double-segment spacing seen in the protein pattern at germ-band elongation. Thus the entire *caudal* sequence of increasingly complex, multipeaked waveforms may well begin with a homogeneous pattern, then pass to a monotonic gradient, then to a single peak in the posterior third of the egg, then to a four-segment spacing, and finally to a double-segment spacing.

The Middle-Wavelength System: The Gap Genes

Kruppel, hunchback, and probably *knirps, giant,* and *tailless* appear to be members of a gene system which exhibits an increasingly multipeaked pattern during early embryogenesis but which has a shorter natural wavelength than does *caudal.*

As remarked above, gap genes denote a class of genes which cause at least one deletion of a number of adjacent segments (Nüsslein-Volhard and Wieschaus 1980; Preiss, Rosenberg, et al. 1985). As we saw on page 603, *Kruppel* mutants have a single extensive domain of deletion extending from T1 to A5. However, the sequences of RNA transcript patterns is much more complex and goes through an apparent bifurcation series closely analogous to that expected on a reaction-diffusion model (Figure 14.24). At the tenth cleavage division, *Kruppel* transcript is first observed as a single broad band in the middle of the egg. By the fourteenth division, just prior to cellularization, two new zones of transcription arise in *both the anterior and the posterior poles of the egg* (Harding and Levine 1988). Thus at this stage, *Kruppel* has passed from an initial pattern very close to the second longitudinal eigen function mode (low in the poles and high mid-egg) to the more multipeaked four mode, with two full cycles along the longitudinal axis: peak trough peak trough peak. By germ-band extenstion, *Kruppel,* like *caudal,* is expressed along the entire anterior–posterior axis but waxes and wanes in a double-segment periodically which gives the appearance of a pair-rule pattern (Harding and Levine 1988). Thus *Kruppel,* like *caudal,* appears to pass through a bifurcation sequence in which increasingly multipeaked patterns arise along the anterior–posterior axis.

Note that the *Kruppel* pattern becomes multipeaked faster than the *caudal* pattern does. Thus the *caudal* protein passes from uniform to monotonic at the ninth division, attains a pattern with a single zone in the posterior third of the egg only at the cellular blastoderm, and achieves a double-segment pattern by germ-band extension. In contrast, *Kruppel* starts later, appearing first in the tenth division passing to a full four-segment mode with peaks at both poles and mid-egg by the time of cellular blastoderm, and then going on to a double-segment pattern by germ-band extension.

A second gap gene, *hunchback,* also follows a harmonic sequence of increasingly complex waveforms (Reinerts, personal communication). This sequence starts with a uniform distribution of transcript which develops into a monotonic gradient having an anterior maximum by the eighth division. By the twelfth, a second peak appears posteriorly, giving a full period of gene product. During the fourteenth division, this pattern progresses through first two and then three full spatial periods. There is not yet evidence of higher harmonics during gastrulation and germ-band extension.

The mutants *knirps, tailless,* and *giant* are also members of the gap gene system. Based on defect patterns, these genes appear likely to exhibit multipeaked transcript and protein waveforms.

Bifurcation Sequences in the Pair-Rule Genes

The pair-rule genes are those whose mutation affects the primary axis either by deletion alone or by deletion plus mirror-symmetric duplications on *double-segment spacings.* Two of the pair-rule genes, *fushi tarazu* and *even-skipped,* give exquisite evidence for a spatial frequency-doubling bifurcation sequence.

Fushi tarazu is located in the Antennapedia complex. The gene has been cloned (Hafen, Levine, et al. 1983; Hafen, Kurowa, and Gehring 1984) and its pattern of transcription in the early embryo probed by *in situ* hybridization (see, for example, MacDonald, Ingham, and Struhl 1986). In addition, the protein coded by the gene has been cloned, and monoclonal antibodies to it have been raised. All this work has allowed the pattern of protein expression in the early embryo to be assessed using the monoclonal antibodies against that protein to stain sections or whole embryos (Carroll and Scott 1986). The results are striking.

Figure 14.26 shows the protein patterns at cellular blastoderm. Note the following features. First, there are seven stripes around the embryo, rather evenly spaced with nonstaining stripes. Second, notice that these stripes clearly splay toward the anterior and posterior poles. The anterior several stripes curve anteriorly as each passes from dorsal to ventral. In contrast, the posterior several stripes curve posteriorly as each passes from dorsal to ventral. The mid-egg stripes are nearly vertical. This spatial pattern is extremely reminiscent of an eigen function of the Laplacian diffusion operator on the deformed ellipsoidal geometry of the *Drosophila* egg. The egg is bent such that its ventral midline is longer than its dorsal midline arc. It is characteristic of eigen functions of the Laplacian that, where the spatial domain is slightly larger, spacings among stripes spread out slightly. On a deformed ellipsoid like the egg, longitudinal eigen function modes will therefore splay toward the poles ventrally. This similarity may be accidental and explained on other grounds, but it is rather hard to suppose so.

The only difference between the expected eigen function patterns on a full egg shape and the pattern seen in *fushi tarazu* is that bands should occur throughout the domain, but the embryonic pattern has a large empty domain anteriorly and a smaller one posteriorly. However, Edgar, Weir, et al. (1986) showed that two extra

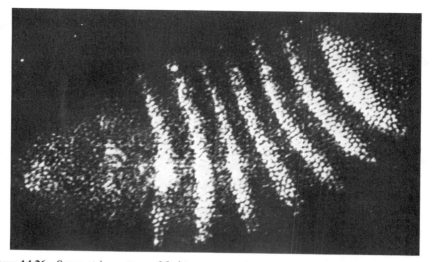

Figure 14.26   Seven-stripe pattern of *fushi tarazu* protein at cellular blastoderm in double-segment intervals. Note splay of stripes toward poles along ventral arc of embryo. (From Carroll and Scott 1986)

bands of gene transcript are observed in the anterior domain if protein synthesis is inhibited by cycloheximide one hour prior to the normal period of *fushi tarazu* banding at the fourteenth division. This suggests that there is a global periodic pattern in the normal embryo, but *fushi tarazu* transcription is repressed in the anterior (and presumably posterior) domain by a local signal. I discuss this point more fully below, for it suggests that some gene products act as components of an *internal positional framework* which confines the activity regions of other genes and couples "pattern-generating engines" such that periodic spatial patterns are generated within defined domains along the egg. If so, this confining framework may be part of the mechanism which allows a certain invariance to the resulting transcript patterns with respect to modest variations in overall size and shape of the egg.

Like the longer-wavelength systems described above, *fushi tarazu* and some other pair-rule genes pass through the now familiar bifurcation sequence. Here, however, the evidence for spatial frequency-doubling bifurcations is the clearest. The *fushi tarazu* transcript can first be found in mid-cleavage embryos, diffusely present in a wide zone ranging from 15 to 85 percent of egg length measured from the posterior pole. By the late syncytial blastoderm stage, when the embryo is still a syncytium and so has a connected common cytoplasm, this pattern sharpens by passing through a transient stage in which two alternative bands of high and low intensity occur on an eight-segment spacing interval. Then the number of bands doubles to four alternative stripes of high and low intensity on a four-segment spacing interval. Thereafter, the anterior three bands split into two bands of high intensity with a dark band between (Hafen, Kurowa, and Gehring 1984; MacDonald, Ingham, and Struhl 1986). Therefore, before cellularization, *fushi tarazu* forms seven alternative stripes of high intensity separated by stripes of low intensity. The spacing corresponds to a two-segment interval (Figure 14.27). Thus, in a syncytium, an initial broad pattern, which is nevertheless confined away from the ends of the egg, sharpens via a two-band and then a four-band pattern to alternative peaks and troughs in a sinusoidal pattern on a two-segment periodicity. As noted above, viewed from the side, these stripes of high intensity are not vertical bars running dorsal to ventral on the embryo; instead, they definitively splay toward the poles as they course ventrally.

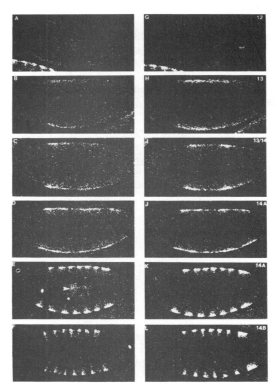

Figure 14.27 Sequential establishment of *fushi tarazu* and *even-skipped* transcription patterns in late syncytial blastoderm. Each transcript establishes a broad pattern which splits first to two bands and then to four. The anterior three bands then split again to yield the ultimate seven stripes. Careful analysis of *fushi tarazu (right panels)* and *even-skipped (left panels)* shows that they are phase-shifted with respect to one another along the anterior–posterior axis. (From MacDonald, Ingham, and Struhl 1986)

Even-skipped exhibits precisely the same spatial frequency-doubling bifurcation sequence (MacDonald, Ingham, and Struhl 1986) but is slightly phase-shifted with respect to the *fushi tarazu* bands (Figure 14.27). Again, bands splay toward the poles along the ventral margin in patterns extremely reminiscent of eigen functions on the deformed ellipsoidal shape of the egg.

The mutant *hairy* is another pair-rule gene, and its deletion yields an embryo having half the normal number of segments. Transcripts of *hairy* are first detectable at the eleventh or twelfth cleavage division. Its transcripts are uniformly distributed throughout the egg (Ingham, Howard, and Ish-Horowicz 1985). Then two domains of high transcript abundance can be seen, one dorsally in the anterior region from 85 to 95 percent of egg length, the other a broad band around the entire circumference of the embryo and extending from 75 to 20 percent of egg length. By the fourteenth cleavage division, the wide band becomes discontinuous and forms eight distinct bands along the length of that region. The *hairy* bands, like those of *fushi tarazu,* are broader ventrally (six to seven nuclei) than dorsally (three to four nuclei). A careful examination of the onset of *fushi tarazu* and *even-skipped* transcripts revealed a transient eight-segment, then four-segment periodicity. It is not clear that the same periodicity occurs in *hairy* because the proper analysis has not yet been done.

The spatial phase relationships between *hairy, fushi tarazu,* and *even-skipped* have been analyzed, however (reviewed in Akam 1987). At blastoderm all three have double-segment periodicities but are slightly out of phase with respect to one another. (That is, one reaches its peaks and troughs of activity slightly anterior to the other.) The three genes, however, are not 180 degrees out of phase in the two-segment spatial period. Thus different regions of the double-segment interval have different combinations, or, more precisely, *different ratios,* of the three transcripts and might use these differences to code for position with the double-segment intervals. I return to this point below and use it as the basis for measuring position by measuring phase defined by these ratios.

Finally, *paired* is another member of the pair-rule class. It, too, has been cloned and its pattern of transcription assessed. Localized *paired* transcripts are first observed at the twelfth division as a band six nuclei wide, stronger ventrally than dorsally, about 63 to 77 percent of egg length. The band widens at the next cleavage division to contain between 12 and 16 nuclei. During the fourteenth division, five additional bands arise spaced evenly to 20 percent of egg length, with a double-segment spacing. Just prior to completion of cellularization, two major events occur. An additional band appears posterior to band 7, at 13 percent of egg length, and bands 2 through 7 begin to split into an anterior and posterior band yielding a banding pattern on a single-segment spacing. This latter process progresses from the anterior to the posterior pole. In addition, a ventral-to-dorsal progression is seen (Kilcherr, Baumgartner, et al. 1986; Akam 1987).

The Segment-Polarity Genes

Among the segment-polarity mutants; whose deletion deletes parts of segments and yields mirror-symmetric duplicates of the remainder, the *engrailed* pattern has been investigated. The overarching phenomena are that *engrailed* is expressed first in a four-segment and then in a two-segment periodicity, which finally transforms to a single-segment spacing, by accumulation of first transcript and then protein in the zone between double-segment stripes (Weir and Kornberg 1985). In more detail, the following is seen. Transcription begins at the fourteenth cleavage division. The initial pattern is diffuse. Next, stripes begin to form but do so in a complex series which has an overall anterior–posterior gradient as well as superimposed complexity. The first two prominent bands are 2 and 8. Bands 4 and 8 become prominent before band 6 does, and band 12 becomes prominent before band 10 does. Thus, the onset witnesses a transient four-segment spacing. Thereafter, bands are present in double-segment invervals corresponding to the even-numbered bands, which remain transiently stronger than the newly arising odd-numbered segments. The final pattern has stripes a single cell wide in *each segment.*

These data offer direct evidence on a number of critical issues:
1. Within the syncytium, regionalized gene transcription or accumulation of transcripts occurs. So does protein synthesis. The regionalization can be on a very fine spatial scale, waxing and waning from maxima to minima to maxima over the distances separating perhaps eight to ten nuclei, counting three to four nuclei (and later cells) per segment.
2. The evidence strongly supports the idea that a spatial frequency-doubling bifurcation sequence or, more generally, a bifurcation sequence occurs. The best evidence for near frequency-doubling derives from *even-skipped* and *fushi tarazu,* where a single broad band alters to two bands, which in turn transform to four bands in a four-segment pattern, which bifurcates to a seven-stripe double-segment pattern and then, in the case of *even-skipped,* ultimately carries on to a 14-

stripe single-segment pattern. Data on *paired, engrailed, Kruppel, hunchback,* and *bicaudal* also give grounds to think that some form of bifurcation sequence to successively more multipeaked patterns is occurring in these gene systems as well. All these genes pass through a sequence from a single broad zone of abundance to at least a double-segment periodicity.

3. The natural wavelengths of the four systems are progressively shorter, and the pace in passing through the bifurcation sequence progressively more rapid. *Caudal* achieves its monotonic pattern at the tenth division, when *Kruppel* is already banded in the mid-egg and the pair-rule and segment-polarity genes have not yet started their bifurcation sequence. *Caudal* achieves the posterior ring pattern, perhaps reflecting the first plus second modes at the fourteenth division, when *Kruppel* has passed to the four mode, with peaks at both ends and at mid-egg. Similarly, the *hunchback* gap gene has achieved either two or three full longitudinal cycles by the cellular blastoderm stage. In the same interval, the *fushi tarazu* and *even-skipped* pair-rule genes have begun and passed through their entire bifurcation sequence from single bands to seven stripes in alternative segments. *Caudal* and *Kruppel,* which begin prior to the pair-rule genes, do not achieve double-segment spacing until germ-band extension.

4. The precise spatial patterns shown by maternal, gap, and pair-rule genes are extremely close to the patterns predicted by the eigen function of the Laplacian operator on the deformed ellipsoidal shape of the egg. Thus *caudal* transcripts form a monotonic gradient flat at both poles and steep in between—just the form predicted by a Turing-like mechanism based on the first longitudinal mode governed by the Laplacian. Even more striking, the precise pattern *fushi tarazu* and *even-skipped* transcripts, splaying ventrally, is exactly the form which eigen functions of the Laplacian operator will take on a deformed ellipsoid such as the *Drosophila* egg.

5. The patterns start diffusely and then sharpen. This is clearest with *caudal, hairy, even-skipped,* and *fushi tarazu* and probably true with *engrailed;* it is not true for *paired.* As in the case of *Polyspondilium,* initiation of a pattern diffusely with subsequent emergence of a few dominant wavelengths is the hallmark of growth of unstable spatial modes from noise.

Taken together, these data are very strong evidence supporting the view that the *Drosophila* embryo has wavelike properties underlying transcription patterns which occur on it early in development, arise in succession with more wavelengths per egg, and are associated with genes playing critical roles in determining segmental properties.

Parametric Couplings Between Gene Systems: An Internal Positional Framework?

In describing these rich phenomena, I have omitted consideration of the known and possible regulatory couplings among these genes, couplings which may in part constitute the molecular machinery which drives pattern generation. This is currently an area of intense research effort (Carroll and Scott 1986; Harding, Rushlow, et al. 1986; Ingham, Ish-Horowicz, and Howard 1986; Jackle, Tautz, et al. 1986; Akam 1987; Frasch and Levine 1987; Scott and Carroll 1987; Harding 1988). For example, it is now clear that deletion of each gap gene alters the banding patterns of pair-rule genes, typically increasing the width of some bands and deleting others. Similarly, products of the pair-rule genes are directly implicated in the generation of the normal segmental *engrailed* pattern (Howard and Ingham 1986; Akam 1987; Martinez-Arias and White 1987).

Several aspects of the problem should be stressed. First, the genes of the maternal-plus-zygotic, gap, pair-rule, and segment-polarity systems generate spatial patterns of very complex, overlapping waveforms. If, in general, these systems are jointly utilized to specify cell position and further development, then the *proper registration* among the diverse waveforms is important. In other words, the proper phase relations must occur. Second, any theory which is to account for the complex patterns must account for the *entire complex* of spatiotemporal patterns. Third, reaction-diffusion or other field models are the natural means to think about the apparent bifurcation sequences in transcript patterns.

If we take seriously the idea that the spatial pattern engine is, in part, some form of reaction-diffusion or other field system governed by partial differential equations and the Laplacian operator, then such a pattern engine must solve the problem of proper registration among different transcript patterns and must ensure that *only some* of the possible eigen function modes occur. In particular, consider the seven-stripe *fushi tarazu* and *even-skipped* patterns at cycle 14. The spacing between these stripes is only a few nuclei wide. However, a Turing-like reaction-diffusion system with such a short wavelength could also generate a striped pattern at right angles to the observed pattern, such that stripes appear along the dorsal–ventral axis. In addition, it might generate a checkerboard pattern. Neither is observed. Thus if a reaction-diffusion or other field model is to account for the onset of longitudinal spatial patterning as eigen functions of the Laplacian operator, it must also account for the failure of such patterns to be established in the dorsal–ventral or left–right axes of the egg.

A quite simple general hypothesis may suffice. The maternal, gap, pair-rule, and segment-polarity gene systems may be coupled such that the longer-wavelength systems act as *bifurcation parameters* to the shorter-wavelength systems. Recall that there are two major parameters in the Turing class of models: β, the ratio of diffusion constants of activator and inhibitor substances, and γ, the size of the domain or, alternatively, the absolute values of the diffusion constants of the substances or, alternatively, the overall rate of the chemical reactions in the system. Thus if components of the longest-wavelength system—*bicoid, oskar,* and *caudal*—act as bifurcation parameters to gap genes such as *Kruppel* or *hunchback,* the former might do so by altering the diffusion ratio of components of the gap gene system, perhaps by binding products of that system or by altering the reaction rate, perhaps by acting as transcriptional or translational regulators. Similarly, the long-wavelength and gap gene systems may act as bifurcation parameters of the pair-rule system. Grounds to think that the longer-wavelength systems may act as bifurcation parameters to the shorter-wavelength systems are based on the fact that deletion of single genes in the longer-wavelength systems typically do not delete or eliminate the basic periodic pattern of transcription in the shorter-wavelength systems but may distort the shorter systems' pattern.

In order to investigate this hypothesis, Axel Hunding in Denmark, in collaboration with B. Goodwin and me, has carried out numerical studies of a nonlinear reaction-diffusion system on the deformed ellipsoidal geometry of the *Drosophila* egg (Hunding, Kauffman, and Goodwin 1990). Hunding finds that, if the *Kruppel* product is assumed to be a bifurcation parameter of the pair-rule system, then the four-mode pattern of *Kruppel*—high at both poles and mid-egg, with troughs between— stabilizes the longitudinal seven-stripe pattern of the *fushi tarazu* and *even-skipped* system. Horizontal and checkerboard patterns are suppressed. Furthermore, on the deformed ellipsoid, the vertical stripes splay poleward along the ventral midline, as observed experimentally. Further, additional faint stripes are present in the anterior and posterior poles, reminiscent of the anterior-pole bands revealed by Edgar, Weir, et al. (1986).

Hunding's results are encouraging. They suggest that the longer-wavelength systems, by serving as bifurcation parameters to the shorter-wavelength systems, can act as successive components, constructing a kind of internal reference framework to build up the spatially fine-grained longitudinal patterns of transcription. The long-wavelength and gap gene systems may control the positions of maxima, the wavelengths, and hence the phase relations of the pair-rule genes by controlling the anterior-posterior boundaries within which pair-rule genes act or by controlling the diffusion constants of the pair-rule system or by acting as inducers or inhibitors of the pair-rule genes and hence controlling the reaction rates of the pair-rule system.

The Turing Mechanism Naturally Generates Phase-Offset Multipeaked Patterns

The observed spatial patterns of the pair-rule genes are phase-offset. It is therefore worth noting that the straightforward Turing model inevitably generates patterns of several variables whose peaks and troughs are phase-offset. In turn, this property offers a way to control spatial phase relations among a number of coupled genes.

Here is the issue. In any such field model, the linearized equations predict that, when a single mode is being amplified on the spatial domain, all the waveforms representing the various underlying chemical variables must be either in phase with one another or exactly 180 degrees out of phase; no other phase offset is possible. The reason is straightforward. In the linearized analysis, a fixed ratio of the underlying variables, measured as deviations from the spatially homogeneous steady state, is amplified at each point in the tissue. Such a fixed ratio is called an *eigen vector*. Thus maxima and minima of all the variables occur at the same position in the domain. However, suppose the domain allows two different modes, with two different wavelengths to be amplified simultaneously. Then typically each mode amplifies a different fixed ratio of the variables. Hence the full pattern, given by the sum of the two modes, typically has maxima and minima of the variables occurring in different spatial positions. Phase offset arises naturally. Furthermore, Turing models generate patterns because only a restricted range of wavelengths, L_1 to L_2, is amplified by the reaction-diffusion system. When the domain is small enough, only a single wavelength from within this range can fit onto the domain and fulfill the boundary conditions. When the domain is large enough, however, it must be the case that N of the shorter wavelengths, L_1, will fit onto the domain while $N - 1$ of the longer wavelengths, L_2, will fit. Therefore, as domain size increases, at first single modes can arise, but ultimately the superposition of more than one mode *must occur*. Thus, eventually, phase-offset patterns can be expected (Kauffman 1984a, 1984c).

The main point to stress is that the Turing model can generate sequentially complex, multipeaked patterns which give rise ultimately to phase-offset patterns. From Arcuri and Murray (1986), we recognize that, in general, the fully nonlinear system of field equations can pass to a spatially inhomogeneous steady state. In that steady state, nonlinear mixing of many modes often occurs; hence stable phase relations among the variables arise on any such stationary spatially inhomogeneous dynamical attractor. In short, nonlinear versions of reaction-diffusion or other field models are very likely to give rise to multiple-wavelength patterns which coexist stably in a tissue domain. Since many of these patterns are nonmonotonic, the question arises as to how cells can utilize such spatially complex patterns to assign positional identities. In a moment, we shall use the observed phase-offset patterns in the four color wheels model to define position in terms of phase angles on each of the different wavelength transcription systems.

THE FOUR COLOR WHEELS MODEL
OF POSITIONAL SPECIFICATION

The Spatial Distribution of Maternal, Gap, Pair-Rule, and Segment-Polarity Genes Products Does Not Yet Dictate How the Embryo Uses Them to Determine Position

We wish to understand how the embryo uses the spatial distribution of the gene products we have discussed, and perhaps others, to specify anterior–posterior pattern elements. Perhaps the first point to stress is that any given spatial distribution of "morphogens" might be used in a variety of ways to specify position. These alternative possible uses, in turn, will predict quite different phenotypes as a consequence of mutations or other perturbations. The simple example shown in Figure 14.11 makes this point. Consider three hypothetical morphogens, X, Y, and Z, arranged in three monotonic gradients in a roughly circular tissue, such as a wing imaginal disc. Let the X gradient be high on one side of the disc and low on the opposite side—on the dorsal and ventral ends of the wing disc, say—while Y is at right angles and high at the anterior margin and low at the posterior margin. Let Z be high in the middle of the wing disc and low at its boundaries, forming a cone-shaped gradient. The wing disc might use the concentrations of the three morphogens to specify position of each cell in the tissue in a Cartesian coordinate system. Instead, however, the *ratios of X and Y above and below some midlevels of their concentration range* might be used to measure an azimuthal *phase,* or *angle,* around a circle of phases, while the concentration *of* Z is used to measure a *radial* positional value from the position of maximum Z. This case corresponds to the simplest molecular interpretation of a polar coordinate system, such as that suggested by French, Bryant, and Bryant (1976) for epimorphic pattern regulation. Finally, the ratio of X and Y might be used to measure a longitudinal angle and the ratio of Z and Y used to measure a latitudinal angle in a spherical coordinate model, such as that proposed by Russell (1978) for epimorphic pattern regulation.

Two points warrant stress: (1) the spatial distribution of the morphogens does not yet tell us how they may be used to specify position and (2) mutant and other effects will differ depending on that use. Hence analysis of the effects of mutants may yield insight into how an organism uses the morphogens to specify pattern.

It Is Unlikely that Longitudinal Position Is Specified by Simple On–Off Combinations of Gene Activity

The pair-rule and segment-polarity genes are expressed in overlapping domains in two-segment or single-segment periodicities. It is extremely attractive to suppose that distinct positional values of each cell within a two-segment period, or within a single-segment interval, are specified by precise combinations of gene-expression patterns. Nevertheless, this effort poses considerable problems even in its simplest form. That simplest form supposes that each position within the double- or single-segment interval is specified by a precise combination of gene-expression patterns, where each gene can be considered to be "on" or "off" in each cell. More exactly, the concentration of each gene's product can be considered to be above or below a precise threshold in each cell. The problem we are concerned with arises when we attempt to apply this idea in the context in which more than two gene products are involved in one periodic domain. For concreteness, we focus on segment-polarity mutants and thus periodic patterns in one-segment spacing domains.

Figure 14.28a shows a concrete case in which four genes arise in a periodic pattern in one domain and in which the phases of the four genes are evenly phase-offset by 45 degrees. It is convenient to recast the phase-offset sine patterns onto a circle whose circumference represents position in a tissue from the beginning to the end of a one-segment interval along the anterior–posterior axis, as shown in Figure 14.28b.

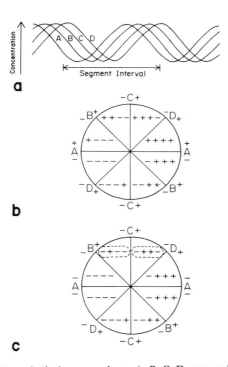

Figure 14.28 (a). Four hypothetical gene products, A, B, C, D, wax and wane sinusoidally along the anterior–posterior axis in single-segment periodicity. Each product is assumed to be phase-offset from the others by 45 degrees. (b). Periodic spatial pattern of variation in a one-segment interval recast onto a circle representing the one-segment interval. Each gene product from (a) divides the circle into two regions separated by a diameter. In one region, the gene is above threshold and hence on (+); in the other, it is off (−). This on–off feature creates a binary combinatorial code word in each sector. (c) The consequence of deleting gene A is that its concentration is below threshold throughout the one-segment interval. This deletion converts all code words with A = on (+) to A = off (−). Rather than yielding mirror-symmetric mutations of the segmental pattern, however, this conversion creates two "illegal" code words, which are circled.

Around this circle, each gene's activity waxes and wanes sinusoidally, passing above and below its threshold level at two defined positions on the circle. Thus each gene's pattern of on and off values is divided by a diameter of the circle. This division yields a pattern of on–off combinations of gene expression transected by four diameters, each demarking the threshold between on and off concentrations of one gene. Mutants of the segment-polarity genes uniformly result in deletion of pattern elements and duplication of the remaining pattern elements. Consider, therefore, the consequence of deleting gene A. Once A is deleted, the concentration of its product falls below threshold in all cells. Rather than yielding a mirror-symmetric pattern, however, this mutation results in a number of "illegal" combinations not present in any sector of the normal pattern (Figure 14.28c). Thus there are no natural reasons to expect these illegal code words to yield pattern elements which are part of a pattern duplication.

The same observation holds for deletion of any of the other genes. In Figure 14.29, we show that the same result arises even for three phase-offset sinusoidal gene products.

A second difficulty with strict on–off binary coding of position is that, if the binary combinations each specify the fate of single cells in a segment and always do so in the same way, then any positions of mirror symmetry due to gene deletion which do arise should be very precise in one mutant among flies, but variability is typically found. We return to this problem below.

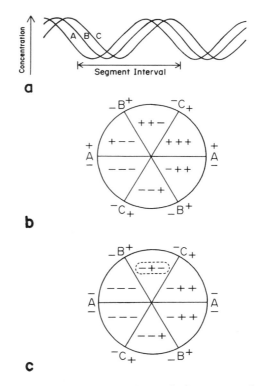

Figure 14.29   Same as Figure 14.28, except that here only three gene products, A, B, C, are used. The result of using three products rather than four is that only one illegal code word is created.

Measuring Position as "Phase": The Color Wheel

We can lead into the concept of measuring position as a "phase" by considering more carefully the on–off model with three genes because its inadequacies point toward a better theory. The linear and circular representations of Figure 14.29 can be made more complete by considering three morphogen axes at right angles, each showing the range of values of one gene from low to high concentrations (Figure 14.30). The idea of a *threshold level* for each gene product at the midlevel of its concentration range now carries over to three orthogonal lines, each representing the level of one gene product in this tissue specificity space (Winfree 1984). The intersection of the three lines occurs at the concentrations which correspond to the threshold level of each morphogen. The concentrations of the three variables along the physical tissue in the segmental domain of the fly (Figure 14.29a) are mapped to a closed cycle of concentrations in tissue specificity space (Figure 14.30). That is, each point on the closed cycle represents the A, B, and C morphogen concentrations at a *single point* in the segmental tissue domain. Traveling around the cycle in tissue specificity space corresponds to mapping the simultaneous concentrations of A, B, and C along with segmental domain from start to end of one period. The same pattern of phase-offset morphogen concentrations then recurs on each segmental repeat.

The hypothesis that each position along the segment is encoded by a specific on–off combination of gene-expression patterns, where "on" and "off" refer to above-threshold and below-threshold concentrations, now amounts to noting that the three orthogonal axes shown in Figure 14.30 intersect at the threshold levels of the three

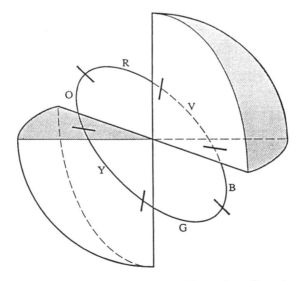

Figure 14.30 Recasting the three genes of Figure 14.29 into a three-dimensional tissue specificity space (TSS), where the concentration of each gene is an independent orthogonal axis, and all three axes intersect at the threshold level of each gene product. Intersection is the origin of the chemical TSS. Each combination of above-threshold and below-threshold levels of the three variables corresponds to one of the $2^3 = 8$ orthants of this TSS. Six of the eight orthants correspond to legal code words from Figure 14.29b and can be thought of as colors of the color wheel arranged in order in a cycle. The two remaining orthants (shaded) correspond to the two illegal code words specifying no meaningful positional information. B, blue; G, green; Y, yellow; O, orange; R, red; V, violet.

morphogens and divide the tissue specificity space into $2^3 = 8$ *orthants*. All points in tissue specificity space within each orthant have the same combination of gene products above and below threshold and hence constitute a combinatorial code word standing for one positional domain.

The first distinction between specifying position in tissue specificity space (Figure 14.30) and the simpler cycle shown in Figure 14.29*a* is that in the former we formally and explicitly extend the definition of the phrase "same code word" to include all parts of the same orthant in tissue specificity space. Thus, we have a concept of the same positional information if the *amplitudes* of the phase-offset wave patterns of A, B, and C decrease or increase slightly or, viewed another way, if the cycle in Figure 14.30 contracts or expands slightly. A visually attractive way of representing the fact that each orthant is meant to be the "same" positional code is to assign each orthant a different color form the color wheel, in which the entire color spectrum is arranged as a ring, with violet next to red. Note that the on–off hypothesis yields a cycle in tissue specificity space which passes through six of the eight possible orthants. Thus these six orthants—colored perhaps red, orange, yellow, green, blue, and violet—have well-defined positional code word values. The two remaining orthants, however, are not traversed by the cycle in tissue specificity space and hence are not assigned any color. This lack of color restates the concept that those quadrants represent illegal combinations of on–off states which have not assigned positional value in the tissue.

Consider the effect of deleting one gene. If A is deleted and B and C exhibit their old waveform, then the cycle of morphogen values remains in the four orthants in tissue specificity space in which A is below threshold. Therefore, the cycle necessarily crosses into at least one of the uncolored orthants representing an illegal code word. This feature restates the difficulty noted above with the on–off combinatorial positional code when three or more gene products are involved: Mirror-symmetric patterns are not produced because illegal code words are generated.

Continuous Color Wheels
and an N − 2 Dimensional Singularity

I believe the familiar on–off combinatorial model is basically on the right track but needs to be generalized to a more realistic theory. Note first that the idea of representing position by *combinations* of several gene products above and below threshold values amounts to representing position of the *ratios* of those gene products, each measured with respect to its threshold. In this context, the use of a single threshold for each gene is equivalent to supposing that the tissue can recognize and respond to only gross differences in gene product concentration—"above threshold" and "below threshold." The combination of all gene values above and below respective thresholds then measures a quantized ratio, or phase angle, around the cycle in tissue specificity space. The quantized angle measured is just the arc in one colored orthant.

Two features of the on–off combinatorial theory are unnecessarily limiting. First, it is reasonable to suppose that cells can respond to only modest differences in concentrations; hence there must be a minimal detectable difference. In terms of distinguishing different "phase," or angle, positions around the cycle in tissue specificity space, there must be a minimum distinguishable angle. However, since cells might be able to discriminate several sufficiently different concentrations of each gene product, the minimum phase angles might be smaller than an orthant.

The second limitation of the theory is more fundamental. Note in Figure 14.30

that the six colored quadrants all *meet at a single point*—the point in tissue specificity space where the three threshold axes intersect. This arrangement is extremely untypical and arises only for quite improbable assignments of colors to tissue specificity space. This leads us to the concept of a singularity (Winfree 1980, 1987). For concreteness, let us quantize the color wheel into a fixed number of distinct colors, each representing a distinct phase-angle sector around the cycle in tissue specificity space. Then each small sector on that cycle is colored appropriately. In order to extend the definition of phase angle to points in tissue specificity space which are not on the normal cycle of morphogen values, we must try to assign a color to each point in tissue specificity space. For example, the most natural way to try to do so is to think of each quantized color as being a colored card of some thickness and then spindle each card in turn by the cycle in TSS. Then the colored cards are spindled in order around the cycle (Figure 14.31) such that, when we pass around the cycle in TSS, we encounter the cards, or phase sectors, in proper order. These cards, each representing one quantized phase angle and called, colorfully, "isochromes," are said to fill, or foliate, in TSS. The critical feature of this image is that each color card extends *inside* the cycle in TSS. Thus it is intuitively clear that all cards might meet somewhere in the middle. In the immediate vicinity of such a locus, all or at least many colors, and hence all or at least many isochrome phase angles, are infinitely close to one another. Indeed, a fundamental topological theorem demonstrates that there must exist at least a *one-dimensional, or line, locus* threading through the cycle in TSS on which all or many such phases abut. Such a line locus is a *phase singularity*. When all isochrome phases terminate on such a singularity, its immediate vicinity contains all possible phase values; hence the singularity is considered a *phaseless locus*. In the

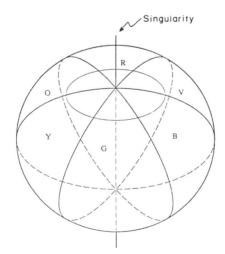

Figure 14.31   Natural generalization of on/off sector model to a color wheel in three-dimensional TSS. Each color, or isochrome, specifies one ratio of the three morphogenetic variables measured from the origin and hence one positional value. All isochromes are two-dimensional surfaces in three-dimensional TSS. Each color is a sector, rather than a surface, and corresponds to a range of ratios. A *singularity* must exist where many or all of the isochrome surfaces meet one another; this is a "phaseless" locus in TSS. The phaseless locus must be one- or two-dimensional in a three-dimensional TSS. The singularity is shown as an axis passing through the sphere from "north" to "south" pole. B, blue; G, green; Y, yellow; O, orange; R, red; V, violet. (From Winfree 1980)

color wheel image, mixing all colors leads to white, which contains all colors but no specific hue. The general theorem states thay, in an N-dimensional TSS in which phase angle is assigned around a cyclic ring in that space, an $N - 1$ or $N - 2$ dimensional phaseless singularity, or "zone," must exist and thread through the cycle (Winfree 1980, 1987). Although the general theorem allows the singularity to be a jump discontinuity where many but not all phases meet, we shall restrict our attention to the simple general case where all phases abut on a phaseless singularity.

Figure 14.31 shows the natural generalization of the on–off model. As before, phase angle is measured by the ratio of concentrations of morphogen variables, where the ratios are taken with respect to an "origin" in TSS rather like the intersection of the three thresholds in Figure 14.30. Unlike the quadrant model, however, we here extend each color to the phaseless singularity which threads the cycle in TSS. Not surprisingly, this phaseless line locus passes through the two quadrants which were left colorless before. Now, however, each color extends from the formerly colored quadrants, or smaller sectors, *well into the formerly uncolored quadrants.* Indeed, each color extends to the one-dimensional phaseless line.

We need one further idea. The supposition that cells cannot distinguish arbitrarily similar concentrations of morphogens implies that cells cannot distinguish positions in TSS which lie very near one another but on opposite sides of the phaseless singularity locus. Therefore, we make the rather natural assumption that a cell which straddles the singularity in TSS and has all or very many phase-angle values literally has *no phase-angle information* and hence makes *no pattern element.* Perhaps the cell dies. Whether it does or not, we shall refer to a "tube" around the singularity as a *dead zone.* Cells in which the concentrations of gene products A, B, and C lie within the dead zone make no pattern elements (Figure 14.32). In effect, we are restricting

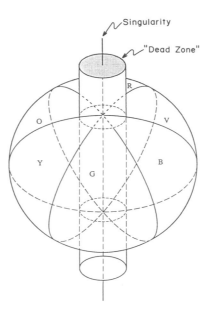

Figure 14.32 A color wheel in three-dimensional TSS having a "dead zone" tube surrounding the one-dimensional singularity where isochromes meet. Cells with positional values in the dead zone are assumed to be phaseless and hence to have no interpretable positional information and make no pattern element. B, blue; G, green; Y, yellow; O, orange; R, red; V, violet.

the illegal quadrants of the on–off model to dead zones which pass through such quadrants but do not occupy them entirely.

Pattern-Element Deletion and Mirror Duplication Are Generic

This color wheel model yields four critical results. First, it naturally predicts pattern-element deletions when the spatial wave patterns of gene products A, B, and C are modified such that part of the cycle in TSS passes through the dead zone. Second, it naturally predicts mirror-symmetric duplications when the wave patterns of A, B, and C in the tissue are modified such that the entire cycle no longer enters or surrounds the dead zone. Third, it predicts that a continuous deformation of the normal cycle in TSS from that which surrounds the dead zone to that which does not enter or surround the dead zone will lead first to pattern-element deletion, then to asymmetric pattern duplication, and finally to mirror-symmetric pattern duplication. Finally, the color wheel model predicts that there will be variability in the position of the mirror symmetry because there is no constraint on the exact manner in which the deformed cycle crosses the singularity.

These generic predictions of a color wheel model containing a singularity and a dead zone are shown in Figure 14.33. Note that, as the waveform of A, B, and C across the tissue is progressively distorted, the cycle in TSS first approaches the dead zone and then enters it but still surrounds the singularity in the center of the dead zone (Figure 14.33a), then no longer surrounds the singularity (Figure 14.33b), then falls entirely on one side of the dead zone and no longer surrounds it (Figure 14.33c). When the cycle passes through the dead zone, the cells having the corresponding phase angles make no corresponding pattern elements; hence those elements are deleted. When the cycle no longer enters or surrounds the dead zone, then, in passing around the cycle (hence from one end of the periodic repeat domains in the physical tissue to the other), each phase angle is encountered *twice, and in reverse order;* hence a mirror-symmetric duplicate is formed. Note further that such duplications require no cell death and intercalary regeneration. As noted by Russell (1985) and Kauffman (1984c), these duplications are a topological consequence of measuring phase by ratios from an origin in a TSS. Finally, note that if the cycle of values does not surround the singularity but does pass through the dead zone, then an *asymmetric* mirror duplicate is formed, with fewer pattern elements in the mirror-duplicated region than in the region in normal orientation.

The first issue to stress is that these properties are precisely those observed in the pattern-element deletions and duplications on all length scales. Weak alleles lead to deletions. Stronger alleles, or more extreme phenotypes, lead to deletion plus duplication—first asymmetric and then symmetric—of remaining pattern elements. The fact that the same phenomena arise on all length scales in the *Drosophila* embryo markedly encourages the belief that similar color wheel principles must underlie the phenomena at all length scales. We return to this idea shortly.

The next feature to emphasize with respect to the generic properties of such color wheel models is that minor differences in how the cycle in TSS is distorted to intersect the dead zone and cross it lead to minor differences in the way in which pattern elements are deleted and in the way in which further pattern elements are duplicated. This, too, matches what is observed on all length scales. In *bicaudal,* the precise position of mirror symmetry can lie at any position in any of several abdominal segments. In *Kruppel,* the position of mirror symmetry varies from embryo to embryo. Even in *runt* and the segment-polarity mutants, precise locations of deletion and mir-

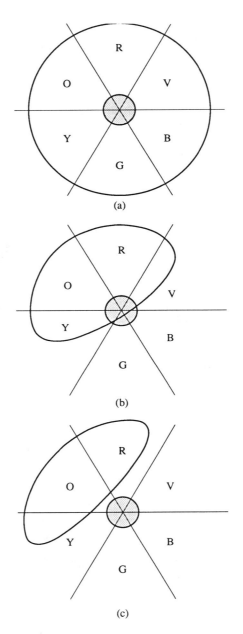

Figure 14.33 Generic predictions of deletion followed by first asymmetric and then mirror-symmetric duplication as the cycle of morphogen values in tissue is displaced first through and then past the dead zone and singularity. (*a*) Cycle distorted so that it enters dead zone but still surround singularity, leading to deletion but no duplication. (*b*) Cycle distorted to enter dead zone but no longer surround singularity. This distortion yields deletion of some pattern elements and asymmetric duplication of others, with fewer duplicated than original pattern elements. (*c*) Cycle no longer surrounds either dead zone or singularity. This conformation yields deletion of some pattern elements and mirror-symmetric duplication of the remainder. B, blue; G, green; Y, yellow; O, orange; R, red; V, violet.

ror symmetry vary. The color wheel model has the capacity to account for such variability naturally.

Recall from page 604 the puzzling phenomena in double mutants of particular alleles of *even-skipped* and *odd-skipped*. Each mutant alone deletes even or odd denticle bands, but the double mutant results not in the deletion of all denticle bands but in the formation of narrow mirror-symmetric denticle band duplicates. The color wheel model on a two-segment spacing can naturally account for this. Each single mutant deforms the cycle of morphogen values such that it encounters the dead zone. One mutant removes even denticles; the other, odd denticles. The joint mutant, we readily suppose, is shifted such that it no longer either enters or surrounds the dead zone and hence yields a duplication.

Gergen, Coulter, and Wieschaus (1986) report a further interesting experiment with *runt*. Flies having different dosages of the normal *runt*+ gene were constructed. Hypomorphs gave familiar weak *runt* phenotypes, but flies with extra doses of *runt*+ yielded *antirunt* phenotypes in which the segments normally deleted were unaffected, and the alternate segments were partially deleted. Thus *antirunt* is similar to *even-skipped*. This phenotype would correspond, on a two-segment color wheel, to assuming that the *runt* and *antirunt* phenotypes correspond to shifting the cycle in TSS in opposite directions with respect to the dead zone, such that pattern elements in the odd or even set are affected.

The Four Color Wheels Model

Evidence presented in the section beginning on page 605 demonstrates that maternal gene products and those of *caudal* are present in long-wavelength patterns—either as monotonic gradients or as a single peak in the posterior third of the egg—at cellular blastoderm. At the same stage, *Kruppel* is a full two-cycle pattern, with three peaks at the anterior and posterior pole and mid-egg and two troughs between the three peaks (Harding and Levine 1988). *Hunchback* also appears to be in a full two-cycle pattern with peaks in the anterior and posterior halves of the egg and three troughs at the two poles and between the two peaks (Akam 1987; Reinerts and Levine 1988). Hence *Kruppel, hunchback,* and other gap genes are arrayed in shorter-wavelength patterns along the egg during the cellular blastoderm stage. At the same time, the pair-rule genes have progressed through their bifurcation sequence and are arrayed with phase differences on a two-segment periodic spacing. Shortly thereafter, the segment-polarity genes are arrayed in presumptive phase-spaced intervals in single-segment repeat units.

Since the basic phenomena of deletion and duplication of pattern elements occur at all length scales along the egg and exhibit fundamentally similar features, Goodwin and I propose that position along the entire anterior–posterior axis of the embryo is specified *simultaneously* by measuring phase angle, or "color," on at least four separate color wheels. The first color wheel, representing the longest wavelengths, gives crude overall information about position along the anterior–posterior axis. The crudeness is a necessary consequence of supposing that cells can measure differences in concentration only beyond some minimal range; hence any color wheel must be quantized into minimum discriminable sectors. Maternal genes such as *bicaudal* (Nüsslein–Volhard 1977) and zygotic-plus-maternal genes such as *caudal* (MacDonald and Struhl 1986) presumably are the constituents of the first color wheel. The second wheel represents information from the gap genes, such as *Kruppel, hunchback, knirps, giant,* and perhaps *tailless*. This wheel passes through about two cycles along the full egg length and hence refines the positional precision yielded by the first wheel. The third wheel is based on the phase-offset patterns of the pair-rule genes, and the fourth is derived from the segment-polarity genes. Jointly, the information tells any cell along the anterior–posterior axis what pattern element to make in the larval cuticle. We note that this formal (1986), model is, in effect, a more precise statement of a concept proposed by Gergen, Coulter, and Wieschaus (1986), in which each cell specifies position by measuring ratios of segment-polarity, pair-rule, and perhaps gap gene products.

This four color wheels model therefore specifies the positional identity of each cell combinatorially, not merely in the sense that each phase angle on each color wheel is itself a ratio combination of many genes but in the further sense that position is specified simultaneously and combinatorially on at least four color wheels.

The four color wheels model is a natural generalization of the binary combinatorial code presented above and preserves its major strengths. Each wheel must have a phase singularity threading through it and hence a dead zone. Thus it is natural to postulate that, if the morphogen cycle on *any* of the four color wheels lies in the dead zone, then that cell will make no pattern element. Thus the model asserts that, in order to make a pattern element, a cell must have a color from each of the four wheels. The combination of four colors constitutes the epigenetic code for that cell. Naturally, the model predicts that a deformation of the cycle of morphogen values on one wheel such that a cell has the "wrong" color from that wheel but correct colors from the remaining wheels will lead to a homeotic-like transformation which alters the pattern made to one appropriate for a different longitudinal region of the embryo. Such transformations are precisely what are observed, although not usually thought of as homeotic, in the duplication phenotypes at all length scales we have described. Further, because the four color wheels each reflect nonmonotonic patterns of gene transcripts on different wavelengths, transformations to a different color across the singularity on a wheel can jump long distances on the fate map. Transformations on the longest-wavelength wheel can convert genitalia to head structures, transformations on the middle-wavelength system can transform eye to wing, and so on. The combinatorial color code in each cell, engendered by the four color wheels, thus maintains the combinatorial features of the binary code model allowing the latter to fit the transdetermination and homeotic data of Tables 12.1 and 12.2.

Bicaudal *Phenotypes*

Bicaudal embryos, as noted earlier, range from mirror-symmetric forms in which typically more ventral segments are present than dorsal, to asymmetric, to headless embryos (Nüsslein-Volhard 1977; Mohler and Wieschaus 1986). The mirror symmetry can be at any position within one of these segments (Gergen, Coulter, and Wieschaus 1986). But *bicaudal* phenotypes can be even more striking. For example, the left side of an embryo may be a mirror-symmetric double abdomen, while the right side is a headless phenotype. Worse, the entire dorsal left and right sides can be mirror-symmetric double abdomen together with the ventral right side, while the ventral left side is a headless phenotype. These embryos juxtapose posterior spiricles with thoracic pattern elements along a dorsal–ventral line around the embryo near the anterior end.

The first color wheel presumably comprises genes such as *bicaudal* and *caudal.* The presumption that *oskar, bicoid,* and other genes are arrayed in monotonic gradients from pole to pole (Nüsslein-Volhard, Frohnhofer, and Lehmann 1987), while *caudal* is a superposition of first and second longitudinal modes with a peak in the posterior third of the egg in the late syncytial blastoderm leads to a color wheel which does not close in a full cycle (Figure 14.34). While there may be other long-wavelength variables which complete the first color wheel cycle, we shall for the present base analysis on the incomplete wheel. The implication is that the range of angular values where the first wheel is incomplete has no meaning and hence forms no pattern elements.

Figure 14.34 models the headless, asymmetric, and mirror-symmetric *bicaudal* phenotypes as continuous deformations of the incomplete cycle in the long wavelength TSS corresponding to the first color wheel. As expected, a continuous deformation of the cycle leads first to loss of pattern elements, than to asymmetric duplication, and then to mirror duplication. Slight differences in the deformation of the cycle lead to different positions of mirror symmetry.

A color wheel model readily accounts for the presence of thoracic and posterior

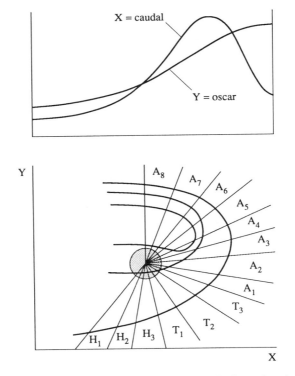

Figure 14.34 Predictions of the *bicaudal* phenotypes based on the four color wheels model via distortions in the long-wavelength color wheel. X, pattern of transcription of *caudal* at cellular blastoderm stage, Y; pattern of *oskar* or other gene in long-wavelength system; H_{1-3}, head segments; T_{1-3}, first, second, and third thoracic segments; A_{1-8}, abdominal segments 1 through 8. The open curve surrounding the dead zone but not entering it yields normal segment pattern. The curve entering the dead zone but still surrounding the singularity yields *headless* phenotype. The curve not entering the dead zone and not surrounding the singularity gives mirror-symmetric double abdomen. A curve which entered the dead zone but did not surround the singularity (not shown) would yield asymmetric double abdomen.

abdominal tissues near one another in a transect around one side of the anterior pole of an embryo. The existence of a phase singularity is a kind of universal mirroring switch. Crossing that locus leads to a jump to a distant phase angle without a large discontinuity in morphogen concentration. Thus, dorsally, the anterior pole can exhibit a symmetric phenotype with posterior abdomen pattern elements, while ventrally nearby morphogen concentrations can yield thoracic pattern elements. By contrast, the supposition that anterior–posterior position along the embryo is specified by a monotonic gradient has a hard time accounting for this phenomenon (Meinhardt 1986).

Bicaudal embryos have more ventral than dorsal segments (previewed in Gergen, Coulter, and Wieschaus 1986). The eigen functions of the Laplacian on a deformed ellipsoid having the ventral arc longer than the dorsal arc exhibit a poleward splay of stripes ventrally, as do the observed pair-rule stripes. Thus formation of a mirror duplicate along a fixed anterior–posterior position of symmetry will delete more dorsal pattern elements than ventral pattern elements.

It is interesting here that strong mutants of the *caudal* gene, mutants which delete maternal and zygotic contributions, lead not only to loss of abdominal segments but to transformation of the posterior telson into mouth-hooks (MacDonald and Struhl 1986). The supposition that *caudal* is a component in the first color wheel and the proximity of posterior abdomen to anterior pattern elements allow the model to explain the transformation rather naturally.

Similar considerations apply to mutants in the second, third, and fourth color wheels. All lead to progressive loss of pattern elements and, in strong alleles, or perhaps double mutants, to mirror-symmetric duplications. The natural expectation for the gap-gene color wheel, since most presumptive genetic components appear to have two or more transcript peaks along the anterior–posterior axis so that the wheel cycles roughly twice from pole to pole, is that mutants in any gene affect more than one area along the axis. This is true for *hunchback, giant,* and *tailless* (reviewed in Akam 1987) and probably for *Kruppel,* whose deletion deletes thorax and proximal abdomen but also affects malphigian tubes and hindgut. A second color wheel which cycles twice in one egg length might be expected to allow two regions of mirror symmetry, but this has not yet been observed. As noted on page 604, double mutants of two offset pair-rule genes can lead to mirror duplication when neither alone causes such phenotypes. It may be that double or triple mutants of the gap-gene color wheel will reveal mirror-symmetric duplications in more than one domain.

The four color wheels model readily accounts for the deletions in different frames in the pair-rule genes and for the onset of mirror duplication in strong alleles or double mutants. The fourth color wheel, representing the segment-polarity genes, as introduced by Russell (1985) and Kauffman (1984d) yields the phenotypes of the segment-polarity genes, deleting part of a segment interval and duplicating the rest.

Finally, note that all four wavelength systems are present simultaneously at the cellular blastoderm stage and are assumed to specify longitudinal positional identities. Thus distortion of the gradient patterns by constriction of the early-cleavage egg and subsequent recovery (Schubiger 1976; Schubiger and Wood 1977) should be interpretable as the restoration of the normal simultaneous patterns of all components after restoration of diffusive continuity.

Cell Autonomy and Its Implications

It is now well established that a number of the segment-polarity genes, some pair-rule genes, and some gap genes are largely cell-autonomous (Gergen and Wieschaus

1986). Thus in *runt*, formation of a homozygous *runt* clone in a heterozygous embryo leads to *runt* deformations in the homozygous clone and perhaps in a few heterozygous neighboring cells at the clone boundary (Gergen and Wieschaus 1985). Cell autonomy suggests that such mutations can occur in single cells or in small groups of cells without distorting the global waveform of morphogens. Thus, presumably, in *runt/runt* clones, the global transcript pattern of *runt* is unaffected but the *runt* product is absent from the mutant cells.

With respect to the four color wheels model, cell autonomy poses this issue: If the concentrations of *runt* and other pair-rule genes or of *Kruppel, giant,* and other gap genes are the cycle of wave patterns in TSS constituting the morphogens measured on the gap gene color wheel, then cell-autonomous loss of one such gene product is directly interpretable on the model. In that deficient cell, the cycle of morphogens is deformed in TSS; hence a pattern element is deleted or one is duplicated. However, the evidence that such homozygous deletion clones—*Kruppel/Kruppel,* for example—do *not* alter the global pattern of *Kruppel* expression in the embryo in wild-type tissue amounts to evidence that *Kruppel* product itself is not necessary in *generating* the wild-type pattern. Then either other gene products perform redundant functions or else nongenetic metabolic variables (such as ions) which can communicate between cells are the generators of the spatially distributed patterns via a Turing-like mechanism, and *Kruppel* and other genes are controlled by that underlying pattern-generating system.

Experiments

The four color wheels model suggests directions for further experiments. Most notably, any one phase on each color wheel is specified by more than one combination of morphogen values. More precisely, each phase angle, or isochrome, is a volume of TSS specifying the same "quantized" color, or angle. Thus perhaps the most important experiments suggested by the model require careful mapping out of the concentrations of presumptive morphogens which constitute the "same" positional value. This task would require assessing the distributions of gene products of at least two genes at a time in normal and mutant embryos, and correlating such alterations with the phenotype of the subsequent larva. Since assessing patterns of transcripts of protein distribution currently requires sacrificing the embryo, such experiments would obviously require correlation between sectioned embryos and larvae of the same genotype and hence quite tight phenotypes. Transformations of embryos with cloned gap, pair-rule, or segment-polarity genes under the control of inducible promoters may allow a fuller investigation of these issues.

The four color wheels model predicts the existence of isochromes, dead zones, and hence of morphogen values at which duplication and deletion occur. Note that, if deletions are correlated with cell death, then finding cells whose morphogen concentratons correspond to a dead zone will not be open to direct demonstration. The boundaries of such dead zones may be definable, however, by analyzing the full set of isochrome values which allow patterns and pattern duplications. The missing morphogen values would correspond to the dead zones.

The model also predicts that the presence of morphogen values corresponding to the dead zone of any one color wheel cause deletion of pattern elements. This should be testable.

In short, it must be stressed that *in situ* hybridization of normal and mutant embryos with batteries of different gene probes in the set belonging to the four color wheels, together with annotation of which embryonic regions will be deleted and which duplicated, constitutes *mapping the cuticular pattern elements of the larva into a real TSS.* Thus if the mapping is into a 15-dimensional space comprising the concentrations of 15 gene products and if there are four independent dead zones, those zones should appear as four independent tubes in the TSS. Similarly, if isochromes end on the dead zone, that, too, should be visible.

Any attempt to account for cell autonomy by supposing that the level of a given gene produce is low or zero in mutant clones, while the global transcript pattern and protein pattern across the whole embryo are normal in wild-type tissue is directly testable in gynandromorphs or mitotic recombinant clones, given cell-autonomous markers for specific cells which can be made visible at the blastoderm stage. This clearly is an important issue beyond the four color wheels model. If genes whose transcripts occur in complex spatial waveforms can be deleted from specific parts of the pattern, and if the products of these genes are absent from those parts of the pattern while the remaining spatial pattern is normal, then the deleted genes and their products are not necessary to the mechanism which generates the spatial pattern.

The hypothesis that the patterns generated by the different genetic systems are due to Turing-like mechanisms might be tested in a variety of ways. In particular, transplantation of nuclei and cytoplasm from the middle or late syncytial blastoderm to a free spherical droplet under fluorcarbon oil might allow gene-transcription patterns to emerge in this altered geometry. Theory predicts that eigen functions of the sphere would arise as transcript patterns. For example, concentric onion shells of transcript patterns could form. Similar experiments might place such a droplet of nuclei and cytoplasm in the protected environment of an egg whose own contents have been removed.

Drosophilia's Summary

We have now witnessed, in part, *Drosophila* and its puzzling, beautiful, and exciting phenomena. I have pursued several major conceptual avenues which *Drosophila*'s data seem to warrant:

1. With respect to ectodermal tissues such as imaginal discs, it seems very reasonable that the cell-heritable determined state is "remembered" by components, a condition which represents a developmental decision. In the simplest case each decision is between two alternatives, but real life may be more complex. I have supposed that the underlying decision-taking entities are similar to a genetic circuit in *lambda,* where C1 and cro repress each other and thus carry the im + and im − alternative steady states (Neubauer and Calef 1970; Ptashne 1986). I stress again, however, that the real circuitry is likely to be far more complex; indeed, the real circuitry may well be found in the functionally isolated live genetic circuits which are not part of the forcing structure which freezes the activity of many genes into fixed active and inactive states, as described in Chapter 12. Thus a point to stress is that a developmental "choice" may best be thought of as one of a number of alternative attractors of such a functionally isolated circuit. We should not assume such a circuit to be as simple as a single gene which regulates its own activity.

2. I have recapitulated data on metaplasias derived from late-acting homeotic mutants and transdetermination which strongly suggest that distant regions, such as genitalia and antenna, are developmental neighbors and have interpreted the concept of developmental neighbors in terms of a binary combina-

torial epigenetic code. Two tissues differing in a single decision are developmental neighbors.

3. I have focused on the evidence which shows that *Drosophila* imaginal discs and probably the early embryo are sequentially subdivided into compartmental domains which appear to reflect alternative developmental commitments.

4. I have suggested that the monotonic and nonmonotonic eigen function patterns of reaction-diffusion or other field equations might account for the positions, sequences, and symmetries of observed compartmental boundaries. Further, the nonmonotonicity of most eigen function patterns, in which zones above or below a threshold concentration may repeat in nonadjacent regions of the embryo and trigger identical decisions, may account for the fact that distant regions on the fate map can have neighboring developmental programs.

5. I have described maternal, gap, pair-rule, and segment-polarity mutants that produce longitudinal deletions and mirror-symmetric defects on roughly half, quarter, sixteenth, and $\frac{1}{32}$ egg length scales.

6. I have reviewed the data showing that maternal, gap, pair-rule, and segment-polarity gene transcripts all pass through similar sequences in which increasingly complex multiple bands and stripes of expression evolve in the syncytial egg and early embryo.

7. I have suggested that successive bifurcations of a reaction-diffusion Turing mechanism or of other field equations account for the transcription phenomena observed.

8. I have suggested that the longer-wavelength gene systems may control the regions of activity, locations of maxima and minima, and phase offset of the shorter-wavelength systems by acting as bifurcation parameters controlling diffusion constants and reaction rates of the shorter-wavelength systems and hence providing a kind of internal framework.

9. With the four color wheels model, I have suggested that the deletion and duplication seen in maternal, gap, pair-rule, and segment-polarity mutants are most naturally accounted for by supposing that cells specify longitudinal position combinatorially in terms of the ratios of the phase-offset variables splayed out along the embryo in each of at least four different spatial wavelengths.

10. I have suggested that the four color wheels model yields a combinatorial epigenetic color code for each cell. Because of the monotonic and nonmonotonic properties of the morphogen patterns creating the different color wheels, this code yields transdetermination and homeotic jumps to distant regions on the embryo. The color code model generalizes as the binary code model.

I have therefore attempted to show that *Drosophila*'s sequential patterning in space is the result of a coupling of reaction-diffusion instabilities to generate phase-offset transcription patterns on a variety of wavelengths; the various wavelengths are then used to specify positional identities in the four color wheels and color code models. I have sketched a picture in which the sequence of patterns drives a sequence of commitments to alternative heritable fates and generates a combinatorial code, binary but generalizable to a color code, which fits much of the data on metaplasias in the imaginal tissues of *Drosophila*.

The theory is almost certainly not correct in detail. At most, it is sensible to hope that the framework of ideas will provide components useful to an ultimate understanding of pattern generation and developmental commitments in this organism. Other simpler theories, specifically a monotonic gradient model, have had different successes (Sander 1984; Meinhardt 1986) but have difficulty with many of the phenomena covered above. Therefore we should bear in mind that our purposes in exploring the Turing model and *Drosophila* have been not merely to examine a plausible way of picturing *Drosophila* commitment but, more deeply, to find yet another area where Turing's basic idea has a reasonable chance of applying. The general notion of instabilities (and eigen functions) via either reaction-diffusion instabilities or chemomechanical models affords a useful class of developmental mechanisms which generate defined classes of well-formed patterns, each having well-formed neighbors. Well beyhond *Drosophila,* the importance of such mechanisms is their clear hint of natural forms with which selection may tinker but to which it may often be constrained.

TURING AND BEYOND

The arc of this chapter began with a discussion of induction and its relation to the poised states of cell types. We found that simple Boolean models gave rise to simple spatial ordering. Thereafter we turned to control of long-range order as exhibited by epimorphic pattern regulation. We examined the concept of positional information and alternative models of coordinate systems. Inadequacies in a formulation in which pattern regeneration is due merely to smoothing of discontinuities in preexisting positional cues led us to consider reaction-diffusion models of the Turing type, which might set up such spatially graded cues. In this final section, I close the arc by examining first the implications of reaction-diffusion models for the *creation of two-dimensional positional information* and then a final application of reaction-diffusion models to shell patterns. I then emphasize again the homology between predictions of reaction-diffusion models and a class of mechanochemical models, here applied to a small set of other developmental problems. There are, of course, a host of other developmental mechanisms based largely on chemomechanical or simply mechanical features of cells. I describe two briefly. Each yields a family of related forms. Thus I return to our interest in thinking about the relation between the forms such mechanisms generate and the effects of selection. The concept of positional information, advanced by Wolpert (1971), is important but also limiting in an important way. Cells manifestly talk to themselves and to one another in complex ways. What that might imply for the evolutionary onset of maps and spatial organization looms as a large issue. Finally, I ask what a theory of morphology might be.

Two-Dimensional Positional Information

Recall from the section which begins on page 552 the evidence which shows that positional axes in vertebrate limb fields arise sequentially. First the anterior–posterior axis is formed, then the dorsal–ventral axis. It is therefore of interest to realize that this phenomenon arises almost inevitably in the Turing class of models. Recall again that, if a single harmonic mode is present in a spatial domain, then the maxima and minima of the underlying chemical variables are exactly in phase. If two or more modes are present, however, each with a different wavelength, and if each mode amplifies a different eigen vector, or ratio, of the variables, then the full pattern is some sort of linear or nonlinear superposition of these modes. The result is that the

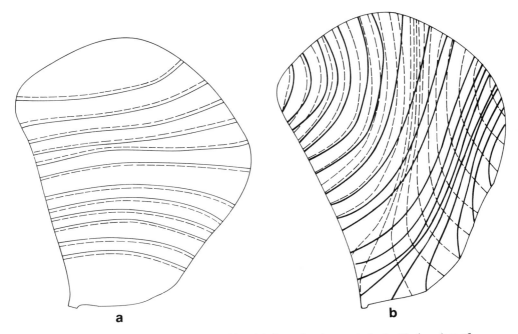

Figure 14.35 Onset of two-dimensional positional information is generic in the Turing class of models. Patterns of X and Y chemical concentrations on an enlarging wing disc. Solid lines are isoconcentration lines of X; dotted lines are isoconcentration lines of Y. (*a*) The first pattern to emerge is monotonic in the long axis of the disc, amplifying a single eigen function. The two chemicals have parallel concentration profiles and hence supply one axis of positional information. (*b*) As the disc enlarges, monotonic gradients in both long and short axes can amplify. The superimposed nonlinear mixed patterns yield stable nonparallel lines of X and Y concentration and hence have formed a second positional axis and two dimensions of positional information. (From Kauffman 1981)

maxima and mimima of one variable need not occur in the same locations as the maxima and minima of the other variable(s). Consequently, in a two-dimensional spatial domain, or tissue, lines of constant concentration of one variable will in general be oblique to lines of constant concentration of another variable. Two dimensions' worth of positional information is present in such transverse gradients. These ideas imply that, as a small tissue which is longer than it is wide grows larger, initially a first mode will form in the longer direction and yield one-dimensional positional information. If a second mode arises in the shorter direction, the superposition of the two modes will cause the underlying morphogens to form transverse gradients. A second axis will form, as shown in Figure 14.35 (Kauffman 1984b). Sequential formation of positional axes in vertebrate limb fields, eye fields, otic fields,, and so forth is at least weak evidence in favor of some form of field model which, like the Turing class of models, predicts sequential axis formation.

Shell Patterns: A Last Turing Model

The Turing class of models has been used to consider many aspects of pattern formation, from hydra and *Dicteostelium discoidum* to the striking stripe and check patterns on the gastropod *Bankivia fasciata* from the Great Barrier Reef in Australia (Figure 14.36). These patterns are laid down in the growing shell by the mantel organ, which extends across the margin of the growing shell. Pigment cells in defined areas along the mantle organ color the shell as it is deposited. Thus the pattern is a temporal record of the positions along the mantle organ where pigment cells are secreting. A constant position on the mantle becomes a stripe paralleling the spiral winding of the shell. If a traveling "wave" of pigment-secreting activity propagates across the mantle in one or the other direction, stripes which are at diagonal angles to the spiral winding axis are developed. Figure 14.37 shows results of a model developed by Meinhardt and Klingler (1987) based on reaction-diffusion instabilities on the mantle. Stripes and checks closely reminiscent of those on the shells are produced by the model. The stripes, checks, and deformations between patterns observed in the model are close to what is seen on the shells.

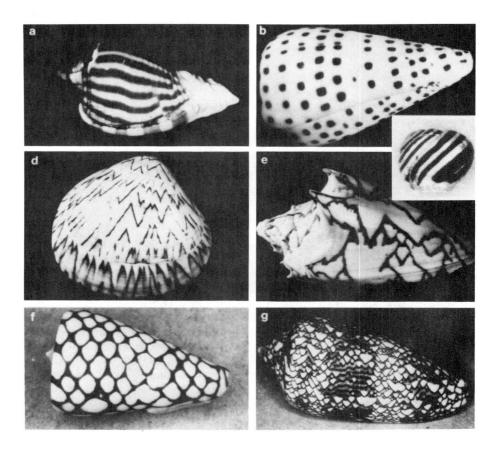

Figure 14.36 The range of shell patterns in *Bamkivia.*

It is important to stress that the Meinhardt–Klingler model based on local reaction-diffusion processes in the mantle organ generates much the same patterns as logically similar models based on neural activities. Thus, as usual, the capacity to fit a macroscopic pattern or set of patterns with a class of models does not come close to identifying the underlying mechanisms, even if it be accepted that such a fit helps identify the logical structure of the mechanism.

I leave for the final section of the chapter an effort to place reaction-diffusion models in a broader context. I emphasize, however, that this class of models has been applied in more areas than I have discussed. Notable examples can be found in Meinhard's (1982) and in Murray's (1981a, 1981b, 1989) examination of the markings on animals from zebras to giraffes, and patterns in Lepidopteran wings.

Mechanochemical Models

I stressed above and do so here again that, as a result of the similarity of the dispersion relation between the eigen value of a mode and its wavelength, properties which show up in the Turing class of models recur in a range of mechanochemical models (Oster and Odell 1984; Oster 1988; Oster, Alberch, et al. 1988; Oster and Murray 1989). Since Chladni figures obtained by allowing powder to gather at the nodal lines of vibrating plates show these kinds of patterns, this recurrence is not a total surprise. The eigen functions for buckling modes or patterns in materials are like those for reaction-diffusion instabilities. For example, Oster and colleagues (1988) have proposed a model to account for the sequential formation of the "canonical" vertebrate limb, where one sees a succession of a single bone (humerus in the arm), splitting to two bones (radius and ulna), splitting to still smaller bones (carpals and metacarpals), and finally formation of the digits. In this view, the sequential formation of increasing numbers of bone elements is again a bifurcation sequence in the growing limb bud. A similar viscoelastic model for the branching pattern of cap elements in *Acetabularia* has been proposed by Goodwin and Trainor (1985). In accounts of neurulation and elsewhere, similar patterns show up in models linking cell motion over substrates which yields folding in epithelial sheets (Odell, Oster, et al. 1981). More recently, Jacobson and colleagues (Jacobson, Odell, and Oster 1985; Jacobson, Oster, et al. 1986) have proposed a novel "cortical tractor" model for the cell motion which induces neurulation. Like other classes of morphogenetic mechanisms, these can be expected to give rise to a defined family of forms as a function of the parameters of the system.

In a somewhat similar spirit, Goodwin and Trainor (1980) attempted to understand the radial and spiral cleavage patterns found in almost all kinds of embryos as the natural harmonic modes which arise on a sphere. The natural modes are given by the Legendre polynomials, or spherical harmonics. These yield successive orthogonal patterns for the first three planes of division, followed by sets of pairs of planes parallel to each axis, yielding radial cleavage. The sequence is close to that observed in radial cleavage. Spiral cleavage patterns, however, do not arise as the natural harmonics of the sphere from the Legendre polynomials but rather, the authors suggest, reflect some other basic form on a sphere. In fact, years ago T. H. Morgan proposed that spiral cleavage reflects high affinity of blastomeres for one another, such that

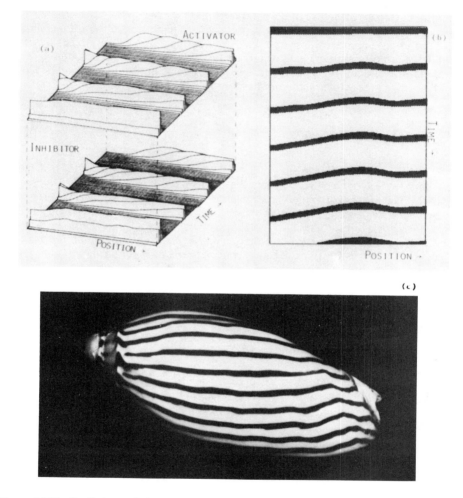

Figure 14.37 Predictions of shell patterns based on Turing class of models. Correspondence between theory-derived patterns and those observed. (From Meinhardt and Klingler 1987)

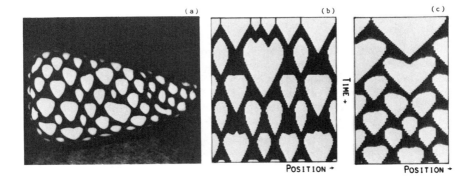

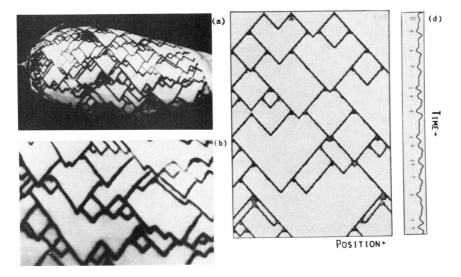

Figure 14.37 *(Continued.)*

each maximized surface contact with the others (Gilbert, personal communication). Thus in contrast to radial cleavage, slight rotation of each tier of cells allows each cell to nestle into the groove between cells in the tier below. Morgan's idea seems quite reasonable. The ideas to focus on the effort of Goodwin and Trainor are (1) that the natural harmonics of the Laplacian or of related operators on a spherical object generate a *limited range* of successive modes and (2) that those modes come quite close to matching observed cleavage patterns. Morgan's ideas lead to similar results, a natural and limited set of forms governed by an energy minimization criterion. Hunding (1987) has advanced a use of Turing systems to explain cleavage plane orientation.

In Chapter 1, I briefly mentioned phyllotaxis (Erikson 1982) and a model by Mitchison (1977) to account for the phenomena. Recall that scale patterns in pine cones, in terms of the number of left-hand and right-hand spirals, are typically adjacent terms in the Fibonnaci series. Mitchison accounted for these patterns on the basis of close packing of scale primordia in the growth region of the apical meristem. More recently, Green (1987) has proposed an elegant model for the different phyllotactic patterns in plants: alternate, whorled, radial, and spiral. His chemomechanical model is based on the dynamics of scale primordia formation on the apical meristem. His aim is to explain the origins of these patterns and the major and minor shifts between them. The idea that the spatial regularity results from some kind of spatial oscillatory activity at the apical meristem dates back at least to Church (1904). Green locates this activity in the processes on the dome of the meristem's apex, and it leads to the formation of an incipient region of hoop reinforcement, by cellulose, over many cells, forming a concentric pattern. This growth pattern leads to a bulge of cells enlarging to form a hoop-reinforced organ which grows outward from the dome. In turn, the organ then acts back on the central area of the dome and stretches it locally, parallel to the organ base. This system generates, via cell responses involving microtubules, a new area of reinforcement on the dome, with lines running tangential to the base of the organ. The combination of parts of several such new areas, from existing organs on the dome, forms the structural potential localizing the next organ. Thus the cycle is complete only if several organs act in concert. Green shows that these processes give rise to only a *limited number of patterns which are self-regenerating.* Thse are the common phyllotactic patterns.

We have now spent considerable time discussing one class developmental mechanisms, based on Turing models or analogous viscoelastic field equations. Each mechanism generates a family of forms. Yet the mechanisms on which we have focused could readily be enlarged. For example, Steinberg (1962, 1970) has for years analyzed the capacity of mixtures of different cell types to *sort out* from one another when agitated together and form concentric or partially concentric onion layers of different cell types. These beautifully ordered phemomena appear to be simple consequences of differential cohesive forces (affinities of each cell type to bind to itself) and adhesive forces (affinities of each cell type to bind to other cell types or to the medium). If cell type A is more cohesive than cell type B, while B adheres to A better than to the medium, A cells will tend to form a central spherical mass, completely or partially surrounded by B cells. Similarly, if B is more cohesive than C and the other conditions are met, C will surround B. But this surface-energy model then predicts that C will also surround A. That is, Steinberg's model predicts a *transitivity relation* in the cell types which will surround one another. And just such a transitivity is found. Three points about this mechanism are important to us. First, it is based only on a generalized "stickiness" which differs in different cells and would be expected to arise almost inevitably in cells having diverse molecules embedded in their lipid membranes. Second, the morphologies which are generated reflect the simple physical principle of falling to a most favorable (lower) energy state; thus the ordering is analogous to that seen in protein folding and self-assembly of viruses. Third, a restricted range of morphologies is again generated by this mechanism. And changing affinities alters which cell type is enclosed by which cell type, or how completely one cell type surrounds another. That is, alteration of parameters predicts deformation to a set of neighboring forms.

The general subject of specific cell adhesion and the molecules which mediate it has received increasing attention in the past decades. A particularly useful discussion is due to Edelman (1988). It seems clear that homopolymer and heteropolymer recognition due to cell surface molecules plays a fundamental role in cell recognition, in the formation of cell clusters, and in the subsequent regulation of cell growth, cell death, and morphogenesis.

Morphogenesis May Be Robust and Adapt on Smooth Fitness Landscapes

We now turn to the relation between morphogenesis and selection. I shall suggest a heterodox possibility: Morphogenesis may be inherently robuts. For each developmental mechanism and, more surprising, for *integrated combinations of developmental mechanisms,* a small number of the morphologies that can be generated by those mechanisms may emerge as natural forms. These natural forms are those most readily generated by the developmental mechanisms in the precise sense that the forms are generated whenever the initial states and the parameters of the mechanisms are located anywhere within very large volumes of state space and of parameter space. The natural forms are thus the simplest forms for the mechanisms to generate and are robust to minor variations in initial state or parameter values. In turn, within those large volumes of state and parameter space, morphology changes slowly as initial state or parameters shift; hence fitness landscapes are smooth.

Our discussion in this chapter suggests that any developmental mechanism generates a set of forms as the parameters of the mechanism are altered. The striking feature about all the attempts we have discussed—from fruit flies and slime molds to limb bones, shell patterns, and plant morphology—is that a concerted effort is being made to link the natural geometry of biological systems with the expected "natural" behavior of nonlinear dynamical systems operating on them. In all cases, any such developmental mechanism has a limited number of alternative modes of behavior. The effort is to find that view under which the alternative behavior modes span the biological forms observed. Among other critical possibilities, this effort should eventually allow us to understand which morphologies are evolutionary neighbors, in the sense that small parameter alternations shift the morphogenetic system from one mode to another. To state the issue in slightly different terms, the developmental mechanisms we have noted each give rise to a family of neighboring forms. This is the structuralist theme sounded by Webster and Goodwin (1982) and Goodwin (1990). Its aim is to understand the set of possible forms generated by any specific mechanism, and the rules of transformation between those forms.

In the present context, transformations among possible forms are due to shifts in the parameters underlying the developmental mechanism. After all, when we truly understand the coordinate mechanisms generating the sequence of bone elements in the vertebrate limb, we shall understand the neighboring forms which were available to selection in the branching tree of vertebrate phylogeny.

If developmental mechanisms give rise to families of forms, and if transitions among these forms are due to shifts in the parameters of the underlying mechanisms, then we confront the central question: How can we begin to think about selection acting on organisms which deploy specific developmental mechanisms in their ontogeny? As is now clear, for any of these proposed developmental mechanisms, each specific pattern in the family will arise only if the initial state of the system lies in the appropriate basin-of-attraction volume in state space, and only if the parameters of the developmental system are held within some defined bounded "bifurcation" volume in parameter space. It is now obvious that, for any mechanism, some basin and bifurcation volumes are large, while others are very small. Some volumes are smooth, like inflated balloons squeezed into a room; others may be tortuous, labyrinthine volumes which interleave with yet other volumes. How will selection bear on the morphology seen? Suppose that many genes influence each variable in the mechanism as well as each parameter and that there are many variables and parameters for a given mechanism. If the volumes associated with a specific pattern in state and parameter space are very small, or very labyrinthine, then selection is unlikely to be able to bring or to maintain the initial state or parameters within that volume. Thus selection will be hard pressed to achieve and maintain the morphology corresponding to that particular volume in state and parameter space. That is, the second form of the complexity catastrophe (the error catastrophe) indicates that, as systems become more complex—with more genes, variables, and parameters—in the face of a constant mutation rate, selection cannot hold a population at rare optima. It follows that we are likely to see the patterns which correspond to the large basins of attraction and bifurcation volumes. These are the "easy," or natural, patterns which any mechanism can generate.

Morphological fitness landscapes might be smooth or rugged. If basin volumes in state space and bifurcation volumes in parameter space are like balloons, then small changes in state or parameters leads to small changes in morphology. If basins of attraction and parameter volumes are tortuous and intertwine, then small genetic changes yielding small smooth changes in initial states, or in parameters, would cross boundaries separating sharply different morphologies. Almost certainly, the resulting fitness landscape will be rugged. Hence trapping of adaptive walks on local optima will occur readily. The first complexity catastrophe suggests that, as systems become more complex, the attainable optima via adaptive walks on rugged fitness landscapes tend to become trapped in local regions of state and parameter space, which tend to be the large basins of attraction in state space, and large bifurcation volumes in parameter space. Thus, again, selection may be limited to those natural patterns which are easy for the developmental mechanism in question.

The same issues come up in *rhythmic phenomena,* which I shall only point to in this book since excellent books covering this material are available (Winfree 1980, 1987; Glass and Mackey 1988). A single example will make the point. Heart cells in culture beat rhythmically, with different individual cells having different rhythms. When coupled, such oscillatory systems can mutually entrain one another, such that the collection beats as a whole with a common rhythm—in 1:1 phase-locking, say or in 2:1 phase-locking, where some cells beat twice while others beat once, or in 3:1 phase-locking, and so on. In addition to these specific, stable phase relations between the phase-locked cells, chaotic behavior can arise instead. Which pattern emerges is governed by a set of parameters. Among the largest volumes in parameter space are those corresponding to 1:1 phase-locking. By contrast, achieving 23:27 phase-locking requires occupancy of a tiny volume in parameter space. Our hearts, made up of two atria and two ventricles, are linked in 1:1 phase-locking. In part this linkage is mediated via sophisticated Purkinje signaling fibers from the atria to the ventricles. In part it undoubtedly reflects efficient pumping design. In some part, however, it probably reflects the fact that 1:1 phase-locking is easy. It is conceivable that a cardiac system might work on the basis of a 23:27 phase-locking. I know of none which does. All utilize 1:1 phase locking. Design? Probably. But easier as well? It seems probable. In short, real developmental mechanisms do some things easily. Those things are, by and large, what we are likely to see.

With my colleagues Goodwin and Murray (Goodwin, Kauffman, and Murray 1992), I now suggest that there are clues that, contrary to intuition, morphogenesis may be deeply robust. Organisms, rather than being tinkered-together contraptions, may exhibit a nearly inevitable and stable order. Real morphogenesis is due not to the unfolding of any single developmental mechanism but to the beautifully ordered unfolding in time and space of some richly integrated combination of simpler mechanisms such as cell-sorting, sheet-folding, positional discontinuities, and reaction-diffusion mechanisms. One's naive intuition might be that reliable occurrence of an ordered morphology from a richly integrated developmental mechanism would require exquisite control of all the variables and parameters of the subsystems making up the integrated system. Since each subsystem by itself can give rise to a family of forms, the set of mechanisms would seem to allow the *product* of all the patterns generated by the diverse subsystems. High genetic precision would appear to be required to choose reliably among these many forms. The suggestion is that intuition is quite wrong.

There are good grounds to think that, when a variety of different developmental mechanisms are integrated into a compound mechanism, the integrated mechanism will *constrain* the morphologies which emerge to a small subset, each of which occupies a large volume of state space and parameter space. Rather than causing complexity, integration of developmental mechanisms may generically yield simplicity and order. The reason is simple: Each submechanism generates a family of forms by breaking symmetries. Each time a symmetry is broken by the mechanism, *that symmetry might be broken in two or more ways.* For example, for the Turing class of reaction-diffusion mechanisms on an ellipsoidal egg, the first longitudinal mode might be high at the anterior pole and low at the posterior pole, or low at the anterior and high at the posterior. The simple new insight is that, when mechanisms are coupled, the currently established mode(s) of one mechanism *choose or determine* how the breaking symmetry of another mechanism will actually be broken. Thus each submechanism literally constrains the alternative choices open to other submechanisms. The overall result is that a few of the conceivable morphologies in the family of forms come to occupy large basins of attraction in state space and large bifurcation volumes. In short, and rather wonderfully, complex mechanisms seem likely to generate a few morphologies with extreme robustness.

We have already seen a harbinger of this general suggestion. Recall the studies of Arcuri and Murray (1986) described in the discussion beginning on page 591. A fully nonlinear version of the Turing model was simulated on a bounded linear domain such that the first and second modes, and hence other multiple modes, could simultaneously amplify. In this condition, generically each mode amplifies a different eigen vector of the underlying variables and the two modes span the state space. These properties imply that the two coupled, approximately linear modes of one mechanism, if simultaneously amplified, should mutually break one another's symmetries. And, in fact, Arcuri and Murray found that the presence of the first mode and its nonlinear coupling to the second mode ensured that the final pattern was independent of the initial conditions and of the orientations of early modes. Each successive mode broke the symmetries of the other.

A complex mechanism that integrates several submechanisms may generically have the property that the nearly linear modes of the compound system will couple nonlinearly such that, when multiple modes are present, each will tend to break the symmetries of the others in preferred ways. For example, Shaw and Murray (1990) have considered a model uniting a reaction-diffusion mechanism operating on one length scale with a chemomechanical model of epidermal behavior. To the authors' surprise, the combined mechanisms sharply restricted the choices each alone could make. Goodwin, Kauffman, and Murray (1992) report that an *Acetabularia* model, which coupled a mechanochemical model of cytoskeletal behavior with a model of elastic cell wall deformation, yielded a highly robust sequence of morphogenetic events, including formation of an elongating tip which generated radial whorl patterns. In short, we confront another bold hypothesis: Morphogenesis may inherently be robust and thus also may adapt on smooth fitness landscapes. If so, selection is likely to be constrained to those robust morphologies engendered by the integrated mechanism in question. In short, we shall tend to see the natural robust forms such mechanisms can create.

Selection often achieves well-wrought morphologies, from sleek shapes for reducing drag when swimming to well-crafted wings. In this regard, it is of interest to note that mutation selection procedures have been used successfully in engineering design by Rechenberg (1973). That Rechenberg often succeeds suggests that many engineering problems confront simple landscapes. In contrast, it is clear that fitness landscapes in *sequence spaces* are likely to be rugged and multipeaked. Our bold hypothesis suggests that morphological evolution typically confronts smooth landscapes. Morphology may be both robust and within the limits of each natural form, smoothly tunable to fit versions.

Thus the direction of inquiry, it would appear, requires us to identify distinct developmental mechanisms. For each mechanism and for the compound integrated mechanism each mechanism is part of, we must identify the forms which that mechanism generates. Within that family of forms, we must identify those which are easy for the mechanism in the sense that a large volume of parameter space yields them. Then we must identify how many genes influence the cellular mechanisms which constitute the variables and parameters of that system. Then we can begin to understand how hard it may be for selection to "pull" the set of variables to any basin of attraction and the parameters to any specific volume of the corresponding parameter space, or to hold an adapting population with the initial state and parameters within those volumes. Further, we can see whether trapping on local optima does in fact occur, either because "good" volumes are tortuous or because mutations cannot "move" smoothly in parameter space but can occupy only distinct subregions. All this must be undertaken, in conjunction with the familiar analysis of design requirements in organisms, to unpack the question D'Arcy Thompson failed to raise.

Maps and Spatial Organization as Generic Self-Organized Properties of Coupled Cells

Is it astonishing that multicellular organisms exhibit spatial organization? The word "astonishing" is meant to convey, imprecisely, the contemporary sense that organisms are accidental contraptions—Rube Goldberg affairs—and the sense that selection had an imponderably difficult task in finding means to generate spatial order. The question cannot, of course, be precise. What we are really asking includes the question of whether the occurrence of something like positional "maps" and ordered spatial heterogeneity can be understood as generic self-organized properties in biological systems. Perhaps the answer is "yes."

We have seen that genomic cybernetic systems generically have multiple attractors, which we may take as alternative cell types. Almost any complex dynamical system will have multiple attractors. Thus achieving multiple cell types "lies to hand."

Given diverse cell types, differential cohesive and adhesive properties are probably almost inevitable; thus mechanical processes of cell-sorting and layer formation also lie to hand. So, too, do certain aspects of mechanical deformation of tissue masses, yielding expected patterns of folding and pleating in two-dimensional tissues.

Induction of novel cell types and of simple spatial ordering of those cell types also probably lies to hand. We have seen that, in the Boolean idealization, if an *arbitrary* subset of gene variables have products which diffuse between cells, then these products generate inductive interactions, generate new cell types as a result of persistent inductive interactions, and suffice to generate such primitive spatial ordering as alternative patterns of one gene's activity along a line of cells. Here gene products can diffuse to only neighboring cells, but the onset of some spatial order is already virtually inevitable.

Given a genomic system which can generate a number of cell types, and given inductive interactions among neighboring cells to generate still further cell types, and given diffusion or its analogues allowing molecules to spread to numbers of adjacent cells, then the formation of gradients is almost inevitable, even without the elegant diffusive instabilities of the Turing class of models. The Boolean idealization allows a picture of genes as only active or inactive and products as present or absent. This is, as stressed repeatedly, an idealization allowing examination of complex systems but also an injustice. The transcriptional activities of genes, translational activities of RNA, and catalytic activities of enzymes all have graded responses to their molecular control inputs; equally important, two different genes controlled by the same molecular variable typically respond to different concentrations of that variable. The generic consequence is that, in a line of cells where there is a spatial gradient in the concentration of a molecular input, the cells near the top of the gradient will activate both downstream genes to a high level, while farther away only the more sensitive downstream gene will be activated. This trivial observation implies that, wherever spatial gradients over many cells exist, the patterns of gene expression induced are likely to vary with distance. Otherwise stated, a graded "morphogen" is likely to induce different "interpretations" at different levels down its concentration gradient. This is just the old idea of gradients and a series of threshold levels, so beloved to all developmental biologists. The further consequence, however, is that all the properties of positional smoothing, duplication, and regeneration which flow from something as simple as two-dimensional "cross gradients" are incipient in this simple, general picture. Given a spatially heterogeneous sheet of cells and locally monotonic gradients of regulatory molecules which are oblique to one another, the familiar pattern regulation properties lie to hand within each monotonic cross gradient domain. Local maps may be easy.

Are the Turing class of models or other more general models capable of breaking a homogeneous state hard to find? Probably not. A Turing model requires rather simple forms of self- and cross-excitation and -inhibition, and greater inhibitor diffusion.

Given these requirements, the volume of parameter space which allows diffusive instabilities is modestly large. Indeed, Vastano, Pearson, et al. (1987) have shown that chemical patterns can form in bistable systems having equal rather than unequal diffusion coefficients. And these two-variable models are the simplest members of what might be a very large class of systems which can break the homogeneous state. It is perhaps a happy fact that even a system comprising dozens of chemical variables will, in the linear analysis, go through essentially the same sequence of eigen functions as will a two-variable system based on the same familiar dispersion relation, $\lambda = f(k)$. Thus eigen functions of the appropriate operators, Laplacian and otherwise, emerge as basic schemata of possible forms, each function defining a family of related patterns which can deform into one another continuously or discontinuously by tuning parameters smoothly. Selection is then constrained by the normal behaviors of the mechanisms it has achieved.

What is a "map"? Perhaps maps are nearly inevitable. Consider *Drosophila* again. The number of distinct RNA sequences transcribed in each nucleus may be on the order of 15 000. Most of these and other molecular variables remain in the cell which synthesizes them, with only a small fraction passing to other cells. In addition, physical forces, such as mechanical deformations, pass between cells. In the context of dynamical systems, a tissue is a system of coupled ordinary and partial differential equations. The variables which are confined to remain in cells constitute the "ordinary differential equations" part of the system, while the variables which can move from one cell to another constitute the "partial differential equations" component. The spatially inhomogeneous behavior of the partial differential equations portion then constitutes the positional information in the tissue. The attractors to which individual cells flow, given their location in the tissue and hence in the positional information field, then constitute the interpretation each cell makes of its position. In an inaccurate analogy to the wave–particle duality of quantum physics, the spatially inhomogeneous patterns—for example, the patterns of maternal, gap, pair-rule, and segment-polarity transcripts—are the waves, and the attractors attained in individual cells are the particles. The duality points to the fact that the entire genomic system is, in reality, a single coupled system whose attractors constitute both map and interpretation at once.

To behave in a unique way then, a growing tissue needs to begin in a well-defined state—probably small—grow, and generate spatially inhomogeneous distributions of inductive agents which drive differentiation, mitosis, cell deformation, and matrix secretion—all of which form a linked, spatially distributed dynamical system which flows to a unique final spatially inhomogeneous state which is that tissue. In this case, the system generates a map which has a unique interpretation. Each cell does one thing and does it properly.

If the linked system has a number of distinct final states that it can go to, each spatially inhomogeneous and each with a unique distribution of inductive agents, then each tissue has a unique map and interprets it in a unique way.

There is yet another sense of "map," however. Suppose that, for the *same* spatial distribution of inducer substances, positional-information signals, or external network variables, the cells each have two different attractors available to them. Then the cells might interpret the same positional information to form wing or leg. Here is "map" in a different sense, and Wolpert's (1969, 1971) idea of the same positional system open to multiple interpretations arises.

Is the attainment of such ordered behavior difficult in coupled nonlinear dynamical systems? The full answer, of course, is unknown, but my own intuition is that such behavior is probably not extremely difficult. Nonlinear dynamical systems of ordinary and partial differential equations are graced by spatially inhomogeneous attractors. All our growing experience in physics and chemistry where spatial ordering occurs indicates that, given modest control of parameters, the onset of specific spatial order is often attainable. Maps and their interpretation may be easy after all.

A Theory of Morphology?

Could we hope to have a theory of morphology which, were evolution to tap-dance again its erratic course, would give us a basis for predicting features of organisms? One is always aware of the intellectual error, whatever it was, which allowed James Mill, John Stuart's father, to deduce from first principles that a constitutional monarchy remarkably like that in England was the highest form of government. We stand in considerable danger of falsely deducing the inevitable existence of what we observe. What might we want in our theory? Design principles, of course. Selection more or less works. An account of the probable developmental mechanisms which would readily have been first found and then presumably maintained by selection. An account of the easy and hard forms each mechanism alone generated. An ontogeny is obtained by the coordinated linking of several or many different developmental mechanisms. We must understand how integrated combinations of mechanisms robustly give rise to more complex but well-defined superfamilies of forms on which selection could carry out its further sifting. Perhaps we may come to think of such morphologies as kinds of morphogrammars and to think of life cycles as stable, repetitive, recursive forms within those grammars.

Our ultimate aims are broad, for it seems proper that those aims are not merely to identify developmental mechanisms in ontogeny and their evolutionary history but also to expand evolutionary theory to embrace the mechanisms found, the ease of finding them, and the natural forms they generate. We want the constraints those normal forms place on selection to achieve forms which are difficult for the mechanisms available. Yet we also need to understand the capacity of selection to find novel mechanisms opening new families of forms. Still again, this effort must be integrated with random drift and with historical contingency. Biology is surely harder than physics. Yet our broad aims constitute a kind of physics of biology.

SUMMARY

This chapter has taken up the overwhelming topic of morphogenesis. No single chapter can hope to do justice to morphology. No book can yet comprehend its scale. A deep theory of morphology would justly reckon with the ancient sense of inherent order in organisms caught by the rational morphologists in their impulse to find laws for form. Darwinsim and evolutionism, branching morphologies flowering under the aegis of natural selection and genetic drift, have focused our attention on the tempo and mode of evolution. We rightly want to understand which organisms branched from which, why, and what features of each reflect adaptations wrought and maintained by selection. In our proper reductionist mode, we properly seek developmental mechanics, the unrolling machinery of genetic interactions and morphogenetic mechanisms which generate any specific ontogeny. Simultaneously, we suspect that the morphologies we see are expressions of a modest number of fundamental mechanisms, each yielding a well-defined family of forms. To suspect this is inevitably to confront two basic unanswered questions: What are those fundamental mechanisms? To what extent can and has selection modified the forms which we see from those naturally generated by the underlying mechanisms?

I have sought to sketch some of the fundamental mechanisms which underlie morphogenesis. Each appears rather easy for evolution to have found. Each is powerful and engenders ordered features of ontogeny. Among the mechanisms we have considered are

- Induction of differentiation by cell–cell interactions
- Evidence for long-range order and positional information expressed in duplication and regeneration phenomena in many systems
- The beautiful Turing-like patterns of standing waves which appear to underwrite pattern formation in many organisms
- Similar mechanochemical models
- Theories of phyllotaxis
- Theories of cell-sorting based on simple surface-energy considerations

It may be that none of these contemporary theories is correct. That issue, however, is far less important than the conceptual one. Each of these theories exemplifies a class of developmental mechanisms which carves out a family of related forms as the parameters of the mechanism are altered. It is not foolish to suppose that any adequate account of any developmental mechanism will show that it, too, generates some family of forms. In any such mechanism, some forms require precise tuning of many parameters into tiny regions of parameter space. Other forms arise when the parameters are located anywhere in large volumes of parameter space. Given that different forms in the same family may have different fitness values, selection will attempt to climb to forms and parameter values of high fitness. Yet where such optimal forms require all parameters to remain in tiny volumes of parameter space, the complexity catastrophe on rugged landscapes or the mutation/selection error catastrophe on smooth landscapes can be expected to arise. Selection typically cannot reach and hold an adapting population in arbitrarily located or overprecise volumes of parameter space. Thus it is not foolish to suppose that the forms we see are largely those which are easily generated by the underlying developmental mechanisms.

Ontogeny is the result of the unfolding of many developmental mechanisms which have become integrated with one another. If our bold hypothesis is right, such integrated complexity generically yields simplicity: Morphogenesis may be inherently robust, not exquisitely fragile.

To characterize the natural forms individual and integrated mechanisms generate is the first, perhaps the outstanding, step to take in elucidating the immanent order expressed in ontogeny. Whether the current models of morphogenesis are correct or not, they surely suffice to persuade us that simple mechanisms yield large domains of order. Selection has not struggled alone these many eons.

Epilogue

Immanuel Kant, writing in the eighteenth century, saw organisms in a clear way. Organisms, he wrote, are self-organized systems in which the parts exist both as means and as ends. Organisms are the mechanisms by which the whole is maintained, and the whole is the ordering such that the parts are maintained. In the nineteenth century, William Paley argued in his "Natural Theology": When we find a watch, we cannot imagine a watch arising without a watchmaker. Because organisms are vastly more complicated than watches, it is even harder to imagine them arising without a divine organism-maker. Darwin, in Richard Dawkins's apt phrase, proposed a blind watchmaker. Darwin turned the argument for God's existence from design on its head: Nature produces contraptions.

The creationists so animating one another, the lay public, and our contemporary court system today rest uneasy with Darwin's heritage. Natural selection, operating on variations which are random with respect to usefulness, appears a slim force for order in a chaotic world. Yet the creationists' impulse is not merely misplaced religion. Science consists in discovering that point of view under which what did occur is what we have good grounds to expect might have occurred. Our legacy from Darwin, powerful as it is, has fractures as its foundations. We do not understand the sources of order on which natural selection was privileged to work. As long as our deepest theory of living entities is the geneology of contraptions and as long as biology is the laying bare of the ad hoc, the intellectually honorable motivation to understand partially lying behind the creationist impulse will persist.

D'Arcy Thompson's famous and elegant book *On Growth and Form* stands as one of the best efforts to find aspects of organismic order which can be understood as aspects which we might, on good grounds, expect. His enquiry, which led him to consider minimal energy surfaces, transformations of coordinate systems as a function of differential growth, and a whole beautiful panoply of phenomena, has stood as a persistent spring for a small trickling of intellectual tradition down through contemporary biology.

Thompson applied classical physics to biology. It has been said that a weakness of some biologists is persistent physics-envy: the seeking of a deep structure to biology. Rest content, is the sensible refrain, with middle-level theories capturing parts of how organisms work. Understand how a genetic cascade works, how sodium transport

643

across a membrane is mediated. Surely we should, have, and will. Yet there is a new physics aborning, and it is time to again fall open victim to physics-envy. For want of a better name, the area which is emerging is something like a theory of complex systems. The trend grew out of statistical mechanics initially and now is clearest in solid-state physics. Study of strongly disordered systems, such as spin-glasses, where many elementary units interact with one another in randomly chosen but specified ways, has already revaled strikingly ordered properties in apparently chaotic systems. Indeed, even the passage of a fluid from laminar flow to turbulence is beginning to reveal hidden order. In short, physics is beginning to discover ways in which very complex systems nevertheless exhibit remarkable order. No reflective biologist can view these developments without wondering whether the origins of order in nonliving systems augurs new insights for the origins of order in living ones as well.

This book is an effort to continue in Thompson's tradition with the spirit now animating parts of physics. It seeks origins of order in the generic properties of complex systems. Those properties discussed have ranged from the origin of life as a connected autocatalytic metabolism in sufficiently complex sets of polymer catalysts and organic molecules, to grammar models of functional integration, to models of the origins of ordered dynamics in genomic regulatory systems as a function of the specificity of molecular interactions and hence as a function of the connectivity of the cybernetic systems, to the origins of spatial integration in multicellular systems when products of individual cells can reach their neighbors.

Unlike physics, biology is the domain of natural selection. If we begin to glimpse origins of order in the inherent properties of complex systems, then we must come to understand the mutual interactions of such self-ordered properties and the actions of selection. Even the outline of an adequate theory is not available but must be found, for the proper marriage of these conceptual realms constitutes the proper evolution of our theory of life.

In this book, therefore, I have tried to take steps toward characterizing the interaction of selection and self-organization. To some great extent, evolution is a complex combinatorial optimization process in each of the coevolving species in a linked ecosystem, where the landscape of each actor deforms as the other actors move. Within each organism, conflicting constraints yield a rugged fitness landscape graced with many peaks, ridges, and valleys. Two major alternative limitations to selection exist. First, selection is limited by the structure of the fitness landscape it acts on; in many landscapes, as the organisms under selection grow more complex, the attainable optima fall toward the average features of the class of systems on which selection is acting. Second, on any landscape, a mutation–selection balance is struck; beyond some level of complexity, selection cannot hold an adapting population at the high peaks of the landscape, and the population then falls toward the average properties of the underlying class of systems. Both limitations suggest that, in sufficiently complex systems, much of the order found is that spontaneously present in the class of systems under selection. Therefore, I have made bold to suggest that much of the order seen in organisms is precisely the spontaneous order in the systems of which we are composed. Such order has beauty and elegance, casting an image of permanence and underlying law over biology. Evolution is not just "chance caught on the wing." It is not just a tinkering of the ad hoc, of bricolage, of contraption. It is emergent order honored and honed by selection.

It is no rhetorical apology to recognize the inadequacy of the efforts undertaken here. The themes are large and new and remain incompletely articulated. Yet they are provoking. Wherever complex systems exhibited order spontaneously, there selection had such order to work with. Wherever the consequent fitness landscape was rugged, there selection was limited and there the self-ordered properties persisted for want of adequate means of escaping them. Such properties may be true biological universals, worthy of our efforts of discovery. Moreover, we expect that selection can mold the entitites it acts on to improve the character of the landscapes these entities explore and to improve the character of the internal model each entity builds of its world, thereby permitting those systems to coadapt better. It follows that we must seek to understand the construction laws that allow complex systems to adapt on properly correlated landscapes and to understand how the couplings between land-scapes evolve. In short, the capacity to evolve is itself subject to evolution and may have its own lawful properties. The construction principles permitting adaptation, too, may emerge as universals. Adaptation to the edge of chaos is just such a candi-date construction principle.

Thus for all the known organisms on this branching river we call life, biology should aim ultimately to account for those essential features which we would expect to find in any recurrence of such a river. To suppose, as I do, that such an intellectual task may one day be achieved is, among other things, to suspect with quiet passion that below the particular teeming molecular traffic in each cell lie fundamental prin-ciples of order any life would reexpress.

Bibliography

Aarts, E., and van Laarhoven, P. (1985). Statistical cooling: A general approach to combinatorial optimization problems. *Phillips J. Res.* **40**:193.

Abramowitz, M., and Stegun, I. A. (1972). *Handbook of Mathematical Functions.* Government Printing Office, Washington, D.C.

Adams, J. M., Harris, A. W., Pinkert, C. A., Corcoran, L. M., Alexander, W. S., Cory, S., Palmiter, R. D., and Brinster, R. L. (1985). The c-*myc* oncogene driven by immunoglobulin enhancers induces lymphoid malignancy in transgenic mice. *Nature* **318**:533.

Agrell, I. (1964). Natural division synchrony and mitotic gradients in metazoan tissues. In *Synchrony in Cell Division and Growth* (E. Zeuthen, ed.). Wiley-Interscience, New York.

Akam M. (1987). The molecular basis for metameric patterns in the *Drosophila* embryo. *Development* **101**:1.

Alberch, P. (1980). Ontogenesis and morphological diversification. *Am. Zool.* **20**:653.

Alberch, P. (1981). Convergence and parallelism in foot morphology in the neotropical salamander *Bolitoglossa. Evolution* **35**:84.

Alberch, P. (1982). Developmental constraints in evolutionary processes. In *Evolution and Development* (J. T. Bonner, ed.). Springer-Verlag, Berlin.

Alberch, P., and Alberch, J. (1981). Heterochronic mechanisms of morphological diversification and evolutionary change in the neotropical salamander *Bolitoglossa occidentalis. J. Morph.* **167**:249.

Alberch, P., and Gale, E. (1983). Size dependence during the development of the amphibian foot: Colchicine-induced digital loss and reduction. *J. Embryol. Exp. Morph.* **76**:177.

Alberch, P., Gould, S. J., Oster, G. F., and Wake, D. B. (1979). Size and shape in ontogeny and phylogeny. *Paleobiology* **5**:296.

Alberts, A., Bray, D., Lewis, J., Raff, M., Roberts, K., and Watson, J. D. (1983). *Molecular Biology of the Cell.* Garland, New York.

Albery, W. J., and Knowles, J. R. (1976). Evolution of enzyme function and the development of catalytic efficiency. *Biochemistry* **15**:5631.

Aleksander, I. (1973). Random logic nets: Stability and adaptation. *Int. J. Man/Machine Studies* **5**:115.

Allen, D., Simon, T., Sablitzky, F., Rajewsky, K., and Cumano, A. (1988). Antibody engineering for the analysis of affinity maturation of an anti-hapten response. *EMBO J.* **7**:1995.

Allen, M. K., and Yanofsky, C. (1963). A biochemical and genetic study of reversion with the A-gene A-protein system of *Escherichia coli* tryptophan synthetase. *Genetics* **48**:1065.

Allison, D. S., and Schatz, G. (1986). Artificial mitochondrial presequences. *Proc. Natl. Acad. Sci. USA* **83**:9011.

Amara, S. G., Jonas, V., Rosenfeld, M. G., Ong, E. S., and Evans, R. M. (1982). Alternative RNA processing in calcitonin gene expression generates mRNAs encoding different polypeptide products. *Nature* **298**:240.

Amitrano, C., Peliti, L., and Saber, M. (1991). A spin-glass model of evolution I: Molecular evolution on rugged landscapes: Proteins, RNA, and the immune system. In *Molecular Evolution on Rugged Landscapes: Santa Fe Institute Studies in the Sciences of Complexity* (S. A. Kauffman and A. S. Perelson, eds.), vol. 9. Addison-Wesley, Reading, Mass.

Anderson, K., and Nüsslein-Volhard, C. (1984). Genetic analysis of dorsal-ventral embryonic pattern in *Drosophila*. In *Pattern Formation* (G. M. Malacinski and S. W. Bryant, eds.). Macmillan, New York.

Anderson, P. W. (1983). Suggested model for prebiotic evolution: The use of chaos. *Proc. Natl. Acad. Sci. USA* **80**:3386.

Anderson, P. W. (1985). Spin glass Hamiltonians: A bridge between biology, statistical mechanics, and computer science. In *Emerging Synthesis in Science: Proceedings of the Founding Workshops of the Santa Fe Institute* (D. Pines, ed.). Santa Fe Institute, Santa Fe, N.M.

Andrews, R. C. (1921). A remarkable case of external hind limbs in a humpback whale. *Am. Mus. Novitates* **9**:1.

Ansari, A., Berendzen, J., Bowne, S. F., Frauenfelder, H., Iben, I.E.T., Sauke, T. B., Shyamsunder, E., and Young, R. D. (1985). Protein states and protein quakes. *Proc. Natl. Acad. Sci. USA* **82**:5000.

Appel, T. (1987). *The Cuvier-Geoffroy Debate: French Biology in the Decades Before Darwin.* Oxford University Press, Oxford.

Apter, M. J. (1966). *Cybernetics and Development.* Pergamon Press, Oxford.

Archer, R. (1971). The neurohypophyseal hormones: An example of molecular evolution. In *Molecular Evolution* (E. Schoffeniels, ed.), vol. 2. North Holland, Amsterdam.

Arcuri, P., and Murray, J. D. (1986). Pattern sensitivity to boundary and initial conditions in reaction-diffusion models. *J. Math. Biol.* **22**:141.

Argyle, E. (1980). A similarity ring for amino acids based on their evolutionary substitution rates. *Origins of Life* **10**:357.

Arthur, C. G., Weide, C. M., Vincent, W. S., and Goldstein, E. S. (1979). mRNA sequence diversity during early embryogenesis in *Drosophila melanogaster. Exp. Cell Res.* **121**:87.

Ashburner, M. (1970). Puffing patterns in *Drosophila melanogaster* and related species. In *Developmental Studies on Giant Chromosomes: Results and Problems in Cell Differentiation* (W. Beerman, ed.), vol. 4. Springer, New York.

Ashburner, M. (1972). Patterns of puffing activity in the salivary glands of *Drosophila:* VI. Induction by ecdysone in salivary glands of *D. melanogaster* cultured in vitro. *Chromosoma* **38**:255.

Ashburner, M. (1974). Sequential gene activation by ecdysone in polytene chromosomes of *Drosophila melanogaster:* II. Effects of inhibitors of protein synthesis. *Dev. Biol.* **39**:141.

Ashburner, M., and Berendes, H. D. (1978). Puffing of polytene chromosomes. In *The Genetics and Biology of Drosophila* (M. Ashburner and T.R.F. Wright, eds.), vol 2B. Academic Press, New York.

Ashburner, M., Chihara, C., Meltzer, P., and Richards, G. (1974). Temporal control of puffing activity in polytene chromosomes. In *Cold Spring Harbor Symposia on Quantitative Biology,* vol. 38. Cold Spring Harbor Laboratory, Cold Spring Harbor, N.Y.

Ashby, W. R. (1960). *Design for a Brain,* 2d ed. Wiley, New York.

Atassi, M. (1980). Precise determination of protein antigenic structures has revealed the molecular immune recognition of proteins and provided a prototype for synthetic mimicking of other protein binding sites. *Mol. and Cell Biol.* **32**:21.

Atlan, H., Fogelman-Soulie, F., Salomon, J., and Weisbuch, G. (1981). Random boolean networks. *Cybernetics and Systems* **12**:103.

Axel, R., Feigelson, P., and Shultz, B. (1976). Analysis of the complexity and diversity of mRNA from chicken liver and oviduct. *Cell* **7**:247.

Axelrod, R. (1984). *The Evolution of Cooperation.* Basic Books, New York.

Axelrod, R. (1987). The evolution of strategies in the iterated prisoner's dilemma. In *Genetic Algorithms and Simulated Annealing* (L. Davis, ed.). Morgan Kaufmann, Los Altos, Calif.

Babcock, A. K. (1976). Logical probability models and representation theorems on the stable dynamics of the genetic net. Ph.D. diss., State University of New York, Buffalo.

Babloyantz, A., and Hiernaux, J. (1975). Models for cell differentiation and generation of polarity in diffusion-governed morphogenetic fields. *Bull. Math. Biol.* **37**:637.

Bachmann, P. A., Luisi, P. L., and Lang, J. (1992). Auto catalytic self-replicating micelles as models for prebiotic structures. *Nature* **357**:57.

Bagley, R. J. (1991). The functional self-organization of autocatalytic networks in a model of the evolution of biogenesis. Ph.D. diss., University of California, San Diego.

Bagley, R. J., Farmer, J. D., and Fontana, W. (1992). Evolution of a metabolism. In *Artificial Life II: A Proceedings Volume in the Santa Fe Institute Studies in the Sciences of Complexity* (C. G. Langton, J. D. Farmer, S. Rasmussen, and C. Taylor, eds.), vol. 10. Addison-Wesley, Reading, Mass.

Bagley, R. J., Farmer, J. D., Kauffman, S. A., Packard, N. H., Perelson, A. S., and Stadnyk, I. M. (1989). Modeling adaptive biological systems. *Biosystems.*

Bak, P., Tang, C., and Wiesenfeld, K. (1988). Self-organized criticality. *Phys. Rev. A* **38**:364.

Baker, A., and Schatz, G. (1987). Sequences from a prokaryotic genome or the mouse dihydrofolate reductase gene can restore the import of a truncated precursor protein into yeast mitochondria. *Proc. Natl. Acad. Sci. USA* **84**:3117.

Baker, B., Nagoshi, R. N., and Burtin, K. C. (1987). Molecular genetic aspects of sex determination in *Drosophila. BioEssays* **6**:66.

Baker, W. M. (1978). A clonal analysis reveals early developmental restrictions in the *Drosophila* head. *Dev. Biol.* **62**:447.

Baldwin, H. H., and Rusch, A. P. (1965). The chemistry of differentiation in lower organisms. *Ann. Rev. Biochem.* **34**:565.

Bambach, R. K. (1985). Classes and adaptive variety: The ecology of diversification in marine faunas through the Phanerozoic. In *Phanerozoic Diversity Patterns: Profiles in Macroevolution* (J. W. Valentine, ed.). Princeton University Press, Princeton, N.J.

Bantle, J. A., and Hahn, W. E. (1976). Complexity and characterization of polyadenylated RNA in the mouse brain. *Cell* **8**:139.

Barbas, C. F., Bain, J. D., Hockstra, D. M., and Lerner, R. A. (1992). Semisynthetic combinatorial antibody libraries: A chemical solution to the diversity problem. *Proc. Natl. Acad. Sci. USA* **89**:4457.

Bars, S., Green, R., and Wells, J. A. (1990). Hormone phage: An enrichment method for variant protein with altered binding properties. *Protein* **8**:309.

Bartlett, P. A., Marlowe, C. K., Giannousis, P. P., and Hanson, J. E. (1987). Phosphorus-containing peptide analogs as peptidase inhibitors. In *Cold Spring Harbor Symposia on Quantitative Biology.* vol. 52. Cold Spring Harbor Laboratory, Cold Spring Harbor, N.Y.

Bateson, W. (1894). *Materials for the Study of Variation.* Cambridge University Press, Cambridge.

Becker, H. J. (1959). Die puffs der speicheldrüsenchromosomen von *Drosophila melanogaster.* I. Beobachtungen zum verhalten des puffmusters in normalstamm und bei zwei mutanten, giant-und lethal-giant larvae. *Chromosoma* **10**:654.

Becker, H. J. (1978). Mitotic recombination and position effect variegation. In *Genetic Mosaics and Cell Differentiation* (W. Gehring, ed.). Springer, New York.

Bedard, P-A., and Brandhorst, B. P. (1983). Patterns of protein synthesis and metabolism during sea urchin embryogenesis. *Dev. Biol.* **96:**74.

Bedian, V. (1982). The possible role of assignment catalysts in the origin of the genetic code. *Origins of Life* **12:**181.

Bedian, V., Oliver, C. E., McCoon, P., and Kauffman, S. A. (1986). A cell surface differentiation antigen of *Drosophila. Dev. Biol.* **115:**105.

Been, M. D., Barfod, J. M., Burke, J. M., Price, J. V., Tanner, N. K., Zaug, A. J., and Cech, T. R. (1987). Structures involved in *Tetrahymena* rRNA self-splicing and RNA enzyme activity. In *Cold Spring Harbor Symposia on Quantitative Biology*, vol. 52. Cold Spring Harbor Laboratory, Cold Spring Harbor, N.Y.

Beerman, W. (1952). Chromomeren konstanz und spezifische modifikationen der chromosomenstruktur in der entwicklung und organdifferenzierung von *Chironomous tentans. Chromosoma* **5:**139.

Behmoaras, T., Toulme, J. J., and Helene, C. (1981a). A tryptophan-containing peptide recognizes and cleaves DNA at apurinic sites. *Nature* **292:**858.

Behmoaras, T., Toulme, J. J., and Helene, C. (1981b). Specific recognition of apurinic sites in DNA by a tryptophan-containing peptide. *Proc. Natl. Acad. Sci. USA* **78:**926.

Bell, G. (1970). Mathematical model of clonal selection and antibody production. *J. Theoret. Biol.* **29:**191.

Bell, G. (1971). Mathematical model of clonal selection and antibody production II. *J. Theoret. Biol.* **33:**339.

Benner, S. A., Allemann, R. K., Ellington, A. D., Ge, L., Glasfeld, A., Leanz, G. F., Krauch, T., MacPherson, L. J., Moroney, S., Piccirilli, J. A., and Weinhold, E. (1987). Natural selection, protein engineering, and the last riborganism: Rational model building in biochemistry. In *Cold Spring Harbor Symposia on Quantitative Biology*, vol. 52. Cold Spring Harbor Laboratory, Cold Spring Harbor, N.Y.

Bennett, D. (1975). The T-locus of the mouse. *Cell* **6:**441.

Benton, W. D., and Davis, R. W. (1977). Screening lambda-gt recombinant clones by hybridization to single plaques *in situ. Science* **196:**180.

Berek, C., Griffiths, G. M., and Milstein, C. (1985). Molecular events during maturation of the immune response to oxazolone. *Nature* **316:**412.

Berek, C., Jarvis, J. M., and Milstein, C. (1987). Activation of memory and virgin b cell clones in hyperimmune animals. *Eur. J. Immunol.* **17:**1121.

Berendes, H. D. (1966). Gene activities in the malphigian tubules of *Drosophila hydei* at different developmental stages. *J. Exp. Zool.* **162:**209.

Berg, D. E. (1980). Control of gene expression by a mobile recombinational switch. *Proc. Natl. Acad. Sci. USA.* **77:**4880.

Berge, C. (1962). *The Theory of Graphs and Its Applications.* Methusena, London.

Betz, J. L., Brown, P. R., Smyth, M. J., and Clarke, P. H. (1974). Evolution in action. *Nature* **247:**261.

Bialock, J. E., and Bost, K. L. (1986). Binding of peptides that are specified by complementary RNAs. *Biochem. J.* **234:**679.

Bialock, J. E., and Smith, E. M. (1984). Hydropathic anti-complementarity of amino acids based on the genetic codes. *Biochem. Biophys. Res. Commun.* **121:**203.

Biebricher, C. K. (1987). Replication and evolution of short-chained RNA species replicated by B replicase. In *Cold Spring Harbor Symposia on Quantitative Biology*, vol. 52. Cold Spring Harbor Laboratory, Cold Spring Harbor, N.Y.

Biebricher, C., and Eigen, M. (1988). *RNA Genetics* (E. Domingo, P. Ahlquist, and J. Holland, eds.), vol. 1. CRC Press, Boca Raton, Fla.

Bienenstock, E., Fogelman-Soulie, F., and Weisbuch, G., eds. (1986). *Disordered Systems and Biological Organization.* Series F: *Computer and Systems Sciences,* vol. 20. Springer, New York.

Binder, K., and Young, A. (1986). Spin-glasses: Experimental facts, theoretical concepts and open questions. *Rev. Mod. Phys.* **54**:801.

Bishop, J. O. (1974). The gene numbers game. *Cell* **2**:81.

Bluestone, J. A., Sharrow, S. O., Epstein, S. L., Ozato, K., and Sachs, D. H. (1981). Induction of anti-H-2 antibodies in the absence of alloantigen exposure by *in vivo* administration of anti-idiotype. *Nature* **219**:233.

Bodner, M., and Karin, M. (1987). A pituitary-specific *trans*-acting factor can stimulate transcription from the growth hormone promoter in extracts of non-expressing cells. *Cell* **50**:267.

Boggs, R. T., Gregor, P., Idriss, S., Belote, J. M., and McKeown, M. (1987). Regulation of sexual differentiation in *D. melanogaster* via alternative splicing of RNA from the *transformer* gene. *Cell* **50**:739.

Bohn, H. (1972). The origin of the epidermis in the supernumerary regenerates of triple legs in cockroaches *(Blattaria). J. Embryol. Exp. Morph.* **28**:185.

Bonner, J. T. (1958). *The Evolution of Development.* Cambridge University Press, Cambridge.

Bonner, J. T., and Pardue, M. L. (1977). Ecdysone-stimulted RNA synthesis in salivary glands of *Drosophila melanogaster:* Assay by in situ hybridization. *Cell* **12**:219.

Borradaile, L. A., Potts, F. A., Eastham, L.E.S., and Saunders, J. T. (1958). *The Invertebrata,* 3d ed. Cambridge University Press, Cambridge.

Borstnik, B., Pumpernik, D., and Hofacker, G. L. (1987). Point mutations as an optimal search process in biological evolution. *J. Theoret. Biol.* **125**:249.

Bost, K. L., Smith, E. M., and Bialock, E. J. (1985a). Similarity between the corticotropin (ACTH) receptor and a peptide encoded by an RNA that is complementary to ACTH mRNA. *Proc. Natl. Acad. Sci. USA* **82**:1372.

Bost, K. L., Smith, E. M., and Bialock, J. E. (1985b). Regions of complementarity between the messenger RNAs for epidermal growth factor, transferrin, interleukin-2 and their respective receptors. *Biochem. Biophys. Res. Commun.* **128**:1373.

Bothwell, A.L.M., Paskind, M., Reth, M., Imanishi-Kari, T., Rajewsky, K., and Baltimore, D. (1982). Somatic variants of murine immunoglobulin lambda light chains. *Nature* **298**:380.

Boulet, A. M., Erwin, C. R., and Rutter, W. J. (1986). Cell-specific enhancers in the rat exocrine pancreas. *Proc. Natl. Acad. Sci. USA* **83**:3599.

Bownes, M. (1975). A photographic study of development in the living embryo of *Drosophila melanogaster. J. Embryol. Exp. Morph.* **33**:789.

Bownes, M., and Sang, J. H. (1974a). Experimental manipulations of early *Drosophila* embryos: I. Adult and embryonic defects resulting from microcautery at nuclear multiplication and blastoderm stages. *J. Embryol. Exp. Morph.* **32**:253.

Bownes, M., and Sang, J. H. (1974b). Experimental manipultion of early *Drosophila* embryos: II. Adult and embryological defects resulting from the removal of blastoderm cells by pricking. *J. Embryol. Exp. Morph.* **32**:273.

Brady, R. M. (1986). Optimization strategies gleaned from biological evolution. *Nature* **317**:804.

Brandhorst, B. P. (1976). Two-dimensional gel patterns of protein synthesis before and after fertilization of sea urchin eggs. *Dev. Biol.* **52**:310.

Braun, A. C. (1968). *The Stability of the Differentiated State.* Springer, New York.

Braun, A. C. (1969). *The Cancer Problem.* Columbia University Press, New York.

Breckenridge, R. J. (1991). Molecular recognition: Models for drug design. *Experientia* **47**:1148.

Breiman, L. (1968). *Probability.* Addison-Wesley, Reading, Mass.

Breitbart, R. E., Andreadis, A., and Nadal-Ginard, B. (1987). Alternative splicing: A ubiquitous mechanism for the generation of multiple protein isoforms from single genes. *Ann. Rev. Biochem.* **56**:467.

Breslow, R. (1987). Artificial enzymes. In *Cold Spring Harbor Symposia on Quantitative Biology,* vol. 52. Cold Spring Harbor Laboratory, Cold Spring Harbor, N.Y.

Bretscher, M. S. (1968). How repressor molecules function. *Nature* **217**:509.

Britten, R. J., and Davidson, E. H. (1969). Gene regulation for higher cells: A theory. *Science* **165**:349.

Britten, R. J., and Kohne, D. E. (1968). Repeated sequences in DNA. *Science* **161**:529.

Britten, R. J., and Kohne, D. E. (1970). Repeated segments of DNA. *Sci. Am.* **222**:24.

Browder, L. W. (1984). *Developmental Biology.* Saunders, Philadelphia.

Brower, D. L., Piovant, M., and Reger, L. A. (1985). Developmental analysis of *Drosophila* position-specific antigens. *Dev. Biol.* **108**:120.

Brown, D. D. (1981). Gene expression in eukaryotes. *Science* **211**:667.

Brown, D. D., and Dawid, I. B. (1968). Specific gene amplification in oocytes. *Science* **160**:272.

Bruck, C., Co, M. S., Slaoui, M., Gaulton, G. N., Smith, T., Fields, B. N., Mullins, J. I., and Greene, M. I. (1986). Nucleic acid sequence of an internal image-bearing monoclonal anti-idiotype and its comparison to the sequence of the external antigen. *Proc. Natl. Acad. Sci. USA* **83**:6578.

Bryant, P. J. (1975). Pattern formation in the imaginal wing disc of *Drosophila melanogaster:* Fate map, regeneration and duplication. *J. Exp. Biol.* **193**:49.

Bryant, P. J. (1978). Pattern formation, growth control and cell interactions in *Drosophila* imaginal discs. In *Determinants of Spatial Organization* (S. Subtelny and I. R. Konigsberg, eds.). Academic Press, New York.

Bryant, S. W., French, V., and Bryant, P. (1980). Distal regeneration and symmetry. *Science* **212**:993.

Bryant, S. W., and Iten, L. E. (1976). Supernumerary limbs in amphibians: Experimental production in *Notophthalmus viridescens* and a new interpretation of their formation. *Dev. Biol.* **50**:212.

Bryngelson, J. D. (1988). Glassy models of protein folding. Ph.D. diss., University of Illinois, Urbana-Champaign.

Bryngelson, J. D., and Wolynes, P. G. (1987). Spin glasses and the statistical mechanics of protein folding. *Proc. Natl. Acad. Sci. USA* **84**:7524.

Bull, A. L. (1966). *Bicaudal,* a genetic factor which affects the polarity of the embryo in *Drosophila melanogaster. J. Exp. Zool.* **161**:221.

Bulliere, D. (1970). Interpretation des regenerats multiples chez les insectes. *J. Embryol. Exp. Morph.* **23**:337.

Bunow, B., Kernevez, P., Joly, G., and Thomas, D. (1980). Pattern formation by reaction-diffusion instabilities: Application to morphogenesis in *Drosophila. J. Theoret Biol.* **84**:629.

Burch, J. B., and Weintraub, H. (1983). Temporal order of chromatin structural changes associated with activation of the major chicken vitellogenin gene. *Cell* **33**:65.

Burnet, M. (1959). *The Clonal Selection Theory of Acquired Immunity.* Cambridge University Press, Cambridge.

Burnette, W. N. (1981). "Western blotting": Electrophoretic transfer of proteins from sodium dodecyl sulfate-polyacrylamide gels to unmodified nitrocellulose and radiographic detection with antibody and radioiodinated protein A. *Anal. Biochem.* **112**:195.

Burstein, C., Cohn, M., Kepes, A., and Monod, J. (1965). Role du lactose et de ses produits metaboliques dans l'induction de l'operon lactose chez *Escherichia coli. Biochim. Biophys. Acta* **95**:634.

Bush, G. L. (1981). Stasipatric speciation and rapid evolution in animals. In *Evolution and Speciation: Essays in Honor of M.J.D. White* (W. R. Ashby and D. S. Woodruff, eds.). Cambridge University Press, Cambridge.

Bush, G. L., Case, S. M., Wilson, A. C., and Patton, J. L. (1977). Rapid speciation and chromosomal evolution in mammals. *Proc. Natl. Acad. Sci. USA* **74**:3942.

Buss, L. (1987). *Evolution of Individuality.* Princeton University Press, Princeton, N.J.

Buzayan, J. M., Gerlach, W. L., and Bruening, G. (1986). Nonenzymatic cleavage and ligation of RNAs complementary to a plant virus satellite RNA. *Nature* **323**:349.

Buzin, C. H., and Seacof, L. (1981). Developmental modulation of protein synthesis in *Drosophila* primary embryonic cell cultures. *Dev. Genet.* **2**:237.

Byrne, G., and Cox, E. C. (1986). Spatial patterning in *Polysphondylium:* Monoclonal antibodies specific for whorl prepatterns. *Dev. Biol.* **117**:442.

Byrne, G., and Cox, E. C. (1987). Genesis of a spatial pattern in the cellular slime mold *Polysphondylium pallildum. Proc. Natl. Acad. Sci. USA* **84**:4140.

Cairns-Smith, A. G. (1982). *Genetic Takeover and the Mineral Origins of Life.* Cambridge University Press, Cambridge.

Calvin, M. (1969). *Chemical Evolution.* Oxford University Press, New York.

Cambou, B., and Klibanov, A. M. (1984). Unusual catalytic properties of usual enzymes. *Ann. N.Y. Acad. Sci.* **434**.

Cameron, J. R., Loh, E. Y., and Davis, R. W. (1979). Evidence for transposition of dispersed repetitive DNA families in yeast. *Cell* **16**:739.

Campbell, J. H. (1982). In *Perspectives on Evolution* (R. Milkman, ed.). Sinauer, Sunderland, Mass.

Campbell, J. H. (1985). An organizational interpretation of evolution. In *Evolution at a Crossroads* (D. J. Depew and B. H. Weber, eds.). MIT Press, Cambridge, Mass.

Cancro, M. P., Gerhard, W., and Klinman, N. H. (1978). The diversity of the influenza-specific primary b-cell repertoire in Balb/c mice. *J. Exp. Med.* **147**:776.

Capdevila, M. P., and Garcia-Bellido, A. (1974). A development and genetic analysis of *bithorax* phenocopies in *Drosophila. Nature* **250**:500.

Carrell, R. W., Pemberton, P. A., and Boswell, D. R. (1987). The serpins: Evolution and adaptation in a family of protease inhibitors. In *Cold Spring Harbor Symposia on Quantitative Biology,* vol. 52. Cold Spring Harbor Laboratory, Cold Spring Harbor, N.Y.

Carroll, S. B., and Scott, M. P. (1986). Zygotically active genes that affect the spatial expression of the *fushi-tarazu* segmentation gene during early *Drosophila* embryogenesis. *Cell* **45**:113.

Cassier, E. (1950). *The Problem of Knowledge.* Yale University Press, New Haven, Conn.

Cavadore, J. C. (1971). Polycondensation d'alpha-amino acids en milieu aqueux. Ph.D. diss., Université des Sciences et Techniques du Languedoc Academie de Montepellier,

Cavalier-Smith, T. (1987). The origin of cells: A symbiosis between genes, catalysts and membranes. In *Cold Spring Harbor Symposia on Quantitative Biology,* vol. 52. Cold Spring Harbor Laboratory, Cold Spring Harbor, N.Y.

Cavender, J. A. (1977). Kauffman's square root law: Possible correlations. *J. Theoret. Biol.* **65**:791.

Cavener, D. R. (1987). Combinatorial control of structural genes in *Drosophila:* Solutions that work for the animal. *BioEssays* **7**:103.

Cech, T. R., (1986a). The generality of self-splicing RNA: Relationship to nuclear RNA splicing. *Cell* **44**:207.

Cech, T. R., (1986b). A model for the RNA-catalyzed replication of RNA. *Proc. Natl. Acad. Sci. USA* **83**:4360.

Cech, T. R. (1986c). RNA as an enzyme. *Sci. Am.* **255**:64.

Cech, T. R. (1987). The chemistry of self-splicing RNA and RNA enzymes. *Science* **236**:1532.

Chaleff, D. T., and Fink, G. R. (1980). Genetic events associated with an insertion mutation in yeast. *Cell* **21**:227.

Chalfie, M., Horovitz, H. R., and Sulston, J. E. (1981). Mutations that lead to reiterations in the cell lineages of *C. elegans. Cell* **24**:59.

Chan, L. N., and Gehring, W. (1971). Determination of blastoderm cells in *Drosophila melanogaster. Proc. Natl. Acad. Sci. USA* **68**:2217.

Chang, C.-Y., and Kohler, H. (1986). A novel chimeric antibody with circular network characteristics. *Ann. N.Y. Acad. Sci.* **475**:114.

Chao, A. T., and Guild, G. M. (1986). Molecular analysis of the ecdysterone-inducible 2B5 "early" puff in *Drosphila melanogaster. EMBO J.* **5**:143.

Chaplin, M. P., and Bucke, C. (1990). *Enzyme Technology.* Cambridge University Press, Cambridge.

Charbit, A., Boulain, J. C., Ryter, A., and Hofnung, M. (1986). Probing the topology of a bacterial membrane protein by genetic insertion of a foreign epitope: Expression at the cell surface. *EMBO J.* **5**:3029.

Charlesworth, B., Lande, R., and Slatkin, M. (1982). A neo-Darwinian commentary on macroevolution. *Evolution* **36**:474.

Chen, C. B., Inoue, T., and Orgel, L. E. (1985). Template-directed synthesis on oligodeoxycytidylate and polydeoxycytidylate templates. *J. Mol. Biol.* **181**:271.

Chikarraishi, D. M., Deeb, S. S., and Sueoka, N. (1978). Sequence complexity of nuclear RNAs in adult rat tissues. *Cell* **13**:111.

Childs, J., Villanueva, K., Barrick, D., Schneider, T. D., Stormo, G., Gold, L., Leitner, M., and Caruthers, M. (1985). Ribosome binding site sequences and function. *UCLA Symp. Mol. Biol.* **30**:341.

Chovnick, A., Gelbart, W., McCarron, M., Osmond, B., Candido, E.P.M., and Baillie, D. L. (1976). Organization of the *rosy* locus in *Drosophila melanogaster:* Evidence for a control element adjacent to the xanthine dehydrogenase structural element. *Genetics* **84**:233.

Christensen, H., and Pain, R. H. (1991). Molten globule intermediates and protein folding. *Eur. Biophys. J.* **19**:221.

Church, A. H. (1904). *On the Relation of Phyllotaxis to Mechanical Laws.* Williams and Nargate, London.

Clark, S. H., Huppi, K., Ruezinsky, D., Staudt, L., Gerhard, W., and Weigert, M. (1985). Inter- and intraclonal diversity in the antibody response to influenza hemagglutinin. *J. Exp. Med.* **161**:687.

Clarke, P. H. (1976). Genes and enzymes. *FEBS Lett.* **62**:E37.

Clothia, C., and Lesk, A. M. (1987). The evolution of protein structures. In *Cold Spring Harbor Symposia on Quantitative Biology,* vol 52. Cold Spring Harbor Laboratory, Cold Spring Harbor, N.Y.

Cohen, J. E. (1988). Threshold phenomena in random structures. *Disc. Appl. Math.* **19**:113.

Coleclough, C. (1983). Chance, necessity and antibody gene dynamics. *Nature* **303**:23.

Comings, D. E. (1973). A general theory of carcinogenesis. *Proc. Natl. Acad. Sci. USA* **70**:3324.

Conklin, E. G. (1905). Organization and cell-lineage of the ascidian egg. *J. Natl. Acad. Sci., Phil.* **13**:1.

Corces, V., Pellicer, A., Axel, R., and Meselson, M. (1981). Integration, transcription and control of a *Drosophila* heat shock gene in mouse cells. *Proc. Natl. Acad. Sci. USA* **78**:7038.

Corsten, M., and Poole, P. (1987). *J. Statis. Phys.*

Costa, U. M. (1987). Spreading of "damage" in a three-dimensional Ising model. *J. Phys. A: Math. Gen.* **20**:L583.

Coste, J., and Henon, H. (1986). In *Disordered Systems and Biological Organization* (E. Bienenstock, F. Fogelman-Soulie, and G. Weisbuch, eds.), Series F: *Computer and Systems Sciences,* vol. 20. Springer, New York.

Counce, S. J., and Waddington, C. H., eds. (1972). *Developmental Systems: Insects.* Academic Press, London.

Cox, T., and McNally, J. G. (1989). Spots and stripes patterning spectrum in the cellular slime mold: *Polysphonylium pallium. Development* **105**:323.

Crenshaw, E. B. III, Russo, A. F., Swanson, L. W., and Rosenfeld, M. G. (1987). Neuron-specific alternative RNA processing in transgenic mice expressing a metallothionein-calcitonin fusion gene. *Cell* **49**:389.

Crews, S., Griffin, J., Huang, H., Calame, K., and Hood, L. (1981). A single V_h gene segment encodes the immune response to phosphorylcholine: Somatic mutation is correlated with the class of the antibody. *Cell* **25**:59.

Crick, F.H.C. (1968). The origin of the genetic code. *J. Mol. Biol.* **38**:367.

Crick, F.H.C., and Lawrence, P. (1975). Compartments and polyclones in insect development. *Science* **189**:49.

Croce, C. M. (1985). Chromosomal translocations, oncogenes, and B-cell tumors. *Hosp. Pract.* **20**(1):41.

Croce, C. M. (1987). Role of chromosome translocations in human neoplasia. *Cell* **49**:155.

Crow, J. F. (1987). Neural models in molecular evolution. In *Neutral Models in Biology* (M. H. Nitecki and A. Hoffman, eds.). Oxford University Press, New York.

Crow, J. R., and Kimura, M. (1965). Evolution in sexual and asexual populations. *Am. Nat.* **99**:439.

Crow, J. R., and Kimura, M. (1970). *An Introduction to Population Genetics Theory.* Harper & Row, New York.

Cull, M. G., Miller, J. F., and Schatz, P. J. (1992). Screening for receptor ligands using large libraries of peptides linked to the C terminus of the *lac* repressor. *Proc. Natl. Acad. Sci. USA* **89**:1865.

Cull, P. (1971). Linear analysis of switching nets. *Kybernetik* **8**:31.

Cummins, F. W., and Prothero, J. W. (1978). A model of pattern formation in multicellular organisms. *Coll. Phenom.* **3**:41.

Cupp, J. E., Leary, J. F., Cernichiari, E., Wood, J.C.S., and Doherty, R. A. (1984). Rare-event analysis methods for detection of fetal blood cells in maternal blood. *Cytometry* **5**:138.

Cwirla et al. (1990). Peptides on shape: A vast library of peptides for identifying ligands. *Proc. Natl. Acad. Sci. USA.*

Darnell, J. E., and Doolittle, W. F. (1986). Speculations on the early course of evolution. *Proc. Natl. Acad. Sci. USA* **83**:1271.

Darnell, J., Lodish, H., and Baltimore, D. (1986). *Molecular Cell Biology.* Scientific American Books, New York.

Darwin, C. (1966). *On the Origin of Species.* Facsimile ed. Harvard University Press, Cambridge, Mass.

Davidson, E. H. (1986). *Gene Activity in Early Development,* 3d ed. Academic Press, New York.

Davidson, E. H., and Britten, R. J. (1976). Regulation of gene expression: Possible role of repetitive sequences. *Science* **204**:1052.

Davidson, E. H., Hough, B. R., Amenson, C. S., and Britten, R. J. (1973). General interspersion of repetitive with non-repetitive sequence elements in the DNA of *Xenopus. J. Mol. Biol.* **77**:1.

Davies, R. W., Waring, R. B., and Towner, P. (1987). Internal guide sequence and reaction specificity of group 1 self-splicing introns. In *Cold Spring Harbor Symposia on Quantitative Biology,* vol. 52. Cold Spring Harbor Laboratory, Cold Spring Harbor, N.Y.

Davison, B. L., Edgly, J. M., Mulvihill, E. R., and Chambon, P. (1983). Formation of stable preinitiation complexes between eukaryotic class B transcription factors and promoter sequences. *Nature* **301**:680.

Deamer, D. W., and Barchfeld, G. L. (1982). Encapsulation of macromolecules by lipid vesicles under simulated prebiotic conditions. *J. Mol. Evol.* **18**:203.

de Arcangelis, L. (1987). Fractal dimensions in three-dimensional Kauffman cellular automata. *J. Phys. A. Lett.* **20**:L369.

de Arcangelis, L., and Coniglio, A. (1988). Critical temperature in Kauffman cellular automata. *Europhys. Lett.* **7**:113.

De Beer, G. R. (1958). *Embryos and Ancestors,* 3d ed. Clarendon Press, Oxford.

De Boer, R. J. (1988). Symmetric idiotypic networks: Connectance and switching, stability and suppression. In *Theoretical Immunology II: Santa Fe Institute Studies in the Sciences of Complexity* (A. S. Perelson, ed). Addison-Wesley, Reading, Mass.

Debrunner, P. G., and Frauenfelder, H. (1982). Dynamics of proteins. *Ann. Rev. Phys. Chem.* **33**:283.

Dent, J. N. (1954). A study of regenerates emanating from limb transplants with reversed proximodistal polarity in the adult newt. *Anat. Rec.* **118**:841.

Department of Trade and Industry (1989). Method for obtaining DNA, RNA, peptides, polypeptides or proteins by means of a DNA recombination technique. English patent no. 2183661, issued to M. Ballivet and S. A. Kauffman.

Derrida, B.(1980). Limit of a family of disordered models. *Phys. Rev. Lett.* **45**:79.

Derrida, B. (1981). Random energy model: An exactly solvable model of disordered systems. *Phys. Rev. B* **24**:2613.

Derrida, B. (1987a). Dynamics of automata, spin glasses and neural network models. From lectures presented at the School for Nonlinear Evolution and Chaotic Phenomena. Noto, Sicily, June 8–19.

Derrida, B. (1987b). Valleys and overlaps in Kauffman's model. *Philos. Mag. B* **56**:917.

Derrida, B. (1987c). Dynamical phase transitions in nonsymmetric spin glasses. *J. Phys. A. Lett.* **20**:L721.

Derrida, B., and Bessis, D. (1988). Statistical properties of valleys in the annealed random map model. *J. Phys. A. Lett.* **21**:L509.

Derrida, B., and Flyvbjerg, H. (1986). Multivalley structure in Kauffman's model: Analogy with spin glasses. *J. Phys. A: Math. Gen.* **19**:L1003.

Derrida, B., and Flyvbjerg, H. (1987a). Distribution of local magnetizations in random networks of automata. *J. Phys. A. Lett.* **20**:L1107.

Derrida, B., and Flyvbjerg, H. (1987b). The random map model: A disordered model with deterministic dynamics. *J. Physique* **48**:971.

Derrida, B., and Pomeau, Y. (1986). Random networks of automata: A simple annealed approximation. *Europhys. Lett.* **1**:45.

Derrida, B., and Stauffer, D. (1986a). Phase transitions in two-dimensional Kauffman cellular automata. *Europhys. Lett.* **2**:739.

Derrida, B., and Stauffer, D. (1986b). *Europhys. Lett.* **1**:45.

Derrida, B., and Weisbuch, G. (1986). Evolution of overlaps between configurations in random Boolean networks. *J. Physique* **47**:1297.

Derrida, B., and Weisbuch, G. (1987). Dynamical phase transitions in three-dimensional spin glasses. *Europhys. Lett.* **4**:657.

Derrida, B., Gardner, E., and Zippelius, A. (1987). An exactly soluble asymmetric neural network model. *Europhys. Lett.* **4**:167.

Desplan, C., Theis, J., and O'Farrell, P. H. (1985). A *Drosophila* developmental gene, *engrailed,* encodes a sequence-specific DNA binding activity. *Nature* **318**:630.

Deutsches Patentent (1990). Verfahren zur herstellung von peptiden, polypeptiden oder proteinen. German patent no. 3,590,766.5-41, issued to M. Ballivet and S. A. Kauffman.

Devlin, J. J., Panganiban, L. C., and Devlin, P. E. (1990). Random peptide libraries: A source of specific protein binding molecules. *Science* **249**:404.

Dickerson, R. E. (1977). Energy and evolution in the folding of proteins. In *Molecular Evolution and Polymorphism* (M. Kimura, ed.). National Institute of Genetics, Mishima, Japan.

Dickinson, W. J. (1975). A genetic locus affecting the developmental expression of an enzyme in *Drosophila melanogaster. Dev. Biol.* **42**:131.

Dickinson, W. J. (1978). Genetic control of enzyme expression in *Drosophila:* A locus influencing tissue specificity of aldehyde oxidase. *J. Exp. Zool.* **206**:333.

Dickinson, W. J. (1980a). Complex *cis*-acting regulatory genes demonstrated in *Drosophila* hybrids. *Dev. Genet.* **1**:229.

Dickinson, W. J. (1980b). Evolution of patterns of gene expression in Hawaiian picture-winged *Drosophila. J. Mol. Evol.* **16**:73.

Dickinson, W. J. (1980c). Tissue specificity of enzyme expression regulated by diffusible factors: Evidence in *Drosophila* hybrids. *Science* **207**:995.

Dickinson, W. J. (1988). Problems and paradigms: On the architecture of regulatory systems: Evolutionary insights and implications. *BioEssays* **8**:204.

Dixon, M., and Webb, E. C. (1960). *Enzymes.* Academic Press, New York.

Dobzhansky, T. (1937). *Genetics and the Origin of Species.* Columbia University Press, New York.

Dobzhansky, T. (1970). *Genetics of the Evolutionary Process.* Columbia University Press, New York.

Dohmen, M. R., and Verdonk, N. H. (1979). The ultrastructure and role of the polar lobe in development of molluscs. In *Determinants of Spatial Organization* (S. Subtelny and I. R. Konigsberg, eds.). Academic Press, New York.

Doolittle, W. F., and Sapienza, C. (1980). Selfish genes, the phenotype paradigm and genome evolution. *Nature* **284**:601.

Doria, G. (1982). Immunoregulatory implications of changes in antibody affinity. In *Regulation of Immune Response Dynamics* (C. DeLisi and J.R.J. Hiernaux, eds.), vol. 2. CRC Press, Boca Raton, Fla.

Doudna, J. A., and Szostak, J. W. (1989). RNA-catalyzed synthesis of complementary strand RNA. *Nature* **339**:519.

Dover, G. A. (1970). The evolution of DNA sequences common to closely related insect genomes. In *Insect Cytogenetics* (R. L. Blackman, G. M. Hewett, and M. Ashburner, eds.), Royal Entomological Society, Symposium, no. 9. Blackwell, Oxford.

Dover, G. A., and Flavell, R. B. (eds.) (1982). *Genome Evolution.* Academic Press, London.

Dover, G., Brown, S., Coen, E., Dallas, J., Strachan, T., and Trick, M. (1982). The dynamics of genome evolution and species differentiation. In *Genome Evolution.* (G. A. Dover and R. B. Flavel, eds.). Academic Press, London.

Dower, W. J., and Foder, S.P.A. (1991). The search for molecular diversity II: Recombinant and synthetic randomized peptide libraries. *Ann. Rep. Med. Chem.* **26**:271.

Driesch, H. (1892/1964). The potency of the first two cleavage cells in the development of echinoderms. In *Foundations of Experimental Embryology* (B. H. Willier and J. M. Oppenheimer, eds.). Prentice-Hall, Englewood Cliffs, N.J.

Driesch, H. (1894). *Analytische Theorie der Organischen Entwicklung.* Engelmann, Leipzig.

Driesch, H. (1908). *Science and Philosophy of the Organism.* A. and C. Black, London.

Driesch, H. (1914). *The History and Theory of Vitalism.* Macmillan, London.

Dube, D. K., and Loeb, L. A. (1989). Mutants generated by the insertion of random oligonucleotides into the active site of the β-lactamase gene. *Biochemistry* **28**:5703.

Dubendorfer, K. (1977). Ph.D. diss., University of Zurich.

Dugundji, J., Gillespie, P., Marquarding, D., Ugi, I., and Ramirez, F. (1976). Metric spaces and graphs representing the logical structure of chemistry. In *Chemical Applications of Graph Theory* (A. T. Balaban, ed.). Academic Press, London.

Dutta, S. K., and Chaudhuri, R. K. (1975). Differential transcription of non-repeated DNA during development of *Neurospora crassa*. *Dev. Biol.* **43**:35.

Dwyer, D. S., Vakil, M., and Kearney, J. F. (1986). Idiotypic network connectivity and a possible cause of myasthenia gravis. *J. Exp. Med.* **164**:1310.

Dyson, F. (1982). A model for the origin of life. *J. Mol. Evol.* **18**:344.

Dyson, F. (1985). *Origins of Life*. Cambridge University Press, London.

Ebling, W., Engel, A., Esser, B., and Feistel, R. (1984). Diffusion and reaction in random media and models of evolution processes. *J. Statis. Phys.* **37**:369.

Edelman, G. M. (1988). *Topobiology: An Introduction to Molecular Embryology*. Basic Books, New York.

Eden, M. (1967). In *Mathematical Challenges to the Neo-Darwinian Interpretation of Evolution* (P. S. Moorhead and M. M. Kaplan, eds.), Wistar Institute Press, Philadelphia.

Edgar, B. A., Weir, M. P., Schubiger, G., and Kornberg, T. (1986). Repression and turnover pattern of *fushi tarazu* RNA in early *Drosophila* embryo. *Cell* **47**:747.

Edwards, S. F., and Anderson, P. W. (1975). Theory of spin glasses. *J. Phys. F* **5**:965.

Efstratiadis, A., Pasakony, J. W., Maniatis, T., Lawn, R. M., O'Connell, C., Spritz, R. A., DeRiel, J. K., Forget, B. G., Weissman, S. M., Slightom, J. L., Biechl, A. E., Smithies, O., Barelle, F. E., Shoulders, C. C., and Proudfoot, N. J., (1980). The structure and evolution of the human β-globin gene family. *Cell* **21**:653.

Eigen, M. (1971). Self-organization of matter and the evolution of biological macromolecules. *Naturwissenschaften* **58**:465.

Eigen, M. (1985). Macromolecular evolution: Dynamical ordering in sequence space. In *Emerging Synthesis in Science: Proceedings of the Founding Workshops of the Santa Fe Institute* (David Pines, ed.). Santa Fe Institute, Santa Fe, N.M.

Eigen, M. (1986). The physics of molecular evolution. *Chemica Scripta* **26B**:13.

Eigen, M. (1987). New concepts for dealing with the evolution of nucleic acids. In *Cold Spring Harbor Symposia on Quantitative Biology*, vol. 52. Cold Spring Harbor Laboratory, Cold Spring Harbor, N.Y.

Eigen, M., and Schuster, P. (1977). The hypercycle: A principle of natural self-organization, Part A: Emergence of the hypercycle. *Naturwissenschaften* **64**:541.

Eigen, M., and Schuster, P. (1978a). The hypercycle: A principle of natural self-organization, Part B: The abstract hypercycle. *Naturwissenschaften* **65**:7.

Eigen, M., and Schuster, P. (1978b). The hypercycle: A principle of natural self-organization, Part C: The realistic hypercycle. *Naturwissenschaften* **65**:341.

Eigen, M., and Schuster, P. (1979). *The Hypercycle: A Principle of Natural Self-Organization*. Springer, New York.

Eigen, M., McCaskill, J., and Schuster P. (1988). The molecular quasi-species. *J. Phys. Chem.* **92**:6881.

Eigen, M., Gardiner, W., Schuster, P., and Winkler-Oswatitsch, R. (1981). *Sci. Am.* **244**:88.

Eisen, H., and Ptashne, M. (1971). The bacteriophage lambda. In *Cold Spring Harbor Symposia on Quantitative Biology*, vol. 36. Cold Spring Harbor Laboratory, Cold Spring Harbor, N.Y.

Eisen, H. N., and Siskind, G. W. (1964). Variations in affinities of antibodies during the immune response. *Biochemistry* **3**:996.

Elber, R., and Karplus, M. (1987). Multiple conformational states of proteins: A molecular dynamics analysis of myoglobin. *Science* **235**:318.

Eldredge, N., and Gould, S. J. (1972). Punctuated equilibria: An alternative to phyletic gradualism. In *Models in Paleobiology* (T.J.M. Schopf, ed.). Freeman, Cooper, San Francisco.

Ellington, A. D., and Szostak, J. W. (1990). *In vitro* selection of RNA molecules that bind specific ligands. *Nature* **346**:818.

Engelke, D. R., Ng, S.-Y., Shastry, B. S, and Roeder, R. G. (1980). Specific interaction of a purified transcription factor with an internal control region of 5 S RNA genes. *Cell* **19**:717.

Erdos, P., and Renyi, A. (1959). *On the Random Graphs 1,* vol. 6. Institute of Mathematics, University of Debreceniens, Debrecar, Hungary.

Erdos, P., and Renyi, A. (1960). *On the Evolution of Random Graphs,* Institute of Mathematics, Hungarian Academy of Sciences, publication no. 5.

Erikson, R. O. (1982). The geometry of phyllotaxis. In *The Growth and Functioning of Leaves* (J. E. Dele and Milthrope, eds.). Cambridge University Press, Cambridge.

Erneux, T., and Hiernaux, J. (1980). Transition from polar to duplicate patterns. *J. Math. Biol.* **9**:193.

Errede, B., Cardillo, T. S., Sherman, F., Dubois, E., Deschamps, J., and Wiame, J. M. (1980). Mating signals control expression of mutations resulting from insertion of a transposable repetitive element adjacent to diverse yeast genes. *Cell* **25**:427.

Erwin, D. H., Valentine, J. W., and Sepkoski, J. J., Jr. (1987). A comparative study of diversification events: The early Paleozoic versus the Mesozoic. *Evolution* **41**:1177.

Ewens, W. (1979). *Mathematical Population Genetics.* Springer, New York.

Fahrlander, P. D., Sumergi, J., Yang, J. Q., Weiner, F., Marcu, K. B., and Klein, G. (1985). Activation of the c-*myc* oncogene by the immunoglobulin heavy chain gene enhancer after multiple switch region-mediated chromosome rearrangements in a murine plasmacytoma. *Proc. Natl. Acad. Sci. USA* **82**:3746.

Fainsod, A., Bogarad, L. D., Ruusala, T., Lubin, M., Crothers, D. M., and Ruddle, F. H. (1986). The homeo domain of a murine protein binds 5′ to its own homeo box. *Proc. Natl. Acad. Sci. USA* **33**:9532.

Farmer, J. D. (1982). Chaotic attractors of an infinite dimensional dynamical system. *Physica* **4D**:366.

Farmer, J. D., Kauffman, S. A., and Packard, N. H. (1986). Autocatalytic replication of polymers. *Physica* **22D**:50.

Farmer, J. D., Ott, E., and Yorke, J. A. (1983). The dimension of chaotic attractors. *Physica* **7D**:153.

Farmer, J. D., Packard, N., and Perelson, A. (1986). The immune system, adaptation, and machine learning. *Physica* **22D**:187.

Feigenbaum, M. J. (1978). Quantitative universality for a class of nonlinear transformations. *J. Statis. Phys.* **19**:25.

Feller, W. (1971). *Introduction to Probability Theory and Its Applications,* vol. 2. 2d ed. Wiley, New York.

Felsenstein, J. (1988). Phylogenies and quantitative characters. *Ann. Rev. Ecol. Syst.* **19**:445.

Ferris, J. P. (1987). Prebiotic synthesis: Problems and challenges. In *Cold Spring Harbor Symposia on Quantitative Biology,* vol. 52. Cold Spring Harbor Laboratory, Cold Spring Harbor, N.Y.

Ferris, J. P., Sanchez, R. A., and Orgel, L. E. (1968). Studies in prebiotic synthesis. III: Synthesis of pyrimidines from cyanoacetylene and cyanate. *J. Mol. Biol.* **33**:693.

Fersht, A. R. (1974). Catalysis binding and enzyme-substrate complementarity. *Proc. Roy. Soc. Lond. B* **187**:397.

Fersht, A. R. (1985). *Enzyme Structure and Mechanism.* Freeman, New York.

Fierke, C. A., Kuchta, R. D., Johnson, K. A., and Benkovic, S. J. (1987). Implications for enzymic catalysis from free-energy reaction coordinate profiles. In *Cold Spring Harbor Symposia on Quantitative Biology,* vol. 52. Cold Spring Harbor Laboratory, Cold Spring Harbor, N.Y.

Fieser, T. M., Tainer, J. A., Geysen, H. M., Houghten, R. A., and Lerner, R. A. (1987). Influence of protein flexibility and peptide conformation on reactivity of monoclonal antipeptide antibodies with a protein alpha-helix. *Proc. Natl. Acad. Sci. USA* **84**:8568.

Finnegan, D. J., Will, B. H., Bayev, A. A., Bowcock, C., and Brown, L. (1982). Transposable DNA sequences in eukaryotes. In *Genome Evolution* (G. A. Dover and R. B. Flavell, eds.). Academic Press, London.

Firtel, R. A. (1972). Changes in the expression of single-copy DNA during development of the cellular slime mold *Dictyostelium discoideum. J. Mol. Biol.* **66**:363.

Fish, S., and Manser, T. (1987). Influence of the macromolecular form of a B cell epitope on the expression of antibody variable and constant region structure. *J. Exp. Med.* **166**:711.

Fisher, R. A. (1930). *The Genetical Theory of Natural Selection.* Oxford University Press, Oxford.

Flavell, R. (1982). Sequence amplification, deletion and rearrangement: Major sources of variation during species diversity. In *Genome Evolution* (G. A. Dover and R. B. Flavell, eds.). Academic Press, London.

Flesselles, J. M., Campbell, I. A., Jullien R., and Botet, R. (1988). *Percolation on a Hypercube.* Orsay Preprints.

Flory, J. (1953). *Principles of Polymer Chemistry,* Cornell University Press, Ithaca, N.Y.

Flyvberg, H., and Kjaer, N. J. (1988). Exact solution of Kauffman's model with connectivity one. *J. Phys. A* **21**:1695.

Foe, V. E., and Alberts, B. M. (1983). Studies of nuclear and cytoplasmic behavior in the five mitotic cycles that precede gastrulation in *Drosophila* embryogenesis. *J. Cell Sci.* **61**:31.

Fogel, L. J., Owens, A. J., and Walsh, M. J. (1966). *Artificial Intelligence Through Simulated Evolution.* Wiley, New York.

Fogelman-Soulie, F. (1984). Frustration and stability in random Boolean networks. *Disc. Appl. Math.* **9**:139.

Fogelman-Soulie, F. (1985a). Ph.D. diss., Université Scientifique et Medical de Grenoble, Grenoble, France.

Fogelman-Soulie, F. (1985b). Parallel and sequential computation on Boolean networks. In *Theoretical Computer Science 40,* North Holland, Amsterdam.

Fogelman-Soulie, F., Goles Chacc, E., and Weisbuch, G. (1982). Specific roles of the different Boolean mappings in random networks. *Bull. Math. Biol.* **44**:715.

Fontana, W. (1992). Algorithmic chemistry. In *Artificial Life II: A Proceedings Volume in the Sante Fe Institute Studies in the Sciences of Complexity* (C. G. Langton, J. D. Farmer, S. Rasmussen, and C. Taylor, eds.), vol. 10. Addison-Wesley, Reading, Mass.

Fontana, W., and Schuster, P. (1987). A computer model of evolutionary optimization. *Biophys. Chem.* **26**:123.

Fontana, W., Schnabl, W., and Schuster P. (1989). Physical aspects of evolutionary optimization and adaptation. *Phys. Rev. A* **40**:3301.

Fontana, W., Griesmacher, T., Schnabl, W., Stadler, P. F., and Schuster, P. (1991). Statistics of landscapes based on free energies' replication and degradation rate constants of RNA secondary structures. *Methods Chem.* **122**:795.

Fontana, W., Stadler, P. F., Bornberg-Bauer, E. G., Griesmacher, T., Hofacker, I. L., Tacher, M., Weinberger, E. D., and Schuster, P. (1992). RNA folding and combinatory landscapes. *Phys. Rev. E.,* submitted.

Fox, R. F. (1988). *Energy and the Evolution of Life.* Freeman, New York.

Fox, S. W. (1980). Metabolic microspheres: Origins and evolution. *Naturwissenschaften* **67**:373.

Fox, S. W. (1981). Origins of the protein-synthesis cycle. *Int. J. Quantum Chem.: Quant. Biol. Symp.* **S8**:441.

Fox, S. W., and Dose, H. (1977). *Molecular Evolution and the Origin of Life*. Academic Press, New York.

Fox, S. W., and Nakashima, T. (1980). The assembly and properties of protobiological structures: The beginnings of cellular peptide synthesis. *Biosystems* **12**:155.

Fox, S. W., and Wang, C. T. (1968). Melanocyte-stimulating hormone: Activity in thermal polymers of alpha-amino acids. *Science* **160**:547.

Fox, S. W., Jungck, J. R., and Nakashima, E. (1974). From proteinoid microsphere to contemporary cell: Formation of internucleotide and peptide bonds by proteinoid particles. *Origins of Life* **5**:227.

Fox, S. W., Nakashima, T., Przybylski, A., and Syren, R. M. (1982). The updated experimental proteinoid model. *Intl. J. Quantum Chem.: Quant. Biol. Symp.* **S9**:195.

Franklin, I., and Lewontin, R. C. (197). Is the gene the unit of selection? *Genetics* **65**:707.

Frasch, M., and Levine, M. (1987). Complementary patterns of *even-skipped* and *fushi tarazu* expression involve their differential regulation by a common set of segmentation genes in *Drosophila*. *Genes and Dev.* **1**:981.

French, V., Bryant, P., and Bryant, S (1976). A theory of pattern regulation in epimorphic fields. *Science* **193**:969.

French Patent Office (1987). Procédé d'obtention d'ADN, ARN peptides, polypeptides ou protéines par une technique de recombinaison d'ADN. French patent no. 863683, issued to M. Ballivet and S. A. Kauffman and registered as 2,579,518.

Freudl, R., McIntyre, S., Degen, M., and Henning, U. (1986). Cell surface exposure of the outer membrane protein Omp A of *Escherichia coli*. *J. Mol. Biol.* **188**:491.

Fruton, J. S. (1982). Proteinase-catalyzed synthesis of peptide bonds. *Adv. Enzymol.* **53**:239.

Fullilove, S. L., Jacobsen, A. G., and Turner, F. R. (1978). Embryonic development: Descriptive. In *The Genetics and Biology of Drosophila* (M. Ashburner and T.R.F. Wright, eds.). Academic Press, New York.

Galau, G. A., Klein, W. H., Davis, M. M., Wold, B. J., Britten, R. J., and Davidson, E. H. (1976). Structural gene sets active in embryos and adult tissues of the sea urchin. *Cell* **7**:487.

Garabedian, M. J., Shepherd, B. M., and Wensink, P. C. (1986). A tissue-specific transcription enhancer from the *Drosophila* yolk protein 1 gene. *Cell* **45**:859.

Garcia-Bellido, A. (1975). Genetic control of wing disc development in *Drosophila*. In *Cell Patterning* (R. Porter and J. Rivers, eds.), *CIBA Foundation Symposium*, vol. 29. Associated Scientific, Amsterdam.

Garcia-Bellido, A. (1983). Comparative anatomy of cuticular patterns in the genus *Drosophila*. In *Development and Evolution* (B. Goodwin and C. Wylie, eds.). Cambridge University Press, Cambridge.

Garcia-Bellido, A., and Merriam, J. R. (1969). Cell lineage in gynandromorphs of *Drosophila*. *J. Exp. Zool.* **170**:1.

Garcia-Bellido, A., and Santamaria, P. (1972). Developmental analysis of the wing disc in the mutant engrailed of *Drosophila melanotaster*. *Genetics* **72**:87.

Garcia-Bellido, A., Ripoll, P., and Morata, G. (1973). Developmental compartmentalization of the wing disk of *Drosophila*. *Nature New Biol.* **245**:251.

Garcia-Bellido, A., Ripoll, P., and Morata, G. (1976). Develomental compartmentalization in the dorsal mesothoracic disc of *Drosophila*. *Dev. Biol.* **48**:132.

Gardner, R. L. (1978). The relationship between cell lineage and differentiation in the early mouse embryo. In *Results and Problems in Cell Differentiation* (W. J. Gehring, ed.), vol. 9. Springer-Verlag, Berlin.

Garrels, J. I. (1979). Two-dimensional gel electrophoresis and computer analysis of proteins synthesized by clonal cell lines. *J. Biol. Chem* **254**:7961.

Garrels, J. I. (1983). Quantitative two-dimensional gel electrophoresis of proteins. *Methods Enzymol.* **100**:411.

Gaulton, G., and Greene, M. (1986). Idiotypic mimicry of biological receptors. *Ann. Rev. Immunol.* **4**:253.

Gehring, W. J. (1966a). In *The Stability of the Determined State.* Springer, New York.

Gehring, W. J. (1966b). *Arch. Julius Klaus-Stift. Vererbungsforsch. Sozialanthropol. Rassenhyg.* **41**:44.

Gehring, W. J. (1973). Genetic control of determination in the *Drosophila* embryo. In *Genetic Mechanisms of Development* (F. H. Ruddle, ed.). Academic Press, New York.

Gehring, W. J. (1975). Determination und transdetermination von imaginal scheiben. *Verh. Dtsch. Zool. Ges.* **1974**:70.

Gehring, W. J. (1976). Developmental genetics of *Drosophila. Ann. Rev. Genet.* **10**:209.

Gehring, W. J., and Nothiger, R. (1973). The imiginal disc of *Drosophila.* In *Developmental Systems: Insects* (S. J. Counce and C. H. Waddington, eds.), vol. 1. Academic Press, London.

Gelfand, A. E., and Walker, C. C. (1982). On the character of and distance between states in a binary switching net. *Biol. Cybernetics* **43**:79.

Gelfand, A. E., and Walker, C. C. (1984). *Ensemble Modeling.* Marcel Dekker, New York.

Gergen, J. P., and Wieschaus, E. F. (1985). The localized requirements for a gene affecting segmentation in *Drosophila:* Analysis of larvae mosaic for *runt. Dev. Biol.* **109**:321.

Gergen, J. P., and Wieschaus, E. F. (1986). Localized requirements for gene activity in segmentation of *Drosophila* embryos: Analysis of *armadillo, fused, giant* and *unpaired* mutations in mosaic embryos. *Roux's Arch. Dev. Biol.* **195**:49.

Gergen, J. P., Coulter, D., and Wieschaus, E. (1986). Segmental pattern and blastoderm cell identities. In *Gametogenesis and the Early Embryo* (S. Subtelny, ed.). *Symp. Soc. Devl. Biol.* **43**:195.

Getsoff, E. D., Geysen, H. M., Rodda, S. J., Alexander, H., Tainer, J. A., and Lerner, R. A. (1987). Mechanisms of antibody binding to a protein. *Science* **235**:1191.

Geysen, H. M., Barteling, S. J., and Meloen, R. H. (1985). Small peptides induce antibodies with a sequence and structural requirement for binding antigen comparable to antibodies raised against the native protein. *Proc. Natl. Acad. Sci. USA* **82**:178.

Geysen, H. M., Rodda, S. J., and Mason, T. J. (1986). The delineation of peptides able to mimic assembled epitopes. *CIBA Foundation Symposium* **119**:130.

Geysen, H. M., Rodda, S. J., and Mason, T. J. (1987). Strategies for epitope analysis using peptide synthesis. *J. Immunol. Methods* **102**:259.

Gierer, A., and Meinhardt, H. (1972). A theory of biological pattern formation. *Kybernetik* **12**:30.

Gilbert, S. F. (1988). *Developmental Biology.* 2d ed. Sinauer, Sunderland, Mass.

Gilbert, W. (1978). Why genes in pieces? *Nature* **271**:501.

Gilbert, W. (1979). Introns and exons: Playgrounds of evolution. In *Eucaryotic Gene Regulation* (R. Axel, T. Maniatis, and C. F. Fox, eds.). Academic Press, New York.

Gilbert, W. (1985). Genes in pieces revisited. *Science* **228**:823.

Gilbert, W. (1986). The RNA world. *Nature* **319**:618.

Gilbert, W. (1987). The exon theory of genes. In *Cold Spring Harbor Symposia on Quantitative Biology,* vol. 52. Cold Spring Harbor Laboratory, Cold Spring Harbor, N.Y.

Gilbert, W., Marchionni, M., and McKnight, G. (1986). On the antiquity of introns. *Cell* **46**:151.

Gill, G., Sadowski, I., and Ptashne, M. (1990). Mutations that increase the activity of a transcriptional activator in yeast and mammalian cells. *Proc. Natl. Acad. Sci. USA* **87**:2127.

Gillespie, J. H. (1983). A simple stochastic gene substitution model. *Theoret. Pop. Biol.* **23**:202.

Gillespie, J. H. (1984). Molecular evolution over the mutational landscape. *Evolution* **38**:1116.

Gillies, S. D., Morrison, S. L., Oi, V. T., and Tonegawa, S. (1983). A tissue-specific transcription enhancer element is located in the major intron of a rearranged immunoglobulin heavy-chain gene. *Cell* **33**:717.

Ginger, E., and Ptashne, M. (1987). Transcription in yeast activated by a putative amphipathic alpha helix linked to a DNA binding unit. *Nature* **330**:670.

Glass, L. (1975). Classification of biological networks by their qualitative dynamics. *J. Theoret. Biol.* **54**:85.

Glass, L. (1977a). Patterns of supernumerary limb regeneration. *Science* **198**:321.

Glass, L. (1977b). Combinatorial aspects of dynamics in biological systems. In *Statistical Mechanics and Statistical Methods in Theory and Application* (U. Landman, ed.). Plenum, New York.

Glass, L. (1985). Boolean and continuous models for the generation of biological rhythms. In *Dynamical Systems and Cellular Automata* (J. Demongeot, E. Goles, and M. Tchuente, eds.). Academic Press, London

Glass, L., and Kauffman, S. A. (1972). Co-operative components, spatial localization and oscillatory cellular dynamics. *J. Theoret. Biol.* **34**:219.

Glass, L., and Kauffman, S. A. (1973). The logical analysis of continuous non-linear biochemical control networks. *J. Theoret. Biol.* **39**:103.

Glass, L., and Mackey, M. C. (1988). *From Clocks to Chaos: The Rhythm of Life.* Princeton University Press, Princeton, N.J.

Glass, L., and Pasternack, J. S. (1978a). Prediction of limit cycles in mathematical models of biolocial oscillations. *Bull. Math. Biol.* **40**:27.

Glass, L. and Pasternack, J. S. (1978b). Stable oscillations in mathematical models of biological control systems. *J. Math. Biol.* **6**:207.

Glass, L., and Perez, R. (1982). The fine structure of phase locking. *Phys. Rev. Lett.* **48**:1772.

Gloor, H. (1947). Phanokopie-veruche mit ather an *Drosophila. Rev. Suisse Zool.* **27**:637.

Gmitro, J. E., and Scriven, L. E. (1966). A physico-chemical basis for pattern formation. In *Intracellular Transport* (K. K. Warren, ed.). Academic Press, New York.

Goldberg, D. E. (1989). *Genetic Algorithms in Search, Optimization, and Machine Learning.* Addison-Wesley, Reading, Mass.

Goldberg, R. B., Galau, G. A., Britten, R. J., and Davidson, E. H. (1973). Nonrepetitive DNA sequence representation in sea urchin embryo messenger RNA. *Proc. Natl. Acad. Sci. USA* **70**:3516.

Goldschmidt, R. (1938). *Physiological Genetics.* McGraw-Hill, New York.

Goldschmidt, R. (1940). *The Material Basis of Evolution.* Yale University Press, New Haven, Conn.

Goodwin, B. C. (1963) *Temporal Organization in Cells.* Academic Press, London.

Goodwin, B. C. (1976). *Analytical Physiology of Cells and Developing Organisms.* Academic Press, New York.

Goodwin, B. C (1990). Structuralism in biology. *Sci. Prog.* **74**:227.

Goodwin, B. C., and Kauffman, S. A. (1989). Bifurcation, harmonics and the four color wheel model of *Drosophila* development. In *Cell-to-cell signalling: From experiment to theoretical models* (A. Goldbeter, ed.). Academic Press, San Diego, Calif.

Goodwin, B. C., and Kauffman, S. A. (1990). Spatial harmonics and pattern specification in early *Drosophila* development. Part I: Bifurcation sequences and gene expression. *J. Theoret. Biol.* **144**:303.

Goodwin, B. C., and Trainor, L.E.H. (1980). A field description of the cleavage process in embryogenesis. *J. Theoret. Biol.* **85**:757.

Goodwin, B. C., and Trainor, L.E.H. (1983). The ontogeny and phylogeny of the pentadactyl limb. In *Development and Evolution* (B. C. Goodwin, N. Holder, and C. C. Wylie, eds.),

British Society for Developmental Biology, Symposium, no. 6. Cambridge University Press, Cambridge.

Goodwin, B. C., and Trainor, L.E.H. (1985). Tip and whorl morphogenesis in *Acetabularia* by calcium-regulated strain fields. *J. Theoret. Biol.* **117**:79.

Goodwin, B. C., Kauffman, S. A., and Murray, J. D. (1992). Is morphogenesis inherently robust? *J. Theoret. Biol.*, submitted.

Gordon, M., and Temple, W. B. (1976). The graph-like state of matter and polymer science. In *Chemical Applications of Graph Theory* (A. T. Balaban, ed.). Academic Press, London.

Gould, S. J. (1977). *Ontogeny and Phylogeny.* Harvard University Press, Cambridge, Mass.

Gould, S. J. (1989). *Wonderful Life.* Harvard University Press, Cambridge, Mass.

Gould, S. J., and Lewontin, R. C. (1979). The spandrels of San Marco and the panglossian paradigm: A critique of the adaptationist programme. *Proc. Roy. Soc. Lond. B* **205**:581.

Gould, S. J., Gilinsky, N. L., and German, R. Z. (1987). Asymmetry of lineages and the direction of evolutionary time. *Science* **236**:1437.

Grassberger, P., and Procaccia, I. (1983). Measuring the strangeness of strange attractors. *Physica* **9D**:189.

Grayeb, J., Kimura, H., Takahara, M., Hsiung, H., Masui, Y., and Inouye, M. (1984). Secretion cloning vectors in *Escherichia coli. EMBO J.* **3**:2437.

Green, M. M. (1980). Transposable elements in *Drosophila* and other diptera. *Ann. Rev. Genet.* **14**:109.

Green, P. B. (1987). Inheritance of pattern: Analysis from phenotype to gene. *Am. Zool.* **27**:657.

Greenberg, R. M., and Adler, P. N. (1982). Protein synthesis and accumulation in *Drosophila melanogaster* imaginal discs: Identification of a protein with a nonrandom spatial distribution. *Dev. Biol.* **89**:273.

Greene, M. I., and Nisonoff, A., eds. (1984). *The Biology of Idiotypes.* Plenum, New York.

Greenleaf, A. L., Weebs, J. R., Voelker, R. A., Ohnishi, S., and Kickson, B. (1980). Genetic and biochemical characterization of mutants at an RNA polymerase II locus in *Drosophila melanogaster. Cell* **21**:785.

Greenspan, N. S., and Roux, K. H. (1988). Categories of idiotype overlap and anti-idiotypic mimicry of antigen. In *Theoretical Immunology II: Santa Fe Institute Studies in the Sciences of Complexity* (A. S. Perelson, ed.). Addison-Wesley, Reading, Mass.

Greenwald, I. (1987). The *lin-12* locus of *Caenorhabditis elegans. BioEssays* **6**:70.

Grossberg, S. (1987). *The Adaptive Brain.* Elsevier, Amsterdam.

Grosschedl, R., and Birnstiel, M. L. (1980). Spacer DNA upstream from the TATAATA sequence is essential for promotion of H2A histone gene transcription *in vivo. Proc. Natl. Acad. Sci. USA* **77**:7102.

Groudine, M., and Weintraub, H. (1981). Activation of globin genes during chick development. *Cell* **24**:393.

Grunstein, M., and Hogness, D. S. (1975). Colony hybridization: A method for the isolation of cloned DNAs that contain a specific gene. *Proc. Natl. Acad. Sci. USA* **72**:3961.

Gurdon, J. B. (1974). *The Control of Gene Expression in Animal Development.* Harvard University Press, Cambridge, Mass.

Gutzeit, H. O., and Gehring, W. J. (1979). Localized synthesis of specific proteins during oogenesis and early embryogenesis in *Drosophila melanogaster. Roux's Arch. Dev. Biol.* **187**:151.

Haas, M. J., and Davies, J. D. (1981). *Molecular Basis of the Drug Action,* North Holland, Amsterdam.

Hadorn, E. (1966). In *Major Problems in Developmental Biology* (M. Locke, ed.). Academic Press, New York.

Hadorn, E. (1967). Dynamics of determination. *Symp. Dev. Biol.* **25**:85.

Hadorn, E. (1978). Transdetermination. In *The Genetics and Biology of Drosophila* (M. Ashburner and T.R.F. Wright, eds.), vol 2C. Academic Press, New York.

Hafen, E., Kurowa, A., and Gehring, W. J. (1984). Spatial distribution of transcripts from the segmentation gene *fushi tarazu* during *Drosophila* embryonic development. *Cell* **37**:833.

Hafen, E., Levine, M., Garber, R. L., and Gehring, W. J. (1983). An improved *in situ* hybridization method for the detection of cellular RNAs in *Drosophila* tissue sections, and its application for localizing transcripts of the homeotic *Antennapedia* gene complex. *EMBO J.* **2**:617.

Haldane, J.B.S. (1932). *The Forces of Evolution.* Longmans Green, London.

Hall, B. G. (1976). Experimental evolution of a new enzymatic function: Kinetic analysis of the ascestral (*ebg°*) and evolved (*ebg⁺*) enzymes. *J. Mol. Biol.* **107**:71.

Hall, B. G. (1978). Experimental evolution of a new enzymatic function: II. Evolution of multiple functions for *EBG* enzyme in *E. coli. Genetics* **89**:453.

Hall, B. G., and Zuzel, T. (1980). Evolution of a new enzymatic function by recombination within a gene. *Proc. Natl. Acad. Sci. USA* **77**:3529.

Hamming, R. W. (1986). *Coding and Information Theory.* 2d ed. Prentice-Hall, Englewood Cliffs, N.J.

Harary, F. (1969). *Graph Theory.* Addison-Wesley, Reading, Mass.

Harary, F., Norman, R. Z., and Cartwright, D. (1975). *Structural Models: An Introduction to the Theory of Directed Graphs.* Wiley, New York.

Harding, K. (1988). Spatial regulation of homeobox genes of *Drosophila.* Ph.D. diss., Columbia University, New York.

Harding, K., and Levine, M. (1988). Gap genes define the limits of *Antennapedia* and *Bithorax* gene expression during early development in *Drosophila. EMBO J.* **7**:205.

Harding, K., Rushlow, C., Doyle, H. J., Hoey, T., and Levine, M. (1986). Cross-regulatory interactions among pair-rule genes in *Drosophila. Science* **233**:953.

Hargreaves, W. R., and Deamer, D. W. (1978). Origin and early evolution of bilayer membranes. In *Light Transducing Membranes: Structure and Function* (D. W. Deamer, ed.). Academic Press, New York.

Hargreaves, W. R., Mulvihill, S. J., and Deamer, D. W. (1977). Synthesis of phospopholipids and membranes in prebiotic conditions. *Nature* **266**:78.

Harrison, R. G. (1918). Experiments on the development of the forelimb of *Amblystoma,* a self-differentiating, equipotential system. *J. Exp. Zool.* **25**:413.

Harrison, R. G. (1921). On relations of symmetry in transplanted limbs. *J. Exp. Zool.* **32**:1.

Hartley, I., Altosaar, J. M., Dothie, J. M., and Neuberger, M. S. (1976). Experimental evolution of a xylitol dehydrogenase. In *Structure–Function Relationships of Proteins* (R. Markham and R. W. Horne, eds.). North Holland, Amsterdam.

Hartman, H., and Vichniac, G. Y. (1986). Inhomogeneous cellular automata. In *Disordered Systems and Biological Organization* (E. Bienenstock, F. Fogelman-Soulie, and G. Weisbuch, eds.). Series F: *Computer and Systems Sciences,* vol. 20. Springer, New York.

Hartwell, L., Culotti, J., Pringle, J., and Reid, B. (1974). Genetic control of the cell division cycle in yeast. *Science* **183**:46.

Hastie, N. D., and Bishop, J. O. (1976). The expression of three abundance classes of messenger RNA in mouse tissues. *Cell* **9**:761.

Haynie, J. L., and Bryant, P. J. (1976). Intercalary regeneration in imaginal wing disk of *Drosophila melanogaster. Nature* **259**:659.

Hayward, W., Neel, B. G., and Astrin, S. (1981). Activation of a cellulr *onc* gene by promoter insertion on ALV-induced lymphoid leukosis. *Nature* **290**:475.

Hebb, D. O. (1949). *The Organization of Behavior.* Wiley, New York.

Heinrich, G., Traunecker, A., and Tonegawa, S. (1984). Somatic mutation creates diversity in the major group of mouse immunogolulin K light chains. *J. Exp. Med.* **159**:417.

Hennig, W. (1966). *Phylogenetic Systematics* (D. D. Davis and R. Zangerl, trans.). University of Illinois Press, Urbana.

Hermes, J. D., and Kowles, J. R. (1988). The search for catalytic improvement in an enzyme: The generation of and the selection from pools of random mutants, unpublished manuscript.

Hermes, J. D., Blacklow, S. C., and Knowles, J. R. (1987). The development of enzyme catalytic efficiency: An experimental approach. In *Cold Spring Harbor Symposia on Quantitative Biology,* vol. 52. Cold Spring Harbor Laboratory, Cold Spring Harbor, N.Y.

Hermes, J. D., Blacklow, S. C., and Knowles, J. R. (1990). Searching sequence space by definably random mutagenesis: Improving the catalytic potency of an enzyme. *Proc. Natl. Acad. Sci. USA* **87**:696.

Hermes, J. D., Blacklow, S. C., Gallo, K. A., Bauer, A. J., and Knowles, J. R., (1987). The evolution of enzyme function. In *Protein Structure and Design* (D. Opender, ed.). Liss, New York.

Hershey, A. D., ed. (1971). *The Bacteriophage Lambda,* Cold Spring Harbor Laboratory. Cold Spring Harbor, N.Y.

Hilhorst, H. J., and Nijmeijer, M. (1987). On the approach of the stationary state in Kauffman's random Boolean network. *J. Physique* **48**:185.

Hinton, G. E., and Sejnowski, T. J. (1986). Learning and relearning in Boltzmann machines.

Hinton, G., Sejnowski, T., and Ackley, D. (1984). Boltzmann machines: Constraint satisfaction networks that learn. Report CMU-CS-84-119, Department of Computer Science, Carnegie-Mellon University, Pittsburgh.

Hirsh, M. W., and Smale, S. (1974). *Differential Equations, Dynamical Systems and Linear Algebra.* Academic Press, New York.

Ho, M. W., Bolton, E., and Saunders, P. T. (1983). The bithorax phenocopy and pattern formation. I: Spatiotemporal characteristics of the phenocopy response. *Exp. Cell Biol.* **51**:282.

Ho, M. W., Matheson, A., Saunders, P. T., Goodwin, B. C., and Smallcombe, A. (1987). Ether-induced segmentation disturbances in *Drosophila melanogaster. Roux's Arch. Dev. Biol.* **196**:511.

Ho, M. W., Tucker, C., Keeley, D., and Saunders, P. T. (1983). Effects of successive generations of ether treatment on penetrance and expression of bithorax phenocopy in *Drosophila melanogaster. J. Exp. Zool.* **225**:357.

Hoffman, G. W., Kion, T. A., Forsyth, R. B., Saga, K. G., and Cooper-Willis, A. (1988). The *N*-dimensional network. In *Theoretical Immunology II: Santa Fe Institute Studies in the Sciences of Complexity* (A. S. Perelson, ed.). Addison-Wesley, Reading, Mass.

Holden, A. V. (ed.) (1986). *Chaos.* Princeton University Press, Princeton, N.J.

Holder, N., Tank, P., and Bryant, S. V. (1980). Regeneration of symmetrical forelimbs in the axolotl, *Ambystoma mexicanum. Dev. Biol.* **74**:302.

Holland, J. H. (1960). Cycles in logical nets. *J. Franklin Institute* **270**:202.

Holland, J. H. (1981). Genetic algorithms and adaptation. Technical report, no. 34. Department of Cognitive Sciences, University of Michigan, Ann Arbor.

Holland, J. H. (1986). Escaping brittleness: The possibilities of general purpose learning algorithms applied to parallel rule-based systems. In *Machine Learning,* (R. S. Michalski, J. G. Carbonell, and T. M. Mitchell, eds.), vol. 2. Morgan Kaufmann, Los Altos, Calif.

Holland, P.W.H., and Hogan, B.L.M. (1986). Phylogenetic distribution of *Antennapedia*-like homeoboxes. *Nature* **321**:251.

Holm, D. M., and Cram, L. S. (1973). An improved flow microfluorometer for rapid measurements of cell flourescence. *Exp. Cell. Res.* **80**:105.

Honjo, T. (1983). Immunoglobulin genes. *Ann. Rev. Immunol.* **1**:499.

Hopfield, J. J. (1982a). Neural networks and physical systems with emergent collective computational abilities. *Proc. Natl. Acad. Sci USA* **79**:2554.

Hopfield, J. J. (1982b). Neural networks and physical systems with emerging collective computational ability. *Proc. Natl. Acad Sci. USA* **83**:1847.

Hopfield, J. J. (1984). Neurons with graded responses have collective computational properties like those of two-state neurons. *Proc. Natl. Acad. Sci. USA* **81**:3088.

Hopfield, J. J., and Tank. D. W. (1986a). Computing with neural circuits: A model. *Science* **233**:625.

Hopfield, J. J., and Tank, D. W. (1986b). Collective computation with continuous variables. In *Disordered Systems and Biological Organization* (E. Bienenstock, F. Fogelman-Soulie, and G. Weisbuch, eds.), Series F: *Computer and Systems Sciences*, vol. 20. Springer, New York.

Horowitz, N. H. (1945). On the evolution of biochemical synthesis. *Proc. Natl. Acad. Sci USA* **31**:453.

Horowitz, N. H., and Metzenberg, R. L. (1965). Biochemical aspects of genetics. *Ann. Rev. Biochem.* **34**:527.

Horwitz, M.S.Z., and Loeb, L.A. (1986). Promoters selected from random DNA sequences. *Proc. Natl. Acad. Sci. USA* **83**:7405.

Horwitz, M.S.Z., and Loeb, L. A. (1988). DNA sequences of random origin as probes of *E. coli* promoter architecture. *J. Biol. Chem.* **263**:14724.

Horwitz, M.S.Z., Dube, D. K., and Loeb, L. A. (1991). Studies in the evolution of biological activity from random sequences. In *Molecular Evolution on Rugged Landscapes: Santa Fe Institute Studies in the Sciences of Complexity* (S. A. Kauffman and A. S. Perelson, eds.), vol. 9. Addison-Wesley, Reading, Mass.

Hough-Evans, B. R., Ernst, S. G., Britten, R. J., and Davidson, E. H. (1979). RNA complexity in developing sea urchin oocytes. *Dev. Biol.* **69**:258.

Hough-Evans, B. R., Smith, J. J., Britten, R. J., and Davidson, E. H. (1975). Sequence complexity of heterogeneous nuclear RNA in sea urchin embryos. *Cell* **5**:291.

Houghten, R. A. (1985). General method for the rapid solid-phase synthesis of large numbers of peptides: Specificity of antigen-antibody interaction at the level of individual amino acids. *Proc. Natl. Acad. Sci. USA* **82**:5131.

Houghton, R. A., Pinilla, C., Blondelle, S. E., Appel, J. R., Dooley, C. T., and Cuervo, J. H. (1991). Generation and use of synthetic peptide combinatorial libraries for basic research and drug discovery. *Nature* **354**:84.

Howard, K., and Ingham, P. (1986). Regulatory interactions between the segmentation genes *fushi-tarazu, hairy,* and *engrailed* in the *Drosophila* blastoderm. *Cell* **44**:949.

Hoyle, F., and Wickramasinghe, N. C. (1981). *Evolution from Space.* Dent, London.

Hubby, J. L., and Lewontin, R. C. (1966). A molecular approach to the study of genic heterozygosity in natural populations. I: The number of alleles at different loci of *Drosophila pseudoobscura. Genetics* **54**:577.

Hudson, R. R. (1983). Testing the constant-rate neutral allele model with protein sequence data. *Evolution* **37**:203.

Hunding, A. (1987). Bifurcations in Turing systems of the second kind may explain blastula cleavage plane orientation. *J. Math. Biol.* **25**:109.

Hunding, A., Kauffman, S. A., and Goodwin, B. C. (1990). *Drosophila* segmentation: Supercomputer simulation of prepattern hierarchy. *J. Theoret. Biol.* **145**:369.

Hunt, R. K. (1975). Developmental programming for retinotectal patterns. In *Cell Patterning* (R. Porter and J. Rivers, eds.), *CIBA Foundation Symposium*, vol. 29. Associated Scientific, Amsterdam.

Hurt, E. C., and van Loon, A.P.G.M. (1986). How proteins find mitochondria and intramitochondrial compartments. *Trends Biochem. Sci.* **11**:204.

Huse, W. D., Sastry, L., Iverson, S. A., Kang, A. S., Alting-Mees, M., Burton, D. R., Benkovic, S. J., and Lerner, R. A. (1989). Generation of a large combinatorial library of the immunoglobulin repertoire in phage lambda. *Science* **246**:1275.

Hutchins, C. J., Rathjen, P. D., Forster, A. C., and Symons, R. H. (1986). Self-cleavage of plus and minus RNA transcripts of avocado sunblotch viroid. *Nucl. Acid Res.* **14**:3627.

Huxley, J. S. (1932). *Problems of Relative Growth.* Methuen, London.

Huxley, J. S. (1972). *Problems of Relative Growth.* 2d ed. Dover, New York.

Illmensee, K. (1972). Developmental potencies of nuclei from cleavage, preblastoderm and syncytial blastoderm transplanted into unfertilized eggs. *Roux's Arch. Dev. Biol.* **170**:267.

Illmensee, K. (1976). Nuclear and cytoplasmic transplantation in *Drosophila.* In *Insect Development.* Royal Entomological Society Symposium, no. 8. Blackwell, Oxford.

Illmensee, K., and Mahowald, A. P. (1974). Transplantation of posterior polar plasm in *Drosophila:* Induction of germ cells at the anterior pole of the egg. *Proc. Natl. Acad. Sci. USA* **71**:1016.

Illmensee, K., and Mahowald, A. P. (1976). The autonomous function of germ plasm in a somtic region of the *Drosophila* egg. *Exp. Cell Res.* **97**:127.

Ingham, P. W., Howard, K. R., and Ish-Horowicz, D. (1985). Transcription pattern of the *Drosophila* segmentation gene *hairy. Nature* **318**:439.

Ingham, P. W., Ish-Horowicz, D., and Howard, K. R. (1986). Correlative changes in homoeotic and segmentation gene expression in *Kruppel* mutant embryos of *Drosophila. EMBO J.* **5**:1659.

Inoue, T., and Orgel, L. E. (1983). A nonenzymatic RNA polymerase model. *Science* **219**:859.

Irvine, D., Tuerk, C., and Gold L. (1991). SELEXION: Systematic evolution of ligands by exponential enrichment with integrated optimization by non-linear analysis. *J. Mol. Biol.* **222**:739.

Jablonski, D., and Bottjer, D. J. (1988). The ecology of evolutionary innovation. In *Evolutionary Innovations* (M. H. Nitecki, ed.). University of Chicago Press, Chicago.

Jackle, H., Tautz, D., Schuh, R., Seifert, E., and Lehmann, R. (1986). Cross-regulatory interactions among the gap genes of *Drosophila. Nature* **324**:668.

Jackson, E. R., Johnson, D., and Nash, W. G. (1986). Gene networks in development. *J. Theoret. Biol.* **119**:379.

Jacob, F. (1977a). Evolution and tinkering. *Science* **196**:1161.

Jacob, F. (1977b). Mouse teratocarcinoma and embryonic antigens. *Immunological Rev.* **33**:7.

Jacob, F. (1982). *The Possible and the Actual.* Pantheon, New York.

Jacob, F. (1983). Molecular tinkering in evolution. In *Evolution from Molecules to Men* (D. S. Rondall, ed.). Cambridge University Press, Cambridge.

Jacob, F., and Monod, J. (1961). On the regulation of gene activity. In *Cold Spring Harbor Symposia on Quantitative Biology,* vol. 26, Cold Spring Harbor Laboratory, Cold Spring Harbor, N.Y.

Jacob, F., and Monod, J. (1963). Genetic repression, allosteric inhibition and cellular differentiation. In *Cytodifferentiation and Macromolecular Synthesis* (M. Locke, ed.). Academic Press, New York.

Jacobs, J., Schultz, P.G., Sugasawara, R., and Powell, M. (1987). Catalytic antibodies. *J. Am. Chem. Soc.* **109**:2174.

Jacobson, A. G. (1966). Inductive processes in embryonic development. *Science* **152**:25.

Jacobson, A. G., Odell, G. M., and Oster, G. F. (1985). The cortical tractor model for epithelial folding: Application to the neural plate. In *Molecular Determinants of Animal Form* (G. M. Edelman, ed.). Liss, New York.

Jacobson, A. G., Oster, G. F., Odell, G. M., and Cheng, L. Y. (1986). Neurulation and the cortical tractor model for epithelial folding. *J. Embryol. Exp. Morph.* **96**:19.

Jaffe, L. F. (1979). Control of development by ionic currents. In *Membrane Transduction Mechanisms* (R. A. Cone and J. E. Dowling, eds.). Raven Press, New York.

Jaffe, S. (1988). Kauffman networks: Cycle structure of random clocked boolean networks. Ph.D. diss., New York University.

Jahnke, E., Emde, F., and Lusch, F. (1960). *Table of Higher Mathematical Functions.* McGraw-Hill, New York.

Janning, W. (1974). Entwicklungs-genetische untersuchungen an gynandern von *Drosophila melanogaster.* I: Deinneren organe der imago. *Roux's Arch. Dev. Biol.* **174**:313.

Janning, W. (1978). Gynandromorph fate maps in *Drosophila.* In *Genetic Mosaics and Cell Differentiation* (W. Gehring, ed.). Springer, New York.

Jeffries, C. (1974). Qualitative stability and digraphs in model eco-systems. *Ecology* **55**:1415.

Jencks, W. P. (1987). Economics of enzyme catalysis. In *Cold Spring Harbor Symposia on Quantitative Biology,* vol. 52. Cold Spring Harbor Laboratory, Cold Spring Harbor, N.Y.

Jerne, N. K. (1974). Toward a network theory of the immune system. *Ann. Immunol. (Inst. Pasteur)* **125C**:373.

Jerne, N. K. (1984). Idiotypic networks and other preconceived ideas. *Immunol. Rev.* **79**:5.

Johnson, D. S., and Papadimitriou, C. H. (1985). Computational complexity. In *The Traveling Salesman Problem* (E. L. Lawler, J. K. Lenstra, A. H. Rinnoykan, and D. B. Shimoys, eds.). Wiley-Interscience, Chicester, England.

Joyce, G. F. (1987). Nonenzymatic template-directed synthesis of informational macromolecules. In *Cold Spring Harbor Symposia on Quantitative Biology,* vol. 52. Cold Spring Harbor Laboratory, Cold Spring Harbor, N.Y.

Joyce, G. (1989). RNA evolution and the origins of life. *Nature* **338**:217.

Joyce, G. F., and Orgel, L. E. (1986). Non-enzymatic template-directed synthesis on RNA random copolymers: Poly-(C,G) templates. *J. Mol. Biol.* **188**:433.

Joyce, G. F., Inoue, T., and Orgel, L. E. (1984). Non-enzymatic template-directed synthesis on RNA random copolymers: Poly(C,U) templates. *J. Mol. Biol.* **176**:279.

Joyce, G. F., Schwartz, A. W., Miller, S. 1., and Orgel, L. E. (1987). The case for an ancestral genetic system involving simple analogues of the nucleotides. *Proc. Natl. Acad. Sci. USA* **84**:4398.

Jungck, J. R., and Fox, S. W. (1973). Synthesis of oligonucleotides by proteinoid microspheres acting on ATP. *Naturwissenschaften* **60**:425.

Jurka, J., and Smith, T. F. (1987). B turns in early evolution: Chirality, genetic code, and biosynthetic pathways. In *Cold Spring Harbor Symposia on Quantitative Biology,* vol. 52. Cold Spring Harbor Laboratory, Cold Spring Harbor, N.Y.

Kaartinen, M., Griffiths, G. M., Markham, A. F., and Milstein, C. (1983). mRNA sequences define an unusually restricted IgG response to 2-phenyloxazolone and its early diversification. *Nature* **304**:320.

Kabat, E. E. (1976). *Structural Concepts in Immunology and Immunochemistry.* Holt, Rinehart and Winston, New York.

Kabat, E. A., Wu, T. T., and Biolofsky, H. (1979). *Sequence of Immunoglobulin Chains.* National Institutes of Health, Bethesda, Md.

Kacser, H., and Beeby, R. (1984). Evolution of catalytic proteins or on the origin of enzyme species by means of natural selection. *J. Mol. Evol.* **20**:38.

Kafatos, F. C. (1983). Structure, evolution, and developmental expression of the chorion multigene families in silkmoths and *Drosophila.* In *Gene Structure in Regulation and in Development* (S. Subtelny and F. C. Kafatos, eds.). Liss, New York.

Kaiser, C. A., Preuss, D., Grisafi, P., and Botstein, D. (1987). Many random sequences functionally replace the secretion signal sequence of yeast invertase. *Science* **235**:312.

Kaltohoff, K. (1979). In *Determinants of Spatial Organization* (W. Gehring, ed.). Springer, New York.

Kalthoff, K. (1983). Cytoplasmic determinants in dipteran eggs. In *Time, Space and Pattern in Embryonic Development* (R. A. Raff and W. Jeffreys, eds.). Liss, New York.

Kalthoff, K., and Sander, K. (1968). Der Entwicklungsgang der Missbildung "Doppelabdomen" impartiell UV-bestrahlten Ei von *Smitia parthenogenetica* (Dipt. Chironomidae). *Roux's Arch. Dev. Biol.* **161**:129.

Kamalay, J. C., and Goldberg, R. B. (1980). Regulation of structural gene expression in tobacco. *Cell* **19**:935.

Kandler-Singer, I., and Kalthoff, K. (1976). RNAase sensitivity of an anterior morphogenetic determinant in an insect egg (*Smithia* sp., Chironomidae, Diptera). *Proc. Natl. Acad. Sci. USA* **73**:3739.

Karlsson, J. (1984). Morphogenesis and compartments in *Drosophila*. In *Pattern Formation* (G. M. Malacinski and S. V. Bryant, eds.). Macmillan, New York.

Karpen, G. H., and Shubiger, G. (1981). Extensive regulatory capabilities of a *Drosophila* imaginal disk blastema. *Nature* **294**:744.

Karplus, M., and Kushick, J. N. (1983). Dynamics of proteins: Elements and function. *Ann. Rev. Biochem.* **53**:263.

Karplus, M., Brunger, A. T., Elber, R., and Kuriyan, J. (1987). Molecular dynamics: Applications to proteins. In *Cold Spring Harbor Symposia on Quantitative Biology*, vol. 52. Cold Spring Harbor Laboratory, Cold Spring Harbor, N.Y.

Kauffman, S. A. (1969). Metabolic stability and epigenesis in randomly connected nets. *J. Theoret. Biol.* **22**:437.

Kauffman, S. A. (1971a). Gene regulation networks: A theory for their global structure and behavior. *Current Topics in Dev. Biol.* **6**:145.

Kauffman, S. A. (1971b). Cellular homeostasis, epigenesis and replication in randomly aggregated macromolecular systems. *J. Cybernetics* **1**:71.

Kauffman, S. A. (1971c). Differentiation of malignant to benign cells. *J. Theoret. Biol.* **31**:429.

Kauffman, S. A. (1973). Control circuits for determination and transdetermination. *Science* **181**:310.

Kauffman, S. A. (1974). The large-scale structure and dynamics of gene control circuits: An emsemble approach. *J. Theoret. Biol.* **44**:167.

Kauffman, S. A. (1975). Control circuits for determination and transdermination: Interpreting positional information in a binary epigenetic code. In *Cell Patterning*. (R. Porter and J. Rivers, eds.), *CIBA Foundation Symposium*, vol. 29. Associated Scientific, Amsterdam.

Kauffman, S. A. (1978). A cartesian coordinate model of positional information in imaginal discs of *Drosophila*. Presented at the 20th Annual *Drosophila* Conference, Chicago.

Kauffman, S. A. (1979). The compartmental and combinatorial code hypothesis in *Drosophila* development. *Bioscience* **29**:581.

Kauffman, S. A. (1980). Heterotopic transplantation in the syncytial blastoderm of *Drosophila*: Evidence for anterior and posterior nuclear commitments. *Roux's Arch. Dev. Biol.* **189**:135.

Kauffman, S. A. (1981). Pattern formation in the *Drosophila* embryo. *Philos. Trans. Roy. Soc. London B* **295**:567.

Kauffman, S. A. (1983). Developmental constraints: Internal factors in evolution. In *Development and Evolution* (B. C. Goodwin, N. Holder, and C. G. Wylie, eds.). Cambridge University Press, Cambridge.

Kauffman, S. A. (1984a). Emergent properties in random complex automata. *Physica* **10D**:145.

Kauffman, S. A. (1984b). Pattern generation and regeneration. In *Pattern Formation* (G. M. Malacinski and S. V. Bryant, eds.). Macmillan, New York.

Kauffman, S. A. (1984c). Bifurcations in insect morphogenesis. In *Aspects of Chemical Evolution* (G. Nicolis, ed.). Wiley, New York.

Kauffman, S. A. (1985a). Self-organization, selective adaptation and its limits: A new pattern of inference in evolution and development. In *Evolution at the Crossroads* (J. DePew and B. H. Weber, eds.). MIT Press, Cambridge, Mass.

Kauffman, S. A. (1985b). New questions in genetics and evolution. *Cladistics* **1**:247.

Kauffman, S. A. (1986c) Developmental logic and its evolution. *BioEssays* **6**:82.

Kauffman, S. A. (1986b). Autocatalytic sets of proteins, *J. Theoret. Biol.* **119**:1.

Kauffman, S. A. (1986c). A framework to think about regulatory systems. In *Integrating Scientific Disciplines* (W. Bechtel, ed.). Martinus Nijhoff, Dordrecht.

Kauffman, S. A. (1986d). Boolean systems, adaptive automata, evolution. In *Disordered Systems and Biological Organization* (E. Bienenstock, F. Fogelman-Soulie, and G. Weisbuch, eds.), Series F: *Computer and Systems Sciences,* vol. 20. Springer, New York.

Kauffman, S. A. (1987b). Problems and paradigms: Developmental logic and its evolution. *BioEssays* **6**:82.

Kauffman, S. A. (1988). The evolution of economic webs. In *The Economy as an Evolving Complex System: Santa Fe Institute Studies in the Sciences of Complexity* (P. W. Anderson, J. Arrow, and D. Pines, eds.), vol. 5. Addison-Wesley, Reading, Mass.

Kauffman, S. A. (1989a). Adaptation on rugged fitness landscapes. In *Lectures in the Sciences of Complexity* (E. Stein, ed.). Addison-Wesley, Reading, Mass.

Kauffman, S. A. (1989b). Principles of adaptation in complex systems. In *Lectures in the Science of Complexity* (E. Stein, ed.). Addison-Wesley, Reading, Mass.

Kauffman, S. A. (1991). Antichaos and adaptation. *Scientific American* **265**:78.

Kauffman, S. A. (1992). Applied molecular evolution. *J. Theoret. Biol.* **157**:1.

Kauffman, S. A. (1993). Whispers from Carnot. In *Integrative Themes: A Proceedings Volume in the Sante Fe Institute Studies in the Sciences of Complexity* (G. Cowan and D. Pines, eds.), vol. 19. Addison-Wesley, Reading, Mass.

Kauffman, S. A., and Ballivet M. (1985). Process for obtaining DNA, RNA, peptides, polypeptides, or proteins by recombinant DNA techniques. International patent applications.

Kauffman, S. A., and Goodwin, B. C. (1990). Spatial harmonics and pattern specification in early *Drosophila* development. Part II: The four color wheel model. *J. Theoret. Biol.* **144**:321.

Kauffman, S. A., and Johnsen, S. (1991). Coevolution to the edge of chaos: Coupled fitness landscapes, poised states, and coevolutionary avalanches. *J. Theoret. Biol.* **149**:467.

Kauffman, S. A., and Johnsen, S. (1992). Coevolution to the edge of chaos: Coupled fitness landscapes, poised states, and coevolutionary avalanches. In *Artificial Life II: A Proceedings Volume in the Santa Fe Institute Studies in the Sciences of Complexity* (C. G. Langton, J. D. Farmer, S. Rasmussen, and C. Taylor, eds.), vol. 10. Addison-Wesley, Reading, Mass.

Kauffman, S. A., and Levin, S. (1987). Towards a general theory of adaptative walks on rugged landscapes. *J. Theoret. Biol.* **128**:11.

Kauffman, S. A., and Ling, E. (1980). Timing and heritability of the *nasobemia* transformation in *Drosophila. Roux's Arch. Dev. Biol.* **189**:147.

Kauffman, S. A. and Ling, E. (1981). Regeneration by complementary wing disc fragments of *Drosophila malanogaster. Dev. Biol.* **82**:238.

Kauffman, S. A., and Shymko, R. M. (1982). Mitotic delay following inhibition by 5'-fluorodeoxyuridine of S-phase in *Physarum* is not due to delay in termination of s-phase. *J. Cell Sci.* **53**:143.

Kauffman, S. A., and Smith, R. G. (1986). Adaptive automata based on Darwinian selection. *Physica* **22D**:68.

Kauffman, S. A., and Stein, D. (1989). Application of the *NK* model of rugged landscapes to protein evolution and protein folding. Abstract of meeting of the American Association for the Advancement of Science on protein folding..

Kauffmn, S. A., and Weinberger, E. D. (1989). The *NK* model of rugged fitness landscapes and its application to maturation of the immune response. *J. Theoret. Biol.* **141**:211.

Kauffman, S. A., and Weinberger, E. D. (1991). The *NK* model of rugged fitness landscapes and its application to the maturation of the immune response. *J. Theoret. Biol.* **141**:211.

Kauffman, S. A., and Wille, J. J. (1975). The mitotic oscillator in *Physarum polycephalum. J. Theoret. Biol.* **55**:47.

Kauffman, S. A., Shymko, R., and Trabert, K. (1978). Control of sequential compartment formation in *Drosophila. Science* **199**:259.

Kauffman, S. A., Weinberger, E. D., and Perelson, A. S. (1988). Maturation of the immune response via adaptive walks on affinity landscapes. In *Theoretical Immunology I: Santa Fe Institute Studies in the Sciences of Complexity* (A. S. Perelson, ed.). Addison-Wesley, Reading, Mass.

Kaufman, M. (1988). Role of multistability in an immune response model: A combined discrete and continuous approach. In *Theoretical Immunology I: Sante Fe Institute Studies in the Sciences of Complexity* (A. S. Perelson, ed.). Addison-Wesley, Reading, Mass.

Kaufman, M., and Thomas, R. (1987). Model analysis of the bases of multistationarity in the humoral immune response. *J. Theoret. Biol.* **129**:141.

Kaufman, M., Urbain, J., and Thomas, R. (1985). Towards a logical analysis of the immune response, *J. Theoret. Biol.* **114**:527.

Kaufman, T. C. (1983). The genetic regulation of segmentation in *Drosophila melanogaster.* In *Time, Space and Pattern in Embryonic Development* (R. A. Raff and W. Jeffreys, eds.). Liss, New York.

Kaufman, T. C., and Abbot, M. K. (1984). Homeotic genes and the specification of segment identity in the embryo and adult thorax of *Drosophila melanogaster.* In *Molecular Aspects of Development* (G. M. Malacinski and W. H. Klein, eds.). Plenum, New York.

Kay, P. S., and Inoue, T. (1987) Implications of intermolecularly catalyzed reactions by the *Tetrahymena* ribozyme. In *Cold Spring Harbor Symposia on Quantitative Biology,* vol. 52. Cold Spring Harbor Laboratory, Cold Spring Harbor, N.Y.

Kearney, J. F., and Vakil, M. (1986). Functional idiotype networks during B-cell ontogeny. *Ann. Immunol. (Inst. Pasteur)* **137C**:77.

Kennett, R. H., Denis, K. A., Tung, A. S., and Kleinman, N. R. (1978). Hybrid plasmacytoma production: Fusions with adult spleen cells, monoclonal spleen fragments, neonatal spleen cells and human spleen cells. *Current Topics Microbiol. Immunol.* **81**:77.

Kenyon, D. H., and Steinman, G. (1969). *Biochemical Predestination.* McGraw-Hill, New York.

Kernevez, J. P. (1980). *Enzyme Mathematics.* North Holland, Amsterdam.

Kettler, M. K., Ghent, A. W., and Whitt, G. S. (1986). A comparison of phylogenies based on structural and tissue-expressional differences of enzymes in a family of teleost fishes (Salmoniformes: Umbridae). *Mol. Biol. Evol.* **3**:485.

Kilcherr, R., Baumgartner, S., Bopp, D., Frei, E., and Noll, M. (1986). Isolation of the *paired* gene of *Drosophila* and its spatial expression during early embryogenesis. *Nature* **321**:493.

Kimble, J., and White, J. (1981). On the control of germ cell development in *Caenorhabdites elegans. Dev. Biol.* **81**:208.

Kimura, M. (1983). *The Neutral Theory of Molecular Evolution.* Cambridge University Press, New York.

Kimura, M., and Ohta, T. (1971) *Theoretical Aspects of Population Genetics.* Princeton University Press, Princeton, N.J.

King, M. J., and Wiltse, J. C. (1958). Derivative zeros and other data pertaining to mathieu functions. Johns Hopkins Radiation Laboratory technical report, no. AF57. Johns Hopkins University, Baltimore.

Kirkpatrick, S., and Toulouse, G. (1985). Configuration space analysis of the traveling salesman problem. *J. Physique* **46**:1277.

Kirkpatrick, S., Gelatt, C. D., and Vecci, H. P. (1983). Optimization by simulated annealing. *Science* **220**:671.

Kleene, K. C., and Humphreys, T. (1977). Similarity of hRNA sequences in blastula and pluteus stage sea urchin embryos. *Cell* **12**:143.

Kleene, K. C., and Humphreys, T. (1985). Transcription of similar sets of rare maternal RNAs and rare nuclear RNAs in sea urchin blastulae and adult coelomocytes. *J. Embryol. Exp. Morph.* **85**:131.

Klein, G. (1987). The approaching era of the tumor suppressor genes. *Science* **238**:1539.

Klibanov, A. M., Alberti, B. N., and Marletta, M. A. (1982). Stereospecific oxidation of aliphatic alcohols catalyzed by galactos oxidase. *Biochem. Biophys. Res. Commun.* **10B**:804.

Klibanov, A. M., Berman, Z., and Alberti, B. N. (1981). Preparative hydroxylation of aromatic compounds catalyzed by perosicase. *J. Am. Chem. Soc.* **103**:6263.

Knipple, D. C., Seifert, E., Rosenberg, U. B., Preiss, A., and Jackle, H. (1985). Spatial and temporal patterns of *Kruppel* gene expression in early *Drosophila* embryos. *Nature* **317**:40.

Knowles, J. R. (1987). Tinkering with enzymes: What are we learning? *Science* **236**:1252.

Knudson, A. G. (1987). *Adv. Viral Oncol.* **7**:1.

Kohler, G., and Milstein, C. (1976). Derivation of specific antibody-producing tissue culture and tumor lines by cell fusion. *Eur. J. Immunol.* **6**:511.

Kolakofsky, D., Bellocq, R., and Raju, R. (1987). The translational requirement for la cross virus S-mRNA synthesis. In *Cold Spring Harbor Symposia on Quantitative Biology,* vol. 52. Cold Spring Harbor Laboratory, Cold Spring Harbor, N.Y.

Kollar, E. J., and Fisher, C. (1980). Tooth induction in chick epithelium: Expression of quiescent genes for enamel synthesis. *Science* **207**:993.

Korn, L. (1982). Transcription of *Xenopus* 5 S ribosomal RNA genes. *Nature* **295**:101.

Kornher, J. S., and Brutlag, D. L. (1986). Proximity-dependent enhancement of Sgs-4 gene expression in *D. melanogaster. Cell* **44**:879.

Koshland, D. E. Jr. (1987). Evolution of catalytic function. In *Cold Spring Harbor Symposia on Quantitative Biology,* vol. 52. Cold Spring Harbor Laboratory, Cold Spring Harbor, N.Y.

Krieg, P. A., and Melton, D. A. (1987). An enhancer responsible for activating transcription at the mid-blastula transition in *Xeonopus* development. *Proc. Natl. Acad. Sci. USA* **84**:2331.

Kruger, K., Grabowski, P. J., Zaug, A. J., Sands, J., Gottschling, D. E., and Cech, T. R. (1982). Self-splicing RNA autoexcision and autocyclization of the ribosomal RNA intervening sequences of *Tetrahymena. Cell* **31**:147.

Kuhn, H., and Kuhn, C. (1978). Evolution of a genetic code simulated with a computer. *Origins of Life* **9**:135.

Kuhn, T. S. (1970). *The Structure of Scientific Revolutions.* University of Chicago Press, Chicago.

Kumar, S., Calef, E.,and Szybalski, W. (1970). Regulation of the transcription of *Escherichia coli* phage lambda by its early genes N and *tof.* In *Cold Spring Harbor Symposia on Quantitative Biology,* vol. 35. Cold Spring Harbor Laboratory, Cold Spring Harbor, N.Y.

Kurten, B. (1963). Return of a lost structure in the evolution of the felid dentition. *Soc. Scient. Fenn. Comm. Biol.* **26**:1.

Kurten, K. E. (1988a). Correspondence between neural threshold networds and Kauffman Boolean cellular automata. *J. Phys. A: Math. Gen.* **21**:615.

Kurten, K. E. (1988b). Critical phenomena in model neural networks. *Phys. Lett. A* **129**:157.

LaBean, T. H., Butt, T. R., and Kauffman, S. A. (1992). Evidence of significant secondary structure and collapsed conformations in a random sample of protein sequence space. Presented at the Sixth Symposium of the Protein Society, San Diego, Calif.

LaBean, T. H., Kauffman, S. A., and Butt, T. R. (1992). Design, expression, and characterization of random sequence polypeptides as fusions with ubiquitin. *FASEB J.* **6**:A471.

Lam, K. S., Salmon, S. E., Hersh, E. M., Hruby, V. J., Kazmierski, W. M., and Knapp, R. J. (1991). A new type of synthetic peptide library for identifying ligand-binding activity. *Nature* **354**:82.

Lam, P. M. (1988). A percolation approach to the Kauffman model. *J. Statis. Phys.* **50**:1263.

Lamb, M. M., and Daneholt, B. (1979). Characterization of active transcription units in Balbiani rings of *Chironomus tentans. Cell* **17**:835.

Land, H., Chen, A. C. Morgenstern, J. P., Parada, L. F., and Weinberg, R. A. (1986). Behavior of *myc* and *ras* oncogenes in transformation of rat embryo fibroblasts. *Mol. and Cell Biol.* **6**:1917.

Lande, R. (1978). Evolutionary mechanisms of limb loss in tetrapods. *Evolution* **32**:73.

Langton, C. (1986). Studying artificial life with cellular automata. *Physica* **22D**:120.

Langton, C. (1990). Adaptation to the edge of chaos. Presented at the Second Artificial Life Conference, Santa Fe, N.M.

Langton, C. G. (1990). Computation to the edge of chaos: Phase transitions and emergent computation. *Physica* **42D**:12.

Langton, C. G. (1992). Adaptation to the edge of chaos. In *Artificial Life II: A Proceedings Volume in the Santa Fe Institute Studies in the Sciences of Complexity* (C. G. Langton, J. D. Farmer, S. Rasmussen, and C. Taylor, eds.), vol. 10. Addison-Wesley, Reading, Mass.

Lawless, J. G., and Boynton, C. G. (1973). Thermal synthesis of amino acids from a simulated primitive atmosphere. *Nature* **243**:405.

Lawrence, P. A., and Morata, G. (1976a). The compartment hypothesis. In *Insect Development.* Royal Entomological Society Symposium, no. 8. Blackwell, Oxford.

Lawrence, P. A., and Morata, G. (1976b). Compartments in the wing of *Drosophila:* A study of the engrailed gene. *Dev. Biol.* **50**:321.

Lawrence, P. A., and Morata, G. (1977). The early development of mesothoracic compartments in *Drosophila. Dev. Biol.* **56**:40.

Lawrence, P. A., and Morata, G. (1979). In *Determinants of Spatial Organization* (S. Subtelny and I. R. Konigsberg, eds.). Academic Press, New York.

Leder, P., Battery, J., Lenoir, G., Moulding, C., Murphy, W., Potter, H., Stewart, T., and Taub, R. (1983). Translocations among antibody genes in human cancer. *Science* **222**:765.

Lehman, R., and Nüsslein-Volhard C. (1987). *Hunchback,* a gene required for segmentation of an anterior and posterior region of the *Drosophila* embryo. *Dev. Biol.* **119**:402.

Lenat, D. B. (1977). Automated theory formation in mathematics. In *Proceedings of the Fifth International Joint Conference on Artificial Intelligence.* Cambridge, Mass.

Lenat, D. B. (1980). The heuristics of nature: The plausible mutation of DNA. Report HPP-80-27, Department of Computer Science, Stanford University.

Lenoir, T. (1982). *The Strategy of Life.* Reidel, Dordrecht.

Lesk, A. M., and Clothia, C. (1980). How different amino acid sequences determine similar protein structures: The structure and evolutionary dynamics of the globins. *J. Mol. Biol.* **136**:225.

Leuthausser, I. (1986). An exact correspondence between Eigen's evolution model and a two-dimensional ising system. *J. Chem. Phys.* **94**:1884.

Levin, S. A. (1978). On the evolution of ecological parameters. In *Ecological Genetics: The Interface* (P. F. Brussard, ed.). Springer, New York.

Levin, Y., Berger, A., and Katchalski, E. (1956). Hydrolysis and transpeptidation of lysine peptides by trypsin. *Biochem. J.* **63**:308.

Levine, M., and Harding, K. (1987). Spatial regulation of homeo box gene expression in *Drosophila.* In *Oxford Surveys on Eukaryotic Genes* (N. MacLean, ed.). Oxford University Press, Oxford.

Levins, R. (1977). Qualitative analysis of complex systems. In *Mathmatics and the Life Sciences* (D. E. Mathews, ed.), Lecture Notes in Biomathematics, vol. 18. Springer, New York.

Levinton, J. (1988). *Genetics, Paleontology, and Macroevolution.* Cambridge University Press, Cambridge.

Levy, W. B., and McCarthy, B. J. (1975). Messenger RNA complexity in *Drosophila melanogaster. Biochemistry* **14**:2440.

Lewin, B. (1980). *Gene Expression 2,* 2d ed. Wiley-Interscience, New York.

Lewin, B. (1987). *Genes,* 3d ed. Wiley, New York.

Lewin, R. (1988). A lopsided view of evolution. *Science* **241**:291.

Lewis, E. B. (1978). A gene complex controlling segmentation in *Drosophila. Nature* **276**:565.

Lewis, E. B. (1981). Developmental genetics of the bithorax complex in *Drosophila.* In *Development Biology Using Purified Genes* (D. D. Brown and C. F. Fox, eds.). Academic Press, New York.

Lewis, J. (1981). Simpler rules of epimorphic regeneration: The polar coordinate model without polar coordinates. *J. Theoret. Biol.* **88**:371.

Lewis, J. (1982). Continuity and discontinuity in pattern formation. In *Developmental Order: Its Origin and Regulation* (S. Subtelny, ed.). Liss, New York.

Lewis, R., Wakimoto, B., Denell, R., and Kaufman, T. (1980). Genetic analysis of the *Antennapedia* gene complex (ANT-C) and adjacent chromosomal regions of *Drosophila melanogaster.* II: Polytene chromosome segments 84A-84B1, 2. *Genetics* **95**:383.

Lewontin, R. C. (1974). *The Genetic Basis of Evolutionary Change.* Columbia University Press, New York.

Lewontin, R. C., and Hubby, J. L. (1966). A molecular approach to the study of genic heterozygosity in natural populations. II: Amount of variation and degree of heterozygosity in natural populations of *Drosophila pseudoobscura. Genetics* **54**:595.

Li, W. H. (1983). Evolution of duplicate genes and pseudogenes. In *Evolution of Genes and Proteins.* (M. Nei and R. K. Koehn, eds.). Sinauer, Sunderland, Mass.

Li, W. W. (1989). Mutual information functions of natural language texts. Working paper 89-008, Santa Fe Institute, Santa Fe, N.M.

Lieberman, D., Hoffman-Lieberman, B., and Sachs, L. (1980). Molecular dissection of differentiation in normal and leukemic meyloblasts: Separately programmed pathways of gene expression. *Dev. Biol.* **79**:46.

Lienhrd, G. E. (1973). Enzymatic catalysis and transition-state theory. *Science* **180**:149.

Lin, S., and Kernighan, B. W. (1973). An effective heuristic algorithm for the traveling salesman problem. *Oper. Res.* **21**:498.

Lindsley, D. L., and Grell, E. H. (1968). Genetic variations of *Drosophila melanogaster.* Carnegie Institute, publication no. 627.

Lipmann, F. (1941). Metabolic generation and utilization of phosphate bond energy. *Adv. Enzymol.* **1**:100.

Little, W. A. (1974). The existence of persistent states in the brain. *Math Biosci.* **19**:101.

Liu, C. P., Tucker, P.W., Mushinski, F., and Blattner, F. R. (1980). Mapping of heavy gene chains for mouse immunoglobins. *Science* **209**:1348.

Lohrmann, R., and Orgel, L. E. (1978). Preferential formation of (2'-5')-linked internucleotide bonds in nonenzymatic reactions. *Tetrahedron* **34**:853.

Lohs-Schardin, M., Cremer, C., and Nüsslein-Volhard, C. (1979). A fate map for the larval epidermis of *Drosophila melanogaster:* Localized cuticle defects following irradiation of the blastoderm with an ultraviolet laser microbeam. *Dev. Biol.* **73**:239.

Longberg, N., and Gilbert, W. (1985). Intron/exon structure of the chicken pyruvate kinase gene. *Cell* **4**:81.

Lorenz, E. N. (1963). Deterministic nonperiodic flow. *J. Atmos. Sci.* **20**:130.

Lovejoy, A. O. (1936). *The Great Chain of Being.* Harvard University Press, Cambridge, Mass.

Lowe, C. U., Rees, M. W., and Markham, R. M. (1963). Synthesis of complex organic compounds from simple precursors: Formation of amino acids, amino-acid polymers, fatty acids and purines from ammonium cyanide. *Nature* 199:219.

Lyon, M. F. (1961). Gene activation in the X-chromosome of the mouse (*Mus musculus* L.). *Nature* 190:372.

Ma, J., and Ptashne, M. (1987). A new class of yeast transcriptional activators. *Cell* 51:113.

Maas, A. (1948). Uber die auslosbarkeit von temperaturmodifikationen wahrend der embryoanalentwicklung vor *Drosophila melanogaster. Roux's Arch. Dev. Biol.* 143:515.

MacArthur, R. H., and Wilson, E. O. (1967). *The Theory of Island Biogeography.* Princeton University Press, Princeton, N.J.

MacDonald, N. (1983). *Trees and Networks in Biological Models.* Wiley, New York.

MacDonald, P. M., and Struhl, G. (1986). A molecular gradient in early *Drosophila* embryos and its role in specifying the body pattern. *Nature* 324:537.

MacDonald, P. M., Ingham, P., and Struhl, G. (1986). Isolation, structure and expression of *even-skipped:* A second pair-rule gene of *Drosophila* containing a homeo box. *Cell* 47:721.

MacIntyre, R. J. (1982). Regulatory genes and adaptation: Past, present and future. *Evol. Biol.* 15:247.

Macken, C. A., and Perelson, A. S. (1989). Protein evolution of rugged landscapes. *Proc. Natl. Acad. Sci. USA* 86:6191.

Macklin, M. (1968). Biological phase relationships. *Nature* 217:622.

Maizels, N., and Weiner, A. M. (1987). Peptide-specific ribosomes, genomic tags, and the origin of the genetic code. In *Cold Spring Harbor Symposia on Quantitative Biology,* vol. 52. Cold Spring Harbor Laboratory, Cold Spring Harbor, N.Y.

Maki, R., Roeder, W., Traunecker, A., Sidman, C., Wabl, M., Rasjhke, W., and Tonegawa, S. (1981). The role of DNA rearrangement and alternate mRNA processing in the expression of immunoglobulin delta genes. *Cell* 24:353.

Mandecki, W. (1990). A method for construction of long randomized open reading frames and polypeptides. *Prot. Engin.* 3:221.

Mandel, J. L., and Chambon, P. (1979). DNA methylation differences: Organ-specific variations in methylation pattern within and around ovalbumin and other chick genes. *Nucl. Acid Res.* 7:2081.

Mandelbrot, B. (1977). *The Fractal Geometry of Nature.* Freeman, San Francisco.

Maniatis, T., Goodbuourn, S., and Fischer, J. A. (1987). Regulation of inducible and tissue-specific gene expression. *Science* 236:1237.

Manser, T., Wysocki, L. J., Gridley, T., Near, R. I., and Gefter, M. L. (1985). The molecular evolution of the immune response. *Immunol. Today* 6:94.

Margulis, L. (1970). *Origin of Eucaryotic Cells.* Yale University Press, New Haven, Conn.

Margulis, L. (1981). *Symbiosis in Cell Evolution.* Freeman, San Francisco.

Markert, C. L., Shaklee, J. B., and Whitt, G. S. (1975). Evolution of a gene. *Science* 189:102.

Markland, W., Roberts, B. L., Saxena, M. J., Guterman, S. K., and Ladner, R. C. (1991). Design, construction and function of a multicopy display vector using fusions to the major coat protein of bacteriophage M13. *Gene* 109:13.

Martinez-Arias, A., and White, R.A.H. (1987). *Ultrabithorax* and *engrailed* expression in *Drosophila* embryos mutant for segmentation genes of the pair-rule class. *Development* 102:325.

Mavrothalassitis, G., Beal, G., and Papas, T. S. (1990). Defining target sequences of DNA-binding proteins by random selection and PCR: Determination of the GCN4 binding sequence repertoire. *DNA and Cell Biol.* 9:783.

May, R. M. (1973). *Stability and Complexity in Ecosystems.* Princeton University Press, Princeton, N.J.

May, R. M., ed. (1976). *Theoretical Ecology: Principles and Applications*. Saunders, Philadelphia.

May, R. M., and Oster, G. F. (1976). Bifurcations and dynamic complexity in simple ecological models. *Am. Nat.* **110**:573.

Mayer-Kress, G., ed. (1986). *Dimensions and Entropies in Chaotic Systems: Quantification of Complex Behavior*. Springer-Verlag, Berlin.

Maynard Smith, J. (1970). Natural selection and the concept of a protein space. *Nature* **225**:563.

Maynard Smith, J. (1974). The theory of games and the evolution of animal conflicts. *J. Theoret. Biol.* **47**:209.

Maynard Smith, J. (1976). What determines the rate of evolution? *Am. Nat.* **110**:331.

Maynard Smith, J. (1982). *Evolution and the Theory of Games*. Cambridge University Press, Cambridge.

Maynard Smith, J., and Price, G. R. (1973). The logic of animal conflict. *Nature* **246**:15.

Maynard Smith, J., and Sondhi, K. C. (1960). The genetics of a pattern. *Genetics* **45**:1039.

Maynard Smith, J., Burian, R., Kauffman, S., Alberch, P., Campbell, J., Goodwin, B., Lande, R., Raup, D., and Wolpert, L. (1985). Developmental constraints and evolution. *Qu. Rev. Biol.* **60**:265.

Mayr, E. (1942). *Systematics and the Origin of Species*. Columbia University Press, New York.

Mayr, E. (1982). *The Growth of Biological Thought*. Belknap Press, Cambridge, Mass.

Mayr, E., and Provine, W. B., eds. (1980). *The Evolutionary Synthesis: Perspectives on the Unification of Biology*. Harvard University Press, Cambridge, Mass.

McCafferty, J., Jackson, R. H., and Chiswell, D. J. (1991). Phage enzymes: Expression and affinity chromatography of functional alkaline phosphates on the surface of bacteriophage. *Prot. Engin.* **4**:955.

McClintock, B. (1956). Controlling elements and the gene. In *Cold Spring Harbor Symposia on Quantitative Biology*, vol. 21. Cold Spring Harbor Laboratory, Cold Spring Harbor, N.Y.

McCulloch, W. S., and Pitts, W. (1943). A logical calculus of the ideas immanent in nervous activity. *Bull. Math. Biophys.* **5**:115.

McGinnis, W., Garber, R. L., Wirz, J. Kurioiwa, A., and Gehring, W. J. (1984). A homologous protein-coding sequence in *Drosophila* homeotic genes and its conservation in other metazoans. *Cell* **37**:403.

McKean, D., Huppi, K., Bell, M., Staudt, L., Gerhard, W., and Weigert, M. (1984). Generation of antibody diversity in the immune response of BALB/c mice to influenza virus hemaglutinin. *Proc. Natl. Acad. Sci. USA* **81**:3180.

McKnight, S., and Tjian, R. (1986). Transcriptional selectivity of viral genes in mammalian cells. *Cell* **46**:795.

McLachlan, N. W. (1947). *Theory and Applications of Mathieu Functions*. Clarendon Press, Oxford.

McNamara, M. K., Ward, R. E., and Kohler, H. (1984). Monoclonal idiotape vaccine against *Streptococcus pneumoniae* infection. *Science* **226**:1325.

Meijlink, F., Curran, T., Miller, A. D., and Verma, I. M. (1985). Removal of a 67-base-pair sequence in the noncoding region of protooncogene *fos* converts it to a tranforming gene. *Proc. Natl. Acad. Sci. USA* **82**:4987.

Meinhardt, H. (1977). A model of pattern formation in insect embryogenesis. *J. Cell Sci.* **23**:117.

Meinhardt, H. (1982). *Models of Biological Pattern Formation*. Academic Press, London.

Meinhardt, H. (1986). Hierarchal inductions of cell states: A model for segmentation in *Drosophila. J. Cell Sci. Suppl.* **4**:357.

Meinhardt, H., and Gierer, A. (1974). Applications of a theory of biological pattern formation based on lateral inhibition. *J. Cell Sci.* **15**:1.

Meinhardt, H., and Klingler, M. (1987). A model for pattern formation on the shells of molluscs. *J. Theoret. Biol.* **126**:63.

Menger, F. M. (1991). Groups of organic molecules that operate collectively. *Agnew. Chem. Int. Ed. Eng.* **30**:1086.

Meyer, A. Y., and Richards, W. G. (1991). Similarity of molecular shape. *J. Computer-Aided Mol. Des.* **5**:427.

Miesfeld, R., Rusconi, S., Godowski, P. J. Maler, B. A., Okret, S., Wikstrom, A.-C., Gustaffson, J. A., and Yamamoto, K. R. (1986). Genetic complementation of a glucocorticoid receptor deficiency by expression of cloned receptor cDNA. *Cell* **46**:389.

Miller, J. H. (1988). The evolution of automata in the repeated prisoner's dilemma. Ph.D. diss., University of Michigan, Ann Arbor.

Miller, J. H., Coulondre, C., Hofer, M., Schmeissner, U., Sommer, H., and Schmitz, A. (1979). Genetic studies of the *lac* repressor. IX: Generation of altered proteins by the suppression of nonsense mutations. *J. Mol. Biol.* **131**:191.

Miller, J. H., and Forrest, S. (1989). The dynamical behavior of classifier systems. In *Proceedings of the International Conference on Genetic Algorithms* (J. D. Shaffer, ed.). Morgan Kaufmann, Los Altos, Calif.

Miller, J. H., and Forrest, S. (1990). Emergent behaviors of classifier systems. *Physica* **42D**:213.

Miller, S. L. (1953). Production of amino acids under possible primitive earth conditions. *Science* **117**:528.

Miller, S. L. (1955). Production of some organic compounds under possible primitive earth conditions. *J. Am. Chem. Soc.* **77**:2351.

Miller, S. L. (1957). The formation of organic compounds on the earth. *Ann. N.Y. Acad. Sci.* **69**:260.

Miller, S. L. (1987). Which organic compounds could have occurred on the prebiotic earth? In *Cold Spring Harbor Symposia on Quantitative Biology,* vol. 52. Cold Spring Harbor Laboratory, Cold Spring Harbor, N.Y.

Miller, S. L., Urey, H. C., and Oro, J. (1976). Origin of organic compounds of the primitive earth and in meteorites. *J. Mol. Evol.* **9**:59.

Milstein, C. (1986). From antibody strucutre ot immunological diversification of the immune response. *Science* **231**:1261.

Minsky, M. L. (1967). *Computation: Finite and Infinite Machines.* Prentice-Hall, Englewood Cliffs, N.J.

Mintz, B., and Illmensee, K. (1975). Normal genetically mosaic mice produced from malignant teratocarcinoma cells. *Proc. Natl. Acad. Sci. USA* **72**:3585.

Mirsky, A. E., and Osawa, S. (1961). In *The Cell,* vol. 4. Academic Press, New York.

Mitchison, G. J. (1977). Phylotaxis and the fibonacci series. *Science* **196**:270.

Mitchison, J. M. (1971). *The Biology of the Cell Cycle.* Cambridge University Press, Cambridge.

Mittenthal, J. E. (1981). The rule of normal neighbors: A hypothesis for morphogenetic pattern regulation. *Dev. Biol.* **88**:15.

Mohler, J., and Wieschaus, E. F. (1986). Dominant maternal-effect mutations of *Drosophila melanogaster* causing the production of double-abdomen embryos. *Genetics* **1122**:802.

Monia, B. P. (1989). Structure and function of ubiquitin and its carboxyl extension proteins. Ph.D. diss., University of Pennsylvania.

Monod, J. (1971). *Chance and Necessity* (A. Wainhouse, trans.). Knopf, New York.

Monod, J. Changeux, J. P., and Jacob, F. (1963). Allosteric proteins and cellular control mechanisms. *J. Mol. Biol.* **6**:306.

Morata, G. (1975). Analysis of gene expression during development in the homeotic mutant *Contrabithorax* of *Drosophila melanogaster. J Embryol. Exp. Morph.* **34**:19.

Morata, G., and Lawrence, P. A. (1975). Control of compartment development by the *engrailed* gene in *Drosophila. Nature* **255**:614.

Morata, G., and Lawrence, P. A. (1977). Homeotic genes control compartments and cell determination in *Drosophila. Nature* **265**:211.

Morata, G., and Lawrence, P. A. (1978). Anterior and posterior compartments in the head of *Drosophila. Nature* **274**:473.

Morowitz, H. (1968). *Energy Flow and Biology: Biological Organization as a Problem in Thermal Physics.* Academic Press, New York.

Morse, P. M., and Feshbach, H. (1953). *Methods of Theoretical Physics.* McGraw-Hill, New York.

Muldoon, J. J., Evans, T. E., Nygaard, O. F., and Evans, H. H. (1971). Control of DNA replication by protein synthesis at defined times during the s period in *physarum polycephalum. Biochem. Biophys. Acta* **247**:310.

Muller, H. J. (1932). Some genetic aspects of sex. *Am. Nat.* **66**:118.

Muller, H. J. (1950a). Our load of mutations. *Am. J. Human Genet.* **2**:111.

Muller, H. J. (1950b). Evidence of the precision of genetic adaptation. In *The Harvey Lectures,* series 43, 1947–1948. Thomas, Springfield, Ill.

Muller-Hill, B., Rickenberg, H. V., and Wallenfels, K. (1964). Specificity of the induction of the enzymes of the *lac* operon in *Escherichia coli. J. Mol. Biol.* **10**:303.

Murray, J. D. (1981a). On pattern formation for lepidopteran wing patterns and mammalian coat markings. *Philos. Trans. Roy. Soc. London B* **295**:473.

Murray, J. D. (1981b). A pre-pattern formation mechanism for animal coat markings. *J. Theoret. Biol.* **88**:161.

Murray, J. D. (1989). *Mathematical Biology.* Springer, New York

Nanney, D. L., and Rudzinska, M. A. (1960), In *The Cell,* vol. 4. Academic Press, New York.

Nash, J. F. (1951). Noncooperative games. *Ann. Math.* **54**:289.

Nei, M. (1975). *Molecular Population Genetics and Evolution.* North Holland, Amsterdam.

Neubauer, Z., and Calef, E. (1970). Immunity phase shift in defective lysogens: Nonmutational hereditary change of early regulation of prophage. *J. Mol. Biol.* **51**:1.

Neumann, H., Levin, Y., Berger, A., and Katchalski, F. (1959). Pepsin-catalyzed transpeptidation of the amino transfer type. *Biochemistry* **73**:33.

Neurath, H. (1984). Evolution of proteolytic enzymes. *Science* **224**:350.

Newell, A., and Rosenbloom, P. (1981). *Cognitive Skills and Their Cognition.* Erlbaum, Hillsdale, N.J.

Newman, S. A., and Rice, S. A. (1971). Model for constraint and control in biochemical networks. *Proc. Natl. Acad. Sci. USA* **68**:92.

Newman, S. M. Jr., and Schubiger, G. (1980). A morphological and developmental study of *Drosophila* embryos ligated during nuclear mutliplication. *Dev. Biol.* **79**:128.

Ng, S.-Y., Parker, C. S., and Roeder, R. G. (1979). Transcription of cloned *Xenopus laervis* RNA polymerase III in reconstituted systems. *Proc. Natl. Acad. Sci. USA* **76**:136.

Nicolis, G., and Prigogine, I. (1977). *Self-Organization in Nonequilibrium Systems.* Wiley-Interscience, New York.

Niesert, U., Harnasch, D., and Bresch, C. (1981). Origin of life between Scylla and Charybdis. *J. Mol. Evol.* **17**:348.

Ninio, J. (1970). *Approaches moleculaires de l'evolution, collection de biologie evolutive.* Masson, Paris.

Noda, M., Selinger, Z., Scolnick, E. M., and Bassin, R. H. (1983). Flat revertants isolated from

Kirsten sarcoma virus-transformed cells are resistant to the action of specific oncogenes. *Proc. Natl. Acad. Sci. USA* **80:**5602.

North, J., Chu, N. R., Chan, A., Steele, J. K., Singhai, R., Strammers, A. T., and Levy, J. G. (1988). Idiotypic control of the immune response to ferredoxin. In *Theoretical Immunology II: Santa Fe Institute Studies in the Sciences of Complexity* (A. S. Perelson, ed.). Addison-Wesley, Reading, Mass.

Nothiger, R. (1972). The larval development of imaginal discs. In *Results and Problems in Cell Differentiation* (V.C.H. Urspung and R. Nothiger, eds.). Springer-Verlag, Berlin.

Nothiger, R., and Gehring, W. (1973). In *Development Systems II: Insects* (S. Counce and C. H. Waddington, eds.). Academic Press, New York.

Nüsslein-Volhard, C. (1977). Genetic analysis of pattern formation in the embryo of *Drosophila melanogaster:* Characterization of the maternal-effect mutant *bicaudal. Roux's Arch. Dev. Biol.* **183:**249.

Nüsslein-Volhard, C., and Wieschaus, E. (1980). Mutations affecting segment number and polarity in *Drosophila. Nature* **287:**795.

Nüsslein-Volhard, C., Frohnhofer, H. G., and Lehmann, R. (1987). Determination of antero-posterior polarity in *Drosophila. Science* **238:**1676.

Odell, G. Oster, G., Burnside, B., and Alberch, P. (1981). The mechanical basis of morphogenesis. I: Epithelial folding and invagination. *Dev. Biol.* **85:**446.

O'Farrell, P. H. (1975). High-resolution two-dimensional gel electrophoresis of proteins. *J. Biol. Chem.* **250:**4007.

O'Farrell, P. H., Desplan, C., DiNardo. S., Kassis, J. A., Kuner, J. M., Sher, E., Theis, J., and Wright, D. (1985). Embryonic pattern in *Drosophila:* The spatial distribution and sequence-specific DNA binding of *engrailed* protein. In *Cold Spring Harbor Symposia on Quantitative Biology,* vol. 50. Cold Spring Harbor Laboratory. Cold Spring Harbor, N.Y.

Ohta, T. (1983). On the evolution of multigene families. *Theoret. Pop. Biol.* **23:**216.

Ohta, T., and Dover, G. (1983). Population genetics of multigene families that are dispersed into two or more chromosomes. *Proc. Natl. Acad. Sci. USA* **80:**4079.

Ohta, T., and Kimura, M. (1971). On the constancy of the evolutionary rates of cistrons. *J. Mol. Evol.* **1:**18.

Okada, M., Kleinman, I. A., and Schneiderman, H. A. (1974). Chimeric *Drosophila* adults produced by transplantation of nuclei into specific regions of fertilized eggs. *Dev. Biol.* **39:**286.

Olby, R. C. (1979). Mendel no Mendelian? *His. Sci* **17:**53.

Oldenburg, K. R., Loganathan, D., Goldstein, I. J., Schultz, P. G., and Gallop, M. A. (1992). Peptide ligands for a sugar-binding protein isolated from a random peptide library. *Proc. Natl. Acad. Sci. USA* **89:**5393.

Oliphant, A. R., and Struhl, K. (1987). The use of random-sequence oligonculeotides for determining consensus sequences. *Methods Enzymol.* **155:**568.

Oliphant, A. R., and Struhl, K. (1989). An efficient method for generating proteins with altered enzymatic properties: Application to β-lactamase. *Proc. Natl. Acad. Sci. USA* **86:**9094.

Oliphant, A., Nussbaum, A. I., and Stuhl, K. (1986). Cloning of random-sequence oligodeoxynucleotides. *Gene* **44:**177.

Oparin, A. I. (1957). *The Origin of Life on Earth.* Academic Press, New York.

Oparin, A. I. (1966). *The Origin and Initial Development of Life.* Meditsina, Moscow.

Oparin, A. I. (1971). Routes for the origin of the first forms of life. *Sub. Cell. Biochem.* **1:**75.

Orgel, L. E. (1963). The maintenance of the accuracy of protein synthesis and its relevance to aging. *Proc. Natl. Acad. Sci. USA* **49:**517.

Orgel, L. E. (1986). RNA catalysis and the origins of life. *J. Theoret. Biol.* **123:**127.

Orgel, L. E. (1987). Evolution of the genetic apparatus: A review. In *Cold Spring Harbor Sym-*

posia on Quantitative Biology, vol. 52. Cold Spring Harbor Laboratory, Cold Spring Harbor, New York.

Orgel, L. E., and Miller, S. (1974). *The Origins of Life on the Earth.* Prentice-Hall, Englewood Cliffs, N.J.

Oro, J., and Kimball, A. P. (1961). Synthesis of purines under possible primitive earth conditions. I: Adenine from hydrogen cyanide. *Arch. Biochem. Biophys.* **94**:221.

Oro, J., and Kimball, A. P. (1962). Synthesis of purines under possible primitive earth conditions. II: Purine intermediates from hydrogen cyanide. *Arch. Biochem. Biophys.* **96**:293.

Oster, G. (1988). Lateral inhibition models of developmental processes. *Math. Biosci.* **90**:265.

Oster, G., and Alberch, P. (1982). Evolution and bifurcation of developmental patterns. *Evolution* **36**:444.

Oster, G., and Murray, J. D. (1989). Pattern formation models and developmental constraints. *J. Exp. Zool.* **251**:186.

Oster, G., and Odell, G. (1984). The mechanochemistry of cytogenesis. *Physica* **12D**:333.

Oster, G., Alberch, P., Murray, J. D., and Shubin, N. (1988). Evolution and morphogenetic rules: The shape of the vertebrate limb in ontogeny and phylogeny. *Evolution* **42**:862.

Ouweneel, W. J. (1976). Developmental genetics of homeosis. *Adv. Genet.* **18**:179.

Overall, R., and Jaffe, L. F. (1985). Patterns of ionic current through *Drosophila* follicles and eggs. *Dev. Biol.* **108**:102.

Packard, N. (1988). Adaptation to the edge of chaos. In *Complexity in Biologic Modelling* (S. Kelso and M. Shlesinger, eds.).

Packard, N. H., Crutchfield, J. P., Farmer, J. D., and Shaw, R. S. (1980). Geometry from a time series. *Phys. Rev. Lett.* **45**:712.

Paecht-Horowitz, M., and Katchalsky, A. (1973). Synthesis of amino acyl-adenylates under prebiotic conditions. *J. Mol. Evol.* **2**:91.

Paecht-Horowitz, M., Berger, J., and Katchalsky, A. (1970). Prebiotic synthesis of polypeptides by heterogeneous polycondensation of amino-acid adenylates. *Nature* **228**:636.

Paigen, K. (1979). In *Physiological Genetics* (J. G. Scandalias, ed.). Academic Press, New York.

Palmer, E. M. (1985). *Graphical Evolution: An Introduction to the Theory of Random Graphs.* Wiley, New York.

Palmer, R. (1988). Statistical mechanics approaches to complex optimization problems. In *The Economy as an Evolving Complex System: Santa Fe Institute Studies in the Sciences of Complexity* (P. W. Anderson, J. Arrow, and D. Pines, eds.), vol. 5. Addison-Wesley, Reading, Mass.

Palmer, R., and Pond, C. M. (1979). Internal field distributions in model spin-glasses. *J. Phys. F-Metal Physics* **9**:1451.

Parker, C. S., and Topol, J. (1984). A *Drosophila* RNA polymerase II transcription factor contains a promoter region-specific DNA-binding activity. *Cell* **36**:357.

Parmley, S. F., and Smith, G. (1988). Antibody-selectable filamentous fd phage vectors: Affinity purification of target genes. *Gene* **73**:305.

Patterson, C., ed. (1987). *Molecules and Morphology in Evolution: Conflict or Compromise?* Cambridge University Press, Cambridge.

Paul, J. (1988). In *Theories of Carcinogenesis* (O. H. Iverson, ed.). Hemisphere, Washington, D. C.

Pei, D., Ulrich, H. D., and Schultz, P. G. (1991). A combinatorial approach toward DNA recognition. *Science* **253**:1408.

Perelson, A. S. (1988). Toward a realistic model of the immune system.In *Theoretical Immunology II: Santa Fe Institute Studies in the Sciences of Complexity* (A. S. Perelson, ed.). Addison-Wesley, Reading, Mass.

Perelson, A. S., and Oster, G. F. (1979). Theoretical studies of clonal selection: Minimal antibody repertoire size and reliability of self–nonself discrimination. *J. Theoret. Biol.* **81**:645.

Perlmutter, R. M. (1984). The molecular genetics of phosphocholine-binding antibodies. In *The Biology of Idiotypes* (M. I. Greene and A. Nisonoff, eds.). Plenum, New York.

Peterson, P. A. (1981). Diverse expression of controlling element components in maize: Test of a model. In *Cold Spring Harbor Symposia on Quantitative Biology,* vol. 45. Cold Spring Harbor Laboratory, Cold Spring Harbor, N.Y..

Pierre, J., and Laval, J. (1981). Specific nicking of DNA at apurinic sites by peptides containing aromatic residues. *J. Biol. Chem.* **256**:10217.

Pimm, S. L. (1982). *Food Webs.* Chapman and Hall, London.

Pineda, F. J. (1987). Generalization of back-propagation to recurrent neural networks. *Phys. Rev. Lett.* **59**:2229.

Plückhun, A., and Ge, L. (1991). The rationality of random screening—efficient methods of selection of peptides and oligonucleotide ligands. *Agnew. Chem. Int. Ed. Eng.* **30**:296.

Pollack, S. J., and Schultz, P. G. (1987). Antibody catalysis by transition state stabilization. In *Cold Spring Harbor Symposia on Quantitative Biology,* vol. 52. Cold Spring Harbor Laboratory, Cold Spring Harbor, N.Y.

Pollack, S. J., Jacobs, J. W., and Schultz, P. G. (1986). Selective chemical catalysis by an antibody. *Science* **234**:1570.

Ponder, J. W., and Richards, F. M. (1987). Internal packing and protein structrual classes. In *Cold Spring Harbor Symposia on Quantitative Biology,* vol. 52. Cold Spring Harbor Laboratory, Cold Spring Harbor, N.Y.

Popper, K. R. (1963). *Conjectures and Refutations: The Growth of Scientific Knowledge.* Harper & Row, New York.

Postlethwait, J. H. (1974). Development of the temperature-sensitive homeotic mutant *opthalmoptera* of *Drosophila melanogaster. Dev. Biol.* **36**:212.

Postlethwait, J. H., and Schneiderman, H. A. (1971). Pattern formation and determination in the antenna of the homeotic mutant *Antennapedia of Drosophilia melanogaster. Dev. Biol.* **25**:606.

Postlethwait, J. H., and Schneiderman, H. A. (1974). Developmental genetics of *Drosophila* imaginal discs. *Ann. Rev. Genet.* **7**:381.

Postlethwait, J. H., Bryant, P., and Schubiger, G. (1972). The homeotic effect of *tumerus head* in *Drosophila melanogaster. Dev. Biol.* **29**:337.

Potter, M. (1971). Myeloma proteins (M-components) with antibody-like activity. *N. Engl. J. Med.* **284**:831.

Potter, H., Weir, L., and Leder, P. (1984). Enhancer-dependent expression of human K-immunoglobulin genes introduced into mouse pre-B lymphocytes by electroporation. *Proc. Natl. Acad. Sci. USA* **81**:7161.

Poulson, D. F. (1950a). Histogenesis, organogenesis, and differentiation in the embryo of *Drosophila melanogaster* Meigen. In *The Biology of Drosophila* (M. Demerec, ed.). Hafner, New York.

Poulson, D. F. (1950b). Histogenesis, organogenesis and differentiation in the embryo of *Drosophila melanogaster. Dev. Biol.* **74**:286.

Preiss, A., Rosenberg, U. B., Kienlin, A., Seifert, E., and Jackle, H. (1985). Molecular genetics of *Kruppel,* a gene required for segmentation of the *Drosophila* embryo. *Nature* **313**:27.

Press, J. L., and Klinman, N. R. (1974). Frequency of hapten-specific B cells in neonatal and adult mouse spleen. *Eur. J. Immunol.* **4**:155.

Provine, W. B. (1971). *The Origins of Theoretical Population Genetics.* University of Chicago Press, Chicago.

Provine, W. B. (1986). *Sewall Wright and Evolutionary Biology.* University of Chicago Press, Chicago.

Ptashne, M. (1986). *A Genetic Switch*. Cell Press and Blackwell Scientific, Cambridge, Mass., and Palo Alto, Calif.

Quine, W. (1961). *From a Logical Point of View*. Harvard University Press, Cambridge, Mass.

Rabinowitz, Z., and Sachs, L. (1970). Control of the reversion properties in transformed cells. *Nature* **225**:136.

Raff, R. A., and Kaufman, T. C. (1983). *Embryos, Genes and Evolution*. Macmillan, New York.

Raines, R. T., Straus, D. R., Gilbert, W., and Knowles, F.R.S. (1986). The kinetic consequences of altering the catalytic residues of triosephosphate isomerase. *Philos. Trans. Roy. Soc. London A* **317**:371.

Raines, R. T., Sutton, E. T., Straus, D. R., Wilber, W., and Knowles, J. R. (1986). Reaction energetics of a mutant triosephosophate isomerase in which the active-site glutamate has been changed to aspartate. *Biochemistry* **25**:7142.

Rasmussen, S. (1989). Toward a quantitative theory of the origin of life. In *Artificial Life: Santa Fe Institute Studies in the Sciences of Complexity* (C. G. Langton, ed.). Addison-Wesley, Reading, Mass.

Raup, D. M. (1968). Theoretical morphology of echinoid growth. *J. Paleont.* **42**:50.

Raup, D. M. (1972). Taxonomic diversity during the Phanerozoic. *Science* **177**:1065.

Raup, D. M. (1983). On the early origins of major biologic groups. *Paleobiology* **9**:107.

Raup, D. M. (1986). Biological extinction in earth history. *Science* **231**:1528.

Raup, D. M., and Gould, S. J. (1974). Stochastic simulation and evolution of morphology: Towards a nomothetic paleontology. *Syst. Zool.* **23**:305.

Raup, D. M., Gould, S. J., Schopf, T.J.M., and Simberloff, D. (1973). Stochastic models of phylogeny and the evolution of diversity. *J. Geol.* **81**:525.

Read, R. C. (1976). The enumeration of acyclic chemical compounds. In *Chemical Applications of Graph Theory* (A. T. Balaban, ed.). Academic Press, London.

Rebek, J., Jr. (1991). Molecular recognition and the development of self-replicating systems. *Experimentia* **47**:1096.

Rechenberg, I. (1973). *Evolutionsstrategie*. Friedrich Frommann Verlag, Stuttgart.

Reinitz, J., and Levine, M. (1990). Control of the initiation of homeotic gene expression by the gap genes: *Giant* and *tailless* in *Drosophila*. *Dev. Biol.* **140**:57.

Rendel, J. M. (1967). *Canalization and Gene Control*. Academic Press, New York.

Rendel, J. M. (1979). Canalization and selection. In *Quantitative Genetic Variation* (J. N. Thompson and J. M. Thoday, eds.). Academic Press, New York.

Richardson, J. S. (1981). Anatomy and taxonomy of protein structure. *Adv. Protein Chem.* **34**:167.

Richter, R. H. (1978). The network idea and the immune response. In *Theoretical Immunology* (G. I. Bell, A. S. Perelson, and G. H. Pimbley, eds.). Marcel Dekker, New York.

Riedl, R. (1978). *Order in Living Organisms* (R.P.S. Jeffries, trans.). Wiley, London.

Rigby, P.W.J., Burleigh, B. D., Jr., and Hartley, B. S. (1974). Gene duplication in experimental enzyme evolution. *Nature* **251**:200.

Roberts, B. L., Markland, W., Ley, A. C., Kent, R. B., White, D. W., Guterman, S. K., and Ladner, R. C. (1992). Directed evolution of a protein: Selection of potent neutrophil elastase inhibitors displayed on MI3 fusion phage. *Proc. Natl. Acad. Sci. USA* **89**:2429.

Robertson, D. L, and Joyce, G. F. (1990). Selection *in vivo* of an RNA enzyme that specifically cleaves single-stranded DNA. *Nature* **344**:467.

Roder, H., Elove, G. A., and Englander, S. W. (1988). Structural characterization of folding intermediates in cytochrome *c* by H-exchange labelling and proton NMR. *Nature* **335**:700.

Rodgers, M., and Shearn, A. (1977). Patterns of protein synthesis in imaginal discs of *Drosophila melanogaster*. *Cell* **12**:915.

Rohlfing, D. L., and Fox, S. W. (1969). Catalytic activities of thermal poly-anhydro-α-amino acids. *Adv. Catal.* **20**:373.

Rokshar, D. S., Anderson, P. W., and Stein, D. L. (1986). Self-organization in prebiological systems: Simulation of a model for the origin of genetic information. *J. Mol. Evol.* **23**:110.

Root-Bernstein, R. S. (1983). Protein replication by amino acid pairing. *J. Theoret. Biol.* **100**:99.

Rose, S. M. (1962). Tissue-arch control of regeneration in the amphibian limb. *Symp. Soc. Study Dev. Growth* **20**:153.

Rosenzweig, M. L., and Taylor, J. A. (1980). Speciation and diversity in Ordovician invertebrates: Filling niches quickly and carefully. *Oikos* **35**:236.

Rosenzweig, M. L., Brown, J. S., and Vincent, T. L. (1987). Red queens and ESS: The coevolution of evolutionary rates. *Evol. Ecol.* **1**:59.

Rossler, O. (1971). A system-theoretic model of biogenesis. *Z. Naturforsch* **B266**:741.

Rossler, O. (1974). Chemical automata in homogeneous and reaction-diffusion kinetics. *Notes Biomath.* **B4**:399.

Rossler, O. (1983). Deductive prebiology. In *Molecular Evolution and the Prebiological Paradigm* (K. L. Rolfing, ed.). Plenum, New York.

Roughgarden, J. (1979). *Theory of Population Genetics and Evolutionary Ecology: An Introduction.* Macmillan, New York.

Rubin, H., and Sitgreave, R. (1954). Probability distributions related to random transformations on a finite set. Technical report, no. 19A, Applied Mathematics and Statistics Laboratory, Stanford University.

Ruelle, D. (1979). Sensitive dependence on initial condition and turbulent behavior of dynamical systems. *Ann. N.Y. Acad. Sci.* **316**:408.

Rummelhart, D. E., McClelland, J. L., and the PDP research group (1986). *Parallel Distributed Processing: Explorations in the Microstructure of Cognition,* vols. 1 and 2. Bradford, Cambridge, Mass.

Russell, M. A. (1978). A spherical coordinate model of positional information. Presented at the 20th Annual *Drosophila* Conference, Chicago.

Russell, M. A. (1985). Positional information in insect segments. *Dev. Biol.* **108**:269.

Sablitzky, F., Weisbaum, D., and Rajewsky, K. (1985). Sequence analysis of non-expressed immunoglobulin heavy-chain loci in clonally related, somatically mutated hybridoma cells. *EMBO J.* **4**:3435.

Sablitzky, F., Wildner, G., and Rajewsky, K. (1985). Somatic mutation and clonal expansion of B cells in an antigen-driven immune response. *EMBO J.* **4**:345.

Sacks, D. L., Kirchhoff, L. V., Hieny, S., and Sher, A. (1985). Molecular mimicry of a carbohydrate epitope on a major surface glycoprotein of *trypanosoma cruzi* by using anti-idiotypic antibodies. *J. Immunol.* **135**:4155.

Sakoyama, Y., and Okubo, S. (1981). Two-dimensional gel patterns of protein species during development of *Drosophila* embryos. *Dev. Biol.* **81**:361.

Sampsell, B. M., and Held, W. A. (1985). Variation in the major urinary protein multigene family in wild-derived mice. *Genetics* **109**:549.

Sander, K. (1975). Pattern specification in the insect embryo. In *Cell Patterning* (R. Porter and J. Rivers, eds.), *CIBA Foundation Symposium,* vol. 29. Associated Scientific, Amsterdam.

Sander, K. (1977). Current understanding of cytoplasmic control centers. In *Insect Embryology* (S. W. Vischer, ed.). Montana State University Press,

Sander, K. (1980). Pattern generation and pattern conservation in insect ontogenesis: Problems, data, and models. In *Progress in Developmental Biology* (H. W. Sauer, ed.). Fischer-Verlag, New York.

Sander, K. (1984). Embryonic pattern formation in insects: Basic concepts and their experimental foundations. In *Pattern Formation* (G. M. Malacinski and S. V. Bryant, eds.). Macmillan, New York.

Sanderson, M. J., and Donoghue, M. J. (1989). Patterns of variation in levels of homoplasy. *Evolution* **43**:1781.

Sang, J. H., and McDonald, J. M. (1954). Production of phenocopies in *Drosophila* using salts, particularly sodium metaborate. *J. Genet.* **52**:392.

Saxen, L., and Toivonen, S. (1962). *Primary Embryonic Induction.* Logos, London.

Schaffer, W. M., and Kott, M. (1986). Differential systems in ecology and epidemiology. In *Chaos* (A. V. Holden, ed.). Princeton University Press, Princeton, N.J.

Schlissel, M. S., and Brown, D. D. (1984). The transcriptional regulation of *Xenopus* 5 S RNA genes in chromatin: The roles of active stable transcription complex and histone H1. *Cell* **37**:903.

Schlomchik, M. J., Pisetsky, D. S., and Weigert, M. G. (1987). The structure and function of anti-DNA autoantibodies derived from a single autoimmune mouse. *Proc. Natl. Acad. Sci. USA* **84**:9150.

Schrieber, A. B., Courand, P. O., Andre, C., Vray, B., and Strosberg, D. A. (1980). Anti-alprenolol anti-idiotypic antibodies bind to B-andrenergic receptors and modulate catecholamine-sensitive adrenylate cyclase. *Proc. Natl. Acad. Sci. USA* **77**:7385.

Schrödinger, E. (1944). *What Is Life? The Physical Aspect of the Living Cell.* Cambridge University Press, Cambridge.

Schubiger, G. (1976). Adult differentiation from partial *Drosophila* embryos after egg ligation during stages of nuclear multiplication and cellular blastoderm. *Dev. Biol.* **50**:476.

Schubiger, G., and Karpen, G. (1981). Extensive regulatory capabilities of a *Drosophila* iminginal disk blastema. *Nature* **294**:744.

Schubiger, G., and Schubiger, M. (1978). Distal transformation in *Drosophila* leg imaginal disc fragments. *Dev. Biol.* **67**:286.

Schubiger, G., and Wood, W. J. (1977). Determination during early embryogenesis in *Drosophila melanogaster. Am. Zool.* **17**:565.

Schulz, G. E., and Schirmer, R. H. (1979). *Principles of Protein Structure.* Springer, New York.

Schuster, P. (1986). The physical basis of molecular evolution. *Chemica Scripta* **26B**:27.

Schuster, P. (1987). Structure and dynamics of replication-mutation systems. *Physica Scripta* **35**:402.

Schutzenberger, M. P. (1967). Algorithms and the Neo-Darwinian theory of evolution. In *Mathematical Challenges to the Neo-Darwinian Interpretation of Evolution* (P. S. Moorhead and M. M. Caplan, eds.). Wistar Institute Press, Philadelphia.

Schwartz, A. W., and Chittenden, G.J.F. (1977). Synthesis of uracil and thymine under simulated prebiotic conditions. *Biosystems* **9**:87.

Schwartz, A. W., and Orgel, L. E. (1985). Template-directed synthesis of novel nucleic acid-like structures. *Science* **228**:585.

Scott, J. K., and Smith, G. P. (1990). Searching for peptide ligands with an epitome library. *Science* **249**:386.

Scott, J. K., Loganathan, D., Easley, R. B., Gong, X. Y., and Goldstein, I. J. (1992). A family of concanavalin A-binding peptides from a hexapeptide epitope library. *Proc. Natl. Acad. Sci. USA* **89**:5398.

Scott, M. P., and Carroll, S. B. (1987). The segmentation and homeotic gene network in early *Drosophila* development. *Cell* **51**:689.

Scott, M. P., and Weiner, A. J. (1984). Structural relationships among genes that control development: Sequence nomology between the *Antennapedia, Ultrabithorax,* and *Fushi tarazu* loci of *Drosophila. Proc. Natl. Acad. Sci. USA* **81**:4115.

Sege, K., and Peterson, P. A. (1983). Use of antiidiotypic antibodies as cell-surface receptor probes. *Proc. Natl. Acad. Sci. USA* **75:**2443.

Segel, L. A., and Perelson, A. S. (1988). Computations in shape space: A new approach to immune network theory. In *Theoretical Immunology II: Santa Fe Institute Studies in the Sciences of Complexity* (A. S. Perelson, ed.). Addison-Wesley, Reading, Mass.

Sellers, P. H. (1970). Combinatorial analysis of a chemical network. *J. Franklin Inst.* **290:**113.

Sellers, P. H. (1971). An introduction to a mathematical theory of chemical reaction networks, I. *Arch. Rat. Mech. Anal.* **44:**23.

Sellers, P. H. (1972). An introduction to a mathematical theory of chemical reaction networks, II. *Arch. Rat. Mech. Anal.* **44:**376.

Sellers, P. H. (1984). Combinatorial classification of chemical mechanisms. *SIAM J. Appl. Math.* **44:**784.

Shai, Y., Flashner, M., and Chaiken, I. M. (1987). Antisense peptide recognition of sense peptides: Direct quantitative characterization with the ribonuclease S-peptide system using analytical high-performance affinity chromatography. *Biochemistry* **26:**669.

Shapiro, R. (1986). *Origins: A Skeptic's Guide to the Creation of Life on Earth.* Summit Books, New York.

Shearn, A., Rice, T., Garen, A., and Gehring, W. (1971). Imaginal disc abnormalities in lethal mutants of *Drosophila. Proc. Natl. Acad. Sci. USA* **68:**2594.

Shenkin, P. S., Yarmush, D. L., Fine, R. M., Wang, H., and Levinthal, C. (1987). Predicting antibody hypervariable loop conformation. I: Ensembles of random conformations for ring-like structures. *Biopolymers* **26:**2053.

Sherlock, R. A. (1979a). Analysis of the behavior of Kauffman binary networks. I: State space description of limit cycle lengths. *Bull. Math. Biol.* **41:**687.

Sherlock, R. A. (1979b). Analysis of the behavior of Kauffman binary networks. II: The state cycle fraction for networks of different connectivities. *Bull. Math. Biol.* **41:**707.

Sherman, F., and Helms, C. (1978). A chromosomal translocation causing overproduction of iso-2-cytochrome *c* in yeast. *Genetics* **88:**689.

Sherrington, D., and Kirkpatrick, S. (1975). Solvable model of a spin glass. *Phys. Rev. Lett.* **35:**1792.

Shih, C., and Weinberg, R. A. (1982). Isolation of a transforming sequence from a human bladder carcinoma cell line. *Cell* **29:**161.

Shoemaker, K. R., Fairman, R., Kim, P. S., York, E. J., Stewart, J. M., and Baldwin, R. L. (1987). The C-peptide helix from ribonuclease A considered as an autonomous folding unit. In *Cold Spring Harbor Symposia on Quantitative Biology,* vol. 52. Cold Spring Harbor Laboratory, Cold Spring Harbor, N.Y.

Shokat, K. M., and Schultz, P. G. (1991). Catalytic antibodies. *Methods Enzymol.* **203:**327.

Shvedova, T. A., Korneeva, G. A., Otroshchenko, V. A., and Venkstern, T. V. (1987). Catalytic activity of the nucleic acid component of the 1,4-alpha-glucan branching enzyme from rabbit muscles. *Nucl. Acid. Res.* **15:**1745.

Shymko, R. S., Klevecz, R. R., and Kauffman, S. A. (1984). The cell cycle as an oscillatory system. In *Cell Cycle Clocks* (L. N. Edmunds Jr., ed.). Marcel Dekker, New York.

Sieburg, H. B. (1988). A logical dynamic systems approach to the regulation of antigen-driven lymphocyte stimulation. In *Theoretical Immunology I: Santa Fe Institute Studies in the Sciences of Complexity* (A. S. Perelson, ed.). Addison-Wesley, Reading, Mass.

Siegel, S. (1956). *Non-Parametric Statistics for the Behavioral Sciences.* McGraw-Hill, New York.

Siekevitz, M., Cocks, C., Jajewsky, K., and Dildrop, R. (1987). Analysis of somatic mutation and class switching in naive and memory B cells generating adaptive primary and secondary responses. *Cell* **48:**757.

Silver, M. S., and James, S.L.T. (1980). Mechanistic features of pepsin-catalyzed amino transfer reactions. *J. Biol. Chem.* **255**:555.

Silver, M. S., and James, S.L.T. (1981a). Enzyme-catalyzed condensation reactions which initiate rapid peptic cleavage of substrates. 1: How the structure of an activating peptide determines its efficiency. *Biochemistry* **20**:3177.

Silver, M. A., and James. S.L.T. (1981b). Enzyme-catalyzed condensation reactions which initiate rapid peptic cleavage of substrates. 2: Proof of mechanism for three examples. *Biochemistry* **20**:3183.

Simon, H. (1962). The architecture of complexity. *Proc. Am. Phil. Soc.* **106**:467.

Simpson, G. G. (1944). *Tempo and Mode in Evolution.* Columbia University Press, New York.

Simpson, G. G. (1950). *The Meaning of Evolution.* Oxford University Press, Oxford.

Simpson, G. G. (1953). *The Major Features of Evolution.* Columbia University Press, New York.

Siskind, G. W., and Benaceraf, B. (1969). Cell selection by antigen in the immune response. *Immunol. Rev.* **10**:1.

Siskind, G. W., Dunn, P., and Walker, J. G. (1968). Studies on the control of antibody synthesis IrI: The effect of antigen dose and of suppression by passive antibody on the affinity of antibody synthesized. *J. Exp. Med.* **127**:55.

Slack, J. (1980). A serial threshold theory of regeneration. *J. Theoret. Biol.* **82**:105.

Slack, J. (1985). Homeotic transformations in man: Implications for the mechanism of embryonic development and for the organization of epithelia. *J. Theoret. Biol.* **114**:463.

Slack, J., and Savage, S. (1978). Regeneration of reduplicated limbs in contravention of the complete circle rule. *Nature* **271**:760.

Slaughter, C. A., and Capra, J. D. (1984). Structural and genetic basis of the major corss-reactive idiotype of the A strain mouse. In *The Biology of Idiotypes* (M. I. Greene and A. Nisonoff, eds.). Plenum, New York.

Smith, G. P. (1974). Unequal crossover and the evolution of multigene familes. In *Cold Spring Harbor Symposia on Quantitative Biology,* vol. 38. Cold Spring Harbor Laboratory, Cold Spring Harbor, N.Y.

Sober, E. (1984). *The Nature of Selection.* MIT Press, Cambridge, Mass.

Solla, S. A. Sorkin, G. B., and White, S. R. (1986). Configuration space analysis for optimization problems. In *Disordered Systems and Biological Organization* (E. Bienenstock, F. Fogelman-Soulie, and G. Weisbuch, eds.), Series F: *Computer and Systems Sciences,* vol. 20. Springer, New York.

Sorkin, G. (1988). Combinatorial opitimization, simulated annealing and fractals. IBM research report RC13674 (No. 61253).

Sparrow, A. H., and Evans, H. J. (1961). *Brookhaven Symp. Biol.* **14**:76.

Spemann, H. (1938). *Embryonic Development and Induction.* Yale University Press, New Haven, Conn.

Spemann, H., and Mangold, H. (1924). Uber Induktion von Embryonalanlagen durch Implantation artfremder Organisatoren. *Roux's Arch. Dev. Biol.* **100**:599.

Sporn, M. B., and Todaro, G. J. (1980). Autocrine secretion and malignant transformation of cells. *N. Engl. J. Med.* **303**:878.

Spradling, A. C., and Mahowald, A. P. (1980). Amplification of genes for chorion proteins during cogenesis in *Drosophila melanogaster. Proc. Natl. Acad. Sci. USA* **77**:1096.

Spradling, A. C., and Rubin, G. M. (1981). *Drosophila* genome organization: Conserved and dynamic aspects. *Ann. Rev. Genet.* **15**:219.

Stadler, P. F. (1992). Correlation of landscapes of combinatorial optimization problems. *Europhys. Lett.,* submitted.

Stadler, P. F., and Happel, R. (1992). Correlation structure of the graph bipartition problem. *J. Phys. A: Math. Gen.* **25**:3103.

Stadler, P. F., and Schnabl, W. (1992). The landscape of the traveling saleman problem. *Phys. Lett. A,* **161**:337.

Stanley, H. E., Stauffer, D., Kertesz, J., and Herrman, H. J. (1987). *Phys. Rev. Lett.* **59**:114.

Stanley, S. M. (1979). *Macroevolution: Pattern and Process.* Freeman, San Francisco.

Stassen, A.P.M., Zaman, G.J.R., van Deursen, J.M.A., Schoenmakers, J.G.G., and Konings, R.N.H. (1992). Selection and characterization of randomly produced mutants of gene V protein of bacteriophage MI3. *Eur. J. Biochem.* **204**:1003.

Stauffer, D. (1985). *Introduction to Percolation Theory.* Taylor and Francis, London.

Stauffer, D. (1987a). Random Boolean networks: Analogy with percolation. *Philos. Mag. B* **56**:901.

Stauffer, D. (1987b). On forcing functions in Kauffman's random Boolean networks. *J. Statis. Phys.* **40**:789.

Stauffer, D. (1989). Percolation thresholds in square-lattice Kauffman model. *J. Theoret. Biol.*

Stebbins, G. L. (1950). *Variation and Evolution in Plants.* Columbia University Press, New York.

Stein, D. L. (1985). A model of protein conformational substrates. *Proc. Natl. Acd. Sci. USA* **82**:3670.

Stein, D. L. (1987). In *Statistics and Dynamics of Complex Systems* (J. Souletie, J. Vanni-menus, and R. Story, eds.). Elsevier, Amsterdam.

Stein, D. L., and Anderson, P. W. (1984). A model for the origin of biological catalysis. *Proc. Natl. Acad. Sci. USA* **81**:1751.

Stein, D., and Ogielski, A. (1985). Dynamics on ultrametric spaces. *Phys. Rev. Lett.* **55**:1634.

Stein, D., Baskaran, G., Liang, S., and Barber, M. (1987). Ground-state structure of short-range ising spin glasses in two and three dimensions. *Phys. Rev. B* **36**.

Stein, D., Palmer, R. G., Abrahams, E., and Anderson, P. W. (1984). Models of hierarchically constrained dynamics for glassy relaxation. *Phys. Rev. Lett.* **53**:958.

Stein, K. E., and Soderstrom, T. (1984). Neonatal administration of idiotype or antiidiotype primes for protection against *Escherichia coli* K13 infection in mice. *J. Exp. Med.* **160**:1001.

Stein, R., Gruenbaum, Y., Pollack, Y., Razin, A., and Cedar, H. (1982). Clonal inheritance of the pattern of methylation in mouse cells. *Proc. Natl. Acad. Sci. USA* **79**:61.

Steinberg, M. S. (1962). On the mechanism of tissue reconstruction by dissociated cells. III: Free energy relations and the reorganization of fused heteronomic tissue fragments. *Proc. Natl. Acad. Sci. USA* **98**:1769.

Steinberg, M. S. (1970). Does differential adhesion govern self-assembly processes in histogenesis? Equilibrium configurations and the emergence of a hierarchy among populations of embryonic cells. *J. Exp. Zool.* **173**:395.

Steiner, E. (1976). Establishment of compartments in the developing leg imaginal discs of *Drosophila melanogaster. Roux's Arch. Dev. Biol.* **180**:9.

Stenseth, N. C. (1985). Darwinian evolution in ecosystems: The Red Queen view. In *Evolution: Essays in Honor of John Maynard Smith* (P. J. Greenwood, P. H. Harvey, and M. Slatkin, eds.). Cambridge University Press, Cambridge.

Stenseth, N. C., and Maynard Smith, J. (1984). Coevolution in ecosystems: Red Queen evolution or stasis? *Evolution* **38**:870.

Stepshin, V. P., and Ginter, E. K. (1972). A study of the homeotic genes *Antennopedia* and *Nasobemia* in *Drosophila melanogaster.* III: Influence of temperature on the penetrance and expressivity of the *Apx* and *Ns* genes. *Genetika* **8**:55.

Stern, C. (1968). Developmental genetics of pattern. In *Genetic Mosaics and Other Essays.* Harvard University Press, Cambridge, Mass.

Stern, D. F., Hare, D. L., Cecchini, M. A., and Weinberg, R. A. (1987). Construction of a novel oncogene based on synthetic sequences encoding epidermal growth factor. *Science* **235**:321.

Sternberg, P. W. (1986). Control of cell type and cell lineage in *Saccharomyces cerevisiae*. In *Genetic Regulation of Development* (W. F. Loomis, ed.). Liss, New York.

Steward, F. C. (1958). From cultured cells to whole plants: The induction and control of their growth and morphogenesis. *Proc. Roy. Soc. Lond. B* **175**:1.

Steward, F. C., Mapes, M. O., Kent, A. E., and Holsten, R. D. (1964). Growth and development of cultured plant cells. *Science* **143**:20.

Stocum, D. L. (1980). Regeneration of symmetrical hindlimbs in larval salamanders. *Science* **200**:790.

Strand, M., and August, J. T. (1977). Polypeptides of cells transformed by RNA or DNA tumor viruses. *Proc. Natl. Acad. Sci. USA* **74**:2729.

Strand, M., and August, J. T. (1978). Polypeptide maps of cells infected with murine type C leukemia or sarcoma oncovirus. *Cell* **13**:399.

Straus, D., Raines, R., Kawashima, E., Knowles, J. R., and Gilbert, W. (1985). Active site of triosephosphate isomerase: *In vitro* mutagenesis and characterization of an altered enzyme. *Proc. Natl. Acad. Sci. USA* **82**:2272.

Struhl, G. (1981a). Anterior and posterior compartments in the proboscis of *Drosophila. Dev. Biol.* **84**:372.

Struhl, G. (1981b). A gene product required for correct initiation of segmental determination in *Drosophila. Nature* **293**:36.

Struhl, G. (1983). Role of the esc^+ gene product in ensuring the selective expression of segment specific homeotic genes in *Drosophila. J. Embryol. Exp. Morph.* **76**:297.

Suckling, C. J. (1991a). Molecular recognition: A universal molecular science? *Experientia* **47**:1093.

Suckling, C. J. (1991b). Molecular recognition in applied enzyme chemistry. *Experientia* **47**:1139.

Sugihara, G., Schoenly, K., and Trombla, A. (1989). Scale invariance in food web properties. *Science* **245**:48.

Sugita, M. (1963). Functional analysis of chemical systems *in vivo* using a logical circuit equivalent. II: The idea of a molecular automaton. *J. Theoret. Biol.* **4**:179.

Summers, M. C., Bedian, V., and Kauffman, S. A. (1986). An analysis of stage-specific protein synthesis in the early *Drosophila* embryo using high-resolution, two-dimensional gel electrophoresis. *Dev. Biol.* **113**:49.

Swenson, R. (1987). Optimization in spin glasses. Presented at Conference on Chaos and Complexity, Turin, Italy.

Szostack, J. W. (1986). Enzymatic activity of the conserved core of a group I self-splicing intron. *Nature* **322**:82.

Szybalski, W., Bovre, K., Fiandt, M. Hayes, S., Hradecna, Z. Kumar, S., Lozeron, H. A., Nijkamp, H.J.J., and Stevens, W. F. (1970). Transcriptional units and their controls in *Escherichia coli* phage lambda: Operons and scriptons. In *Cold Spring Harbor Symposia on Quantitative Biology*, vol. 38. Cold Spring Harbor Laboratory, Cold Spring Harbor, N.Y.

Tang, Y., Hicks, J. B., and Hilvert, D. (1991). *In vivo* catalysis of a metabolically essential reaction by an antibody. *Proc. Natl. Acad. Sci. USA* **88**:8784.

Tautz, D., Lehman, R., Schnurch, H., Schuh, R., Seifert, E., Kienlin, A., Jones, K., and Jackle, H. (1987). Finger protein of novel structure encoded by *hunchback*, a second member of the gap class of *Drosophila* segmentation genes. *Nature* **327**:383.

Templeton, A. R. (1982). Genetic architecture of speciation. In *Mechanisms of Speciation* (C. Barigozzi, ed.). Liss, New York.

Thom, R. (1970). Topological models in biology. In *Towards a Theoretical Biology*, vol. 3: *Drafts* (C. H. Waddington, ed.). Aldine, Chicago.

Thom, R. (1972). *Stabilité structurelle et morphogenese: Essai d'une theorie general des modeles.* Benjamin, Reading.

Thomas, R. (1979). Kinetic logic: A Boolean analysis of the dynamic behavior of control circuits. In *Kinetic Logic* (R. Thomas, ed.), Lecture Notes in Biomathematics, vol. 29. Springer, New York.

Thomas, R. (1984). Logical description, analysis and synthesis of biological and other networks comprising feedback loops. *Adv. Chem. Phys.* **55**:247.

Thompson, D. (1966). *On Growth and Form,* 2d ed. Cambridge University Press, Cambridge.

Tonegawa, S. (1983). Somatic generation of antibody diversity. *Nature* **302**:575.

Toneguzzo, F., Hayday, A. C., and Keating, A. (1986). Electric-field-mediated DNA transfer: Transient and stable gene expression in human and mouse lymphoid cells. *Mol. and Cell Biol.* **6**:703.

Totafurno, J., and Trainor, L.E.H. (1987). A non-linear vector field model of supernumerary limb production in salamanders. *J. Theoret. Biol.* **124**:415.

Toulouse, G. (1987). Progrès récent dans la physique des systèmes desordonnés. *Helv. Physiol. Pharmacol. Acta* **57**:459.

Toulouse, G., Dehaene, S., and Changeux, J. P. (1986). Spin-glass model of learning by selection. *Proc. Natl. Acad. Sci. USA* **83**:1695.

Trainor, L.E.H., and Goodwin, B. C. (1986). Stability analysis on a set of calcium-regulated viscoelastic equations. *Physica* **21D**:137.

Tramontano, A., Janda, K. D., and Lerner, R. A. (1986a). Catalytic antibodies. *Science* **234**:1566.

Tramontano, A. Janda, K. D., and Lerner, R. A. (1986b). Chemical reactivity at an antibody-bonding site elicited by mechanistic design of a synthetic antigen. *Proc. Natl. Acad. Sci. USA* **83**:6736.

Tramontano, A., Janda, K., Napper, A. D., Benkovick, S. J., and Lerner, R. A. (1987). Catalytic antibodies. In *Cold Spring Harbor Symposia on Quantitative Biology,* vol. 52. Cold Spring Harbor Laboratory, Cold Spring Harbor, N.Y.

Tuerk, C., and Gold, L. (1990). Systematic evolution of ligands by exponential enrichment: RNA ligands to bacteriophage T4 DNA polymerase. *Science* **249**:505.

Tufano, F., and Brandhorst, B. P. (1979). Similarity of proteins synthesized by isolated blastomeres of early sea urchin embryos, *Dev. Biol.* **72**:390.

Turing, A. M. (1952). The chemical basis of morphogenesis. *Philos. Trans. Roy. Soc. London B* **237**:37.

Turing, A. M. (1959). In *Alan M. Turing, Mathematician and Scientist* (S. Turing, ed.). Heffer, Cambridge.

Turner, F. R., and Mahowald, A. P. (1976). Scanning electron microsocpy of *Drosophila* embryogenesis. I: The structure of the egg envelope and the formation of the cellular blastoderm. *Dev. Biol.* **50**:95.

Turner, F. R., and Mahowald, A. P. (1977). Scanning electron microscopy of *Drosophila* embryogenesis. II: Gastrulation and segmentation. *Devl. Biol.* **57**:403.

Turner, F. R., and Mahowald, A. P. (1979). Scanning electron microscopy of *Drosophila* embryogenesis. III: Formation of the head and caudal segments. *Devl. Biol.* **68**:96.

Underwood, E. M., Turner, F. R., and Mahowald, A. P. (1980). Analysis of cell movements and fate mapping during early embryogenesis in *Drosophila melanogaster. Dev. Biol.* **74**:286.

Urbain, J., VanAcker, A., De Vos-Cloetens, C., and Urbain-Vasanten, G. (1972). Increase and decrease in binding affinity of antibodies during the immune response. *Immunology* **9**:121.

Valentine, J. W. (1977). General patterns of metazoan evolution. In *Patterns of Evolution as Illustrated by the Fossil Record* (A. Hallam, ed.). Elsevier, Amsterdam.

Valentine, J. W. (1980). Determinants of diversity in higher taxonomic categories. *Paleobiology* **6**:444.

Valentine, J. W. (1986). Fossil record of the origin of Bauplane and its implications. In *Patterns and Processes in the History of Life* (D. M. Raup and D. Jablonski, eds.). Springer, New York.

Valentine, J. W., and Campbell, C. A. (1975). Genetic regulation and the fossil record. *Am. Sci.* **63**:673.

Valentine, J. W., and Erwin, D. H. (1986). Interpreting great developmental experiments: The fossil record. In *Development as an Evolutionary Process.* (R. A. Raff and E. C. Raff, eds.). Liss, New York.

Valentine, J. W., and Walker, T. D. (1986). Diversity trends within a model taxonomic hierarchy. *Physica* **22D**:31.

Van der Meer, J. M. (1984). Parameters influencing reversal of segment sequence in posterior egg fragments of *Callosobruchus (Coleoptera). Roux's Arch. Dev. Biol.* **193**:339.

Van der Meer, J. M., and Ouweneel, W. (1974). Differentiation capacities of the dorsal mesothoracic *(haltere)* disc of *Drosophila melanogaster.* II: Regeneration and duplication. *Roux's Arch. Dev. Biol.* **174**:361.

Van Valen, L. (1973). A new evolutionary theory. *Evol. Theory* **1**:1.

Vastano, J. A., Pearson, J. E., Horsthemke, W., and Swinney, H. L. (1987). Chemical pattern formation with equal diffusion coefficients. *Phys. Lett. A* **124**:320.

Vendrely, R. (1955). In *The Nucleic Acids,* vol. 2. Academic Press, New York.

Vermeu, G. J. (1973a). Adaptation, versatility ane evolution. *Syst. Zool.* **22**:466.

Vermeu, G. J. (1973b). Biological versatility and earth history. *Proc. Natl. Acad. Sci. USA* **70**:1936.

Villee, C. A. (1943). Phenogenetic studies of the homeotic mutants of *Drosophila melanogaster.* I: The effects of temperature on the expression of *aristepedia. J. Exp. Zool.* **93**:75.

Villee, C. A. (1944). Phenogenetic studies of the homeotic mutants of *Drosophila melanogaster.* II: The effects of temperature on the expression of *proboscipedia. J. Exp. Zool.* **96**:85.

Villee, C. A. (1945). Phenogenetic studies of the homeotic mutants of *Drosophila melanogaster.* III: The effects of temperature on the expression of *biothorax-34e. Amer. Nat.* **79**:246.

Vogel, H. J., ed. (1971). *Metabolic Pathways,* vol. 5. Academic Press, New York.

von Kiedrowski, G. (1986). A self-replicating hexadeoxynucleotide. *Agnew. Chem. Int. Ed. Eng.* **25**:982.

Voss, R. (1985). In *Random Fractals: Characterization and Measurement, Scaling Phenomena in Disordered Systems* (R. Pynn and A. Skjeltorp, eds.). Plenum, New York.

Wabl, M., Burrows, P. D., vonGabain, A., and Steinberg, C. (1985). Hypermutation at the immunoglobin heavy chanin locus in a pre-b-cell line. *Proc. Natl. Acad. Sci, USA* **82**:479.

Waddington, C. H. (1940). *Organizers and Genes.* Cambridge University Press, Cambridge.

Waddington, C. H. (1942). Canalization of development and inheritance of acquired characters. *Nature* **150**:563.

Waddington, C. H. (1956). Genetic assimilation of the bithorax phenotype. *Evolution* **10**:1.

Waddington, C. H. (1957). *The Strategy of the Genes.* George Allen and Unwin, London.

Waddington, C. H. (1962). *New Patterns in Genetics and Development.* Columbia University Press, New York.

Waddington, C. H. (1966). *Principles of Development and Differentiation.* Macmillan, New York.

Wald, G. (1954). The origin of life. *Sci. Am.,* August.

Walker, C. C., and Gelfand, A. E. (1979). A system theoretical approach to the management of complex organizations: Management by exception, priority and input span in a class of fixed-structure models. *Behav. Sci.* **24**:112.

Walker, M. D., Edlund, T., Boulet, A. M., and Rutter, W. J. (1983). Cell-specific expression controlled by the 5′ flanking region of the insulin and chymotrypsin genes. *Nature* **306**:557.

Walter, C., Parker, R., and Ycas, M. (1967). A model for binary logic in biochemical systems. *J. Theoret. Biol.* **15**:208.

Wand, A. J., Roder, H., and Englander, S. W. (1986). Two-dimensional 'HNMR of cytochrome C: Hydrogen exchange in the N-terminal helix. *Biochemstry* **25**:1107.

Wang, E. A., Kallen, R., and Walsh, C. (1981). Mechanism-based inactivation of Serine Transhydroxymethylaxes by d-Fluoroalanine and related amino acids. *J. Biol. Chem.* **256**:6917.

Wang, T. T., Dorrington, K. J., and Hoffman, T. (1974). Activation of the action of penicillopepsin on leucyl-tyrosyl-amide by a non-substrate peptide and evidence for a conformational change associated with a secondary binding site. *Biochem. Biophys. Res. Commun.* **57**::865.

Wasteneys, H., and Borsook, H. (1930). The enzymatic synthesis of protein. *Physiol. Rev.* **10**:110.

Watson, J. D. (1965). *The Molecular Biology of the Gene.* Benjamin, New York.

Webster, G., and Goodwin, B. C. (1982). The origin of species: A structuralist approach. *J. Soc. Biol. Struc.* **5**:15.

Weigert, M. (1987). Personal communication.

Weinberger, E. D. (1987). A stochastic generalization of Eigen's model of natural selection. Ph.D. diss., Courant Institute of Mathematical Sciences, New York.

Weinberger, E. D. (1990). Correlated and uncorrelated fitness landscapes and how to tell the difference. *Biol. Cybernet.* **63**:325.

Weinberger, E. D. (1991a). A more rigorous derivation of some properties of uncorrelated fitness landscapes. *J. Theoret. Biol.* **134**:125.

Weinberger, E. D. (1991b). Local properties of the *NK* model, a tuneably rugged energy landscape. *Phys. Rev. A* **44**:6399.

Weinberger, E. D., and Stadler, P. F. (1992). Why some fitness landscapes are fractal. *J. Statis. Phys.*, submitted.

Weintraub, H. (1985). Assembly and propagation of repressed and depressed chromosomal states. *Cell* **42**:705.

Weir, M. P., and Kornberg, T. (1985). Patterns of *engrailed* and *Fushi-tarazu* transcripts reveal novel intermediate stages in *Drosophila* segmentation. *Nature* **318**:433.

Weir, M. P., and Lo, C. (1982). Gap junction communication compartments in the *Drosophila* wing disc. *Proc. Natl. Acad. Sci. USA* **79**:3232.

Weisbuch, G. (1984). *C. R. Acad. Sci.* **298**:375.

Weisbuch, G. (1989a). Dynamical behavior of discrete models of Jerne's network. In *Theories of Immune Networks* (H. Atlan, ed.). Springer, New York.

Weisbuch, G. (1989b). *Dynamics of Complex Systems: An Introduction to Networks of Automata.* Interedition, Paris.

Weisbuch, G., and Stauffer, D. (1987). Phase transition in cellular random Boolean nets. *J. Physique* **48**:11.

Weischaus, E., Nüsslein-Volhard, C., and Kluding, H. (1984). *Kruppel,* a gene whose activity is required early in the zygotic genome for normal embryonic segmentation. *Dev. Biol.* **104**:172.

Weismann, A. (1885). The continuity of the germ-plasm as the foundation of a theory of heredity. In *Readings in Heredity and Development* (J. A. Moore, ed.). Oxford University Press, New York.

Weismann, A. (1904). *The Evolution Theory.* Edward Arnold, London.

Weiss, P. (1939). *Principles of Development: A Text in Experimental Embryology.*

Weiss, P. A. (1968). *Dynamics of Development: Experiments and Inferences,* Academic Press, New York.

Werblin, T. P., Kim, Y. T., Quagliata, F., and Siskind, G. W. (1973). Studies on the control of antibody synthesis. III: Changes in heterogeneity of antibody affinity during the course of an immune response. *Immunochemistry* 24:477.

Wetzel, R. (1991). Learning from the immune system: Laboratory methods for creating and refining molecular diversity in polypeptides. *Mol. Engin.* 4:371.

Whyte, L. L. (1965). *Internal Factors in Evolution.* Braziller, New York.

Wigner, E. (1961). The probability of the existence of a self-reproducing unit. In *The Logic of Personal Knowledge* (E. Shils, ed.). Free Press, Glencoe, Ill.

Wille, J. J., Scheffey, C., and Kauffman, S. A. (1977). Novel behaviour of the mitotic clock in *Physarum. J. Cell Sci.* 27:91.

Wilmer, E. N. (1970). *Cytology and Evolution,* 2d ed. Academic Press, New York.

Wilson, A. C., Carlson, S. S., and White, T. J. (1977). Biochemical evolution. *Ann. Rev. Biochem.* 46:573.

Wilson, A. C., Sarich, V. M., and Maxson, L. R. (1974). The importance of gene rearrangement in evolution: Evidence from studies on rates of chromosomal, protein, and anatomic evolution. *Proc. Natl. Acad. Sci. USA* 71:3028.

Wimsatt, W. C. (1986a). Developmental constraints, generative entrenchment, and the innate-acquired distinction. In *Science and Philosophy, Integrating Scientific Disciplines* (W. Bechtel, ed.). Martinus Nijhoff, Dordrecht.

Wimsatt, W. C. (1986b). In *Interdisciplinary Relations Among the Sciences* (P. W. Bechtel, ed.). North Holland, Amsterdam.

Winfree, A. T. (1980). *The Geometry of Biological Time,* Lecture Notes in Biomathematics, vol. 8. Springer, New York.

Winfree, A. T. (1984). A continuity principle for regeneration. In *Pattern Formation* (G. M. Malacinski and S. V. Bryant, eds.). Macmillan, New York.

Winfree, A. T. (1987). *When Time Breaks Down.* Princeton University Press, Princeton, N.J.

Woese, C. R. (1972). *The Genetic Code: The Molecular Basis for Genetic Expression.* Harper & Row, New York.

Wolfenden, R. (1972). Analog approaches to structure of transition state in enzyme reactions. *Accts. Chem. Res.* 5:10.

Wolfram, S. (1983). Statistical mechanics of cellular automata. *Rev. Mod. Phys.* 55:601.

Wolfram, S. (1984). Universality and complexity in cellular automata. *Physica* 10D:1.

Wolpert, L. (1969). Positional information and the spatial pattern of cellular differentiation. *J. Theoret. Biol.* 25:1.

Wolpert, L. (1971). Positional information and pattern formation. In *Current Topics in Developmental Biology* (A. A. Moscana and A. Monroy, eds.), vol. 6. Academic Press, New York.

Woolf, C. M. (1949). The effect of temperature treatments on an early developmental period of *tumerous head* in *Drosophila melanogaster. Utah Acad. Sci. Arts Lett. Proc.* 26:139.

Wright, S. (1931). Evolution in Mendelian populations. *Genetics* 16:97.

Wright, S. (1932). The roles of mutation, inbreeding, crossbreeding and selection in evolution. *Proceedings of the Sixth International Congress on Genetics* 1:356.

Wuensche, A. (1992). The ghost in the machine: Basin of attraction fields of disordered cellular automata networks. Working paper 92-14-017, Santa Fe Institute, Santa Fe, N.M.

Wuensche, A., and Lesser, M. J. (1992). *The Global Dynamics of Cellular Automata: An Atlas of Basin of Attraction Fields of One-Dimensional Cellular Automata.* Santa Fe Institute Studies in the Sciences of Complexity, ref. vol. 1. Addison-Wesley, Reading, Mass.

Wysocki, L., Manser, T., and Gefter, M. L. (1986). Somatic evolution of variable region structures during an immune response. *Proc. Natl. Acad. Sci. USA* 83:1847.

Yancouplos, G. D., and Alt, F. W. (1986). Regulation of the assembly and expression of variable-region genes. *Ann. Rev. Immunol.* 4:339.

Yeo, Y., Rota K., and Rechsteiner, M. (1989). Synthesis of peptides as cloned ubiquitin extensions. *J. Biol. Chem.* **264:**17078.

Young, M. W. (1979). Middle-repetitive DNA: A fluid component of the *Drosophila* genome. *Proc. Natl. Acad. Sci. USA* **76:**6274.

Young, R. A., and Davis, R. W. (1983). Yeast RNA polymerase II genes: Isolation with antibody probes. *Science* **222:**778.

Yourno, J., Kohno, T., and Roth, J. R. (1970). Enzyme evolution: Generation of a bifunctional enzyme by infusion of adjacent genes. *Nature* **228:**820.

Zaret, K. S., and Yamamoto, K. R. (1984). Reversible and persistent changes in chromatin structure accompany activation of a glucocorticoid-dependent enhancer element. *Cell* **38:**29.

Zaug, A. J., and Cech, T. R. (1985). Oligomerization of intervening sequence RNA molecules in the absence of proteins. *Science* **229:**1060.

Zaug, A. J., and Cech, T. R. (1986). The intervening sequence RNA of *Tetrahymena* is an enzyme. *Science* **231:**470.

Zebedee, S. L., Barbas, C. F., Hom, Y. L., Caothien, R. H., Graff, R., Degraw, J., Pyati, J., Lapolla, R., Burton, D. R., Lerner, R. A., and Thornton, G. B. (1992). Human conbinatorial antibody libraries on hepatitis-B surface-antigen. *Proc. Natl. Acad. Sci. USA* **89:**3175.

Zhimulev, I. F., Vlassova, I. E., and Belyaeva, E. S. (1982). Cytogenetic analysis of the 2B3-4 to 2B11 region of the X chromosome of *Drosophila melanogaster*. *Chromosoma* **85:**659.

Zielinski, W. S., and Orgel, L. E. (1985). Oligomerization of activated derivatives of 3'-amino-3'-deoxyguanosine on poly(c) and poly(dC) templates. *Nucl. Acid Res.* **13:**2469.

Zubay, G., and Chambers, D. A. (1971). Regulating the *lac* operon. In *Metabolic Pathways* (H. J. Vogel, ed.), vol. 5. Academic Press, New York.

Zubay, G., Gielow, L., and Englesberg, E. (1971). *Nature New Biol.* **233:**164.

Zuckerkandl, E., and Pauling, L. (1965). Evolutionary divergence and convergence in proteins. In *Evolving Genes and Proteins* (V. Bryson and H. J. Vogel, eds.). Academic Press, New York.

Index